Student Solutions M

Chemistry

Seventh Edition

Raymond Chang
Williams College

by
Brandon Cruickshank
Northern Arizona University

Boston   Burr Ridge, IL   Dubuque, IA   Madison, WI   New York   San Francisco   St. Louis
Bangkok   Bogotá   Caracas   Kuala Lumpur   Lisbon   London   Madrid   Mexico City
Milan   Montreal   New Delhi   Santiago   Seoul   Singapore   Sydney   Taipei   Toronto

McGraw-Hill Higher Education

*A Division of The **McGraw-Hill** Companies*

Student Solutions Manual by Brandon Cruickshank
CHEMISTRY, SEVENTH EDITION
RAYMOND CHANG

Published by McGraw-Hill Higher Education, an imprint of The McGraw-Hill Companies, Inc.,
1221 Avenue of the Americas, New York, NY 10020. Copyright © The McGraw-Hill Companies,
Inc., 2002, 1998. All rights reserved.

This book is printed on acid-free paper.

1 2 3 4 5 6 7 8 9 0 QPD/QPD 0 3 2 1

ISBN 0-07-231802-3

www.mhhe.com

FOR THE STUDENT

The *Student Solutions Manual* is intended for use with the seventh edition of Raymond Chang's *Chemistry*. For this manual to be of maximum assistance, you should incorporate it into your overall plan for studying general chemistry. The *Student Solutions Manual* contains material to help you develop problem-solving skills, material to practice your problem-solving skills, and solutions to the even-numbered problems in the text. Each chapter in the manual corresponds to one in the text. Most chapters of the manual contain the following main features:

Problem-Solving Strategies

You will encounter several different types of problems in each chapter. To make your problem-solving task easier, we will break down each chapter into the most common problem types. We will present step-by-step methods for solving each problem type.

Example Problems

Incorporated into each problem type are example problems with detailed step-by-step tutorial solutions. These example problems and tutorial solutions are intended to reinforce the problem-solving strategy presented with each problem type. The solutions often contain further explanation of important concepts.

Practice Exercises

Throughout all chapters are numerous practice exercises to allow you to test your problem-solving skills and your knowledge of the material. Answers to the practice exercises are found at the end of the problem-solving strategy section of each chapter.

Text Solutions

Solutions to all the even-numbered text problems are presented following the problem-solving strategies for each chapter. Usually one text problem of each problem type is solved in the step-by-step tutorial manner presented in the problem-solving strategy section. The problems with tutorial solutions refer back to the problem type presented in the strategy section. This will allow you to refer back to the appropriate material if you are having difficulty solving the problem.

Similar Problems

Following each problem type is a list of similar even-numbered problems in the text. Problems in bold-face type are solved in the step-by-step tutorial manner presented in the problem-solving strategy section. Similar problems are listed to allow you to practice more problems that follow a comparable strategy.

I hope that the *Student Solutions Manual* helps you succeed in general chemistry. I have attempted to present detailed problem-solving strategies to help you become a better problem-solver. Improved problem-solving will not only help you succeed in general chemistry, but also in many other courses throughout your college career. Probably one of the best ways to learn general chemistry is to work through, and sometimes struggle through, the problems assigned by your instructor. It always looks easy when your instructor solves a problem in class, but you really do not learn the material until you prove to yourself that you can solve the problem.

When solving problems, always try to be flexible. Please do not think that all problems that you encounter in general chemistry will fall into one of the problem types presented in each chapter. However, take the knowledge that you gain from the tutorial solutions and learn to apply that knowledge to different and perhaps more difficult problems. Also, try to work a few problems *every* day. (OK, you can take Friday off!) You cannot afford to fall behind in a general chemistry course.

If you have any comments or suggestions regarding the *Student Solutions Manual*, I would like to hear from you. My e-mail address is Brandon.Cruickshank@nau.edu or you can write me at the following address:

Dr. Brandon Cruickshank
Northern Arizona University
Department of Chemistry
P.O. Box 5698
Flagstaff, AZ 86011-5698

Good luck in all your college endeavors! And try to enjoy chemistry. You might be pleasantly surprised.

ACKNOWLEDGMENTS
My most sincere thanks go to Dr. Raymond Chang who provided solutions for all the new problems in this edition and reviewed the entire manuscript. His insights, helpful suggestions, and corrections greatly improved this manual.

My deepest thanks go to Ms Suzanne Guinn, Senior Developmental Editor at McGraw-Hill, for directing and organizing this project.

I also wish to thank Mr. Kent Peterson, Executive Editor at McGraw-Hill, for developing and running an excellent "Chemistry Team" at McGraw-Hill.

Finally, I express my love and thanks to my partner, Liz Hobbs, for her support and encouragement during work on this project.

Brandon J. Cruickshank

CONTENTS

CHAPTER 1
CHEMISTRY: THE STUDY OF CHANGE

PROBLEM-SOLVING STRATEGIES AND TUTORIAL SOLUTIONS

TYPES OF PROBLEMS

Problem Type 1: Density Calculations.

Problem Type 2: Temperature Conversions.
- (a) °C → °F
- (b) °F → °C

Problem Type 3: Scientific Notation.
- (a) Expressing a number in scientific notation.
- (b) Addition and subtraction.
- (c) Multiplication and division.

Problem Type 4: Significant Figures.
- (a) Addition and subtraction.
- (b) Multiplication and division.

Problem Type 5: The Factor-Label Method of Solving Problems.

PROBLEM TYPE 1: DENSITY CALCULATIONS

Density is the mass of an object divided by its volume.

$$\text{density} = \frac{\text{mass}}{\text{volume}}$$

$$d = \frac{m}{V}$$

Densities of solids and liquids are typically expressed in units of grams per cubic centimeter (g/cm^3) or equivalently grams per milliliter (g/mL). Because gases are much less dense than solids and liquids, typical units are grams per liter (g/L).

EXAMPLE 1.1
A lead brick with dimensions of 5.08 cm by 10.2 cm by 20.3 cm has a mass of 11,950 g. What is the density of lead in g/cm^3?

Step 1: You are given the mass of the lead brick in the problem. You need to calculate the volume of the lead brick to solve for the density. The volume of a rectangular object is equal to the length × width × height.

$$\text{Volume} = \text{length} \times \text{width} \times \text{height}$$
$$\text{Volume} = 5.08 \text{ cm} \times 10.2 \text{ cm} \times 20.3 \text{ cm} = \textbf{1052 cm}^3$$

Step 2: Calculate the density by substituting the mass and the volume into the equation.

$$d = \frac{m}{V} = \frac{11,950 \text{ g}}{1052 \text{ cm}^3} = \textbf{11.4 g/cm}^3$$

PRACTICE EXERCISE

1.   Platinum has a density of 21.4 g/cm^3. What is the mass of a small piece of platinum that has a volume of 7.50 cm^3?

| **Text Problem: 1.22** |

PROBLEM TYPE 2: TEMPERATURE CONVERSIONS

To convert between the Fahrenheit scale and the Celsius scale, you must account for two differences between the two scales.

(1) The Fahrenheit scale defines the normal freezing point of water to be exactly 32°F, whereas the Celsius scale defines it to be exactly 0°C.

(2) A Fahrenheit degree is 5/9 the size of a Celsius degree.

A.   Converting degrees Fahrenheit to degrees Celsius

The equation needed to complete a conversion from degrees Fahrenheit to degrees Celsius is:

$$? \, ^\circ C = (^\circ F - 32 ^\circ F) \times \frac{5 ^\circ C}{9 ^\circ F}$$

32°F is subtracted to compensate for the normal freezing point of water being 32°F, compared to 0° on the Celsius scale. We multiply by (5/9) because a Fahrenheit degree is 5/9 the size of a Celsius degree.

EXAMPLE 1.2
Convert 20°F to degrees Celsius.

$$? \, ^\circ C = (^\circ F - 32 ^\circ F) \times \frac{5 ^\circ C}{9 ^\circ F}$$

$$? \, ^\circ C = (20 ^\circ F - 32 ^\circ F) \times \frac{5 ^\circ C}{9 ^\circ F} = -6.7 ^\circ C$$

B.   Converting degrees Celsius to degrees Fahrenheit

The equation needed to complete a conversion from degrees Celsius to degrees Fahrenheit is:

$$? \, ^\circ F = \left(^\circ C \times \frac{9 ^\circ F}{5 ^\circ C} \right) + 32 ^\circ F$$

°C is multiplied by (9/5) because a Celsius degree is 9/5 the size of a Fahrenheit degree. 32°F is then added to compensate for the normal freezing point of water being 32°F, compared to 0° on the Celsius scale.

EXAMPLE 1.3
Normal human body temperature on the Celsius scale is 37.0°C. Convert this to the Fahrenheit scale.

$$? \, ^\circ F = \left(^\circ C \times \frac{9 ^\circ F}{5 ^\circ C} \right) + 32 ^\circ F$$

$$? \, ^\circ F = \left(37.0 ^\circ C \times \frac{9 ^\circ F}{5 ^\circ C} \right) + 32 ^\circ F = 98.6 ^\circ F$$

PRACTICE EXERCISE

2.   Convert –40°F to degrees Celsius.

Text Problems: **1.24**, 1.26, 1.48 a,b

PROBLEM TYPE 3: SCIENTIFIC NOTATION

Scientific notation is typically used when working with small or large numbers. All numbers can be expressed in the form

$$N \times 10^{n}$$

where N is a number between 1 and 10 and n is an exponent that can be a positive or negative integer, or zero.

A.   Expressing a number in scientific notation

Step 1: Determine n by counting the number of places that the decimal point must be moved to give N, a number between 1 and 10.

Step 2: If the decimal point is moved to the left, n is a positive integer, the number you are working with is larger than 10. If the decimal point is moved to the right, n is a negative integer. The number you are working with is smaller than 1.

EXAMPLE 1.4
Express 0.000105 in scientific notation.

Step 1: The decimal point must be moved four places to the right to give N, a number between 1 and 10. In this case,
$$N = \mathbf{1.05}$$

Step 2: Since 0.000105 is a number less than one, n is a negative integer. In this case, $n = -4$ (The decimal point was moved four places to the right to give $N = 1.05$).

Combining the above two steps:
$$0.000105 = \mathbf{1.05 \times 10^{-4}}$$

Tip: The notation 1.05×10^{-4} means the following: Take 1.05 and multiply by 10^{-4} (0.0001).

$$1.05 \times 0.0001 = 0.000105$$

EXAMPLE 1.5
Express 4224 in scientific notation.

Step 1: The decimal point must be moved three places to the left to give N, a number between 1 and 10. In this case,
$$N = \mathbf{4.224}$$

Step 2: Since 4,224 is a number greater than one, n is a positive integer. In this case, $n = 3$ (the decimal point was moved three places to the left to give $N = 4.224$).

Combining the above two steps: $4224 = \mathbf{4.224 \times 10^{3}}$

Tip: The notation 4.224×10^{3} means the following: Take 4.224 and multiply by 10^{3} (1000).

$$4.224 \times 1000 = 4,224$$

PRACTICE EXERCISE

3.   Express the following numbers in scientific notation:

   (a)   45,781   **(b)**   0.0000430

> **Text Problem: 1.30**

B. Addition and subtraction using scientific notation

Step 1: Write each quantity with the same exponent, n.

Step 2: Add or subtract the N parts of the numbers, keeping the exponent, n, the same.

EXAMPLE 1.6
Express the answer to the following calculation in scientific notation.

$$2.43 \times 10^1 + 5.955 \times 10^2 = ?$$

Step 1: Write each quantity with the same exponent, n. Let's write 2.43×10^1 in such a way that $n = 2$.

> **Tip:** We are *increasing* 10^n by a factor of 10, so we must *decrease* N by a factor of 10. We move the decimal point one place to the left.

$$2.43 \times 10^1 = \mathbf{0.243 \times 10^2}$$
(n was increased by 1. Move the decimal point one place to the left.)

Step 2: Add or subtract, as required, the N parts of the numbers, keeping the exponent, n, the same. In this example, the process is addition.

$$
\begin{array}{r}
0.243 \times 10^2 \\
+ \; 5.955 \times 10^2 \\
\hline
\mathbf{6.198 \times 10^2}
\end{array}
$$

C. Multiplication and division using scientific notation

Step 1: Multiply or divide the N parts of the numbers in the usual way.

Step 2: When multiplying, *add* the exponents. When dividing, *subtract* the exponents.

EXAMPLE 1.7
Divide 4.2×10^{-7} by 5.0×10^{-5}.

Step 1: Divide the N parts of the numbers in the usual way.

$$4.2 \div 5.0 = 0.84$$

Step 2: When dividing the 10^n parts, *subtract* the exponents.

$$0.84 \times 10^{-7 - (-5)} = 0.84 \times 10^{-7 + 5} = \mathbf{0.84 \times 10^{-2}}$$

The usual practice is to express N as a number between 1 and 10. Therefore, it is more appropriate to move the decimal point of the above number one place to the right, decreasing the exponent by 1.

$$0.84 \times 10^{-2} = \mathbf{8.4 \times 10^{-3}}$$

> **Tip:** In the answer above, we moved the decimal point to the right, *increasing N* by a factor of 10. Therefore, we must *decrease* 10^n by a factor of 10. The exponent, *n*, is changed from −2 to −3.

EXAMPLE 1.8
Multiply 2.2×10^{-3} by 1.4×10^{6}.

Step 1: Multiply the *N* parts of the numbers in the usual way.

$$2.2 \times 1.4 = 3.1$$

Step 2: When multiplying the 10^n parts, *add* the exponents.

$$3.1 \times 10^{-3 + 6} = \mathbf{3.1 \times 10^{3}}$$

PRACTICE EXERCISE

4. Express the answer to the following calculations in scientific notation. Try these without using a calculator.

    (a) $2.20 \times 10^{3} - 4.54 \times 10^{2} =$

    (b) $4.78 \times 10^{5} \div 6.332 \times 10^{-7} =$

Text Problem: 1.32

PROBLEM TYPE 4: SIGNIFICANT FIGURES

See Section 1.8 of your text for guidelines for using significant figures.

A. Addition and subtraction

The number of significant figures to the right of the decimal point in the answer is determined by the lowest number of significant figures to the right of the decimal point in any of the original numbers.

EXAMPLE 1.9
Carry out the following operations and express the answer to the correct number of significant figures.

$$\mathbf{102.226 + 2.51 + 736.0 =}$$

$$
\begin{array}{r}
102.226 \\
2.51 \\
+ \; \underline{736.0} \quad \leftarrow \text{fewest digits to the right of the decimal point} \\
840.736
\end{array}
$$

The 3 and 6 are nonsignificant digits, since 736.0 only has one digit to the right of the decimal point. The answer should only have one digit to the right of the decimal point.

The correct answer rounded off to the correct number of significant figures is **840.7**

> **Tip:** To round off a number at a certain point, simply drop the digits that follow if the first of them is less than 5. If the first digit following the point of rounding off is equal to or greater than 5, add 1 to the preceding digit.

B. Multiplying and dividing

The number of significant figures in the answer is determined by the original number having the smallest number of significant figures.

EXAMPLE 1.10

Carry out the following operations and express the answer to the correct number of significant figures.

$$12 \times 2143.1 \div 3.11 = ?$$

$$12 \times 2143.1 \div 3.11 = 8269.2 = 8.2692 \times 10^3$$

The 6, 9, and 2 (bolded) are nonsignificant digits because the original number 12 only has two significant figures. Therefore, the answer has only two significant figures.

The correct answer rounded off to the correct number of significant figures is $\mathbf{8.3 \times 10^3}$

PRACTICE EXERCISE

5.   Carry out the following operations and express the answer to the correct number of significant figures.

   (a) $90.25 - 83 + 1.0015 =$
   (b) $55.6 \times 3.482 \div 505.34 =$

Text Problem: 1.36

PROBLEM TYPE 5: THE FACTOR-LABEL METHOD OF SOLVING PROBLEMS

In order to convert from one unit to another, you need to be proficient at the factor-label method. See Section 1.9 of your text. Unit conversions can seem daunting, but if you keep track of the units, making sure that the appropriate units cancel, your effort will be rewarded.

Step 1: Map out a strategy to proceed from initial units to final units based on available conversion factors.

Step 2: Use the following method as many times as is necessary to ensure that you obtain the desired unit.

$$\text{Given unit} \times \left(\frac{\text{desired unit}}{\text{given unit}} \right) = \text{desired unit}$$

EXAMPLE 1.11

How long will it take to fly from Denver to New York, a distance of 1631 miles, at a speed of 815 km/hr?

Step 1: One conversion factor is given in the problem, 815 km/hr. This conversion factor can be used to convert from distance (in km) to time (in hr). If you can convert the distance of 1631 miles to km, then you can use the conversion factor (815 km/hr) to convert to time in hours. Another conversion factor that you can look up is

   1 mi = 1.61 km

   You should come up with the following strategy.

   miles → km → hours

Step 2: Carry out the necessary conversions, making sure that units cancel.

$$\textbf{? hours} = 1631 \text{ mi (given)} \times \frac{1.61 \text{ km (desired)}}{1 \text{ mi (given)}} \times \frac{1 \text{ hr (desired)}}{815 \text{ km (given)}} = \textbf{3.22 hr}$$

> **Tip**: In the first conversion factor (km/mi), km is the desired unit. When moving on to the next conversion factor (hr/km), km is now given, and the desired unit is hr.

EXAMPLE 1.12

The *Voyager II* mission to the outer planets of our solar system transmitted by radio signals many spectacular photographs of Neptune. Radio waves, like light waves, travel at a speed of 3.00×10^8 m/s. If Neptune was 2.75 billion miles from Earth during these transmissions, how many hours were required for radio signals to travel from Neptune to Earth?

Step 1: One conversion factor is given in the problem, 3.00×10^8 m/s. This conversion factor will allow you to convert from distance (in m) to time (in seconds). If you can convert the distance of 2.75 billion miles to meters, then the speed of light (3.00×10^8 m/s) can be used to convert to time in seconds. Other conversion factors that you can look up are:

$$\text{billion} = 10^9 \qquad\qquad 60 \text{ s} = 1 \text{ min}$$
$$1 \text{ mi} = 1.61 \text{ km} \qquad\qquad 60 \text{ min} = 1 \text{ hr}$$
$$1 \text{ km} = 1000 \text{ m}$$

You should come up with the following strategy.

$$\text{miles} \rightarrow \text{km} \rightarrow \text{meters} \rightarrow \text{seconds} \rightarrow \text{min} \rightarrow \text{hours}$$

Step 2: Carry out the necessary conversions, making sure that units cancel.

$$? \text{ hr} = 2.75 \times 10^9 \text{ mi} \times \frac{1.61 \text{ km}}{1 \text{ mi}} \times \frac{1000 \text{ m}}{1 \text{ km}} \times \frac{1 \text{ s}}{3.00 \times 10^8 \text{ m}} \times \frac{1 \text{ min}}{60 \text{ s}} \times \frac{1 \text{ hr}}{60 \text{ min}} = \mathbf{4.10 \text{ hr}}$$

PRACTICE EXERCISES

6. On a certain day, the concentration of carbon monoxide, CO, in the air over Denver reached 1.8×10^{-5} g/L. Convert this concentration to mg/m^3.

7. Copper (Cu) is a trace element that is essential for nutrition. Newborn infants require 80 μg of Cu per kilogram of body mass per day. The Cu content of a popular baby formula is 0.48 μg of Cu per milliliter. How many milliliters should a 7.0 lb baby consume per day to obtain the minimum daily Cu requirement?

> **Text Problems: 1.38, 1.40**, 1.44, 1.46, 1.48 c,d, 1.50

ANSWERS TO PRACTICE EXERCISES

1. 161 g Pt

2. −40°C

3. (a) 4.5781×10^4
   (b) 4.30×10^{-5}

4. (a) 1.75×10^3
   (b) 7.55×10^{11}

5. (a) 8
   (b) 0.383

6. 18 mg/m^3

7. 530 mL/day

SOLUTIONS TO SELECTED TEXT PROBLEMS

1.4   **(a)** hypothesis   **(b)** law   **(c)** theory

1.12   **(a)** Physical change. The helium isn't changed in any way by leaking out of the balloon.

      (b) Chemical change in the battery.

      (c) Physical change. The orange juice concentrate can be regenerated by evaporation of the water.

      (d) Chemical change. Photosynthesis changes water, carbon dioxide, etc., into complex organic matter.

      (e) Physical change. The salt can be recovered unchanged by evaporation.

1.14   **(a)** K   **(b)** Sn   **(c)** Cr   **(d)** B   **(e)** Ba
      (f) Pu   **(g)** S   **(h)** Ar   **(i)** Hg

1.16   **(a)** homogeneous mixture   **(b)** element   **(c)** compound
      (d) homogeneous mixture   **(e)** heterogeneous mixture   **(f)** homogeneous mixture
      (g) heterogeneous mixture

1.22   Density Calculation, Problem Type 1.

$$\text{density} = \frac{\text{mass}}{\text{volume}}$$

Step 1: You are asked to solve for the mass of mercury. Therefore, solve the above equation algebraically for the mass.

$$\textbf{mass} = \text{density} \times \text{volume}$$

Step 2: Calculate the mass of mercury by substituting the known values into the equation.

$$\textbf{mass of Hg} = 13.6 \; \frac{\text{g}}{\text{mL}} \times 95.8 \; \text{mL} = \textbf{1.30} \times \textbf{10}^3 \; \textbf{g}$$

1.24   Temperature Conversion, Problem Type 2.

    (a) Conversion from Fahrenheit to Celsius.

$$? \, ^\circ C = (^\circ F - 32 \, ^\circ F) \times \frac{5 \, ^\circ C}{9 \, ^\circ F}$$

$$? \, ^\circ C = (105 \, ^\circ F - 32 \, ^\circ F) \times \frac{5 \, ^\circ C}{9 \, ^\circ F} = \textbf{41} \, ^\circ\textbf{C}$$

    (b) Conversion from Celsius to Fahrenheit.

$$? \, ^\circ F = \left(^\circ C \times \frac{9 \, ^\circ F}{5 \, ^\circ C} \right) + 32 \, ^\circ F$$

$$? \, ^\circ F = \left(-11.5 \, ^\circ C \times \frac{9 \, ^\circ F}{5 \, ^\circ C} \right) + 32 \, ^\circ F = \textbf{11.3} \, ^\circ\textbf{F}$$

(c)   Conversion from Celsius to Fahrenheit.

$$? \,^{\circ}F = \left(^{\circ}C \times \frac{9\,^{\circ}F}{5\,^{\circ}C} \right) + 32\,^{\circ}F$$

$$? \,^{\circ}F = \left(6.3 \times 10^3 \,^{\circ}\!\!\!\!/C \times \frac{9\,^{\circ}F}{5\,^{\circ}\!\!\!\!/C} \right) + 32\,^{\circ}F = \mathbf{1.1 \times 10^4 \,^{\circ}F}$$

1.26   **(a)** $K = (^{\circ}C + 273\,^{\circ}C)\dfrac{1\,K}{1\,^{\circ}C}$

$^{\circ}C = K - 273 = 77\,K - 273 = \mathbf{-196\,^{\circ}C}$

(b) $^{\circ}C = 4.2\,K - 273 = \mathbf{-269\,^{\circ}C}$

(c) $^{\circ}C = 601\,K - 273 = \mathbf{328\,^{\circ}C}$

1.30   **(a)** 10^{-2} indicates that the decimal point must be moved two places to the left.

$$1.52 \times 10^{-2} = \mathbf{0.0152}$$

(b) 10^{-8} indicates that the decimal point must be moved 8 places to the left.

$$7.78 \times 10^{-8} = \mathbf{0.0000000778}$$

1.32   Scientific Notation, Problem Types 3B and 3C

(a)   Addition using scientific notation.

Step 1: Write each quantity with the same exponent, n.

Let's write 0.0095 in such a way that $n = -3$. This means to move the decimal point three places to the right. To *decrease* 10^n by 10^3, you must *increase* N by 10^3.

$$0.0095 = \mathbf{9.5 \times 10^{-3}}$$

Step 2: Add the N parts of the numbers, keeping the exponent, n, the same.

$$\begin{array}{r} 9.5 \times 10^{-3} \\ +\ \ 8.5 \times 10^{-3} \\ \hline \mathbf{18.0 \times 10^{-3}} \end{array}$$

The usual practice is to express N as a number between 1 and 10. Since we must *decrease* N by a factor of 10 to express N between 1 and 10 (1.8), we must *increase* 10^n by a factor of 10. The exponent, n, is increased by 1 from -3 to -2.

$$18.0 \times 10^{-3} = \mathbf{1.8 \times 10^{-2}}$$

(b)   Division using scientific notation.

Step 1: Make sure that all numbers are expressed in scientific notation.

$$653 = \mathbf{6.53 \times 10^2}$$

Step 2: Divide the N parts of the numbers in the usual way.

$$6.53 \div 5.75 = 1.14$$

Step 3: *Subtract* the exponents, n.

$$1.14 \times 10^{+2-(-8)} = 1.14 \times 10^{+2+8} = \mathbf{1.14 \times 10^{10}}$$

(c) Subtraction using scientific notation.

Step 1: Write each quantity with the same exponent, n.

Let's write 850,000 in such a way that $n = 5$. This means to move the decimal point five places to the left.

$$850,000 = \mathbf{8.5 \times 10^5}$$

Step 2: Subtract the N parts of the numbers, keeping the exponent, n, the same.

$$\begin{array}{r} 8.5 \times 10^5 \\ -\ 9.0 \times 10^5 \\ \hline \mathbf{-0.5 \times 10^5} \end{array}$$

The usual practice is to express N as a number between 1 and 10. Since we must *increase* N by a factor of 10 to express N between 1 and 10 (5), we must *decrease* 10^n by a factor of 10. The exponent, n, is decreased by 1 from 5 to 4.

$$-0.5 \times 10^5 = \mathbf{-5 \times 10^4}$$

(d) Multiplication using scientific notation.

Step 1: Make sure that all numbers are expressed in scientific notation.

Step 2: Multiply the N parts of the numbers in the usual way.

$$3.6 \times 3.6 = 13$$

Step 3: *Add* the exponents, n.

$$13 \times 10^{-4+(+6)} = \mathbf{13 \times 10^2}$$

The usual practice is to express N as a number between 1 and 10. Since we must *decrease* N by a factor of 10 to express N between 1 and 10 (1.3), we must *increase* 10^n by a factor of 10. The exponent, n, is increased by 1 from 2 to 3.

$$\mathbf{13 \times 10^2 = 1.3 \times 10^3}$$

1.34 **(a)** one **(b)** three **(c)** three **(d)** four
        (e) two or three **(f)** one **(g)** one or two

1.36 Significant Figures, Problem Types 4B and 4C

(a) Division

The number of significant figures in the answer is determined by the original number having the smallest number of significant figures.

$$\frac{7.310 \text{ km}}{5.70 \text{ km}} = 1.28\mathbf{3}$$

The 3 (bolded) is a nonsignificant digit because the original number 5.70 only has three significant digits. Therefore, the answer has only three significant digits.

The correct answer rounded off to the correct number of significant figures is:

$$1.28 \quad \text{(Why are there no units?)}$$

(b)   Subtraction

The number of significant figures to the right of the decimal point in the answer is determined by the lowest number of digits to the right of the decimal point in any of the original numbers. Writing both numbers with exponents = −3, we have

$$3.26 \times 10^{-3} \text{ mg} - 0.0788 \times 10^{-3} \text{ mg} = \mathbf{3.18 \times 10^{-3} \text{ mg}}$$

Since 3.26×10^{-3} has only two digits to the right of the decimal point, two digits are carried to the right of the decimal point in the final answer.

(c)   Addition

The number of significant figures to the right of the decimal point in the answer is determined by the lowest number of digits to the right of the decimal point in any of the original numbers. Writing both numbers with exponents = +7, we have

$$0.402 \times 10^{7} \text{ dm} + 7.74 \times 10^{7} \text{ dm} = \mathbf{8.14 \times 10^{7} \text{ dm}}$$

Since 7.74×10^{7} has only two digits to the right of the decimal point, two digits are carried to the right of the decimal point in the final answer.

1.38   Factor-Label Method, Problem Type 5.

(a)

Step 1: You can look up the following conversion factors.

> 1 lb = 453.6 grams
> 1 gram = 1000 mg

You should come up with the following strategy.

> lb → grams → mg

Step 2: Carry out the necessary conversions, making sure that units cancel.

$$\mathbf{?\ mg} = 242 \text{ lb (given)} \times \frac{453.6 \text{ g (desired)}}{1 \text{ lb (given)}} \times \frac{1 \text{ mg (desired)}}{0.001 \text{ g (given)}} = \mathbf{1.10 \times 10^{8} \text{ mg}}$$

> **Tip**: In the first conversion factor (g/lb), g is the desired unit. When moving on to the next conversion factor (mg/g), g is now given, and the desired unit is mg.

(b)

Step 1: You can look up or you should know the following conversion factor.

> 1 m = 100 cm

You should come up with the following strategy.

> $cm^3 \rightarrow m^3$

Step 2: Carry out the necessary conversions, making sure that units cancel.

$$\mathbf{?\ m^3} = 68.3 \text{ cm}^3 \text{ (given)} \times \left(\frac{1 \text{ m (desired)}}{100 \text{ cm (given)}} \right)^3 = \mathbf{6.83 \times 10^{-5} \text{ m}^3}$$

> **Tip:** Make sure to cube the unit factor, so that the cm^3 units cancel and the answer has units of m^3.

1.40   Factor-Label Method, Problem Type 5.

Step 1: You should know the following conversion factors.

$1 \text{ day} = 24 \text{ hours}$
$1 \text{ hr} = 60 \text{ minutes}$
$1 \text{ min} = 60 \text{ seconds}$

You should come up with the following strategy.

$$\text{days} \rightarrow \text{hours} \rightarrow \text{minutes} \rightarrow \text{seconds}$$

Step 2: Carry out the necessary conversions, making sure that units cancel.

$$? \text{ s } = 365.24 \text{ day} \times \frac{24 \text{ hr}}{1 \text{ day}} \times \frac{60 \text{ min}}{1 \text{ hr}} \times \frac{60 \text{ s}}{1 \text{ min}} = \mathbf{3.1557 \times 10^7 \text{ s}}$$

1.42   (a) $? \text{ in/s } = \dfrac{1 \text{ mi}}{13 \text{ min}} \times \dfrac{5280 \text{ ft}}{1 \text{ mi}} \times \dfrac{12 \text{ in}}{1 \text{ ft}} \times \dfrac{1 \text{ min}}{60 \text{ s}} = \mathbf{81 \text{ in/s}}$

(b) $? \text{ m/min } = \dfrac{1 \text{ mi}}{13 \text{ min}} \times \dfrac{1609 \text{ m}}{1 \text{ mi}} = \mathbf{1.2 \times 10^2 \text{ m/min}}$

(c) $? \text{ km/hr } = \dfrac{1 \text{ mi}}{13 \text{ min}} \times \dfrac{1609 \text{ m}}{1 \text{ mi}} \times \dfrac{1 \text{ km}}{1000 \text{ m}} \times \dfrac{60 \text{ min}}{1 \text{ hr}} = \mathbf{7.4 \text{ km/hr}}$

1.44 $? \text{ km/hr } = \dfrac{55 \text{ mi}}{1 \text{ hr}} \times \dfrac{1.609 \text{ km}}{1 \text{ mi}} = \mathbf{88 \text{ km/hr}}$

1.46 $0.62 \text{ ppm Pb} = \dfrac{0.62 \text{ g Pb}}{1 \times 10^6 \text{ g blood}}$

$6.0 \times 10^3 \text{ g of blood} \times \dfrac{0.62 \text{ g Pb}}{1 \times 10^6 \text{ g blood}} = \mathbf{3.7 \times 10^{-3} \text{ g Pb}}$

1.48   (a) $?°C = (47.4 - 32.0)°F \left(\dfrac{5°C}{9°F} \right) = \mathbf{8.56°C}$

(b) $?°F = \left(\dfrac{9°F}{5°C} \right) (-273.15°C) + 32.00°F = \mathbf{-459.67°F}$

(c) $? \text{ m}^3 = 71.2 \text{ cm}^3 \left(\dfrac{1 \text{ m}}{10^2 \text{ cm}} \right)^3 = \mathbf{7.12 \times 10^{-5} \text{ m}^3}$

(d) $? \text{ L } = 7.2 \text{ m}^3 \left(\dfrac{10 \text{ dm}}{1 \text{ m}} \right)^3 \left(\dfrac{1 \text{ L}}{1 \text{ dm}^3} \right) = \mathbf{7.2 \times 10^3 \text{ L}}$

1.50 $\text{density} = \dfrac{0.625 \text{ g}}{1 \text{ L}} \times \dfrac{1 \text{ L}}{1000 \text{ mL}} \times \dfrac{1 \text{ mL}}{1 \text{ cm}^3} = \mathbf{6.25 \times 10^{-4} \text{ g/cm}^3}$

1.52   See Section 1.6 of your text for a discussion of these terms.

   (a)   <u>Chemical property</u>. Iron has changed its composition and identity by chemically combining with oxygen and water.

   (b)   <u>Chemical property</u>. The water reacts with chemicals in the air (such as sulfur dioxide) to produce acids, thus changing the composition and identity of the water.

   (c)   <u>Physical property</u>. The color of the hemoglobin can be observed and measured without changing its composition or identity.

   (d)   <u>Physical property</u>. The evaporation of water does not change its chemical properties. Evaporation is a change in matter from the liquid state to the gaseous state.

   (e)   <u>Chemical property</u>. The carbon dioxide is chemically converted into other molecules.

1.54   Volume of rectangular bar = length × width × height

   $\text{density} = \dfrac{m}{V} = \dfrac{52.7064 \text{ g}}{(8.53 \text{ cm})(2.4 \text{ cm})(1.0 \text{ cm})} = \mathbf{2.6 \text{ g/cm}^3}$

1.56   You are asked to solve for the inner diameter of the tube. If you can calculate the volume that the mercury occupies, you can calculate the radius of the cylinder, $V_{\text{cylinder}} = \pi r^2 h$ (r is the inner radius of the cylinder, and h is the height of the cylinder). The cylinder diameter is $2r$.

   $\text{volume of Hg filling cylinder} = \dfrac{\text{mass of Hg}}{\text{density of Hg}}$

   $\mathbf{\text{volume of Hg filling cylinder}} = \dfrac{105.5 \text{ g}}{13.6 \text{ g/cm}^3} = \mathbf{7.76 \text{ cm}^3}$

   Next, solve for the radius of the cylinder.

   $\text{Volume of cylinder} = \pi r^2 h$

   $r = \sqrt{\dfrac{\text{volume}}{\pi \times h}}$

   $r = \sqrt{\dfrac{7.76 \text{ cm}^3}{\pi \times 12.7 \text{ cm}}} = \mathbf{0.441 \text{ cm}}$

   The cylinder diameter equals $2r$.

   $\mathbf{\text{Cylinder diameter}} = 2r = 2(0.441 \text{ cm}) = \mathbf{0.882 \text{ cm}}$

1.58 $\dfrac{343 \text{ m}}{1 \text{ s}} \times \dfrac{1 \text{ mi}}{1609 \text{ m}} \times \dfrac{3600 \text{ s}}{1 \text{ hr}} = \mathbf{767 \text{ mph}}$

1.60   In order to work this problem, you need to understand the physical principles involved in the experiment in Problem 1.59. The volume of the water displaced must equal the volume of the piece of silver. If the silver did not sink, would you have been able to determine the volume of the piece of silver?

The liquid must be *less dense* than the ice in order for the ice to sink. The temperature of the experiment must be maintained at or below *0°C* to prevent the ice from melting.

1.62 $$\text{Volume} = \frac{\text{mass}}{\text{density}}$$

Volume occupied by Li $= \dfrac{1.20 \times 10^3 \text{ g}}{0.53 \text{ g} / \text{cm}^3} = \mathbf{2.3 \times 10^3 \ cm^3}$

1.64   To work this problem, we need to convert from cubic feet to L. Some tables will have a conversion factor of 28.3 L = 1 ft^3, but we can also calculate it using the factor-label method described in Section 1.9.

First, converting from cubic feet to liters:

$$(5.0 \times 10^7 \text{ ft}^3) \times \left(\frac{12 \text{ in}}{1 \text{ ft}}\right)^3 \times \left(\frac{2.54 \text{ cm}}{1 \text{ in}}\right)^3 \times \frac{1 \text{ mL}}{1 \text{ cm}^3} \times \frac{1 \text{ L}}{1000 \text{ mL}} = 1.4 \times 10^9 \text{ L}$$

The mass of vanillin (in g) is:

$$\frac{2.0 \times 10^{-11} \text{ g vanillin}}{1 \text{ L}} \times (1.4 \times 10^9 \text{ L}) = 2.8 \times 10^{-2} \text{ g vanillin}$$

The cost is:

$$(2.8 \times 10^{-2} \text{ g vanillin}) \times \frac{\$ 112}{50 \text{ g vanillin}} = \mathbf{\$0.063} = \mathbf{6.3¢}$$

1.66   There are 78.3 + 117.3 = 195.6 Celsius degrees between 0°S and 100°S. We can write this as a unit factor.

$$\left(\frac{195.6°C}{100°S}\right)$$

Set up the equation like a Celsius to Fahrenheit conversion. We need to subtract 117.3°C, because the zero point on the new scale is 117.3°C lower than the zero point on the Celsius scale.

$$? \ °C = \left(\frac{195.6°C}{100°S}\right)(? \ °S) - 117.3°C$$

Solving for ? °S gives: $$? \ °S = (?°C + 117.3°C)\left(\frac{100°S}{195.6°C}\right)$$

For 25°C we have: $$\mathbf{?} \ °S = (25°C + 117.3°C)\left(\frac{100°S}{195.6°C}\right) = \mathbf{73°S}$$

1.68   **(a)** $\dfrac{6000 \text{ mL of inhaled air}}{1 \text{ min}} \times \dfrac{0.001 \text{ L}}{1 \text{ mL}} \times \dfrac{60 \text{ min}}{1 \text{ hr}} \times \dfrac{24 \text{ hr}}{1 \text{ day}} = \mathbf{8.6 \times 10^3 \ L \ of \ air/day}$

(b) $\dfrac{8.6 \times 10^3 \text{ L of air}}{1 \text{ day}} \times \dfrac{2.1 \times 10^{-6} \text{ L CO}}{1 \text{ L of air}} = \mathbf{0.018 \ L \ CO/day}$

1.70 First, calculate the volume of 1 kg of seawater from the density and the mass. We chose 1 kg of seawater, because the problem gives the amount of Mg in every kg of seawater. The density of seawater is given in Problem 1.69.

$$\text{volume} = \frac{\text{mass}}{\text{density}}$$

$$\text{volume of 1 kg of seawater} = \frac{1000 \text{ g}}{1.03 \text{ g} / \text{mL}} = \textbf{971 mL} = \textbf{0.971 L}$$

In other words, there are 1.3 g of Mg in every 0.971 L of seawater.

Next, let's convert tons of Mg to grams of Mg.

$$(8.0 \times 10^4 \text{ tons Mg}) \times \frac{2000 \text{ lb}}{1 \text{ ton}} \times \frac{453.6 \text{ g}}{1 \text{ lb}} = 7.3 \times 10^{10} \text{ g Mg}$$

Volume of seawater needed to extract 8.0×10^4 ton Mg =

$$(7.3 \times 10^{10} \text{ g Mg}) \times \frac{0.971 \text{ L seawater}}{1.3 \text{ g Mg}} = \textbf{5.5} \times \textbf{10}^{\textbf{10}} \textbf{ L of seawater}$$

1.72 Volume = surface area $\times$ depth

Recall that $1 \text{ L} = 1 \text{ dm}^3$. Let's convert the surface area to units of dm^2 and the depth to units of dm.

$$\text{surface area} = (1.8 \times 10^8 \text{ km}^2) \times \left(\frac{1000 \text{ m}}{1 \text{ km}}\right)^2 \times \left(\frac{1 \text{ dm}}{0.1 \text{ m}}\right)^2 = 1.8 \times 10^{16} \text{ dm}^2$$

$$\text{depth} = (3.9 \times 10^3 \text{ m}) \times \frac{1 \text{ dm}}{0.1 \text{ m}} = 3.9 \times 10^4 \text{ dm}$$

Volume = surface area $\times$ depth = $(1.8 \times 10^{16} \text{ dm}^2)(3.9 \times 10^4 \text{ dm}) = 7.0 \times 10^{20} \text{ dm}^3 = \textbf{7.0} \times \textbf{10}^{\textbf{20}} \textbf{ L}$

1.74 Volume of sphere $= \frac{4}{3}\pi r^3$

$$\text{Volume} = \frac{4}{3}\pi\left(\frac{15 \text{ cm}}{2}\right)^3 = 1.8 \times 10^3 \text{ cm}^3$$

$$\text{mass} = \text{volume} \times \text{density} = (1.8 \times 10^3 \text{ cm}^3) \times \frac{22.57 \text{ g Os}}{1 \text{ cm}^3} \times \frac{1 \text{ kg}}{1000 \text{ g}} = \textbf{41 kg Os}$$

$$41 \text{ kg Os} \times \frac{2.205 \text{ lb}}{1 \text{ kg}} = \textbf{9.0} \times \textbf{10}^{\textbf{1}} \textbf{ lb Os}$$

1.76 62 kg $= 6.2 \times 10^4$ g

O: $(6.2 \times 10^4 \text{ g})(0.65) = \textbf{4.0} \times \textbf{10}^{\textbf{4}} \textbf{ g O}$ N: $(6.2 \times 10^4 \text{ g})(0.03) = \textbf{2} \times \textbf{10}^{\textbf{3}} \textbf{ g N}$

C: $(6.2 \times 10^4 \text{ g})(0.18) = \textbf{1.1} \times \textbf{10}^{\textbf{4}} \textbf{ g C}$ Ca: $(6.2 \times 10^4 \text{ g})(0.016) = \textbf{9.9} \times \textbf{10}^{\textbf{2}} \textbf{ g Ca}$

H: $(6.2 \times 10^4 \text{ g})(0.10) = \textbf{6.2} \times \textbf{10}^{\textbf{3}} \textbf{ g H}$ P: $(6.2 \times 10^4 \text{ g})(0.012) = \textbf{7.4} \times \textbf{10}^{\textbf{2}} \textbf{ g P}$

1.78 $? \,°C = (7.3 \times 10^2 - 273) \, K = \mathbf{4.6 \times 10^2 \,°C}$

$? \,°F = \left((4.6 \times 10^2 \,°\cancel{C}) \times \dfrac{9°F}{5°\cancel{C}} \right) + 32°F = \mathbf{8.6 \times 10^2 \,°F}$

1.80 $(8.0 \times 10^4 \text{ tons Au}) \times \dfrac{2000 \text{ lb Au}}{1 \text{ ton Au}} \times \dfrac{16 \text{ oz Au}}{1 \text{ lb Au}} \times \dfrac{\$350}{1 \text{ oz Au}} = \mathbf{\$9.0 \times 10^{11}} \text{ or } \mathbf{900 \text{ billion dollars}}$

1.82 $? \text{ Fe atoms} = 4.9 \text{ g Fe} \times \dfrac{1.1 \times 10^{22} \text{ Fe atoms}}{1.0 \text{ g Fe}} = \mathbf{5.4 \times 10^{22} \text{ Fe atoms}}$

1.84   10 cm = 0.1 m. We need to find the number of times the 0.1 m wire must be cut in half until the piece left is 1.3×10^{-12} m long. Let n be the number of times we can cut the Cu wire in half. We can write:

$$\left(\frac{1}{2} \right)^n \times 0.1 \text{ m} = 1.3 \times 10^{-12} \text{ m}$$

$$\left(\frac{1}{2} \right)^n = 1.3 \times 10^{-11} \text{ m}$$

Taking the log of both sides of the equation:

$$n \log \left(\frac{1}{2} \right) = \log(1.3 \times 10^{-11})$$

$$\mathbf{n = 36 \text{ times}}$$

1.86   Volume = area × thickness.

From the density, we can calculate the volume of the Al foil.

$$\text{Volume} = \frac{\text{mass}}{\text{density}} = \frac{3.636 \text{ g}}{2.699 \text{ g} / \text{cm}^3} = 1.347 \text{ cm}^3$$

Convert the unit of area from ft^2 to cm^2.

$$1.000 \text{ ft}^2 \times \left(\frac{12 \text{ in}}{1 \text{ ft}} \right)^2 \times \left(\frac{2.54 \text{ cm}}{1 \text{ in}} \right)^2 = 929.0 \text{ cm}^2$$

$$\mathbf{thickness} = \frac{\text{volume}}{\text{area}} = \frac{1.347 \text{ cm}^3}{929.0 \text{ cm}^2} = 1.450 \times 10^{-3} \text{ cm} = \mathbf{1.450 \times 10^{-2} \text{ mm}}$$

1.88   First, let's calculate the mass (in g) of water in the pool. We perform this conversion because we know there is 1 g of chlorine needed per million grams of water.

$$2.0 \times 10^4 \text{ gallons H}_2\text{O} \times \frac{3.79 \text{ L}}{1 \text{ gallon}} \times \frac{1 \text{ mL}}{0.001 \text{ L}} \times \frac{1 \text{ g}}{1 \text{ mL}} = 7.6 \times 10^7 \text{ g H}_2\text{O}$$

Next, let's calculate the mass of chlorine that needs to be added to the pool.

$$7.6 \times 10^7 \text{ g H}_2\text{O} \times \frac{1 \text{ g chlorine}}{1 \times 10^6 \text{ g H}_2\text{O}} = 76 \text{ g chlorine}$$

The chlorine solution is only 6 percent chlorine by mass. We can now calculate the volume of chlorine solution that must be added to the pool.

$$76 \text{ g chlorine} \times \frac{100\% \text{ soln}}{6\% \text{ chlorine}} \times \frac{1 \text{ mL soln}}{1 \text{ g soln}} = \textbf{1.3} \times \textbf{10}^3 \textbf{ mL of chlorine solution}$$

1.90   We assume that the thickness of the oil layer is equivalent to the length of one oil molecule. We can calculate the thickness of the oil layer from the volume and surface area.

$$40 \text{ m}^2 \times \left(\frac{1 \text{ cm}}{0.01 \text{ m}}\right)^2 = 4.0 \times 10^5 \text{ cm}^2$$

$$0.10 \text{ mL} = 0.10 \text{ cm}^3$$

$$\text{Volume} = \text{surface area} \times \text{thickness}$$

$$\text{thickness} = \frac{\text{volume}}{\text{surface area}} = \frac{0.10 \text{ cm}^3}{4.0 \times 10^5 \text{ cm}^2} = 2.5 \times 10^{-7} \text{ cm}$$

Converting to nm:

$$2.5 \times 10^{-7} \text{ cm} \times \frac{0.01 \text{ m}}{1 \text{ cm}} \times \frac{1 \text{ nm}}{1 \times 10^{-9} \text{ m}} = \textbf{2.5 nm}$$

1.92   **(a)** $\dfrac{\$1.30}{15.0 \text{ ft}^3} \times \left(\dfrac{1 \text{ ft}}{12 \text{ in}}\right)^3 \times \left(\dfrac{1 \text{ in}}{2.54 \text{ cm}}\right)^3 \times \dfrac{1 \text{ cm}^3}{1 \text{ mL}} \times \dfrac{1 \text{ mL}}{0.001 \text{ L}} = \textbf{\$3.06} \times \textbf{10}^{-3}\textbf{/L}$

(b) $2.1 \text{ L water} \times \dfrac{0.304 \text{ ft}^3 \text{ gas}}{1 \text{ L water}} \times \dfrac{\$1.30}{15.0 \text{ ft}^3} = \textbf{\$0.055} = \textbf{5.5¢}$

1.94   **(a)**   Data was collected to indicate a high iridium content in clay deposited above sediments formed during the Cretaceous period. A hypothesis was formulated that the iridium came from a large asteroid. When the asteroid impacted Earth, large amounts of dust and debris blocked the sunlight. Plants eventually died, then many plant eating animals gradually perished, then, in turn, meat eating animals starved.

(b)   If the hypothesis is correct, we should find a similarly high iridium content in corresponding rock layers at different locations on Earth. Also, we should expect the simultaneous extinction of other large species in addition to dinosaurs.

(c)   Yes. Hypotheses that survive many experimental tests of their validity may evolve into theories. A *theory* is a unifying principle that explains a body of facts and/or those laws that are based on them.

(d)   Twenty percent of the asteroid's mass turned into dust. First, let's calculate the mass of the dust.

$$? \text{ g dust} = \frac{0.02 \text{ g}}{\text{cm}^2} \times \left(\frac{1 \text{ cm}}{0.01 \text{ m}}\right)^2 \times (5.1 \times 10^{14} \text{ m}^2) = 1.02 \times 10^{17} \text{ g dust}$$

This mass is 20% of the asteroid's mass.

$$\textbf{mass of asteroid} = \frac{1.0 \times 10^{17} \text{ g}}{0.20} = 5.0 \times 10^{17} \text{ g} = \mathbf{5.0 \times 10^{14} \text{ kg}}$$

Converting to tons:

$$5.0 \times 10^{14} \text{ kg} \times \frac{1000 \text{ g}}{1 \text{ kg}} \times \frac{1 \text{ lb}}{453.6 \text{ g}} \times \frac{1 \text{ ton}}{2000 \text{ lb}} = \mathbf{5.5 \times 10^{11} \text{ tons}}$$

From the mass and density of the asteroid, we can calculate its volume. Then, from its volume, we can calculate the radius.

$$\text{Volume} = \frac{\text{mass}}{\text{density}} = \frac{5.0 \times 10^{17} \text{ g}}{2 \text{ g}/\text{cm}^3} = 2.5 \times 10^{17} \text{ cm}^3$$

$$\text{Volume}_{\text{sphere}} = \frac{4}{3}\pi r^3$$

$$2.5 \times 10^{17} \text{ cm}^3 = \frac{4}{3}\pi r^3$$

$$r = 4 \times 10^5 \text{ cm} = \mathbf{4 \times 10^3 \text{ m}}$$

CHAPTER 2
ATOMS, MOLECULES, AND IONS

PROBLEM-SOLVING STRATEGIES AND TUTORIAL SOLUTIONS

TYPES OF PROBLEMS

Problem Type 1: Atomic number, Mass number, and Isotopes.

Problem Type 2: Empirical and Molecular Formulas.

Problem Type 3: Naming Compounds.
      (a) Ionic compounds.
      (b) Molecular compounds.
      (c) Acids.
      (d) Bases.

Problem Type 4: Formulas of Ionic Compounds.

PROBLEM TYPE 1: ATOMIC NUMBER, MASS NUMBER, AND ISOTOPES

The **atomic number** (Z) is the number of protons in the nucleus of each atom of an element.

EXAMPLE 2.1
What is the atomic number of an oxygen atom?

The atomic number is listed above each element in the periodic table. For oxygen, the atomic number is 8, meaning that an oxygen atom has eight protons in the nucleus.

The **mass number** (A) is the total number of neutrons and protons present in the nucleus of an atom of an element.

$$\text{mass number} = \text{number of protons} + \text{number of neutrons}$$
$$\text{mass number} = \text{atomic number} + \text{number of neutrons}$$

EXAMPLE 2.2
A particular oxygen atom has nine neutrons in the nucleus. What is the mass number of this atom?

Looking at a periodic table, you should find that every oxygen atom has an atomic number of 8. Thus,

$$\text{mass number} = \text{atomic number} + \text{number of neutrons}$$

$$\textbf{mass number} = \textbf{8 + 9 = 17}$$

Isotopes are atoms that have the same atomic number, but different mass numbers. For example, there are three isotopes of oxygen found in nature, oxygen-16, oxygen-17, and oxygen-18. The accepted way to denote the atomic number and mass number of an element X is as follows:

$$_{Z}^{A}\text{X}$$

where, A = mass number
                  Z = atomic number

EXAMPLE 2.3
The three isotopes of oxygen found in nature are oxygen-16, -17, and -18. Write their isotopic symbols.

The atomic number of oxygen is 8, so all isotopes of oxygen contain eight protons. The mass numbers are 16, 17, and 18, respectively.

$$^{16}_{8}O \qquad ^{17}_{8}O \qquad ^{18}_{8}O$$

The number of **electrons** in an *atom* is equal to the number of protons.

> **number of electrons (atom)** = number of protons = atomic number

The number of **electrons** in an *ion* is equal to the number of protons minus the charge on the ion.

> **number of electrons (ion)** = number of protons − charge on the ion

EXAMPLE 2.4
What is the total number of fundamental particles (protons, neutrons, and electrons) in
(a) an atom of $^{56}_{26}Fe$ and (b) an $^{56}_{26}Fe^{3+}$ ion?

Both $^{56}_{26}Fe$ and $^{56}_{26}Fe^{3+}$ have the same atomic number and mass number. Thus, for both (a) and (b),

> **number of protons** = atomic number = **26**

and

> **number of neutrons** = mass number − atomic number = 56 − 26 = **30**

However, the number of electrons for the above species differ.

(a) $^{56}_{26}Fe$ is a neutral atom. Therefore,

> **number of electrons** = number of protons = **26**

(b) $^{56}_{26}Fe^{3+}$ is an ion with a +3 charge. Therefore,

> **number of electrons** = number of protons − charge = 26 − (+3) = **23**

PRACTICE EXERCISE

1.   How many protons, neutrons, and electrons are contained in each of the following atoms or ions?

(a) ^{19}F (b) $^{79}Se^{2-}$ (c) ^{40}Ca (d) $^{48}Ti^{4+}$

Text Problems: **2.14**, 2.16, 2.18, 2.34

PROBLEM TYPE 2: EMPIRICAL AND MOLECULAR FORMULAS

A *molecular* formula shows the exact number of atoms of each element in the smallest unit of a substance. An *empirical formula* tells us which elements are present and the simplest whole-number ratio of their atoms. Empirical formulas are therefore the simplest chemical formulas; they are always written so that the subscripts in the molecular formulas are converted to the smallest possible whole numbers.

EXAMPLE 2.5

What is the empirical formula of each of the following compounds: (a) H_2O_2, (b) $C_6H_8O_6$, (c) $MgCl_2$, and (d) C_8H_{18}?

(a) The simplest whole number ratio of the atoms in H_2O_2 is **HO**.

(b) The simplest whole number ratio of the atoms in $C_6H_8O_6$ is **$C_3H_4O_3$**.

(c) The molecular formula as written contains the simplest whole number ratio of the atoms present. In this case, the molecular formula and the empirical formula are the same.

(d) The simplest whole number ratio of the atoms in C_8H_{18} is **C_4H_9**.

PRACTICE EXERCISE

2.   What is the empirical formula of each of the following compounds?

    (a) C_2H_6O **(b)** $C_6H_{12}O_6$ **(c)** CH_3COOH **(d)** $C_{12}H_{22}O_{11}$

Text Problems: 2.42, 2.44

PROBLEM TYPE 3: NAMING COMPOUNDS

A.   Naming ionic compounds

(1)   **Metal cation has only one charge**. You should memorize the metal cations that have only one charge when they form ionic compounds. These include the alkali metals (Group 1A), which always have a +1 charge in ionic compounds, the alkaline earth metals (Group 2A), which always have a +2 charge in ionic compounds, and Al^{3+}, Ag^+, Cd^{2+}, and Zn^{2+}.

Since the metal cation has only one possible charge, we do not need to specify this charge in the compound. Therefore, the name of this type of ionic compound can be written simply by first naming the metal cation as it appears on the periodic table, followed by the nonmetallic anion. Anions from elements are named by changing the suffix to "-ide".

EXAMPLE 2.6

Name the following ionic compounds: (a) AlF_3, (b) Na_3N, (c) $Ba(NO_3)_2$.

(a)   aluminum fluor*ide*
(b)   sodium nitr*ide*
(c)   barium nitrate

> **Tip:** There are a number of ions that contain more than one atom. These ions are called **polyatomic ions**. Nitrate, NO_3^-, is an example. Ask your instructor which polyatomic ions you should know.

(2)   **Metals that form more than one type of cation**. Transition metals typically can form more than one type of cation. For compounds containing these metals, you must specify the charge of the cation. Roman numerals are used for this purpose. The Roman numeral I is used for one positive charge, II for two positive charges, III for three positive charges, and so on.

EXAMPLE 2.7

Name the following compounds: (a) FeO, (b) Fe_2O_3, (c) $HgSO_4$.

(a)   In this compound, the iron cation has a +2 charge, since oxide has a −2 charge (see Table 2.2 of your text). Therefore, the compound is named **iron(II) oxide**.

(b)   In this compound, the iron cation has a +3 charge. Therefore, the compound is named **iron(III) oxide**.

(c)   This compound contains the polyatomic ion sulfate, which has a −2 charge (SO_4^{2-}). Thus, the charge on mercury (Hg) is +2. The compound is named **mercury(II) sulfate**.

B. Naming molecular compounds

Unlike ionic compounds, molecular compounds contain discrete molecular units. They are usually composed of nonmetallic elements. There are two types of molecular compounds to consider.

(1)   Only one compound of the two elements exists. If this is the case, you simply name the first element in the formula as it appears on the periodic table, followed by naming the second element with an "-ide" suffix.

EXAMPLE 2.8
Name the following compounds: (a) HF, (b) SiC.

(a)   hydrogen fluoride
(b)   silicon carbide

(2)   More than one compound composed of the two elements exists. It is quite common for one pair of elements to form several different compounds. Therefore, we must be able to differentiate between the compounds. Greek prefixes are used to denote the number of atoms of each element present. See Table 2.4 in the text for the Greek prefixes used.

EXAMPLE 2.9
Name the following compounds: (a) CO, (b) CO2, (c) N2O5, (d) SF6.

(a)   carbon *mono*xide
(b)   carbon *di*oxide
(c)   *di*nitrogen *pent*oxide
(d)   sulfur *hexa*fluoride

> **Tip:** The prefix "mono-" may be omitted when naming the first element in a molecular compound (see carbon monoxide and carbon dioxide above). The absence of a prefix for the first element usually means that only one atom of that element is present in the molecule.

C. Naming acids

An acid is a substance that yields hydrogen ions (H^+) when dissolved in water. There are two types of acids to consider.

(1)   Acids that do not contain oxygen. This type of acid contains one or more hydrogen atoms as well as an anionic group. To name these acids, you add the prefix "hydro-" to the anion name, change the "-ide" suffix of the anion to "-ic", and then add the word "acid" at the end.

EXAMPLE 2.10
Name the following binary acids: (a) HF (*aq*), (b) HCN (*aq*).

(a)   *hydro*fluor*ic* acid
(b) CN^- is a polyatomic ion called cyanide. In the acid, the "-ide" suffix is changed to "-ic". The correct name is **hydrocyanic acid**.

> **Tip:** The (*aq*) above means that the substance is dissolved in water. HF dissolved in water is an acid and is named hydrofluoric acid. However, HF in its pure state is a molecular compound and is named hydrogen fluoride.

(2) **Oxoacids.** This type of acid contains hydrogen, oxygen, and another element. To name oxoacids, you must look carefully at the anion name. If the suffix of the anion is "-ate", change the suffix to "-ic" and add the word "acid" at the end. If the suffix of the anion is "-ite", change the suffix to "-ous" and add the word "acid" at the end.

EXAMPLE 2.11

Name the following oxoacids: (a) HNO_2 (*aq*), (b) $HClO_3$ (*aq*).

(a) The NO_2^- polyatomic ion is called nitr*ite*. Simply change the suffix to "-ous" and add the word "acid". The correct name is **nitr*ous* acid**.

(b) The ClO_3^- polyatomic ion is called chlor*ate*. Simply change the suffix to "-ic" and add the work "acid". The correct name is **chlor*ic* acid**.

> **Tip:** If one O atom is added to the "-ic" acid, the acid is called "per...ic" acid. For example, $HClO_4$ is named perchloric acid. Compare this acid to chloric acid above. If one O atom is removed from the "-ous" acid, the acid is called "hypo...ous" acid. For example, HClO is named hypochlorous acid. Compare this to chlorous acid, $HClO_2$.

D. Naming Bases

A base is a substance that produces the hydroxide ion (OH^-) when dissolved in water. At this point, for naming purposes, we will only consider bases that contain the hydroxide ion. To name this type of base, simply name the metal cation first as it appears on the periodic table, then add "hydroxide".

EXAMPLE 2.12

Name the following bases: (a) KOH, (b) $Sr(OH)_2$.

(a) potassium hydroxide
(b) strontium hydroxide

PRACTICE EXERCISE

3. Name the following compounds:

    (a) $MgCl_2$ (b) $CuCl_2$ (c) HNO_3 (d) P_2O_5 (e) $Ca(OH)_2$

Text Problems: **2.54**, 2.56

PROBLEM TYPE 4: FORMULAS OF IONIC COMPOUNDS

The formulas of ionic compounds are usually the same as their empirical formulas because ionic compounds do not consist of discrete molecular units. See Section 2.6 of your text if you need further information.

Ionic compounds are electrically neutral. In order for ionic compounds to be electrically neutral, the sum of the charges on the cation and anion in each formula unit must add up to zero. There are two possibilities to consider.

(1) If the charges on the cation and anion are numerically equal, no subscripts are necessary in the formula.
(2) If the charges on the cation and anion are numerically different, the subscript of the cation is numerically equal to the charge on the anion, and the subscript of the anion is numerically equal to the charge on the cation.

You should memorize the metal cations that have only one charge when they form ionic compounds. These include the alkali metals (Group 1A), which always have a +1 charge in ionic compounds, the alkaline earth metals (Group 2A), which always have a +2 charge in ionic compounds, and Al^{3+}, Ag^+, Cd^{2+}, and Zn^{2+}. All other metals can have more than one possible positive charge when forming ionic compounds. This positive charge will be specified using a Roman numeral in the name of the compound.

You should also memorize the charges of polyatomic ions (see Table 2.3 of your text) and the common charges of monatomic anions based on their positions in the periodic table (see Table 2.2 of your text).

EXAMPLE 2.13
Write the formula for the ionic compound, magnesium oxide.

The magnesium cation is an alkaline earth metal cation which always has a +2 charge in an ionic compound, and the oxide anion is –2 in an ionic compound (see Table 2.2 of your text). Mg^{2+} and the oxide anion, O^{2-}, combine to form the ionic compound magnesium oxide. The sum of the charges is +2 + (–2) = 0, so no subscripts are necessary. The formula is **MgO**.

EXAMPLE 2.14
Write the formula for the ionic compound, iron(II) chloride.

An iron cation can either have a +2 or +3 charge in an ionic compound. The Roman numeral, II, specifies that in this compound it is the +2 cation, Fe^{2+}. The chloride anion has a –1 charge in an ionic compound (see Table 2.2 of your text). Fe^{2+} and the chloride anion, Cl^-, combine to form the ionic compound iron(II) chloride. The charges on the cation and anion are numerically different, so make the subscript of the cation (Fe^{2+}) numerically equal to the charge of the anion (subscript = 1). Also, make the subscript of the anion (Cl^-) numerically equal to the charge of the cation (subscript = 2). The formula is **FeCl$_2$**.

> **Tip:** Check to make sure that the compound is electrically neutral by multiplying the charge of each ion by its subscript and then adding them together. The sum should equal zero.

$$(+2)(1) + (-1)(2) = 0$$

PRACTICE EXERCISE

4.   Write the correct formulas for the following ionic compounds:

   (a) Sodium oxide   **(b)** Copper(II) nitrate   **(c)** Aluminum oxide

Text Problems: 2.56 a, b, f, g, i, j

ANSWERS TO PRACTICE EXERCISES

1.   **(a)** 9p, 10n, 9e
     (b) 34p, 45n, 36e
     (c) 20p, 20n, 20e
     (d) 22p, 26n, 18e

2.   **(a)** C_2H_6O
     (b) CH_2O
     (c) CH_2O
     (d) $C_{12}H_{22}O_{11}$

3.   **(a)** magnesium chloride
     (b) copper(II) chloride
     (c) nitric acid
     (d) diphosphorus pentoxide
     (e) calcium hydroxide

4.   **(a)** Na_2O
     (b) $Cu(NO_3)_2$
     (c) Al_2O_3

SOLUTIONS TO SELECTED TEXT PROBLEMS

2.8 Note that you are given the information to set up the unit factor relating meters and miles.

$$r_{atom} = 10^4 r_{nucleus} = 10^4 \times 2.0 \text{ cm} \times \frac{1 \text{ m}}{100 \text{ cm}} \times \frac{1 \text{ mi}}{1609 \text{ m}} = \textbf{0.12 mi}$$

2.14 Problem Type 1, Atomic number, Mass number, and Isotopes.

The 239 in Pu-239 is the mass number. The **mass number (A)** is the total number of neutrons and protons present in the nucleus of an atom of an element. You can look up the atomic number (number of protons) on the periodic table.

mass number = number of protons + number of neutrons

number of neutrons = mass number − number of protons = 239 − 94 = **145**

2.16

Isotope	$^{15}_{7}N$	$^{33}_{16}S$	$^{63}_{29}Cu$	$^{84}_{38}Sr$	$^{130}_{56}Ba$	$^{186}_{74}W$	$^{202}_{80}Hg$
No. Protons	7	16	29	38	56	74	80
No. Neutrons	8	17	34	46	74	112	122
No. Electrons	7	16	29	38	56	74	80

2.18 The accepted way to denote the atomic number and mass number of an element X is as follows:

$$^{A}_{Z}X$$

where,
  A = mass number
  Z = atomic number

(a) $^{186}_{74}W$ (b) $^{201}_{80}Hg$

2.24 (a) Metallic character increases as you progress down a group of the periodic table. For example, moving down Group 4A, the nonmetal carbon is at the top and the metal lead is at the bottom of the group.

(b) Metallic character decreases from the left side of the table (where the metals are located) to the right side of the table (where the nonmetals are located).

2.26 F and Cl are Group 7A elements; they should have similar chemical properties. Na and K are both Group 1A elements; they should have similar chemical and physical properties. P and N are both Group 5A elements; they should have similar chemical properties.

2.32 There are more than two correct answers for each part of the problem.

(a) H_2 and F_2 (b) HCl and CO   (c) S_8 and P_4
(d) H_2O and $C_{12}H_{22}O_{11}$ (sucrose)

2.34 The **atomic number (Z)** is the number of protons in the nucleus of each atom of an element. You can find this on a periodic table. The number of **electrons** in an *ion* is equal to the number of protons minus the charge on the ion.

number of electrons (ion) = number of protons − charge on the ion

Ion	K^+	Mg^{2+}	Fe^{3+}	Br^-	Mn^{2+}	C^{4-}	Cu^{2+}
No. protons	19	12	26	35	25	6	29
No. electrons	18	10	23	36	23	10	27

2.42   Problem Type 2, Empirical and Molecular Formulas.

An *empirical formula* tells us which elements are present and the simplest whole-number ratio of their atoms.

(a)   The simplest whole number ratio of the atoms in Al_2Br_6 is **$AlBr_3$**.

(b)   The simplest whole number ratio of the atoms in $Na_2S_2O_4$ is **$NaSO_2$**.

(c)   The molecular formula as written, **N_2O_5**, contains the simplest whole number ratio of the atoms present. In this case, the molecular formula and the empirical formula are the same.

(d)   The molecular formula as written, **$K_2Cr_2O_7$**, contains the simplest whole number ratio of the atoms present. In this case, the molecular formula and the empirical formula are the same.

2.44   The molecular formula of ethanol is **C_2H_6O**.

2.46   Compounds of metals with nonmetals are usually ionic. Nonmetal-nonmetal compounds are usually molecular.

Ionic: $NaBr$, BaF_2, $CsCl$.

Molecular: CH_4, CCl_4, ICl, NF_3

2.54   Problem Type 3, Naming Compounds.

(a)   This is an ionic compound in which the metal cation (K^+) has only one charge (see Problem Type 3A). The correct name is **potassium hypochlorite**. Hypochlorite is a polyatomic ion with one less O atom than the chlorite ion, ClO_2^-

(b)   **silver carbonate**

(c)   This is an oxoacid that contains the nitrite ion, NO_2^-. The "-ite" suffix is changed to "-ous". The correct name is **nitrous acid**.

(d)   **potassium permanganate**   (e)   **cesium chlorate**   (f)   **potassium ammonium sulfate**

(g)   This is an ionic compound in which the metal can form more than one cation (see Problem Type 3A). Use a Roman numeral to specify the charge of the Fe ion. Since the oxide ion has a −2 charge, the Fe ion has a +2 charge. The correct name is **iron(II) oxide**.

(h)   **iron(III) oxide**

(i)   This is an ionic compound in which the metal can form more than one cation (see Problem Type 3A). Use a Roman numeral to specify the charge of the Ti ion. Since each of the four chloride ions has a −1 charge (total of −4), the Ti ion has a +4 charge. The correct name is **titanium(IV) chloride**.

(j)   **sodium hydride**   (k)   **lithium nitride**   (l)   **sodium oxide**

(m)   This is an ionic compound in which the metal cation (Na^+) has only one charge (see Problem Type 3A). The O_2^{2-} ion is called the peroxide ion. Each oxygen has a −1 charge. You can determine that each oxygen only has a −1 charge, because each of the two Na ions has a +1 charge. Compare this to sodium oxide in part (l). The correct name is **sodium peroxide**.

2.56   Problem Types 3 and 4.

    (a)   The Roman numeral I tells you that the Cu cation has a +1 charge. Cyanide has a −1 charge. Since, the charges are numerically equal, no subscripts are necessary in the formula. The correct formula is **CuCN**.

    (b)   Strontium is an alkaline earth metal. It only forms a +2 cation. The polyatomic ion chlorite, ClO_2^-, has a −1 charge. Since the charges on the cation and anion are numerically different, the subscript of the cation is numerically equal to the charge on the anion, and the subscript of the anion is numerically equal to the charge on the cation. The correct formula is $\mathbf{Sr(ClO_2)_2}$.

    (c)   Perbromic tells you that the anion of this oxoacid is perbromate, BrO_4^-. The correct formula is $\mathbf{HBrO_4}$ ***(aq)***. Remember that *(aq)* means that the substance is dissolved in water.

    (d)   Hydroiodic tells you that the anion of this binary acid is iodide, I^-. The correct formula is **HI** ***(aq)***.

    (e)   Na is an alkali metal. It only forms a +1 cation. The polyatomic ion ammonium, NH_4^+, has a +1 charge and the polyatomic ion phosphate, PO_4^{3-}, has a −3 charge. To balance the charge, you need 2 Na^+ cations. The correct formula is $\mathbf{Na_2(NH_4)PO_4}$.

    (f)   The Roman numeral II tells you that the Pb cation has a +2 charge. The polyatomic ion carbonate, CO_3^{2-}, has a −2 charge. Since, the charges are numerically equal, no subscripts are necessary in the formula. The correct formula is $\mathbf{PbCO_3}$.

    (g)   The Roman numeral II tells you that the Sn cation has a +2 charge. Fluoride has a −1 charge. Since the charges on the cation and anion are numerically different, the subscript of the cation is numerically equal to the charge on the anion, and the subscript of the anion is numerically equal to the charge on the cation. The correct formula is $\mathbf{SnF_2}$.

    (h)   This is a molecular compound. The Greek prefixes tell you the number of each type of atom in the molecule. The correct formula is $\mathbf{P_4S_{10}}$.

    (i)   The Roman numeral II tells you that the Hg cation has a +2 charge. Oxide has a −2 charge. Since, the charges are numerically equal, no subscripts are necessary in the formula. The correct formula is **HgO**.

    (j)   The Roman numeral I tells you that the Hg cation has a +1 charge. However, this cation exists as Hg_2^{2+}. Iodide has a −1 charge. You need two iodide ion to balance the +2 charge of Hg_2^{2+}. The correct formula is $\mathbf{Hg_2I_2}$.

    (k)   This is a molecular compound. The Greek prefixes tell you the number of each type of atom in the molecule. The correct formula is $\mathbf{SeF_6}$.

2.58   Changing the electrical charge of an atom usually has a major effect on its chemical properties. The two electrically neutral carbon isotopes should have nearly identical chemical properties.

2.60   Atomic number = 127 − 74 = 53. This anion has 53 protons, so it is an iodide ion. Since there is one more electron than protons, the ion has a −1 charge. The correct symbol is $\mathbf{I^-}$.

2.62   NaCl is an ionic compound; it doesn't form molecules.

2.64   The species and their identification are as follows:

(a)	SO_2	molecule and compound		**(g)**	O_3	element and molecule
(b)	S_8	element and molecule		**(h)**	CH_4	molecule and compound
(c)	Cs	element		**(i)**	KBr	compound
(d)	N_2O_5	molecule and compound		**(j)**	S	element
(e)	O	element		**(k)**	P_4	element and molecule
(f)	O_2	element and molecule		**(l)**	LiF	compound

2.66   **(a)** CO_2 (*s*), solid carbon dioxide   **(b)** NaCl, sodium chloride

   (c) N_2O, nitrous oxide   **(d)** $CaCO_3$, calcium carbonate

   (e) CaO, calcium oxide   **(f)** $Ca(OH)_2$, calcium hydroxide

   (g) $NaHCO_3$, sodium bicarbonate   **(h)** $Na_2CO_3 \cdot 10H_2O$, sodium carbonate decahydrate

   (i) $CaSO_4 \cdot 2H_2O$, calcium sulfate dihydrate   **(j)** $Mg(OH)_2$, magnesium hydroxide

2.68   **(a)** Ionic compounds are typically formed between metallic and nonmetallic elements.

   (b) In general the transition metals, the actinides and lanthanides have variable charges.

2.70   The symbol ^{23}Na provides more information than $_{11}$Na. The mass number plus the chemical symbol identifies a specific isotope of Na (sodium) while combining the atomic number with the chemical symbol tells you nothing new. Can other isotopes of sodium have different atomic numbers?

2.72   Mercury (Hg) and bromine (Br_2)

2.74 H_2, N_2, O_2, F_2, Cl_2, He, Ne, Ar, Kr, Xe, Rn

2.76   They do not have a strong tendency to form compounds. Helium, neon, and argon are chemically inert.

2.78   All isotopes of radium are radioactive. It is a radioactive decay product of uranium-238. Radium itself does *not* occur naturally on Earth.

2.80   Argentina is named after silver (argentum, Ag).

2.82   **(a)** NaH, sodium hydride   **(b)** B_2O_3, diboron trioxide   **(c)** Na_2S, sodium sulfide

   (d) AlF_3, aluminum fluoride   **(e)** OF_2, oxygen difluoride   **(f)** $SrCl_2$, strontium chloride

CHAPTER 3
MASS RELATIONSHIPS IN CHEMICAL REACTIONS

PROBLEM-SOLVING STRATEGIES

TYPES OF PROBLEMS

Problem Type 1:   Calculating Average Atomic Mass.

Problem Type 2:   Calculations Involving Molar Mass of an Element and Avogadro's Number.
     (a) Converting between moles of atoms and mass of atoms.
     (b) Calculating the mass of a single atom.
     (c) Converting mass in grams to number of atoms.

Problem Type 3:   Calculations Involving Molecular Mass.
     (a) Calculating molecular mass.
     (b) Calculating the number of moles in a given amount of a compound.
     (c) Calculating the number of atoms in a given amount of a compound.

Problem Type 4:   Calculations Involving Percent Composition
     (a) Calculating percent composition of a compound.
     (b) Determining empirical formula from percent composition.
     (c) Calculating mass from percent composition.

Problem Type 5:   Experimental Determination of Empirical Formulas.

Problem Type 6:   Determining the Molecular Formula of a Compound.

Problem Type 7:   Calculating the Amounts of Reactants and Products.

Problem Type 8:   Limiting Reagent Calculations.

Problem Type 9:   Calculating the Percent Yield of a Reaction.

PROBLEM TYPE 1: CALCULATING AVERAGE ATOMIC MASS

The atomic mass you look up on a periodic table is an average atomic mass. The reason for this is that most naturally occurring elements have more than one isotope. The average atomic mass can be calculated as follows:

Step 1: Convert the percentage of each isotope to fractions. For example, an isotope that is 69.09 percent abundant becomes 69.09/100 = 0.6909.

Step 2: Multiply the mass of each isotope by its abundance and add them together.

> **average atomic mass** = (fraction of isotope A)(mass of isotope A) + (fraction of isotope B)
> (mass of isotope B) + . . . + (fraction of isotope Z)(mass of isotope Z).

EXAMPLE 3.1

The element lithium has two isotopes that occur in nature: ^6_3Li with 7.5 percent abundance and ^7_3Li with 92.5 percent abundance. The atomic mass of ^6_3Li is 6.01513 amu and that of ^7_3Li is 7.01601 amu. Calculate the average atomic mass of lithium.

Step 1: Convert the percentage of each isotope to fractions.

$$^6_3Li : 7.5/100 = 0.075$$

$$^7_3Li : 92.5/100 = 0.925$$

Step 2: Multiply the mass of each isotope by its abundance and add them together.

average atomic mass = (0.075)(6.01513) + (0.925)(7.01601) = **6.94 amu**

PRACTICE EXERCISE

1. The element boron (B) consists of two stable isotopes with atomic masses of 10.0129 amu and 11.0093 amu. The average atomic mass of B is 10.81 amu. Which isotope is more abundant?

Text Problem: 3.6

PROBLEM TYPE 2: CALCULATIONS INVOLVING MOLAR MASS OF AN ELEMENT AND AVOGADRO'S NUMBER

A. Converting between moles of atoms and mass of atoms

In order to convert from one unit to another, you need to be proficient at the factor-label method. See Section 1.9 of your text and Problem Type 5, chapter 1. Unit conversions can seem daunting, but if you keep track of the units, making sure that the appropriate units cancel, your effort will be rewarded.

Step 1: Map out a strategy to proceed from initial units to final units based on available conversion factors.

Step 2: Use the following method to ensure that you obtain the desired unit.

$$Given\ unit \times \left(\frac{desired\ unit}{given\ unit} \right) = desired\ unit$$

To convert between moles and mass, you need to use the molar mass of the element as a conversion factor.

$$mol \times \frac{g}{mol} = g$$

Also, going in the opposite direction

$$g \times \frac{mol}{g} = mol$$

Tip: Whether you are converting from g → mol or from mol → g, you will need to use the molar mass as the conversion factor. The molar mass of an element can be found directly on the periodic table.

EXAMPLE 3.2
How many grams are there in 0.130 mole of Cu?

Step 1: The conversion factor needed is the molar mass of Cu.

$$\frac{63.55 \text{ g Cu}}{1 \text{ mol Cu}}$$

You should come up with the following strategy.

$$\text{moles} \rightarrow \text{grams}$$

Step 2:

$$\textbf{? g Cu} = 0.130 \text{ mol Cu} \times \frac{63.55 \text{ g Cu}}{1 \text{ mol Cu}} = \textbf{8.26 g Cu}$$

PRACTICE EXERCISE
2.   How many moles of Cu are in 125 g of Cu?

Text Problem: 3.16

B.   Calculating the mass of a single atom

To calculate the mass of a single atom, you must use Avogadro's number. The conversion factor is

$$\frac{1 \text{ mol}}{6.022 \times 10^{23} \text{ atoms}}$$

EXAMPLE 3.3
Copper is a minor component of pennies minted since 1981, and it is also used in electrical cables. Calculate the mass (in grams) of a single Cu atom.

Step 1: One conversion factor needed is given above. Another necessary conversion factor is the molar mass of Cu.

$$\frac{63.55 \text{ g Cu}}{1 \text{ mol Cu}}$$

You should come up with the following strategy.

$$\frac{\text{grams}}{\text{mol of Cu atoms}} \rightarrow \frac{\text{grams}}{\text{single Cu atom}}$$

Step 2:

$$\textbf{? g/Cu atom} = \frac{63.55 \text{ g Cu}}{1 \text{ mol Cu}} \times \frac{1 \text{ mol Cu}}{6.022 \times 10^{23} \text{ Cu atoms}}$$

$$= \textbf{1.055} \times \textbf{10}^{-22} \textbf{ g/Cu atom}$$

PRACTICE EXERCISE
3.   Titanium (Ti) is a transition metal with a very high strength-to-weight ratio. For this reason, titanium is used in the construction of aircraft. What is the mass (in grams) of one Ti atom?

Text Problem: 3.18

C. Converting mass in grams to number of atoms

To complete the following conversion, you need to use both molar mass and Avogadro's number as conversion factors.

EXAMPLE 3.4
Zinc is the main component of pennies minted after 1981. How many zinc atoms are present in 20.0 g of Zn?

Step 1: The two conversion factors needed are:

$$\frac{1 \text{ mol Zn}}{65.39 \text{ g Zn}} \qquad \frac{6.022 \times 10^{23} \text{ Zn atoms}}{1 \text{ mol Zn}}$$

You should come up with the following strategy.

$$\text{grams of Zn} \rightarrow \text{moles of Zn} \rightarrow \text{atoms of Zn}$$

Step 2:

$$? \text{ atoms of Zn} = 20.0 \text{ g Zn} \times \frac{1 \text{ mol Zn}}{65.39 \text{ g Zn}} \times \frac{6.022 \times 10^{23} \text{ Zn atoms}}{1 \text{ mol Zn}}$$

$$= 1.84 \times 10^{23} \text{ Zn atoms}$$

PRACTICE EXERCISE
4.   What is the mass (in grams) of 9.09×10^{23} atoms of Zn?

Text Problem: 3.20

PROBLEM TYPE 3: CALCULATIONS INVOLVING MOLECULAR MASS

A. Calculating molecular mass

The molecular mass is simply the sum of the atomic masses (in amu) of all the atoms in the molecule.

EXAMPLE 3.5
Calculate the molecular mass of carbon tetrachloride (CCl_4).

$$\text{molecular mass } CCl_4 = (\text{mass of C}) + 4(\text{mass of Cl})$$

$$\textbf{molecular mass } CCl_4 = (12.01 \text{ amu}) + 4(35.45 \text{ amu}) = \textbf{153.8 amu}$$

PRACTICE EXERCISE
5.   Bananas owe their characteristic smell and flavor to the ester, isopentyl acetate [$CH_3COOCH_2CH_2CH(CH_3)_2$].
     Calculate the molecular mass of isopentyl acetate.

Text Problem: 3.24

B. Calculating the number of moles in a given amount of a compound

To complete this conversion, the only conversion factor needed is the molar mass in units of g/mol. Remember, the molar mass of a compound (in grams) is numerically equal to its molecular mass (in atomic mass units). For example, the molar mass of CCl_4 is 153.8 g/mol, compared to its molecular mass of 153.8 amu.

EXAMPLE 3.6

How many moles of ethane (C_2H_6) are present in 50.3 g of ethane?

Step 1: First, the molar mass of ethane must be calculated in order to have the appropriate conversion factor for the problem.

$$\text{molar mass of } C_2H_6 = 2(12.01 \text{ g}) + 6(1.008 \text{ g}) = 30.07 \text{ g}$$

Hence, the conversion factor is

$$\frac{1 \text{ mol } C_2H_6}{30.07 \text{ g } C_2H_6}$$

Step 2: Based on the conversion factor above, you should come up with the following strategy.

$$\text{grams } C_2H_6 \rightarrow \text{moles of } C_2H_6$$

Step 3:

$$? \text{ moles of } \mathbf{C_2H_6} = 50.3 \text{ g } C_2H_6 \times \frac{1 \text{ mol } C_2H_6}{30.07 \text{ g } C_2H_6} = \mathbf{1.67 \text{ moles}}$$

PRACTICE EXERCISE

6.   What is the mass (in grams) of 0.436 moles of ethane (C_2H_6)?

Text Problem: 3.26

C.   Calculating the number of atoms in a given amount of a compound

Again, this is a unit conversion problem. This calculation is more difficult than the conversions above, because you must convert from *grams of compound* to *moles of compound* to *moles of a particular atom* to *number of atoms*. Sound tough? Let's try an example.

EXAMPLE 3.7

How many carbon atoms are present in 50.3 g of ethane (C_2H_6)?

We started this problem in Example 3.6 when we calculated the moles of ethane in 50.3 g ethane. To continue, we need two additional conversion factors. One should represent the mole ratio between moles of C atoms and moles of ethane molecules. The other conversion factor needed is Avogadro's number.

Step 1: The two conversion factors needed are:

$$\frac{2 \text{ mol C}}{1 \text{ mol } C_2H_6} \qquad \frac{6.022 \times 10^{23} \text{ C atoms}}{1 \text{ mol C}}$$

You should come up with the following strategy.

$$\text{grams of } C_2H_6 \rightarrow \text{moles of } C_2H_6 \rightarrow \text{moles of C} \rightarrow \text{atoms of C}$$

Step 2:

$$? \text{ C atoms} = 50.3 \text{ g } C_2H_6 \times \frac{1 \text{ mol } C_2H_6}{30.07 \text{ g } C_2H_6} \times \frac{2 \text{ mol C}}{1 \text{ mol } C_2H_6} \times \frac{6.022 \times 10^{23} \text{ C atoms}}{1 \text{ mol C}}$$

$$= \mathbf{2.01 \times 10^{24} \text{ C atoms}}$$

PRACTICE EXERCISE

7.   Glucose, the sugar used by the cells of our bodies for energy, has the molecular formula, $C_6H_{12}O_6$. How many atoms of *carbon* are present in a 3.50 g sample of glucose?

Text Problem: 3.28

PROBLEM TYPE 4: CALCULATIONS INVOLVING PERCENT COMPOSITION

A.   Calculating percent composition of a compound

The *percent composition by mass* is the percent by mass of each element the compound contains. Percent composition is obtained by dividing the mass of each element in 1 mole of the compound by the molar mass of the compound, then multiplying by 100 percent.

$$\textbf{percent by mass of each element} = \frac{\text{mass of element in 1 mol of compound}}{\text{molar mass of compound}} \times 100\%$$

EXAMPLE 3.8

Calculate the percent composition by mass of all the elements in sodium bicarbonate, $NaHCO_3$.

First, calculate the molar mass of sodium bicarbonate.

   molar mass sodium bicarbonate = 22.99 g + 1.008 g + 12.01 g + 3(16.00 g) = 84.01 g

$$\%Na = \frac{22.99 \text{ g}}{84.01 \text{ g}} \times 100\% = \textbf{27.37\%}$$

$$\%H = \frac{1.008 \text{ g}}{84.01 \text{ g}} \times 100\% = \textbf{1.200\%}$$

$$\%C = \frac{12.01 \text{ g}}{84.01 \text{ g}} \times 100\% = \textbf{14.30\%}$$

$$\%O = \frac{3(16.00 \text{ g})}{84.01 \text{ g}} \times 100\% = \textbf{57.14\%}$$

> **Tip:** You can check your work by making sure that the mass percents of all the elements added together equals 100%. Checking above, 27.37% + 1.200% + 14.30% + 57.14% = 100.01% ≈ 100%.

PRACTICE EXERCISE

8.   Cinnamic alcohol is used mainly in perfumes, particularly for soaps and cosmetics. Its molecular formula is $C_9H_{10}O$. Calculate the percent composition by mass of *hydrogen* in cinnamic alcohol.

Text Problems: 3.40, 3.42

B.   Determining empirical formula from percent composition

The procedure used above to calculate the percent composition of a compound can be reversed. Given the percent composition by mass of a compound, you can determine the empirical formula of the compound.

EXAMPLE 3.9

Dieldrin, like DDT, is an insecticide that contains only C, H, Cl, and O. It is composed of 37.84 percent C, 2.12 percent H, 55.84 percent Cl, and 4.20 percent O. Determine its empirical formula.

Step 1: Assume you have exactly 100 g of substance. 100 g is a convenient amount, because all the percentages sum to 100 percent. In 100 g of dieldrin there will be 37.84 g C, 2.12 g H, 55.84 g Cl, and 4.20 g O.

Step 2: Calculate the number of moles of each element in 100 g of the compound. Remember, an *empirical formula* tells us which elements are present and the simplest whole-number ratio of their atoms. This ratio is also a mole ratio. Let n_C, n_H, n_{Cl}, and n_O be the number of moles of elements present. Use the molar masses of these elements as conversion factors to convert to moles

$$n_C = 37.84 \text{ g C} \times \frac{1 \text{ mol C}}{12.01 \text{ g C}} = \textbf{3.151 mol C}$$

$$n_H = 2.12 \text{ g H} \times \frac{1 \text{ mol H}}{1.008 \text{ g H}} = \textbf{2.10 mol H}$$

$$n_{Cl} = 55.84 \text{ g Cl} \times \frac{1 \text{ mol Cl}}{35.45 \text{ g Cl}} = \textbf{1.575 mol Cl}$$

$$n_O = 4.20 \text{ g O} \times \frac{1 \text{ mol O}}{16.00 \text{ g O}} = \textbf{0.263 mol O}$$

Thus, we arrive at the formula $C_{3.151}H_{2.10}Cl_{1.575}O_{0.263}$, which gives the identity and the ratios of atoms present. However, chemical formulas are written with whole numbers.

Step 3: Try to convert to whole numbers by dividing all the subscripts by the smallest subscript.

$$\text{C: } \frac{3.151}{0.263} = 12.0 \qquad \text{H: } \frac{2.10}{0.263} = 7.98 \approx 8 \qquad \text{Cl: } \frac{1.575}{0.263} = 5.99 \approx 6 \qquad \text{O: } \frac{0.263}{0.263} = 1$$

This gives us the empirical for dieldrin, $C_{12}H_8Cl_6O$.

> **Tip:** It's not always this easy. Step 3 often does not give all whole numbers. If this is the case, you must multiply all the subscripts by some *integer* to come up with whole number subscripts. Try the practice exercise below.

PRACTICE EXERCISE

9. The substance responsible for the green color on the yolk of a boiled egg is composed of 53.58 percent Fe and 46.42 percent S. Determine its empirical formula.

> **Text Problems:** **3.44**, 3.50, 3.52

C. Calculating mass from percent composition

Step 1: Convert the mass percentage to a fraction. For example, if the mass percent of an element in a compound were 54.73 percent, you would convert this to 54.73/100 = 0.5473.

Step 2: Multiply the fraction by the total mass of the compound. This gives the mass of the particular element in the compound.

EXAMPLE 3.10

Calculate the mass of carbon in exactly 10 g of glucose ($C_6H_{12}O_6$).

Step 1: Calculate the percentage by mass of carbon in glucose, then convert the percentage to a fraction.

$$\text{mass \% C} = \frac{\text{mass of C in 1 mol of glucose}}{\text{molar mass of glucose}} \times 100\%$$

$$\textbf{mass \% C} = \frac{6(12.01) \text{ g}}{180.16 \text{ g}} \times 100\% = \textbf{40.00\% C}$$

Converting this percentage to a fraction, we obtain 40.00/100 = **0.4000**

Step 2: Multiply the fraction by the total mass of the compound.

$$\textbf{? g C in 10 g glucose} = (0.4000)(10 \text{ g}) = \textbf{4.000 g C}$$

PRACTICE EXERCISE

10. Calculate the mass of hydrogen in exactly 10 grams of glucose ($C_6H_{12}O_6$).

Text Problems: 3.46, **3.48**

PROBLEM TYPE 5: EXPERIMENTAL DETERMINATION OF EMPIRICAL FORMULAS

See Section 3.6 of your text for a description of the experimental setup. To solve this type of problem, you must recognize that all of the carbon in the sample is converted to CO_2 and all the hydrogen in the sample is converted to H_2O. Then, you can calculate the mass of C in CO_2 and the mass of H in H_2O. Finally, you can calculate the mass of oxygen by difference, if necessary.

EXAMPLE 3.11
When a 0.761-g sample of a compound containing only carbon and hydrogen is burned in an apparatus with CO_2 and H_2O absorbers, 2.23 g CO_2 and 1.37 g H_2O are collected. Determine the empirical formula of the compound.

Step 1: Calculate the mass of C in 2.23 g CO_2, and the mass of H in 1.37 g H_2O. This is a factor-label problem. To calculate the mass of each component, you need the molar masses and the correct mole ratio.

You should come up with the following strategy.

$$\text{g } CO_2 \rightarrow \text{mol } CO_2 \rightarrow \text{mol C} \rightarrow \text{g C}$$

Step 2:
$$? \text{ g C} = 2.23 \text{ g } CO_2 \times \frac{1 \text{ mol } CO_2}{44.01 \text{ g } CO_2} \times \frac{1 \text{ mol C}}{1 \text{ mol } CO_2} \times \frac{12.01 \text{ g C}}{1 \text{ mol C}} = 0.609 \text{ g C}$$

Similarly,

$$? \text{ g H} = 1.37 \text{ g } H_2O \times \frac{1 \text{ mol } H_2O}{18.02 \text{ g } H_2O} \times \frac{2 \text{ mol H}}{1 \text{ mol } H_2O} \times \frac{1.008 \text{ g H}}{1 \text{ mol H}} = 0.153 \text{ g H}$$

Step 3: Calculate the number of moles of each element present in the 0.761 g sample. This is also a factor-label problem, using the molar mass as a conversion factor. You should come up with the following strategy. See Problem Type 2A.

$$\text{g C} \rightarrow \text{mol C}$$

$$\textbf{? mol C} = 0.609 \text{ g C} \times \frac{1 \text{ mol C}}{12.01 \text{ g C}} = \textbf{0.0507 mol C}$$

Similarly,

$$\textbf{? mol H} = 0.153 \text{ g H} \times \frac{1 \text{ mol H}}{1.008 \text{ g H}} = \textbf{0.152 mol H}$$

Thus, we arrive at the formula $C_{0.0507}H_{0.152}$, which gives the identity and the ratios of atoms present. However, chemical formulas are written with whole numbers.

Step 4: Try to convert to whole numbers by dividing all the subscripts by the smallest subscript.

$$\text{C: } \frac{0.0507}{0.0507} = 1.00 \qquad \text{H: } \frac{0.152}{0.0507} = 3.00$$

This gives us the empirical formula, **CH$_3$**.

PRACTICE EXERCISE

11. Diethyl ether, commonly known as "ether", was used as an anesthetic for many years. Diethyl ether contains C, H, and O. When a 1.45 g sample of ether is burned in an apparatus such as that shown in Figure 3.5 of the text, 2.77 g of CO_2 and 1.70 g of H_2O are collected. Determine the empirical formula of diethyl ether.

Text Problem: 3.124

PROBLEM TYPE 6: DETERMINING THE MOLECULAR FORMULA OF A COMPOUND

To determine the molecular formula of a compound, we must know both the *approximate* molar mass and the empirical formula of the compound. The molecular formula will either be equal to the empirical formula or be some integral multiple of it. Thus, the molar mass divided by the empirical mass will be an integer greater than or equal to one.

$$\frac{\text{molar mass}}{\text{empirical molar mass}} \geq 1 \text{ (integer values)}$$

EXAMPLE 3.12

A mass spectrum obtained on the compound in Example 3.11, shows its molecular mass to be about 31 g/mol. What is its molecular formula?

Step 1: Determine the empirical formula. This was already done in the previous example and was determined to be **CH$_3$**.

Step 2: Calculate the empirical molar mass.

$$\text{empirical molar mass} = 12.01 \text{ g} + 3(1.008 \text{ g}) = 15.03 \text{ g/mol}$$

Step 3: Determine the number of (CH$_3$) units present in the molecular formula. This number is found by taking the ratio

$$\frac{\text{molar mass}}{\text{empirical molar mass}} = \frac{31 \text{ g}}{15.03 \text{ g}} = 2.1 \approx 2$$

Thus, there are two CH$_3$ units in each molecule of the compound, so the molecular formula is (CH$_3$)$_2$, or **C$_2$H$_6$**.

PRACTICE EXERCISE

12. In Example 3.9, the empirical formula of dieldrin was determined to be $C_{12}H_8Cl_6O$. If the molar mass of dieldrin is 381 ± 10 g/mol, what is the molecular formula of dieldrin?

Text Problem: 3.52

PROBLEM TYPE 7: CALCULATING THE AMOUNTS OF REACTANTS AND PRODUCTS

These types of problems are factor-label problems. You must always remember to start this type of problem with a balanced chemical equation. The typical approach is given below. See Section 3.8 of your text for a step-by-step method.

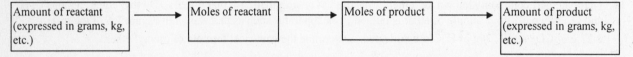

> **Tip:** Always try to be flexible when solving problems. Most problems of this type will follow an approach similar to the one above, but you may have to modify it sometimes.

EXAMPLE 3.13
Sulfur dioxide can be removed from stack gases by reaction with quicklime (CaO):

$$SO_2 \, (g) \; + \; CaO \, (s) \; \longrightarrow \; CaSO_3 \, (s)$$

If 975 kg of SO_2 are to be removed from stack gases by the above reaction, how many kilograms of CaO are required?

Step 1: Start with a balanced chemical equation. It's given in the problem.

$$SO_2 \, (g) \; + \; CaO \, (s) \; \longrightarrow \; CaSO_3 \, (s)$$

Step 2: Map out the following strategy to solve this problem.

$$kg \, SO_2 \; \rightarrow \; g \, SO_2 \; \rightarrow \; moles \, SO_2 \; \rightarrow \; moles \, CaO \; \rightarrow \; g \, CaO \; \rightarrow \; kg \, CaO$$

Step 3: Calculate the mass of CaO in kg using the strategy above.

$$? \, \textbf{kg CaO} = 975 \, kg \, SO_2 \times \frac{1000 \, g \, SO_2}{1 \, kg \, SO_2} \times \frac{1 \, mol \, SO_2}{64.07 \, g \, SO_2} \times \frac{1 \, mol \, CaO}{1 \, mol \, SO_2} \times \frac{56.08 \, g \, CaO}{1 \, mol \, CaO} \times \frac{1 \, kg \, CaO}{1000 \, g \, CaO}$$

$$= \textbf{853 kg CaO}$$

> **Tip:** Notice that the approach followed was a slight modification of the one given above. We went from mass of one reactant, to moles of that reactant, to moles of a second reactant, and finally to mass of second reactant.

PRACTICE EXERCISE

13. Carbon dioxide in the air of a spacecraft can be removed by its reaction with a lithium hydroxide solution.

$$CO_2 \, (g) \; + \; 2 \, LiOH \, (aq) \; \longrightarrow \; Li_2CO_3 \, (aq) \; + \; H_2O \, (l)$$

On average, a person will exhale about 1 kg of CO_2/day. How many kilograms of LiOH are required to react with 1.0 kg of CO_2?

Text Problems: **3.62**, 3.64, 3.66, 3.68, 3.70, 3.72

PROBLEM TYPE 8: LIMITING REAGENT CALCULATIONS

When a chemist carries out a reaction, the reactants are usually not present in exact **stoichiometric amounts**, that is, in the proportions indicated by the balanced equation. The reactant used up first in a reaction is called the **limiting reagent**. When this reactant is used up, no more product can be formed.

Typically, the only difference between this type of problem and Problem Type 7, Calculating the Amounts of Reactants and Products, is that you must first determine which reactant is the limiting reagent.

EXAMPLE 3.14

Phosphine (PH$_3$) burns in oxygen (O$_2$) to produce phosphorus pentoxide and water.

$$2 \text{ PH}_3 \ (g) \ + \ 4 \text{ O}_2 \ (g) \ \longrightarrow \ \text{P}_2\text{O}_5 \ (s) \ + \ 3 \text{ H}_2\text{O} \ (l)$$

How many grams of P$_2$O$_5$ will be produced when 17.0 g of phosphine are reacted with 16.0 g of O$_2$?

Step 1: **Determine which reactant is the limiting reagent.** There are two different methods to determine the limiting reagent. One is given in Section 3.9 of your text. A second method involves treating each reactant separately. Calculate the amount of product formed if that reactant reacts completely, assuming the other reactant is in excess. Repeat the procedure for the second reactant. The reactant that produces the *least* amount of product is the *limiting reagent*; it *limits* the amount of product that can be formed.

You should follow the strategy below to calculate the amount of product formed

g reactant $\rightarrow$ mol reactant $\rightarrow$ mol product

Calculate moles of P$_2$O$_5$ formed if all PH$_3$ reacts.

$$\text{? mol P}_2\text{O}_5 = 17.0 \text{ g PH}_3 \ \times \ \frac{1 \text{ mol PH}_3}{33.99 \text{ g PH}_3} \ \times \ \frac{1 \text{ mol P}_2\text{O}_5}{2 \text{ mol PH}_3} = \textbf{0.250 mol P}_2\textbf{O}_5$$

Calculate moles of P$_2$O$_5$ formed if all O$_2$ reacts.

$$\text{? mol P}_2\text{O}_5 = 16.0 \text{ g O}_2 \ \times \ \frac{1 \text{ mol O}_2}{32.0 \text{ g O}_2} \ \times \ \frac{1 \text{ mol P}_2\text{O}_5}{4 \text{ mol O}_2} = \textbf{0.125 mol P}_2\textbf{O}_5$$

The initial amount of O$_2$ limits the amount of product that can be formed; therefore, it is the limiting reagent.

Step 2: The problem asks for grams of P$_2$O$_5$ produced. We already know the moles of P$_2$O$_5$ produced, 0.125 moles. Use the molar mass of P$_2$O$_5$ as a conversion factor to convert to grams.

$$\text{? g P}_2\text{O}_5 = 0.125 \text{ mol P}_2\text{O}_5 \ \times \ \frac{141.94 \text{ g P}_2\text{O}_5}{1 \text{ mol P}_2\text{O}_5} = \textbf{17.7 g P}_2\textbf{O}_5$$

PRACTICE EXERCISE

14. Iron can be produced by reacting iron ore with carbon. The iron produced can then be used to make steel. The reaction is

$$2 \text{ Fe}_2\text{O}_3 \ (s) + \ 3 \text{ C} \ (s) \ \xrightarrow{\text{heat}} \ 4 \text{ Fe} \ (l) \ + \ 3 \text{ CO}_2 \ (g)$$

(a) How many grams of Fe can be produced from a mixture of 200.0 g of Fe$_2$O$_3$ and 300.0 g C?
(b) How many grams of excess reagent will remain after the reaction ceases?

Text Problems: 3.76, 3.78

PROBLEM TYPE 9: CALCULATING THE PERCENT YIELD OF A REACTION

The **theoretical yield** is the amount of product that would result if all the limiting reagent reacted. This is the maximum obtainable yield predicted by the balanced equation. However, the amount of product obtained is almost always less than the theoretical yield. The **actual yield** is the quantity of product that actually results from a reaction.

To determine the efficiency of a reaction, chemists often calculate the **percent yield**, which describes the proportion of the actual yield to the theoretical yield. The percent yield is calculated as follows:

$$\% \text{ yield } = \frac{\text{actual yield}}{\text{theoretical yield}} \times 100\%$$

EXAMPLE 3.15

In Example 3.14, the theoretical yield of P_2O_5 was determined to be 17.7 g. If only 12.6 g of P_2O_5 are actually obtained, what is the percent yield of the reaction?

$$\% \text{ yield } = \frac{\text{actual yield}}{\text{theoretical yield}} \times 100\%$$

$$\textbf{\% yield } = \frac{12.6 \text{ g}}{17.7 \text{ g}} \times 100\% = \textbf{71.2\%}$$

PRACTICE EXERCISE

15. Refer back to Practice Exercise 14 to answer this question. If the actual yield of Fe is 110 g, what is the percent yield of Fe?

Text Problems: **3.82**, 3.84

ANSWERS TO PRACTICE EXERCISES

1. ^{11}B

2. 1.97 moles Cu

3. 7.951×10^{-23} g/Ti atom

4. 98.7 g Zn

5. 130.18 amu

6. 13.1 g ethane

7. 7.02×10^{22} C atoms

8. 7.513 percent H by mass

9. Fe_2S_3

10. 0.67 g H

11. C_2H_6O

12. $C_{12}H_8Cl_6O$

13. 1.1 kg LiOH

14. **(a)** 139.9 g Fe
     (b) 277 g C

15. 78.6 percent yield

SOLUTIONS TO SELECTED TEXT PROBLEMS

3.6   This is a variation of Problem Type 1, Calculating Average Atomic Mass.

It would seem that there are two unknowns in this problem, the natural abundance of ^{6}Li and the natural abundance of ^{7}Li. However, these two quantities are not independent of each other; they are related by the fact that they must sum to 1. Start by letting x be the natural abundance of ^{6}Li. Since the sum of the two abundance's must be 1, we can write

$$\text{Abundance } ^7\text{Li} = (1 - x)$$

Remember,

average atomic mass =   (fraction of isotope A)(mass of isotope A) + (fraction of isotope B)(mass of isotope B) + . . . + (fraction of isotope Z)(mass of isotope Z).

$$
\begin{aligned}
\textbf{Average atomic mass of Li} = \; 6.941 \text{ amu} &= x(6.0151 \text{ amu}) + (1 - x)(7.0160 \text{ amu}) \\
6.941 &= 1.0009x + 7.0160 \\
1.0009x &= 0.075 \\
\boldsymbol{x} &\boldsymbol{= 0.075}
\end{aligned}
$$

$x = 0.075$ corresponds to a natural abundance of ^{6}Li of **7.5 percent**. The natural abundance of ^{7}Li is $(1 - x) = 0.925$ or **92.5 percent**.

3.8   The unit factor required is $\left(\dfrac{6.022 \times 10^{23} \text{ amu}}{1 \text{ g}} \right)$

$$? \; \textbf{amu} = (8.4 \text{ g}) \times \left(\frac{6.022 \times 10^{23} \text{ amu}}{1 \text{ g}} \right) = \mathbf{5.1 \times 10^{24} \; amu}$$

3.12   The thickness of the book in miles would be:

$$\frac{3.6 \times 10^{-3} \text{ in}}{1 \text{ page}} \times \frac{1 \text{ ft}}{12 \text{ in}} \times \frac{1 \text{ mi}}{5280 \text{ ft}} \times 6.022 \times 10^{23} \text{ pages} = 3.42 \times 10^{16} \text{ mi}$$

The distance, in miles, traveled by light in one year is:

$$1.00 \text{ yr} \times \frac{365 \text{ day}}{1 \text{ yr}} \times \frac{24 \text{ hr}}{1 \text{ day}} \times \frac{3600 \text{ s}}{1 \text{ hr}} \times \frac{3.00 \times 10^8 \text{ m}}{1 \text{ s}} \times \frac{1 \text{ mi}}{1609 \text{ m}} = \mathbf{5.88 \times 10^{12} \; mi}$$

The thickness of the book in light-years is:

$$3.42 \times 10^{16} \text{ mi} \times \frac{1 \text{ light} - \text{yr}}{5.88 \times 10^{12} \text{ mi}} = \mathbf{5.82 \times 10^3 \; light - yr}$$

It will take light 5.82×10^3 years to travel from the first page to the last one!

3.14 $\left(6.00 \times 10^9 \text{ Co atoms} \right) \times \left(\dfrac{1 \text{ mol Co}}{6.022 \times 10^{23} \text{ Co atoms}} \right) = \mathbf{9.96 \times 10^{-15} \; mol \; Co}$

3.16   Converting between moles of atoms and mass of atoms, Problem Type 2A.

Step 1: The conversion factor needed is the molar mass of Au.

$$\frac{197.0 \text{ g Au}}{1 \text{ mol Au}}$$

You should come up with the following strategy.

$$\text{moles} \rightarrow \text{grams}$$

Step 2:

$$? \text{ g Au} = 15.3 \text{ mol Au} \times \frac{197.0 \text{ g Au}}{1 \text{ mol Au}} = \mathbf{3.01 \times 10^3 \text{ g Au}}$$

3.18   Calculating the mass of a single atom, Problem Type 2B.

(a)

Step 1: One conversion factor needed is Avogadro's number. Another necessary conversion factor is the molar mass of As.

$$\frac{74.92 \text{ g As}}{1 \text{ mol As}}$$

You should come up with the following strategy.

$$\text{grams/mole of As} \rightarrow \text{grams/single As atom}$$

Step 2:

$$? \text{g/As atom} = \frac{74.92 \text{ g As}}{1 \text{ mol As}} \times \frac{1 \text{ mol As}}{6.022 \times 10^{23} \text{ As atoms}} = \mathbf{1.244 \times 10^{-22} \text{ g/As atom}}$$

(b)   Follow same method as part (a).

$$? \text{ g/Ni atom} = \frac{58.69 \text{ g Ni}}{1 \text{ mol Ni}} \times \frac{1 \text{ mol Ni}}{6.022 \times 10^{23} \text{ Ni atoms}} = \mathbf{9.746 \times 10^{-23} \text{ g/Ni atom}}$$

3.20   Converting mass in grams to number of atoms, Problem Type 2a.

Step 1: The two conversion factors needed for this problem are the molar mass of Cu and Avogadro's number.

$$\frac{1 \text{ mol Cu}}{63.55 \text{ g Cu}} \qquad \frac{6.022 \times 10^{23} \text{ Cu atoms}}{1 \text{ mol Cu}}$$

You should come up with the following strategy.

$$\text{grams of Cu} \rightarrow \text{moles of Cu} \rightarrow \text{atoms of Cu}$$

Step 2:

$$? \text{ atoms of Cu} = 3.14 \text{ g Cu} \times \frac{1 \text{ mol Cu}}{63.55 \text{ g Cu}} \times \frac{6.022 \times 10^{23} \text{ Cu atoms}}{1 \text{ mol Cu}} = \mathbf{2.98 \times 10^{22} \text{ Cu atoms}}$$

3.22 $(2 \text{ Pb atoms}) \times \left(\dfrac{1 \text{ mol Pb}}{6.022 \times 10^{23} \text{ Pb atoms}} \right) \times \left(\dfrac{207.2 \text{ g Pb}}{1 \text{ mol Pb}} \right) = \mathbf{6.881 \times 10^{-22} \text{ g Pb}}$

$\left(5.1 \times 10^{-23} \text{ mol He} \right) \left(\dfrac{4.003 \text{ g He}}{1 \text{ mol He}} \right) = \mathbf{2.0 \times 10^{-22} \text{ g He}}$

2 atoms of lead has a greater mass than 5.1×10^{-23} mol of helium.

3.24   Calculating molar mass, modification of Problem Type 3A.

    (a)   **molar mass Li_2CO_3** $= 2(6.941 \text{ g}) + 12.01 \text{ g} + 3(16.00 \text{ g}) = \mathbf{73.89 \text{ g}}$

    (b)   **molar mass CS_2** $= 12.01 \text{ g} + 2(32.07 \text{ g}) = \mathbf{76.15 \text{ g}}$

    (c)   **molar mass $CHCl_3$** $= 12.01 \text{ g} + 1.008 \text{ g} + 3(35.45 \text{ g}) = \mathbf{119.37 \text{ g}}$

    (d)   **molar mass $C_6H_8O_6$** $= 6(12.01 \text{ g}) + 8(1.008 \text{ g}) + 6(16.00 \text{ g}) = \mathbf{176.12 \text{ g}}$

    (e)   **molar mass KNO_3** $= 39.10 \text{ g} + 14.01 \text{ g} + 3(16.00 \text{ g}) = \mathbf{101.11 \text{ g}}$

    (f)   **molar mass Mg_3N_2** $= 3(24.31 \text{ g}) + 2(14.01 \text{ g}) = \mathbf{100.95 \text{ g}}$

3.26   Calculating the number of molecules in a given amount of compound, similar to Problem Type 3B.

Step 1: First, the molar mass of ethane must be calculated in order to have the appropriate conversion factor for the problem.

     molar mass of $C_2H_6 = 2(12.01 \text{ g}) + 6(1.008 \text{ g}) = 30.07 \text{ g}$

Hence, the conversion factor is

$$\frac{1 \text{ mol } C_2H_6}{30.07 \text{ g } C_2H_6}$$

Step 2: Based on the conversion factor above, you should come up with the following strategy.

     grams C_2H_6 $\rightarrow$ moles of C_2H_6

Step 3:

$$\textbf{? molecules of } C_2H_6 = 0.334 \text{ g } C_2H_6 \times \frac{1 \text{ mol } C_2H_6}{30.07 \text{ g } C_2H_6} \times \frac{6.022 \times 10^{23} \text{ } C_2H_6 \text{ molecules}}{1 \text{ mol } C_2H_6}$$

$$= \mathbf{6.69 \times 10^{21} \text{ } C_2H_6 \text{ molecules}}$$

3.28   Calculating the number of atoms in a given amount of a compound, Problem Type 3C.

This is a unit conversion problem. You must convert from *grams of compound* to *moles of compound* to *moles of a particular atom* to *number of atoms*.

Step 1: The conversion factors needed are:

$$\frac{1 \text{ mol urea}}{60.06 \text{ g urea}} \qquad \frac{2 \text{ mol N}}{1 \text{ mol urea}} \qquad \frac{6.022 \times 10^{23} \text{ N atoms}}{1 \text{ mol N}}$$

You should come up with the following strategy.

     grams of urea $\rightarrow$ moles of urea $\rightarrow$ moles of N $\rightarrow$ atoms of N

Step 2:

$$? \text{ atoms of N} = 1.68 \times 10^4 \text{ g urea} \times \frac{1 \text{ mol urea}}{60.06 \text{ g urea}} \times \frac{2 \text{ mol N}}{1 \text{ mol urea}} \times \frac{6.022 \times 10^{23} \text{ N atoms}}{1 \text{ mol N}}$$

$$= \mathbf{3.37 \times 10^{26} \text{ N atoms}}$$

We could calculate the number of atoms of the remaining elements in the same manner, or we can use the atom ratios from the molecular formula. The carbon atom to nitrogen atom ratio in a urea molecule is 1:2, the oxygen atom to nitrogen atom ratio is 1:2, and the hydrogen atom to nitrogen atom ration is 4:2.

$$? \text{ atoms of C} = 3.37 \times 10^{26} \text{ N atoms} \times \frac{1 \text{ C atom}}{2 \text{ N atoms}} = \mathbf{1.69 \times 10^{26} \text{ C atoms}}$$

$$? \text{ atoms of O} = 3.37 \times 10^{26} \text{ N atoms} \times \frac{1 \text{ O atom}}{2 \text{ N atoms}} = \mathbf{1.69 \times 10^{26} \text{ O atoms}}$$

$$? \text{ atoms of H} = 3.37 \times 10^{26} \text{ N atoms} \times \frac{4 \text{ H atoms}}{2 \text{ N atoms}} = \mathbf{6.74 \times 10^{26} \text{ H atoms}}$$

3.30   Mass of water $= (2.56 \text{ mL})\left(\dfrac{1.00 \text{ g}}{1.00 \text{ mL}}\right) = 2.56 \text{ g}$

Molar mass of $H_2O = (16.00 \text{ g}) + 2(1.008 \text{ g}) = 18.02 \text{ g/mol}$

$$? \, H_2O \text{ molecules} = 2.56 \text{ g } H_2O \times \left(\frac{1 \text{ mol } H_2O}{18.02 \text{ g } H_2O}\right) \times \left(\frac{6.022 \times 10^{23} \text{ molecules } H_2O}{1 \text{ mol } H_2O}\right)$$

$$= \mathbf{8.56 \times 10^{22} \text{ molecules}}$$

3.34   Since there are two hydrogen isotopes, they can be paired in three ways: ^1H–^1H, ^1H–^2H, and ^2H–^2H. There will then be three choices for each sulfur isotope. We can make a table showing all the possibilities (masses in amu):

	^{32}S	^{33}S	^{34}S	^{36}S
$^1H_2$	34	35	36	38
$^1H^2H$	35	36	37	39
$^2H_2$	36	37	38	40

There will be seven peaks of the following mass numbers: 34, 35, 36, 37, 38, 39, and 40.

Very accurate (and expensive!) mass spectrometers can detect the mass difference between two ^1H and one ^2H. How many peaks would be detected in such a "high resolution" mass spectrum?

3.40   The molar mass of $CHCl_3 = 12.01 \text{ g/mol} + 1.008 \text{ g/mol} + 3(35.45 \text{ g/mol}) = 119.4 \text{ g/mol}$

$$\%C = \frac{12.01 \text{ g/mol}}{119.4 \text{ g/mol}} \times 100\% = \mathbf{10.06\%}$$

$$\% H = \frac{1.008 \text{ g/mol}}{119.4 \text{ g/mol}} \times 100\% = \mathbf{0.8442\%}$$

$$\%Cl = \frac{3(35.45) \text{ g/mol}}{119.4 \text{ g/mol}} \times 100\% = \mathbf{89.07\%}$$

3.42

	Compound	Molar mass (g)	N% by mass
(a)	$(NH_2)_2CO$	60.06	$\dfrac{2(14.01\text{ g})}{60.06\text{ g}} \times 100\% = 46.65\%$
(b)	NH_4NO_3	80.05	$\dfrac{2(14.01\text{ g})}{80.05\text{ g}} \times 100\% = 35.00\%$
(c)	$HNC(NH_2)_2$	59.08	$\dfrac{3(14.01\text{ g})}{59.08\text{ g}} \times 100\% = 71.14\%$
(d)	NH_3	17.03	$\dfrac{14.01\text{ g}}{17.03\text{ g}} \times 100\% = 82.27\%$

Ammonia, NH_3, is the richest source of nitrogen on a mass percentage basis.

3.44   It is convenient to assume that you have 100 g of sample.

The percentage of oxygen is found by difference:

$$100\% - (19.8\% + 2.50\% + 11.6\%) = 66.1\%$$

$$\text{Moles C} = 19.8\text{ g C}\left(\frac{1\text{ mol C}}{12.01\text{ g C}}\right) = 1.65\text{ mol C}$$

$$\text{Moles H} = 2.50\text{ g H}\left(\frac{1\text{ mol H}}{1.008\text{ g H}}\right) = 2.48\text{ mol H}$$

$$\text{Moles N} = 11.6\text{ g N}\left(\frac{1\text{ mol N}}{14.01\text{ g N}}\right) = 0.828\text{ mol N}$$

$$\text{Moles O} = 66.1\text{ g O}\left(\frac{1\text{ mol O}}{16.00\text{ g O}}\right) = 4.13\text{ mol O}$$

The formula is $C_{1.65}H_{2.48}N_{0.828}O_{4.13}$. Dividing the subscripts by 0.828 gives the empirical formula, **$C_2H_3NO_5$**. Would you get a different answer if you had assumed an amount of sample different than 100 g?

3.46   Using unit factors we convert:

$$\text{g of Hg} \rightarrow \text{mol Hg} \rightarrow \text{mol S} \rightarrow \text{g S}$$

$$\textbf{? g S} = 246\text{ g Hg} \times \left(\frac{1\text{ mol Hg}}{200.6\text{ g Hg}}\right) \times \left(\frac{1\text{ mol S}}{1\text{ mol Hg}}\right) \times \left(\frac{32.07\text{ g S}}{1\text{ mol S}}\right) = \textbf{39.3 g S}$$

3.48   Calculating mass from percent composition, Problem Type 4C.

Step 1: Calculate the percentage by mass of fluorine in tin(II) fluoride, then convert the percentage to a fraction.

$$\text{mass \% F} = \frac{\text{mass of F in 1 mol SnF}_2}{\text{molar mass of SnF}_2} \times 100\%$$

$$= \frac{2(19.00\text{ g})}{156.7\text{ g}} \times 100\% = 24.25\%\text{ F}$$

Converting this percentage to a fraction, we obtain $24.25/100 = 0.2425$.

Step 2: Multiply the fraction by the total mass of the compound.

$$\text{? g F in 24.6 g SnF}_2 = (0.2425)(24.6 \text{ g}) = \textbf{5.97 g F}$$

> **Note:** This problem could have been worked in a manner similar to Problem 3.46. You could complete the following conversions:
>
> g of SnF_2 → mol of SnF_2 → mol of F → g of F

3.50   Assuming 100 g of compound:

(a)
$$\text{mol C} = 40.1 \text{ g C} \times \frac{1 \text{ mol C}}{12.01 \text{ g C}} = 3.34 \text{ mol C}$$

$$\text{mol H} = 6.6 \text{ g H} \times \frac{1 \text{ mol H}}{1.008 \text{ g H}} = 6.5 \text{ mol H}$$

$$\text{mol O} = 53.3 \text{ g O} \times \frac{1 \text{ mol O}}{16.00 \text{ g O}} = 3.33 \text{ mol O}$$

Dividing by the smallest number of moles (3.33 mol) gives the empirical formula, $\textbf{CH}_2\textbf{O}$.

(b)
$$\text{mol C} = 18.4 \text{ g C} \times \frac{1 \text{ mol C}}{12.01 \text{ g C}} = 1.53 \text{ mol C}$$

$$\text{mol N} = 21.5 \text{ g N} \times \frac{1 \text{ mol N}}{14.01 \text{ g N}} = 1.53 \text{ mol N}$$

$$\text{mol K} = 60.1 \text{ g K} \times \frac{1 \text{ mol K}}{39.10 \text{ g K}} = 1.54 \text{ mol K}$$

Dividing by the smallest number of moles (1.53 mol) gives the empirical formula, **KCN**.

3.52   Determining empirical formula from percent composition, Problem Type 4B.

METHOD 1:

Step 1: Assume you have exactly 100 g of substance. 100 g is a convenient amount, because all the percentages sum to 100%. In 100 g of MSG there will be 35.51 g C, 4.77 g H, 37.85 g O, 8.29 g N, and 13.60 g Na.

Step 2: Calculate the number of moles of each element in the compound. Remember, an *empirical formula* tells us which elements are present and the simplest whole-number ratio of their atoms. This ratio is also a mole ratio. Let n_C, n_H, n_O, n_N, and n_{Na} be the number of moles of elements present. Use the molar masses of these elements as conversion factors to convert to moles.

$$n_C = 35.51 \text{ g C} \times \frac{1 \text{ mol C}}{12.01 \text{ g C}} = \textbf{2.957 mol C}$$

$$n_H = 4.77 \text{ g H} \times \frac{1 \text{ mol H}}{1.008 \text{ g H}} = \textbf{4.73 mol H}$$

$$n_O = 37.85 \text{ g O} \times \frac{1 \text{ mol O}}{16.00 \text{ g O}} = \textbf{2.366 mol O}$$

$$n_N = 8.29 \text{ g N} \times \frac{1 \text{ mol N}}{14.01 \text{ g N}} = \textbf{0.592 mol N}$$

$$n_{Na} = 13.60 \text{ g Na} \times \frac{1 \text{ mol Na}}{22.99 \text{ g Na}} = \textbf{0.5916 mol Na}$$

Thus, we arrive at the formula $C_{2.957}H_{4.73}O_{2.366}N_{0.592}Na_{0.5916}$, which gives the identity and the ratios of atoms present. However, chemical formulas are written with whole numbers.

Step 3: Try to convert to whole numbers by dividing all the subscripts by the smallest subscript.

$$\textbf{C:} \ \frac{2.957}{0.5916} = 4.998 \approx 5 \qquad \textbf{H:} \ \frac{4.73}{0.5916} = 8.00 \qquad \textbf{O:} \ \frac{2.366}{0.5916} = 3.999 \approx 4$$

$$\textbf{N:} \ \frac{0.592}{0.5916} = 1.00 \qquad \textbf{Na:} \ \frac{0.5916}{0.5916} = 1$$

This gives us the empirical formula for MSG, $\textbf{C}_5\textbf{H}_8\textbf{O}_4\textbf{NNa}$.

To determine the molecular formula, remember that the molar mass/empirical mass will be an integer greater than or equal to one.

$$\frac{\text{molar mass}}{\text{empirical molar mass}} \geq 1 \text{ (integer values)}$$

In this case,

$$\frac{\text{molar mass}}{\text{empirical molar mass}} = \frac{169 \text{ g}}{169.11 \text{ g}} \approx 1$$

Hence, the molecular formula and the empirical formula are the same, $\textbf{C}_5\textbf{H}_8\textbf{O}_4\textbf{NNa}$. It should come as no surprise that the empirical and molecular formulas are the same since MSG stands for *monosodium*glutamate.

METHOD 2:

Step 1: Multiply the mass % (converted to a decimal) of each element by the molar mass to convert to grams of each element. Then, use the molar mass to convert to moles of each element.

$$n_C = (0.3551) \times (169 \text{ g}) \times \frac{1 \text{ mol C}}{12.01 \text{ g C}} = \textbf{5.00 mol C}$$

$$n_H = (0.0477) \times (169 \text{ g}) \times \frac{1 \text{ mol H}}{1.008 \text{ g H}} = \textbf{8.00 mol H}$$

$$n_O = (0.3785) \times (169 \text{ g}) \times \frac{1 \text{ mol O}}{16.00 \text{ g O}} = \textbf{4.00 mol O}$$

$$n_N = (0.0829) \times (169 \text{ g}) \times \frac{1 \text{ mol N}}{14.01 \text{ g N}} = \textbf{1.00 mol N}$$

$$n_{Na} = (0.1360) \times (169 \text{ g}) \times \frac{1 \text{ mol Na}}{22.99 \text{ g Na}} = \textbf{1.00 mol Na}$$

Step 2: Since we used the molar mass to calculate the moles of each element present in the compound, this method directly gives the molecular formula. The formula is $\textbf{C}_5\textbf{H}_8\textbf{O}_4\textbf{NNa}$.

3.58 The balanced equations are as follows:

(a) $2N_2O_5 \rightarrow 2N_2O_4 + O_2$

(b) $2KNO_3 \rightarrow 2KNO_2 + O_2$

(c) $NH_4NO_3 \rightarrow N_2O + 2H_2O$

(d) $NH_4NO_2 \rightarrow N_2 + 2H_2O$

(e) $2NaHCO_3 \rightarrow Na_2CO_3 + H_2O + CO_2$

(f) $P_4O_{10} + 6H_2O \rightarrow 4H_3PO_4$

(g) $2HCl + CaCO_3 \rightarrow CaCl_2 + H_2O + CO_2$ (k) $Be_2C + 4H_2O \rightarrow 2Be(OH)_2 + CH_4$

(h) $2Al + 3H_2SO_4 \rightarrow Al_2(SO_4)_3 + 3H_2$ (l) $3Cu + 8HNO_3 \rightarrow 3Cu(NO_3)_2 + 2NO + 4H_2O$

(i) $CO_2 + 2KOH \rightarrow K_2CO_3 + H_2O$ (m) $S + 6HNO_3 \rightarrow H_2SO_4 + 6NO_2 + 2H_2O$

(j) $CH_4 + 2O_2 \rightarrow CO_2 + 2H_2O$ (n) $2NH_3 + 3CuO \rightarrow 3Cu + N_2 + 3H_2O$

3.62   Calculating the Amounts of Reactants and Products, Problem Type 7.

Step 1: Start with a balanced chemical equation. It's given in the problem.

$$Si\,(s)\, + 2\,Cl_2\,(g) \longrightarrow SiCl_4\,(l)$$

Step 2: Map out the following strategy to solve this problem.

$$\text{moles } SiCl_4 \rightarrow \text{mol } Cl_2$$

Step 3: Calculate the moles of Cl_2 using the stoichiometric ratio from the balanced equation.

$$\textbf{? mol Cl}_2 = 0.507 \text{ mol } SiCl_4 \times \frac{2 \text{ mol } Cl_2}{1 \text{ mol } SiCl_4} = \textbf{1.01 mol Cl}_2$$

3.64   Molar mass $NaHCO_3$ = 22.99 g + 1.008 g + 12.01 g + 3(16.00 g) = 84.01 g
Molar mass CO_2 = 12.01 g + 2(16.00 g) = 44.01 g

(a) $2\,NaHCO_3 \longrightarrow Na_2CO_3 + H_2O + CO_2$

(b)   The balanced equation shows one mole of CO_2 formed from two moles of $NaHCO_3$.

$$\textbf{mass NaHCO}_3 = 20.5 \text{ g } CO_2 \left(\frac{1 \text{ mol } CO_2}{44.01 \text{ g } CO_2} \right)\left(\frac{2 \text{ mol } NaHCO_3}{1 \text{ mol } CO_2} \right)\left(\frac{84.01 \text{ g } NaHCO_3}{1 \text{ mol } NaHCO_3} \right)$$

$$= \textbf{78.3 g NaHCO}_3$$

3.66   The grams of ethanol that can be obtained are:

$$\textbf{? g C}_2\textbf{H}_5\textbf{OH} = 500.4 \text{ g } C_6H_{12}O_6 \times \frac{1 \text{ mol } C_6H_{12}O_6}{180.16 \text{ g } C_6H_{12}O_6} \times \frac{2 \text{ mol } C_2H_5OH}{1 \text{ mol } C_6H_{12}O_6} \times \frac{46.07 \text{ g } C_2H_5OH}{1 \text{ mol } C_2H_5OH}$$

$$= \textbf{255.9 g C}_2\textbf{H}_5\textbf{OH}$$

The liters of ethanol can be calculated from the density and the mass of ethanol.

$$\text{volume} = \frac{\text{mass}}{\text{density}}$$

$$\textbf{Volume of ethanol obtained} = \frac{255.9 \text{ g}}{0.789 \text{ g} / \text{mL}} = 324 \text{ mL} = \textbf{0.324 L}$$

3.68   The balanced equation shows that eight moles of KCN are needed to combine with four moles of Au.

$$\textbf{? mol KCN} = 29.0 \text{ g } Au \times \left(\frac{1 \text{ mol } Au}{197.0 \text{ g } Au} \right) \times \left(\frac{8 \text{ mol } KCN}{4 \text{ mol } Au} \right) = \textbf{0.294 mol KCN}$$

3.70   Calculating the Amounts of Reactants and Products, Problem Type 7.

(a) $NH_4NO_3 (s) \longrightarrow N_2O (g) + 2 H_2O (g)$

(b)
Step 1: Map out the following strategy to solve this problem.

$$\text{mol } NH_4NO_3 \rightarrow \text{mol } N_2O \rightarrow \text{g } N_2O$$

Step 2: The molar mass of nitrous oxide is needed as a conversion factor.

$$\frac{44.02 \text{ g } N_2O}{1 \text{ mol } N_2O}$$

Step 3: Calculate the grams of N_2O using the strategy in step 1.

$$\textbf{? g } N_2O = 0.46 \text{ mol } NH_4NO_3 \times \frac{1 \text{ mol } N_2O}{1 \text{ mol } NH_4NO_3} \times \frac{44.02 \text{ g } N_2O}{1 \text{ mol } N_2O} = \textbf{2.0} \times \textbf{10}^{1} \textbf{ g } N_2O$$

3.72   The balanced equation for the decomposition is :

$$2 KClO_3 (s) \longrightarrow 2 KCl (s) + 3 O_2 (g)$$

$$\textbf{? g } O_2 = 46.0 \text{ g } KClO_3 \times \frac{1 \text{ mol } KClO_3}{122.6 \text{ g } KClO_3} \times \frac{3 \text{ mol } O_2}{2 \text{ mol } KClO_3} \times \frac{32.00 \text{ g } O_2}{1 \text{ mol } O_2} = \textbf{18.0 g } O_2$$

3.76   Limiting Reagent Calculation, Problem Type 8.

Step 1: Determine which reactant is the limiting reagent.

You should follow the strategy below to calculate the amount of product formed.

$$\text{g reactant} \rightarrow \text{mol reactant} \rightarrow \text{mol product}$$

Calculate moles of NO_2 formed if all O_3 reacts.

$$\text{? mol } NO_2 = 0.740 \text{ g } O_3 \times \frac{1 \text{ mol } O_3}{48.00 \text{ g } O_3} \times \frac{1 \text{ mol } NO_2}{1 \text{ mol } O_3} = 0.0154 \text{ mol } NO_2$$

Calculate moles of NO_2 formed if all NO reacts.

$$\text{? mol } NO_2 = 0.670 \text{ g } NO \times \frac{1 \text{ mol } NO}{30.01 \text{ g } NO} \times \frac{1 \text{ mol } NO_2}{1 \text{ mol } NO} = 0.0223 \text{ mol } NO_2$$

The initial amount of O_3 limits the amount of product that can be formed; therefore, it is the **limiting reagent**.

Step 2: The problem asks for grams of NO_2 produced. We already know the moles of NO_2 produced, 0.0154 moles. Use the molar mass of NO_2 as a conversion factor to convert to grams.

$$\textbf{? mol } NO_2 = 0.0154 \text{ mol } NO_2 \times \frac{46.01 \text{ g } NO_2}{1 \text{ mol } NO_2} = \textbf{0.709 g } NO_2$$

Step 3: The number of moles of NO (excess reagent) remaining at the end of the reaction is

mol NO remaining = mol NO initial − mol NO reacted.

$$\text{mol NO initial} = 0.670 \text{ g NO} \times \frac{1 \text{ mol NO}}{30.01 \text{ g NO}} = 0.0223 \text{ mol NO}$$

mol NO reacted $=$ mol NO_2 formed $= 0.0154$ mol NO (1:1 mol ratio between NO and NO_2)

mol NO remaining $= 0.0223$ mol NO $- 0.0154$ mol NO $=$ **0.0069 mol NO**

3.78   As in all limiting reagent problems, we first must ascertain which reagent is limiting.

(a)   We first convert the mass of HCl given to moles

$$48.2 \text{ g HCl} \left(\frac{1 \text{ mol HCl}}{36.46 \text{ g HCl}} \right) = 1.32 \text{ mol HCl}$$

Since one mole of MnO_2 reacts with four moles of HCl, the number of moles of HCl needed to combine with 0.86 mole of MnO_2 is

$$0.86 \text{ mol } MnO_2 \left(\frac{4 \text{ mol HCl}}{1 \text{ mol } MnO_2} \right) = 3.4 \text{ mol HCl}$$

Since we do not have that much HCl present, HCl must be the limiting reagent and MnO_2 the excess reagent.

(b)   The mass of Cl_2 formed is then:

$$? \text{ g } \textbf{Cl}_2 = 1.32 \text{ mol HCl} \times \frac{1 \text{ mol } Cl_2}{4 \text{ mol HCl}} \times \frac{70.90 \text{ g } Cl_2}{1 \text{ mol } Cl_2} = \textbf{23.4 g Cl}_2$$

3.82   This is a combination of Problem Type 7, Calculating the Amounts of Reactants and Products, and Problem Type 9, Calculating the Percent Yield of a Reaction.

(a)
Step 1: Start with a balanced chemical equation. It's given in the problem. We use NG as an abbreviation for nitroglycerin. The molar mass of NG $= 227.1$ g/mol.

$$4 \, C_3H_5N_3O_9 \longrightarrow 6 \, N_2 + 12 \, CO_2 + 10 \, H_2O + O_2$$

Step 2: Map out the following strategy to solve this problem.

$$\text{g NG} \rightarrow \text{mol NG} \rightarrow \text{mol } O_2 \rightarrow \text{g } O_2$$

Step 3: Calculate the grams of O_2 using the strategy above.

$$? \text{ g } \textbf{O}_2 = 2.00 \times 10^2 \text{ g NG} \times \frac{1 \text{ mol NG}}{227.1 \text{ g NG}} \times \frac{1 \text{ mol } O_2}{4 \text{ mol NG}} \times \frac{32.00 \text{ g } O_2}{1 \text{ mol } O_2}$$

$$= \textbf{7.05 g O}_2$$

(b)   Calculate the percent yield

$$\% \text{ yield} = \frac{\text{actual yield}}{\text{theoretical yield}} \times 100\%$$

$$\% \textbf{ yield} = \frac{6.55 \text{ g } O_2}{7.05 \text{ g } O_2} \times 100\% = \textbf{92.9\%}$$

3.84   The actual yield of ethylene is 481 g. Let's calculate the yield of ethylene if the reaction is 100 percent efficient. We can calculate this from the definition of percent yield. We can then calculate the mass of hexane that must be reacted.

$$\% \text{ yield } = \frac{\text{actual yield}}{\text{theoretical yield}} \times 100\%$$

$$42.5 \% \text{ yield } = \frac{481 \text{ g } C_2H_4}{\text{theoretical yield}} \times 100\%$$

$$\text{theoretical yield } C_2H_4 = 1.13 \times 10^3 \text{ g } C_2H_4$$

The mass of hexane that must be reacted is:

$$1.13 \times 10^3 \text{ g } C_2H_4 \times \frac{1 \text{ mol } C_2H_4}{28.05 \text{ g } C_2H_4} \times \frac{1 \text{ mol } C_6H_{14}}{1 \text{ mol } C_2H_4} \times \frac{86.15 \text{ g } C_6H_{14}}{1 \text{ mol } C_6H_{14}} = \mathbf{3.47 \times 10^3 \text{ g } C_6H_{14}}$$

3.86   We assume that all the Cl in the compound ends up as HCl and all the O ends up as H_2O. Therefore, we need to find the number of moles of Cl in HCl and the number of moles of O in H_2O.

$$\text{mol Cl } = 0.233 \text{ g HCl} \times \frac{1 \text{ mol HCl}}{36.46 \text{ g HCl}} \times \frac{1 \text{ mol Cl}}{1 \text{ mol HCl}} = 0.00639 \text{ mol Cl}$$

$$\text{mol O } = 0.403 \text{ g } H_2O \times \frac{1 \text{ mol } H_2O}{18.02 \text{ g } H_2O} \times \frac{1 \text{ mol O}}{1 \text{ mol } H_2O} = 0.0224 \text{ mol O}$$

Dividing by the smallest number of moles (0.00639 mol) gives the formula, $ClO_{3.5}$. Multiplying both subscripts by two gives the empirical formula, $\mathbf{Cl_2O_7}$.

3.88   The symbol "O" refers to moles of oxygen atoms, not oxygen molecule (O_2). Look at the molecular formulas given in parts (a) and (b). What do they tell you about the relative amounts of carbon and oxygen?

(a) $0.212 \text{ mol C} \times \left(\dfrac{1 \text{ mol O}}{1 \text{ mol C}} \right) = \mathbf{0.212 \text{ mol O}}$

(b) $0.212 \text{ mol C} \times \left(\dfrac{2 \text{ mol O}}{1 \text{ mol C}} \right) = \mathbf{0.424 \text{ mol O}}$

3.90   This is a calculation involving percent composition. Remember,

$$\textbf{percent by mass of each element } = \frac{\text{mass of element in 1 mol of compound}}{\text{molar mass of compound}} \times 100\%$$

The molar masses are: Al, 26.98 g/mol; $Al_2(SO_4)_3$, 342.2 g/mol; H_2O, 18.02 g/mol. Thus, using x as the number of H_2O molecules,

$$\text{mass } \% \text{ Al } = \left(\frac{2(\text{molar mass of Al})}{\text{molar mass of } Al_2(SO_4)_3 + x(\text{molar mass of } H_2O)} \right) \times 100\%$$

$$8.20 \% = \left(\frac{2(26.98 \text{ g})}{342.2 \text{ g} + x(18.02 \text{ g})} \right) \times 100\%$$

$$x = \mathbf{17.53}$$

Rounding off to a whole number of water molecules, $x = \mathbf{18}$. Therefore the formula is $\mathbf{Al_2(SO_4)_3 \cdot 18 \, H_2O}$.

3.92   The number of carbon atoms in a 24-carat diamond is:

$$(24 \text{ carat})\left(\frac{200 \text{ mg C}}{1 \text{ carat}}\right)\left(\frac{1 \text{ g C}}{1000 \text{ mg C}}\right)\left(\frac{1 \text{ mol C}}{12.01 \text{ g C}}\right)\left(\frac{6.022 \times 10^{23} \text{ atoms C}}{1 \text{ mol C}}\right) = \textbf{2.4} \times \textbf{10}^{\textbf{23}} \textbf{ atoms C}$$

3.94   The mass of oxygen in MO is 39.46 g − 31.70 g = 7.76 g O. Therefore, for every 31.70 g of M, there is 7.76 g of O in the compound MO. The molecular formula shows a mole ratio of 1 mol M : 1 mol O. First, calculate moles of M that react with 7.76 g O.

$$\text{mol M} = 7.76 \text{ g O} \times \frac{1 \text{ mol O}}{16.00 \text{ g O}} \times \frac{1 \text{ mol M}}{1 \text{ mol O}} = 0.485 \text{ mol M}$$

$$\textbf{molar mass M} = \frac{31.70 \text{ g M}}{0.485 \text{ mol M}} = \textbf{65.4 g/mol}$$

Thus, the atomic mass of M is **65.4 amu**. The metal is most likely **Zn**.

3.96   The wording of the problem suggests that the actual yield is less than the theoretical yield. The percent yield will be equal to the percent purity of the iron(III) oxide. We find the theoretical yield :

$$2.62 \times 10^3 \text{ kg Fe}_2\text{O}_3 \left(\frac{1000 \text{ g Fe}_2\text{O}_3}{1 \text{ kg Fe}_2\text{O}_3}\right)\left(\frac{1 \text{ mol Fe}_2\text{O}_3}{159.7 \text{ g Fe}_2\text{O}_3}\right)\left(\frac{2 \text{ mol Fe}}{1 \text{ mol Fe}_2\text{O}_3}\right)\left(\frac{55.85 \text{ g Fe}}{1 \text{ mol Fe}}\right)\left(\frac{1 \text{ kg Fe}}{1000 \text{ g Fe}}\right)$$

$$= \textbf{1.83} \times \textbf{10}^{\textbf{3}} \textbf{ kg Fe}$$

$$\text{percent yield} = \frac{\text{actual yield}}{\text{theoretical yield}} \times 100\%$$

$$\textbf{percent yield} = \frac{1.64 \times 10^3 \text{ kg Fe}}{1.83 \times 10^3 \text{ kg Fe}} \times 100\% = \textbf{89.6\%} = \textbf{purity of Fe}_2\textbf{O}_3$$

3.98   The carbohydrate contains 40 percent carbon; therefore, the remaining 60 percent is hydrogen and oxygen. The problem states that the hydrogen to oxygen ratio is 2:1. We can write this 2:1 ratio as H_2O.

Assume 100 g of compound.

$$40.0 \text{ g C} \times \frac{1 \text{ mol C}}{12.01 \text{ g C}} = 3.33 \text{ mol C}$$

$$60.0 \text{ g H}_2\text{O} \times \frac{1 \text{ mol H}_2\text{O}}{18.02 \text{ g H}_2\text{O}} = 3.33 \text{ mol H}_2\text{O}$$

Dividing by 3.33 gives **CH$_2$O** for the empirical formula.

To find the molecular formula, divide the molar mass by the empirical mass.

$$\frac{\text{molar mass}}{\text{empirical mass}} = \frac{178 \text{ g}}{30.03 \text{ g}} \approx 6$$

Thus, there are six CH_2O units in each molecule of the compound, so the molecular formula is $(CH_2O)_6$, or **C$_6$H$_{12}$O$_6$**.

3.100   If we assume 100 g of compound, the masses of Cl and X are 67.2 g and 32.8 g, respectively. We can calculate the moles of Cl.

$$67.2 \text{ g Cl} \times \frac{1 \text{ mol Cl}}{35.45 \text{ g Cl}} = 1.90 \text{ mol Cl}$$

Then, using the mole ratio from the chemical formula (XCl_3), we can calculate the moles of X contained in 32.8 g.

$$1.90 \text{ mol Cl} \times \frac{1 \text{ mol X}}{3 \text{ mol Cl}} = 0.633 \text{ mol X}$$

0.633 moles of X has a mass of 32.8 g. Calculating the molar mass of X:

$$\frac{32.8 \text{ g X}}{0.633 \text{ mol X}} = \textbf{51.8 g/mol}$$

The element is most likely **chromium** (molar mass = 52.00 g/mol).

3.102   A 100 g sample of myoglobin contains 0.34 g of iron (0.34% Fe). The number of moles of Fe is:

$$0.34 \text{ g Fe} \times \frac{1 \text{ mol Fe}}{55.85 \text{ g Fe}} = 6.1 \times 10^{-3} \text{ mol Fe}$$

Since there is one Fe atom in a molecule of myoglobin, the moles of myoglobin also equals 6.1×10^{-3} moles. The molar mass of myoglobin can be calculated.

$$\textbf{molar mass myoglobin} = \frac{100 \text{ g myoglobin}}{6.1 \times 10^{-3} \text{ mol myoglobin}} = \textbf{1.6} \times \textbf{10}^{\textbf{4}} \textbf{ g/mol}$$

3.104   Assume 100 g of sample. Then,

$$\text{mol Na} = 32.08 \text{ g Na} \times \frac{1 \text{ mol Na}}{22.99 \text{ g Na}} = 1.395 \text{ mol Na}$$

$$\text{mol O} = 36.01 \text{ g O} \times \frac{1 \text{ mol O}}{16.00 \text{ g O}} = 2.251 \text{ mol O}$$

$$\text{mol Cl} = 19.51 \text{ g Cl} \times \frac{1 \text{ mol Cl}}{35.45 \text{ g Cl}} = 0.5504 \text{ mol Cl}$$

Since Cl is only contained in NaCl, the moles of Cl equals the moles of Na contained in NaCl.

mol Na (in NaCl) = **0.5504 mol**

The number of moles of Na in the remaining two compounds is: 1.395 mol − 0.5504 mol = 0.8446 mol Na.

To solve for moles of the remaining two compounds, let

x = moles of Na_2SO_4

y = moles of $NaNO_3$

Then, from the mole ratio of Na and O in each compound, we can write

$2x + y$ = mol Na = 0.8446 mol

$4x + 3y$ = mol O = 2.251 mol

Solving two equations with two unknowns gives

$x = 0.1414$ = moles Na_2SO_4 and $y = 0.5618$ = moles $NaNO_3$

Finally, we convert to mass of each compound to calculate the mass percent of each compound in the sample. Remember, the sample size is 100 g.

$$\text{mass \% NaCl} = 0.5504 \text{ mol NaCl} \times \frac{58.44 \text{ g NaCl}}{1 \text{ mol NaCl}} \times \frac{1}{100 \text{ g sample}} \times 100\% = \textbf{32.17\% NaCl}$$

$$\text{mass \% Na}_2\text{SO}_4 = 0.1414 \text{ mol Na}_2\text{SO}_4 \times \frac{142.1 \text{ g Na}_2\text{SO}_4}{1 \text{ mol Na}_2\text{SO}_4} \times \frac{1}{100 \text{ g sample}} \times 100\% = \textbf{20.09\% Na}_2\textbf{SO}_4$$

$$\text{mass \% NaNO}_3 = 0.5618 \text{ mol NaNO}_3 \times \frac{85.00 \text{ g NaNO}_3}{1 \text{ mol NaNO}_3} \times \frac{1}{100 \text{ g sample}} \times 100\% = \textbf{47.75\% NaNO}_3$$

3.106   The mass percent of an element in a compound can be calculated as follows:

$$\textbf{percent by mass of each element} = \frac{\text{mass of element in 1 mol of compound}}{\text{molar mass of compound}} \times 100\%$$

The molar mass of $Ca_3(PO_4)_2 = 310.18$ g/mol

$$\%\text{Ca} = \frac{3 \times 40.08 \text{ g}}{310.18 \text{ g}} \times 100\% = \textbf{38.76\% Ca}$$

$$\%\text{P} = \frac{2 \times 30.97 \text{ g}}{310.18 \text{ g}} \times 100\% = \textbf{19.97\% P}$$

$$\%\text{O} = \frac{8 \times 16.00 \text{ g}}{310.18 \text{ g}} \times 100\% = \textbf{41.27\% O}$$

3.108   **Yes.** The number of hydrogen atoms in one gram of hydrogen molecules is the same as the number in one gram of hydrogen atoms. There is no difference in mass, only in the way that the particles are arranged.

Would the mass of 100 dimes be the same if they were stuck together in pairs instead of separated?

3.110   Since we assume that water exists as either H_2O or D_2O, the natural abundances are 99.985 percent and 0.015 percent, respectively. If we convert to molecules of water (both H_2O or D_2O), we can calculate the molecules that are D_2O from the natural abundance (0.015%).

The necessary conversions are:

mL water $\rightarrow$ g water $\rightarrow$ mol water $\rightarrow$ molecules water $\rightarrow$ molecules D_2O

$$400 \text{ mL water} \times \frac{1 \text{ g water}}{1 \text{ mL water}} \times \frac{1 \text{ mol water}}{18.02 \text{ g water}} \times \frac{6.022 \times 10^{23} \text{ molecules}}{1 \text{ mol water}} \times \frac{0.015\% \text{ molecules D}_2\text{O}}{100\% \text{ molecules water}}$$

$$= \textbf{2.01} \times \textbf{10}^{21} \textbf{ molecules D}_2\textbf{O}$$

3.112   First, we can calculate the moles of oxygen.

$$2.445 \text{ g C} \times \frac{1 \text{ mol C}}{12.01 \text{ g C}} \times \frac{1 \text{ mol O}}{1 \text{ mol C}} = 0.2036 \text{ mol O}$$

Next, we can calculate the molar mass of oxygen.

$$\text{molar mass O} = \frac{3.257 \text{ g O}}{0.2036 \text{ mol O}} = 16.00 \text{ g/mol}$$

If 1 mole of oxygen atoms has a mass of 16.00 g, then 1 atom of oxygen has an **atomic mass of 16.00 amu**.

3.114   **(a)**   The mass of chlorine is **5.0 g**.

(b)   From the percent by mass of Cl, we can calculate the mass of chlorine in 60.0 g of $NaClO_3$.

$$\text{mass \% Cl} = \frac{35.45 \text{ g Cl}}{106.44 \text{ g compound}} \times 100\% = 33.31\% \text{ Cl}$$

mass Cl $= 60.0 \text{ g} \times 0.3331 = $ **20.0 g Cl**

(c)   0.10 mol of KCl contains 0.10 mol of Cl.

$$0.10 \text{ mol Cl} \times \frac{35.45 \text{ g Cl}}{1 \text{ mol Cl}} = \textbf{3.5 g Cl}$$

(d)   From the percent by mass of Cl, we can calculate the mass of chlorine in 30.0 g of $MgCl_2$.

$$\text{mass \% Cl} = \frac{2 \times 35.45 \text{ g Cl}}{95.21 \text{ g compound}} \times 100\% = 74.47\% \text{ Cl}$$

mass Cl $= 30.0 \text{ g} \times 0.7447 = $ **22.3 g Cl**

(e)   The mass of Cl can be calculated from the molar mass of Cl_2.

$$0.50 \text{ mol Cl}_2 \times \frac{70.90 \text{ g Cl}}{1 \text{ mol Cl}_2} = \textbf{35.45 g Cl}$$

Thus, **0.50 mol Cl_2** contains the greatest mass of chlorine.

3.116   Both compounds contain only Pt and Cl. The percent by mass of Pt can be calculated by subtracting the percent Cl from 100 percent.

Compound A: Assume 100 g of compound.

$$26.7 \text{ g Cl} \times \frac{1 \text{ mol Cl}}{35.45 \text{ g Cl}} = 0.753 \text{ mol Cl}$$

$$73.3 \text{ g Pt} \times \frac{1 \text{ mol Pt}}{195.1 \text{ g Pt}} = 0.376 \text{ mol Pt}$$

Dividing by the smallest number of moles (0.376 mol) gives the empirical formula, **$PtCl_2$**.

Compound B: Assume 100 g of compound.

$$42.1 \text{ g Cl} \times \frac{1 \text{ mol Cl}}{35.45 \text{ g Cl}} = 1.19 \text{ mol Cl}$$

$$57.9 \text{ g Pt} \times \frac{1 \text{ mol Pt}}{195.1 \text{ g Pt}} = 0.297 \text{ mol Pt}$$

Dividing by the smallest number of moles (0.297 mol) gives the empirical formula, **$PtCl_4$**.

3.118   Both compounds contain only Mn and O. When the first compound is heated, oxygen gas is evolved. Let's calculate the empirical formulas for the two compounds, then we can write a balanced equation.

(a) Compound X: Assume 100 g of compound.

$$63.3 \text{ g Mn} \times \frac{1 \text{ mol Mn}}{54.94 \text{ g Mn}} = 1.15 \text{ mol Mn}$$

$$36.7 \text{ g O} \times \frac{1 \text{ mol O}}{16.00 \text{ g O}} = 2.29 \text{ mol O}$$

Dividing by the smallest number of moles (1.15 mol) gives the empirical formula, **MnO_2**.

Compound Y: Assume 100 g of compound.

$$72.0 \text{ g Mn} \times \frac{1 \text{ mol Mn}}{54.94 \text{ g Mn}} = 1.31 \text{ mol Mn}$$

$$28.0 \text{ g O} \times \frac{1 \text{ mol O}}{16.00 \text{ g O}} = 1.75 \text{ mol O}$$

Dividing by the smallest number of moles gives $MnO_{1.33}$. Recall that an empirical formula must have whole number coefficients. Multiplying by a factor of 3 gives the empirical formula **Mn_3O_4**.

(b) The unbalanced equation is: $MnO_2 \longrightarrow Mn_3O_4 + O_2$

Balancing by inspection gives: $3 \text{ } MnO_2 \longrightarrow Mn_3O_4 + O_2$

3.120 SO_2 is converted to H_2SO_4 by reaction with water. The mole ratio between SO_2 and H_2SO_4 is 1:1.

This is a unit conversion problem. You should come up with the following strategy to solve the problem.

tons SO_2 → ton-mol SO_2 → ton-mol H_2SO_4 → tons H_2SO_4

? tons H_2SO_4 $= 4.0 \times 10^5 \text{ tons } SO_2 \times \dfrac{1 \text{ ton-mol } SO_2}{64.07 \text{ tons } SO_2} \times \dfrac{1 \text{ ton-mol } H_2SO_4}{1 \text{ ton-mol } SO_2} \times \dfrac{98.09 \text{ tons } H_2SO_4}{1 \text{ ton-mol } H_2SO_4}$

$= \mathbf{6.1 \times 10^5}$ **tons H_2SO_4**

> **Tip:** You probably won't come across a ton-mol that often in chemistry. However, it was convenient to use in this problem. We normally use a g-mol. 1 g-mol SO_2 has a mass of 64.07 g. In a similar manner, 1 ton-mol of SO_2 has a mass of 64.07 tons.

3.122   We assume that the increase in mass results from the element nitrogen. The mass of nitrogen is:

$$0.378 \text{ g} - 0.273 \text{ g} = 0.105 \text{ g N}$$

The empirical formula can now be calculated. Assume 100 g of compound.

$$0.273 \text{ g Mg} \times \frac{1 \text{ mol Mg}}{24.31 \text{ g Mg}} = 0.0112 \text{ mol Mg}$$

$$0.105 \text{ g N} \times \frac{1 \text{ mol N}}{14.01 \text{ g N}} = 0.00749 \text{ mol N}$$

Dividing by the smallest number of moles gives $Mg_{1.5}N$. Recall that an empirical formula must have whole number coefficients. Multiplying by a factor of 2 gives the empirical formula **Mg_3N_2**. The name of this compound is **magnesium nitride**.

3.124 This problem involves the experimental determination of empirical formulas, Problem Type 5.

Step 1: Calculate the mass of C in 55.90 g CO_2, and the mass of H in 28.61 g H_2O. This is a factor-label problem. To calculate the mass of each component, you need the molar masses and the correct mole ratio.

You should come up with the following strategy:

$$g\ CO_2 \rightarrow mol\ CO_2 \rightarrow mol\ C \rightarrow g\ C$$

Step 2: $?\ g\ C = 55.90\ g\ CO_2 \times \dfrac{1\ mol\ CO_2}{44.01\ g\ CO_2} \times \dfrac{1\ mol\ C}{1\ mol\ CO_2} \times \dfrac{12.01\ g\ C}{1\ mol\ C} = \textbf{15.25 g C}$

Similarly,

$?\ g\ H = 28.61\ g\ H_2O \times \dfrac{1\ mol\ H_2O}{18.02\ g\ H_2O} \times \dfrac{2\ mol\ H}{1\ mol\ H_2O} \times \dfrac{1.008\ g\ H}{1\ mol\ H} = \textbf{3.201 g H}$

Since the compound contains C, H, and Pb, we can calculate the mass of Pb by difference.

$$51.36\ g = mass\ C + mass\ H + mass\ Pb$$

$$51.36\ g = 15.25\ g + 3.201\ g + mass\ Pb$$

$$mass\ Pb = \textbf{32.91 g Pb}$$

Step 3: Calculate the number of moles of each element present in the sample. This is also a factor-label problem, using the molar mass as a conversion factor. You should come up with the following strategy (see Problem Type 2A):

$$g\ C \rightarrow mol\ C$$

$?\ mol\ C = 15.25\ g\ C \times \dfrac{1\ mol\ C}{12.01\ g\ C} = \textbf{1.270 mol C}$

Similarly,

$?\ mol\ H = 3.201\ g\ H \times \dfrac{1\ mol\ H}{1.008\ g\ H} = \textbf{3.176 mol H}$

$?\ mol\ Pb = 32.91\ g\ Pb \times \dfrac{1\ mol\ Pb}{207.2\ g\ Pb} = \textbf{0.1588 mol Pb}$

Thus, we arrive at the formula $Pb_{0.1588}C_{1.270}H_{3.176}$, which gives the identity and the ratios of atoms present. However, chemical formulas are written with whole numbers.

Step 4: Try to convert to whole numbers by dividing all the subscripts by the smallest subscript.

Pb: $\dfrac{0.1588}{0.1588} = 1.00$ $\qquad$ **C:** $\dfrac{1.270}{0.1588} \approx 8$ $\qquad$ **H:** $\dfrac{3.176}{0.1588} \approx 20$

This gives the empirical formula, **PbC_8H_{20}**.

3.126   **(a)**   The following strategy can be used to convert from the volume of the Mg cube to the number of Mg atoms.

$$cm^3 \rightarrow grams \rightarrow moles \rightarrow atoms$$

$$1.0 \; cm^3 \times \frac{1.74 \; g \; Mg}{1 \; cm^3} \times \frac{1 \; mol \; Mg}{24.31 \; g \; Mg} \times \frac{6.022 \times 10^{23} \; Mg \; atoms}{1 \; mol \; Mg} = \textbf{4.3} \times \textbf{10}^{\textbf{22}} \; \textbf{Mg atoms}$$

(b)   Since 74 percent of the available space is taken up by Mg atoms, 4.3×10^{22} atoms occupy the following volume:

$$0.74 \times 1.0 \; cm^3 = 0.74 \; cm^3$$

We are trying to calculate the radius of a single Mg atom, so we need the volume occupied by a single Mg atom.

$$volume \; Mg \; atom = \frac{0.74 \; cm^3}{4.3 \times 10^{22} \; Mg \; atoms} = 1.7 \times 10^{-23} \; cm^3/Mg \; atom$$

The volume of a sphere is $(4/3)\pi r^3$. Solving for the radius:

$$V = 1.7 \times 10^{-23} \; cm^3 = \frac{4}{3}\pi r^3$$

$$r^3 = 4.1 \times 10^{-24} \; cm^3$$

$$r = 1.6 \times 10^{-8} \; cm$$

Converting to picometers:

$$\textbf{radius Mg atom} = 1.6 \times 10^{-8} \; cm \times \frac{0.01 \; m}{1 \; cm} \times \frac{1 \; pm}{1 \times 10^{-12} \; m} = \textbf{1.6} \times \textbf{10}^{\textbf{2}} \; \textbf{pm}$$

3.128   The molar mass of air can be calculated by multiplying the mass of each component by its abundance and adding them together. Recall that nitrogen gas and oxygen gas are diatomic.

molar mass air $= (0.7808)(28.02 \; g/mol) + (0.2095)(32.00 \; g/mol) + (0.0097)(39.95 \; g/mol) = \textbf{28.97 g/mol}$

3.130   The surface area of the water can be calculated assuming that the dish is circular.

$$surface \; area \; of \; water = \pi r^2 = \pi (10 \; cm)^2 = 3.1 \times 10^2 \; cm^2$$

The cross-sectional area of one stearic acid molecule in cm^2 is:

$$0.21 \; nm^2 \times \left(\frac{1 \times 10^{-9} \; m}{1 \; nm} \right)^2 \times \left(\frac{1 \; cm}{0.01 \; m} \right)^2 = 2.1 \times 10^{-15} \; cm^2/molecule$$

Assuming that there is no empty space between molecules, we can calculate the number of stearic acid molecules that will fit in an area of $3.1 \times 10^2 \; cm^2$.

$$3.1 \times 10^2 \; cm^2 \times \frac{1 \; molecule}{2.1 \times 10^{-15} \; cm^2} = 1.5 \times 10^{17} \; molecules$$

Next, we can calculate the moles of stearic acid in the 1.4×10^{-4} g sample. Then, we can calculate Avogadro's number – the number of molecules per mole.

$$1.4 \times 10^{-4} \text{ g stearic acid} \times \frac{1 \text{ mol stearic acid}}{284.5 \text{ g stearic acid}} = 4.9 \times 10^{-7} \text{ mol stearic acid.}$$

$$\textbf{Avogadro's number } (N_A) = \frac{1.5 \times 10^{17} \text{ molecules}}{4.9 \times 10^{-7} \text{ mol}} = \textbf{3.1} \times \textbf{10}^{\textbf{23}} \textbf{ molecules/mol}$$

3.132 **(a)** The balanced chemical equation is:

$$C_3H_8 \ (g) \ + \ 3 \ H_2O \ (g) \ \longrightarrow \ 3 \ CO \ (g) \ + \ 7 \ H_2 \ (g)$$

(b) You should come up with the following strategy to solve this problem. In this problem, we use kg-mol to save a couple of steps.

$$\text{kg } C_3H_8 \ \rightarrow \ \text{mol } C_3H_8 \ \rightarrow \ \text{mol } H_2 \ \rightarrow \ \text{kg } H_2$$

$$\textbf{? kg } H_2 = 2.84 \times 10^3 \text{ kg } C_3H_8 \times \frac{1 \text{ kg-mol } C_3H_8}{44.09 \text{ kg } C_3H_8} \times \frac{7 \text{ kg-mol } H_2}{1 \text{ kg-mol } C_3H_8} \times \frac{2.016 \text{ kg } H_2}{1 \text{ kg-mol } H_2}$$

$$= \textbf{9.09} \times \textbf{10}^{\textbf{2}} \textbf{ kg } H_2$$

CHAPTER 4
REACTIONS IN AQUEOUS SOLUTION

PROBLEM-SOLVING STRATEGIES AND TUTORIAL SOLUTIONS

TYPES OF PROBLEMS

Problem Type 1: Applying Solubility Rules.

Problem Type 2: Writing Molecular, Ionic, and Net Ionic Equations.

Problem Type 3: Acid-Base Reactions.
  (a) Identifying Brønsted acids and bases.
  (b) Writing acid/base reactions.

Problem Type 4: Oxidation-Reduction Reactions.
  (a) Assigning oxidation numbers.
  (b) Writing oxidation/reduction half-reactions.
  (c) Using an activity series.

Problem Type 5: Concentration of Solutions.

Problem Type 6: Dilution of Solutions.

Problem Type 7: Gravimetric Analysis.

Problem Type 8: Acid-Base Titrations.

Problem Type 9: Redox Titrations.

PROBLEM TYPE 1: APPLYING SOLUBILITY RULES

Ionic compounds are classified as "soluble", "slightly soluble", or "insoluble". Table 4.2 of your text provides solubility rules that will help you determine how a given compound behaves in aqueous solution.

EXAMPLE 4.1
According to the solubility rules, which of the following compounds are soluble in water?
(a) $MgCO_3$ (b) $AgNO_3$ (c) $MgCl_2$ (d) $Ca_3(PO_4)_2$ (e) KOH

Refer to Table 4.2 of your text to answer this question.

(a) $MgCO_3$ is *insoluble* (Most ionic compounds containing carbonate ions are *insoluble*).

(b) $AgNO_3$ is *soluble* (All ionic compounds containing nitrate ions are *soluble*).

(c) $MgCl_2$ is *soluble* (Most ionic compounds containing chloride ions are *soluble*).

(d) $Ca_3(PO_4)_2$ is *insoluble* (Most ionic compounds containing phosphate ions are *insoluble*).

(e) KOH is *soluble* (All ionic compounds containing alkali metal ions are *soluble*).

PRACTICE EXERCISE

1. Predict whether the following ionic compounds are soluble or insoluble in water.

   (a) $NaNO_3$ (b) AgCl   (c) $Ba(OH)_2$ (d) $CaCO_3$

Text Problems: **4.16**, 4.20

PROBLEM TYPE 2: WRITING MOLECULAR, IONIC, AND NET IONIC EQUATIONS

In a *molecular equation*, the formulas are written as though all species existed as molecules or whole units. However, a molecular equation does not accurately describe what actually happens at the microscopic level. To better describe the reaction in solution, the equation should show the dissociation of dissolved ionic compounds into ions. An *ionic equation* shows dissolved ionic compounds in terms of their free ions. A *net ionic equation* shows only the species that actually take part in the reaction.

EXAMPLE 4.2

Write balanced molecular, ionic, and net ionic equations for the reaction that occurs when a $BaCl_2$ solution is mixed with a Na_2SO_4 solution.

Both ionic solids given in the problem are soluble, because the problem states that both ionic solids are in solution. You could also check the solubility rules. The combined solution will contain Na^+, SO_4^{2-}, Ba^{2+}, and Cl^- ions. The two possible ionic products are NaCl and $BaSO_4$. If one (or both) is insoluble, it (they) will precipitate. According to Table 4.2, NaCl is soluble. On the other hand, $BaSO_4$ is insoluble. The *molecular equation* should show $BaSO_4$ (*s*) as a product along with NaCl (*aq*).

$$BaCl_2\ (aq)\ +\ Na_2SO_4\ (aq)\ \longrightarrow\ BaSO_4\ (s)\ +\ 2\ NaCl\ (aq)$$

The *ionic equation* should show dissolved ionic compounds in terms of their free ions.

$$Ba^{2+}\ (aq)\ +\ 2\ Cl^-\ (aq)\ +\ 2\ Na^+\ (aq)\ +\ SO_4^{2-}\ (aq)\ \longrightarrow\ BaSO_4\ (s)\ +\ 2\ Na^+\ (aq)\ +\ 2\ Cl^-\ (aq)$$

As you write out the ionic equation above, you should notice that some ions (Na^+ and Cl^-) are not involved in the overall reaction. These ions are called *spectator ions*. Since the spectator ions appear on both sides of the equation and are unchanged in the chemical reaction, they can be canceled from both sides of the equation. A *net ionic equation* shows only the species that actually take part in the reaction.

Cancel the spectator ions to write the *net ionic equation*.

$$Ba^{2+}\ (aq)\ +\ SO_4^{2-}\ (aq)\ \longrightarrow\ BaSO_4\ (s)$$

> **Tip:** To help pick out the spectator ions, think about spectators at a sporting event. The spectators are at the stadium, watching the action, but they do *not* participate in the game.

PRACTICE EXERCISE

2.   Write the balanced molecular, ionic, and net ionic equations for the following reaction:

$$CaCl_2\ (aq)\ +\ Na_2CO_3\ (aq)\ \longrightarrow$$

Text Problem: **4.18**

PROBLEM TYPE 3: ACID-BASE REACTIONS

A. Identifying Brønsted acids and bases

A **Brønsted acid** is a proton donor, and a **Brønsted base** is a proton acceptor. To identify a Brønsted acid, you should look for a substance that contains hydrogen. The formula of inorganic acids will begin with H. For example, HCl (hydrochloric acid), HNO_2 (nitrous acid), H_3PO_4 (phosphoric acid). Carboxylic acids contain the carboxyl group, –COOH. The hydrogen from the carboxyl group can be donated. Examples of carboxylic acids are CH_3COOH (acetic acid) and HCOOH (formic acid).

To identify a Brønsted base, you should look for soluble hydroxide salts. The hydroxide ion (OH^-) will accept a proton to form H_2O. Also look for weak bases, which are amines. Ammonia (NH_3) is an example. Finally, look for anions from acids. These negatives ions can accept a proton (H^+). Some examples, are $H_2PO_4^{2-}$, NO_2^-, and HCO_3^-.

EXAMPLE 4.3

Identify each of the following species as a Brønsted acid, base, or both: (a) HNO_3, (b) $Ba(OH)_2$, (c) SO_4^{2-}, (d) $CH_3CH_2CH_2COOH$, (e) HPO_4^{2-}.

(a)   Brønsted acid. The formula of this compound starts with H; this indicates that it is probably an acid.
(b)   Brønsted base. This is a soluble hydroxide salt.
(c)   Brønsted base. This negative ion can accept a proton; therefore, it is a base.
(d)   Brønsted acid. This is a carboxylic acid. It contains a carboxyl group, –COOH.
(e)   Both a Brønsted acid and base. This ion has a proton (H^+) that it can donate. It also has a negative charge and therefore can accept a proton.

PRACTICE EXERCISE

3.   Identify each of the following species as a Brønsted acid, base, or both.

    (a) CH_3CH_2COOH **(b)** HF   **(c)** KOH   **(d)** HCO_3^-

> **Text Problem: 4.28**

B. Writing Acid-Base Reactions

An acid-base reaction is called a **neutralization reaction**. The typical products of an acid-base reaction are a salt and water.

$$acid \; + \; base \; \longrightarrow \; salt \; + \; water$$

Let's consider a generic acid, HA, reacted with a generic base, MOH.

$$HA \; + \; MOH \; \longrightarrow \; MA \; (salt) \; + \; H_2O$$

The H^+ from the acid combines with OH^- from the base to produce water. The anion from the acid, A^-, combines with the metal cation from the base, M^+, to form the salt, MA.

EXAMPLE 4.4

Complete and balance the following equations and write the corresponding ionic and net ionic equations:

(a) HBr (aq) + $Ba(OH)_2$ (aq) $\longrightarrow$

(b) $HCOOH$ (aq) + $NaOH$ (aq) $\longrightarrow$

(a) $2\ HBr\ (aq)\ +\ Ba(OH)_2\ (aq)\ \longrightarrow\ BaBr_2\ (aq)\ +\ 2\ H_2O\ (l)$

HBr (aq) is a strong acid and therefore it completely ionizes in solution. Ba(OH)$_2$ (aq) is a strong base and therefore it also completely ionizes in solution. The balanced *ionic equation* is:

$$2H^+(aq) + 2Br^-(aq) + Ba^{2+}(aq) + 2OH^-(aq) \longrightarrow Ba^{2+}(aq) + 2Br^-(aq) + 2H_2O(l)$$

Cancel the spectator ions to write the *net ionic equation*.

$$2\ H^+\ (aq)\ +\ 2\ OH^-\ (aq)\ \longrightarrow\ 2\ H_2O\ (l)\quad \text{or}\quad H^+\ (aq)\ +\ OH^-\ (aq)\ \longrightarrow\ H_2O\ (l)$$

(b) $HCOOH\ (aq)\ +\ NaOH\ (aq)\ \longrightarrow\ HCOONa\ (aq)\ +\ H_2O\ (l)$

Formic acid, HCOOH (aq), is a weak acid and therefore it only ionizes to a small extent in water. In solution, most formic acid molecules remain as HCOOH. NaOH is a strong base and therefore it completely ionizes in solution. The balanced *ionic equation* is:

$$HCOOH\ (aq)\ +\ Na^+\ (aq)\ +\ OH^-\ (aq)\ \longrightarrow\ Na^+\ (aq)\ +\ HCOO^-\ (aq)\ +\ H_2O\ (l)$$

Cancel the spectator ion (Na$^+$) to write the *net ionic equation*.

$$HCOOH\ (aq)\ +\ OH^-\ (aq)\ \longrightarrow\ HCOO^-\ (aq)\ +\ H_2O\ (l)$$

PRACTICE EXERCISE

4.   Write the balanced molecular, ionic, and net ionic equations for the following acid-base reaction:

$$HNO_2\ (aq)\ +\ Ba(OH)_2\ (aq)\ \longrightarrow$$

Text Problem: 4.30

PROBLEM TYPE 4: OXIDATION-REDUCTION REACTIONS

A. Assigning oxidation numbers

Oxidation numbers are assigned to reactants and products in oxidation-reduction (redox) reactions to keep track of electrons. An oxidation number refers to the number of charges an atom would have in a molecule (or an ionic compound) if electrons were transferred completely.

Rules for assigning oxidation numbers are in Section 4.4 of your text. These rules will be used in the following example.

To assign oxidation numbers you should refer to the following *two* steps:

Step 1: Use the rules in Section 4.4 to assign oxidation numbers to as many atoms as possible.

Step 2: Often times, one atom does not follow any rules outlined in Section 4.4. To assign an oxidation number to this atom, follow rule 6 of the text. In a neutral molecule, the sum of the oxidation numbers of all the atoms must be zero. In a polyatomic ion, the sum of the oxidation numbers of all the elements in the ion must be equal to the net charge of the ion.

EXAMPLE 4.5
Assign oxidation numbers to all the atoms in the following compounds and ion:
(a) Na_2SO_4, (b) $CuCl$, (c) SO_3^{2-}

(a)

Step 1: Na always has an oxidation number of +1 (Rule 2). The oxidation number of oxygen in most compounds is –2 (Rule 3).

Step 2: You can now assign an oxidation number to S based on Na having a +1 oxidation number and O having a –2 oxidation number. This is a neutral ionic compound, so the sum of the oxidation numbers of all the atoms must be zero.

$$2(\text{oxi. no. Na}) + (\text{oxi. no. S}) + 4(\text{oxi. no. O}) = 0$$

$$2(+1) + (\text{oxi. no. S}) + 4(-2) = 0$$

$$\textbf{(oxi. no. S)} = \textbf{8} - \textbf{2} = \textbf{+6}$$

(b)

Step 1: An oxidation number of –1 can be assigned to Cl (Rule 5).

Step 2: You can now assign an oxidation number to Cu. This is a neutral ionic compound.

$$(\text{oxi. no. Cu}) + (\text{oxi. no. Cl}) = 0$$
$$(\text{oxi. no. Cu}) + (-1) = 0$$

$$\textbf{(oxi. no. Cu)} = \textbf{+1}$$

(c)

Step 1: An oxidation number of –2 can be assigned to oxygen (Rule 3).

Step 2: You can now assign an oxidation number to S. SO_3^{2-} is a polyatomic ion. The sum of the oxidation numbers of all elements in the ion must be equal to the net charge of the ion, in this case –2.

$$(\text{oxi. no. S}) + 3(\text{oxi. no. O}) = -2$$
$$(\text{oxi. no. S}) + 3(-2) = -2$$

$$\textbf{(oxi. no. S)} = \textbf{-2} + \textbf{6} = \textbf{+4}$$

PRACTICE EXERCISE

5.   Assign oxidation numbers to the underlined atoms in the following molecules or ions:

(a) $\underline{C}O_3^{2-}$ (b) $\underline{Cu}Cl_2$ (c) $\underline{Ti}O_2$ (d) $\underline{N}O_3^-$

Text Problems: **4.42**, 4.44, 4.46

B. Writing oxidation-reduction half-reactions

In order to break a redox reaction down into an oxidation half-reaction and a reduction half-reaction, you must first assign oxidation numbers to all the atoms in the reaction. In this way, you can determine which element is oxidized (loses electrons) and which element is reduced (gains electrons).

EXAMPLE 4.6
For the following redox reaction, break down the reaction into its half-reactions.

$$2\ \textbf{Al} + \textbf{Fe}_2\textbf{O}_3 \longrightarrow \textbf{Al}_2\textbf{O}_3 + 2\ \textbf{Fe}$$

Reactants
Step 1: The oxidation number of Al is 0 (Rule 1), and the oxidation number of O in a compound is –2 (Rule 3).

Step 2: Solve for the oxidation number of Fe in Fe_2O_3. This is a neutral ionic compound.

$$2(\text{oxi. no. Fe}) + 3(\text{oxi. no. O}) = 0$$
$$2(\text{oxi. no. Fe}) + 3(-2) = 0$$
$$2(\text{oxi. no. Fe}) = +6$$

(oxi. no. Fe) = +3

Products
Step 1: The oxidation number of Fe is 0 (Rule 1), and the oxidation number of O in a compound is –2 (Rule 3).

Step 2: Solve for the oxidation number of Al in Al_2O_3. This is a neutral ionic compound.

$$2(\text{oxi. no. Al}) + 3(\text{oxi. no. O}) = 0$$
$$2(\text{oxi. no. Al}) + 3(-2) = 0$$
$$2(\text{oxi. no. Al}) = +6$$

(oxi. no. Al) = +3

$$Al \longrightarrow Al^{3+} + 3\,e^- \quad (\text{oxidation half-reaction, 3 } e^- \text{ lost})$$
$$Fe^{3+} + 3\,e^- \longrightarrow Fe \quad (\text{reduction half-reaction, 3 } e^- \text{ gained})$$

Tip: When a species is oxidized, the oxidation number will *increase*. In this example, the oxidation number of Al *increased* from 0 to +3. When a species is reduced, the oxidation number will *decrease*. In this example, the oxidation number of Fe *decreased* from +3 to 0.

PRACTICE EXERCISE

6. The nickel-cadmium (nicad) battery, a popular rechargeable "dry cell" used in battery-operated tools, uses the following redox reaction to generate electricity:

$$Cd\,(s) + NiO_2\,(s) + 2\,H_2O\,(l) \longrightarrow Cd(OH)_2\,(s) + Ni(OH)_2\,(s)$$

Assign oxidation numbers to all the atoms and ions, identify the substances that are oxidized and reduced, and write oxidation and reduction half-reactions.

Text Problem: 4.40

C. Using an activity series

An activity series is used to predict whether a metal or hydrogen displacement reaction will occur (See Figure 4.15). An activity series can be described as a convenient summary of the results of many possible displacement reactions.

(1) Hydrogen displacement. Any metal above hydrogen in the activity series will displace it from water or from an acid. Metals below hydrogen will *not* react with either water or an acid.

(2) Metal displacement. Any metal will react with a compound containing any metal ion listed below it.

EXAMPLE 4.7
Predict the outcome of the reactions represented by the following equations by using the activity series, and balance the equations.

(a) $Mg\,(s) + HCl\,(aq) \longrightarrow$

Since Mg is above hydrogen in the activity series, it will displace hydrogen from the acid.

$$Mg\,(s) + 2\,HCl\,(aq) \longrightarrow MgCl_2\,(aq) + H_2\,(g)$$

(b) Au (s) + H$_2$O (l) $\longrightarrow$

Since Au (gold) is below hydrogen in the activity series, it will *not* react with water. You probably already knew this.

$$\text{Au } (s) \text{ + H}_2\text{O } (l) \longrightarrow \text{ No reaction}$$

(c) Cu (s) + NiCl$_2$ (aq) $\longrightarrow$

Any metal will react with a compound containing any metal ion listed *below* it. In this case Ni^{2+} is listed *above* Cu in the activity series. No reaction will occur.

$$\text{Cu } (s) \text{ + NiCl}_2 \text{ } (aq) \longrightarrow \text{ No reaction}$$

PRACTICE EXERCISE

7.   Predict the outcome of the following reactions using the activity series. If a reaction occurs, balance the equation.

    (a) Al (s) + HCl (aq) $\longrightarrow$
    (b) Au (s) + KCl (aq) $\longrightarrow$

Text Problems: **4.48**, 4.50

PROBLEM TYPE 5: CONCENTRATION OF SOLUTIONS

Solutions are characterized by their concentration, that is, the amount of solute dissolved in a given quantity of solvent. One of the most common units of concentration in chemistry is **molarity (M)**. Molarity is the number of moles of solute in 1 liter of solution:

$$M = \text{molarity} = \frac{\text{moles of solute}}{\text{liters of solution}}$$

Sometimes, it is useful to rearrange the above equation to the following form:

$$\text{moles solute} = \text{molarity} \times \text{liters of solution}$$

EXAMPLE 4.8

What is the molarity of a solution made by dissolving 32.1 g of KNO$_3$ in enough water to make 500 mL of solution?

Step 1: Based on the definition of molarity, grams of solute must be converted to moles of solute, and mL of solution must be converted to L of solution. For discussion of unit conversions, see Problem Type 5, chapter 1.

$$\mathcal{M}\,(\text{KNO}_3) = 101.1 \text{ g/mol}$$

$$? \text{ moles solute} = 32.1 \text{ g KNO}_3 \times \frac{1 \text{ mol KNO}_3}{101.1 \text{ g KNO}_3} = 0.318 \text{ mol KNO}_3$$

$$? \text{ liters of solution} = 500 \text{ mL solution} \times \frac{1 \text{ L}}{1000 \text{ mL}} = 0.500 \text{ L solution}$$

Step 2: Substitute the above values into the molarity equation.

$$M = \frac{\text{moles of solute}}{\text{liters of solution}}$$

$$M = \frac{0.318 \text{ mol KNO}_3}{0.500 \text{ L solution}} = 0.636 \, M$$

This is normally written 0.636 M KNO$_3$.

PRACTICE EXERCISE

8.   An aqueous nutrient solution is prepared by adding 50.23 g of KNO$_3$ (molar mass = 101.1 g/mol) to enough water to fill a 40.0 L container. What is the molarity of the KNO$_3$ solution?

EXAMPLE 4.9
How many moles of solute are in 2.50 × 10^2 mL of 0.100 M KCl?

Step 1: Since the problem asks for moles of solute, you must solve the equation algebraically for moles of solute.

moles solute = molarity × liters of solution

Step 2: Substitute the molarity and liters of solution into the above equation to solve for moles solute.

2.50 × 10^2 mL = 0.250 L

$$\text{? moles KCl solute} = \frac{0.100 \text{ moles solute}}{1 \text{ L solution}} \times 0.250 \text{ L solution} = \textbf{0.0250 mol}$$

PRACTICE EXERCISE

9.   You need to prepare 1.00 L of a 0.500 M NaCl solution. What mass of NaCl (in g) must you weigh out to prepare this solution?

Text Problems: **4.56**, 4.58, 4.60, 4.62

PROBLEM TYPE 6: DILUTION OF SOLUTIONS

Dilution refers to the procedure for preparing a less-concentrated solution from a more-concentrated one. The key to solving a dilution problem is to realize that

moles of solute *before* dilution = moles of solute *after* dilution

In Problem Type 5 above, we discussed how to calculate moles of solute from the molarity and the volume of solution.

moles solute = molarity × volume of solution (in L)

Thus,

moles of solute *before* dilution (initial) = moles of solute *after* dilution (final)

$$M_{initial}V_{initial} = M_{final}V_{final}$$

EXAMPLE 4.10
What volume of a concentrated (12.0 M) hydrochloric acid stock solution is needed to prepare 8.00 × 10^2 mL of 0.120 M HCl?

Recognize that the problem asks for the initial volume of stock solution needed to prepare the dilute solution. Solve the above equation algebraically for $V_{initial}$, then substitute in the appropriate values from the problem.

$$V_{initial} = \frac{M_{final}V_{final}}{M_{initial}}$$

$$V_{\text{initial}} = \frac{(0.120\ M)(8.00 \times 10^2\ mL)}{12.0\ M} = \textbf{8.00 mL}$$

> **Tip:** The units of V_{initial} and V_{final} can be milliliters or liters for a dilution problem as long as they are the same. Be consistent. Also, make sure to check whether your results seem reasonable. Be sure that $M_{\text{initial}} > M_{\text{final}}$ and $V_{\text{final}} > V_{\text{initial}}$.

PRACTICE EXERCISE

10. A 20.0 mL sample of 0.127 M Ca(NO$_3$)$_2$ is diluted to 5.00 L. What is the molarity of the resulting solution?

Text Problems: 4.66, 4.68, 4.70

PROBLEM TYPE 7: GRAVIMETRIC ANALYSIS

Gravimetric analysis is an analytical technique based on the measurement of mass. The type of gravimetric analysis discussed in your text involves the formation, isolation, and mass determination of a precipitate. This procedure is applicable only to reactions that go to completion, or have nearly a 100 percent yield. Thus, the precipitate must be insoluble rather than slightly soluble. See Section 4.6 of your text for further discussion of this technique.

The typical problem involves determining the mass percent of an element in one of the reactants. The element (ion) of interest is completely precipitated from solution. Use the following approach to solve this type of problem.

Step 1: From the measured mass of precipitate, calculate the mass of the element of interest in the precipitate. This is the same amount of the element that was present in the original sample.

Step 2: Calculate the mass percent of the element of interest in the original sample. See Problem Type 4A, Chapter 3.

> **Tip:** Try to be flexible when solving problems. Some gravimetric analysis problems may ask you to calculate the *molar concentration* of the component of interest in the sample, rather than mass percent. For this type of problem, you must modify your approach by converting grams of the component of interest to moles, then dividing by the volume of solution in liters.

EXAMPLE 4.11
A 0.7469 g sample of an ionic compound containing Pb ions is dissolved in water and treated with excess Na$_2$SO$_4$. If the mass of PbSO$_4$ that precipitates is 0.6839 g, what is the percent by mass of Pb in the original sample?

Step 1: Calculate the mass of Pb in 0.6839 g of the PbSO$_4$ precipitate.

$$\textbf{? mass of Pb} = 0.6839\ g\ PbSO_4 \times \frac{1\ mol\ PbSO_4}{303.27\ g\ PbSO_4} \times \frac{1\ mol\ Pb}{1\ mol\ PbSO_4} \times \frac{207.2\ g\ Pb}{1\ mol\ Pb}$$

$$= \textbf{0.4673 g Pb}$$

Step 2: Calculate the mass percent of Pb in the unknown sample.

$$\%Pb\ by\ mass = \frac{mass\ of\ Pb}{mass\ of\ sample} \times 100\%$$

$$\textbf{\%Pb by mass} = \frac{0.4673\ g}{0.7469\ g} \times 100\% = \textbf{62.57\%}$$

PRACTICE EXERCISE

11.   The concentration of Pb^{2+} ions in tap water could be determined by adding excess sodium sulfate solution to water. Excess sodium sulfate solution is added to 0.250 L of tap water. Write the net ionic equation and calculate the molar concentration of Pb^{2+} in the water sample if 0.01685 g of solid $PbSO_4$ is formed.

Text Problems: 4.74, 4.76

PROBLEM TYPE 8: ACID-BASE TITRATIONS

You must try to convince yourself that a titration problem follows the same thought process as the stoichiometry problems discussed in Chapter 3. The difference is that for acid and base solutions, you will typically be given the molarity rather than grams of substance.

Remember, you cannot directly compare grams of one substance to grams of another. Similarly, you typically cannot compare molarities or volume of one substance to that of another. Therefore, you must convert to *moles* of one substance, then apply the correct mole ratio from the balanced chemical equation to convert to *moles* of the other substance.

A typical approach to a stoichiometry problem is outlined below.

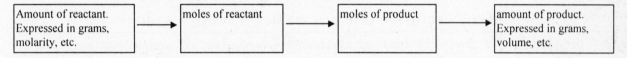

Amount of reactant. Expressed in grams, molarity, etc. → moles of reactant → moles of product → amount of product. Expressed in grams, volume, etc.

EXAMPLE 4.12

What volume of 0.900 *M* HCl is required to completely neutralize 50.0 mL of a 0.500 *M* $Ba(OH)_2$ solution?

Step 1: In order to have the correct mole ratio to solve the problem, you must start with a balanced chemical equation.

$$2 \, HCl \, (aq) \; + \; Ba(OH)_2 \, (aq) \; \longrightarrow \; BaCl_2 \, (aq) \; + \; 2 \, H_2O \, (l)$$

Step 2: From the molarity and volume of the $Ba(OH)_2$ solution, you can calculate moles of $Ba(OH)_2$. Then, using the mole ratio from the balanced equation above, you can calculate moles of HCl.

50.0 mL = 0.0500 L

$$\textbf{? mol HCl} \; = \; 0.0500 \, L \times \frac{0.500 \text{ mol } Ba(OH)_2}{1 \, L \text{ of solution}} \times \frac{2 \text{ mol HCl}}{1 \text{ mol } Ba(OH)_2}$$

$$= \; \textbf{0.0500 mol HCl}$$

Thus, 0.0500 mol of HCl are required to neutralize 50.0 mL of 0.500 *M* $Ba(OH)_2$.

Step 3: Solve the molarity equation algebraically for liters of solution. Then, substitute in the moles of HCl and molarity of HCl to solve for volume of HCl.

$$M = \frac{\text{moles of solute}}{\text{liters of solution}}$$

$$\text{liters of solution} = \frac{\text{moles of solute}}{M}$$

$$\textbf{volume of HCl} = \frac{0.0500 \text{ mol HCl}}{0.900 \text{ mol} / L} = \textbf{0.0556 L}$$

PRACTICE EXERCISE

12.   The distinctive odor of vinegar is due to acetic acid, CH_3COOH. Acetic acid reacts with sodium hydroxide in the following fashion:

$$CH_3COOH\ (aq)\ +\ NaOH\ (aq)\ \longrightarrow\ H_2O\ (l)\ +\ CH_3COONa\ (aq)$$

If 2.50 mL of vinegar requires 34.9 mL of 0.0960 M NaOH to reach the equivalence point in a titration, how many grams of acetic acid are in the 2.50 mL sample?

Text Problem: **4.80**

PROBLEM TYPE 9: REDOX TITRATIONS

Redox titration problems are solved in a similar manner to acid-base titration problems. A redox reaction is an *electron* transfer reaction; whereas an acid-base reaction is typically a *proton* transfer reaction. In a redox titration, an oxidizing agent is titrated against a reducing agent.

EXAMPLE 4.13

A 20.32 mL volume of 0.2002 M $KMnO_4$ solution is needed to oxidize 10.00 mL of an oxalic acid ($H_2C_2O_4$) solution. What is the concentration of the oxalic acid solution? The net ionic equation is:

$$2\ MnO_4^-\ +5\ C_2O_4^{2-}\ +\ 16\ H^+\ \longrightarrow\ 2\ Mn^{2+}\ +\ 10\ CO_2\ +\ 8\ H_2O$$

Step 1: In order to have the correct mole ratio to solve the problem, you must start with a balanced chemical equation. The balanced net ionic equation is given in the problem.

Step 2: From the molarity and volume of the $KMnO_4$ solution, you can calculate moles of MnO_4^-. Then, using the mole ratio from the balanced equation above, you can calculate moles of $C_2O_4^{2-}$.

20.32 mL = 0.02032 L

$$0.02032\ L\ soln\ \times\ \frac{0.2002\ mol\ MnO_4^{2-}}{1\ L\ soln}\ \times\ \frac{5\ mol\ C_2O_4^{2-}}{2\ mol\ MnO_4^{2-}}\ =\ \textbf{0.01017 mol } C_2O_4^{2-}$$

The mole ratio between $C_2O_4^{2-}$ and oxalic acid, $H_2C_2O_4$, is 1:1; therefore, the number of moles of oxalic acid is 0.01017 mole.

Step 3: We can now calculate the molarity of the oxalic acid solution from the moles of oxalic acid and the volume of the solution.

$$M\ =\ \frac{moles\ of\ solute}{liters\ of\ solution}\ =\ \frac{0.01017\ mol\ oxalic\ acid}{10.00\times10^{-3}\ L\ soln}\ =\ \textbf{1.017 } M \textbf{ oxalic acid}$$

PRACTICE EXERCISE

13.   Fe metal reacts with hydrochloric acid to produce Fe^{2+} ions and hydrogen gas. It takes 55.6 mL of 1.15 M HCl to completely react with a piece of Fe. What is the mass of the Fe? The balanced equation is:

$$Fe\ (s)\ +\ 2\ H^+\ (aq)\ +\ 2\ Cl^-\ (aq)\ \longrightarrow\ Fe^{2+}\ (aq)\ +\ H_2\ (g)\ +\ 2\ Cl^-\ (aq)$$

Text Problem: **4.84**, 4.86, 4.88, 4.90

ANSWERS TO PRACTICE EXERCISES

1. **(a)** soluble   **(b)** insoluble   **(c)** soluble   **(d)** insoluble

2. $CaCl_2\ (aq)\ +\ Na_2CO_3\ (aq)\ \longrightarrow\ 2\ NaCl\ (aq)\ +\ CaCO_3\ (s)$

    $Ca^{2+}(aq) + 2\ Cl^-(aq) + 2\ Na^+(aq) + CO_3^{2-}(aq)\ \longrightarrow\ 2\ Na^+(aq) + CO_3^{2-}(aq) + CaCO_3(s)$

    $Ca^{2+}\ (aq)\ +\ CO_3^{2-}\ (aq)\ \longrightarrow\ CaCO_3\ (s)$

3. **(a)** acid   **(b)** acid   **(c)** base   **(d)** both

4. $2\ HNO_2\ (aq)\ +\ Ba(OH)_2\ (aq)\ \longrightarrow\ Ba(NO_2)_2\ (aq)\ +\ 2\ H_2O\ (l)$

    $2\ HNO_2\ (aq)\ +\ Ba^{2+}\ (aq)\ +\ 2\ OH^-\ (aq)\ \longrightarrow\ Ba^{2+}\ (aq)\ +\ 2\ NO_2^-\ (aq)\ +\ 2\ H_2O\ (l)$

    $2\ HNO_2\ (aq)\ +\ 2\ OH^-\ (aq)\ \longrightarrow\ 2\ NO_2^-\ (aq)\ +\ 2\ H_2O\ (l)$

5. **(a)** +4   **(b)** +2   **(c)** +4   **(d)** +5

6. The oxidation numbers of the atoms and ions are:

    $$Cd^0\ (s)\ +\ Ni^{4+}O_2^{\ 2-}\ (s)\ +\ 2\ H_2^+O^{2-}\ (l)\ \longrightarrow\ Cd^{2+}(OH)_2^{\ -}\ (s)\ +\ Ni^{2+}(OH)_2^{\ -}\ (s)$$

    $Cd^0\ (s)$ is oxidized, and Ni^{4+} is reduced. The oxidation and reduction half-reactions are:

    $$Cd\ \longrightarrow\ Cd^{2+}\ +\ 2e^-$$
    $$Ni^{4+}\ +\ 2e^-\ \longrightarrow\ Ni^{2+}$$

7. **(a)** $2\ Al\ (s)\ +\ 6\ HCl\ (aq)\ \longrightarrow\ 2\ AlCl_3\ (aq)\ +\ 3\ H_2\ (g)$ **(b)** No reaction

8. $0.0124\ M$ **9.** 29.22 g NaCl   **10.** $5.08 \times 10^{-4}\ M$

11. $Pb^{2+}\ (aq)\ +\ SO_4^{\ 2-}\ (aq)\ \longrightarrow\ PbSO_4\ (s)$

    $[Pb^{2+}]\ =\ 2.222 \times 10^{-4}\ M$

12. 0.201 g acetic acid   **13.** 1.79 g Fe

SOLUTIONS TO SELECTED TEXT PROBLEMS

4.8   Ionic compounds, strong acids, and strong bases (metal hydroxides) are strong electrolytes (completely broken up into ions of the compound). Weak acids and weak bases are weak electrolytes. Molecular substances other than acids or bases are nonelectrolytes.

   (a) strong electrolyte (ionic)   **(b)**   nonelectrolyte

   (c) weak electrolyte (weak base)   **(d)**   strong electrolyte (strong base)

4.10   **(a)** Solid NaCl does not conduct. The ions are locked in a rigid lattice structure.

   (b) Molten NaCl conducts. The ions can move around in the liquid state.

   (c) Aqueous NaCl conducts. NaCl ionizes completely to Na^+ (aq) and Cl^- (aq) in water.

4.12   Since HCl dissolved in water conducts electricity, then HCl (aq) must actually exists as H^+ (aq) cations and Cl^- (aq) anions. Since HCl dissolved in benzene solvent does not conduct electricity, then we must assume that the HCl molecules in benzene solvent do not ionize, but rather exist as un-ionized molecules.

4.16   Applying solubility rules, Problem Type 1. Refer to Table 4.2 of your text to solve this problem.

   (a) $CaCO_3$ is <u>insoluble</u>.
   (b) $ZnSO_4$ is <u>soluble</u>.
   (c) $Hg(NO_3)_2$ is <u>soluble</u>.
   (d) $HgSO_4$ is <u>insoluble</u>.
   (e) NH_4ClO_4 is <u>soluble</u>.

4.18   Writing Molecular, Ionic, and Net Ionic Equations, Problem Type 2.

   Recall that an <u>ionic equation</u> shows dissolved ionic compounds in terms of their free ions. A <u>net ionic equation</u> shows only the species that actually take part in the reaction.

   (a) <u>Ionic:</u> $2\,Na^+\,(aq) + S^{2-}\,(aq) + Zn^{2+}\,(aq) + 2\,Cl^-\,(aq) \longrightarrow ZnS\,(s) + 2\,Na^+\,(aq) + 2\,Cl^-\,(aq)$

   <u>Net ionic:</u> $Zn^{2+}\,(aq) + S^{2-}\,(aq) \longrightarrow ZnS\,(s)$

   Table 4.2 indicates that ZnS is <u>insoluble</u>.

   (b) <u>Ionic:</u> $6K^+(aq) + 2PO_4^{3-}(aq) + 3Sr^{2+}(aq) + 6NO_3^-(aq) \longrightarrow Sr_3(PO_4)_2(s) + 6K^+(aq) + 6NO_3^-(aq)$

   <u>Net ionic:</u> $3\,Sr^{2+}\,(aq) + 2\,PO_4^{3-}\,(aq) \longrightarrow Sr_3(PO_4)_2\,(s)$

   Table 4.2 indicates that $Sr_3(PO_4)_2$ is <u>insoluble</u>.

   (c) <u>Ionic:</u> $Mg^{2+}(aq) + 2NO_3^-\,(aq) + 2Na^+\,(aq) + 2OH^-\,(aq) \longrightarrow Mg(OH)_2\,(s) + 2Na^+(aq) + 2NO_3^-(aq)$

   <u>Net ionic:</u> $Mg^{2+}\,(aq) + 2\,OH^-\,(aq) \longrightarrow Mg(OH)_2\,(s)$

   Table 4.2 indicates that $Mg(OH)_2$ is <u>insoluble</u>.

4.20 **(a)** Add chloride ions. KCl is soluble, but AgCl is not.

  (b) Add sulfate ions. Ag_2SO_4 is moderately soluble, but $PbSO_4$ is insoluble.

  (c) Add carbonate ions. $(NH_4)_2CO_3$ is soluble, but $CaCO_3$ is insoluble.

  (d) Add sulfate ions. $CuSO_4$ is soluble, but $BaSO_4$ is insoluble.

4.28 Identifying Brønsted acids and bases, Problem Type 3A.

  (a) PO_4^{3-} in water can accept a proton to become HPO_4^{2-}, and is thus a Brønsted base.

  (b) ClO_2^- in water can accept a proton to become $HClO_2$, and is thus a Brønsted base.

  (c) NH_4^+ dissolved in water can donate a proton H^+, thus behaving as a Brønsted acid.

  (d) HCO_3^- can either accept a proton to become H_2CO_3, thus behaving as a Brønsted base. Or, HCO_3^- can donate a proton to yield H^+ and CO_3^{2-}, thus behaving as a Brønsted acid.

4.30 Writing acid-base reactions, Problem Type 3B.

  Recall that strong acids and strong bases are strong electrolytes. They are completely ionized in solution. An <u>ionic equation</u> will show strong acids and strong bases in terms of their free ions. A <u>net ionic equation</u> shows only the species that actually take part in the reaction.

  (a) <u>Ionic:</u> $CH_3COOH\ (aq) + K^+\ (aq) + OH^-\ (aq) \longrightarrow K^+\ (aq) + CH_3COO^-\ (aq) + H_2O\ (l)$

  <u>Net ionic:</u> $CH_3COOH\ (aq) + OH^-\ (aq) \longrightarrow CH_3COO^-\ (aq) + H_2O\ (l)$

  (b) <u>Ionic:</u> $H_2CO_3\ (aq) + 2\ Na^+\ (aq) + 2\ OH^-\ (aq) \longrightarrow 2\ Na^+\ (aq) + CO_3^{2-}\ (aq) + 2\ H_2O\ (l)$

  <u>Net ionic:</u> $H_2CO_3\ (aq) + 2\ OH^-\ (aq) \longrightarrow CO_3^{2-}\ (aq) + 2\ H_2O\ (l)$

  (c) <u>Ionic:</u> $2H^+\ (aq) + 2\ NO_3^-\ (aq) + Ba^{2+}\ (aq) + 2OH^-\ (aq) \longrightarrow Ba^{2+}\ (aq) + 2NO_3^-\ (aq) + 2H_2O\ (l)$

  <u>Net ionic:</u> $2\ H^+\ (aq) + 2\ OH^-\ (aq) \longrightarrow 2\ H_2O\ (l)$

4.40 Writing oxidation/reduction half-reactions, Problem Type 4B.

  In each part, the reducing agent is the reactant in the first half-reaction and the oxidizing agent is the reactant in the second half-reaction. The coefficients in each half-reaction have been reduced to smallest whole numbers.

  (a) The product is an ionic compound whose ions are Fe^{3+} and O^{2-}.

$$Fe \longrightarrow Fe^{3+} + 3e^-$$
$$O_2 + 4e^- \longrightarrow 2\ O^{2-}$$

  O_2 is the oxidizing agent; Fe is the reducing agent.

  (b) Na^+ does not change in this reaction. It is a "spectator ion."

$$2\ Br^- \longrightarrow Br_2 + 2e^-$$
$$Cl_2 + 2e^- \longrightarrow 2\ Cl^-$$

  Cl_2 is the oxidizing agent; Br^- is the reducing agent.

(c)   Assume SiF_4 is made up of Si^{4+} and F^-.

$$Si \longrightarrow Si^{4+} + 4e^-$$
$$F_2 + 2e^- \longrightarrow 2 F^-$$

F_2 is the oxidizing agent; Si is the reducing agent.

(d)   Assume HCl is made up of H^+ and Cl^-.

$$H_2 \longrightarrow 2 H^+ + 2e^-$$
$$Cl_2 + 2e^- \longrightarrow 2 Cl^-$$

Cl_2 is the oxidizing agent; H_2 is the reducing agent.

4.42   Assigning oxidation numbers, Problem Type 4A.

Step 1: In each case the oxidation number of oxygen is −2 and that of hydrogen is +1 (rules 3 and 4, Section 4).

Step 2: All the compounds listed are neutral compounds, so the oxidation numbers must sum to zero
(Rule 6, Section 4).

Let the oxidation number of $P = x$.

(a) $x + 1 + (3)(-2) = 0$, $x = +5$ **(d)** $x + (3)(+1) + (4)(-2) = 0$, $x = +5$
(b) $x + (3)(+1) + (2)(-2) = 0$, $x = +1$ **(e)** $2x + (4)(+1) + (7)(-2) = 0$, $2x = 10$, $x = +5$
(c) $x + (3)(+1) + (3)(-2) = 0$, $x = +3$ **(f)** $3x + (5)(+1) + (10)(-2) = 0$, $3x = 15$, $x = +5$

The molecules in part (a), (e), and (f) can be made by strongly heating the compound in part (d). Are these oxidation-reduction reactions?

4.44   All are free elements, so all have an oxidation number of **zero**.

4.46   **(a)**   N: −3   **(b)**   O: −1/2   **(c)**   C: −1   **(d)**   C: +4

(e)   C: +3   **(f)**   O: −2   **(g)**   B: +3   **(h)**   W: +6

4.48   Using an activity series, Problem Type 4C.

Hydrogen displacement. Any metal above hydrogen in the activity series will displace it from water or from an acid. Metals below hydrogen will <u>not</u> react with either water or an acid.

Only (b) Li and (d) Ca are above hydrogen in the activity series, so they are the only metals in this problem that will react with water.

4.50   **(a)** $Cu\ (s) + HCl\ (aq) \rightarrow$ no reaction, since Cu(s) is less reactive than the hydrogen from acids.

(b) $I_2\ (s) + NaBr\ (aq) \rightarrow$ no reaction, since $I_2\ (s)$ is less reactive than $Br_2\ (l)$.

(c) $Mg\ (s) + CuSO_4\ (aq) \rightarrow MgSO_4\ (aq) + Cu\ (s)$, since Mg (s) is more reactive than Cu (s).

Net ionic equation: $Mg\ (s) + Cu^{2+}\ (aq) \rightarrow Mg^{2+}\ (aq) + Cu\ (s)$

(d) $Cl_2\ (g) + 2\ KBr\ (aq) \rightarrow Br_2\ (l) + 2\ KCl\ (aq)$, since $Cl_2\ (g)$ is more reactive than $Br_2\ (l)$

Net ionic equation: $Cl_2\ (g) + 2\ Br^-\ (aq) \rightarrow 2\ Cl^-\ (aq) + Br_2\ (l)$

4.52   **(a)**   Combination reaction
        (b)   Decomposition reaction
        (c)   Displacement reaction
        (d)   Disproportionation reaction

4.56   Concentration of Solutions, Problem Type 5.

Step 1: From the molarity (0.707 M), we can calculate the moles of $NaNO_3$ needed to prepare 250 mL of solution.

$$\text{Moles } NaNO_3 = \frac{0.707 \text{ mol } NaNO_3}{1 \text{ L soln}} \times \frac{1 \text{ L soln}}{1000 \text{ mL soln}} \times 250 \text{ mL soln} = 0.177 \text{ mol}$$

Step 2: Use the molar mass of $NaNO_3$ as a conversion factor to convert from moles to grams.

$\mathcal{M}(NaNO_3) = 85.00$ g/mol.

$$0.177 \text{ mol } NaNO_3 \times \frac{85.00 \text{ g } NaNO_3}{1 \text{ mol } NaNO_3} = \textbf{15.0 g NaNO}_3$$

4.58   Since the problem asks for grams of solute (KOH), you should be thinking that you can calculate moles of solute from the molarity and volume of solution. Then, you can convert moles of solute to grams of solute.

$$? \text{ moles KOH solute} = \frac{5.50 \text{ moles solute}}{1 \text{ L solution}} \times 0.0350 \text{ L solution} = 0.193 \text{ moles KOH}$$

The molar mass of KOH is 56.11 g/mol. Use this conversion factor to calculate grams of KOH.

$$\textbf{? grams KOH} = 0.193 \text{ mol KOH} \times \frac{56.11 \text{ g KOH}}{1 \text{ mol KOH}} = \textbf{10.8 g KOH}$$

4.60   **(a)** $? \text{ mol } CH_3OH = 6.57 \text{ g } CH_3OH \times \dfrac{1 \text{ mol } CH_3OH}{32.04 \text{ g } CH_3OH} = 0.205 \text{ mol } CH_3OH$

$$M = \frac{0.205 \text{ mol } CH_3OH}{0.150 \text{ L}} = \textbf{1.37 } M$$

(b) $? \text{ mol } CaCl_2 = 10.4 \text{ g } CaCl_2 \times \dfrac{1 \text{ mol } CaCl_2}{111.0 \text{ g } CaCl_2} = 0.0937 \text{ mol } CaCl_2$

$$M = \frac{0.0937 \text{ mol } CaCl_2}{0.220 \text{ L}} = \textbf{0.426 } M$$

(c) $? \text{ mol } C_{10}H_8 = 7.82 \text{ g } C_{10}H_8 \times \dfrac{1 \text{ mol } C_{10}H_8}{128.2 \text{ g } C_{10}H_8} = 0.0610 \text{ mol } C_{10}H_8$

$$M = \frac{0.0610 \text{ mol } C_{10}H_8}{0.0852 \text{ L}} = \textbf{0.716 } M$$

4.62   A 250 mL sample of 0.100 M solution contains 0.0250 mol of solute (mol = $M \times$ L). The computation in each case is the same:

(a) $0.0250 \text{ mol CsI} \times \dfrac{259.8 \text{ g CsI}}{1 \text{ mol CsI}} = \textbf{6.50 g CsI}$

(b) $0.0250 \text{ mol H}_2\text{SO}_4 \times \dfrac{98.09 \text{ g H}_2\text{SO}_4}{1 \text{ mol H}_2\text{SO}_4} = \textbf{2.45 g H}_2\textbf{SO}_4$

(c) $0.0250 \text{ mol Na}_2\text{CO}_3 \times \dfrac{106.0 \text{ g Na}_2\text{CO}_3}{1 \text{ mol Na}_2\text{CO}_3} = \textbf{2.65 g Na}_2\textbf{CO}_3$

(d) $0.0250 \text{ mol K}_2\text{Cr}_2\text{O}_7 \times \dfrac{294.2 \text{ g K}_2\text{Cr}_2\text{O}_7}{1 \text{ mol K}_2\text{Cr}_2\text{O}_7} = \textbf{7.36 g K}_2\textbf{Cr}_2\textbf{O}_7$

(e) $0.0250 \text{ mol KMnO}_4 \times \dfrac{158.0 \text{ g KMnO}_4}{1 \text{ mol KMnO}_4} = \textbf{3.95 g KMnO}_4$

4.66   Dilution of Solutions, Problem Type 6.

You should be able to come up with the following equation.

$$M_{\text{initial}}V_{\text{initial}} = M_{\text{final}}V_{\text{final}}$$

You can solve the equation algebraically for M_{final}. Then substitute in the given quantities to solve the problem.

$$M_{\textbf{final}} = \frac{M_{\text{initial}} \times V_{\text{initial}}}{V_{\text{final}}} = \frac{0.866 \text{ M} \times 25.0 \text{ mL}}{500 \text{ mL}} = \textbf{0.0433 } \textbf{\textit{M}}$$

4.68   You need to calculate the final volume of the dilute solution. Then, you can subtract 505 mL from this volume to calculate the amount of water that should be added.

$$V_{\text{final}} = \frac{M_{\text{initial}}V_{\text{initial}}}{M_{\text{final}}} = \frac{(0.125 \text{ } M)(505 \text{ mL})}{(0.100 \text{ } M)} = 631 \text{ mL}$$

Add (631 − 505) mL = **126 mL of water**.

4.70   Moles of calcium nitrate in the first solution:

$$\frac{0.568 \text{ mol}}{1000 \text{ mL soln}} \times 46.2 \text{ mL soln} = 0.0262 \text{ mol Ca(NO}_3)_2$$

Moles of calcium nitrate in the second solution:

$$\frac{1.396 \text{ mol}}{1000 \text{ mL soln}} \times 80.5 \text{ mL soln} = 0.112 \text{ mol Ca(NO}_3)_2$$

The volume of the combined solutions = 46.2 mL + 80.5 mL = 126.7 mL. The concentration of the final solution is:

$$M = \frac{(0.0262 + 0.112) \text{ mol}}{0.1267 \text{ L}} = \textbf{1.09 } \textbf{\textit{M}}$$

4.74   Gravimetric Analysis, Problem Type 7.

We assume the precipitation is quantitative, that is, that all of the barium in the sample has been precipitated as barium sulfate.

Step 1: Calculate the mass of Ba in 0.4105 g of the $BaSO_4$ precipitate.

$$? \text{ mass of Ba } = 0.4105 \text{ g BaSO}_4 \times \frac{1 \text{ mol BaSO}_4}{233.4 \text{ g BaSO}_4} \times \frac{1 \text{ mol Ba}}{1 \text{ mol BaSO}_4} \times \frac{137.33 \text{ g Ba}}{1 \text{ mol Ba}}$$

$$= 0.2415 \text{ g Ba}$$

Step 2: Calculate the mass percent of Ba in the unknown compounds.

$$\% \text{ Ba by mass } = \frac{\text{mass of Ba}}{\text{mass of sample}} \times 100\%$$

$$\textbf{\% Ba by mass } = \frac{0.2415 \text{ g}}{0.6760 \text{ g}} \times 100\% = \textbf{35.72 \%}$$

4.76   The net ionic equation is: $Cu^{2+}(aq) + S^{2-}(aq) \longrightarrow CuS(s)$

The answer sought is the molar concentration of Cu^{2+}, that is, moles of Cu^{2+} ions per liter of solution. The factor-label method is used to convert, in order:

$$\text{g of CuS} \rightarrow \text{moles CuS} \rightarrow \text{moles } Cu^{2+} \rightarrow \text{moles } Cu^{2+} \text{ per liter soln}$$

$$[Cu^{2+}] = 0.0177 \text{ g CuS} \times \frac{1 \text{ mol CuS}}{95.62 \text{ g CuS}} \times \frac{1 \text{ mol } Cu^{2+}}{1 \text{ mol CuS}} \times \frac{1}{0.800 \text{ L}} = \textbf{2.31} \times \textbf{10}^{-4} \textbf{ M}$$

4.80   Acid-Base Titrations, Problem Type 8.

(a)   In order to have the correct mole ratio to solve the problem, you must start with a balanced chemical equation.

$$HCl(aq) + NaOH(aq) \longrightarrow NaCl(aq) + H_2O(l)$$

From the molarity and volume of the NaOH solution, you can calculate moles of NaOH. Then, using the mole ratio from the balanced equation above, you can calculate moles of HCl.

10.0 mL = 0.0100 L

$$? \text{ mol HCl} = 0.0100 \text{ L} \times \frac{0.300 \text{ mol NaOH}}{1 \text{ L of solution}} \times \frac{1 \text{ mol HCl}}{1 \text{ mol NaOH}} = 3.00 \times 10^{-3} \text{ mol HCl}$$

Solve the molarity equation algebraically for liters of solution. Then, substitute in the moles of HCl and molarity of HCl to solve for volume of the HCl solution.

$$M = \frac{\text{moles of solute}}{\text{liters of solution}}$$

$$\text{liters of solution } = \frac{\text{moles of solute}}{M}$$

$$\textbf{volume of HCl} = \frac{3.00 \times 10^{-3} \text{ mol HCl}}{0.500 \text{ mol/L}} = \textbf{6.00} \times \textbf{10}^{-3} \textbf{ L} = \textbf{6.00 mL}$$

(b)   This problem is similar to part (a). The difference is that the mole ratio between acid and base is 2:1.

$$2\, HCl\,(aq)\,+\,Ba(OH)_2\,(aq)\,\longrightarrow\,BaCl_2\,(aq)\,+2\,H_2O\,(l)$$

$$?\,mol\,HCl\,=\,0.0100\,L\,\times\,\frac{0.200\,mol\,Ba(OH)_2}{1\,L\,of\,solution}\,\times\,\frac{2\,mol\,HCl}{1\,mol\,Ba(OH)_2}\,=\,4.00\times10^{-3}\,mol\,HCl$$

$$\textbf{volume of HCl}\,=\,\frac{4.00\times10^{-3}\,mol\,HCl}{0.500\,mol/L}\,=\,\textbf{8.00}\times\textbf{10}^{-3}\,\textbf{L}\,=\,\textbf{8.00 mL}$$

4.84   Redox Titrations, Problem Type 9.

The balanced equation is given in the problem.

$$5\,SO_2\,+\,2\,MnO_4^-\,+\,2\,H_2O\,\longrightarrow\,5\,SO_4^{2-}\,+\,2\,Mn^{2+}\,+\,4\,H^+$$

The mass of SO_2 in the air sample can be calculated as follows:

$$\frac{0.00800\,mol\,KMnO_4}{1\,L\,soln}\,\times\,0.00737\,L\,=\,5.89\times10^{-5}\,mol\,KMnO_4$$

$$5.89\times10^{-5}\,mol\,KMnO_4\,\times\,\frac{5\,mol\,SO_2}{2\,mol\,KMnO_4}\,\times\,\frac{64.07\,g\,SO_2}{1\,mol\,SO_2}\,=\,\textbf{9.43}\times\textbf{10}^{-3}\,\textbf{g SO}_2$$

4.86   The balanced equation is given in the problem.

$$2\,MnO_4^-\,+\,5\,H_2O_2\,+\,6\,H^+\,\longrightarrow\,5\,O_2\,+\,2\,Mn^{2+}\,+\,8\,H_2O$$

First, calculate the moles of potassium permanganate in 36.44 mL (0.03644 L) of solution.

$$\frac{0.01652\,mol\,KMnO_4}{1\,L\,soln}\,\times\,0.03644\,L\,=\,6.019\times10^{-4}\,mol\,KMnO_4$$

Next, calculate the moles of hydrogen peroxide using the mole ratio from the balanced equation.

$$6.019\times10^{-4}\,mol\,KMnO_4\,\times\,\frac{5\,mol\,H_2O_2}{2\,mol\,KMnO_4}\,=\,1.505\times10^{-3}\,mol\,H_2O_2$$

Finally, calculate the molarity of the H_2O_2 solution. The volume of the solution is 0.02500 L.

$$\textbf{Molarity of H}_2\textbf{O}_2\,=\,\frac{1.505\times10^{-3}\,mol\,H_2O_2}{0.02500\,L}\,=\,\textbf{0.06020}\,\textbf{\textit{M}}$$

4.88   First, calculate the moles of $KMnO_4$ in 24.0 mL (0.0240 L) of solution.

$$\frac{0.0100\,mol\,KMnO_4}{1\,L\,soln}\,\times\,0.0240\,L\,=\,2.40\times10^{-4}\,mol\,KMnO_4$$

Next, calculate the mass of oxalic acid needed to react with 2.40×10^{-4} mol $KMnO_4$. Use the mole ratio from the balanced equation.

$$2.40\times10^{-4}\,mol\,KMnO_4\,\times\,\frac{5\,mol\,H_2C_2O_4}{2\,mol\,KMnO_4}\,\times\,\frac{90.04\,g\,H_2C_2O_4}{1\,mol\,H_2C_2O_4}\,=\,0.0540\,g\,H_2C_2O_4$$

The original sample had a mass of 1.00 g. The mass percent of $H_2C_2O_4$ in the sample is:

$$\text{mass \%} = \frac{0.0540 \text{ g}}{1.00 \text{ g}} \times 100\% = \textbf{5.40\% } \mathbf{H_2C_2O_4}$$

4.90   The balanced equation is:

$$2 \text{ MnO}_4^- + 16 \text{ H}^+ + 5 \text{ C}_2\text{O}_4^{2-} \longrightarrow 2 \text{ Mn}^{2+} + 10 \text{ CO}_2 + 8 \text{ H}_2\text{O}$$

24.2 mL = 0.0242 L

$$\text{mol MnO}_4^- = \frac{9.56 \times 10^{-4} \text{ mol MnO}_4^-}{1 \text{ L of soln}} \times 0.0242 \text{ L} = 2.31 \times 10^{-5} \text{ mol MnO}_4^-$$

Using the mole ratio from the balanced equation, we can calculate the mass of Ca^{2+} in the 10.0 mL sample of blood.

$$2.31 \times 10^{-5} \text{ mol MnO}_4^- \times \frac{5 \text{ mol C}_2\text{O}_4^{2-}}{2 \text{ mol MnO}_4^-} \times \frac{1 \text{ mol Ca}^{2+}}{1 \text{ mol C}_2\text{O}_4^{2-}} \times \frac{40.08 \text{ g Ca}^{2+}}{1 \text{ mol Ca}^{2+}} = 2.31 \times 10^{-3} \text{ g Ca}^{2+}$$

Converting to mg/mL:

$$\frac{2.31 \times 10^{-3} \text{ g Ca}^{2+}}{10.0 \text{ mL of blood}} \times \frac{1 \text{ mg}}{0.001 \text{ g}} = \textbf{0.231 mg Ca}^{2+}\textbf{/mL of blood}$$

4.92   First, the gases could be tested to see if they supported combustion. O_2 would support combustion, CO_2 would not. Second, if CO_2 is bubbled through a solution of calcium hydroxide [$Ca(OH)_2$], a white precipitate of $CaCO_3$ forms. No reaction occurs when O_2 is bubbled through a calcium hydroxide solution.

4.94   Starting with a balanced chemical equation:

$$\text{Mg } (s) + 2 \text{ HCl } (aq) \longrightarrow \text{ MgCl}_2 (aq) + \text{H}_2 (g)$$

From the mass of Mg, you can calculate moles of Mg. Then, using the mole ratio from the balanced equation above, you can calculate moles of HCl reacted.

$$4.47 \text{ g Mg} \times \frac{1 \text{ mol Mg}}{24.31 \text{ g Mg}} \times \frac{2 \text{ mol HCl}}{1 \text{ mol Mg}} = 0.368 \text{ mol HCl reacted}$$

Next we can calculate the number of moles of HCl in the original solution.

5.00×10^2 mL = 0.500 L

$$\frac{2.00 \text{ mol HCl}}{1 \text{ L soln}} \times 0.500 \text{ L soln} = 1.00 \text{ mol HCl}$$

Moles HCl remaining = 1.00 mol − 0.368 mol = 0.632 mol HCl

$$\textbf{conc. of HCl after reaction} = \frac{\text{mol HCl}}{\text{L soln}} = \frac{0.632 \text{ mol HCl}}{0.500 \text{ L}} = \textbf{1.26 mol/L} = \textbf{1.26 } \textbf{\textit{M}}$$

4.96   The balanced equation is:

$$2 \, HCl \, (aq) \; + \; Na_2CO_3 \, (s) \; \longrightarrow \; CO_2 \, (g) \; + \; H_2O \, (l) \; + \; 2 \, NaCl \, (aq)$$

The mole ratio from the balanced equation is 2 mol HCl : 1 mol Na_2CO_3. The moles of HCl needed to react with 0.256 g of Na_2CO_3 are:

$$0.256 \, g \, Na_2CO_3 \; \times \; \frac{1 \, mol \, Na_2CO_3}{106.0 \, g \, Na_2CO_3} \; \times \; \frac{2 \, mol \, HCl}{1 \, mol \, Na_2CO_3} \; = \; 4.83 \times 10^{-3} \, mol \; HCl$$

Molarity HCl $= \dfrac{moles \, HCl}{L \, soln} = \dfrac{4.83 \times 10^{-3} \, mol \, HCl}{0.0283 \, L \, soln} = \textbf{0.171 mol/L} = \textbf{0.171 } \boldsymbol{M}$

4.98   Starting with a balanced chemical equation:

$$CH_3COOH \, (aq) \; + \; NaOH \, (aq) \; \longrightarrow \; CH_3COONa \, (aq) \; + \; H_2O \, (l)$$

From the molarity and volume of the NaOH solution, you can calculate moles of NaOH. Then, using the mole ratio from the balanced equation above, you can calculate moles of CH_3COOH.

$$5.75 \times 10^{-3} \, L \, solution \times \; \frac{1.00 \, mol \, NaOH}{1 \, L \, of \, solution} \; \times \; \frac{1 \, mol \, CH_3COOH}{1 \, mol \, NaOH} \; = \; 5.75 \times 10^{-3} \, mol \, CH_3COOH$$

Molarity CH$_3$COOH $= \dfrac{5.75 \times 10^{-3} \, mol \, CH_3COOH}{0.0500 \, L} = \textbf{0.115 } \boldsymbol{M}$

4.100   The balanced equation is:

$$Zn \, (s) \; + \; 2 \, AgNO_3 \, (aq) \; \longrightarrow \; Zn(NO_3)_2 \, (aq) \; + \; 2 \, Ag \, (s)$$

Let x = mass of Ag produced. We can find the mass of Zn reacted in terms of the amount of Ag produced.

$$x \, g \, Ag \; \times \; \frac{1 \, mol \, Ag}{107.9 \, g \, Ag} \; \times \; \frac{1 \, mol \, Zn}{2 \, mol \, Ag} \; \times \; \frac{65.39 \, g \, Zn}{1 \, mol \, Zn} \; = \; 0.303x \; g \, Zn \, reacted.$$

The mass of Zn remaining will be:

$$2.50 \, g \; - \; amount \, of \, Zn \, reacted \; = \; 2.50 \, g \, Zn - 0.303x \, g \, Zn$$

The final mass of the strip, 3.37 g, equals the mass of Ag produced + the mass of Zn remaining.

$$3.37 \, g \; = \; x \, g \, Ag \; + \; 2.50 \, g \, Zn - 0.303 \, x \, g \, Zn$$

$$x = \textbf{1.25 g} = \textbf{mass of Ag produced}$$

mass of Zn remaining $= \; 3.37 \, g - 1.25 \, g = \textbf{2.12 g Zn}$

4.102   The balanced equation is: $HNO_3 \, (aq) \; + \; NaOH \, (aq) \; \longrightarrow \; NaNO_3 \, (aq) \; + \; H_2O \, (l)$

$$mol \, HNO_3 \; = \; \frac{0.211 \, mol \, HNO_3}{1 \, L \, soln} \; \times \; 0.0107 \, L \, soln \; = \; 2.26 \times 10^{-3} \, mol \, HNO_3$$

$$mol \, NaOH \; = \; \frac{0.258 \, mol \, NaOH}{1 \, L \, soln} \; \times \; 0.0163 \, L \, soln \; = \; 4.21 \times 10^{-3} \, mol \, NaOH$$

Since the mole ratio from the balanced equation is 1 mol NaOH : 1 mol HNO_3, then 2.26×10^{-3} mol HNO_3 will react with 2.26×10^{-3} mol NaOH.

$$\text{mol NaOH remaining} = 4.21 \times 10^{-3} \text{ mol} - 2.26 \times 10^{-3} \text{ mol} = 1.95 \times 10^{-3} \text{ mol NaOH}$$

$$\textbf{molarity NaOH} = \frac{1.95 \times 10^{-3} \text{ mol NaOH}}{(10.7 + 16.3)\text{mL}} \times \frac{1 \text{ mL}}{0.001 \text{ L}} = \textbf{0.0722 } \textbf{\textit{M}}$$

4.104   The balanced equations for the two reactions are:

$$X\ (s)\ +\ H_2SO_4\ (aq) \longrightarrow XSO_4\ (aq)\ +\ H_2\ (g)$$

$$H_2SO_4\ (aq)\ +\ 2\ NaOH\ (aq) \longrightarrow Na_2SO_4\ (aq)\ +\ 2\ H_2O\ (l)$$

First, let's find the number of moles of excess acid from the reaction with NaOH.

$$0.0334 \text{ L} \times \frac{0.500 \text{ mol NaOH}}{1 \text{ L soln}} \times \frac{1 \text{ mol } H_2SO_4}{2 \text{ mol NaOH}} = 8.35 \times 10^{-3} \text{ mol } H_2SO_4$$

The original number of moles of acid was:

$$0.100 \text{ L} \times \frac{0.500 \text{ mol } H_2SO_4}{1 \text{ L soln}} = 0.0500 \text{ mol } H_2SO_4$$

The amount of sulfuric acid that reacted with the metal, X, is

$$0.0500 \text{ mol } H_2SO_4 - 8.35 \times 10^{-3} \text{ mol } H_2SO_4 = 0.0417 \text{ mol } H_2SO_4.$$

Since the mole ratio from the balanced equation is 1 mol X : 1 mol H_2SO_4, then the amount of X that reacted is 0.0417 mol X.

$$\textbf{molar mass X} = \frac{1.00 \text{ g X}}{0.0417 \text{ mol X}} = \textbf{24.0 g/mol}$$

The element is **magnesium**.

4.106   First, calculate the number of moles of glucose present.

$$\frac{0.513 \text{ mol glucose}}{1 \text{ L soln}} \times 60.0 \times 10^{-3} \text{ L} = 0.0308 \text{ mol glucose}$$

$$\frac{2.33 \text{ mol glucose}}{1 \text{ L soln}} \times 120.0 \times 10^{-3} \text{ L} = 0.280 \text{ mol glucose}$$

Add the moles of glucose, then divide by the total volume of the combined solutions to calculate the molarity.

$$\textbf{Molarity of final solution} = \frac{0.0308 \text{ mol} + 0.280 \text{ mol glucose}}{(60.0 + 120.0) \text{ mL}} \times \frac{1 \text{ mL}}{0.001 \text{ L}} = 1.73 \text{ mol/L} = \textbf{1.73 } \textbf{\textit{M}}$$

4.108   Iron(II) compounds can be oxidized to iron(III) compounds. The sample could be tested with a small amount of a strongly colored oxidizing agent like a $KMnO_4$ solution, which is a deep purple color. A loss of color would imply the presence of an oxidizable substance like an iron(II) salt.

4.110   Since both of the original solutions were strong electrolytes, you would expect a mixture of the two solutions to also be a strong electrolyte. However, since the light dims, the mixture must contain fewer ions than the original solution. Indeed, H^+ from the sulfuric acid reacts with the OH^- from the barium hydroxide to form water. The barium cations react with the sulfate anions to form insoluble barium sulfate.

$$2\,H^+\,(aq)\,+\,SO_4^{2-}\,(aq)\,+\,Ba^{2+}\,(aq)\,+\,2\,OH^-\,(aq)\,\longrightarrow\,2\,H_2O\,(l)\,+\,BaSO_4\,(s)$$

Thus, the reaction depletes the solution of ions and the conductivity decreases.

4.112   You could test the conductivity of the solutions. Sugar is a nonelectrolyte and an aqueous sugar solution will not conduct electricity; whereas, NaCl is a strong electrolyte when dissolved in water. Silver nitrate could be added to the solutions to see if silver chloride precipitated. In this particular case, the solutions could also be tasted.

4.114   In a redox reaction, the oxidizing agent gains one or more electrons. In doing so, the oxidation number of the element gaining the electrons must become more negative. In the case of chlorine, the -1 oxidation number is already the most negative state possible. The chloride ion *cannot* accept any more electrons; therefore, hydrochloric acid is *not* an oxidizing agent.

4.116   The reaction is too violent. This could cause the hydrogen gas produced to ignite, and an explosion could result.

4.118   The solid sodium bicarbonate would be the better choice. The hydrogen carbonate ion, HCO_3^-, behaves as a Brønsted base to accept a proton from the acid.

$$HCO_3^-\,(aq)\,+\,H^+\,(aq)\,\longrightarrow\,H_2CO_3\,(aq)\,\longrightarrow\,H_2O\,(l)\,+\,CO_2\,(g)$$

The heat generated during the reaction of hydrogen carbonate with the acid causes the carbonic acid, H_2CO_3, that was formed to decompose to water and carbon dioxide.

The reaction of the spilled sulfuric acid with sodium hydroxide would produce sodium sulfate, Na_2SO_4, and water. There is a possibility that the Na_2SO_4 could precipitate. Also, the sulfate ion, SO_4^{2-} is a weak base; therefore, the "neutralized" solution would actually be *basic*.

$$H_2SO_4\,(aq)\,+\,2\,NaOH\,(aq)\,\longrightarrow\,Na_2SO_4\,(aq)\,+\,2\,H_2O\,(l)$$

Also, NaOH is a caustic substance and therefore is not safe to use in this manner.

4.120   **(a)**   Table salt, NaCl, is very soluble in water and is a strong electrolyte. Addition of $AgNO_3$ will precipitate AgCl.

   (b)   Table sugar or sucrose, $C_{12}H_{22}O_{11}$, is soluble in water and is a nonelectrolyte.

   (c)   Aqueous acetic acid, CH_3COOH, the primary ingredient of vinegar, is a weak electrolyte. It exhibits all of the properties of acids (Section 4.3).

   (d)   Baking soda, $NaHCO_3$, is a water-soluble strong electrolyte. It reacts with acid to release CO_2 gas. Addition of $Ca(OH)_2$ results in the precipitation of $CaCO_3$.

   (e)   Washing soda, $Na_2CO_3 \cdot 10H_2O$, is a water-soluble strong electrolyte. It reacts with acids to release CO_2 gas. Addition of a soluble alkaline-earth salt will precipitate the alkaline-earth carbonate. Aqueous washing soda is also slightly basic (Section 4.3).

   (f)   Boric acid, H_3BO_3, is weak electrolyte and a weak acid.

   (g)   Epsom salt, $MgSO_4 \cdot 7H_2O$, is a water-soluble strong electrolyte. Addition of $Ba(NO_3)_2$ results in the precipitation of $BaSO_4$. Addition of hydroxide precipitates $Mg(OH)_2$.

(h)   Sodium hydroxide, NaOH, is a strong electrolyte and a strong base. Addition of $Ca(NO_3)_2$ results in the precipitation of $Ca(OH)_2$.

(i)   Ammonia, NH_3, is a sharp-odored gas that when dissolved in water is a weak electrolyte and a weak base. NH_3 in the gas phase reacts with HCl gas to produce solid NH_4Cl.

(j)   Milk of magnesia, $Mg(OH)_2$, is an insoluble, strong base that reacts with acids. The resulting magnesium salt may be soluble or insoluble.

(k) $CaCO_3$ is an insoluble salt that reacts with acid to release CO_2 gas. $CaCO_3$ is discussed in the Chemistry in Action essays entitled, "An Undesirable Precipitation Reaction" and "Metal from the Sea" in Chapter 4.

With the exception of NH_3 and vinegar, all the compounds in this problem are white solids.

4.122   The balanced equation for the reaction is:

$$XCl\,(aq) + AgNO_3\,(aq) \longrightarrow AgCl\,(s) + XNO_3\,(aq) \quad \text{where X = Na, or K}$$

From the amount of AgCl produced, we can calculate the moles of XCl reacted (X = Na, or K).

$$1.913\ g\ AgCl \times \frac{1\ mol\ AgCl}{143.4\ g\ AgCl} \times \frac{1\ mol\ XCl}{1\ mol\ AgCl} = 0.01334\ mol\ XCl.$$

Let x = number of moles NaCl. Then, the number of moles of KCl = 0.01334 mol $- x$. The sum of the NaCl and KCl masses must equal the mass of the mixture, 0.8870 g. We can write:

mass NaCl + mass KCl = 0.8870 g

$$\left[x\ mol\ NaCl \times \frac{58.44\ g\ NaCl}{1\ mol\ NaCl} \right] + \left[(0.01334 - x)mol\ KCl \times \frac{74.55\ g\ KCl}{1\ mol\ KCl} \right] = 0.8870\ g$$

$$x = 6.673 \times 10^{-3} = \text{moles NaCl}$$

mol KCl = $0.01334 - x$ = 0.01334 mol $- 6.673 \times 10^{-3}$ mol = 6.667×10^{-3} mol KCl

Converting moles to grams:

$$\text{mass NaCl} = 6.673 \times 10^{-3}\ mol\ NaCl \times \frac{58.44\ g\ NaCl}{1\ mol\ NaCl} = 0.3900\ g\ NaCl$$

$$\text{mass KCl} = 6.667 \times 10^{-3}\ mol\ KCl \times \frac{74.55\ g\ KCl}{1\ mol\ KCl} = 0.4970\ g\ KCl$$

The percentages by mass for each compound are:

$$\textbf{\% NaCl} = \frac{0.3900\ g}{0.8870\ g} \times 100\% = \textbf{43.97\% NaCl}$$

$$\textbf{\% KCl} = \frac{0.4970\ g}{0.8870\ g} \times 100\% = \textbf{56.03\% KCl}$$

4.124   The number of moles of oxalic acid in 500 mL (0.500 L) is:

$$\frac{0.100\ mol\ H_2C_2O_4}{1\ L\ soln} \times 0.500\ L = 0.0500\ mol\ H_2C_2O_4$$

The balanced equation shows a mole ratio of 1 mol Fe_2O_3 : 6 mol $H_2C_2O_4$. The mass of rust that can be removed is:

$$0.0500 \text{ mol } H_2C_2O_4 \times \frac{1 \text{ mol } Fe_2O_3}{6 \text{ mol } H_2C_2O_4} \times \frac{159.7 \text{ g } Fe_2O_3}{1 \text{ mol } Fe_2O_3} = \textbf{1.33 g } Fe_2O_3$$

4.126   The precipitation reaction is: $Ag^+ (aq) + Br^- (aq) \longrightarrow AgBr (s)$

In this problem, the relative amounts of NaBr and $CaBr_2$ are not known. However, the total amount of Br^- in the mixture can be determined from the amount of AgBr produced. Let's find the number of moles of Br^-.

$$1.6930 \text{ g AgBr} \times \frac{1 \text{ mol AgBr}}{187.8 \text{ g AgBr}} \times \frac{1 \text{ mol } Br^-}{1 \text{ mol AgBr}} = 9.015 \times 10^{-3} \text{ mol } Br^-$$

The amount of Br^- comes from both NaBr and $CaBr_2$. Let x = number of moles NaBr. Then, the number of moles of $CaBr_2 = \dfrac{9.015 \times 10^{-3} \text{ mol} - x}{2}$. The moles of $CaBr_2$ are divided by 2, because 1 mol of $CaBr_2$ produces 2 moles of Br^-. The sum of the NaBr and $CaBr_2$ masses must equal the mass of the mixture, 0.9157 g. We can write:

mass NaBr + mass $CaBr_2$ = 0.9157 g

$$\left[x \text{ mol NaBr} \times \frac{102.9 \text{ g NaBr}}{1 \text{ mol NaBr}} \right] + \left[\left(\frac{9.015 \times 10^{-3} - x}{2} \right) \text{mol } CaBr_2 \times \frac{199.9 \text{ g } CaBr_2}{1 \text{ mol } CaBr_2} \right] = 0.9157 \text{ g}$$

$2.95 x = 0.01465$

$x = 4.966 \times 10^{-3}$ = moles NaBr

Converting moles to grams:

$$\text{mass NaBr} = 4.966 \times 10^{-3} \text{ mol NaBr} \times \frac{102.9 \text{ g NaBr}}{1 \text{ mol NaBr}} = 0.5110 \text{ g NaBr}$$

The percentage by mass of NaBr in the mixture is:

$$\textbf{\% NaBr} = \frac{0.5110 \text{ g}}{0.9157 \text{ g}} \times 100\% = \textbf{55.80\% NaBr}$$

4.128   There are two moles of Cl^- per one mole of $CaCl_2$.

(a) $25.3 \text{ g } CaCl_2 \times \dfrac{1 \text{ mol } CaCl_2}{111.0 \text{ g } CaCl_2} \times \dfrac{2 \text{ mol } Cl^-}{1 \text{ mol } CaCl_2} = 0.456 \text{ mol } Cl^-$

Molarity $Cl^- = \dfrac{0.456 \text{ mol } Cl^-}{0.325 \text{ L soln}} = 1.40 \text{ mol/L} = \textbf{1.40 } M$

(b) We need to convert from mol/L to g/0.100 L.

$$\frac{1.40 \text{ mol Cl}^-}{1 \text{ L soln}} \times \frac{35.45 \text{ g Cl}}{1 \text{ mol Cl}^-} \times 0.100 \text{ L soln} = \textbf{4.96 g Cl}^-$$

4.130 (a) The precipitation reaction is: $Mg^{2+}(aq) + 2 OH^-(aq) \longrightarrow Mg(OH)_2(s)$

The acid-base reaction is: $Mg(OH)_2(s) + 2 HCl(aq) \longrightarrow MgCl_2(aq) + 2 H_2O(l)$

The redox reactions are: $Mg^{2+} + 2 e^- \longrightarrow Mg(s)$

$2 Cl^- \longrightarrow Cl_2(g) + 2 e^-$

(b) NaOH is much more expensive than CaO.

(c) Dolomite has the advantage of being an additional source of magnesium that can also be recovered.

4.132 (a) $NH_4^+(aq) + OH^-(aq) \longrightarrow NH_3(aq) + H_2O(l)$

(b) From the amount of NaOH needed to neutralize the 0.2041 g sample, we can find the amount of the 0.2041 g sample that is NH_4NO_3.

First, calculate the moles of NaOH.

$$\frac{0.1023 \text{ mol NaOH}}{1 \text{ L of soln}} \times 0.02442 \text{ L soln} = 2.498 \times 10^{-3} \text{ mol NaOH}$$

Using the mole ratio from the balanced equation, we can calculate the amount of NH_4NO_3 that reacted.

$$2.498 \times 10^{-3} \text{ mol NaOH} \times \frac{1 \text{ mol NH}_4\text{NO}_3}{1 \text{ mol NaOH}} \times \frac{80.05 \text{ g NH}_4\text{NO}_3}{1 \text{ mol NH}_4\text{NO}_3} = 0.2000 \text{ g NH}_4\text{NO}_3$$

The purity of the NH_4NO_3 sample is:

$$\textbf{\% purity} = \frac{0.2000 \text{ g}}{0.2041 \text{ g}} \times 100\% = \textbf{97.99\%}$$

4.134 Using the rules for assigning oxidation numbers given in Section 4.4, H is +1, F is −1, so the oxidation number of O must be **zero**.

4.136 The balanced equation is:

$3 CH_3CH_2OH + 2 K_2Cr_2O_7 + 8 H_2SO_4 \longrightarrow 3 CH_3COOH + 2 Cr_2(SO_4)_3 + 2 K_2SO_4 + 11 H_2O$

From the amount of $K_2Cr_2O_7$ required to react with the blood sample, we can calculate the mass of ethanol (CH_3CH_2OH) in the 10.0 g sample of blood.

First, calculate the moles of $K_2Cr_2O_7$ reacted.

$$\frac{0.07654 \text{ mol K}_2\text{Cr}_2\text{O}_7}{1 \text{ L soln}} \times 0.00423 \text{ L} = 3.24 \times 10^{-4} \text{ mol K}_2\text{Cr}_2\text{O}_7$$

Next, using the mole ratio from the balanced equation, we can calculate the mass of ethanol that reacted.

$$3.24 \times 10^{-4} \text{ mol K}_2\text{Cr}_2\text{O}_7 \times \frac{3 \text{ mol ethanol}}{2 \text{ mol K}_2\text{Cr}_2\text{O}_7} \times \frac{46.07 \text{ g ethanol}}{1 \text{ mol ethanol}} = 0.0224 \text{ g ethanol}$$

The percent ethanol by mass is:

$$\text{\% by mass ethanol} = \frac{0.0224 \text{ g}}{10.0 \text{ g}} \times 100\% = \textbf{0.224\%}$$

This is well above the legal limit of 0.1 percent by mass ethanol in the blood. The individual should be prosecuted for drunk driving.

4.138 **(a)** $\text{Zn } (s) + \text{H}_2\text{SO}_4 \, (aq) \longrightarrow \text{ZnSO}_4 \, (aq) + \text{H}_2 \, (g)$

   (b) $2 \text{ KClO}_3 \, (s) \longrightarrow 2 \text{ KCl} \, (s) + 3 \text{ O}_2 \, (g)$

   (c) $\text{Na}_2\text{CO}_3 \, (s) + 2 \text{ HCl} \, (aq) \longrightarrow 2 \text{ NaCl} \, (aq) + \text{CO}_2 \, (g) + \text{H}_2\text{O} \, (l)$

   (d) $\text{NH}_4\text{NO}_2 \, (s) \xrightarrow{\text{heat}} \text{N}_2 \, (g) + 2 \text{ H}_2\text{O} \, (g)$

4.140 Because the volume of the solution changes (increases or decreases) when the solid dissolves.

4.142 NH_4Cl exists as NH_4^+ and Cl^-. To form NH_3 and HCl, a proton (H^+) is transferred from NH_4^+ to Cl^-. Therefore, this is a Brønsted acid-base reaction.

CHAPTER 5
GASES

PROBLEM-SOLVING STRATEGIES AND TUTORIAL SOLUTIONS

TYPES OF PROBLEMS

Problem Type 1: Pressure Conversions (Also see Problem Type 5, Chapter 1).

Problem Type 2: Calculations Using the Ideal Gas Law.
    (a) Given three of the four variable quantities in the equation (P, V, n, and T).
    (b) Only one or two variable quantities have fixed values.
    (c) Calculations using density or molar mass (Also see Density calculations, Problem Type 1, Chapter 1).
    (d) Stoichiometry involving gases (Also see Stoichiometry problems, Problem Type 7, Chapter 3).
    (e) Collecting a gas over water (Also see Dalton's Law of Partial Pressures, below).

Problem Type 3: Dalton's Law of Partial Pressures.

Problem Type 4: Root-Mean-Square (RMS) Speed.

Problem Type 5: Deviations from Ideal Behavior.

PROBLEM TYPE 1: PRESSURE CONVERSIONS

(Also see Problem Type 5, Chapter 1)

In order to convert from one unit of pressure to another, you need to be proficient at the factor-label method. See Section 1.9 of your text. Unit conversions can seem daunting, but if you keep track of the units, making sure that the appropriate units cancel, your effort will be rewarded.

EXAMPLE 5.1
Convert 555 mmHg to kPa.

Step 1: Map out a strategy to proceed from initial units to final units based on available conversion factors. Looking in Section 5.2 of your text, you should find the following conversions.

$$1 \text{ atm} = 760 \text{ mmHg}$$
$$1 \text{ atm} = 1.01325 \times 10^2 \text{ kPa}$$

You should come up with the following strategy:

$$\text{mm Hg} \rightarrow \text{atm} \rightarrow \text{kPa}$$

Step 2: Use the following method to ensure that you obtain the desired unit.

$$\text{Given unit} \times \left(\frac{\text{desired unit}}{\text{given unit}} \right) = \text{desired unit}$$

$$\textbf{? kPa} = 555 \text{ mmHg (given)} \times \left(\frac{1 \text{ atm (desired)}}{760 \text{ mmHg (given)}} \right) \times \left(\frac{1.01325 \times 10^2 \text{ kPa (desired)}}{1 \text{ atm (given)}} \right)$$

$$= \textbf{74.0 kPa}$$

> **Note**: In the first conversion factor (atm/mmHg), atm is the desired unit. When moving on to the next conversion factor (kPa/atm), atm is now given, and the desired unit is kPa.

PRACTICE EXERCISE

1.   Convert 5.32 atm to mmHg.

Text Problem: 5.14

PROBLEM TYPE 2: CALCULATIONS USING THE IDEAL GAS LAW

You will encounter five different types of problems in this chapter that use the Ideal Gas Law. Each type will be addressed individually below. Remember that R (the gas constant) has the units $L \cdot atm/mol \cdot K$. Therefore, in problems that contain R, you *must* use the following units: volume in liters (L), pressure in atmospheres (atm), and temperature in Kelvin (K).

A.   Given three of the four variable quantities in the equation (P, V, n, and T)

If you know the values of any three of the variable quantities, you can solve $PV = nRT$ algebraically for the fourth one. Then, you can calculate its value by substituting in the three known quantities.

EXAMPLE 5.2
Carbon monoxide gas, CO, stored in a 2.00 L container at 25.0°C exerts a pressure of 15.5 atm. How many moles of CO (g) are in the container?

Step 1: You are given P, V, and T. Solve $PV = nRT$ algebraically for n.

$$n = \frac{PV}{RT}$$

Step 2: Check that the three known quantities (P, V, and T) have the appropriate units.

P and V have the correct units, but T must be converted to units of K.

$$T(K) = °C + 273°$$
$$T(K) = 25.0° + 273° = 298 \text{ K}$$

Step 3: Calculate the value of n by substituting the three known quantities into the equation.

$$n = \frac{PV}{RT}$$

$$n = \frac{(15.5 \text{ atm})(2.00 \text{ L})}{298 \text{ K}} \times \frac{\text{mol} \cdot \text{K}}{0.0821 \text{ L} \cdot \text{atm}} = \textbf{1.27 mol CO}$$

PRACTICE EXERCISE

2.   Carbon dioxide gas, CO_2, and water vapor are the two species primarily responsible for the greenhouse effect. Combustion of fossil fuels adds 22 billion tons of carbon dioxide to the atmosphere each year. How many liters of CO_2 is this at 25°C and 1.0 atm pressure?

Text Problems: 5.30, 5.40

B. Only one or two variable quantities have fixed values.

Let's look at the case where n and P are constant (Charles' law).

Step 1: Rearrange the ideal gas equation so that the constants are all on one side.

$$\frac{V}{T} = \frac{nR}{P} \quad \left(\frac{nR}{P} = \text{constant}\right)$$

$$\text{so,} \quad \frac{V}{T} = \textbf{constant}$$

Step 2: The individual values of volume and temperature can vary greatly for a given sample of gas. However, as long as the pressure and the amount of gas (n) do *not* change, $\frac{V}{T}$ always equals the same constant. Thus,

$$\frac{V_1}{T_1} = \text{constant} = \frac{V_2}{T_2}$$

$$\text{and,} \quad \frac{V_1}{T_1} = \frac{V_2}{T_2}$$

This is Charles' law.

Step 3: Solve the equation algebraically for the missing quantity. Then, you can calculate its value by substituting in the known quantities.

EXAMPLE 5.3
Given 10.0 L of neon gas at 5.0°C and 630 mmHg, calculate the new volume at 400°C and 2.5 atm.

Step 1: In this problem, only n is constant. As long as gas is not added or removed, the number of moles of gas (n) is constant. Therefore,

$$\frac{PV}{T} = nR = \text{constant}$$

Step 2: Volume, temperature, and pressure change.

$$\frac{P_1V_1}{T_1} = \text{constant} = \frac{P_2V_2}{T_2}$$

$$\frac{P_1V_1}{T_1} = \frac{P_2V_2}{T_2}$$

Step 3: Solve the equation algebraically for V_2, then substitute in the known quantities to solve for V_2.

$$P_2 = 2.5 \text{ atm} \times \frac{760 \text{ mmHg}}{1 \text{ atm}} = 1.9 \times 10^3 \text{ mmHg}$$

$$V_2 = \frac{P_1V_1T_2}{P_2T_1}$$

$$V_2 = \frac{(630 \text{ mmHg})(10.0 \text{ L})(673 \text{ K})}{(1.9 \times 10^3 \text{ mmHg})(278 \text{ K})} = \textbf{8.0 L}$$

> **Tip:** Since R is not in this equation, the units of pressure and volume do *not* have to be atm and liters, respectively. However, temperature must *always* be in units of Kelvin for gas law calculations.

PRACTICE EXERCISE

3. After flying in an airplane, a passenger finds that his sticky hair-gel has opened in his bag making a complete mess. This happens because the cabin is pressurized to about 8000 feet above sea level. During flight, the volume of air inside the bottle increases due to the decreased pressure. Sometimes, this can cause the flip-top cap to open. If the volume of air inside the bottle is 125 mL at sea level ($P = 760$ mmHg), what volume does the air in the bottle occupy in the airplane ($P = 595$ mmHg)? Assume that the temperature is kept constant.

Text Problems: 5.18, **5.20**, **5.22**, 5.32, 5.34, 5.36, 5.38

C. Calculations involving density (*d*) or molar mass ($\mathcal{M}$)

(Also see Problem Type 1, Chapter 1)

Step 1: Algebraically solve the ideal gas equation for either density $\left(\dfrac{m}{V}\right)$ or molar mass ($\mathcal{M}$).

$$PV = nRT$$

$$\frac{n}{V} = \frac{P}{RT} \quad \text{and} \quad n = \frac{m}{\mathcal{M}}$$

Substituting for n,

$$\frac{m}{\mathcal{M}V} = \frac{P}{RT}$$

and,

$$d = \frac{m}{V} = \frac{P\mathcal{M}}{RT} \qquad\qquad (5.1)$$

Furthermore,

$$\mathcal{M} = \frac{dRT}{P}$$

Step 2: Calculate the value of density or molar mass by substituting in the known quantities.

EXAMPLE 5.4

What is the density of methane gas, CH_4, at STP.

This problem is simplified if you recall that 1 mole of an ideal gas occupies a volume of 22.4 L at STP.

Step 1: Let's assume that we have 1 mole of methane gas. At STP, 1 mole of methane will occupy a volume of 22.4 L. First, we calculate the molar mass of methane.

$$\mathcal{M}(CH_4) = 12.01 \text{ g} + 4(1.008 \text{ g}) = 16.04 \text{ g}$$

Step 2: Knowing the mass of 1 mole of methane and the volume it occupies, we can calculate the density.

$$d = \frac{\text{mass}}{\text{volume}}$$

$$d = \frac{16.04 \text{ g}}{22.4 \text{ L}} = \textbf{0.716 g/L}$$

This problem could also be solved using Equation (5.1) derived above.

Step 1: The necessary equation to derive is:

$$d = \frac{P\mathcal{M}}{RT}$$

Step 2: Calculate the density by substituting in the known quantities.

STP: $P = 1.00$ atm
$T = 273$ K

$\mathcal{M} = 16.04$ g/mol

$$d = \frac{(1.00 \text{ atm})\left(16.04 \ \frac{g}{mol}\right)}{273 \text{ K}} \times \frac{mol \cdot K}{0.0821 \text{ L} \cdot atm} = \textbf{0.716 g/L}$$

PRACTICE EXERCISE

4.   An element that exists as a diatomic gas at room temperature has a density of 1.553 g/L at 25°C and 1 atm pressure. Identify the unknown gas.

Text Problems: **5.42**, 5.46, 5.48

D. Stoichiometry Involving Gases

(Also see Problem Type 7, Chapter 3)

A typical gas stoichiometry problem involves the following approach:

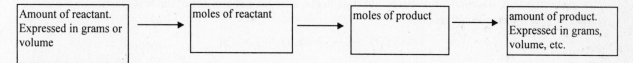

EXAMPLE 5.5
Oxygen gas was discovered by decomposing mercury (II) oxide.

$$2 \text{ HgO } (s) \longrightarrow 2 \text{ Hg } (l) + O_2 (g)$$

What volume of oxygen gas would be produced by the reaction of 35.2 g of the oxide if the gas is collected at STP?

Step 1: Since stoichiometry uses mole ratios, make sure that you start with a balanced equation. In this problem, you are given the balanced equation.

Step 2: Use the following strategy to solve this problem.

grams HgO → moles HgO → moles O_2 → volume O_2

First, calculate the moles of O_2.

$$? \text{ mol } O_2 = 35.2 \text{ g HgO} \times \frac{1 \text{ mol HgO}}{216.6 \text{ g HgO}} \times \frac{1 \text{ mol } O_2}{2 \text{ mol HgO}} = 0.0813 \text{ mol } O_2$$

Step 3: Solve the ideal gas equation algebraically for V_{O_2}. Then, calculate the volume by substituting the known quantities into the equation.

STP: $P = 1.00$ atm
$T = 273$ K

$$V_{O_2} = \frac{n_{O_2} RT}{P}$$

$$V_{O_2} = \frac{(0.0813 \text{ mol})(273 \text{ K})}{1.00 \text{ atm}} \times \frac{0.0821 \text{ L} \cdot atm}{mol \cdot K} = \textbf{1.82 L}$$

EXAMPLE 5.6

Calculate the volume of methane, CH$_4$, at STP required to completely consume 3.50 L of oxygen at STP.

Step 1: Write the balanced chemical equation.

$$CH_4 \, (g) + 2 \, O_2 \, (g) \longrightarrow CO_2 \, (g) + 2 \, H_2O \, (l)$$

Step 2: You could follow the strategy used in the previous example to solve this problem. You could find the moles of oxygen in 3.50 L, apply the correct mole ratio to find moles of methane, and then use the ideal gas equation to find volume of methane. However, there is a short cut if you remember Avogadro's Law. Avogadro's Law states that the volume of a gas is directly proportional to the number of moles of gas at constant temperature and pressure.

Thus, the stoichiometric ratio, $\dfrac{1 \, mol \, CH_4}{2 \, mol \, O_2}$, can be written in terms of volume, $\dfrac{1 \, L \, CH_4}{2 \, L \, O_2}$.

Step 3: Calculate the volume of methane using the above conversion factor.

$$? \, L \, CH_4 = 3.50 \, \cancel{L} \, O_2 \times \frac{1 \, L \, CH_4}{2 \, \cancel{L} \, O_2} = 1.75 \, L \, CH_4$$

PRACTICE EXERCISE

5.   The decomposition of sodium azide (NaN$_3$) to sodium and nitrogen gas was one of the first reactions used to inflate air-bag systems in automobiles.

$$2 \, NaN_3 \, (s) \longrightarrow 2 \, Na \, (l) + 3 \, N_2 \, (g)$$

What mass of NaN$_3$ in grams would be needed to inflate a 100 L air bag with nitrogen at 25.0°C and 755 mmHg?

Text Problems: 5.50, 5.52, 5.54

E.   Collecting a gas over water

In this type of problem, you must account for the fact that

$$P_{TOTAL} = P_{SAMPLE} + P_{H_2O(g)}$$

The pressure of the gas sample of interest is:

$$P_{SAMPLE} = P_{TOTAL} - P_{H_2O(g)}$$

Also see Dalton's law of partial pressures, Problem Type 3.

EXAMPLE 5.7

When heated, calcium carbonate decomposes forming solid calcium oxide and carbon dioxide gas. A sample of calcium carbonate is completely decomposed by heating and 50.5 mL of gas is collected at 731.2 torr and 25.0°C. How much did the sample of calcium carbonate weigh?

Step 1: Write the balanced chemical equation.

$$CaCO_3 \, (s) \longrightarrow CaO \, (s) + CO_2 \, (g)$$

Step 2: Map out the following strategy to solve this problem.

volume CO$_2$ → moles CO$_2$ → moles CaCO$_3$ → grams CaCO$_3$

Step 3: Solve the ideal gas equation algebraically for moles of CO_2. Then, substitute in the known values.

$$P_{CO_2} = 731.2 \text{ torr} \times \frac{1 \text{ atm}}{760 \text{ torr}} = 0.9621 \text{ atm}$$

$$T(\text{K}) = 25.0° + 273° = 298 \text{ K}$$

$$V = 50.5 \text{ mL} \times \frac{1 \text{ L}}{1000 \text{ mL}} = 0.0505 \text{ L}$$

$$n_{CO_2} = \frac{P_{CO_2}V}{RT}$$

$$n_{CO_2} = \frac{(0.9621 \text{ atm})(0.0505 \text{ L})}{298 \text{ K}} \times \frac{\text{mol} \cdot \text{K}}{0.0821 \text{ L} \cdot \text{atm}} = 1.99 \times 10^{-3} \text{ mol } CO_2$$

Step 4: Calculate the mass of $CaCO_3$ using the strategy in Step 2.

$$? \text{ g } CaCO_3 = 1.99 \times 10^{-3} \text{ mol } CO_2 \times \frac{1 \text{ mol } CaCO_3}{1 \text{ mol } CO_2} \times \frac{100.09 \text{ g } CaCO_3}{1 \text{ mol } CaCO_3}$$

$$= 0.199 \text{ g } CaCO_3$$

PRACTICE EXERCISE

6. Fermentation is one of the oldest studied biochemical processes. The fermentation process is a catabolic pathway, which involves the use of yeast to convert sugars into ethanol and carbon dioxide. In brewing beer, malt sugar is fermented to form ethanol (CH_3CH_2OH) and carbon dioxide (CO_2) according to the following overall reaction:

$$C_6H_{12}O_6 (aq) \longrightarrow 2 CH_3CH_2OH (aq) + 2 CO_2 (g)$$

A micro-brewery is making a batch of beer, and they wish to determine the mass percentage of ethanol in their brew by collecting the CO_2 (g) evolved. They start the process by mixing malt sugar, hops, and yeast in water. The mixture has a total mass of 800 lbs. During fermentation, 10,000 L of CO_2 is collected over water at 25.0°C and 750 mmHg. What is the mass percentage of ethanol in the beer? (The vapor pressure of water at 25°C is 23.76 mm Hg.)

Text Problems: 5.60, 5.62

PROBLEM TYPE 3: DALTON'S LAW OF PARTIAL PRESSURES

In order to understand how to approach problems with partial pressures, it is important to know (1) the derivation of Dalton's law of partial pressures and (2) the relationship between partial pressure and the total pressure.

(1) Dalton's law for two gases, **A** and **B**, in a container of volume, *V*.

$$P_{\text{Total}} = \frac{n_{\text{Total}}RT}{V}$$

and, $n_{\text{Total}} = n_A + n_B$

$$P_{\text{Total}} = \frac{(n_A + n_B)RT}{V}$$

$$P_{\text{Total}} = \frac{n_A RT}{V} + \frac{n_B RT}{V}$$

$$\textbf{\textit{P}}_{\textbf{Total}} = \textbf{\textit{P}}_{\textbf{A}} + \textbf{\textit{P}}_{\textbf{B}}$$

In general, $P_{Total} = P_1 + P_2 + P_3 + \ldots + P_n$, where $P_1, P_2, P_3, \ldots$ are the partial pressures of components 1, 2, 3, $\ldots$

(2)   The relationship between partial pressure and P_{Total} for two gases, **A** and **B**, in a container of volume, V.

$$\frac{P_A}{P_T} = \frac{\dfrac{n_A RT}{V}}{\dfrac{(n_A + n_B)RT}{V}}$$

$$\frac{P_A}{P_T} = \frac{n_A}{(n_A + n_B)} \quad \text{and} \quad \frac{n_A}{(n_A + n_B)} = X_A$$

where X_A = mole fraction of component A.

$$\frac{P_A}{P_T} = X_A$$

$$P_A = X_A P_T$$

Similarly,

$$P_B = X_B P_T$$

In general, $P_i = X_i P_T$, where X_i is the mole fraction of substance i.

EXAMPLE 5.8

A 2.00 L container at 22.0°C contains a mixture of 1.00 g H_2 (g) and 1.00 g He (g).

(a)   What are the partial pressures of H_2 and He? What is the total pressure?

This is a mixture of two gases that obeys Dalton's law of partial pressures.

$$P_T = P_{H_2(g)} + P_{He(g)}$$

Step 1: Calculate the moles of H_2 and He.

$$n_{H_2} = 1.00 \text{ g} \times \frac{1 \text{ mol } H_2}{2.016 \text{ g } H_2} = 0.496 \text{ mol } H_2$$

$$n_{He} = 1.00 \text{ g} \times \frac{1 \text{ mol He}}{4.003 \text{ g He}} = 0.250 \text{ mol He}$$

Step 2: Solve the ideal gas equation algebraically for P, then substitute in the known quantities to solve for P_{H_2} or P_{He}.

$T(K) = 22.0° + 273° = 295 \text{ K}$

$$P_{H_2} = \frac{n_{H_2} RT}{V} \qquad\qquad P_{He} = \frac{n_{He} RT}{V}$$

$$P_{H_2} = \frac{(0.496 \text{ mol})(295 \text{ K})}{2.00 \text{ L}} \times \frac{0.0821 \text{ L·atm}}{\text{mol·K}} \qquad P_{He} = \frac{(0.250 \text{ mol})(295 \text{ K})}{2.00 \text{ L}} \times \frac{0.0821 \text{ L·atm}}{\text{mol·K}}$$

$$P_{H_2} = \textbf{6.01 atm} \qquad\qquad P_{He} = \textbf{3.03 atm}$$

$$P_{Total} = P_{H_2} + P_{He}$$

$$P_{Total} = 6.01 \text{ atm} + 3.03 \text{ atm} = \textbf{9.04 atm}$$

(b) What are the mole fractions of H_2 and He?

Remember that $P_i = X_i P_T$.

$$P_{H_2} = X_{H_2} P_T \qquad\qquad P_{He} = X_{He} P_T$$

$$X_{H_2} = \frac{P_{H_2}}{P_T} \qquad\qquad X_{He} = \frac{P_{He}}{P_T}$$

$$X_{H_2} = \frac{6.01 \text{ atm}}{9.04 \text{ atm}} = \mathbf{0.665} \qquad\qquad X_{He} = \frac{3.03 \text{ atm}}{9.04 \text{ atm}} = \mathbf{0.335}$$

PRACTICE EXERCISE

7. About two-thirds of the carbon monoxide (CO) emissions in the United States come from automobiles. CO is extremely toxic to humans because it binds 210 times more strongly to hemoglobin than does O_2. Hemoglobin is the iron-containing protein responsible for oxygen transport in blood. In a typical urban environment, the CO concentration is 10 parts per million (ppm). Assuming an atmospheric pressure of 750 torr, calculate the partial pressure of CO.

> **Hint:** 1 ppm of a gas refers to one part by volume in 1 million volume units of the whole (volume fraction).

Text Problems: 5.56, 5.58, 5.64

PROBLEM TYPE 4: ROOT-MEAN-SQUARE (RMS) SPEED

This type of problem typically involves substituting known values into the following equation:

$$u_{rms} = \sqrt{\frac{3RT}{\mathcal{M}}}$$

You must remember to use R in units of $\dfrac{J}{mol \cdot K}$ and $\mathcal{M}$ in units of kg/mol. 1 joule $= 1 \dfrac{kg \cdot m^2}{s^2}$. If these units are used, the units of u_{rms} are m/s.

$$u_{rms} \text{ (units)} = \sqrt{\frac{\left(\dfrac{J}{mol \cdot K}\right)(K)}{\left(\dfrac{kg}{mol}\right)}}$$

$$u_{rms} \text{ (units)} = \sqrt{\frac{\left(\dfrac{kg \cdot m^2}{s^2} \times \dfrac{1}{mol \cdot K}\right)(K)}{\left(\dfrac{kg}{mol}\right)}}$$

$$u_{rms} \text{ (units)} = \sqrt{\frac{m^2}{s^2}}$$

$$u_{rms} \text{ (units)} = \frac{m}{s}$$

EXAMPLE 5.9

Which has a higher root mean square velocity, H_2 (g) at 150 K or He (g) at 650K?

Step 1: The molar mass ($\mathcal{M}$) must be in units of kg/mol for the units to work out correctly. Convert g/mol to kg/mol.

$$\mathcal{M}_{H_2} = \frac{2.016 \text{ g } H_2}{1 \text{ mol } H_2} \times \frac{1 \text{ kg}}{1000 \text{ g}} = 2.016 \times 10^{-3} \text{ kg/mol}$$

$$\mathcal{M}_{He} = \frac{4.003 \text{ g He}}{1 \text{ mol He}} \times \frac{1 \text{ kg}}{1000 \text{ g}} = 4.003 \times 10^{-3} \text{ kg/mol}$$

Step 2: Substitute the appropriate values into the equation to solve for u_{rms}.

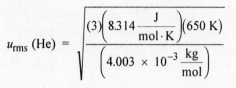

$$u_{rms} (H_2) = \sqrt{\frac{(3)\left(8.314 \frac{J}{mol \cdot K}\right)(150 \text{ K})}{\left(2.016 \times 10^{-3} \frac{kg}{mol}\right)}} \qquad u_{rms} (He) = \sqrt{\frac{(3)\left(8.314 \frac{J}{mol \cdot K}\right)(650 \text{ K})}{\left(4.003 \times 10^{-3} \frac{kg}{mol}\right)}}$$

$$u_{rms} (H_2) = 1.36 \times 10^3 \text{ m/s} \qquad\qquad u_{rms} (He) = 2.01 \times 10^3 \text{ m/s}$$

He (g) has the higher root mean square velocity.

PRACTICE EXERCISE

8.   What is the root-mean-square velocity of F_2 (g) at 298 K?

Text Problems: 5.72, 5.74

PROBLEM TYPE 5: DEVIATIONS FROM IDEAL BEHAVIOR

Van der Waals' equation is a simple modification of the ideal gas equation. It takes into account intermolecular forces and finite molecular volumes.

Starting with the ideal gas equation:

$$P_{ideal} V_{ideal} = nRT$$

van der Waals made two corrections.

(1) $P_{ideal} = P_{real}$ + correction for attraction between particles

$$P_{ideal} = P_{real} + \frac{an^2}{V_{real}^2} \quad \text{(see Section 5.8 of your text for discussion)}$$

(2) $V_{ideal} = V_{real}$ − correction for finite volume of particles

$$V_{ideal} = V_{real} - nb \quad \text{(see Section 5.8 of your text for discussion)}$$

where,

  a and b are constants for a particular gas (See Table 5.4)
  n = moles of gas

Substituting these corrections into the ideal gas equation leads to the **van der Waals equation.**

$$\left(P_{real} + \frac{an^2}{V_{real}^2}\right)(V_{real} - nb) = nRT$$

EXAMPLE 5.10

The molar volume of isopentane (C_5H_{12}) is 1.00 L at 503 K and 30.0 atm.

(a) Does isopentane behave like an ideal gas?

Calculate the pressure of 1 mol of an ideal gas at 503 K that occupies a volume of 1.00 L.

Step 1: Solve the ideal gas equation algebraically for P, and then substitute in the known values.

$$P = \frac{nRT}{V} = \frac{(1.00 \text{ mol})(503 \text{ K})}{1.00 \text{ L}} \times \frac{0.0821 \text{ L} \cdot \text{atm}}{\text{mol} \cdot \text{K}}$$

$$P = 41.3 \text{ atm}$$

Step 2: Compare the ideal pressure to the actual pressure. Calculating the percentage error would be helpful. Percentage of error is the difference between the two values divided by the actual value.

$$\% \text{ error} = \frac{41.3 \text{ atm} - 30.0 \text{ atm}}{30.0 \text{ atm}} \times 100 = \textbf{37.7 \%}$$

Because of the large percent error, we conclude that under these conditions, C_5H_{12} behaves in a nonideal manner.

(b) Given that $a = 17.0 \text{ L}^2 \cdot \text{atm/mol}^2$ and $b = 0.136$ L/mol, calculate the pressure of isopentane as predicted by the van der Waals' equation.

Calculate the pressure of 1 mol of gas at 503 K that occupies 1.00 L using van der Waals' equation.

Step 1: Calculate the correction terms.

$$\frac{an^2}{V_{obs}^2} = \frac{\left(17.0 \frac{L^2 \cdot \text{atm}}{\text{mol}^2}\right)(1.00 \text{ mol})^2}{1.00 \text{ L}^2} = 17.0 \text{ atm}$$

$$nb = (1.00 \text{ mol})\left(0.136 \frac{L}{\text{mol}}\right) = 0.136 \text{ L}$$

Step 2: Substitute the values into the van der Waals' equation and solve for P_{real}.

$$\left(P_{real} + \frac{an^2}{V_{real}^2}\right)(V_{real} - nb) = nRT$$

$$(P_{real} + 17.0 \text{ atm})(1.00 \text{ L} - 0.136 \text{ L}) = (1.00 \text{ mol})\left(\frac{0.0821 \text{ L} \cdot \text{atm}}{\text{mol} \cdot \text{K}}\right)(503 \text{ K})$$

$$(P_{real} + 17.0 \text{ atm})(0.864) = 41.3 \text{ atm}$$

$$P_{real} + 17.0 \text{ atm} = 47.8 \text{ atm}$$

$$P_{real} = 47.8 \text{ atm} - 17.0 \text{ atm} = \textbf{30.8 atm}$$

The pressure calculated for 1 mole of this "real" gas at 503 K and 1.00 L using van der Waals' equation is much closer to the actual value of 30.0 atm. The percent error is only 2.7 percent.

PRACTICE EXERCISE

9.   Using van der Waals' equation, calculate the pressure exerted by 1.00 mol of water vapor that occupies a volume of 5.55 L at 150°C.

Text Problem: 5.80

ANSWERS TO PRACTICE EXERCISES

1. 4.04×10^3 mmHg

2. $V_{CO_2} = 1.1 \times 10^{16}$ L

3. $V_{air} = 160$ mL

4.   The gas is fluorine, F_2.

5.   176 g NaN_3

6.   mass % ethanol = 5.0 %

7. $P_{CO} = 7.5 \times 10^{-3}$ torr

8. $u_{rms} = 442$ m/s

9. $P_{real} = 6.11$ atm

SOLUTIONS TO SELECTED TEXT PROBLEMS

5.14   Pressure Conversions, Problem Type 1.

The conversion needed to solve this problem are:

$$1 \text{ atm } = 760 \text{ mmHg} \qquad 1 \text{ atm} = 101.325 \text{ kPa}$$

$$\textbf{? atm } = 606 \text{ mmHg} \times \frac{1 \text{ atm}}{760 \text{ mmHg}} = \textbf{0.797 atm}$$

$$\textbf{? kPa } = 606 \text{ mmHg} \times \frac{101.325 \text{ kPa}}{760 \text{ mmHg}} = \textbf{80.8 kPa}$$

5.18   Temperature does not change in this problem. Pressure and volume are the variables; it is a Boyle's law problem.

$$V_2 = 0.10 \, V_1$$

$$P_2 = \frac{P_1 V_1}{V_2}$$

$$P_2 = \frac{(5.3 \text{ atm})V_1}{0.10 V_1} = \textbf{53 atm}$$

5.20   Only one or two variable quantities have fixed values, Problem Type 2B. This is a Boyle's law problem.

(a)
Step 1: n and T are constant. The ideal gas equation is already arranged with all constants on one side.

$$PV = nRT$$

$$PV = \text{constant}$$

Step 2: Pressure and volume change.

$$P_1 V_1 = \text{constant} = P_2 V_2$$

and,

$$P_1 V_1 = P_2 V_2$$

(Initial Pressure)(Initial Volume) = (Final Pressure)(Final Volume)

Step 3: Solve the equation algebraically for the missing quantity (V_2). Then, calculate its value by substituting in the known quantities.

$$V_2 = \frac{P_1 V_1}{V_2}$$

$$V_2 = \frac{(1.2 \text{ atm})(3.8 \text{ L})}{(6.6 \text{ atm})} = \textbf{0.69 L}$$

(b)   On this part, you should also come up with $P_1 V_1 = P_2 V_2$.

Solve the equation algebraically for the missing quantity (P_2). Then, calculate its value by substituting in the known quantities.

$$P_2 = \frac{P_1 V_1}{V_2}$$

$$P_2 = \frac{(1.2 \text{ atm})(3.8 \text{ L})}{(0.075 \text{ L})} = \textbf{61 atm}$$

5.22   Only one or two variable quantities have fixed values, Problem Type 2B. This is a Charles' law problem.

Step 1: P and n are constant. Remember, as long as gas is not added or removed, the number of moles of gas (n) is constant. Rearrange the ideal gas equation so that the constants are all on one side.

$$\frac{V}{T} = \frac{nR}{P}$$

$$\frac{V}{T} = \text{constant}$$

Step 2: Volume and temperature change.

$$\frac{V_1}{T_1} = \text{constant} = \frac{V_2}{T_2}$$

and,

$$\frac{V_1}{T_1} = \frac{V_2}{T_2}$$

$$\frac{\text{Initial Volume}}{\text{Initial Temperature}} = \frac{\text{Final Volume}}{\text{Final Temperature}}$$

Step 3: Solve the equation algebraically for the missing quantity (T_2). Then, calculate its value by substituting in the known quantities.

$$T_1(\text{K}) = 88° + 273° = 361 \text{ K}$$

$$T_2 = \frac{T_1 V_2}{V_1}$$

$$T_2 = \frac{(361 \text{ K})(3.4 \text{ L})}{(9.6 \text{ L})} = \textbf{1.3} \times \textbf{10}^2 \text{ K}$$

5.24   This is a gas stoichiometry problem that requires a knowledge of Avogadro's law to solve. Avogadro's law states that the volume of a gas is directly proportional to the number of moles of gas at constant temperature and pressure.

The volume ratio, 1 vol. Cl_2 : 3 vol. F_2 : 2 vol. product, can be written as a mole ratio, 1 mol Cl_2 : 3 mol F_2 : 2 mol product.

Attempt to write a balanced chemical equation. The subscript of F in the product will be three times the Cl subscript, because there are three times as many F atoms reacted as Cl atoms.

$$1 \, Cl_2 \, (g) + 3 \, F_2 \, (g) \longrightarrow 2 \, Cl_x F_{3x} \, (g)$$

Balance the equation. The x must equal one so that there are two Cl atoms on each side of the equation. If x = 1, the subscript on F is 3.

$$Cl_2 \, (g) + 3 \, F_2 \, (g) \longrightarrow 2 \, ClF_3 \, (g)$$

The formula of the product is **ClF_3**.

5.30   Given three of the four variable quantities in the equation (P, V, n, and T), Problem Type 2A.

Step 1: You are given n, V, and T. Solve $PV = nRT$ algebraically for P.

$$P = \frac{nRT}{V}$$

Step 2: Check that the three known quantities (n, V, and T) have the appropriate units.

T must be converted to units of K

$$T(\text{K}) = °\text{C} + 273°$$

$$T(\text{K}) = 62° + 273° = 335 \text{ K}$$

Step 3: Calculate the pressure by substituting the known values into the equation.

$$P = \frac{(6.9 \text{ mol})\left(0.0821 \dfrac{\text{L} \cdot \text{atm}}{\text{mol} \cdot \text{K}}\right)(335 \text{ K})}{30.4 \text{ L}} = \mathbf{6.2 \text{ atm}}$$

5.32   In this problem, the moles of gas and the volume the gas occupies are constant. Temperature and pressure change.

$$\frac{P_1}{T_1} = \frac{P_2}{T_2}$$

$$T_2 = \frac{T_1 P_2}{P_1}$$

$$T_2 = \frac{(298 \text{ K})(2.00 \text{ atm})}{(0.800 \text{ atm})} = \mathbf{745 \text{ K}} = \mathbf{472°C}$$

5.34 $P_1 = 1.0 \text{ atm}$ $P_2 = ?$
         $T_1 = 273 \text{ K}$ $T_2 = (250 + 273)\text{K} = 523 \text{ K}$

$$P_2 = \frac{P_1 T_2}{T_1}$$

$$P_2 = \frac{(1.0 \text{ atm})(523 \text{ K})}{273 \text{ K}} = \mathbf{1.9 \text{ atm}}$$

5.36 $$V_2 = \frac{V_1 T_2}{T_1}$$

$$V_2 = \frac{(0.78 \text{ L})(36.5 + 273) \text{ K}}{(20.1 + 273) \text{ K}} = \mathbf{0.82 \text{ L}}$$

5.38   In this problem, the number of moles of gas is constant. Pressure, volume, and temperature change.

$$\frac{P_1 V_1}{T_1} = \frac{P_2 V_2}{T_2}$$

$$T_2(\text{K}) = 150° + 273° = 423 \text{ K}$$

$$V_2 = \frac{P_1 V_1 T_2}{P_2 T_1}$$

$$V_2 = \frac{(1\ \text{atm})(488\ \text{mL})(423\ \text{K})}{(22.5\ \text{atm})(273\ \text{K})} = \textbf{33.6 mL}$$

Note: Since R is not in this equation, the units of pressure and volume do *not* have to be atm and liters, respectively. However, temperature must *always* be in units of Kelvin for gas law calculations.

5.40   The molar mass of CO_2 = 44.01 g/mol. Since $PV = nRT$, we write:

$$P = \frac{nRT}{V}$$

$$P = \frac{\left(0.050\ \text{g} \times \dfrac{1\ \text{mol}}{44.01\ \text{g}}\right)\left(0.0821\ \dfrac{\text{L} \cdot \text{atm}}{\text{mol} \cdot \text{K}}\right)(30 + 273)\ \text{K}}{4.6\ \text{L}} = \textbf{6.1} \times \textbf{10}^{-3}\ \textbf{atm}$$

5.42   Calculations involving molar mass, Problem Type 2C.

Step 1: Algebraically solve the ideal gas equation for molar mass. You should come up with

$$\mathcal{M} = \frac{dRT}{P}$$

Step 2: Calculate the density, and check that the other known quantities (P and T) have the appropriate units.

$$d = \frac{7.10\ \text{g}}{5.40\ \text{L}} = 1.31\ \text{g} / \text{L}$$

$$T = 44° + 273° = 317\ \text{K}$$
$$P = 741\ \text{torr} \times \frac{1\ \text{atm}}{760\ \text{torr}} = 0.975\ \text{atm}$$

Step 3: Calculate the molar mass by substituting in the known quantities.

$$\mathcal{M} = \frac{\left(1.31\ \dfrac{\text{g}}{\text{L}}\right)\left(0.0821\ \dfrac{\text{L} \cdot \text{atm}}{\text{mol} \cdot \text{K}}\right)(317\ \text{K})}{0.975\ \text{atm}} = \textbf{35.0 g/mol}$$

5.44   The number of particles in 1 L of gas at STP is:

$$\text{Number of particles} = 1.0\ \text{L} \times \frac{1\ \text{mol}}{22.414\ \text{L}} \times \frac{6.022 \times 10^{23}\ \text{particles}}{1\ \text{mol}} = 2.7 \times 10^{22}\ \text{particles}$$

$$\textbf{Number of N}_2\ \textbf{molecules} = \left(\frac{78\%}{100\%}\right)(2.7 \times 10^{22}\ \text{particles}) = \textbf{2.1} \times \textbf{10}^{22}\ \textbf{N}_2\ \textbf{molecules}$$

$$\textbf{Number of O}_2\ \textbf{molecules} = \left(\frac{21\%}{100\%}\right)(2.7 \times 10^{22}\ \text{particles}) = \textbf{5.7} \times \textbf{10}^{21}\ \textbf{O}_2\ \textbf{molecules}$$

$$\textbf{Number of Ar atoms} = \left(\frac{1\%}{100\%}\right)(2.7 \times 10^{22}\ \text{particles}) = \textbf{3} \times \textbf{10}^{20}\ \textbf{Ar atoms}$$

5.46   The density can be calculated from the ideal gas equation.

$$d = \frac{P\mathcal{M}}{RT}$$

$\mathcal{M} = 1.008 \text{ g/mol} + 79.90 \text{ g/mol} = 80.91 \text{ g/mol}$
$T = 46° + 273° = 319 \text{ K}$
$P = 733 \text{ mmHg} \times \dfrac{1 \text{ atm}}{760 \text{ mmHg}} = 0.964 \text{ atm}$

$$d = \frac{(0.964 \text{ atm})\left(\dfrac{80.91 \text{ g}}{\text{mol}}\right)}{319 \text{ K}} \times \frac{\text{mol} \cdot \text{K}}{0.0821 \text{ L} \cdot \text{atm}} = \mathbf{2.98 \text{ g/L}}$$

5.48   This is an extension of an ideal gas law calculation involving molar mass. If you determine the molar mass of the gas, you will be able to determine the molecular formula from the empirical formula (see Determining the Molecular Formula of a Compound, Problem Type 6, Chapter 3).

$$\mathcal{M} = \frac{dRT}{P}$$

Calculate the density, then substitute its value into the equation above.

$$d = \frac{0.100 \text{ g}}{22.1 \text{ mL}} \times \frac{1000 \text{ mL}}{1 \text{ L}} = 4.52 \text{ g / L}$$

$T(\text{K}) = 20° + 273° = 293 \text{ K}$

$$\mathcal{M} = \frac{\left(4.52 \dfrac{\text{g}}{\text{L}}\right)\left(0.0821 \dfrac{\text{L} \cdot \text{atm}}{\text{mol} \cdot \text{K}}\right)(293 \text{ K})}{1.02 \text{ atm}} = 107 \text{ g/mol}$$

Compare the empirical mass to the molar mass.

$$\text{empirical mass} = 32.07 \text{ g/mol} + 4(19.00 \text{ g/mol}) = 108.07 \text{ g/mol}$$

Remember, the molar mass will be a whole number multiple of the empirical mass. In this case, the $\dfrac{\text{molar mass}}{\text{empirical mass}} \approx 1$. Therefore, the molecular formula is the same as the empirical formula, **SF$_4$**.

5.50   Gas stoichiometry, Problem Type 2D.

We can calculate the moles of M reacted, and the moles of H$_2$ gas produced. By comparing the number of moles of M reacted to the number of moles H$_2$ produced, we can determine the mole ratio in the balanced equation.

Step 1: First let's calculate the moles of the metal (M) reacted.

$$\text{mol M} = 0.225 \text{ g M} \times \frac{1 \text{ mol M}}{27.0 \text{ g M}} = 8.33 \times 10^{-3} \text{ mol M}$$

Step 2: Solve the ideal gas equation algebraically for n_{H_2}. Then, calculate the moles of H_2 by substituting the known quantities into the equation.

$$P = 741 \text{ mmHg} \times \frac{1 \text{ atm}}{760 \text{ mmHg}} = 0.975 \text{ atm}$$

$$T = 17° + 273° = 290 \text{ K}$$

$$n_{H_2} = \frac{PV_{H_2}}{RT}$$

$$n_{H_2} = \frac{(0.975 \text{ atm})(0.303 \text{ L})}{\left(0.0821 \frac{\text{L} \cdot \text{atm}}{\text{mol} \cdot \text{K}}\right)(290 \text{ K})} = 1.24 \times 10^{-2} \text{ mol } H_2$$

Step 3: Compare the number moles of H_2 produced to the number of moles of M reacted.

$$\frac{1.24 \times 10^{-2} \text{ mol } H_2}{8.33 \times 10^{-3} \text{ mol M}} \approx 1.5$$

This means that the mole ratio of H_2 to M is 1.5 : 1.

Step 4: We can now write the balanced equation since we know the mole ratio between H_2 and M.

The unbalanced equation is:

$$M (s) + HCl (aq) \longrightarrow 1.5 \, H_2 (g) + M_xCl_y (aq)$$

We have 3 atoms of H on the products side of the reaction, so a 3 must be placed in front of HCl. The ratio of M to Cl on the reactants side is now 1 : 3. Therefore the formula of the metal chloride must be MCl_3. The balanced equation is:

$$\textbf{M (s) + 3 HCl (aq)} \longrightarrow \textbf{1.5 } H_2 \textbf{ (g) + } MCl_3 \textbf{ (aq)}$$

From the formula of the metal chloride, we determine that the charge of the metal is +3. Therefore, the formula of the metal oxide and the metal sulfate are $\textbf{M}_2\textbf{O}_3$ and $\textbf{M}_2\textbf{(SO}_4\textbf{)}_3$, respectively.

5.52   From the moles of CO_2 produced, we can calculate the amount of calcium carbonate that must have reacted. We can then determine the percent by mass of $CaCO_3$ in the 3.00 g sample.

The balanced equation is:

$$CaCO_3 (s) + 2 HCl (aq) \longrightarrow CO_2 (g) + CaCl_2 (aq) + H_2O (l)$$

The moles of CO_2 produced can be calculated using the ideal gas equation.

$$n_{CO_2} = \frac{PV_{CO_2}}{RT}$$

$$n_{CO_2} = \frac{\left(792 \text{ mmHg} \times \frac{1 \text{ atm}}{760 \text{ mmHg}}\right)(0.656 \text{ L})}{\left(0.0821 \frac{\text{L} \cdot \text{atm}}{\text{mol} \cdot \text{K}}\right)(20 + 273 \text{ K})} = \textbf{2.84} \times \textbf{10}^{-2} \textbf{ mol } \textbf{CO}_2$$

The balanced equation shows a 1:1 mole ratio between CO_2 and $CaCO_3$. Therefore, 2.84×10^{-2} mol of $CaCO_3$ must have reacted.

$$? \text{ g } CaCO_3 \text{ reacted} = 2.84 \times 10^{-2} \text{ mol } CaCO_3 \times \frac{100.1 \text{ g } CaCO_3}{1 \text{ mol } CaCO_3} = 2.84 \text{ g } CaCO_3$$

The percent by mass of the $CaCO_3$ sample is:

$$\% \text{ } CaCO_3 = \frac{2.84 \text{ g}}{3.00 \text{ g}} \times 100\% = \mathbf{94.7\%}$$

Assumption: The impurity (or impurities) must not react with HCl to produce CO_2 gas.

5.54   The balanced equation is:

$$C_2H_5OH \text{ } (l) + 3 \text{ } O_2 \text{ } (g) \longrightarrow 2 \text{ } CO_2 \text{ } (g) + 3 \text{ } H_2O \text{ } (l)$$

The moles of O_2 needed to react with 227 g ethanol are:

$$227 \text{ g } C_2H_5OH \times \frac{1 \text{ mol } C_2H_5OH}{46.07 \text{ g } C_2H_5OH} \times \frac{3 \text{ mol } O_2}{1 \text{ mol } C_2H_5OH} = 14.8 \text{ mol } O_2$$

14.8 mol of O_2 corresponds to a volume of:

$$V_{O_2} = \frac{n_{O_2}RT}{P} = \frac{(14.8 \text{ mol } O_2)\left(0.0821 \dfrac{L \cdot atm}{mol \cdot K}\right)(35 + 273 \text{ K})}{\left(790 \text{ mmHg} \times \dfrac{1 \text{ atm}}{760 \text{ mmHg}}\right)} = 3.60 \times 10^2 \text{ L } O_2$$

Since air is 21.0 percent O_2 by volume, we can write:

$$V_{air} = V_{O_2}\left(\frac{100\% \text{ air}}{21\% \text{ } O_2}\right) = (3.60 \times 10^2 \text{ L } O_2)\left(\frac{100\% \text{ air}}{21\% \text{ } O_2}\right) = \mathbf{1.71 \times 10^3 \text{ L air}}$$

5.58   Dalton's law states that the total pressure of the mixture is the sum of the partial pressures.

(a) $P_{total} = 0.32 \text{ atm} + 0.15 \text{ atm} + 0.42 \text{ atm} = \mathbf{0.89 \text{ atm}}$

(b)   We know: $P_1 = 0.15 \text{ atm} + 0.42 \text{ atm} = 0.57 \text{ atm}$ $P_2 = 1.0 \text{ atm}$

$T_1 = (15 + 273) \text{ K} = 288 \text{ K}$ $T_2 = 273 \text{ K}$

$V_1 = 2.5 \text{L}$ $V_2 = ?$

$$V_2 = \frac{V_1 P_1 T_2}{P_2 T_1}$$

$$V_2 = \frac{(2.5 \text{ L})(0.57 \text{ atm})(273 \text{ K})}{(1.0 \text{ atm})(288 \text{ K})} = \mathbf{1.4 \text{ L at STP}}$$

5.60 $P_{\text{Total}} = P_1 + P_2 + P_3 + \ldots + P_n$

In this case,

$$P_{\text{Total}} = P_{\text{Ne}} + P_{\text{He}} + P_{\text{H}_2\text{O}}$$

$$P_{\text{Ne}} = P_{\text{Total}} - P_{\text{He}} - P_{\text{H}_2\text{O}}$$

$$\mathbf{P_{\text{Ne}}} = 745 \text{ mm Hg} - 368 \text{ mmHg} - 28.3 \text{ mmHg} = \mathbf{349 \text{ mmHg}}$$

5.62   Collecting a gas over water, Problem Type 2E.

Step 1: Dalton's law of partial pressure states that

$$\mathbf{P_{\text{Total}} = P_1 + P_2 + P_3 + \ldots + P_n}$$

In this case,

$$P_{\text{Total}} = P_{\text{H}_2} + P_{\text{H}_2\text{O}}$$

Step 2: Solve the equation algebraically for P_{H_2}. Then, calculate its value by substituting in the known quantities.

$$P_{\text{H}_2} = P_{\text{Total}} - P_{\text{H}_2\text{O}}$$

$$P_{\text{H}_2} = 0.980 \text{ atm} - (23.8 \text{ mmHg})\left(\frac{1 \text{ atm}}{760 \text{ mmHg}}\right) = 0.949 \text{ atm}$$

Step 3: Map out the following strategy to calculate the mass of Zn (s) reacted.

volume H_2 → moles H_2 → moles Zn → grams Zn

Step 4: Solve the ideal gas equation algebraically for moles of H_2. Then, calculate its value by substituting in the known quantities.

$$T(\text{K}) = 25° + 273° = 298 \text{ K}$$

$$n_{\text{H}_2} = \frac{P_{\text{H}_2}V}{RT}$$

$$n_{\text{H}_2} = \frac{(0.949 \text{ atm})(7.80 \text{ L})}{298 \text{ K}} \times \frac{\text{mol} \cdot \text{K}}{0.0821 \text{ L} \cdot \text{atm}} = 0.303 \text{ mol } \text{H}_2$$

Step 5: Calculate the mass of Zn metal using the strategy in Step 3.

$$\mathbf{? \text{ g Zn}} = 0.303 \text{ mol } \text{H}_2 \times \frac{1 \text{ mol Zn}}{1 \text{ mol } \text{H}_2} \times \frac{65.38 \text{ g Zn}}{1 \text{ mol Zn}} = \mathbf{19.8 \text{ g Zn}}$$

5.64 $P_i = X_i P_T$

We need to determine the mole fractions of each component in order to determine their partial pressures. To calculate mole fraction, write the balanced chemical equation to determine the correct mole ratio.

$$2 \text{ NH}_3 \text{ }(g) \longrightarrow \text{ N}_2 \text{ }(g) + 3 \text{ H}_2 \text{ }(g)$$

The mole fractions of H_2 and N_2 are:

$$X_{H_2} = \frac{3 \text{ mol}}{3 \text{ mol} + 1 \text{ mol}} = 0.750$$

$$X_{N_2} = \frac{1 \text{ mol}}{3 \text{ mol} + 1 \text{ mol}} = 0.250$$

The partial pressures of H_2 and N_2 are:

$$P_{H_2} = X_{H_2} P_T = (0.750)(866 \text{ mm Hg}) = \textbf{650 mmHg}$$

$$P_{N_2} = X_{N_2} P_T = (0.250)(866 \text{ mm Hg}) = \textbf{217 mmHg}$$

5.72 Root-Mean-Square (RMS) Speed, Problem Type 4.

Step 1: Recognize that you need to use the following equation.

$$u_{rms} = \sqrt{\frac{3RT}{\mathcal{M}}}$$

Step 2: Calculate the molar masses ($\mathcal{M}$) of N_2, O_2, and O_3. Remember, $\mathcal{M}$ must be in units of kg/mol.

$$\mathcal{M}_{N_2} = 2(14.01 \text{ g/mol}) = 28.02 \; \frac{g}{mol} \times \frac{1 \text{ kg}}{1000 \text{ g}} = 0.02802 \text{ kg/mol}$$

$$\mathcal{M}_{O_2} = 2(16.00 \text{ g/mol}) = 32.00 \; \frac{g}{mol} \times \frac{1 \text{ kg}}{1000 \text{ g}} = 0.03200 \text{ kg/mol}$$

$$\mathcal{M}_{O_3} = 3(16.00 \text{ g/mol}) = 48.00 \; \frac{g}{mol} \times \frac{1 \text{ kg}}{1000 \text{ g}} = 0.04800 \text{ kg/mol}$$

Step 3: Substitute the known quantities into the equation and solve for u_{rms}.

$$T(K) = -23° + 273° = 250 \text{ K}$$

$$u_{rms} (N_2) = \sqrt{\frac{(3)\left(8.314 \dfrac{J}{mol \cdot K}\right)(250 \text{ K})}{\left(0.02802 \dfrac{kg}{mol}\right)}}$$

$$u_{rms} (N_2) = \textbf{472 m/s}$$

Similarly,

$$u_{rms} (O_2) = \textbf{441 m/s} \qquad u_{rms} (O_3) = \textbf{360 m/s}$$

5.74 **Average speed** $= \dfrac{(2.0 + 2.2 + 2.6 + 2.7 + 3.3 + 3.5) \text{m} / \text{s}}{6} = \textbf{2.7 m/s}$

RMS speed $= \sqrt{\dfrac{\left(2.0^2 + 2.2^2 + 2.6^2 + 2.7^2 + 3.3^2 + 3.5^2\right)(\text{m}/\text{s})^2}{6}} = \textbf{2.8 m/s}$

The root-mean-square value is always greater than the average value, because squaring favors the larger values compared to just taking the average value.

5.80   Deviations from Ideal Behavior, Problem Type 5. In this problem we wish to determine if the gas deviates from ideal behavior, by comparing the ideal pressure with the actual pressure.

Step 1: Calculate the pressure of 10.0 moles of an ideal gas at 27°C in a 1.50 L container. Solve the ideal gas equation algebraically for P, then substitute in the known values.

$$T(K) = 27°C + 273° = 300 \text{ K}$$

$$P = \frac{nRT}{V} = \frac{(10.0 \text{ mol})(0.0821 \frac{L \cdot atm}{mol \cdot K})(300 \text{ K})}{1.50 \text{ L}} = \textbf{164 atm}$$

Step 2: Compare the ideal pressure to the actual pressure by calculating the percent error.

$$\% \text{ error} = \frac{164 \text{ atm} - 130 \text{ atm}}{130 \text{ atm}} \times 100 = \textbf{26.2 \%}$$

Based on the large percent error, we conclude that under this condition of high pressure, the gas behaves in a non-ideal manner.

5.82   When a and b are zero, the van der Waals equation simply becomes the ideal gas equation. In other words, an ideal gas has zero for the a and b values of the van der Waals equation. It therefore stands to reason that the gas with the smallest values of a and b will behave most like an ideal gas under a specific set of pressure and temperature conditions. Of the choices given in the problem, the gas with the smallest a and b values is **Ne** (see Table 5.4).

5.84   We need to determine the molar mass of the gas. Comparing the molar mass to the empirical mass will allow us to determine the molecular formula.

$$n = \frac{PV}{RT} = \frac{(0.74 \text{ atm})\left(97.2 \text{ mL} \times \frac{1 \text{ mL}}{0.001 \text{ L}}\right)}{\left(0.0821 \frac{L \cdot atm}{mol \cdot K}\right)(473 \text{ K})} = 1.85 \times 10^{-3} \text{ mol}$$

$$\text{molar mass} = \frac{0.145 \text{ g}}{1.85 \times 10^{-3} \text{ mol}} = 78.4 \text{ g/mol}$$

The empirical mass of $CH = 13.02$ g/mol

Since $\dfrac{78.4 \text{ g/mol}}{13.02 \text{ g/mol}} = 6.02 \approx 6$, the molecular formula is $(CH)_6$ or $\textbf{C}_6\textbf{H}_6$.

5.86   The reaction is: $HCO_3^- (aq) + H^+ (aq) \longrightarrow H_2O (l) + CO_2 (g)$

The mass of HCO_3^- reacted is:

$$3.29 \text{ g tablet} \times \frac{32.5\% \ HCO_3^-}{100\% \text{ tablet}} = 1.07 \text{ g } HCO_3^-$$

$$\text{mol } CO_2 \text{ produced} = 1.07 \text{ g } HCO_3^- \times \frac{1 \text{ mol } HCO_3^-}{61.02 \text{ g } HCO_3^-} \times \frac{1 \text{ mol } CO_2}{1 \text{ mol } HCO_3^-} = 0.0175 \text{ mol } CO_2$$

$$V_{CO_2} = \frac{n_{CO_2}RT}{P} = \frac{(0.0175 \text{ mol } CO_2)\left(0.0821 \frac{L \cdot atm}{mol \cdot K}\right)(37 + 273)K}{(1.00 \text{ atm})} = 0.445 \text{ L} = \textbf{445 mL}$$

5.88   (a)   The number of moles of $Ni(CO)_4$ formed is:

$$86.4 \text{ g Ni} \times \frac{1 \text{ mol Ni}}{58.69 \text{ g Ni}} \times \frac{1 \text{ mol Ni(CO)}_4}{1 \text{ mol Ni}} = 1.47 \text{ mol Ni(CO)}_4$$

The pressure of $Ni(CO)_4$ is:

$$P = \frac{nRT}{V} = \frac{(1.47 \text{ mol})\left(0.0821 \frac{L \cdot atm}{mol \cdot K}\right)(43 + 273)K}{4.00 \text{ L}} = \textbf{9.53 atm}$$

(b) $Ni(CO)_4$ decomposes to produce more moles of gas (CO), which increases the pressure.

$$Ni(CO)_4 \,(g) \longrightarrow Ni\,(s) + 4\,CO\,(g)$$

5.90   Using the ideal gas equation, we can calculate the moles of gas.

$$n = \frac{PV}{RT} = \frac{(1.1 \text{ atm})\left(5.0 \times 10^2 \text{ mL} \times \frac{0.001 \text{ L}}{1 \text{ mL}}\right)}{\left(0.0821 \frac{L \cdot atm}{mol \cdot K}\right)(37 + 273)K} = 0.0216 \text{ mol gas}$$

Next, use Avogadro's number to convert to molecules of gas.

$$0.0216 \text{ mol gas} \times \frac{6.022 \times 10^{23} \text{ molecules}}{1 \text{ mol gas}} = \textbf{1.30} \times \textbf{10}^{\textbf{22}} \textbf{ molecules of gas}$$

The most common gases present in exhaled air are: CO_2, O_2, N_2, and H_2O.

5.92   Mass of the Earth's atmosphere = (surface area of the earth in cm^2) × (mass per 1 cm^2 column)

Mass of a single column of air with a surface area of 1 cm^2 area is:

$$76.0 \text{ cm} \times 13.6 \text{ g/cm}^3 = 1.03 \times 10^3 \text{ g/cm}^2$$

The surface area of the Earth in cm^2 is:

$$4\pi r^2 = 4\pi(6.371 \times 10^8 \text{ cm})^2 = 5.10 \times 10^{18} \text{ cm}^2$$

Mass of atmosphere = $(5.10 \times 10^{18} \text{ cm}^2)(1.03 \times 10^3 \text{ g/cm}^2) = 5.25 \times 10^{21} \text{ g} = \textbf{5.25} \times \textbf{10}^{\textbf{18}} \textbf{ kg}$

5.94   To calculate the molarity of NaOH we need moles of NaOH and volume of the NaOH solution. The volume is given in the problem; therefore, we need to calculate the moles of NaOH. The moles of NaOH can be calculated from the reaction of NaOH with HCl. The balanced equation is:

$$NaOH\,(aq) + HCl\,(aq) \longrightarrow H_2O\,(l) + NaCl\,(aq)$$

The number of moles of HCl gas is found from the ideal gas equation. $V = 0.189$ L, $T = (25 + 273)$ K $= 298$ K, and $P = 108$ mmHg $\left(\dfrac{1 \text{ atm}}{760 \text{ mmHg}} \right) = 0.142$ atm.

$$n_{HCl} = \frac{PV_{HCl}}{RT} = \frac{(0.142 \text{ atm})(0.189 \text{ L})}{\left(0.0821 \dfrac{\text{L} \cdot \text{atm}}{\text{mol} \cdot \text{K}} \right)(298 \text{ K})} = 1.10 \times 10^{-3} \text{ mol HCl}$$

The moles of NaOH can be calculated using the mole ratio from the balanced equation.

$$1.10 \times 10^{-3} \text{ mol HCl} \times \frac{1 \text{ mol NaOH}}{1 \text{ mol HCl}} = 1.10 \times 10^{-3} \text{ mol NaOH}$$

The molarity of the NaOH solution is:

$$M = \frac{\text{mol NaOH}}{\text{L of soln}} = \frac{1.10 \times 10^{-3} \text{ mol NaOH}}{0.0157 \text{ L soln}} = 0.0701 \text{ mol/L} = \textbf{0.0701 } \boldsymbol{M}$$

5.96   To calculate the partial pressures of He and Ne, the total pressure of the mixture is needed. To calculate the total pressure of the mixture, we need the total number of moles of gas in the mixture (mol He + mol Ne).

$$n_{He} = \frac{PV}{RT} = \frac{(0.63 \text{ atm})(1.2 \text{ L})}{\left(0.0821 \dfrac{\text{L} \cdot \text{atm}}{\text{mol} \cdot \text{K}} \right)(16 + 273) \text{K}} = 0.032 \text{ mol He}$$

$$n_{Ne} = \frac{PV}{RT} = \frac{(2.8 \text{ atm})(3.4 \text{ L})}{\left(0.0821 \dfrac{\text{L} \cdot \text{atm}}{\text{mol} \cdot \text{K}} \right)(16 + 273) \text{K}} = 0.40 \text{ mol Ne}$$

The total pressure is:

$$P_{Total} = \frac{(n_{He} + n_{Ne})RT}{V_{Total}} = \frac{(0.032 + 0.40) \text{mol} \left(0.0821 \dfrac{\text{L} \cdot \text{atm}}{\text{mol} \cdot \text{K}} \right)(16 + 273) \text{K}}{(1.2 + 3.4) \text{L}} = 2.2 \text{ atm}$$

$P_i = X_i P_T$. The partial pressures of He and Ne are:

$$P_{He} = \frac{0.032 \text{ mol}}{(0.032 + 0.40) \text{mol}} \times 2.2 \text{ atm} = \textbf{0.16 atm}$$

$$P_{Ne} = \frac{0.40 \text{ mol}}{(0.032 + 0.40) \text{mol}} \times 2.2 \text{ atm} = \textbf{2.0 atm}$$

5.98   When the water enters the flask from the dropper, some hydrogen chloride dissolves, creating a partial vacuum. Pressure from the atmosphere forces more water up the vertical tube.

5.100   Use the ideal gas equation to calculate the moles of water produced.

$$n_{H_2O} = \frac{PV}{RT} = \frac{(24.8 \text{ atm})(2.00 \text{ L})}{\left(0.0821 \dfrac{\text{L} \cdot \text{atm}}{\text{mol} \cdot \text{K}} \right)(120 + 273) \text{K}} = 1.54 \text{ mol H}_2\text{O}$$

Next, we can determine the mass of H_2O in the 54.2 g sample. Subtracting the mass of H_2O from 54.2 g will give the mass of $MgSO_4$ in the sample.

$$1.54 \text{ mol } H_2O \times \frac{18.02 \text{ g } H_2O}{1 \text{ mol } H_2O} = 27.8 \text{ g } H_2O$$

$$\text{Mass } MgSO_4 = 54.2 \text{ g sample } - 27.8 \text{ g } H_2O = 26.4 \text{ g } MgSO_4$$

Finally, we can calculate the moles of $MgSO_4$ in the sample. Comparing moles of $MgSO_4$ to moles of H_2O will allow us to determine the correct mole ratio in the formula.

$$26.4 \text{ g } MgSO_4 \times \frac{1 \text{ mol } MgSO_4}{120.4 \text{ g } MgSO_4} = 0.219 \text{ mol } MgSO_4$$

$$\frac{\text{mol } H_2O}{\text{mol } MgSO_4} = \frac{1.54 \text{ mol}}{0.219 \text{ mol}} = 7.03$$

Therefore, the mole ratio between H_2O and $MgSO_4$ in the compound is 7 : 1. Thus, the value of $x = 7$, and the formula is **$MgSO_4 \cdot 7H_2O$**.

5.102 The circumference of the cylinder is $= 2\pi r = 2\pi \left(\dfrac{15.0 \text{ cm}}{2} \right) = 47.1 \text{ cm}$

(a) The speed at which the target is moving equals:

$$\text{speed of target } = \text{ circumference } \times \text{ revolutions/sec}$$

$$\textbf{speed of target } = \frac{47.1 \text{ cm}}{1 \text{ revolution}} \times \frac{130 \text{ revolutions}}{1 \text{ s}} \times \frac{0.01 \text{ m}}{1 \text{ cm}} = \textbf{61.2 m/s}$$

(b) $2.80 \text{ cm} \times \dfrac{0.01 \text{ m}}{1 \text{ cm}} \times \dfrac{1 \text{ s}}{61.2 \text{ m}} = \textbf{4.58} \times \textbf{10}^{-4} \textbf{ s}$

(c) The Bi atoms must travel across the cylinder to hit the target. This distance is the diameter of the cylinder, which is 15.0 cm. The Bi atoms travel this distance in 4.58×10^{-4} s.

$$\frac{15.0 \text{ cm}}{4.58 \times 10^{-4} \text{ s}} \times \frac{0.01 \text{ m}}{1 \text{ cm}} = \textbf{328 m/s}$$

$$u_{rms} = \sqrt{\frac{3RT}{\mathcal{M}}} = \sqrt{\frac{3(8.314 \text{ J/K} \cdot \text{mol})(850 + 273)\text{K}}{209.0 \times 10^{-3} \text{ kg/mol}}} = \textbf{366.1 m/s}$$

The magnitudes of the speeds are comparable, but not identical. This is not surprising since 328 m/s is the velocity of a particular Bi atom, and u_{rms} is an average value.

5.104 The moles of O_2 can be calculated from the ideal gas equation. The mass of O_2 can then be calculated using the molar mass as a conversion factor.

$$n_{O_2} = \frac{PV}{RT} = \frac{(132 \text{ atm})(120 \text{ L})}{\left(0.0821 \dfrac{\text{L} \cdot \text{atm}}{\text{mol} \cdot \text{K}} \right)(22 + 273)\text{K}} = 654 \text{ mol } O_2$$

$$\text{? g } O_2 = 654 \text{ mol } O_2 \times \frac{32.00 \text{ g } O_2}{1 \text{ mol } O_2} = \textbf{2.09} \times \textbf{10}^4 \textbf{ g } O_2$$

The volume of O_2 gas under conditions of 1.00 atm pressure and a temperature of 22°C can be calculated using the ideal gas equation. The moles of O_2 = 654 moles.

$$V_{O_2} = \frac{n_{O_2} RT}{P} = \frac{(654 \text{ mol})\left(0.0821 \frac{\text{L} \cdot \text{atm}}{\text{mol} \cdot \text{K}}\right)(22 + 273)\text{K}}{1.00 \text{ atm}} = 1.58 \times 10^4 \text{ L O}_2$$

5.106   The fruit ripens more rapidly because the quantity (partial pressure) of ethylene gas inside the bag increases.

5.108   As the pen is used the amount of ink decreases, increasing the volume inside the pen. As the volume increases, the pressure inside the pen decreases. The hole is needed to equalize the pressure as the volume inside the pen increases.

5.110   (a) $NH_4NO_3 (s) \longrightarrow N_2O (g) + 2 H_2O (l)$

(b) $R = \dfrac{PV}{nT} = \dfrac{\left(718 \text{ mmHg} \times \dfrac{1 \text{ atm}}{760 \text{ mmHg}}\right)(0.340 \text{ L})}{\left(0.580 \text{ g N}_2\text{O} \times \dfrac{1 \text{ mol N}_2\text{O}}{44.02 \text{ g N}_2\text{O}}\right)(24 + 273)\text{K}} = 0.0821 \dfrac{\text{L} \cdot \text{atm}}{\text{mol} \cdot \text{K}}$

5.112   The value of a indicates how strongly molecules of a given type of gas attract one anther. C_6H_6 has the greatest intermolecular attractions due to its larger size compared to the other choices. Therefore, it has the largest a value.

5.114   The gases inside the mine were a mixture of carbon dioxide, carbon monoxide, methane, and other harmful compounds. The low atmospheric pressure caused the gases to flow out of the mine (the gases in the mine were at a higher pressure), and the man suffocated.

5.116   (a)   This is a Boyle's law problem.

$$P_{\text{tire}} V_{\text{tire}} = P_{\text{air}} V_{\text{air}}$$

$$(5.0 \text{ atm})(0.98 \text{ L}) = (1.0 \text{ atm}) V_{\text{air}}$$

$$V_{\text{air}} = \textbf{4.90 L}$$

(b)   Pressure in the tire − atmospheric pressure = gauge pressure

Pressure in the tire − 1.0 atm = 5.0 atm

Pressure in the tire = 6.0 atm

(c)   Again, this is a Boyle's law problem.

$$P_{\text{pump}} V_{\text{pump}} = P_{\text{gauge}} V_{\text{gauge}}$$

$$(1 \text{ atm})(0.33 V_{\text{tire}}) = P_{\text{gauge}} V_{\text{gauge}}$$

$$P_{\text{gauge}} = 0.33 \text{ atm}$$

This is the gauge pressure after one pump stroke. After three strokes, the gauge pressure will be $(3 \times 0.33 \text{ atm})$, or approximately **1 atm**. This is assuming that the initial gauge pressure was zero.

5.118 **(a)** First, let's convert the concentration of hydrogen from atoms/cm^3 to mol/L. The concentration in mol/L can be substituted into the ideal gas equation to calculate the pressure of hydrogen.

$$\frac{1 \text{ H atom}}{1 \text{ cm}^3} \times \frac{1 \text{ mol H}}{6.022 \times 10^{23} \text{ H atoms}} \times \frac{1 \text{ cm}^3}{1 \text{ mL}} \times \frac{1 \text{ mL}}{0.001 \text{ L}} = \frac{2 \times 10^{-21} \text{ mol H}}{\text{L}}$$

The pressure of H is:

$$P = \left(\frac{n}{V}\right)RT = \left(\frac{2 \times 10^{-21} \text{ mol}}{1 \text{ L}}\right)\left(0.0821 \frac{\text{L} \cdot \text{atm}}{\text{mol} \cdot \text{K}}\right)(3 \text{ K}) = \mathbf{5 \times 10^{-22} \text{ atm}}$$

(b) From part (a), we know that 1 L contains 2×10^{-21} mole of H atoms. We convert to the volume that contains 1.0 g of H atoms.

$$\frac{1 \text{ L}}{2 \times 10^{-21} \text{ mol H}} \times \frac{1 \text{ mol H}}{1.008 \text{ g H}} = \mathbf{5 \times 10^{20} \text{ L/g of H}}$$

Note: This volume is about that of all the water on Earth!

5.120 From Table 5.3, the equilibrium vapor pressure at 30°C is 31.82 mmHg.

Converting 3.9×10^3 Pa to units of mmHg:

$$3.9 \times 10^3 \text{ Pa} \times \frac{760 \text{ mmHg}}{1.01325 \times 10^5 \text{ Pa}} = 29 \text{ mmHg}$$

$$\textbf{Relative Humidity} = \frac{\text{partial pressure of water vapor}}{\text{equilibrium vapor pressure}} \times 100\% = \frac{29 \text{ mmHg}}{31.82 \text{ mmHg}} \times 100\% = \mathbf{91\%}$$

5.122 The volume of one alveoli is:

$$V = \frac{4}{3}\pi r^3 = \frac{4}{3}\pi(0.0050 \text{ cm})^3 = 5.2 \times 10^{-7} \text{ cm}^3 \times \frac{1 \text{ mL}}{1 \text{ cm}^3} \times \frac{0.001 \text{ L}}{1 \text{ mL}} = 5.2 \times 10^{-10} \text{ L}$$

The number of moles of air in one alveoli can be calculated using the ideal gas equation.

$$n = \frac{PV}{RT} = \frac{(1.0 \text{ atm})(5.2 \times 10^{-10} \text{ L})}{\left(0.0821 \frac{\text{L} \cdot \text{atm}}{\text{mol} \cdot \text{K}}\right)(37 + 273)\text{K}} = 2.0 \times 10^{-11} \text{ mol of air}$$

Since the air inside the alveoli is 14 percent oxygen, the moles of oxygen in one alveoli equals:

$$2.0 \times 10^{-11} \text{ mol of air} \times \frac{14\% \text{ oxygen}}{100\% \text{ air}} = 2.8 \times 10^{-12} \text{ mol O}_2$$

Converting to O_2 molecules:

$$2.8 \times 10^{-12} \text{ mol O}_2 \times \frac{6.022 \times 10^{23} \text{ O}_2 \text{ molecules}}{1 \text{ mol O}_2} = \mathbf{1.7 \times 10^{12} \text{ O}_2 \text{ molecules}}$$

5.124   The molar mass of a gas can be calculated using Equation (5.10) from the text.

$$\mathcal{M} = \frac{dRT}{P} = \frac{\left(1.33 \ \frac{g}{L}\right)\left(0.0821 \frac{L \cdot atm}{mol \cdot K}\right)(150 + 273)K}{\left(764 \ mmHg \times \frac{1 \ atm}{760 \ mmHg}\right)} = 46 \ g/mol$$

Some nitrogen oxides and their molar masses are:

NO   30 g/mol   N$_2$O   44 g/mol   NO$_2$ 46 g/mol

The nitrogen oxide is most likely **NO$_2$**, although N$_2$O cannot be completely ruled out.

5.126   When calculating root-mean-square speed, remember that the molar mass must be in units of kg/mol.

$$u_{rms} = \sqrt{\frac{3RT}{\mathcal{M}}} = \sqrt{\frac{3(8.314 \ J/mol \cdot K)(1.7 \times 10^{-7} \ K)}{85.47 \times 10^{-3} \ kg/mol}} = \mathbf{7.0 \times 10^{-3} \ m/s}$$

The mass of one Rb atom in kg is:

$$\frac{85.47 \ g \ Rb}{1 \ mol \ Rb} \times \frac{1 \ mol \ Rb}{6.022 \times 10^{23} \ Rb \ atoms} \times \frac{1 \ kg}{1000 \ g} = 1.419 \times 10^{-25} \ kg/Rb \ atom$$

$$\overline{KE} = \frac{1}{2}\overline{mu^2} = \frac{1}{2}(1.419 \times 10^{-25} \ kg)(7.0 \times 10^{-3} \ m/s)^2 = \mathbf{3.5 \times 10^{-30} \ J}$$

5.128   The molar volume is the volume of 1 mole of gas under the specified conditions.

$$V = \frac{nRT}{P} = \frac{(1 \ mol)\left(0.0821 \frac{L \cdot atm}{mol \cdot K}\right)(220 \ K)}{\left(6.0 \ mmHg \times \frac{1 \ atm}{760 \ mmHg}\right)} = \mathbf{2.3 \times 10^3 \ L}$$

5.130   The volume of the bulb can be calculated using the ideal gas equation. Pressure and temperature are given in the problem. Moles of air must be calculated before the volume can be determined.

Mass of air = 91.6843 g − 91.4715 g = 0.2128 g air

Molar mass of air = (0.78 × 28.02 g/mol) + (0.21 × 32.00 g/mol) + (0.01 × 39.95 g/mol) = 29 g/mol

moles air = 0.2128 g air × $\frac{1 \ mol \ air}{29 \ g \ air}$ = 7.3 × 10^{-3} mol air

Now, we can calculate the volume of the bulb.

$$V_{bulb} = \frac{nRT}{P} = \frac{(7.3 \times 10^{-3} \ mol)\left(0.0821 \frac{L \cdot atm}{mol \cdot K}\right)(23 + 273)K}{\left(744 \ mmHg \times \frac{1 \ atm}{760 \ mmHg}\right)} = 0.18 \ L = \mathbf{1.8 \times 10^2 \ mL}$$

5.132   In Problem 5.92, the mass of the Earth's atmosphere was determined to be 5.25×10^{18} kg. Assuming that the molar mass of air is 29.0 g/mol, we can calculate the number of molecules in the atmosphere.

(a) 5.25×10^{18} kg air $\times \dfrac{1000 \text{ g}}{1 \text{ kg}} \times \dfrac{1 \text{ mol air}}{29.0 \text{ g air}} \times \dfrac{6.022 \times 10^{23} \text{ molecules air}}{1 \text{ mol air}} = \mathbf{1.09 \times 10^{44} \text{ molecules}}$

(b)   First, calculate the moles of air exhaled in every breath. (500 mL = 0.500 L)

$$n = \frac{PV}{RT} = \frac{(1 \text{ atm})(0.500 \text{ L})}{\left(0.0821 \dfrac{\text{L} \cdot \text{atm}}{\text{mol} \cdot \text{K}}\right)(37 + 273)\text{K}} = 1.96 \times 10^{-2} \text{ mol air/breath}$$

Next, convert to molecules of air per breath.

$$1.96 \times 10^{-2} \text{ mol air/breath} \times \frac{6.022 \times 10^{23} \text{ molecules air}}{1 \text{ mol air}} = \mathbf{1.18 \times 10^{22} \text{ molecules/breath}}$$

(c) $\dfrac{1.18 \times 10^{22} \text{ molecules}}{1 \text{ breath}} \times \dfrac{12 \text{ breaths}}{1 \text{ min}} \times \dfrac{60 \text{ min}}{1 \text{ hr}} \times \dfrac{24 \text{ hr}}{1 \text{ day}} \times \dfrac{365 \text{ days}}{1 \text{ yr}} \times 35 \text{ yr} = \mathbf{2.60 \times 10^{30} \text{ molecules}}$

(d)   Fraction of molecules in the atmosphere breathed out by Mozart is:

$$\frac{2.60 \times 10^{30} \text{ molecules}}{1.09 \times 10^{44} \text{ molecules}} = 2.39 \times 10^{-14}$$

Or,

$$\frac{1}{2.39 \times 10^{-14}} = 4.18 \times 10^{13}$$

Thus, about 1 molecule of air in every 4×10^{13} molecules was breathed out by Mozart.

In a single breath containing 1.18×10^{22} molecules, we would breath in on average:

$$1.18 \times 10^{22} \text{ molecules} \times \frac{1 \text{ Mozart air molecule}}{4 \times 10^{13} \text{ air molecules}} = \mathbf{3 \times 10^{8} \text{ molecules that Mozart breathed out}}$$

(e)   We made the following assumptions:

1. Complete mixing of air in the atmosphere.
2. That no molecules escaped to the outer atmosphere.
3. That no molecules were used up during metabolism, nitrogen fixation, and so on.

5.134   The ideal gas law can be used to calculate the moles of water vapor per liter.

$$\frac{n}{V} = \frac{P}{RT} = \frac{1.0 \text{ atm}}{(0.0821 \dfrac{\text{L} \cdot \text{atm}}{\text{mol} \cdot \text{K}})(100 + 273)\text{K}} = 0.033 \frac{\text{mol}}{\text{L}}$$

We eventually want to find the distance between molecules. Therefore, let's convert moles to molecules, and convert liters to a volume unit that will allow us to get to distance (m^3).

$$\left(0.033\ \frac{mol}{L}\right)\left(6.022\times10^{23}\ \frac{molecules}{mol}\right)\left(\frac{1000\ L}{1\ m^3}\right)=2.0\times10^{25}\ \frac{molecules}{m^3}$$

This is the number of ideal gas molecules in a cube that is 1 meter on each side. Assuming an equal distribution of molecules along the three mutually perpendicular directions defined by the cube, a linear density in one direction may be found:

$$\left(2.0\times10^{25}\ \frac{molecules}{m^3}\right)^{\frac{1}{3}}=2.7\times10^{8}\ \frac{molecules}{m}$$

This is the number of molecules on a line *one* meter in length. The distance between each molecule is given by:

$$\frac{1\ m}{2.70\times10^{8}}=3.7\times10^{-9}\ m=\textbf{3.7 nm}$$

Assuming a water molecule to be a sphere with a diameter of 0.3 nm, the water molecules are separated by over 12 times their diameter: $\frac{3.7\ nm}{0.3\ nm}\approx12$ times.

A similar calculation is done for liquid water. Starting with density, we convert to molecules per cubic meter.

$$\frac{0.96\ g}{cm^3}\times\frac{1\ mol\ H_2O}{18.02\ g\ H_2O}\times\frac{6.022\times10^{23}\ molecules}{1\ mol\ H_2O}\times\left(\frac{100\ cm}{1\ m}\right)^3=3.2\times10^{28}\ \frac{molecules}{m^3}$$

This is the number of liquid water molecules in *one* cubic meter. From this point, the calculation is the same as that for water vapor, and the space between molecules is found using the same assumptions.

$$\left(3.2\times10^{28}\ \frac{molecules}{m^3}\right)^{\frac{1}{3}}=3.2\times10^{9}\ \frac{molecules}{m}$$

$$\frac{1\ m}{3.2\times10^{9}}=3.1\times10^{-10}\ m=\textbf{0.31 nm}$$

Asuming a water molecule to be a sphere with a diameter of 0.3 nm, to one significant figure, the water molecules are touching each other in the liquid phase.

CHAPTER 6
THERMOCHEMISTRY

PROBLEM-SOLVING STRATEGIES AND TUTORIAL SOLUTIONS

TYPES OF PROBLEMS

Problem Type 1: Thermochemical Equations.

Problem Type 2: Calculating Heat Absorbed or Released Using Specific Heat Data.

Problem Type 3: Calorimetry.
    (a) Constant-volume calorimetry.
    (b) Constant-pressure calorimetry.

Problem Type 4: Standard Enthalpy of Formation and Reaction.
    (a) Calculating the standard enthalpy of reaction.
    (b) Direct method of calculating the standard enthalpy of formation.
    (c) Indirect method of calculating the standard enthalpy of formation, Hess's law.

Problem Type 5: The First Law of Thermodynamics.
    (a) Applying the First Law of Thermodynamics.
    (b) Calculating the work done in gas expansion.
    (c) Enthalpy and the First Law of Thermodynamics. Calculating the internal energy change of a gaseous reaction.

PROBLEM TYPE 1: THERMOCHEMICAL EQUATIONS

Equations showing both the mass and enthalpy relations are called **thermochemical equations**. The following guidelines are helpful in writing and interpreting thermochemical equations:

1. The stoichiometric coefficients always refer to the number of moles of each substance.
2. When an equation is reversed, the roles of reactants and products change. Consequently, the magnitude of ΔH for the equation remains the same, but its sign changes.
3. If both sides of a thermochemical equation are multiplied by a factor n, then ΔH must also be multiplied by the same factor.
4. When writing thermochemical equations, the physical states of all reactants and products must be specified, because they help determine the actual enthalpy changes.

See Section 6.3 of your text for further discussion.

EXAMPLE 6.1
Given the thermochemical equation

$$SO_2\ (g)\ +\ 1/2\ O_2\ (g) \longrightarrow SO_3\ (g) \quad \Delta H = -99\ kJ$$

how much heat is evolved when (a) 1/2 mol of SO_2 reacts and (b) 3 mol of SO_2 reacts?

(a) Guideline 3 above states that if both sides of a thermochemical equation are multiplied by a factor n, then ΔH must also be multiplied by the same factor. If 1/2 mole of SO_2 reacts, that means that we are multiplying the equation by 1/2. Therefore, we must multiply ΔH by 1/2.

$$\text{heat evolved} = 1/2(99\ kJ) = \mathbf{5.0 \times 10^1\ kJ}$$

(b)   Following the same argument as in part a:

$$\text{heat evolved} = 3(99 \text{ kJ}) = \mathbf{3.0 \times 10^2 \text{ kJ}}$$

Why isn't there a negative sign in our answer? The sign convention for an exothermic reaction (energy, as heat, is released by the system) is negative ($-\Delta H$). However, in the above example, we state that heat is evolved, so a negative sign is unnecessary.

> **Tip:** Remember that the heat evolved or absorbed (q) by a reaction carried out under constant-pressure conditions is equal to the enthalpy change of the system, ΔH.

PRACTICE EXERCISE

1.   Given the thermochemical equation

$$SO_2 \, (g) + 1/2 \, O_2 \, (g) \longrightarrow SO_3 \, (g) \quad \Delta H = -99 \text{ kJ}$$

how much heat is evolved when 75 g of SO_2 is combusted?

Text Problem: 6.42

PROBLEM TYPE 2: CALCULATING HEAT ABSORBED OR RELEASED USING SPECIFIC HEAT DATA

If the specific heat (s) and the amount of substance is known, then the change in the sample's temperature (Δt) will tell us the amount of heat (q) that has been absorbed or released in a particular process. The equation for calculating the heat change is given by:

$$q = ms\Delta t \qquad\qquad (6.1)$$

or

$$q = C\Delta t$$

where m is the mass of the sample and Δt is the temperature change.

$$\Delta t = t_{\text{final}} - t_{\text{initial}}$$

EXAMPLE 6.2

How much heat is absorbed by 80.0 g of iron (Fe) when its temperature is raised from 25°C to 500°C? The specific heat of iron is 0.444 J/g·°C.

Notice that the specific heat of iron is given in the problem. Therefore, we should use Equation (6.1) to solve this problem.

Step 1: Calculate Δt.

$$\Delta t = t_{\text{final}} - t_{\text{initial}}$$
$$\Delta t = 500°C - 25°C = \mathbf{475°C}$$

Step 2: Substitute the known values into Equation (6.1).

$$q = m_{\text{Fe}}s_{\text{Fe}}\Delta t$$

$$q = (80.0 \text{ g})(0.444 \text{ J/g·°C})(475°C) = \mathbf{1.69 \times 10^4 \text{ J}}$$

PRACTICE EXERCISE

2. A piece of iron initially at a temperature of 25°C absorbs 10.0 kJ of heat. If its mass is 50.0 g, calculate the final temperature of the piece of iron. The specific heat of iron is 0.444 J/g·°C.

Text Problems: 6.18, **6.20**, 6.82

PROBLEM TYPE 3: CALORIMETRY

A. Constant-volume calorimetry

For a discussion of constant-volume calorimetry, see Section 6.4 of your text. Heat of combustion is usually measured in a constant-volume calorimeter. The heat released during combustion is absorbed both by the water in the calorimeter and the calorimeter itself. Because no heat enters or leaves the system throughout the process, we can write:

$$q_{system} = 0 = q_{water} + q_{bomb} + q_{rxn}$$

or,

$$q_{rxn} = -(q_{water} + q_{bomb})$$

The heat absorbed by the water can be calculated using the equation

$$q_{water} = m_{water}s_{water}\Delta t$$

The heat absorbed by the bomb calorimeter can be calculated using the heat capacity of the bomb calorimeter.

$$q_{bomb} = C_{bomb}\Delta t$$

> **Note:** The negative sign for q_{rxn} indicates that heat was released during the combustion. You should expect this, because all combustion processes are exothermic. Thermal energy is transferred from the system to the surroundings.

EXAMPLE 6.3

0.500 g of ethanol [CH_3CH_2OH (*l*)] was burned in a bomb calorimeter containing 2.000×10^3 g of water. The heat capacity of the bomb calorimeter was 950 J/°C, and the temperature rise was found to be 1.6°C.
(a) Write a balanced equation for the combustion of ethanol.
(b) Calculate the molar heat of combustion of ethanol.

(a) Recall that a **combustion reaction** is typically a vigorous and exothermic reaction that takes place between certain substances and oxygen. If the reactant contains only C, H, and/or O, then the products are CO_2 and H_2O. Therefore, the balanced equation for the combustion of ethanol is:

$$CH_3CH_2OH\ (l)\ +\ 3\ O_2\ (g)\ \longrightarrow\ 2\ CO_2\ (g) +\ 3\ H_2O\ (g)$$

(b)
Step 1: First, we must calculate the heat absorbed by both the water and the bomb calorimeter.

$$q_{water} = m_{water}s_{water}\Delta t$$
$$q_{water} = (2.000 \times 10^3\ g)(4.184\ J/g\cdot°C)(1.6°C) = 1.3 \times 10^4\ J$$

$$q_{bomb} = C_{bomb}\Delta t$$
$$q_{bomb} = (950\ J/°C)(1.6°C) = 1.5 \times 10^3\ J$$

Step 2: Calculate the heat of the reaction.

$$q_{rxn} = -(q_{water} + q_{bomb})$$
$$q_{rxn} = -(1.3 \times 10^4 \text{ J} + 1.5 \times 10^3 \text{ J}) = -1.5 \times 10^4 \text{ J}$$

Step 3: The problem asks for the **molar heat of combustion** of ethanol. The heat of reaction calculated in Step 2 is for the combustion of 0.500 g of ethanol. We can use the molar mass of ethanol as a conversion factor to convert from J/0.500 g ethanol to J/mol ethanol.

$$\mathcal{M}_{ethanol} = 2(12.01 \text{ g}) + 6(1.008 \text{ g}) + 16.00 \text{ g} = 46.07 \text{ g}$$

$$\textbf{Molar heat of combustion} = \frac{-1.5 \times 10^4 \text{ J}}{0.500 \text{ g ethanol}} \times \frac{46.07 \text{ g ethanol}}{1 \text{ mol ethanol}} = \textbf{-1.4} \times \textbf{10}^6 \textbf{ J/mol ethanol}$$

PRACTICE EXERCISE

3.   The combustion of benzoic acid is often used as a standard source of heat for calibrating combustion bomb calorimeters. The heat of combustion of benzoic acid has been accurately determined to be 26.42 kJ/g. When 0.8000 g of benzoic acid was burned in a calorimeter containing 9.50×10^2 g of water, a temperature rise of 4.08°C was observed. What is the heat capacity of the bomb calorimeter?

Text Problem: 6.94

B. Constant-pressure calorimetry

A simpler device than the constant-volume calorimeter is the constant-pressure calorimeter that is used to determine the heat changes for noncombustion reactions. The reactions usually occur in solution. Because the measurements are carried out under constant atmospheric pressure conditions, the heat change for the process (q_{rxn}) is equal to the enthalpy change (ΔH).

The heat released during reaction is absorbed both by the solution in the calorimeter and the calorimeter itself. Because no heat enters or leaves the system throughout the process, we can write:

$$q_{system} = 0 = q_{soln} + q_{calorimeter} + q_{rxn}$$

or,

$$q_{rxn} = -(q_{soln} + q_{calorimeter})$$

The heat absorbed by the solution can be calculated using the equation

$$q_{soln} = m_{soln}s_{soln}\Delta t$$

The heat absorbed by the calorimeter can be calculated using the heat capacity of the calorimeter

$$q_{calorimeter} = C_{calorimeter}\Delta t$$

EXAMPLE 6.4

The heat of neutralization for the following reaction is −56.2 kJ/mol.

$$\text{NaOH } (aq) + \text{HCl } (aq) \longrightarrow \text{NaCl } (aq) + \text{H}_2\text{O } (l)$$

1.00×10^2 **mL of 1.50** *M* **HCl is mixed with** 1.00×10^2 **mL of 1.50** *M* **NaOH in a constant-pressure calorimeter having a heat capacity of 15.2 J/°C. The initial temperature of the HCl and NaOH solutions is the same, 23.2°C. Calculate the final temperature of the mixed solution. Assume that the density and specific heat of the mixed solution is the same as for water (1.00 g/mL and 4.184 J/g·°C, respectively).**

Step 1: We know that:

$$q_{rxn} = -(q_{soln} + q_{calorimeter})$$

and

$$q_{rxn} = -(m_{soln}s_{soln}\Delta t + C_{calorimeter}\Delta t)$$

We can calculate the mass of the solution; s_{soln} is given; $C_{calorimeter}$ is given. If we can calculate q_{rxn}, then we can solve for Δt. Once we solve for Δt, we can calculate t_{final}, since we know $t_{initial}$.

Step 2: Calculate q_{rxn}. We are given the heat of neutralization in units of kJ/mol. If we can calculate the moles of HCl and NaOH that are reacted, we can calculate the heat of neutralization (q_{rxn}) for this particular reaction.

$$1.00 \times 10^2 \text{ mL} = 1.00 \times 10^{-1} \text{ L} = 0.100 \text{ L}$$

$$\text{mol HCl} = \text{mol NaOH} = M_{soln} \times L_{soln}$$

$$\textbf{mol HCl} = \left(1.50 \frac{\text{mol}}{L}\right) \times 0.100 \; L = \textbf{0.150 mol HCl}$$

$$\textbf{\textit{q}}_{\textbf{rxn}} = \left(-56.2 \frac{\text{kJ}}{\text{mol}}\right) \times 0.150 \text{ mol} = \textbf{--8.43 kJ} = \textbf{--8.43} \times \textbf{10}^3 \textbf{ J}$$

Step 3: Calculate the mass of the solution. You are told to assume that the density of the solution is the same as the density of water, 1.00 g/mL.

$$V_{soln} = V_{HCl} + V_{NaOH} = 1.00 \times 10^2 \text{ mL} + 1.00 \times 10^2 \text{ mL} = 2.00 \times 10^2 \text{ mL}$$

$$\textbf{Mass}_{\textbf{soln}} = (2.00 \times 10^2 \text{ mL}) \times \frac{1.00 \text{ g}}{1 \text{ mL}} = \textbf{2.00} \times \textbf{10}^2 \textbf{ g}$$

Step 4: Substitute the known values into the equation derived in Step 1, then solve for t_{final}.

$$q_{rxn} = -(m_{soln}s_{soln}\Delta t + C_{calorimeter}\Delta t)$$

$$-8.43 \times 10^3 \text{ J} = -[(2.00 \times 10^2 \text{ g})(4.184 \text{ J/g·°C})\Delta t + (15.2 \text{ J/°C})\Delta t]$$

$$-8.43 \times 10^3 \text{ J} = -[837 \Delta t + 15.2 \Delta t]\text{J/°C}$$

$$-8.43 \times 10^3 \text{ J} = -852 \Delta t \text{ (J/°C)}$$

$$\Delta t = 9.89°C$$

$$\Delta t = t_{final} - t_{initial}$$

$$\textbf{\textit{t}}_{\textbf{final}} = \Delta t + t_{initial} = (9.89 + 23.2)°C = \textbf{33.1°C}$$

PRACTICE EXERCISE

4. A 10.4 g sample of an unknown metal at 99.0°C was placed in a constant-pressure calorimeter containing 75.0 g of water at 23.5°C. The final temperature of the system was found to be 25.6°C. Calculate the specific heat of the metal, then use Table 6.1 of your text to predict the identity of the metal. (The heat capacity of the calorimeter is 12.6 J/°C.)

Text Problem: 6.22

PROBLEM TYPE 4: STANDARD ENTHALPY OF FORMATION AND REACTION

The **standard enthalpy of formation** (ΔH_f°) is defined as the heat change that results when one mole of a compound is formed from its elements at a pressure of 1 atm. The standard enthalpy of formation of any element in its most stable form is zero.

A.   Calculating the standard enthalpy of reaction

From standard enthalpies of formation, we can calculate the **standard enthalpy of reaction, ΔH_{rxn}°**.

Consider the hypothetical reaction

$$a\text{A} + b\text{B} \longrightarrow c\text{C} + d\text{D}$$

where a, b, c, and d are stoichiometric coefficients.

The standard enthalpy of reaction is given by

$$\Delta H_{rxn}^\circ = [c\,\Delta H_f^\circ(\text{C}) + d\,\Delta H_f^\circ(\text{D})] - [a\,\Delta H_f^\circ(\text{A}) + b\,\Delta H_f^\circ(\text{B})]$$

where a, b, c, and d all have the unit mol.

The equation can be written in the general form:

$$\Delta H_{rxn}^\circ = \Sigma n\,\Delta H_f^\circ(\text{products}) - \Sigma m\,\Delta H_f^\circ(\text{reactants})$$

where m and n denote the stoichiometric coefficients for the reactants and products, and Σ (sigma) means "the sum of".

EXAMPLE 6.5
A reaction used for rocket engines is

$$\text{N}_2\text{H}_4\,(l) + 2\,\text{H}_2\text{O}_2\,(l) \longrightarrow \text{N}_2\,(g) + 4\,\text{H}_2\text{O}\,(l)$$

What is the standard enthalpy of reaction in kilojoules? The standard enthalpies of formation are
$\Delta H_f^\circ[\text{N}_2\text{H}_4\,(l)] = 95.1$ kJ, $\Delta H_f^\circ[\text{H}_2\text{O}_2\,(l)] = -187.8$ kJ, and $\Delta H_f^\circ[\text{H}_2\text{O}\,(l)] = -285.8$ kJ.

$$\Delta H_{rxn}^\circ = \Sigma n\,\Delta H_f^\circ(\text{products}) - \Sigma m\,\Delta H_f^\circ(\text{reactants})$$

$$\Delta H_{rxn}^\circ = \Delta H_f^\circ[\text{N}_2\,(g)] + 4\,\Delta H_f^\circ[\text{H}_2\text{O}\,(l)] - \{\Delta H_f^\circ[\text{N}_2\text{H}_4\,(l)] + 2\,\Delta H_f^\circ[\text{H}_2\text{O}_2\,(l)]\}$$

Remember, the standard enthalpy of formation of any element in its most stable form is zero. Therefore, $\Delta H_f^\circ[\text{N}_2(g)] = 0$.

$$\Delta H_{rxn}^\circ = [0 + 4(-285.8\text{ kJ})] - [95.1\text{ kJ} + 2(-187.8\text{ kJ})] = \mathbf{-862.7\ kJ}$$

PRACTICE EXERCISE
5.   The combustion of methane, the main component of natural gas, occurs according to the equation

$$\text{CH}_4\,(g) + 2\,\text{O}_2\,(g) \longrightarrow \text{CO}_2\,(g) + 2\,\text{H}_2\text{O}\,(l) \qquad \Delta H_{rxn}^\circ = -890\text{ kJ}$$

Use standard enthalpies of formation for CO_2 and H_2O to determine the standard enthalpy of formation of methane.

Text Problems: **6.36**, 6.40, 6.44

B. Direct method of calculating the standard enthalpy of formation

This method of measuring ΔH_f° applies to compounds that can be readily synthesized from their elements. The best way to describe this direct method is to look at an example.

EXAMPLE 6.6

The combustion of sulfur occurs according to the following thermochemical equation:

$$S \text{ (rhombic)} + O_2 (g) \longrightarrow SO_2 (g) \qquad \Delta H_{rxn}^\circ = -296 \text{ kJ}$$

What is the enthalpy of formation of SO_2?

Step 1: Recall that the enthalpy of reaction carried out under standard-state conditions is given by

$$\Delta H_{rxn}^\circ = \Sigma n \Delta H_f^\circ \text{ (products)} - \Sigma m \Delta H_f^\circ \text{ (reactants)}$$

$$\Delta H_{rxn}^\circ = [\Delta H_f^\circ (SO_2)] - [\Delta H_f^\circ (S) + \Delta H_f^\circ (O_2)]$$

Step 2: Recall that the standard enthalpy of formation of any element in its most stable form is zero. Therefore, ΔH_f° [S(rhombic)] = 0 and $\Delta H_f^\circ [O_2(g)] = 0$.

$$\Delta H_{rxn}^\circ = [\Delta H_f^\circ (SO_2)] - [\Delta H_f^\circ (S) + \Delta H_f^\circ (O_2)]$$

$$-296 \text{ kJ} = [\Delta H_f^\circ (SO_2)] - [0 + 0]$$

$$\boldsymbol{\Delta H_f^\circ (SO_2) = -296 \text{ kJ/mol } SO_2}$$

> **Note:** You should recognize that this chemical equation as written meets the definition of a *formation* reaction. Thus, ΔH_{rxn} *is* ΔH_f° of SO_2 (g).

PRACTICE EXERCISE

6.   Hydrogen bromide (HBr) can be produced according to the following equation:

$$H_2 (g) + Br_2 (l) \longrightarrow 2 HBr (g) \qquad \Delta H_{rxn}^\circ = -72.4 \text{ kJ/mol}$$

What is the enthalpy of formation (ΔH_f°) of HBr?

Text Problem: 6.34

C. Indirect method of calculating the standard enthalpy of formation, Hess's law

Many compounds cannot be directly synthesized from their elements. In these cases, ΔH_f° can be determined by an indirect approach using **Hess's law**. Hess's law states that when reactants are converted to products, the change in enthalpy is the same whether the reaction takes place in one step or in a series of steps. This means that if we can break down the reaction of interest into a series of reactions for which ΔH_{rxn}° can be measured, we can calculate ΔH_{rxn}° for the overall reaction. Let's look at an example.

EXAMPLE 6.7

From the following heats of combustion with fluorine, calculate the enthalpy of formation of methane, CH_4.

(a) $CH_4 (g) + 4 F_2 (g) \longrightarrow CF_4 (g) + 4 HF (g)$ $\qquad \Delta H_{rxn}^\circ = -1942 \text{ kJ}$

(b) $C \text{ (graphite)} + 2 F_2 (g) \longrightarrow CF_4 (g)$ $\qquad \Delta H_{rxn}^\circ = -933 \text{ kJ}$

(c) $H_2 (g) + F_2 (g) \longrightarrow 2 HF (g)$ $\qquad \Delta H_{rxn}^\circ = -542 \text{ kJ}$

Step 1: The enthalpy of formation of methane can be determined from the following equation.

$$C \text{ (graphite)} + 2 H_2 (g) \longrightarrow CH_4 (g) \qquad \Delta H^\circ_{rxn} = ?$$

If we can calculate ΔH°_{rxn} for this reaction, we can calculate $\Delta H^\circ_f (CH_4)$ because the enthalpies of formation of C and H_2 are zero. Why? Remember that the standard enthalpy of formation of any element in its most stable form is zero. See Example 6.6.

Step 2: Since we want to obtain one equation containing only C, H_2, and CH_4, we need to eliminate F_2, CF_4, and HF from the first three equations (a–c). First, we note that equation (a) contains methane, CH_4, on the reactant side. Let's reverse (a) to get CH_4 on the product side.

$$CF_4 (g) + 4 HF (g) \longrightarrow CH_4 (g) + 4 F_2 (g) \qquad \Delta H^\circ_{rxn} = +1942 \text{ kJ}$$

Note: ΔH°_{rxn} changed sign when reversing the direction of the reaction. See Problem Type 1 for discussion.

Next, we notice that equation (b) has C (graphite) on the reactant side so let's keep that equation as written. Also, notice that equation (c) has H_2 on the reactant side. We need two moles of H_2 in the balanced equation for the formation of methane, so let's multiply this equation by two. If we add these three equations together, we end up with the following:

$$CF_4 (g) + 4 HF (g) \longrightarrow CH_4 (g) + 4 F_2 (g) \qquad \Delta H^\circ_{rxn} = +1942 \text{ kJ}$$
$$C \text{ (graphite)} + 2 F_2 (g) \longrightarrow CF_4 (g) \qquad \Delta H^\circ_{rxn} = -933 \text{ kJ}$$
$$2 H_2 (g) + 2 F_2 (g) \longrightarrow 4 HF (g) \qquad \Delta H^\circ_{rxn} = -1084 \text{ kJ}$$

$$C \text{ (graphite)} + 2 H_2 (g) \longrightarrow CH_4 (g) \qquad \Delta H^\circ_{rxn} = -75 \text{ kJ}$$

Step 3: Since the above equation represents the synthesis of CH_4 from its elements, we have

$$\Delta H^\circ_{rxn} = [\Delta H^\circ_f (CH_4)] - [\Delta H^\circ_f (C) + 2 \Delta H^\circ_f (H_2)]$$

$$-75 \text{ kJ} = [\Delta H^\circ_f (CH_4)] - [0 + 0]$$

$$\mathbf{\Delta H^\circ_f (CH_4) = -75 \text{ kJ/mol } CH_4}$$

PRACTICE EXERCISE

7.   From the following enthalpies of reaction, calculate the enthalpy of combustion of methane (CH_4) with F_2:

$$CH_4 (g) + 4 F_2 (g) \longrightarrow CF_4 (g) + 4 HF(g) \qquad \Delta H^\circ_{rxn} = ?$$

$$C \text{ (graphite)} + 2 H_2 (g) \longrightarrow CH_4 (g) \qquad \Delta H^\circ_{rxn} = -75 \text{ kJ}$$
$$C \text{ (graphite)} + 2 F_2 (g) \longrightarrow CF_4 (g) \qquad \Delta H^\circ_{rxn} = -933 \text{ kJ}$$
$$H_2 (g) + F_2 (g) \longrightarrow 2 HF (g) \qquad \Delta H^\circ_{rxn} = -542 \text{ kJ}$$

Text Problems: 6.46, 6.48

PROBLEM TYPE 5: THE FIRST LAW OF THERMODYNAMICS

A. Applying the First Law of Thermodynamics

The **First Law of Thermodynamics** states that energy can be converted from one form to another, but cannot be created or destroyed. Another way of stating the first law is that the energy of the universe is constant. The universe is composed of both the system and the surroundings.

$$\Delta E_{sys} + \Delta E_{surr} = 0$$

where,

the subscripts "sys" and "surr" denote system and surroundings, respectively.

However, in chemistry, we are normally interested in the changes associated with the *system* (which may be a flask containing reactants and products), not with its surroundings. Therefore, a more useful form of the first law is

$$\Delta E = q + w$$

where,

ΔE is the change in the internal energy of the system
q is the heat exchange between the system and surroundings
w is the work done on (or by) the system

Using the sign convention for thermochemical processes (see Section 6.3 of your text for discussion), q is positive for an endothermic process and negative for an exothermic process. For work, w is positive for work done *on* the system *by* the surroundings and negative for work done *by* the system *on* the surroundings. Try to understand the sign convention in this manner. If a *system* loses heat to the surroundings or does work on the surroundings, we expect its internal energy to decrease since both processes are energy depleting. Conversely, if heat is added to the *system* or if work is done on the *system*, then the internal energy of the system would increase.

EXAMPLE 6.8
A system does 975 kJ of work on its surroundings while at the same time it absorbs 625 kJ of heat. What is the change in energy, ΔE, for the system?

To solve this problem, you must make sure to get the sign convention correct. The system does work on the surroundings; this is an energy-depleting process.

$$w = -975 \text{ kJ}$$

The system absorbs 625 kJ of heat. Therefore, the internal energy of the system would increase.

$$q = +625 \text{ kJ}$$

Finally,

$$\Delta E = q + w = 625 \text{ kJ} + (-975 \text{ kJ}) = -350 \text{ kJ}$$

PRACTICE EXERCISE
8.   The surroundings do 455 kJ of work on the system while at the same time the system releases 253 kJ of heat. What is the change in energy, ΔE, for the system?

Text Problem: 6.62

B. Calculating the work done in gas expansion

A useful example of mechanical work is the expansion of a gas. Picture a gas-filled cylinder that is fitted with a weightless, frictionless, movable piston, at a certain temperature, pressure, and volume. As the gas expands, it pushes the piston upward against a constant, opposing, external atmospheric pressure, P. The gas (system) is doing work on the surroundings. The work can be calculated as follows:

$$w = -P\Delta V \tag{6.2}$$

where,

P is the external pressure

ΔV is the change in volume $(V_f - V_i)$

> **Note:** The minus sign in the equation takes care of the sign convention for w. For gas expansion, $\Delta V > 0$, so $-P\Delta V$ is a negative quantity. When a gas expands, it's doing work on the surroundings; the internal energy of the system decreases. For gas compression, $\Delta V < 0$, so $-P\Delta V$ is a positive quantity. When a gas is compressed, the surroundings are doing work on the system, increasing the internal energy.

EXAMPLE 6.9

A gas initially at a pressure of 10.0 atm and occupying a volume of 5.0 L is allowed to expand at constant temperature against a constant external pressure of 4.0 atm. After expansion, the gas occupies a volume of 12.5 L. Calculate the work done by the gas on the surroundings.

Step 1: We are given the external pressure in the problem, but we must calculate ΔV.

$$\Delta V = V_f - V_i = 12.5\ L - 5.0\ L = \textbf{7.5 L}$$

Step 2: Substitute P and ΔV into Equation (6.2) and solve for w.

$$w = -P\Delta V = -(4.0\ \text{atm})(7.5\ L) = \mathbf{-3.0 \times 10^1\ L \cdot atm}$$

It would be more convenient to express w in units of joules. The following conversion factor can be obtained from Appendix 1:

$$1\ L \cdot atm = 101.3\ J$$

Thus, we can write:

$$w = -3.0 \times 10^1\ \cancel{L \cdot atm} \times \frac{101.3\ J}{1\ \cancel{L \cdot atm}} = \mathbf{-3.0 \times 10^3\ J}$$

PRACTICE EXERCISE

9. Calculate the work done on the system when 6.0 L of a gas is compressed to 1.0 L by a constant external pressure of 2.0 atm.

Text Problem: 6.60

C. Enthalpy and the First Law of Thermodynamics—Calculating the internal energy change of a gaseous reaction

Let's return to the following form of the first law of thermodynamics.

$$\Delta E_{sys} = q + w$$

Under constant-pressure conditions we can write:

$$\Delta E = q_p + w$$

Recall that the heat evolved or absorbed (q) by a reaction carried out under constant-pressure conditions is equal to the enthalpy change of the system, ΔH.

Thus,

$$\Delta E = \Delta H + w \tag{6.3}$$

Also, we know that for gas expansion or compression under a constant external pressure, $w = -P\Delta V$. Substituting into Equation (6.3), we have:

$$\Delta E = \Delta H - P\Delta V \tag{6.4}$$

Also, for an ideal gas at constant pressure,

$$P\Delta V = \Delta(nRT)$$

$$\Delta V = \frac{\Delta(nRT)}{P}$$

Substituting into Equation (6.4),

$$\Delta E = \Delta H - \Delta(nRT)$$

Finally, at constant temperature,

$$\Delta E = \Delta H - RT\Delta n \tag{6.5}$$

where Δn is defined as

Δn = number of moles of product gases − number of moles of reactant gases

EXAMPLE 6.10

Calculate the change in internal energy when 1 mole of H_2 and 1/2 mole of O_2 are converted to 1 mole of H_2O at 1 atm and 25°C.

$$H_2\,(g) + 1/2\,O_2\,(g) \longrightarrow H_2O\,(l) \quad \Delta H° = -286 \text{ kJ}$$

Step 1: Calculate the total change in the number of moles of **gas**. Note that the product is a liquid.

Δn = 0 mol − (1 mol + 1/2 mol) = −1.5 mol

Step 2: Substitute the values for $\Delta H°$ and Δn into Equation (6.5).

$T = 25° + 273° = 298$ K

$$\Delta E° = \Delta H° - RT\Delta n$$

$$\Delta E° = -286 \text{ kJ} - \left(8.314 \; \frac{\cancel{J}}{\text{mol} \cdot \cancel{K}}\right)(298 \,\cancel{K})(-1.5 \; \cancel{\text{mol}})\left(\frac{1 \text{ kJ}}{1000 \,\cancel{J}}\right) = -282 \text{ kJ}$$

PRACTICE EXERCISE

10. Calculate the change in the internal energy when 1.0 mole of water vaporizes at 1.0 atm and 100°C. Assume that water vapor is an ideal gas and that the volume of liquid water is negligible compared with that of steam at 100°C. [$\Delta H_{vap}(H_2O)$ = 40.67 kJ/mol at 100°C].

Text Problem: 6.68

ANSWERS TO PRACTICE EXERCISES

1. heat evolved $= 1.2 \times 10^2$ kJ

2. 475°C

3. $C_{bomb} = \dfrac{4.9 \text{ kJ}}{4.08\ ^\circ\text{C}} = 1.2$ kJ/°C

4. $s_{metal} = 0.898$ J/g°C. The metal is probably aluminum.

5. ΔH_f° (CH$_4$) $= -75.1$ kJ

6. ΔH_f° (HBr) $= -36.2$ kJ/mol

7. $\Delta H_{rxn}^\circ = -1942$ kJ

8. $\Delta E = 202$ kJ

9. $w = 1.0 \times 10^3$ J

10. $\Delta E = 37.57$ kJ

SOLUTIONS TO SELECTED TEXT PROBLEMS

6.18 $q = m_{Cu}s_{Cu}\Delta t = (6.22 \times 10^3 \text{ g})(0.385 \text{ J/g·°C})(324.3°C - 20.5°C) = \textbf{728 kJ}$

6.20   Calculating Heat Absorbed Using Specific Heat Data, Problem Type 2.

The hint in this problem gives you some important information. The heat gained by the gold must be equal to the heat lost by the iron.

$$q_{Au} = -q_{Fe}$$

If we can calculate the heat gained by the gold and the heat lost by the iron, we can calculate the final temperature of the combined metals.

Step 1: Calculate the heat gained by the gold and the heat lost by the iron.

Heat gained by the gold = $q_{Au} = m_{Au}s_{Au}\Delta t = (10.0 \text{ g})(0.129 \text{ J/g·°C})(t_f - 18.0)°C$

Heat lost by the iron = $m_{Fe}s_{Fe}\Delta t = (20.0 \text{ g})(0.444 \text{ J/g·°C})(t_f - 55.6)°C$

Step 2: The heat gained by the gold is the opposite sign of the heat lost by the iron.

$$q_{Au} = -q_{Fe}$$

$(10.0 \text{ g})(0.129 \text{ J/g·°C})(t_f - 18.0)°C = -(20.0 \text{ g})(0.444 \text{ J/g·°C})(t_f - 55.6)°C$

$1.29 \, t_f \, (\text{J/°C}) - 23.2 \text{ J} = -8.88 \, t_f \, (\text{J/°C}) + 494 \text{ J}$

$10.2 \, t_f \, (\text{J/°C}) = 517 \text{ J}$

$t_f = \textbf{50.7 °C}$

Must the final temperature be between the two starting values?

6.22   Constant-pressure calorimetry, Problem Type 3B.

Step 1: We know that

$$q_{rxn} = -(q_{soln} + q_{calorimeter})$$

and

$$q_{rxn} = -(m_{soln}s_{soln}\Delta t + C_{calorimeter}\Delta t)$$

$C_{calorimeter}$ is given. If we assume that the solution has the same density as water, we can calculate the mass of the solution. From the moles of HCl and the heat of neutralization, we can calculate q_{rxn}. Also, we need to assume that the solution has the same specific heat as water. Once we have all this information, we can solve for Δt. Knowing $t_{initial}$, we can solve for t_{final}.

Step 2: Calculate the mass of the solution. If we assume that the density of the solution is the same as the density of water, 1.00 g/mL, we have:

$$V_{soln} = V_{HCl} + V_{Ba(OH)_2} = 2.00 \times 10^2 \text{ mL} + 2.00 \times 10^2 \text{ mL} = 4.00 \times 10^2 \text{ mL}$$

$$\text{Mass}_{soln} = (4.00 \times 10^2 \text{ mL})\left(\frac{1.00 \text{ g}}{1 \text{ mL}}\right) = 4.00 \times 10^2 \text{ g}$$

Step 3: The equation for the neutralization is

$$2\,HCl\,(aq)\;+\;Ba(OH)_2\,(aq)\;\longrightarrow\;2\,H_2O\,(l)\;+\;BaCl_2\,(aq)$$

One mole of water is formed from each mole of HCl. There is exactly enough $Ba(OH)_2$ to neutralize all the HCl. The number of moles of HCl is:

$$200.\;mL\;\times\;\frac{1\,L}{1000\;mL}\;\times\;\frac{0.862\;mol\;HCl}{1\,L}\;=\;0.172\;mol\;HCl$$

The amount of heat liberated during the reaction is

$$q_{rxn}\;=\;0.0172\;mol\;\times\;\frac{-56.2\times10^3\;J}{1\;mol}\;=\;-9.67\times10^3\;J$$

Step 4: Substitute the known values into the equation derived in Step 1, then solve for t_{final}.

$$q_{rxn}\;=\;-(m_{soln}s_{soln}\Delta t\;+\;C_{calorimeter}\Delta t)$$

$$-9.67\times10^3\;J\;=\;-[(4.00\times10^2\;g)(4.184\;J/g\cdot°C)\Delta t\;+\;(453\;J/°C)\Delta t]$$

$$-9.67\times10^3\;J\;=\;-(1.67\times10^3\;+\;453)\Delta t\;(J/°C)$$

$$\Delta t\;=\;4.55\;°C$$

$$\Delta t\;=\;t_{final}\;-\;t_{initial}$$

$$\mathbf{t_{final}\;=\;\Delta t\;+\;t_{initial}\;=\;(4.55\;+\;20.48)°C\;=\;25.03°C}$$

6.30   The standard enthalpy of formation of any element in its most stable form is zero. Therefore, since $\Delta H_f^°\,(O_2)=0$, O_2 is the more stable form of the element oxygen at this temperature.

6.32   **(a)** $Br_2\,(l)$ is the most stable form of bromine at 25 °C; therefore, $\Delta H_f^°\,[Br_2\,(l)]=0$. Since $Br_2\,(g)$ is less stable than $Br_2\,(l)$, $\Delta H_f^°\,[Br_2\,(g)]>0$.

(b) $I_2\,(s)$ is the most stable form of iodine at 25 °C; therefore, $\Delta H_f^°\,[I_2\,(s)]=0$. Since $I_2\,(g)$ is less stable than $I_2\,(s)$, $\Delta H_f^°\,[I_2\,(g)]>0$.

6.34   Direct method of calculating the standard enthalpy of formation, Problem Type 4B.

This method of measuring $\Delta H_f^°$ applies to compounds that can be readily synthesized from their elements.

Step 1: Let's write a balanced equation that shows the formation of $Ag_2O\,(s)$ from its elements.

$$2\,Ag\,(s)\;+\;(1/2)\,O_2\,(g)\;\longrightarrow\;Ag_2O\,(s)$$

Step 2: Recall that the enthalpy of reaction carried out under standard-state conditions is given by

$$\Delta H_{rxn}^°\;=\;\Sigma n\,\Delta H_f^°\,(products)\;-\;\Sigma m\,\Delta H_f^°\,(reactants)$$

thus,

$$\Delta H_{rxn}^°\;=\;[\,\Delta H_f^°\,(Ag_2O)\,]\;-\;[2\,\Delta H_f^°\,(Ag)\;+\;(1/2)\,\Delta H_f^°\,(O_2)]$$

Step 3: Recall that the standard enthalpy of formation of any element in its most stable form is zero.
Therefore, $\Delta H_f^\circ [Ag(s)] = 0$ and $\Delta H_f^\circ [O_2(g)] = 0$.

$$\Delta H_{rxn}^\circ = [\Delta H_f^\circ (Ag_2O)] - [2\,\Delta H_f^\circ (Ag) + (1/2)\,\Delta H_f^\circ (O_2)]$$

$$\Delta H_{rxn}^\circ = [\Delta H_f^\circ (Ag_2O)] - [0 + 0]$$

$$\boldsymbol{\Delta H_f^\circ (Ag_2O) = \Delta H_{rxn}^\circ}$$

In a similar manner, you should be able to show that $\boldsymbol{\Delta H_f^\circ (CaCl_2) = \Delta H_{rxn}^\circ}$ for the reaction

$$Ca\,(s) + Cl_2\,(g) \longrightarrow CaCl_2\,(s)$$

6.36   Calculating the standard enthalpy of reaction, Problem Type 4A.

$$\Delta H_{rxn}^\circ = \Sigma n\,\Delta H_f^\circ (products) - \Sigma m\,\Delta H_f^\circ (reactants)$$

(a) $HCl(g) \rightarrow H^+(aq) + Cl^-(aq)$

$$\Delta H_{rxn}^\circ = \Delta H_f^\circ (H^+) + \Delta H_f^\circ (Cl^-) - \Delta H_f^\circ (HCl)$$

$$-74.9\text{ kJ} = 0 + \Delta H_f^\circ (Cl^-) - (1\text{ mol})(-92.3\text{ kJ/mol})$$

$$\boldsymbol{\Delta H_f^\circ (Cl^-) = -167.2 \text{ kJ/mol}}$$

(b)   The neutralization reaction is:

$$H^+\,(aq) + OH^-\,(aq) \rightarrow H_2O\,(l)$$

and,

$$\Delta H_{rxn}^\circ = \Delta H_f^\circ [H_2O(l)] - [\Delta H_f^\circ (H^+) + \Delta H_f^\circ (OH^-)]$$

$$\Delta H_f^\circ [H_2O(l)] = -285.8 \text{ kJ/mol}\ \ (\text{see Table 6.3 of your text})$$

$$\boldsymbol{\Delta H_{rxn}^\circ = (1\text{ mol})(-285.8\text{ kJ/mol}) - [(1\text{ mol})(0\text{ kJ/mol}) + (1\text{ mol})(-229.6\text{ kJ/mol})] = -56.2 \text{ kJ}}$$

6.38   **(a)** $\Delta H^\circ = [2\,\Delta H_f^\circ (CO_2) + 2\,\Delta H_f^\circ (H_2O)] - [\Delta H_f^\circ (C_2H_4) + 3\,\Delta H_f^\circ (O_2)]$

$$\Delta H^\circ = [(2\text{ mol})(-393.5\text{ kJ/mol}) + (2\text{ mol})(-285.8\text{ kJ/mol})] - [(1\text{ mol})(52.3\text{ kJ/mol}) + (3\text{ mol})(0)]$$

$$\boldsymbol{\Delta H^\circ = -1411 \text{ kJ}}$$

(b) $\Delta H^\circ = [2\,\Delta H_f^\circ (H_2O) + 2\,\Delta H_f^\circ (SO_2)] - [2\,\Delta H_f^\circ (H_2S) + 3\,\Delta H_f^\circ (O_2)]$

$$\Delta H^\circ = [(2\text{ mol})(-285.8\text{ kJ/mol}) + (2\text{ mol})(-296.1\text{ kJ/mol})] - [(2\text{ mol})(-20.15\text{ kJ/mol}) + (3\text{ mol})(0)]$$

$$\boldsymbol{\Delta H^\circ = -1124 \text{ kJ}}$$

6.40 $\Delta H_{rxn}^\circ = \Sigma n\,\Delta H_f^\circ (products) - \Sigma m\,\Delta H_f^\circ (reactants)$

The reaction is:

$$H_2\,(g) \longrightarrow H\,(g) + H\,(g)$$

and,

$$\Delta H_{rxn}^\circ = [\Delta H_f^\circ (H) + \Delta H_f^\circ (H)] - \Delta H_f^\circ (H_2)$$

$$\Delta H_f^\circ (H_2) = 0$$

$$\Delta H_{rxn}^\circ = 436.4 \text{ kJ} = 2 \, \Delta H_f^\circ (H) - (1 \text{ mol})(0 \text{ kJ/mol})$$

$$\Delta H_f^\circ (\textbf{H}) = \frac{436.4 \text{ kJ}}{2 \text{ mol}} = \textbf{218.2 kJ/mol}$$

6.42   Thermochemical Equations, Problem Type 1.

The equation as written shows that 879 kJ of heat is released when two moles of ZnS react. We want to calculate the amount of heat released when 1 g of ZnS reacts.

Let ΔH° be the heat change per gram of ZnS roasted. We write:

$$\Delta H^\circ = \frac{-879 \text{ kJ}}{2 \text{ mol ZnS}} \times \frac{1 \text{ mol ZnS}}{97.46 \text{ g ZnS}} = \textbf{-4.51 kJ/g ZnS}$$

6.44 $\Delta H_{rxn}^\circ = \Sigma n \, \Delta H_f^\circ \text{ (products)} - \Sigma m \, \Delta H_f^\circ \text{ (reactants)}$

The balanced equation for the reaction is:

$$CaCO_3 \, (s) \longrightarrow CaO \, (s) + CO_2 \, (g)$$

$$\Delta H_{rxn}^\circ = [\, \Delta H_f^\circ (CaO) + \Delta H_f^\circ (CO_2)] - \Delta H_f^\circ (CaCO_3)$$

$$\Delta H_{rxn}^\circ = [-635.6 \text{ kJ/mol} + -393.5 \text{ kJ/mol}] - (-1206.9 \text{ kJ/mol})$$

$$\Delta H_{rxn}^\circ = 177.8 \text{ kJ/mol}$$

The enthalpy change calculated above is the enthalpy change if 1 mole of CO_2 is produced. The problem asks for the enthalpy change if 66.8 g of CO_2 are produced. We need to use the molar mass of CO_2 as a conversion factor.

$$\Delta H^\circ = 66.8 \text{ g CO}_2 \times \frac{1 \text{ mol CO}_2}{44.01 \text{ g CO}_2} \times \frac{177.8 \text{ kJ}}{1 \text{ mol CO}_2} = \textbf{2.70} \times \textbf{10}^2 \textbf{ kJ}$$

6.46   Indirect method of calculating the standard enthalpy of formation, Hess's law. Problem Type 4C.

$$2 \text{ C (graphite)} + 3 \text{ H}_2 \, (g) \longrightarrow \text{C}_2\text{H}_6 \, (g) \qquad \Delta H_{rxn}^\circ = ? \qquad (6.1)$$

We want to obtain one equation containing only C and H_2 on the reactant side and C_2H_6 on the product side. First, we note that the first equation contains 1 mole of C (graphite) on the reactant side. Two moles of C are on the reactant side of equation (6.1), so let's multiply the first equation by 2.

$$2 \text{ C (graphite)} + 2 \text{ O}_2 \, (g) \longrightarrow 2 \text{ CO}_2 \, (g) \qquad \Delta H_{rxn}^\circ = 2(-393.5 \text{ kJ}) = -787.0 \text{ kJ}$$

Next, notice that the second equation contains 1 mole of H_2 on the reactant side. Three moles of H_2 are on the reactant side of Equation (6.1), so let's multiply the second equation by 3.

$$3 \text{ H}_2 \, (g) + \frac{3}{2} \text{ O}_2 \, (g) \longrightarrow 3 \text{ H}_2\text{O} \, (l) \qquad \Delta H_{rxn}^\circ = 3(-285.8 \text{ kJ}) = -857.4 \text{ kJ}$$

Next, notice that the third equation contains 2 moles of ethane, C_2H_6, on the reactant side. One mole of ethane is on the product side of equation (6.1), so let's reverse the third equation and multiply by 1/2.

$$2\ CO_2\ (g)\ +\ 3\ H_2O\ (l)\ \longrightarrow\ C_2H_6\ (g)\ +\ \frac{7}{2}\ O_2\ (g) \qquad \Delta H^{\circ}_{rxn}\ =\ 1/2(3119.6\ kJ) = 1559.8\ kJ$$

Putting these three equations together, we have:

Reaction	ΔH° (kJ)
$2\ C\ (graphite)\ +\ 2\ O_2\ (g)\ \longrightarrow\ 2\ CO_2\ (g)$	-787.0
$3\ H_2\ (g)\ +\ \frac{3}{2}\ O_2\ (g)\ \longrightarrow\ 3\ H_2O\ (l)$	-857.4
$2\ CO_2\ (g)\ +\ 3\ H_2O\ (l)\ \longrightarrow\ C_2H_6\ (g)\ +\ \frac{7}{2}\ O_2\ (g)$	1559.8

$2\ C\ (graphite)\ +\ 3\ H_2\ (g)\ \longrightarrow\ C_2H_6\ (g)$	**$\Delta H^{\circ} = -84.6\ kJ$**

6.48   The second and third equations can be combined to give the first equation.

$$2Al\ (s)\ +\ \frac{3}{2}\ O_2\ (g)\ \longrightarrow\ Al_2O_3\ (s) \qquad\qquad \Delta H^{\circ} = -1601\ kJ$$

$$Fe_2O_3\ (s)\ \longrightarrow\ 2\ Fe\ (s)\ +\ \frac{3}{2}\ O_2\ (g) \qquad\qquad \Delta H^{\circ} = 821\ kJ$$

$$2\ Al\ (s)\ +\ Fe_2O_3\ (s)\ \longrightarrow\ 2\ Fe\ (s)\ +\ Al_2O_3\ (s) \qquad \mathbf{\Delta H^{\circ} = -780\ kJ}$$

6.60   Applying the First Law of Thermodynamics, Problem Type 5A.

To solve this problem, you must make sure to get the sign convention correct. The surroundings do work on the system; this will increase the internal energy of the system.

$$w\ =\ +74\ kJ$$

The system gives off 26 kJ of heat. This is an energy-depleting process.

$$q\ =\ -26\ kJ$$

Finally,

$$\Delta E\ =\ q\ +\ w\ =\ -26\ kJ\ +\ 74\ kJ\ =\ \mathbf{48\ kJ}$$

6.62   Calculating the work done in gas expansion, Problem Type 5B.

The work can be calculated as follows:

$$w\ =\ -P\Delta V \qquad\qquad\qquad\qquad\qquad\qquad\qquad\qquad (6.2)$$

Step 1: First, we need to calculate the volume that the water vapor will occupy (V_f).

Using the ideal gas equation:

$$V_{H_2O}\ =\ \frac{n_{H_2O}RT}{P}\ =\ \frac{(1\ mol)\left(0.0821\ \dfrac{L\cdot atm}{mol\cdot K}\right)(373\ K)}{(1.0\ atm)}\ =\ 31\ L$$

Step 2: It is given that the volume occupied by liquid water is negligible. Therefore,

$$\Delta V\ =\ V_f\ -\ V_i\ =\ 31\ L\ -\ 0\ L\ =\ 31\ L$$

Step 3: Substitute P and ΔV into Equation (6.2) and solve for w.

$$w = -P\Delta V = -(1.0 \text{ atm})(31 \text{ L}) = \textbf{-31 L·atm}$$

It would be more convenient to express w in units of joules. The following conversion factor can be obtained from Appendix 1:

$$1 \text{ L·atm} = 101.3 \text{ J}$$

Thus, we can write:

$$w = -31 \text{ L·atm} \times \frac{101.3 \text{ J}}{1 \text{ L·atm}} = \textbf{-3.1} \times \textbf{10}^3 \textbf{ J}$$

Why is the sign negative?

6.64   Rearrange the equations as necessary so they can be added to yield the desired equation.

$$
\begin{array}{lll}
2\text{ B} \longrightarrow \text{A} & & -\Delta H_1 \\
\text{A} \longrightarrow \text{C} & & \Delta H_2 \\
\hline
2\text{B} \longrightarrow \text{C} & & \Delta H^\circ = \Delta H_2 - \Delta H_1
\end{array}
$$

6.66   **(a)** $\Delta H^\circ_{rxn} = \Sigma n\, \Delta H^\circ_f \text{ (products)} - \Sigma m\, \Delta H^\circ_f \text{ (reactants)}$

$\Delta H^\circ_{rxn} = [4\, \Delta H^\circ_f (\text{NH}_3) + \Delta H^\circ_f (\text{N}_2)] - 3\, \Delta H^\circ_f (\text{N}_2\text{H}_4)$

$\Delta H^\circ_{rxn} = [(4 \text{ mol})(-46.3 \text{ kJ/mol}) + (0)] - (3 \text{ mol})(50.42 \text{ kJ/mol}) = \textbf{-336.5 kJ}$

(b)   The balanced equations are:

(1) $\text{N}_2\text{H}_4\,(l) + \text{O}_2\,(g) \longrightarrow \text{N}_2\,(g) + 2\,\text{H}_2\text{O}\,(l)$

(2) $4\,\text{NH}_3\,(g) + 3\,\text{O}_2\,(g) \longrightarrow 2\,\text{N}_2\,(g) + 6\,\text{H}_2\text{O}\,(l)$

The standard enthalpy change for equation (1) is:

$\Delta H^\circ_{rxn} = \Delta H^\circ_f(\text{N}_2) + 2\,\Delta H^\circ_f[\text{H}_2\text{O}\,(l)] - \{\Delta H^\circ_f[\text{N}_2\text{H}_4\,(l)] + \Delta H^\circ_f(\text{O}_2)\}$

$\Delta H^\circ_{rxn} = [(1 \text{ mol})(0) + (2 \text{ mol})(-285.8 \text{ kJ/mol})] - [(1 \text{ mol})(50.442 \text{ kJ/mol}) + (1 \text{ mol})(0)] = -622.0 \text{ kJ}$

The standard enthalpy change for equation (2) is:

$\Delta H^\circ_{rxn} = [2\,\Delta H^\circ_f(\text{N}_2) + 6\,\Delta H^\circ_f(\text{H}_2\text{O})] - [4\,\Delta H^\circ_f(\text{NH}_3) + 3\,\Delta H^\circ_f(\text{O}_2)]$

$\Delta H^\circ_{rxn} = [(2 \text{ mol})(0) + (6 \text{ mol})(-285.8 \text{ kJ/mol})] - [(4 \text{ mol})(-46.3 \text{ kJ/mol}) - (3 \text{ mol})(0)] = -1529.6 \text{ kJ}$

We can now calculate the enthalpy change per kilogram of each substance. ΔH°_{rxn} above is in units of kJ/mol. We need to convert to kJ/kg.

$\text{N}_2\text{H}_4\,(l)$: $\Delta H^\circ_{rxn} = \dfrac{-622.0 \text{ kJ}}{1 \text{ mol N}_2\text{H}_4} \times \dfrac{1 \text{ mol N}_2\text{H}_4}{32.05 \text{ g N}_2\text{H}_4} \times \dfrac{1000 \text{ g}}{1 \text{ kg}} = \textbf{-1.941} \times \textbf{10}^4 \textbf{ kJ/kg N}_2\textbf{H}_4$

$$NH_3(g): \quad \Delta H^\circ_{rxn} = \frac{-1529.6 \text{ kJ}}{4 \text{ mol } NH_3} \times \frac{1 \text{ mol } NH_3}{17.03 \text{ g } NH_3} \times \frac{1000 \text{ g}}{1 \text{ kg}} = \mathbf{-2.245 \times 10^4 \text{ kJ/kg } NH_3}$$

Since ammonia releases more energy per kilogram of substance, it would be a better fuel.

6.68   We initially have 6 moles of gas (3 moles of chlorine and 3 moles of hydrogen). Since our product is 6 moles of hydrogen chloride, there is no change in the number of moles of gas. Therefore there is no volume change; $\Delta V = 0$.

$$w = -P\Delta V = [-1 \text{ atm} \times (0 \text{ L})] = 0$$

$$\Delta E^\circ = \Delta H^\circ - P\Delta V$$

$-P\Delta V = 0$, so

$$\Delta E = \Delta H$$

$$\Delta H = 3 \Delta H^\circ_{rxn} = 3(-184.6 \text{ kJ}) = -553.8 \text{ kJ}$$

We need to multiply ΔH°_{rxn} by three, because the question involves the formation of 6 moles of HCl; whereas, the equation as written only produces 2 moles of HCl.

$$\Delta E^\circ = \Delta H^\circ = \mathbf{-553.8 \text{ kJ}}$$

6.70   The initial and final states of this system are identical. Since enthalpy is a state function, its value depends only upon the state of the system. The enthalpy change is **zero**.

6.72 $H(g) + Br(g) \longrightarrow HBr(g) \qquad \Delta H^\circ_{rxn} = ?$

Rearrange the equations as necessary so they can be added to yield the desired equation.

$H(g) \longrightarrow 1/2\ H_2(g)$	$\Delta H^\circ_{rxn} = 1/2(-436.4 \text{ kJ}) = -218.2 \text{ kJ}$
$Br(g) \longrightarrow 1/2\ Br_2(g)$	$\Delta H^\circ_{rxn} = 1/2(-192.5 \text{ kJ}) = -96.25 \text{ kJ}$
$1/2\ H_2(g) + 1/2\ Br_2(g) \longrightarrow HBr(g)$	$\Delta H^\circ_{rxn} = 1/2(-72.4 \text{ kJ}) = -36.2 \text{ kJ}$

$$H(g) + Br(g) \longrightarrow 2\ HBr(g) \qquad \mathbf{\Delta H^\circ = -350.7 \text{ kJ}}$$

6.74 $q_{system} = 0 = q_{metal} + q_{water} + q_{calorimeter}$

$q_{metal} + q_{water} + q_{calorimeter} = 0$

$m_{metal}s_{metal}(t_{final} - t_{initial}) + m_{water}s_{water}(t_{final} - t_{initial}) + C_{calorimeter}(t_{final} - t_{initial}) = 0$

All the needed values are given in the problem. All you need to do is plug in the values and solve for s_{metal}.

$(44.0 \text{ g})(s_{metal})(28.4 - 99.0)°C + (80.0 \text{ g})(4.184 \text{ J/g·°C})(28.4 - 24.0)°C + (12.4 \text{ J/°C})(28.4 - 24.0)°C = 0$

$(-3.11 \times 10^3)s_{metal} \text{ (g·°C)} = -1.53 \times 10^3 \text{ J}$

$s_{metal} = \mathbf{0.492 \text{ J/g·°C}}$

6.76   A good starting point would be to calculate the standard enthalpy for both reactions.

Calculate the standard enthalpy for the reaction: $\qquad C(s) + \dfrac{1}{2} O_2(g) \longrightarrow CO(g)$

This reaction corresponds to the standard enthalpy of formation of CO, so we use the value of -110.5 kJ.

Calculate the standard enthalpy for the reaction: $C\ (s)\ +\ H_2O\ (g)\ \longrightarrow\ CO\ (g)\ +\ H_2\ (g)$

$$\Delta H^{\circ}_{rxn} = [\ \Delta H^{\circ}_f(CO) + \Delta H^{\circ}_f(H_2)] - [\ \Delta H^{\circ}_f(C) + \Delta H^{\circ}_f(H_2O)]$$

$$\Delta H^{\circ}_{rxn} = [(1\ mol)(-110.5\ kJ/mol) + (1\ mol)(0)] - [(1\ mol)(0) + (1\ mol)(-241.8\ kJ/mol)] = 131.3\ kJ$$

The first reaction, which is exothermic, can be used to promote the second reaction, which is endothermic. Thus, the two gases are produced alternately.

6.78   First, calculate the energy produced by 1 mole of octane, C_8H_{18}.

$$C_8H_{18}\ (l)\ +\ \frac{25}{2}\ O_2\ (g)\ \longrightarrow\ 8CO_2\ (g)\ +\ 9H_2O\ (l)$$

$$\Delta H^{\circ}_{rxn} = 8\ \Delta H^{\circ}_f(CO_2) + 9\ \Delta H^{\circ}_f[H_2O\ (l)] - [\ \Delta H^{\circ}_f(C_8H_{18}) + \frac{25}{2}\ \Delta H^{\circ}_f(O_2)]$$

$$\Delta H^{\circ}_{rxn} = [(8\ mol)(-393.5\ kJ/mol) + (9\ mol)(-285.8\ kJ/mol)] - [(1\ mol)(-249.9\ kJ/mol) + (\frac{25}{2}\ mol)(0)]$$

$$= -5470\ kJ$$

The problem asks for the energy produced by the combustion of 1 gallon of octane. ΔH°_{rxn} above has units of kJ/mol octane. We need to convert from kJ/mol octane to kJ/gallon octane. The heat of combustion for 1 gallon of octane is:

$$\Delta H^{\circ} = \frac{-5470\ kJ}{1\ mol\ octane} \times \frac{1\ mol\ octane}{114.2\ g\ octane} \times \frac{2660\ g}{1\ gal} = -1.274 \times 10^5\ kJ/gal$$

The combustion of hydrogen corresponds to the standard heat of formation of water:

$$H_2\ (g)\ +\ \frac{1}{2}\ O_2\ (g)\ \longrightarrow\ H_2O\ (l)$$

Thus, ΔH°_{rxn} is the same as ΔH°_f for $H_2O\ (l)$, which has a value of -285.8 kJ. The number of moles of hydrogen required to produce 1.274×10^5 kJ of heat is:

$$n_{H_2} = (1.274 \times 10^5\ kJ) \times \frac{1\ mol\ H_2}{285.8\ kJ} = 445.8\ mol\ H_2$$

Finally, use the ideal gas law to calculate the volume of gas corresponding to 445.8 moles of H_2 at 25°C and 1 atm.

$$V_{H_2} = \frac{n_{H_2}RT}{P} = \frac{(445.8\ mol)\left(0.0821\ \dfrac{L \cdot atm}{mol \cdot K}\right)(298\ K)}{(1\ atm)} = 1.09 \times 10^4\ L$$

That is, the volume of hydrogen that is energy-equivalent to 1 gallon of gasoline is over **10,000 liters** at 1 atm and 25°C!

6.80   The combustion reaction is: $C_2H_6\ (l)\ +\ \frac{7}{2}\ O_2\ (g)\ \longrightarrow\ 2\ CO_2\ (g)\ +\ 3\ H_2O\ (l)$

The heat released during the combustion of 1 mole of ethane is:

$$\Delta H^{\circ}_{rxn} = [2\ \Delta H^{\circ}_f(CO_2) + 3\ \Delta H^{\circ}_f(H_2O)] - [\ \Delta H^{\circ}_f(C_2H_6) + \frac{7}{2}\ \Delta H^{\circ}_f(O_2)]$$

$$\Delta H^{\circ}_{rxn} = [(2\ mol)(-393.5\ kJ/mol) + (3\ mol)(-285.8\ kJ/mol)] - [(1\ mol)(-84.7\ kJ/mol + (\tfrac{7}{2}\ mol)(0)]$$

$$= -1560\ kJ$$

The heat required to raise the temperature of the water to 98°C is:

$$q = m_{H_2O}\,s_{H_2O}\,\Delta t = (855\ g)(4.184\ J/g\cdot°C)(98.0 - 25.0)°C = 2.61 \times 10^5\ J = 261\ kJ$$

The combustion of 1 mole of ethane produces 1560 kJ; the number of moles required to produce 261 kJ is:

$$261\ kJ \times \frac{1\ mol\ ethane}{1560\ kJ} = 0.167\ mol\ ethane$$

The volume of ethane is:

$$V_{ethane} = \frac{nRT}{P} = \frac{(0.167\ mol)\left(0.0821\ \dfrac{L\cdot atm}{mol\cdot K}\right)(296\ K)}{(752\,/\,760)(atm)} = \mathbf{4.10\ L}$$

6.82   The heat gained by the liquid nitrogen must be equal to the heat lost by the water.

$$q_{N_2} = -q_{H_2O}$$

If we can calculate the heat lost by the water, we can calculate the heat gained by 60.0 g of the nitrogen.

$$\text{Heat lost by the water} = q_{H_2O} = m_{H_2O}\,s_{H_2O}\,\Delta t$$

$$q_{H_2O} = (2.00 \times 10^2\ g)(4.184\ J/g\cdot°C)(41.0 - 55.3)°C = \mathbf{-1.20 \times 10^4\ J}$$

The heat gained by 60.0 g nitrogen is the opposite sign of the heat lost by the water.

$$q_{N_2} = -q_{H_2O}$$
$$q_{N_2} = 1.20 \times 10^4\ J$$

The problem asks for the molar heat of vaporization of liquid nitrogen. Above, we calculated the amount of heat necessary to vaporize 60.0 g of liquid nitrogen. We need to convert from J/60.0 g N_2 to J/mol N_2.

$$\Delta H_{vap} = \frac{1.20 \times 10^4\ J}{60.0\ g\ N_2} \times \frac{28.02\ g\ N_2}{1\ mol\ N_2} = \mathbf{5.60 \times 10^3\ J/mol = 5.60\ kJ/mol}$$

6.84   Recall that the standard enthalpy of formation (ΔH°_f) is defined as the heat change that results when 1 mole of a compound is formed from its elements at a pressure of 1 atm. Only in choice **(a)** does $\Delta H^{\circ}_{rxn} = \Delta H^{\circ}_f$. In choice **(b)**, C(diamond) is *not* the most stable form of elemental carbon under standard conditions; C(graphite) is the most stable form.

6.86   **(a)**   No work is done by a gas expanding in a vacuum, because the pressure exerted on the gas is zero.

   (b) $w = -P\Delta V$

   $$w = -(0.20\ atm)(0.50 - 0.050)L = -0.090\ L\cdot atm$$

Converting to units of joules:

$$w = -0.090 \text{ L·atm} \times \frac{101.3 \text{ J}}{\text{L·atm}} = \textbf{-9.1 J}$$

(c)   The gas will expand until the pressure is the same as the applied pressure of 0.20 atm. We can calculate its final volume using the ideal gas equation.

$$V = \frac{nRT}{P} = \frac{(0.020 \text{ mol})\left(0.0821 \dfrac{\text{L·atm}}{\text{mol·K}}\right)(273 + 20)\text{K}}{0.20 \text{ atm}} = \textbf{2.4 L}$$

The amount of work done is:

$$w = -P\Delta V = (0.20 \text{ atm})(2.4 - 0.050)\text{L} = -0.47 \text{ L·atm}$$

Converting to units of joules:

$$w = -0.47 \text{ L·atm} \times \frac{101.3 \text{ J}}{\text{L·atm}} = \textbf{-48 J}$$

6.88   **(a)**   The more closely packed, the greater the mass of food. Heat capacity depends on both the mass and specific heat.

$$C = ms$$

The heat capacity of the food is greater than the heat capacity of air; hence, the cold in the freezer will be retained longer.

(b)   Tea and coffee are mostly water; whereas, soup might contain vegetables and meat. Water has a higher heat capacity than the other ingredients in soup; therefore, coffee and tea retain heat longer than soup.

6.90   The equation given in the problem represents twice the standard enthalpy of formation of Fe_2O_3. From Appendix 3, the standard enthalpy of formation of $Fe_2O_3 = -822.2$ kJ/mol. The mole ratio between Fe_2O_3 and Fe in the balanced equation is 2:4.

We can now convert to the amount of heat produced when 250 g of Fe react.

$$\frac{-822.2 \text{ kJ}}{1 \text{ mol Fe}_2\text{O}_3} \times \frac{2 \text{ mol Fe}_2\text{O}_3}{4 \text{ mol Fe}} \times \frac{1 \text{ mol Fe}}{55.85 \text{ g Fe}} \times 250 \text{ g Fe} = \textbf{-1.84} \times \textbf{10}^{\textbf{3}} \textbf{ kJ}$$

6.92   The heat required to raise the temperature of 1 liter of water by 1°C is:

$$4.184 \frac{\text{J}}{\text{g·°C}} \times \frac{1 \text{ g}}{1 \text{ mL}} \times \frac{1000 \text{ mL}}{1 \text{ L}} \times 1°\text{C} = 4184 \text{ J/L}$$

Next, convert the volume of the Pacific Ocean to liters.

$$(7.2 \times 10^8 \text{ km}^3) \times \left(\frac{1000 \text{ m}}{1 \text{ km}}\right)^3 \times \left(\frac{100 \text{ cm}}{1 \text{ m}}\right)^3 \times \left(\frac{1 \text{ L}}{1000 \text{ cm}^3}\right) = 7.2 \times 10^{20} \text{ L}$$

The amount of heat needed to raise the temperature of 7.2×10^{20} L of water is:

$$(7.2 \times 10^{20}\text{ L}) \times \frac{4184\text{ J}}{1\text{ L}} = 3.0 \times 10^{24}\text{ J}$$

Finally, we can calculate the number of atomic bombs needed to produce this much heat.

$$(3.0 \times 10^{24}\text{ J}) \times \frac{1\text{ atomic bomb}}{1.0 \times 10^{15}\text{ J}} = \textbf{3.0} \times \textbf{10}^9 \textbf{ atomic bombs} = \textbf{3.0 billion atomic bombs}$$

6.94   Constant-volume calorimetry, Problem Type 3A.

Step 1: The heat of the reaction (combustion) is absorbed by both the water and the calorimeter.

$$q_{rxn} = -(q_{water} + q_{bomb})$$

First, we must calculate the heat absorbed by the water.

$$q_{water} = m_{water}s_{water}\Delta t$$
$$q_{water} = (2000.\text{ g})(4.184\text{ J/g}\cdot{}^\circ\text{C})(25.67 - 21.84){}^\circ\text{C} = 3.20 \times 10^4\text{ J} = 32.0\text{ kJ}$$

Step 2: ΔH_{rxn} is given in the problem in units of kJ/mol. Let's convert to q_{rxn} in kJ. This is the amount of heat released during the reaction when 1.9862 g of benzoic acid are combusted.

$$q_{rxn} = 1.9862\text{ g benzoic acid} \times \frac{1\text{ mol benzoic acid}}{122.1\text{ g benzoic acid}} \times \frac{-3226.7\text{ kJ}}{1\text{ mol benzoic acid}} = -52.49\text{ kJ}$$

And,

$$q_{bomb} = -q_{rxn} - q_{water}$$
$$q_{bomb} = 52.49\text{ kJ} - 32.0\text{ kJ} = 20.5\text{ kJ}$$

Step 3: To calculate the heat capacity of the bomb calorimeter, we can use the following equation:

$$q_{bomb} = C_{bomb}\Delta T$$
$$C_{bomb} = \frac{q_{bomb}}{\Delta t} = \frac{20.5\text{ kJ}}{(25.67 - 21.84){}^\circ\text{C}} = \textbf{5.35 kJ/}{}^\circ\textbf{C}$$

6.96   First, let's calculate the standard enthalpy of reaction.

$$\Delta H^\circ_{rxn} = 2\,\Delta H^\circ_f\,(CaSO_4) - [2\,\Delta H^\circ_f\,(CaO) + 2\,\Delta H^\circ_f\,(SO_2) + \Delta H^\circ_f\,(O_2)]$$
$$= (2\text{ mol})(-1432.7\text{ kJ/mol}) - [(2\text{ mol})(-635.6\text{ kJ/mol}) + (2\text{ mol})(-296.1\text{ kJ/mol}) + 0]$$
$$= -1002\text{ kJ}$$

This is the enthalpy change for every 2 moles of SO_2 that are removed. The problem asks to calculate the enthalpy change for this process if 6.6×10^5 g of SO_2 are removed.

$$6.6 \times 10^5\text{ g SO}_2 \times \frac{1\text{ mol SO}_2}{64.07\text{ g SO}_2} \times \frac{-1002\text{ kJ}}{2\text{ mol SO}_2} = \textbf{-5.2} \times \textbf{10}^6\textbf{ kJ}$$

6.98   First, we need to calculate the volume of the balloon.

$$V = \frac{4}{3}\pi r^3 = \frac{4}{3}\pi(8 \text{ m})^3 = 2.1 \times 10^3 \text{ m}^3 \times \frac{1000 \text{ L}}{1 \text{ m}^3} = 2.1 \times 10^6 \text{ L}$$

(a)   We can calculate the mass of He in the balloon using the ideal gas equation.

$$n_{He} = \frac{PV}{RT} = \frac{\left(98.7 \text{ kPa} \times \dfrac{1 \text{ atm}}{1.01325 \times 10^2 \text{ kPa}}\right)(2.1 \times 10^6 \text{ L})}{\left(0.0821 \dfrac{\text{L} \cdot \text{atm}}{\text{mol} \cdot \text{K}}\right)(273 + 18)\text{K}} = 8.6 \times 10^4 \text{ mol He}$$

mass He $= 8.6 \times 10^4 \text{ mol He} \times \dfrac{4.003 \text{ g He}}{1 \text{ mol He}} = \mathbf{3.4 \times 10^5 \text{ g He}}$

(b)   Work done $= -P\Delta V$

$$= -\left(98.7 \text{ kPa} \times \frac{1 \text{ atm}}{1.01325 \times 10^2 \text{ kPa}}\right)(2.1 \times 10^6 \text{ L})$$

$$= -2.0 \times 10^6 \text{ L} \cdot \text{atm} \times \frac{101.3 \text{ J}}{1 \text{ L} \cdot \text{atm}}$$

Work done $= \mathbf{-2.0 \times 10^8 \text{ J}}$

6.100   **(a)**   The heat needed to raise the temperature of the water from 3°C to 37°C can be calculated using the equation:

$$q = ms\Delta t$$

First, we need to calculate the mass of the water.

$$4 \text{ glasses of water} \times \frac{2.5 \times 10^2 \text{ mL}}{1 \text{ glass}} \times \frac{1 \text{ g water}}{1 \text{ mL water}} = 1.0 \times 10^3 \text{ g water}$$

The heat needed to raise the temperature of 1.0×10^3 g of water is:

$$q = ms\Delta t = (1.0 \times 10^3 \text{ g})(4.184 \text{ J/g} \cdot °\text{C})(37 - 3)°\text{C} = 1.4 \times 10^5 \text{ J} = \mathbf{1.4 \times 10^2 \text{ kJ}}$$

(b)   We need to calculate both the heat needed to melt the snow and also the heat needed to heat liquid water form 0°C to 37°C (normal body temperature).

The heat needed to melt the snow is:

$$8.0 \times 10^2 \text{ g} \times \frac{1 \text{ mol}}{18.02 \text{ g}} \times \frac{6.01 \text{ kJ}}{1 \text{ mol}} = 2.7 \times 10^2 \text{ kJ}$$

The heat needed to raise the temperature of the water from 0°C to 37°C is:

$$q = ms\Delta t = (8.0 \times 10^2 \text{ g})(4.184 \text{ J/g} \cdot °\text{C})(37 - 0)°\text{C} = 1.2 \times 10^5 \text{ J} = 1.2 \times 10^2 \text{ kJ}$$

The total heat lost by your body is:

$$(2.7 \times 10^2 \text{ kJ}) + (1.2 \times 10^2 \text{ kJ}) = \mathbf{3.9 \times 10^2 \text{ kJ}}$$

6.102 **(a)** $\Delta H° = \Delta H_f°(F^-) + \Delta H_f°(H_2O) - [\Delta H_f°(HF) + \Delta H_f°(OH^-)]$

$\Delta H° = (1 \text{ mol})(-329.1 \text{ kJ/mol}) + (1 \text{ mol})(-285.8 \text{ kJ/mol}) - [(1 \text{ mol})(-320.1 \text{ kJ/mol}) + (1 \text{ mol})(-229.6 \text{ kJ/mol})$

$\Delta H° = -65.2 \text{ kJ}$

(b) We can add the equation given in part (a) to that given in part (b) to end up with the equation we are interested in.

$$HF\ (aq) + OH^-\ (aq) \longrightarrow F^-\ (aq) + H_2O\ (l) \qquad \Delta H° = -65.2 \text{ kJ}$$
$$\underline{H_2O\ (l) \longrightarrow H^+\ (aq) + OH^-\ (aq) \qquad\qquad\qquad \Delta H° = +56.2 \text{ kJ}}$$
$$\textbf{HF\ (aq)} \longrightarrow \textbf{H}^+\ \textbf{(aq)} + \textbf{F}^-\ \textbf{(aq)} \qquad\qquad \Delta H° = -9.0 \text{ kJ}$$

6.104 The equation we are interested in is the formation of CO form its elements.

$$C\ (graphite) + 1/2\ O_2\ (g) \longrightarrow CO\ (g) \qquad \Delta H° = ?$$

Try to add the given equations together to end up with the equation above.

$$C\ (graphite) + O_2\ (g) \longrightarrow CO_2\ (g) \qquad \Delta H° = -393.5 \text{ kJ}$$
$$\underline{CO_2\ (g) \longrightarrow CO\ (g) + 1/2\ O_2\ (g) \qquad\qquad \Delta H° = +283.0 \text{ kJ}}$$
$$\textbf{C\ (graphite)} + \textbf{1/2 O}_2\ \textbf{(g)} \longrightarrow \textbf{CO\ (g)} \qquad \Delta H° = -110.5 \text{ kJ}$$

We cannot obtain $\Delta H_f°$ for CO directly, because burning graphite in oxygen will form both CO and CO_2.

6.106 **(a)** mass $= 0.0010$ kg

Potential energy $= mgh$

$= (0.0010 \text{ kg})(9.8 \text{ m/s}^2)(51 \text{ m})$

Potential energy $= 0.50$ J

(b) Kinetic energy $= 1/2 mu^2 = 0.50$ J

$1/2(0.0010 \text{ kg})u^2 = 0.50$ J

$u^2 = 1.0 \times 10^3 \text{ m}^2/\text{s}^2$

$u = 32$ **m/s**

(c) $q = ms\Delta t$

$0.50 \text{ J} = (1.0 \text{ g})(4.184 \text{ J/g}°\text{C})\Delta t$

$\Delta t = 0.12$ **°C**

6.108 The reaction we are interested in is the formation of ethanol from its elements.

$$2\ C\ (graphite) + 1/2\ O_2\ (g) + 3\ H_2\ (g) \longrightarrow C_2H_5OH\ (l)$$

Along with the reaction for the combustion of ethanol, we can add other reactions together to end up with the above reaction.

Reversing the reaction representing the combustion of ethanol gives:

$$2\ CO_2\ (g)\ +\ 3\ H_2O\ (l)\ \longrightarrow\ C_2H_5OH\ (l)\ +\ 3\ O_2\ (g) \qquad \Delta H° = +1367.4\ kJ$$

We need to add equations to add C (graphite) and remove H_2O from the reactants side of the equation. We write:

$$2\ CO_2\ (g)\ +\ 3\ H_2O\ (l)\ \longrightarrow\ C_2H_5OH\ (l)\ +\ 3\ O_2\ (g) \qquad \Delta H° = +1367.4\ kJ$$

$$2\ C\ (graphite)\ +\ 2\ O_2\ (g)\ \longrightarrow\ 2\ CO_2\ (g) \qquad \Delta H° = 2(-393.5\ kJ)$$

$$\underline{3\ H_2\ (g)\ +\ 3/2\ O_2\ (g)\ \longrightarrow\ 3\ H_2O\ (l) \qquad \Delta H° = 3(-285.8\ kJ)}$$

$$\mathbf{2\ C\ (graphite)\ +\ 1/2\ O_2\ (g)\ +\ 3\ H_2\ (g)\ \longrightarrow\ C_2H_5OH\ (l) \qquad \Delta H_f° = -277.0\ kJ/mol}$$

6.110   Heat gained by ice = Heat lost by the soft drink

$$m_{ice} \times 334\ J/g = -m_{sd}s_{sd}\Delta t$$

$$m_{ice} \times 334\ J/g = -(361\ g)(4.184\ J/g\cdot°C)(0 - 23)°C$$

$$\mathbf{m_{ice} = 104\ g}$$

6.112   From Chapter 5, we saw that the kinetic energy (or internal energy) of 1 mole of a gas is $(3/2)RT$. For 1 mole of an ideal gas, $PV = RT$. We can write:

$$\text{internal energy} = \frac{3}{2}RT = \frac{3}{2}PV$$

$$= \frac{3}{2}(1.2 \times 10^5\ Pa)(5.5 \times 10^3\ m^3)$$

$$= 9.9 \times 10^8\ Pa\cdot m^3$$

$$1\ Pa\cdot m^3 = 1\ \frac{N}{m^2}m^3 = 1\ N\cdot m = 1\ J$$

Therefore, the internal energy is $\mathbf{9.9 \times 10^8\ J}$.

The final temperature of the copper metal can be calculated.

$$q = m_{Cu}s_{Cu}\Delta t$$

$$9.9 \times 10^8\ J = (9.07 \times 10^6\ g)(0.385\ J/g°C)(t_f - 21°C)$$

$$3.49 \times 10^6\ t_f = 1.06 \times 10^9$$

$$\mathbf{t_f = 304°C}$$

6.114   **(a)** $CaC_2\ (s)\ +\ 2\ H_2O\ (l)\ \longrightarrow\ Ca(OH)_2\ (s)\ +\ C_2H_2\ (g)$

(b)   The reaction for the combustion of acetylene is:

$$2\ C_2H_2\ (g)\ +\ 5\ O_2\ (g)\ \longrightarrow\ 4\ CO_2\ (g)\ +\ 2\ H_2O\ (l) \qquad \Delta H° = -2598.8\ kJ$$

First, we need to calculate the number of moles of acetylene produced when 74.6 g of CaC_2 are reacted.

$$74.6 \text{ g CaC}_2 \times \frac{1 \text{ mol CaC}_2}{64.10 \text{ g CaC}_2} \times \frac{1 \text{ mol C}_2\text{H}_2}{1 \text{ mol CaC}_2} = 1.16 \text{ mol C}_2\text{H}_2$$

Now we can calculate the maximum amount of heat obtained from the combustion of 1.16 moles of acetylene.

$$1.16 \text{ mol C}_2\text{H}_2 \times \frac{2598.8 \text{ kJ}}{2 \text{ mol C}_2\text{H}_2} = \mathbf{1.51 \times 10^3 \text{ kJ}}$$

CHAPTER 7
QUANTUM THEORY AND THE ELECTRONIC STRUCTURE OF ATOMS

PROBLEM-SOLVING STRATEGIES AND TUTORIAL SOLUTIONS

TYPES OF PROBLEMS

Problem Type 1: Calculating the Frequency and Wavelength of an Electromagnetic Wave.

Problem Type 2: Calculating the Energy of a Photon.

Problem Type 3: Calculating the Energy, Wavelength, or Frequency in the Emission Spectrum of a Hydrogen Atom.

Problem Type 4: The De Broglie Equation: Calculating the Wavelengths of Particles.

Problem Type 5: Quantum Numbers.
    (a) Labeling an atomic orbital.
    (b) Counting the number of orbitals associated with a principal quantum number.
    (c) Assigning quantum numbers to an electron.
    (d) Counting the number of electrons in a principal level.

Problem Type 6: Writing Electron Configurations and Orbital Diagrams.

PROBLEM TYPE 1: CALCULATING THE FREQUENCY AND WAVELENGTH OF AN ELECTROMAGNETIC WAVE

All types of *electromagnetic radiation* move through a vacuum at a speed of about 3.00×10^8 m/s, which is called the speed of light (c). Speed is an important property of a wave traveling through space and is equal to the product of the wavelength and the frequency of the wave. For electromagnetic waves

$$c = \lambda\nu \qquad (7.1)$$

Equation (7.1) can be rearranged as necessary to solve for either the wavelength (λ) or the frequency (ν).

EXAMPLE 7.1
A certain AM radio station broadcasts at a frequency of 6.00×10^2 kHz. What is the wavelength of these radio waves in meters?

Step 1: Solve Equation (7.1) algebraically for the wavelength (λ).

$$c = \lambda\nu$$

$$\lambda = \frac{c}{\nu} \qquad (7.2)$$

Step 2: Since the speed of light has units of m/s, we must convert the frequency from units of kHz to Hz (s^{-1})

$$6.00 \times 10^2 \text{ kHz} \times \frac{1000 \text{ Hz}}{1 \text{ kHz}} = 6.00 \times 10^5 \text{ Hz} = 6.00 \times 10^5 \text{ s}^{-1}$$

Step 3: Calculate the value of λ by substituting the known quantities into Equation (7.2).

$$\lambda = \frac{3.00 \times 10^8 \frac{m}{s}}{6.00 \times 10^5 \frac{1}{s}} = \mathbf{5.00 \times 10^2 \text{ m}}$$

EXAMPLE 7.2
What is the frequency of light that has a wavelength of 665 nm?

Step 1: Solve Equation (7.1) algebraically for the frequency (ν).

$$c = \lambda\nu$$

$$\nu = \frac{c}{\lambda} \tag{7.3}$$

Step 2: Since the speed of light has units of m/s, we must convert the wavelength from units of nm to m.

$$665 \text{ nm} \times \frac{10^{-9} \text{ m}}{1 \text{ nm}} = 6.65 \times 10^{-7} \text{ m}$$

Step 3: Calculate the value of ν by substituting the known quantities into Equation (7.3).

$$\nu = \frac{3.00 \times 10^8 \frac{m}{s}}{6.65 \times 10^{-7} \text{ m}} = \mathbf{4.51 \times 10^{14} \text{ s}^{-1}} = \mathbf{4.51 \times 10^{14} \text{ Hz}}$$

PRACTICE EXERCISE

1.   Domestic microwave ovens generate microwaves with a frequency of 2.450 GHz. What is the wavelength of this microwave radiation?

| Text Problems: 7.8, 7.12, **7.16** |

PROBLEM TYPE 2: CALCULATING THE ENERGY OF A PHOTON

Max Planck said that atoms and molecules could emit (or absorb) energy only in discrete quantities. Planck gave the name *quantum* to the smallest quantity of energy that can be emitted (or absorbed) in the form of electromagnetic radiation. The energy E of a single quantum of energy is given by

$$E = h\nu \tag{7.4}$$

where,

   h is Planck's constant = 6.63×10^{-34} J·s
   ν is the frequency of radiation

EXAMPLE 7.3
The yellow light given off by a sodium vapor lamp has a wavelength of 589 nm. What is the energy of a single photon of this radiation?

Step 1: You are given wavelength in this problem. Equation (7.4) shows the relationship between energy and frequency. However, there is a relationship between frequency and wavelength [see Equation (7.3)].

$$\nu = \frac{c}{\lambda}$$

Substituting for the frequency in Equation (7.4), we have

$$E = \frac{hc}{\lambda} \tag{7.5}$$

Step 2: Since the speed of light is in units of m/s, we must convert the wavelength from units of nm to m.

$$589 \text{ nm} \times \frac{10^{-9} \text{ m}}{1 \text{ nm}} = 5.89 \times 10^{-7} \text{ m}$$

Step 3: Calculate the value of E by substituting the known quantities into Equation (7.5).

$$E = \frac{(6.63 \times 10^{-34} \text{ J} \cdot \text{s})\left(3.00 \times 10^8 \frac{\text{m}}{\text{s}}\right)}{5.89 \times 10^{-7} \text{ m}} = \mathbf{3.38 \times 10^{-19} \text{ J}}$$

PRACTICE EXERCISE

2.   The red line in the spectrum of lithium occurs at 670.8 nm. What is the energy of a photon of this light? What is the energy of 1 mol of these photons?

3.   The light-sensitive compound in most photographic films is silver bromide (AgBr). When the film is exposed, assume that the light energy absorbed dissociates the molecule into atoms. (The actual process is more complex.) If the energy of dissociation of AgBr is 1.00×10^2 kJ/mol, find the wavelength of light that is just able to dissociate AgBr.

Text Problems: **7.16**, 7.18, 7.20

PROBLEM TYPE 3: CALCULATING THE ENERGY, WAVELENGTH, OR FREQUENCY IN THE EMISSION SPECTRUM OF A HYDROGEN ATOM

Using arguments based on electrostatic interaction and Newton's laws of motion, Neils Bohr showed that the energies that the electron in the hydrogen atom can possess are given by:

$$E_n = -R_H\left(\frac{1}{n^2}\right) \tag{7.6}$$

where,

R_H is the Rydberg constant $= 2.18 \times 10^{-18}$ J
n is the principal quantum number that has integer values

During the emission process in a hydrogen atom, an electron initially in an excited state characterized by the principal quantum number n_i drops to a lower energy state characterized by the principal quantum number n_f. This lower energy state may be either another excited state or the ground state. The difference between the energies of the initial and final states is:

$$\Delta E = E_f - E_i$$

Substituting Equation (7.6) into the above equation gives:

$$\Delta E = \left(\frac{-R_H}{n_f^2}\right) - \left(\frac{-R_H}{n_i^2}\right)$$

$$\Delta E = R_H\left(\frac{1}{n_i^2} - \frac{1}{n_f^2}\right)$$

Furthermore, since this transition results in the emission of a photon of frequency ν and energy $h\nu$ (See Problem Type 2), we can write:

$$\Delta E = h\nu = R_H\left(\frac{1}{n_i^2} - \frac{1}{n_f^2}\right) \tag{7.7}$$

EXAMPLE 7.4

What wavelength of radiation will be emitted during an electron transition from the $n = 5$ state to the $n = 1$ state in the hydrogen atom? What region of the electromagnetic spectrum does this wavelength correspond to?

Step 1: Using Equation (7.7), we can calculate the energy change for the transition.

$$\Delta E = R_H\left(\frac{1}{n_i^2} - \frac{1}{n_f^2}\right)$$

$$\Delta E = (2.18 \times 10^{-18}\text{ J})\left(\frac{1}{5^2} - \frac{1}{1^2}\right)$$

$$\Delta E = -2.09 \times 10^{-18}\text{ J}$$

Step 2: The negative sign for ΔE indicates that energy is released to the surroundings during the emission process. To calculate the wavelength, we omit the minus sign for ΔE because the wavelength of the photon must be positive. We know that

$$\Delta E = h\nu$$

We also know that $\nu = \dfrac{c}{\lambda}$. Substituting into the above equation gives:

$$\Delta E = \frac{hc}{\lambda}$$

Solving the equation algebraically for the wavelength, then substituting in the known values gives:

$$\lambda = \frac{hc}{\Delta E} = \frac{(6.63 \times 10^{-34}\text{ J·s})\left(3.00 \times 10^8\ \dfrac{\text{m}}{\text{s}}\right)}{2.09 \times 10^{-18}\text{ J}} = \mathbf{9.52 \times 10^{-8}\ m}$$

Step 3: To determine the region of the electromagnetic spectrum that this wavelength corresponds to, we should convert the wavelength from units of meters to nanometers, and then compare the value to Figure 7.4 of your text.

$$9.52 \times 10^{-8}\text{ m} \times \frac{1\text{ nm}}{10^{-9}\text{ m}} = 95.2\text{ nm}$$

Checking Figure 7.4, we see that the ultraviolet region of the spectrum is centered at a wavelength of 10 nm. Therefore, this emission is in the ultraviolet (UV) region of the electromagnetic spectrum.

PRACTICE EXERCISE

4.   A hydrogen emission line in the ultraviolet region of the spectrum at 95.2 nm corresponds to a transition from a higher energy level n_i to the $n = 1$ level. What is the value of n_i for the higher energy level?

Text Problems: 7.30, **7.32**, 7.34

PROBLEM TYPE 4: THE DE BROGLIE EQUATION: CALCULATING THE WAVELENGTHS OF PARTICLES

Albert Einstein showed that light (electromagnetic radiation) can possess particle like properties. De Broglie reasoned that if waves can behave like particles, then particles can exhibit wave properties. De Broglie deduced that the particle and wave properties are related by the expression:

$$\lambda = \frac{h}{mu} \tag{7.8}$$

where,

> λ is the wavelength associated with the moving particle
> m is the mass of the particle
> u is the velocity of the particle

Equation (7.8) implies that a particle in motion can be treated as a wave, and a wave can exhibit the properties of a particle.

EXAMPLE 7.5

When an atom of Th-232 undergoes radioactive decay, an alpha particle, which has a mass of 4.0 amu, is ejected from the Th nucleus with a velocity of 1.4×10^7 m/s. What is the De Broglie wavelength of the alpha particle?

Step 1: Because Planck's constant has units of J·s, and $1 \text{ J} = 1 \text{ kg·m}^2 \cdot \text{s}^{-2}$, the mass of the alpha particle must be expressed in kilograms. Since one particle has a mass of 4.0 amu, a mole of alpha particles will have a mass of 4.0 g. A reasonable strategy to complete the conversion is:

g/mol → kg/mol → kg/particle

$$\frac{4.0 \text{ g}}{1 \text{ mol}} \times \frac{1 \text{ kg}}{1000 \text{ g}} \times \frac{1 \text{ mol}}{6.022 \times 10^{23} \text{ particles}} = 6.6 \times 10^{-27} \text{ kg/particle}$$

Step 2: Substitute the known quantities into Equation (7.8).

$$\lambda = \frac{h}{mu} = \frac{(6.63 \times 10^{-34} \text{ J·s}) \times \left(\frac{1 \text{ kg·m}^2/\text{s}^2}{1 \text{ J}} \right)}{(6.6 \times 10^{-27} \text{ kg})\left(1.4 \times 10^7 \frac{\text{m}}{\text{s}} \right)} = 7.2 \times 10^{-15} \text{ m}$$

The wavelength is smaller than the diameter of the thorium nucleus, which is about 2×10^{-14} m. Is this what you would expect?

PRACTICE EXERCISE

5. The average kinetic energy of a neutron at 25°C is 6.2×10^{-21} J. What is the de Broglie wavelength of an average neutron? The mass of a neutron is 1.008 amu. (**Hint:** kinetic energy $= \frac{1}{2} mu^2$)

Text Problems: **7.40**, 7.42

PROBLEM TYPE 5: QUANTUM NUMBERS

See Section 7.6 of your text for a complete discussion. In quantum mechanics, three quantum numbers are required to describe the distribution of electrons in atoms.

(1) **The principal quantum number (n).** In a hydrogen atom, the value of n determines the energy of an orbital. It can have integral values 1, 2, 3, and so forth.

(2) **The angular momentum quantum number (l)** tells us the "shape" of the orbitals. l has possible integral values from 0 to $(n-1)$. The value of l is generally designated by the letters s, p, d, ..., as follows:

l	0	1	2	3	4	5
Name of orbital	s	p	d	f	g	h

(3) **The magnetic quantum number (m_l)** describes the orientation of the orbital in space. For a certain value of l, there are $(2l + 1)$ integral values of m_l, as follows:

$$-l, (-l + 1), \ldots 0, \ldots (+l - 1), +l$$

The number of m_l values indicates the number of orbitals in a subshell with a particular l value.

Finally, there is a fourth quantum number that tells us the spin of the electron. The **electron spin quantum number (m_s)** has values of $+1/2$ or $-1/2$, which correspond to the two spinning motions of the electron.

A. Labeling an atomic orbital

To "label" an atomic orbital, you need to specify the three quantum numbers (n, l, m_l) that give information about the distribution of electrons in orbitals. Remember, m_s tells us the spin of the electron, which tells us nothing about the orbital.

EXAMPLE 7.6

List the values of n, l, and m_l for orbitals in the $2p$ subshell.

The number given in the designation of the subshell is the principal quantum number, so in this case $n = 2$. For p orbitals, $l = 1$. m_l can have integer values from $-l$ to $+l$. Therefore, m_l can be -1, 0, and $+1$. (The three values for m_l correspond to the three p orbitals.)

PRACTICE EXERCISE

6. List the values of n, l, and m_l for orbitals in the $4f$ subshell.

Text Problems: **7.54**, **7.56**, 7.60

B. Counting the number of orbitals associated with a principal quantum number

To work this type of problem, you must take into account the energy level (n), the types of orbitals in that energy level (l), and the number of orbitals in a subshell with a particular l value (m_l).

EXAMPLE 7.7
What is the total number of orbitals associated with the principal quantum number $n = 2$?

For $n = 2$, there are only two possible values of l, 0 and 1. Thus, there is one $2s$ orbital, and there are three $2p$ orbitals. (For $l = 1$, there are three possible m_l values, −1, 0, and +1.)

Therefore, the total number of orbitals in the $n = 2$ energy level is $1 + 3 = $ **4**.

> **Tip:** The total number of orbitals with a given n value is n^2. For Example 7.7, the total number of orbitals in the $n = 2$ level equals $2^2 = $ **4**.

PRACTICE EXERCISE
7.   What is the total number of orbitals associated with the principal quantum number $n = 4$?

Text Problems: 7.56, 7.60

C. Assigning quantum numbers to an electron

In assigning quantum numbers to an electron, you need to specify all four quantum numbers. In most cases, there will be more than one possible set of quantum numbers that can designate an electron.

EXAMPLE 7.8
List the different ways to write the four quantum numbers that designate an electron in a $4s$ orbital.

To begin with, we know that the principal quantum number n is 4, and the angular momentum quantum number l is 0 (s orbital). For $l = 0$, there is only one possible value for m_l, also 0. Since the electron spin quantum number m_s can be either +1/2 or −1/2, we conclude that there are two possible ways to designate the electron:

$$(4, 0, 0, +1/2) \qquad\qquad (4, 0, 0, -1/2)$$

PRACTICE EXERCISE
8.   List the different ways to write the four quantum numbers that designate an electron in a $3d$ orbital.

Text Problem: 7.54

D. Counting the number of electrons in a principal level

To work this type of problem, you need to know that the number of orbitals with a particular l value is $(2l + 1)$. Also, each orbital can accommodate two electrons.

EXAMPLE 7.9
What is the maximum number of electrons that can be present in the principal level for which $n = 4$?

When $n = 4$, $l = 0$, 1, 2, and 3. The number of orbitals for each l value is given by

Value of l	Number of orbitals $(2l + 1)$
0	1
1	3
2	5
3	7

The total number of orbitals in the principal level $n = 4$ is sixteen. Since each orbital can accommodate two electrons, the maximum number of electrons that can reside in the orbitals is $2 \times 16 = \mathbf{32}$.

> **Tip:** The above result can be generalized by the formula $2n^2$. For Example 7.9, we have $n = 4$, so $2(4)^2 = 32$.

PRACTICE EXERCISE

9.   What is the maximum number of electrons that can be present in the principal level for which $n = 2$?

Text Problem: 7.62

PROBLEM TYPE 6: WRITING ELECTRON CONFIGURATIONS AND ORBITAL DIAGRAMS

The electron configuration of an atom tells us how the electrons are distributed among the various atomic orbitals. To write electron configurations, you should follow the four rules or guidelines given below.

(1)   The electron configurations of all elements except hydrogen and helium are represented by a *noble gas core*, which shows (in brackets) the noble gas element that most nearly precedes the element being considered. The noble gas core is followed by the electron configurations of filled or partially filled subshells in the outermost shells.

(2)   *The Aufbau or "building up" principle* states that electrons are added to atomic orbitals starting with the lowest energy orbital and "building up" to higher energy orbitals.

(3)   In many-electron atoms, the subshells are filled in the order shown in Figure 7.24.

(4)   Each orbital can hold only *two* electrons.

The electron configuration can be represented in a more detailed manner called an **orbital diagram** that shows the spin of the electron. For orbital diagrams, you need to follow two additional rules given below.

(5)   The *Pauli exclusion principle* states that no two electrons in an atom can have the same four quantum numbers. This means that electrons occupying the same orbital *cannot* have the same spin.

(6)   *Hund's rule* states that the most stable arrangement of electrons in subshells is the one with the greatest number of parallel spins, without violating the Pauli exclusion principle.

EXAMPLE 7.10
Write the ground-state electron configuration and the orbital diagram for selenium.

Step 1: Selenium has 34 electrons. These electrons need to be placed in atomic orbitals.

Step 2: The noble gas element that most nearly proceeds Se is Ar. Therefore, the *noble gas core* is [Ar]. This core accounts for 18 electrons.

Step 3: See Figure 7.24 to check the order of filling subshells past the Ar noble gas core. You should find that the order of filling is $4s$, $3d$, then $4p$. There are 16 remaining electrons to distribute among these orbitals.

The $4s$ orbital can hold *two* electrons. Each of the five $3d$ orbitals can hold *two* electrons for a total of *ten* electrons. This leaves *four* electrons to fill the $4p$ orbitals.

The electron configuration for Se is:

$$[Ar]\, 4s^2 3d^{10} 4p^4$$

Step 4: To write an *orbital diagram*, we must also specify the spin of the electrons. The 4*s* and 3*d* orbitals are filled, so according to the Pauli exclusion principle, the paired electrons in the 4*s* orbital and in each of the 3*d* orbitals *must* have opposite spins.

$$[Ar] \quad \underset{4s^2}{\uparrow\downarrow} \qquad \underset{3d^{10}}{\uparrow\downarrow \;\; \uparrow\downarrow \;\; \uparrow\downarrow \;\; \uparrow\downarrow \;\; \uparrow\downarrow}$$

Step 5: Now, let's deal with the 4*p* electrons. Hund's rule states that the most stable arrangement of electrons in subshells is the one with the greatest number of parallel spins. In other words, we want to keep electrons unpaired if possible with parallel spins. Since there are three *p* orbitals, three of the *p* electrons can be placed individually in each of the *p* subshells with parallel spins.

$$\underset{4p^3}{\uparrow \;\; \uparrow \;\; \uparrow}$$

Step 6: Finally, the fourth *p* electron must be paired up in one of the 4*p* orbitals. The complete orbital diagram is:

$$[Ar] \quad \underset{4s^2}{\uparrow\downarrow} \qquad \underset{3d^{10}}{\uparrow\downarrow \;\; \uparrow\downarrow \;\; \uparrow\downarrow \;\; \uparrow\downarrow \;\; \uparrow\downarrow} \qquad \underset{4p^4}{\uparrow\downarrow \;\; \uparrow \;\; \uparrow}$$

PRACTICE EXERCISE

10. Write the electron configuration and the orbital diagram for iron (Fe).

> **Text Problems:** **7.80**, 7.82, **7.84**, 7.86

ANSWERS TO PRACTICE EXERCISES

1. $\lambda = 0.1224$ m

2. $E = 2.97 \times 10^{-19}$ J
    $E = 1.79 \times 10^5$ J/mol

3. $\lambda = 1.20 \times 10^{-6}$ m $= 1.20 \times 10^3$ nm

4. $n_i = 5$

5. $\lambda = 1.5 \times 10^{-10}$ m

6. $n = 4, l = 3, m_l = -3, -2, -1, 0, 1, 2, 3$

7. Total number of orbitals $= n^2 = 4^2 = 16$

8. (3, 2, −2, +1/2)   (3, 2, −2, −1/2)
    (3, 2, −1, +1/2)   (3, 2, −1, −1/2)
    (3, 2, 0, +1/2)   (3, 2, 0, −1/2)
    (3, 2, +1, +1/2)   (3, 2, +1, −1/2)
    (3, 2, +2, +1/2)   (3, 2, +2, −1/2)

9. 8 electrons

10. $[Ar]4s^2 3d^6$

$$[Ar] \quad \underset{4s^2}{\uparrow\downarrow} \qquad \underset{3d^6}{\uparrow\downarrow \;\; \uparrow \;\; \uparrow \;\; \uparrow \;\; \uparrow}$$

SOLUTIONS TO SELECTED TEXT PROBLEMS

7.8   Calculating the Frequency and Wavelength of an Electromagnetic Wave, Problem Type 1.

(a)

Step 1: $c = \lambda\nu$. Solve this equation algebraically for the frequency (ν).

$$\nu = \frac{c}{\lambda} \tag{7.1}$$

Step 2: Since the speed of light has units of m/s, we must convert the wavelength from units of nm to m.

$$456 \text{ nm} \times \frac{1 \times 10^{-9} \text{ m}}{1 \text{ nm}} = 4.56 \times 10^{-7} \text{ m}$$

Step 3: Calculate the value of ν by substituting the known quantities into Equation (7.1).

$$\nu = \frac{3.00 \times 10^8 \ \dfrac{\text{m}}{\text{s}}}{4.56 \times 10^{-7} \ \text{m}} = 6.58 \times 10^{14} \text{ s}^{-1} = \mathbf{6.58 \times 10^{14}} \text{ Hz}$$

(b)

Step 1: $c = \lambda\nu$. Solve this equation algebraically for the wavelength (λ).

$$\lambda = \frac{c}{\nu} \tag{7.2}$$

Step 2: Calculate the value of λ by substituting in the known quantities into Equation (7.2).

$$\lambda = \frac{3.00 \times 10^8 \ \dfrac{\text{m}}{\text{s}}}{2.45 \times 10^9 \ \dfrac{1}{\text{s}}} = 0.122 \text{ m}$$

Step 3: Convert the wavelength in units of meters to nanometers.

$$\lambda = 0.122 \text{ m} \times \frac{1 \text{ nm}}{1 \times 10^{-9} \text{ m}} = \mathbf{1.22 \times 10^8} \text{ nm}$$

7.10 $\text{time} = \dfrac{\text{distance}}{\text{speed}}$

A radio wave is an electromagnetic wave, which travels at the speed of light. The speed of light is in units of m/s, so let's convert distance from units of miles to meters.

$$? \text{ distance (m)} = (2.8 \times 10^7 \text{ mi}) \times \frac{1.61 \text{ km}}{1 \text{ mi}} \times \frac{1000 \text{ m}}{1 \text{ km}} = 4.5 \times 10^{10} \text{ m}$$

$$\mathbf{time} = \frac{\text{distance}}{\text{speed}} = \frac{4.5 \times 10^{10} \text{ m}}{3.00 \times 10^8 \text{ m/s}} = \mathbf{1.5 \times 10^2} \text{ s}$$

7.12   The wavelength is:

$$\lambda = \frac{1\ m}{1,650,763.73\ \text{wavelengths}} = 6.057802106 \times 10^{-7}\ m$$

$$\nu = \frac{c}{\lambda} = \frac{3.00 \times 10^8\ m/s}{6.057802106 \times 10^{-7}\ m} = \mathbf{4.95 \times 10^{14}\ s^{-1}}$$

7.16   Calculating the wavelength of electromagnetic radiation and calculating the energy of a photon, Problem Types 1 and 2.

(a)
Step 1: $c = \lambda\nu$. Solve this equation algebraically for the wavelength (λ).

$$\lambda = \frac{c}{\nu} \qquad\qquad\qquad\qquad (7.2)$$

Step 2: Calculate the value of λ by substituting the known quantities into Equation (7.2).

$$\lambda = \frac{3.00 \times 10^8\ \dfrac{m}{s}}{7.5 \times 10^{14}\ \dfrac{1}{s}} = \mathbf{4.0 \times 10^{-7}\ m = 4.0 \times 10^2\ nm}$$

(b) $E = h\nu$. Substitute the frequency (ν) into this equation to solve for the energy of a single photon associated with this frequency.

$$E = h\nu = (6.63 \times 10^{-34}\ J \cdot s)(7.5 \times 10^{14}\ \tfrac{1}{s}) = \mathbf{5.0 \times 10^{-19}\ J}$$

7.18   The energy given in this problem is for *1 mole* of photons. To apply $E = h\nu$, we must divide by Avogadro's number. The energy of one photon is:

$$E = \frac{1.0 \times 10^3\ kJ}{1\ mol} \times \frac{1\ mol}{6.022 \times 10^{23}\ \text{photons}} \times \frac{1000\ J}{1\ kJ} = 1.7 \times 10^{-18}\ J/\text{photon}$$

The wavelength of this photon can be found using the relationship $E = hc/\lambda$.

$$\lambda = \frac{hc}{E} = \frac{(6.63 \times 10^{-34}\ J \cdot s)\left(3.00 \times 10^8\ \dfrac{m}{s}\right)}{1.7 \times 10^{-18}\ J} \times \left(\frac{1\ nm}{1 \times 10^{-9}\ m}\right) = \mathbf{1.2 \times 10^2\ nm}$$

The radiation is in the ultraviolet region (see Figure 7.4 of the text).

7.20   **(a)** $\lambda = \dfrac{c}{\nu}$

$$\lambda = \frac{3.00 \times 10^8\ \dfrac{m}{s}}{8.11 \times 10^{14}\ \dfrac{1}{s}} = 3.70 \times 10^{-7}\ m = \mathbf{3.70 \times 10^2\ nm}$$

(b)   Checking Figure 7.4 of the text, you should find that the visible region of the spectrum runs from 400 to 700 nm. 370 nm is in the **ultraviolet** region of the spectrum.

(c) $E = h\nu$. Substitute the frequency (ν) into this equation to solve for the energy of one quantum associated with this frequency.

$$E = h\nu = (6.63 \times 10^{-34} \text{ J} \cdot \text{s})(8.11 \times 10^{14} \frac{1}{\text{s}}) = 5.38 \times 10^{-19} \text{ J}$$

7.26   The emitted light could be analyzed by passing it through a prism.

7.28   Excited atoms of the chemical elements emit the same characteristic frequencies or lines in a terrestrial laboratory, in the sun, or in a star many light-years distant from earth.

7.30   We use more accurate values of h and c for this problem.

$$E = h\nu = \frac{hc}{\lambda} = \frac{(6.6256 \times 10^{-34} \text{ J} \cdot \text{s})(2.998 \times 10^8 \text{ m/s})}{656.3 \times 10^{-9} \text{ m}} = 3.027 \times 10^{-19} \text{ J}$$

7.32   Calculating the Energy, Wavelength, or Frequency in the Emission Spectrum of a Hydrogen Atom, Problem Type 3.

Step 1: Using the following equation, we can calculate the energy change for the transition.

$$\Delta E = R_H \left(\frac{1}{n_i^2} - \frac{1}{n_f^2} \right)$$

$$\Delta E = (2.18 \times 10^{-18} \text{ J}) \left(\frac{1}{4^2} - \frac{1}{2^2} \right)$$

$$\Delta E = -4.09 \times 10^{-19} \text{ J}$$

Step 2: The negative sign for ΔE indicates that energy is released to the surroundings during the emission process. To calculate the wavelength, we omit the minus sign for ΔE because the frequency of the photon must be positive. We know that

$$\Delta E = h\nu$$

Rearranging the equation and substituting in the known values,

$$\nu = \frac{\Delta E}{h} = \frac{(4.09 \times 10^{-19} \text{ J})}{(6.63 \times 10^{-34} \text{ J} \cdot \text{s})} = 6.17 \times 10^{14} \text{ s}^{-1} \text{ or Hz}$$

Step 3: We also know that $\lambda = \dfrac{c}{\nu}$. Substituting the frequency calculated above into this equation gives:

$$\lambda = \frac{\left(3.00 \times 10^8 \ \frac{\text{m}}{\text{s}} \right)}{\left(6.17 \times 10^{14} \ \frac{1}{\text{s}} \right)} = 4.86 \times 10^{-7} \text{ m} = 486 \text{ nm}$$

7.34 $\Delta E = R_{\mathrm{H}}\left(\dfrac{1}{n_{\mathrm{i}}^{2}} - \dfrac{1}{n_{\mathrm{f}}^{2}}\right)$

n_{f} is given in the problem and R_{H} is a constant, but we need to calculate ΔE. The photon energy is:

$$E = \frac{hc}{\lambda} = \frac{(6.63 \times 10^{-34}\ \mathrm{J \cdot s})(3.00 \times 10^{8}\ \mathrm{m/s})}{434 \times 10^{-9}\ \mathrm{m}} = 4.58 \times 10^{-19}\ \mathrm{J}$$

Since this is an emission process, the energy change ΔE must be negative, or -4.58×10^{-19} J.

Substitute ΔE into the following equation, and solve for n_{i}.

$$\Delta E = R_{\mathrm{H}}\left(\frac{1}{n_{\mathrm{i}}^{2}} - \frac{1}{n_{\mathrm{f}}^{2}}\right)$$

$$-4.58 \times 10^{-19}\ \mathrm{J} = (2.18 \times 10^{-18}\ \mathrm{J})\left(\frac{1}{n_{\mathrm{i}}^{2}} - \frac{1}{2^{2}}\right)$$

$$\frac{1}{n_{\mathrm{i}}^{2}} = \left(\frac{-4.58 \times 10^{-19}\ \mathrm{J}}{2.18 \times 10^{-18}\ \mathrm{J}}\right) + \frac{1}{2^{2}} = -0.210 + 0.250 = 0.040$$

$$n_{\mathrm{i}} = \frac{1}{\sqrt{0.040}} = \mathbf{5}$$

7.40   The De Broglie Equation: Calculating the Wavelengths of Particles, Problem Type 4.

Step 1: The equation needed to solve this problem is the De Broglie equation.

$$\lambda = \frac{h}{mu}$$

Substitute the known values into the equation to solve for the wavelength of the proton.

$$\lambda = \frac{h}{mu} = \frac{(6.63 \times 10^{-34}\ \mathrm{J \cdot s}) \times \left(\dfrac{1\ \mathrm{kg \cdot m^{2}/s^{2}}}{1\ \mathrm{J}}\right)}{(1.673 \times 10^{-27}\ \mathrm{kg})(2.90 \times 10^{8}\ \mathrm{m/s})} = 1.37 \times 10^{-15}\ \mathrm{m}$$

Step 2: Convert wavelength in units of meters to nanometers.

$$\lambda = 1.37 \times 10^{-15}\ \mathrm{m} \times \frac{1\ \mathrm{nm}}{1 \times 10^{-9}\ \mathrm{m}} = \mathbf{1.37 \times 10^{-6}\ nm}$$

7.42   First, we convert mph to m/s.

$$\frac{35\ \mathrm{mi}}{1\ \mathrm{hr}} \times \frac{1.61\ \mathrm{km}}{1\ \mathrm{mi}} \times \frac{1000\ \mathrm{m}}{1\ \mathrm{km}} \times \frac{1\ \mathrm{hr}}{3600\ \mathrm{s}} = 16\ \mathrm{m/s}$$

$$\lambda = \frac{h}{mu} = \frac{(6.63 \times 10^{-34}\ \mathrm{J \cdot s}) \times \left(\dfrac{1\ \mathrm{kg \cdot m^{2}/s^{2}}}{1\ \mathrm{J}}\right)}{(2.5 \times 10^{-3}\ \mathrm{kg})(16\ \mathrm{m/s})} = 1.7 \times 10^{-32}\ \mathrm{m} = \mathbf{1.7 \times 10^{-23}\ nm}$$

7.54   Quantum Numbers, Problem Type 5.

When $n = 3$, there are three possible values of l: 0, 1, and 2, corresponding to the s, p, and d orbitals, respectively. The values of m_l for each l value are:

$l = 0$: $m_l = 0$ $l = 1$: $m_l = -1, 0, 1$ $l = 2$: $m_l = -2, -1, 0, 1, 2$

7.56   Labeling an atomic orbital and counting the number of orbitals associated with a principal quantum number, Problem Types 5A and 5B.

(a)   The number given in the designation of the subshell is the principal quantum number, so in this case $n = 4$. For p orbitals, $l = 1$. m_l can have integer values from $-l$ to $+l$. Therefore, m_l can be -1, 0, and $+1$.

The three values for m_l correspond to the *three p* orbitals.

Following the same reasoning as part **(a)**

(b) $3d$: $n = 3, l = 2, m_l = 2, 1, 0, -1,$ or -2 (5 orbitals)

(c) $3s$: $n = 3, l = 0, m_l = 0$ (1 orbital)

(d) $5f$: $n = 5, l = 3, m_l = 3, 2, 1, 0, -1, -2,$ or -3 (7 orbitals)

7.58   The two orbitals are identical in size, shape, and energy. They differ only in their orientation with respect to each other.

Can you assign a specific value of the magnetic quantum number to these orbitals? What are the allowed values of the magnetic quantum number for the $2p$ subshell?

7.60   For $n = 6$, the allowed values of l are 0, 1, 2, 3, 4, and 5 $[l = 0, ..., (n - 1)$, integer values]. These l values correspond to the $6s$, $6p$, $6d$, $6f$, $6g$, and $6h$ subshells. These subshells each have 1, 3, 5, 7, 9, and 11 orbitals, respectively (number of orbitals $= 2l + 1$).

7.62

n value	orbital sum	total number of electrons
1	1	2
2	$1 + 3 = 4$	8
3	$1 + 3 + 5 = 9$	18
4	$1 + 3 + 5 + 7 = 16$	32
5	$1 + 3 + 5 + 7 + 9 = 25$	50
6	$1 + 3 + 5 + 7 + 9 + 11 = 36$	72

In each case the total number of orbitals is just the square of the n value (n^2). The total number of electrons is $2n^2$.

7.64   The electron configurations for the elements are

(a)   N: $1s^2 2s^2 2p^3$ There are three p-type electrons.

(b)   Si: $1s^2 2s^2 2p^6 3s^2 3p^2$ There are six s-type electrons.

(c)   S: $1s^2 2s^2 2p^6 3s^2 3p^4$ There are no d-type electrons.

7.66   In the many-electron atom, the $3p$ orbital electrons are more effectively shielded by the inner electrons of the atom (that is, the $1s$, $2s$, and $2p$ electrons) than the $3s$ electrons. The $3s$ orbital is said to be more "penetrating" than the $3p$ and $3d$ orbitals. In the hydrogen atom there is only one electron, so the $3s$, $3p$, and $3d$ orbitals have the same energy.

7.68   **(a)** $2s < 2p$ **(b)** $3p < 3d$ **(c)** $3s < 4s$ **(d)** $4d < 5f$

7.80   Writing Electron Configurations, Problem Type 6.

For aluminum, there are not enough electrons in the $2p$ subshell. (The $2p$ subshell holds six electrons.) The number of electrons (13) is correct. The electron configuration should be $1s^2 2s^2 2p^6 3s^2 3p^1$. The configuration shown might be an excited state of an aluminum atom.

For boron, there are too many electrons. (Boron only has five electrons.) The electron configuration should be $1s^2 2s^2 2p^1$. What would be the electric charge of a boron ion with the electron arrangement given in the problem?

For fluorine, there are also too many electrons. (Fluorine only has nine electrons.) The configuration shown is that of the F^- ion. The correct electron configuration is $1s^2 2s^2 2p^5$.

7.82   You should write the electron configurations for each of these elements to answer this question. In some cases, an orbital diagram may be helpful.

B: [He]$2s^2 2p^1$ (1 unpaired electron)   Ne: (0 unpaired electrons, Why?)

P: [Ne]$3s^2 3p^3$ (3 unpaired electrons)   Sc: [Ar]$4s^2 3d^1$ (1 unpaired electron)

Mn: [Ar]$4s^2 3d^5$ (5 unpaired electrons)   Se: [Ar]$4s^2 3d^{10} 4p^4$ (2 unpaired electrons)

Kr: (0 unpaired electrons)   Fe: [Ar]$4s^2 3d^6$ (4 unpaired electrons)

Cd: [Kr]$5s^2 4d^{10}$ (0 unpaired electrons)   I: [Kr]$5s^2 4d^{10} 5p^5$ (1 unpaired electron)

Pb: [Xe]$6s^2 4f^{14} 5d^{10} 6p^2$ (2 unpaired electrons)

7.84   Writing Electron Configurations, Problem Type 6.

Step 1: Germanium (Ge) has 32 electrons. These electrons need to be placed in atomic orbitals.

Step 2: The noble gas element that most nearly proceeds Se is Ar. Therefore, the *noble gas core* is [Ar]. This core accounts for 18 electrons.

Step 3: See Figure 7.24 of your text to check the order of filling subshells past the Ar noble gas core. You should find that the order of filling is $4s$, $3d$, then $4p$. There are 14 remaining electrons to distribute among these orbitals.

The $4s$ orbital can hold *2* electrons. Each of the five $3d$ orbitals can hold *2* electrons for a total of *10* electrons. This leaves *2* electrons to fill the $4p$ orbitals.

The electrons configuration for Ge is:

$$[Ar]\, 4s^2 3d^{10} 4p^2$$

You should follow the same reasoning for the remaining atoms.

Fe: [Ar]$4s^2 3d^6$ Zn: [Ar]$4s^2 3d^{10}$ Ni: [Ar]$4s^2 3d^8$

W: [Xe]$6s^2 4f^{14} 5d^4$ Tl: [Xe]$6s^2 4f^{14} 5d^{10} 6p^1$

7.86 $\underset{3s^2}{\uparrow\downarrow}$ $\underset{3p^3}{\uparrow\;\uparrow\;\uparrow}$ $\underset{3s^2}{\uparrow\downarrow}$ $\underset{3p^4}{\uparrow\downarrow\;\uparrow\;\uparrow}$ $\underset{3s^2}{\uparrow\downarrow}$ $\underset{3p^5}{\uparrow\downarrow\;\uparrow\downarrow\;\uparrow}$

S^+ (5 valence electrons)   S (6 valence electrons) S^- (7 valence electrons)
3 unpaired electrons   2 unpaired electrons   1 unpaired electron

S^+ has the most unpaired electrons

7.92   The ground state electron configuration of Tc is: $[Kr]5s^2 4d^5$.

7.94   Part **(b)** is correct in the view of contemporary quantum theory. Bohr's explanation of emission and absorption line spectra appears to have universal validity. Parts **(a)** and **(c)** are artifacts of Bohr's early planetary model of the hydrogen atom and are *not* considered to be valid today.

7.96   **(a)**   With $n = 2$, there are n^2 orbitals $= 2^2 = 4$. $m_s = +1/2$, specifies *1* electron per orbital, for a total of **4 electrons**.

(b) $n = 4$ and $m_l = +1$, specifies one orbital in each subshell with $l = 1, 2$, or 3 (i.e., a $4p$, $4d$, and $4f$ orbital). Each of the three orbitals holds 2 electrons for a total of **6 electrons**.

(c)   If $n = 3$ and $l = 2$, m_l has the values 2, 1, 0, –1, or –2. Each of the five orbitals can hold 2 electrons for a total of **10 electrons** (2 e^- in each of the five $3d$ orbitals).

(d)   If $n = 2$ and $l = 0$, then m_l can only be zero. $m_s = -1/2$ specifies 1 electron in this orbital for a total of **1 electron** (one e^- in the $2s$ orbital).

(e) $n = 4$, $l = 3$ and $m_l = -2$, specifies one $4f$ orbital. This orbital can hold **2 electrons**.

7.98   The wave properties of electrons are used in the operation of an electron microscope.

7.100   **(a)**   First convert 100 mph to units of m/s.

$$\frac{100 \text{ mi}}{1 \text{ hr}} \times \frac{1 \text{ hr}}{3600 \text{ s}} \times \frac{1.609 \text{ km}}{1 \text{ mi}} \times \frac{1000 \text{ m}}{1 \text{ km}} = 44.7 \text{ m/s}$$

Using the de Broglie equation:

$$\lambda = \frac{h}{mu} = \frac{(6.63 \times 10^{-34} \text{ J} \cdot \text{s}) \times \left(\dfrac{1 \text{ kg} \cdot \text{m}^2/\text{s}^2}{1 \text{ J}}\right)}{(0.141 \text{ kg})(44.7 \text{ m/s})} = 1.05 \times 10^{-34} \text{ m} = \mathbf{1.05 \times 10^{-25} \text{ nm}}$$

(b)   The average mass of a hydrogen atom is:

$$1.008 \frac{\text{g}}{\text{mol}} \times \frac{1 \text{ mol}}{6.022 \times 10^{23} \text{ atoms}} = 1.674 \times 10^{-24} \text{ g/H atom} = 1.674 \times 10^{-27} \text{ kg}$$

$$\lambda = \frac{h}{mu} = \frac{(6.63 \times 10^{-34} \text{ J} \cdot \text{s}) \times \left(\dfrac{1 \text{ kg} \cdot \text{m}^2/\text{s}^2}{1 \text{ J}}\right)}{(1.674 \times 10^{-27} \text{ kg})(44.7 \text{ m/s})} = 8.86 \times 10^{-9} \text{ m} = \mathbf{8.86 \text{ nm}}$$

7.102   **(a)**   First, we can calculate the energy of a single photon with a wavelength of 633 nm.

$$E = \frac{hc}{\lambda} = \frac{(6.63 \times 10^{-34} \text{ J} \cdot \text{s})(3.00 \times 10^8 \text{ m/s})}{633 \times 10^{-9} \text{ m}} = 3.14 \times 10^{-19} \text{ J}$$

The number of photons produced in a 0.376 J pulse is:

$$0.376 \text{ J} \times \frac{1 \text{ photon}}{3.14 \times 10^{-19} \text{ J}} = \mathbf{1.20 \times 10^{18} \text{ photons}}$$

(b)   Since a 1 W = 1 J/s, the power delivered per a 1.00×10^{-9} s pulse is:

$$\frac{0.376 \text{ J}}{1.00 \times 10^{-9} \text{ s}} = 3.76 \times 10^8 \text{ J/s} = \mathbf{3.76 \times 10^8 \text{ W}}$$

Compare this with the power delivered by a 100-W light bulb!

7.104   First, let's find the energy needed to photodissociate one water molecule.

$$\frac{285.8 \text{ kJ}}{1 \text{ mol}} \times \frac{1 \text{ mol}}{6.022 \times 10^{23} \text{ molecules}} = 4.746 \times 10^{-22} \text{ kJ/molecule} = 4.746 \times 10^{-19} \text{ J/molecule}$$

The maximum wavelength of a photon that would provide the above energy is:

$$\lambda = \frac{hc}{E} = \frac{(6.63 \times 10^{-34} \text{ J} \cdot \text{s})(3.00 \times 10^8 \text{ m/s})}{4.746 \times 10^{-19} \text{ J}} = \mathbf{4.19 \times 10^{-7} \text{ m} = 419 \text{ nm}}$$

This wavelength is in the visible region of the electromagnetic spectrum. Since water is continuously being struck by visible radiation *without* decomposition, it seems unlikely that photodissociation of water by this method is feasible.

7.106   Since 1 W = 1 J/s, the energy output of the light bulb in 1 second is 75 J. The actual energy converted to visible light is 15 percent of this value or 11 J.

First, we need to calculate the energy of one 550 nm photon. Then, we can determine how many photons are needed to provide 11 J of energy.

The energy of one 550 nm photon is:

$$E = \frac{hc}{\lambda} = \frac{(6.63 \times 10^{-34} \text{ J} \cdot \text{s})(3.00 \times 10^8 \text{ m/s})}{550 \times 10^{-9} \text{ m}} = 3.62 \times 10^{-19} \text{ J/photon}$$

The number of photons needed to produce 11 J of energy is:

$$11 \text{ J} \times \frac{1 \text{ photon}}{3.62 \times 10^{-19} \text{ J}} = \mathbf{3.0 \times 10^{19} \text{ photons}}$$

7.108   The Balmer series corresponds to transitions to the $n = 2$ level.

For He^+:

$$\Delta E = R_{He^+}\left(\frac{1}{n_i^2} - \frac{1}{n_f^2}\right) \qquad\qquad \lambda = \frac{hc}{\Delta E} = \frac{(6.63 \times 10^{-34} \text{ J} \cdot \text{s})(3.00 \times 10^8 \text{ m/s})}{\Delta E}$$

For the transition, $n = 3 \rightarrow 2$

$$\Delta E = (8.72 \times 10^{-18} \text{ J})\left(\frac{1}{3^2} - \frac{1}{2^2}\right) = -1.21 \times 10^{-18} \text{ J} \qquad \lambda = \frac{1.99 \times 10^{-25} \text{ J} \cdot \text{m}}{1.21 \times 10^{-18} \text{ J}} = 1.64 \times 10^{-7} \text{ m} = \mathbf{164 \text{ nm}}$$

For the transition, $n = 4 \rightarrow 2$, $\Delta E = -1.64 \times 10^{-18}$ J $\qquad \lambda = \mathbf{121 \text{ nm}}$

For the transition, $n = 5 \rightarrow 2$, $\Delta E = -1.83 \times 10^{-18}$ J $\qquad \lambda = \mathbf{109 \text{ nm}}$

For the transition, $n = 6 \rightarrow 2$, $\Delta E = -1.94 \times 10^{-18}$ J $\qquad \lambda = \mathbf{103 \text{ nm}}$

For H, the calculations are identical to those above, except the Rydberg constant for H is 2.18×10^{-18} J.

For the transition, $n = 3 \rightarrow 2$, $\Delta E = -3.03 \times 10^{-19}$ J $\qquad \lambda = \mathbf{657 \text{ nm}}$

For the transition, $n = 4 \rightarrow 2$, $\Delta E = -4.09 \times 10^{-19}$ J $\qquad \lambda = \mathbf{487 \text{ nm}}$

For the transition, $n = 5 \rightarrow 2$, $\Delta E = -4.58 \times 10^{-19}$ J $\qquad \lambda = \mathbf{434 \text{ nm}}$

For the transition, $n = 6 \rightarrow 2$, $\Delta E = -4.84 \times 10^{-19}$ J $\qquad \lambda = \mathbf{411 \text{ nm}}$

All the Balmer transitions for He$^+$ are in the ultraviolet region; whereas, the transitions for H are all in the visible region. Note the negative sign for energy indicating that a photon has been emitted.

7.110   First, we need to calculate the energy of one 600 nm photon. Then, we can determine how many photons are needed to provide 4.0×10^{-17} J of energy.

The energy of one 600 nm photon is:

$$E = \frac{hc}{\lambda} = \frac{(6.63 \times 10^{-34} \text{ J} \cdot \text{s})(3.00 \times 10^8 \text{ m/s})}{600 \times 10^{-9} \text{ m}} = 3.32 \times 10^{-19} \text{ J/photon}$$

The number of photons needed to produce 4.0×10^{-17} J of energy is:

$$4.0 \times 10^{-17} \text{ J} \times \frac{1 \text{ photon}}{3.32 \times 10^{-19} \text{ J}} = \mathbf{1.2 \times 10^2 \text{ photons}}$$

7.112   A "blue" photon (shorter wavelength) is higher energy than a "yellow" photon. For the same amount of energy delivered to the metal surface, there must be fewer "blue" photons than "yellow" photons. Thus, the yellow light would eject more electrons since there are more "yellow" photons. Since the "blue" photons are of higher energy, blue light will eject electrons with greater kinetic energy.

7.114   The excited atoms are still neutral, so the total number of electrons is the same as the atomic number of the element.

(a)   He (2 electrons), $1s^2$ (b)   N (7 electrons), $1s^2 2s^2 2p^3$

(c)   Na (11 electrons), $1s^2 2s^2 2p^6 3s^1$ (d)   As (33 electrons), $[\text{Ar}]4s^2 3d^{10} 4p^3$

(e)   Cl (17 electrons), $[\text{Ne}]3s^2 3p^5$

7.116   Rutherford and his coworkers might have discovered the wave properties of electrons.

7.118   The wavelength of a He atom can be calculated using the de Broglie equation. First, we need to calculate the root-mean-square speed using Equation 5.12 from the text.

$$u_{\text{rms}} = \sqrt{\frac{3\left(8.314\ \dfrac{J}{K \cdot mol}\right)(273 + 20)K}{4.003 \times 10^{-3}\ kg\,/\,mol}} = 1.35 \times 10^3\ m/s$$

To calculate the wavelength, we also need the mass of a He atom in kg.

$$\frac{4.003 \times 10^{-3}\ kg\ He}{1\ mol\ He} \times \frac{1\ mol\ He}{6.022 \times 10^{23}\ He\ atoms} = 6.647 \times 10^{-27}\ kg/atom$$

Finally, the wavelength of a He atom is:

$$\lambda = \frac{h}{mu} = \frac{(6.63 \times 10^{-34}\ J \cdot s)}{(6.647 \times 10^{-27}\ kg)(1.35 \times 10^3\ m/s)} = \mathbf{7.39 \times 10^{-11}\ m} = \mathbf{7.39 \times 10^{-2}\ nm}$$

7.120   **(a)**   **False**. $n = 2$ is the first excited state.

   (b)   **False**. In the $n = 4$ state, the electron is (on average) further from the nucleus and hence easier to remove.

   (c)   **True**.

   (d)   **False**. The $n = 4$ to $n = 1$ transition is a higher energy transition, which corresponds to a *shorter* wavelength.

   (e)   **True**.

7.122   We use Heisenberg's uncertainty principle with the equality sign to calculate the minimum uncertainty.

$$\Delta x \Delta p = \frac{h}{4\pi}$$

The momentum (p) is equal to the mass times the velocity.

$$p = mu \qquad \text{or} \qquad \Delta p = m\Delta u$$

We can write:

$$\Delta p = m\Delta u = \frac{h}{4\pi\Delta x}$$

Finally, the uncertainty in the velocity of the oxygen molecule is:

$$\Delta u = \frac{h}{4\pi m\Delta x} = \frac{(6.63 \times 10^{-34}\ J \cdot s)}{4\pi(5.3 \times 10^{-26}\ kg)(5.0 \times 10^{-5}\ m)} = \mathbf{2.0 \times 10^{-5}\ m/s}$$

7.124   The Pauli exclusion principle states that no two electrons in an atom can have the same four quantum numbers. In other words, only two electrons may exist in the same atomic orbital, and these electrons must have opposite spins. **(a)** and **(f)** violate the Pauli exclusion principle.

   Hund's rule states that the most stable arrangement of electrons in subshells is the one with the greatest number of parallel spins. **(b)**, **(d)**, and **(e)** violate Hund's rule.

7.126 As an estimate, we can equate the energy for ionization ($Fe^{13+} \rightarrow Fe^{14+}$) to the average kinetic energy ($3/2RT$) of the ions.

$$\frac{3.5 \times 10^4 \text{ kJ}}{1 \text{ mol}} \times \frac{1000 \text{ J}}{1 \text{ kJ}} = 3.5 \times 10^7 \text{ J}$$

$$IE = \frac{3}{2}RT$$

$$3.5 \times 10^7 \text{ J/mol} = \frac{3}{2}(8.314 \text{ J/mol} \cdot K)T$$

$$T = 2.8 \times 10^6 \text{ K}$$

The actual temperature can be, and most probably is, higher than this.

7.128 Looking at the deBroglie equation $\lambda = \dfrac{h}{mu}$, the mass of an N_2 molecule (in kg) and the velocity of an N_2 molecule (in m/s) is needed to calculate the deBroglie wavelength of N_2.

First, calculate the root-mean-square velocity of N_2.

$\mathcal{M}(N_2) = 28.02$ g/mol = 0.02802 kg/mol

$$u_{rms} (N_2) = \sqrt{\frac{(3)\left(8.314 \dfrac{\text{J}}{\text{mol} \cdot K}\right)(300 \text{ K})}{\left(0.02802 \dfrac{\text{kg}}{\text{mol}}\right)}} = 516.8 \text{ m/s}$$

Second, calculate the mass of one N_2 molecule in kilograms.

$$\frac{28.02 \text{ g } N_2}{1 \text{ mol } N_2} \times \frac{1 \text{ mol } N_2}{6.022 \times 10^{23} \text{ } N_2 \text{ molecules}} \times \frac{1 \text{ kg}}{1000 \text{ g}} = 4.653 \times 10^{-26} \text{ kg/molecule}$$

Now, substitute the mass of an N_2 molecule and the root-mean-square velocity into the deBroglie equation to solve for the deBroglie wavelength of an N_2 molecule.

$$\lambda = \frac{h}{mu} = \frac{(6.626 \times 10^{-34} \text{ J} \cdot s)}{(4.653 \times 10^{-26} \text{ kg})(516.8 \text{ m/s})} = 2.755 \times 10^{-11} \text{ m}$$

CHAPTER 8
PERIODIC RELATIONSHIPS
AMONG THE ELEMENTS

PROBLEM-SOLVING STRATEGIES AND TUTORIAL SOLUTIONS

TYPES OF PROBLEMS

Problem Type 1: Writing an Electron Configuration and Identifying an Element (see Problem Type 7.6).

Problem Type 2: Electron Configurations of Cations and Anions.

Problem Type 3: Comparing the Sizes of Atoms.

Problem Type 4: Comparing the Sizes of Ions.

Problem Type 5: Comparing the Ionization Energies of Elements.

Problem Type 6: Electron Affinity.

PROBLEM TYPE 1: WRITING AN ELECTRON CONFIGURATION AND IDENTIFYING AN ELEMENT

See Problem Type 7.6 for information on writing electron configurations. When examining the electron configurations of the elements in a particular group, a clear pattern emerges. Elements in the same group have the same electron configuration of their *outer* electrons. The outer electrons of an atom, which are those involved in chemical bonding, are called **valence electrons**. Having the same number of valence electrons accounts for the similarities in chemical behavior among the elements within each of these groups.

In regards to identifying an element, your text considers a number of items. First, according to the type of subshell being filled, the elements can be divided into categories--the representative elements, the noble gases, the transition metals, the lanthanides, and the actinides.

> The **representative elements** are the elements in Groups 1A through 7A, all of which have *incompletely* filled *s* or *p* subshells of the highest principal quantum number.
>
> The **noble gases**, with the exception of helium, all have a *completely* filled *p* subshell.
>
> The **transition metals** are the elements in Groups 1B and 3B through 8B that have *incompletely* filled *d* subshells, or readily produce cations with *incompletely* filled *d* subshells.
>
> The **lanthanides** and **actinides** are sometimes called *f*-block transition elements because they have *incompletely* filled *f* subshells.

Second, you might be able to classify the element as a metal, nonmetal, or metalloid. However, sometimes you need to be careful in making this classification. For example in Group 4A, carbon is a nonmetal, silicon and germanium are metalloids, and tin and lead are metals.

Third, the problem might ask whether the element is paramagnetic or diamagnetic. In Chapter 7, you learned that an element that contains unpaired electrons is *paramagnetic*, and an element in which all electrons are paired is *diamagnetic*.

EXAMPLE 8.1

A neutral atom of a certain element has 19 electrons. Without consulting a periodic table, answer the following questions: (a) What is the electron configuration of the element? (b) How should the element be classified? (c) Are the atoms of this element diamagnetic or paramagnetic?

(a)   The electron configuration is $1s^2 2s^2 2p^6 3s^2 3p^6 4s^1$ or $[Ar]4s^1$. See Problem Type 7.6 if you are having difficulty writing electron configurations.

(b)   Since the $4s$ subshell is not completely filled, this is a *representative element*. Without consulting a periodic table, you might know that the alkali metal family has one valence electron in the s subshell. You could then further classify this element as an *alkali metal*.

(c)   There is *one* unpaired electron in the s subshell. Therefore, the atoms of this element are paramagnetic.

> **Tip:** Any atom that contains an odd number of electrons *must* be paramagnetic. Why?

PRACTICE EXERCISE

1.   A neutral atom of a certain element has 36 electrons. Without consulting a periodic table, answer the following questions: (a) What is the electron configuration of the element? (b) How should the element be classified? (c) Are the atoms of this element diamagnetic or paramagnetic?

> **Text Problems: 8.20**, 8.22, 8.24

PROBLEM TYPE 2: ELECTRON CONFIGURATIONS OF CATIONS AND ANIONS

Writing an electron configuration for a cation or anion requires only a slight extension of the method used for neutral atoms. Let's group the ions in two categories.

(1)   Ions Derived from Representative Elements

(a)   In the formation of a **cation** from the neutral atom of a representative element, one or more electrons are removed from the highest occupied n shell. The number of electrons removed is equal to the charge of the ion.

Examples: Following are the electron configurations of some neutral atoms and their corresponding cations.

Mg: $[Ne]3s^2$ Mg^{2+}: $[Ne]$

K: $[Ar]4s^1$ K^+: $[Ar]$

Note: Each ion has a stable noble gas electron configuration.

(b)   In the formation of an **anion** from the neutral atom of a representative element, one or more electrons are added to the highest partially filled n shell. The number of electrons added is equal to the charge of the ion.

Examples: Following are the electron configurations of some neutral atoms and their corresponding anions.

O: $[He]2s^2 2p^4$ O^{2-}: $[Ne]$

Cl: $[Ne]3s^2 3p^5$ Cl^-: $[Ar]$

Note: Each ion has a stable noble gas electron configuration.

(2) Cations Derived from Transition Metals

In a neutral transition metal atom, the ns orbital is filled before the $(n-1)d$ orbitals. See Figure 7.24 of your text. However, in a transition metal ion, the $(n-1)d$ orbitals are more stable than the ns orbital. Hence, when a cation is formed from an atom of a transition metal, electrons are *always* removed first from the ns orbital and then from the $(n-1)d$ orbitals if necessary. See Section 8.2 of your text for a more complete discussion.

Examples: Following are the electron configurations of some neutral transition metals and their corresponding cations.

Fe: $[Ar]4s^2 3d^6$ Fe^{2+}: $[Ar]3d^6$

Fe: $[Ar]4s^2 3d^6$ Fe^{3+}: $[Ar]3d^5$

Mn: $[Ar]4s^2 3d^5$ Mn^{7+}: $[Ar]$

PRACTICE EXERCISE

2.   Write electron configurations for the following ions: N^{3-}, Ba^{2+}, Zn^{2+}, and, V^{5+}.

Text Problems: 8.26, **8.28**, 8.30, 8.32

PROBLEM TYPE 3: COMPARING THE SIZES OF ATOMS

The general periodic trends in atomic size are:

(1)   Moving from left to right across a row (period) of the periodic table, the atomic radius *decreases* due to an increase in effective nuclear charge.

(2)   Moving down a column (group) of the periodic table, the atomic radius *increases* since the orbital size increases with increasing principal quantum number.

For a more detailed discussion, see Section 8.3 of your text.

EXAMPLE 8.2
Which one of the following has the smallest atomic radius? (a) Li, (b) Na, (c) Be, (d) Mg

Atomic radii *increase* going down a group of the periodic table; therefore, Li atoms have a smaller radius than Na atoms, and Be atoms have a smaller radius than Mg atoms. We have narrowed our choices to Li and Be. Atomic radii decrease when moving from left to right across a row of the periodic table. Thus, **Be atoms** are smaller than Li atoms as well as the other choices given.

PRACTICE EXERCISE

3.   Which atom should have the largest atomic radius? (a) Br, (b) Cl, (c) Se, (d) Ge, (e) C.

Text Problems: **8.38**, 8.40, 8.42

PROBLEM TYPE 4: COMPARING THE SIZES OF IONS

When a neutral atom is converted to an ion, we expect a change in size. When forming an *anion* from an atom, its size (or radius) *increases*, because the nuclear charge remains the same but the repulsion resulting from the additional electron(s) enlarges the domain of the electron cloud. Conversely, when forming a *cation* from an atom, its size *decreases*, because removing one or more electrons reduces electron-electron repulsion but the nuclear charge remains the same. Thus, the electron cloud shrinks.

The general periodic trends in ionic size are:

(1)   Similar to atomic size, ionic radii increase when moving down a column (group) of the periodic table.

(2)   The next trend only applies to **isoelectronic ions**. Isoelectronic ions have the same number of electrons and the same electron configuration.

   (a)   Cations are smaller than anions. For example, Mg^{2+} is smaller than O^{2-}. Both ions have the same number of electrons (10), but Mg (Z = 12) has more protons then O (Z = 8). The larger effective nuclear charge of Mg^{2+} results in a smaller radius.

   (b)   Considering only *isoelectronic cations*, the radius of a tripositive ion is smaller than the radius of a dipositive ion, which is smaller than the radius of a unipositive ion. For example, Al^{3+} is smaller than Mg^{2+}, which is smaller than Na^{+}. Each of the cations has 10 electrons, but Al^{3+} has 13 protons, Mg^{2+} has 12 protons, and Na^{+} has 11 protons. As the effective nuclear charge increases, the ionic radius decreases.

   (c)   Considering only *isoelectronic anions*, the radius increases as we go from an ion with a uninegative charge, to one with a dinegative charge, and so on. For example, N^{3-} is larger than O^{2-}, which is larger than F^{-}. Each of the anions has 10 electrons, but N^{3-} has 7 protons, O^{2-} has 8 protons, and F^{-} has 9 protons. As the effective nuclear charge increases, ionic radius decreases.

EXAMPLE 8.3

In each of the following pairs, choose the ion with the *larger* ionic radius: (a) K^{+} and Na^{+}; (b) K^{+} and Ca^{2+}; (c) K^{+} and Cl^{-}.

(a) K^{+} and Na^{+} are in the same group of the periodic table (Group 1A, alkali metals). As you proceed down a group of the periodic table, ionic radii increase. This occurs because orbital size increases with increasing principal quantum number. Therefore, **K^{+}** has the larger ionic radius.

(b) K^{+} and Ca^{2+} are isoelectronic cations. Both ions have 18 electrons, but Ca^{2+} has one more proton than K^{+}. The greater effective nuclear charge of Ca^{2+}, pulls the 18 electrons more strongly toward the nucleus. Therefore, **K^{+}** has the larger ionic radius. For isoelectronic cations, the radius of a dipositive ion is smaller than the radius of a unipositive ion.

(c) K^{+} and Cl^{-} are isoelectronic species. Both ions have 18 electrons, but K^{+} has the greater effective nuclear charge, with 19 protons compared to 17 protons for Cl^{-}. Therefore, **Cl^{-}** has the larger ionic radius.

PRACTICE EXERCISE

4.   Which of the following has the largest radius: Na^{+}, Mg^{2+}, Al^{3+}, S^{2-}, or Cl^{-}?

Text Problems: **8.44**, 8.46

PROBLEM TYPE 5: COMPARING THE IONIZATION ENERGIES OF ELEMENTS

Ionization energy is the minimum energy required to remove an electron from a gaseous atom in its ground state. An equation that represents this process is

$$\text{energy} + X(g) \longrightarrow X^+(g) + e^-$$

where,

X represents a gaseous atom of any element

e^- is an electron

The more "tightly" the electron is held in the atom, the more difficult it will be to remove the electron. Hence, the more "tightly" the electron is held, the higher the ionization energy.

The general periodic trends for ionization energy are:

(1) Moving from left to right across a row of the periodic table, the ionization energy *increases* due to an increase in effective nuclear charge. As the effective nuclear charge increases, the electrons will be held more "tightly" by the nucleus, making them more difficult to remove.

(2) Moving down a column (group) of the periodic table, the ionization energy *decreases*. As the principal quantum number *n* increases, so does the average distance of a valence electron from the nucleus. A greater separation between the electron and the nucleus results in a weaker attraction; hence, a valence electron becomes increasingly easier to remove as we proceed down a group of the periodic table.

As with most periodic trends, there are exceptions. For a more detailed discussion, see Section 8.4 of the text.

EXAMPLE 8.4
Which of the following has the highest ionization energy: K, Br, Cl, or S?

The ionization energy increases from left to right across a row of the periodic table. Thus, the ionization energy of Cl is greater than for S, and the ionization energy of Br is greater than for K. We have narrowed our choices to Cl and Br. Moving down a group of the periodic table, the ionization decreases. Therefore, **Cl** has a higher ionization energy than Br as well as the other choices given.

PRACTICE EXERCISE
5.   Based on periodic trends, which of the following elements has the greatest ionization energy:
Cl, K, S, Se, or Br?

> **Text Problems:** 8.52, **8.54**, 8.56

PROBLEM TYPE 6: ELECTRON AFFINITY

Electron affinity is the energy change when an electron is accepted by an atom in the gaseous state. An equation that represents this process is:

$$X(g) + e^- \longrightarrow X^-(g)$$

where,

X represents a gaseous atom of any element

e^- is an electron

A positive electron affinity signifies that energy is liberated when an electron is added to an atom. The more positive the electron affinity, the greater the tendency of the atom to accept an electron. Electron affinity is positive if the reaction is exothermic and negative if the reaction is endothermic.

The general periodic trends for electron affinity are:

(1)   The tendency to accept electrons increases (that is, electron affinity values become more positive) as we move from left to right across a period.

(2)   The electron affinities of metals are generally less than those of nonmetals.

(3)   Electron affinity values vary little within a given group.

EXAMPLE 8.5
Explain why the electron affinities of the halogens are all positive.

The outer-shell electron configuration of the halogens is ns^2np^5. For the process

$$X\,(g) + e^- \longrightarrow X^-\,(g)$$

where X denotes a member of the halogen family, the accepted electron would fill the outer shell, giving the halogen ion a stable noble gas electron configuration. This is a favorable process; consequently, halogens have a strong tendency to accept an extra electron (i.e., a highly positive electron affinity).

| Text Problems: **8.60**, 8.62 |

ANSWERS TO PRACTICE EXERCISES

1.   **(a)** $1s^2 2s^2 2p^6 3s^2 3p^6 4s^2 3d^{10} 4p^6$

  (b) Since the $4p$ subshell is completely filled, this is a *noble gas*. All noble gases are *nonmetals*.

  (c) Since the outer shell is completely filled, there are *no* unpaired electrons. Therefore, the atoms of this element are diamagnetic.

2. N^{3-}:   [Ne] Ba^{2+}: [Xe] Zn^{2+}: [Ar]$3d^{10}$ V^{5+}: [Ar]

3.   **Ge** has the largest atomic radius of the group.

4.   The ions are isoelectronic. S^{2-} has the largest radius.

5.   **Cl**

SOLUTIONS TO SELECTED TEXT PROBLEMS

8.20   Writing an Electron Configuration and Identifying an Element, Problem Type 1.

    (a)   The electron configuration is $1s^2 2s^2 2p^6 3s^2 3p^5$ or $[Ne]3s^2 3p^5$. See Problem Type 7.6 if you are having difficulty writing electron configurations.

    (b)   Since the $3p$ subshell is not completely filled, this is a *representative element*. Without consulting a periodic table, you might know that the halogen family has seven valence electrons. You could then further classify this element as a *halogen*. In addition, all halogens are *nonmetals*.

    (c)   If you were to write an orbital diagram for this electron configuration, you would see that there is *one* unpaired electron in the p subshell. Therefore, the atoms of this element are paramagnetic.

8.22   Elements that have the same number of valence electrons will have similarities in chemical behavior. Looking at the periodic table, elements with the same number of valence electrons are in the same group. Therefore, the pairs that would represent similar chemical properties of their atoms are:

    (a) and **(d)**   **(b)** and **(e)**   **(c)** and **(f)**.

8.24   **(a)**   Group 1A   **(b)**   Group 5A   **(c)**   Group 8A   **(d)**   Group 8B

    Identify the elements.

8.26   You should realize that the metal ion in question is a transition metal ion because it has five electrons in the $3d$ subshell. Remember that in a transition metal ion, the $(n-1)d$ orbitals are more stable than the ns orbital. Hence, when a cation is formed from an atom of a transition metal, electrons are *always* removed first from the ns orbital and then from the $(n-1)d$ orbitals if necessary. Since the metal ion has a +3 charge, three electrons have been removed. Since the $4s$ subshell is less stable than the $3d$, two electrons would have been lost from the $4s$ and one electron from the $3d$. Therefore, the electron configuration of the neutral atom is $[Ar]4s^2 3d^6$. This is the electron configuration of iron. Thus, the metal is **iron**.

8.28   Electron Configurations of Cations and Anions, Problem Type 2.

    In the formation of a **cation** from the neutral atom of a representative element, one or more electrons are *removed* from the highest occupied n shell. In the formation of a **anion** from the neutral atom of a representative element, one or more electrons are *added* to the highest partially filled n shell. Representative elements typically gain or lose electrons to achieve a stable noble gas electron configuration. When a cation is formed from an atom of a transition metal, electrons are *always* removed first from the ns orbital and then from the $(n-1)d$ orbitals if necessary.

    (a)   [Ne]   **(e)**   Same as (c)

    (b)   same as (a). Do you see why?   **(f)** $[Ar]3d^6$. Why isn't it $[Ar]4s^2 3d^4$?

    (c)   [Ar]   **(g)** $[Ar]3d^9$. Why not $[Ar]4s^2 3d^7$?

    (d)   Same as (c). Do you see why?   **(h)** $[Ar]3d^{10}$. Why not $[Ar]4s^2 3d^8$?

8.30   **(a)** Cr^{3+} **(b)** Sc^{3+} **(c)** Rh^{3+} **(d)** Ir^{3+}

8.32   Isoelectronic means that the species have the same number of electrons and the same electron configuration.

    Be^{2+} and He ($2\,e^-$) F^- and N^{3-} ($10\,e^-$) Fe^{2+} and Co^{3+} ($24\,e^-$) S^{2-} and Ar ($18\,e^-$)

8.38   Comparing the Sizes of Atoms, Problem Type 3.

Recall that the general periodic trends in atomic size are:

(1)   Moving from left to right across a row (period) of the periodic table, the atomic radius *decreases* due to an increase in effective nuclear charge.

(2)   Moving down a column (group) of the periodic table, the atomic radius *increases* since the orbital size increases with increasing principal quantum number.

The atoms that we are considering are all in the same period of the periodic table. Hence, the atom furthest to the left in the row will have the largest atomic radius, and the atom furthest to the right in the row will have the smallest atomic radius. Arranged in order of decreasing atomic radius, we have,

$$Na > Mg > Al > P > Cl$$

8.40   Fluorine is the smallest atom in Group 7A. Atomic radius increases moving down a group since the orbital size increases with increasing principal quantum number, n.

8.42   The atomic radius is largely determined by how strongly the outer-shell electrons are held by the nucleus. The larger the effective nuclear charge, the more strongly the electrons are held and the smaller the atomic radius. For the second period, the atomic radius of Li is largest because the $2s$ electron is well shielded by the filled $1s$ shell. The effective nuclear charge that the outermost electrons feel increases across the period as a result of incomplete shielding by electrons in the same shell. Consequently, the orbital containing the electrons is compressed and the atomic radius decreases.

8.44   Comparing the Sizes of Ions, Problem Type 4.

The ions listed are all isoelectronic. They each have ten electrons. The ion with the fewest protons will have the largest ionic radius, and the ion with the most protons will have the smallest ionic radius. The effective nuclear charge increases with increasing number of protons. The electrons are attracted more strongly by the nucleus, decreasing the ionic radius. The order of increasing atomic radius is:

$$Mg^{2+} < Na^+ < F^- < O^{2-} < N^{3-}$$

8.46   Both selenium and tellurium are Group 6A elements. Since atomic radius increases going down a column in the periodic table, it follows that Te^{2-} must be larger than Se^{2-}.

8.48   We assume the approximate boiling point of argon is the mean of the boiling points of neon and krypton, based on its position in the periodic table being between Ne and Kr in Group 8A.

$$\textbf{b.p.} = \frac{-245.9°C + (-152.9°C)}{2} = \textbf{-199.4°C}$$

The actual boiling point of argon is −185.7°C.

8.52   The Group 3A elements (such as Al) all have a single electron in the outermost p subshell, which is well shielded from the nuclear charge by the inner electrons and the ns^2 electrons. Therefore, less energy is needed to remove a single p electron than to remove a paired s electron from the same principal energy level (such as for Mg).

8.54   Comparing the Ionization Energies of Elements, Problem Type 5.

A noble gas electron configuration, such as $1s^2 2s^2 2p^6$, is a very stable configuration, making it extremely difficult to remove an electron. The high ionization energy of 2080 kJ/mol would be associated with the element having this noble gas electron configuration.

The lone electron in the $3s$ orbital will be much easier to remove. Therefore, the ionization energy of 496 kJ/mol is paired with the electron configuration $1s^2 2s^2 2p^6 3s^1$.

8.56   The atomic number of mercury is 80. We write:

$$\Delta E = (2.18 \times 10^{-18} \text{ J})(80^2)\left(\frac{1}{1^2} - \frac{1}{\infty^2}\right) = 1.40 \times 10^{-14} \text{ J/ion}$$

$$\Delta E = \left(1.40 \times 10^{-14} \frac{\text{J}}{\text{ion}}\right)\left(\frac{6.022 \times 10^{23} \text{ ions}}{1 \text{ mol}}\right)\left(\frac{1 \text{ kJ}}{1000 \text{ J}}\right) = \mathbf{8.43 \times 10^6 \text{ kJ/mol}}$$

8.60   Electron Affinity, Problem Type 6.

One of the general periodic trends for electron affinity is that the tendency to accept electrons increases (that is, electron affinity values become more positive) as we move from left to right across a period. However, this trend does not include the noble gases. We know that noble gases are extremely stable, and they do not want to gain or lose electrons.

Based on the above periodic trend, Cl would be expected to have the highest electron affinity. Addition of an electron to Cl forms Cl^-, which has a stable noble gas electron configuration.

8.62   Alkali metals have a valence electron configuration of ns^1 so they can accept another electron in the ns orbital. On the other hand, alkaline earth metals have a valence electron configuration of ns^2. Alkaline earth metals have little tendency to accept another electron, as it would have to go into a higher energy p orbital.

8.66   Since ionization energies decrease going down a column in the periodic table, francium should have the lowest first ionization energy of all the alkali metals. As a result, Fr should be the most reactive of all the Group 1A elements toward water and oxygen. The reaction with oxygen would probably be similar to that of K, Rb, or Cs.

What would you expect the formula of the oxide to be? The chloride?

8.68   The Group 1B elements are much less reactive than the Group 1A elements. The 1B elements are more stable because they have much higher ionization energies resulting from incomplete shielding of the nuclear charge by the inner d electrons. The ns^1 electron of a Group 1A element is shielded from the nucleus more effectively by the completely filled noble gas core. Consequently, the outer s electrons of 1B elements are more strongly attracted by the nucleus.

8.70   (a)   Lithium oxide is a basic oxide. It reacts with water to form the metal hydroxide:

$$Li_2O \, (s) + H_2O \, (l) \longrightarrow 2 \, LiOH \, (aq)$$

(b)   Calcium oxide is a basic oxide. It reacts with water to form the metal hydroxide:

$$CaO \, (s) + H_2O \, (l) \longrightarrow Ca(OH)_2 \, (aq)$$

(c)   Sulfur trioxide is an acidic oxide. It reacts with water to form sulfuric acid:

$$SO_3\ (g)\ +\ H_2O\ (l)\ \longrightarrow\ H_2SO_4\ (aq)$$

8.72   As we move down a column, the metallic character of the elements increases. Since magnesium and barium are both Group 2A elements, we expect barium to be more metallic than magnesium and BaO to be more basic than MgO.

8.74   **(a)**   bromine   **(b)**   nitrogen   **(c)**   rubidium   **(d)**   magnesium

8.76   This is an isoelectronic series with ten electrons in each species. The nuclear charge interacting with these ten electrons ranges from +8 for oxygen to +12 for magnesium. Therefore the +12 charge in Mg^{2+} will draw in the ten electrons more tightly than the +11 charge in Na^+, than the +9 charge in F^-, than the +8 charge in O^{2-}. Recall that the largest species will be the *easiest* to ionize.

(a)   increasing ionic radius: $Mg^{2+}\ <\ Na^+\ <\ F^-\ <\ O^{2-}$

(b)   increasing ionization energy: $O^{2-}\ <\ F^-\ <\ Na^+\ <\ Mg^{2+}$

8.78   According to the *Handbook of Chemistry and Physics* (1966-67 edition), potassium metal has a melting point of 63.6°C, bromine is a reddish brown liquid with a melting point of −7.2°C, and potassium bromide (KBr) is a colorless solid with a melting point of 730°C. M is potassium (K) and X is bromine (Br).

8.80 O^+ and N   Ar and S^{2-} Ne and N^{3-} Zn and As^{3+} Cs^+ and Xe

8.82   **(a)** and **(d)**

8.84   Fluorine is a yellow-green gas that attacks glass; chlorine is a pale yellow gas; bromine is a fuming red liquid; and iodine is a dark, metallic-looking solid.

8.86   Fluorine

8.88 H^- and He are isoelectronic species with two electrons. Since H^- has only one proton compared to two protons for He, the nucleus of H^- will attract the two electrons less strongly compared to He. Therefore, H^- is larger.

8.90

Oxide	Name	Property
Li_2O	lithium oxide	basic
BeO	beryllium oxide	amphoteric
B_2O_3	boron oxide	acidic
CO_2	carbon dioxide	acidic
N_2O_5	dinitrogen pentoxide	acidic

Note that only the highest oxidation states are considered.

8.92   In its chemistry, hydrogen can behave like an alkali metal (H^+) and like a halogen (H^-).

8.94   Replacing Z in the equation given in Problem 8.55 with $(Z - \sigma)$ gives:

$$E_n = (2.18 \times 10^{-18} \text{ J})(Z - \sigma)^2 \left(\frac{1}{n^2}\right)$$

For helium, the atomic number (Z) is 2, and in the ground state, its two electrons are in the first energy level, so $n = 1$. Substitute Z, n, and the first ionization energy into the above equation to solve for σ.

$$E_1 = 3.94 \times 10^{-18} \text{ J} = (2.18 \times 10^{-18} \text{ J})(2 - \sigma)^2 \left(\frac{1}{1^2}\right)$$

$$(2 - \sigma)^2 = \frac{3.94 \times 10^{-18} \text{ J}}{2.18 \times 10^{-18} \text{ J}}$$

$$2 - \sigma = \sqrt{1.81}$$

$$\sigma = 2 - 1.35 = \textbf{0.65}$$

8.96   The percentage of volume occupied by K^+ compared to K is:

$$\frac{\text{volume of } K^+ \text{ ion}}{\text{volume of K atom}} \times 100\% = \frac{\frac{4}{3}\pi(133 \text{ pm})^3}{\frac{4}{3}\pi(216 \text{ pm})^3} \times 100\% = 23.3\%$$

Therefore, there is a decrease in volume of $(100 - 23.3)\% = \textbf{76.7\%}$ when K^+ is formed from K.

8.98   Rearrange the given equation to solve for ionization energy.

$$IE = h\nu - \frac{1}{2}mu^2$$

or,

$$IE = \frac{hc}{\lambda} - KE$$

The kinetic energy of the ejected electron is given in the problem. Substitute h, c, and λ into the above equation to solve for the ionization energy.

$$IE = \frac{(6.63 \times 10^{-34} \text{ J} \cdot \text{s})(3.00 \times 10^8 \text{ m/s})}{162 \times 10^{-9} \text{ m}} - (5.34 \times 10^{-19} \text{ J})$$

$$\textbf{IE} = \textbf{6.94} \times \textbf{10}^{-19} \textbf{ J}$$

We might also want to express the ionization energy in kJ/mol.

$$\frac{6.94 \times 10^{-19} \text{ J}}{1 \text{ photon}} \times \frac{6.022 \times 10^{23} \text{ photons}}{1 \text{ mol}} \times \frac{1 \text{ kJ}}{1000 \text{ J}} = \textbf{418 kJ/mol}$$

To ensure that the ejected electron is the valence electron, UV light of the *longest* wavelength (lowest energy) should be used that can still eject electrons.

8.100   We want to determine the second ionization energy of lithium.

$$Li^+ \longrightarrow Li^{2+} + e^- \qquad I_2 = ?$$

The equation given in Problem 8.55 allows us to determine the third ionization energy for Li. Knowing the total energy needed to remove all three electrons from Li, we can calculate the second ionization energy by difference.

$$\text{Energy needed to remove three electrons} = I_1 + I_2 + I_3$$

First, let's calculate I_3. For Li, $Z = 3$, and $n = 1$ because the third electron will come from the $1s$ orbital.

$$I_3 = \Delta E = E_\infty - E_3$$

$$I_3 = -(2.18 \times 10^{-18} \text{ J})(3)^2\left(\frac{1}{\infty^2}\right) + (2.18 \times 10^{-18} \text{ J})(3)^2\left(\frac{1}{1^2}\right)$$

$$I_3 = +1.96 \times 10^{-17} \text{ J}$$

Converting to units of kJ/mol:

$$I_3 = 1.96 \times 10^{-17} \text{ J} \times \frac{6.022 \times 10^{23} \text{ ions}}{1 \text{ mol}} = 1.18 \times 10^7 \text{ J/mol} = 1.18 \times 10^4 \text{ kJ/mol}$$

Energy needed to remove three electrons $= I_1 + I_2 + I_3$

$$1.96 \times 10^4 \text{ kJ/mol} = 520 \text{ kJ/mol} + I_2 + 1.18 \times 10^4 \text{ kJ/mol}$$

$$\mathbf{I_2 = 7.28 \times 10^3 \text{ kJ/mol}}$$

8.102 **X** must belong to Group 4A; it is probably Sn or Pb because it is not a very reactive metal (it is certainly not reactive like an alkali metal).

Y is a nonmetal since it does *not* conduct electricity. Since it is a light yellow solid, it is probably phosphorus (Group 5A).

Z is an alkali metal since it reacts with air to form a basic oxide or peroxide.

8.104 $\text{Na} \longrightarrow \text{Na}^+ + \text{e}^-$ $I_1 = 495.9 \text{ kJ/mol}$

This equation is the reverse of the electron affinity for Na^+. Therefore, the electron affinity of Na^+ is +495.9 kJ/mol. Note that the electron affinity is positive, indicating that energy is liberated when an electron is added to an atom or ion. You should expect this since we are adding an electron to a positive ion.

8.106 The reaction representing the electron affinity of chlorine is:

$$\text{Cl } (g) + \text{e}^- \longrightarrow \text{Cl}^- (g) \qquad \Delta H° = +349 \text{ kJ/mol}$$

It follows that the energy needed for the reverse process is also +349 kJ/mol.

$$\text{Cl}^- (g) + h\nu \longrightarrow \text{Cl } (g) + \text{e}^- \qquad \Delta H° = +349 \text{ kJ/mol}$$

The energy above is the energy of one mole of photons. We need to convert to the energy of one photon in order to calculate the wavelength of the photon.

$$\frac{349 \text{ kJ}}{1 \text{ mol photons}} \times \frac{1 \text{ mol photons}}{6.022 \times 10^{23} \text{ photons}} \times \frac{1000 \text{ J}}{1 \text{ kJ}} = 5.80 \times 10^{-19} \text{ J/photon}$$

Now, we can calculate the wavelength of a photon with this energy.

$$\lambda = \frac{hc}{E} = \frac{(6.63 \times 10^{-34} \text{ J·s})(3.00 \times 10^8 \text{ m/s})}{5.80 \times 10^{-19} \text{ J}} = \mathbf{3.43 \times 10^{-7} \text{ m}} = \mathbf{343 \text{ nm}}$$

The radiation is in the **ultraviolet** region of the electromagnetic spectrum.

8.108   The equation that we want to calculate the energy change for is:

$$\text{Na}\,(s) \longrightarrow \text{Na}^+\,(g) + \text{e}^- \qquad \Delta H° = ?$$

Can we take information given in the problem and other knowledge to end up with the above equation? This is a Hess's law problem (see Problem Type 4C, Chapter 6).

In the problem we are given:	$\text{Na}\,(s) \longrightarrow \text{Na}\,(g)$	$\Delta H° = 108.4 \text{ kJ}$
We also know the ionization energy of Na (g).	$\text{Na}\,(g) \longrightarrow \text{Na}^+\,(g) + \text{e}^-$	$\Delta H° = 495.9 \text{ kJ}$
Adding the two equations:	$\text{Na}\,(s) \longrightarrow \text{Na}^+\,(g) + \text{e}^-$	$\Delta H° = 604.3 \text{ kJ}$

8.110   The electron configuration of titanium is: $[\text{Ar}]4s^2 3d^2$. Titanium has four valence electrons, so the maximum oxidation number it is likely to have in a compound is +4. The compounds followed by the oxidation state of titanium are: $K_3\text{TiF}_6$, +3; $K_2\text{Ti}_2\text{O}_5$, +4; TiCl_3, +3; $K_2\text{TiO}_4$, +6; and $K_2\text{TiF}_6$, +4. $K_2\text{TiO}_4$ is unlikely to exist because of the oxidation state of Ti of +6. Titanium in an oxidation state greater than +4 is unlikely because of the very high ionization energies needed to remove the fifth and sixth electrons.

8.112   The unbalanced ionic equation is: $\text{MnF}_6^{2-} + \text{SbF}_5 \longrightarrow \text{SbF}_6^- + \text{MnF}_3 + \text{F}_2$

In this redox reaction, Mn^{4+} is reduced to Mn^{3+}, and F^- from both MnF_6^{2-} and SbF_5 is oxidized to F_2.

We can simplify the half-reactions. $\text{Mn}^{4+} \xrightarrow{\text{reduction}} \text{Mn}^{3+}$

$$\text{F}^- \xrightarrow{\text{oxidation}} \text{F}_2$$

Balancing the two half-reactions: $\text{Mn}^{4+} + \text{e}^- \longrightarrow \text{Mn}^{3+}$

$$2\,\text{F}^- \longrightarrow \text{F}_2 + 2\,\text{e}^-$$

Adding the two half-reactions: $2\,\text{Mn}^{4+} + 2\,\text{F}^- \longrightarrow 2\,\text{Mn}^{3+} + \text{F}_2$

We can now reconstruct the complete balanced equation. In the balanced equation, we have 2 mol of Mn ions and 1 mol of F_2 on the products side.

$$2\,K_2\text{MnF}_6 + \text{SbF}_5 \longrightarrow \text{KSbF}_6 + 2\,\text{MnF}_3 + 1\,\text{F}_2$$

We can now balance the remainder of the equation by inspection. Notice that there are 4 mol of K^+ on the left, but only 1 mol of K^+ on the right. The balanced equation is:

$$2\,K_2\text{MnF}_6 + 4\,\text{SbF}_5 \longrightarrow 4\,\text{KSbF}_6 + 2\,\text{MnF}_3 + \text{F}_2$$

8.114   To work this problem, assume that the oxidation number of oxygen is −2.

Oxidation number	Chemical formula
+1	N_2O
+2	NO
+3	N_2O_3
+4	NO_2, N_2O_4
+5	N_2O_5

8.116   The larger the effective nuclear charge, the more tightly held are the electrons. Thus, the atomic radius will be small, and the ionization energy will be large. The quantities show an opposite periodic trend.

8.118   We assume that the m.p. and b.p. of bromine will be between those of chlorine and iodine.

Taking the average of the melting points and boiling points:

$$\textbf{m.p.} = \frac{-101.0°C + 113.5°C}{2} = \textbf{6.3°C} \qquad \text{(Handbook: } -7.2°C)$$

$$\textbf{b.p.} = \frac{-34.6°C + 184.4°C}{2} = \textbf{74.9°C} \qquad \text{(Handbook: } 58.8 °C)$$

The estimated values do not agree very closely with the actual values because Cl_2 (g), Br_2 (l), and I_2 (s) are in different physical states. If you were to perform the same calculations for the noble gases, your calculations would be much closer to the actual values.

8.120   The heat generated from the radioactive decay can break bonds; therefore, few radon compounds exist.

8.122   **(a)**   It was determined that the periodic table was based on atomic number, not atomic mass.

   (b)   Argon:

   $(0.00337 \times 35.9675 \text{ amu}) + (0.00063 \times 37.9627 \text{ amu}) + (0.9960 \times 39.9624 \text{ amu}) = \textbf{39.95 amu}$

   Potassium:

   $(0.93258 \times 38.9637 \text{ amu}) + (0.000117 \times 39.9640 \text{ amu}) + (0.0673 \times 40.9618 \text{ amu}) = \textbf{39.10 amu}$

8.124 $Z = 119$

Electron configuration: $[Rn]7s^2 5f^{14} 6d^{10} 7p^6 8s^1$

8.126   There is a large jump from the second to the third ionization energy, indicating a change in the principal quantum number n. In other words, the third electron removed is an inner, noble gas core electron, which is difficult to remove. Therefore, the element is in **Group 2A**.

8.128   **(a)** SiH_4, GeH_4, SnH_4, PbH_4
   (b)   Metallic character increases going down a family of the periodic table. Therefore, RbH would be more ionic than NaH.
   (c)   Since Ra is in Group 2A, we would expect the reaction to be the same as other alkaline earth metals with water.

$$Ra \ (s) \ + \ 2 \ H_2O \ (l) \ \longrightarrow \ Ra(OH)_2 \ (aq) \ + \ H_2 \ (g)$$

   (d)   Beryllium (diagonal relationship)

8.130   The importance and usefulness of the periodic table lie in the fact that we can use our understanding of the general properties and trends within a group or a period to predict with considerable accuracy the properties of any element, even though the element may be unfamiliar to us. For example, elements in the same group or family have the same valence electron configurations. Due to the same number of valence electrons occupying similar orbitals, elements in the same family have similar chemical properties. In addition, trends in properties such as ionization energy, atomic radius, electron affinity, and metallic character can be predicted based on an element's position in the periodic table. Ionization energy typically increases across a period of the period table and decreases down a group. Atomic radius typically decreases across a period and increases down a group. Electron affinity typically increases across a period and decreases down a group. Metallic character typically decreases across a period and increases down a group. The periodic table is an extremely useful tool for a scientist. Without having to look in a reference book for a particular element's properties, one can look at its position in the periodic table and make educated predictions as to its many properties such as those mentioned above.

CHAPTER 9
CHEMICAL BONDING I:
BASIC CONCEPTS

PROBLEM-SOLVING STRATEGIES AND TUTORIAL SOLUTIONS

TYPES OF PROBLEMS

Problem Type 1:   Classifying Chemical Bonds.

Problem Type 2:   Calculating the Lattice Energy of Ionic Compounds.

Problem Type 3:   Writing Lewis Structures.

Problem Type 4:   Formal Charges.
  (a) Assigning formal charges.
  (b) Choosing the most plausible Lewis structure based on formal charges.

Problem Type 5:   Drawing Resonance Structures.

Problem Type 6:   Exceptions to the Octet Rule.
  (a) The incomplete octet.
  (b) Odd-electron molecules.
  (c) The expanded octet.

Problem Type 7:   Using Bond Energies to Estimate the Enthalpy of a Reaction.

PROBLEM TYPE 1: CLASSIFYING CHEMICAL BONDS

We can classify bonds as three different types: ionic, polar covalent, or covalent.

In a **covalent bond**, the electron pair of the bond is shared *equally* by the two atoms.

In a **polar covalent bond**, the electron pair of the bond is shared *unequally* by the two atoms. The electrons spend more time in the vicinity of one atom than the other.

Covalent bonds (polar or nonpolar) are typically formed between two nonmetals.

In an **ionic bond,** the electron or electrons are nearly completely transferred from one atom to another. Ionic bonds are typically formed between a metal cation and a nonmetal anion or a metal cation and a polyatomic anion.

A property that helps us distinguish a nonpolar covalent bond from a polar covalent bond is **electronegativity**, the ability of an atom to attract toward itself the electrons in a chemical bond. Elements with high electronegativities have a greater tendency to attract electrons than elements with low electronegativities. Linus Pauling devised a method for calculating *relative* electronegativities of most elements. These values are shown in Figure 9.5 of the text.

Only atoms of the same element, which have the same electronegativity, can be joined by a pure *covalent bond*. Atoms of elements with similar electronegativities tend to form *polar covalent bonds* with each other because the shift in electron density is usually small. There is no sharp distinction between a polar bond and an ionic bond, but the following rule is helpful in distinguishing them. An *ionic bond* forms when the electronegativity difference between the two bonding atoms is 2.0 or more. This rule applies to most but not all ionic compounds.

EXAMPLE 9.1

Classify the following bonds as ionic, polar covalent, or covalent: (a) the bond in KCl, (b) the OH bond in H_2O, and (c) the OO bond in oxygen gas (O_2).

(a)   In Figure 9.5 of the text, we see that the electronegativity difference between K and Cl is 2.2, above the 2.0 guideline. Therefore, the bond between K and Cl is ionic. Remember that an ionic bond is typically formed between a metal cation and a nonmetal anion or a metal cation and a polyatomic anion.

(b)   The electronegativity difference between O and H is 1.4, which is appreciable but not large enough to qualify H_2O as an ionic compound. Therefore, the bond between O and H is polar covalent. Recall that polar covalent compounds are typically formed between two different nonmetals.

(c)   The two O atoms are identical in every respect. Therefore, the bond between them is purely covalent.

PRACTICE EXERCISE

1.   Classify the following bonds as ionic, polar covalent, or covalent: (a) the NH bond in NH_3, (b) the OO bond in hydrogen peroxide (H_2O_2), and (c) the bond in NaF.

Text Problems: 9.20, **9.36**, 9.38, 9.40

PROBLEM TYPE 2: CALCULATING THE LATTICE ENERGY OF IONIC COMPOUNDS

See Section 9.3 of your text for a complete discussion of lattice energy. Example 9.2 below will illustrate the method used to calculate the lattice energy of an ionic compound.

EXAMPLE 9.2

Calculate the lattice energy of magnesium oxide, MgO, given that the enthalpy of sublimation of Mg is 150 kJ/mol and the electron affinity of O^- is –780 kJ/mol.

We want to calculate the enthalpy change corresponding to the following reaction:

$$Mg^{2+}(g) + O^{2-}(g) \longrightarrow MgO\ (s) \qquad \Delta H° = ?$$

This reaction is the reverse of the reaction for the lattice energy. If we can calculate $\Delta H°$ for the above reaction, we can determine the *lattice energy*. Starting with Mg(s) and $O_2(g)$, we can follow two different pathways to the product, MgO(s). The enthalpy changes for the two pathways will be equal.

Pathway 1:   This is the overall reaction, which is the $\Delta H_f°$ of MgO(s). You can look up the appropriate value in Appendix 3 of the text.

$$Mg(s)\ +\ O_2\ (g) \longrightarrow MgO(s) \qquad \Delta H_f° = -601.8\ kJ/mol$$

Pathway 2:   This is the indirect pathway. Using a series of steps, we can form the product MgO(s). The last step in the process will be:

$$Mg^{2+}(g) + O^{2-}(g) \longrightarrow MgO\ (s)$$

This reaction is the reverse of the reaction for the lattice energy. Knowing the enthalpy changes for the other steps, we can calculate the enthalpy change for the step above.

Step 1: Convert solid magnesium to magnesium vapor.

$$Mg(s) \longrightarrow Mg(g) \qquad\qquad \Delta H_1^\circ = 150 \text{ kJ}$$

This is the energy needed to sublime $Mg(s)$.

Step 2: Dissociate (1/2) mole of O_2 gas into separate gaseous O atoms.

$$(1/2)\, O_2(g) \longrightarrow O(g) \qquad\qquad \Delta H_2^\circ = (1/2)(498.7 \text{ kJ}) = 249.4 \text{ kJ}$$

498.7 kJ is the amount of energy needed to break a mole of O=O bonds. Here we are breaking the bonds in half a mole of O_2. You can find this bond energy in Table 9.4 of the text.

Step 3: Ionize 1 mole of gaseous Mg atoms.

$$Mg(g) \longrightarrow Mg^{2+}(g) + 2e^- \qquad\qquad \Delta H_3^\circ = 738.1 \text{ kJ} + 1450 \text{ kJ} = 2188 \text{ kJ}$$

This process corresponds to the first and second ionization energies of Mg. See Table 8.2 of your text for the ionization energies.

Step 4: Add 2 moles of electrons to 1 mole of gaseous O atoms.

$$O(g) + e^- \longrightarrow O^-(g) \qquad\qquad \Delta H_{4a}^\circ = -142 \text{ kJ}$$
$$O^-(g) + e^- \longrightarrow O^{2-}(aq) \qquad\qquad \Delta H_{4b}^\circ = 780 \text{ kJ}$$

This process corresponds to the opposite of the electron affinity value of O and the electron affinity value of O^-. The electron affinity of O is given in Table 8.3 of the text.

Step 5: Combine 1 mole of gaseous Mg^{2+} and 1 mole of gaseous O^{2-} to form 1 mole of solid MgO.

$$Mg^{2+}(g) + O^{2-}(g) \longrightarrow MgO(s) \qquad \Delta H_5^\circ = ?$$

This is the enthalpy change that we wish to calculate because it is the opposite of the lattice energy. Changing the sign of the value that we calculate will give us the lattice energy.

Step 6: The enthalpy change for the two pathways is the same. We can write:

$$\Delta H^\circ(\text{pathway 1}) = \Delta H^\circ(\text{pathway 2})$$

$$\Delta H_f^\circ [MgO(s)] = \Delta H_1^\circ + \Delta H_2^\circ + \Delta H_3^\circ + \Delta H_{4a}^\circ + \Delta H_{4b}^\circ + \Delta H_5^\circ$$

Substituting the values from above:

$$-601.8 \text{ kJ} = 150 \text{ kJ} + 249.4 \text{ kJ} + 2188 \text{ kJ} - 142 \text{ kJ} + 780 \text{ kJ} + \Delta H_5^\circ$$

$$\Delta H_5^\circ = -3827 \text{ kJ}$$

This enthalpy change corresponds to the reaction that is the reverse of the lattice energy. Therefore, the lattice energy will have the opposite sign of ΔH_5°.

$$MgO(s) \longrightarrow Mg^{2+}(g) + O^{2-}(g) \qquad \textbf{lattice energy} = \textbf{+3827 kJ/mol}$$

PRACTICE EXERCISE

2.   Given that the enthalpy of sublimation of sodium (Na) is 108 kJ/mol and $\Delta H_f^\circ [NaF(s)] = -570$ kJ/mol, calculate the lattice energy of $NaF(s)$. See Tables 8.2, 8.3, and 9.4 for other data.

Text Problem: 9.26

PROBLEM TYPE 3: WRITING LEWIS STRUCTURES

The general rules for writing Lewis structures are given below. For more detailed rules, see Section 9.6 of the text.

1.   Write the skeletal structure of the compound, using chemical symbols and placing bonded atoms next to one another. In general, the least electronegative atom occupies the central position. Hydrogen and fluorine usually occupy the terminal (end) positions in the Lewis structure.

2.   Count the number of valence electrons present. For polyatomic anions, add the number of negative charges to that total. For polyatomic cations, subtract the number of positive charges from the number of valence electrons.

3.   Draw a single covalent bond between the central atom and each of the surrounding atoms. Complete the octets of the atoms bonded to the central atom with lone pairs. The total number of electrons to be used is that determined in step 2.

4.   Sometimes there will not be enough electrons to satisfy the octet rule of the central atom by placing lone pairs. Try adding double or triple bonds between the surrounding atoms and the central atom, using the lone pairs from the surrounding atoms.

EXAMPLE 9.3

Write the Lewis structure for SO_2.

Step 1: Sulfur is less electronegative than oxygen, so it occupies the central position. The skeletal structure is:

$$O \quad S \quad O$$

Step 2: The outer-shell electron configurations of O and S are $2s^2 2p^4$ and $3s^2 3p^4$, respectively. Thus, there are

$$6 + (2 \times 6) = 18 \text{ valence electrons}$$

> **Tip:** For the representative elements (Group 1A – 7A), the number of valence electrons is equal to the group number. For example, both O and S are in Group 6A; thus, they both have 6 valence electrons.

Step 3: We draw a single covalent bond between S and each O, and then complete the octets for the O atoms. We place the remaining two electrons on S.

$$:\ddot{O}—\ddot{S}—\ddot{O}:$$

Step 4: The octet rule is satisfied for the oxygen atoms; however, the S atom does *not* satisfy the octet rule. Let's try making a sulfur-to-oxygen double bond by moving a lone pair from one of the O atoms to form another bond with S.

$$:\ddot{O}—\ddot{S}=\ddot{O}$$

Now, the octet rule is also satisfied for the S atom. As a final check, we verify that there are 18 valence electrons in the Lewis structure of SO_2.

EXAMPLE 9.4

Write the Lewis structure for NO_3^-.

Step 1: Nitrogen is less electronegative than oxygen, so it occupies the central position. The skeletal structure is:

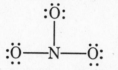

Step 2: Nitrate is a polyatomic anion, so we must add the negative charge to the number of valence electrons. The outer-shell electron configurations of O (Group 4A) and N (Group 5A) are $2s^2 2p^4$ and $2s^2 2p^3$, respectively. Thus, there are

$$5 + (3 \times 6) + 1 = 24 \text{ valence electrons}$$

Step 3: We draw a single covalent bond between N and each O, and then complete the octets for the O atoms.

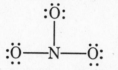

Step 4: The octet rule is satisfied for the oxygen atoms; however, the N atom does *not* satisfy the octet rule. Let's try making a nitrogen-to-oxygen double bond by moving a lone pair from one of the O atoms to form another bond with N.

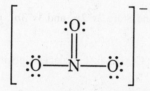

Now, the octet rule is also satisfied for the N atom. As a final check, we verify that there are 24 valence electrons in the Lewis structure of NO_3^-.

Also notice that we draw a bracket with the charge of the polyatomic ion around the Lewis structure. This is to distinguish an ion from a neutral molecule.

PRACTICE EXERCISES

3.   Write the Lewis structure for $AsCl_3$.

4.   Write the Lewis structure for cyanide ion, CN^-.

Text Problems: **9.44**, 9.46

PROBLEM TYPE 4: FORMAL CHARGES

By comparing the number of electrons in an isolated atom (valence electrons) with the number of electrons that are associated with the same atom in a Lewis structure, we can determine the distribution of electrons in the molecule and draw the most plausible Lewis structure. This difference between the valence electrons in an isolated atom and the number of electrons assigned to that atom in a Lewis structure is called the atom's **formal charge**.

We can calculate the formal charge on an atom in a molecule using the equation:

formal charge = (no. of valence e^- in free atom) − [(no. of nonbonding e^-) + 1/2(no. of bonding e^-)]

When you write formal charges, the following rules are helpful.

• For neutral molecules, the sum of the formal charges must add up to zero.

• For cations, the sum of the formal charges must equal the positive charge.

• For anions, the sum of the formal charges must equal the negative charge.

A.   Assigning formal charges

EXAMPLE 9.5
Assign formal charges to the atoms in the following Lewis structures.

(a) $:C\equiv O:$

(b) $:\ddot{O}-\ddot{S}=\ddot{O}$

The formula used to calculate the formal charge of an atom is:

formal charge = (no. of valence e^- in free atom) − [(no. of nonbonding e^-) + 1/2(no. of bonding e^-)]

(a)   For the C atom,
$$\text{formal charge} = 4 - [2 + 1/2(6)] = -1$$

For the O atom,
$$\text{formal charge} = 6 - [2 + 1/2(6)] = +1$$

Some chemists do not approve of this structure for CO because it places a positive formal charge on the more electronegative oxygen atom.

(b)   For the S atom,
$$\text{formal charge} = 6 - [2 + 1/2(6)] = +1$$

For the O atom on the left,
$$\text{formal charge} = 6 - [6 + 1/2(2)] = -1$$

For the O atom on the right,
$$\text{formal charge} = 6 - [4 + 1/2(4)] = 0$$

PRACTICE EXERCISE
5.   Assign formal charges to the atoms in the following Lewis structures.

(a) $\ddot{N}=N=\ddot{O}$

(b) $\left[:\ddot{O}-H \right]^{-1}$

Text Problems: 9.52, 9.54

B. Choosing the most plausible Lewis structure based on formal charges

The following guidelines show how to use formal charges to select a plausible Lewis structure for a given compound.

- For neutral molecules, a Lewis structure in which there are no formal charges is preferable to one in which formal charges are present.

- Lewis structures with large formal charges are less plausible than those with small formal charges.

- When comparing two structures with similar magnitudes of formal charges, the most plausible structure is the one in which negative formal charges are placed on the more electronegative atoms.

EXAMPLE 9.6

Two possible Lewis structures for BF$_3$ are shown below. Which is the more reasonable structure in terms of formal charges?

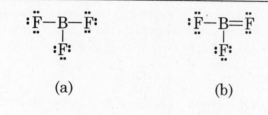

     (a)   (b)

The formal charges in (a) are:

     B atom: formal charge $= 3 - [0 + 1/2(6)] = 0$

     F atoms: formal charge $= 7 - [6 + 1/2(2)] = 0$

The formal charges in (b) are

     B atom: formal charge $= 3 - [0 + 1/2(8)] = -1$

     F atom (double bond):   formal charge $= 7 - [4 + 1/2(4)] = +1$

     F atoms (single bond):   formal charge $= 7 - [6 + 1/2(2)] = 0$

The rule used to establish the most plausible structure is: For neutral molecules, a Lewis structure in which there are no formal charges is preferable to one in which formal charges are present. Thus, structure (a) is preferred over structure (b). We could also rule out structure (b) as the most plausible structure because there is a positive formal charge on the very electronegative F atom.

PRACTICE EXERCISE

6.   Consider the following Lewis structures for the sulfate ion. Which is the more reasonable structure in terms of formal charges?

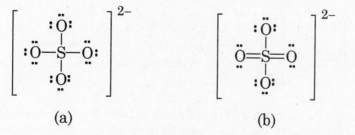

     (a)   (b)

Text Problem: 9.60

PROBLEM TYPE 5: DRAWING RESONANCE STRUCTURES

The Lewis structure for SO_3 is shown below.

We can draw two more equivalent Lewis structures with the double bond between S and a different oxygen atom.

Which is the correct structure? Let's consider experimental data. We would expect the S–O bond to be longer than the S=O bond because double bonds are known to be shorter than single bonds. Yet experimental evidence shows that all three sulfur-to-oxygen bonds are equal in length. Therefore, none of the three structures shown accurately represents the molecule. We resolve this conflict by using all *three* Lewis structures to represent SO_3.

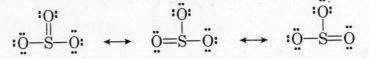

Each of the three structures is called a **resonance structure**. A resonance structure is one of two or more Lewis structures for a single molecule that cannot be described fully with only one Lewis structure. The symbol ⟷ indicates that the structures shown are resonance structures.

EXAMPLE 9.7
We drew the Lewis structure for nitrate ion, NO_3^-, in Example 9.4. However, experimental evidence shows that all N–O bonds are equivalent. Draw resonance structures to indicate the equivalence of the N–O bonds.

The Lewis structure drawn in Example 9.4 was:

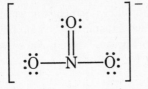

This structure, while a correct Lewis structure, does not show the equivalence of all three N–O bonds. Three contributing resonance structures can be drawn.

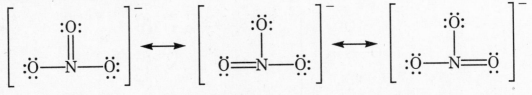

Resonance does not mean that the nitrate ion shifts quickly back and forth from one resonance structure to the other. Keep in mind that *none* of the resonance structures adequately represents the actual molecule, which has its own unique, stable structure. The actual structure is an average or hybrid of the above three structures. Resonance is a human invention, designed to address the limitations in these simple bonding models.

PRACTICE EXERCISE

7.   Draw all the resonance structures for N_2O. The skeletal structure is N–N–O.

Text Problems: **50**, 52, 54

PROBLEM TYPE 6: EXCEPTIONS TO THE OCTET RULE

A. The incomplete octet

In some compounds the number of electrons surrounding the central atom in a stable molecule is fewer than eight. **Be**, **B**, and **Al** tend to form compounds in which they are surrounded by fewer than eight electrons.

EXAMPLE 9.8

Draw the Lewis structure for GaI₃.

Step 1: Gallium is less electronegative than iodine, so it occupies the central position. The skeletal structure is:

$$I$$
$$I \quad Ga \quad I$$

Step 2: The outer-shell electron configurations of Ga (Group IIIA) and I (Group VIIA) are $4s^2 4p^1$ and $5s^2 5p^5$, respectively. Thus, there are

$$3 + (3 \times 7) = 24 \text{ valence electrons}$$

Step 3: We draw a single covalent bond between Ga and each I, and then complete the octets for the I atoms.

The octet rule is satisfied for the iodine atoms; however, the Ga atom does *not* satisfy the octet rule. A resonance structure with a double bond between Ga and I can be drawn that satisfies the octet rule for Ga. However, the properties of GaI₃ are more consistent with a Lewis structure in which there are single bonds between Ga and each I, as shown above.

Based on formal charges, is the structure shown above more plausible than the structure that contains one double bond between Ga and an I atom?

PRACTICE EXERCISE

8.   Write the Lewis structure for BCl₃.

Text Problems: 9.60, 9.64

B. Odd-electron molecules

Some molecules contain an **odd** number of electrons. To satisfy the octet rule, we need an even number of electrons. Therefore, the octet rule cannot be satisfied in a molecule that has an odd number of electrons. When drawing a Lewis structure, if an atom has fewer than eight electrons, make sure that it is the least electronegative atom in the compound.

EXAMPLE 9.9

Draw the Lewis structure for NO$_2$ (all bonds are equivalent).

Step 1: Nitrogen is less electronegative than oxygen, so it occupies the central position. The skeletal structure is:

$$O \quad N \quad O$$

Step 2: The outer-shell electron configurations of N (Group 5A) and O (Group 6A) are $2s^2 2p^3$ and $2s^2 2p^4$, respectively. Thus, there are

$$5 + (2 \times 6) = 17 \text{ valence electrons}$$

Step 3: We have an odd number of valence electrons, so the octet rule cannot be satisfied for at least one of the atoms in the molecule. Either nitrogen or oxygen in the structure will have fewer than eight electrons, because a second-row element cannot exceed an octet of electrons. Since N is less electronegative than O, nitrogen should be electron deficient. We draw a single covalent bond between N and each O, and then complete the octets for the O atoms.

$$:\overset{..}{\underset{..}{O}}\!-\!N\!-\!\overset{..}{\underset{..}{O}}:$$

Step 4: We have one electron left to place on the molecule. Placing the electron on N only gives five electrons around N. However, if we make a nitrogen-to-oxygen double bond by moving a lone pair from one of the O atoms to form another bond with N, nitrogen will be surrounded by seven electrons.

$$:\overset{..}{\underset{..}{O}}\!-\!\overset{\bullet}{N}\!=\!\overset{..}{\underset{..}{O}}$$

We could also draw resonance structures in which oxygen is electron deficient with seven electrons. However, the above structure is the most plausible since the least electronegative element is electron deficient.

C. The expanded octet

A number of compounds contain more than eight electrons around an atom. These **expanded octets** only occur for atoms of elements in and beyond the third period of the periodic table. Atoms from the third period on can accommodate more than eight electrons because they also have *d* orbitals that can be used in bonding.

EXAMPLE 9.10

Draw the Lewis structure for ClF$_3$.

Step 1: Chlorine is less electronegative than fluorine, so it occupies the central position. The skeletal structure is:

$$F$$
$$F \quad Cl \quad F$$

Step 2: The outer-shell electron configurations of F and Cl are $2s^2 2p^5$ and $3s^2 3p^5$, respectively. Thus, there are

$$7 + (3 \times 7) = 28 \text{ valence electrons}$$

Step 3: We draw a single covalent bond between Cl and each F, and then complete the octets for the F atoms.

$$:\overset{..}{\underset{..}{F}}:$$
$$\big|$$
$$:\overset{..}{\underset{..}{F}}\!-\!\overset{..}{Cl}\!-\!\overset{..}{\underset{..}{F}}:$$

Step 4: At this point, we still have two electron pairs (4 e$^-$) to place. Fluorine cannot exceed an octet of electrons, so the electrons must be placed as lone pairs on chlorine. The correct Lewis structure is:

$$:\overset{\cdot\cdot}{\underset{}{F}}:$$
$$\mid$$
$$:\overset{\cdot\cdot}{\underset{\cdot\cdot}{F}}-\overset{\cdot\cdot}{\underset{\cdot\cdot}{Cl}}-\overset{\cdot\cdot}{\underset{\cdot\cdot}{F}}:$$

Chlorine can exceed an octet of electrons. An expanded octet can occur for atoms of elements in and beyond the third period of the periodic table. As a final check, we verify that there are 28 valence electrons in the Lewis structure of ClF_3.

PRACTICE EXERCISE

9.   Write the Lewis structure for PCl_5.

$\boxed{\textbf{Text Problem: } 9.62}$

PROBLEM TYPE 7: USING BOND ENERGIES TO ESTIMATE THE ENTHALPY OF A REACTION

A quantitative measure of the stability of a molecule is its **bond dissociation energy** (or **bond energy**), which is the enthalpy change required to break a particular bond in one mole of gaseous molecules. Table 9.4 of the text lists the average bond energies of a number of bonds found in polyatomic molecules, as well as the bond energies of several diatomic molecules.

In many cases, it is possible to predict the approximate enthalpy of reaction by using the average bond energies. Energy *is always required* to break chemical bonds and chemical bond formation is always accompanied by a *release of energy*. We can estimate the enthalpy of reaction by counting the total number of bonds broken and formed in the reaction and recording all the corresponding energy changes. The enthalpy of reaction in the *gas phase* is given by:

$$\Delta H° = \text{total energy input} - \text{total energy released}$$

$$\Delta H° = \Sigma BE(\text{reactants}) - \Sigma BE(\text{products})$$

Where,

BE is the average bond energy

Σ represents summation

If the total energy input is greater than the total energy released, $\Delta H°$ *is positive* and the reaction is *endothermic*. On the other hand, if more energy is released than absorbed, $\Delta H°$ *is negative* and the reaction is *exothermic*.

EXAMPLE 9.11

Use average bond energies to estimate $\Delta H°$ for the following reaction:

$$Cl_2\ (g)\ +\ I_2\ (g)\ \longrightarrow\ 2\ ICl\ (g)$$

The average I–Cl bond energy is 210 kJ.

Recall that, $\Delta H° = \Sigma BE(\text{reactants}) - \Sigma BE(\text{products})$

$$\Delta H° = BE(\text{Cl–Cl}) + BE(\text{I–I}) - 2BE(\text{I–Cl})$$

$$= 242.7\ \text{kJ} + 151.0\ \text{kJ} - 2(210\ \text{kJ})$$

$$\Delta H° = -26\ \text{kJ}$$

Is the reaction exothermic or endothermic?

PRACTICE EXERCISE

10.   Estimate the enthalpy change for the following reaction:

$$N_2\,(g)\; +\; O_2\,(g)\; \longrightarrow\; 2\,NO\,(g)$$

Hint: NO has a double bond [$BE(N{=}O) = 630$ kJ]

Text Problems: **9.68**, 9.70

ANSWERS TO PRACTICE EXERCISES

1.   **(a)** polar covalent   **(b)** covalent   **(c)** ionic

2.   lattice energy (NaF) = 919 kJ/mol

3.   :C̈l—A̎s—C̈l: 4. $\left[\,:C{\equiv}N:\,\right]^{-}$

5.   **(a)**   Formal charge (left N) = $5 - [4 + 1/2(4)] = -1$

          Formal charge (middle N) = $5 - [0 + 1/2(8)] = +1$

          Formal charge (O) = $6 - [4 + 1/2(4)] = 0$

       (b)   Formal charge (O) = $6 - [6 + 1/2(2)] = -1$

          Formal charge (H) = $1 - [0 + 1/2(2)] = 0$

6.   (b) is the more plausible structure based on formal charges. There is a large positive (+2) formal charge on the S atom in structure (a). The formal charge on the S atom in structure (b) is zero.

7.   :N̈=N=Ö: ⟷   :N≡N—Ö: ⟷   :N̈—N≡O:

8.   :C̈l—B—C̈l:

9.   The lone pairs have been left off the chlorine atoms.

10. $\Delta H° = 180$ kJ

SOLUTIONS TO SELECTED TEXT PROBLEMS

9.16   **(a)** RbI, rubidium iodide   **(b)** Cs_2SO_4, cesium sulfate

       (c) Sr_3N_2, strontium nitride   **(d)** Al_2S_3, aluminum sulfide

9.18   The Lewis representations for the reactions are:

       (a) $\dot{Sr}$ + $\cdot \ddot{Se} \cdot$ $\longrightarrow$ Sr^{2+} $:\ddot{Se}:^{2-}$

       (b) $\dot{Ca}$ + $2\,H\cdot$ $\longrightarrow$ Ca^{2+} $2H:^-$

       (c) $3Li\cdot$ + $\cdot\ddot{N}\cdot$ $\longrightarrow$ $3Li^+$ $:\ddot{N}:^{3-}$

       (d) $2\dot{Al}\cdot$ + $3\cdot\ddot{S}\cdot$ $\longrightarrow$ $2Al^{3+}$ $3:\ddot{S}:^{2-}$

9.20   **(a)** Covalent (BF_3, boron trifluoride)   **(b)** ionic (KBr, potassium bromide)

9.26   (1) $Ca(s) \rightarrow Ca(g)$ $\Delta H_1^\circ = 121$ kJ

       (2) $Cl_2(g) \rightarrow 2Cl(g)$ $\Delta H_2^\circ = 242.8$ kJ

       (3) $Ca(g) \rightarrow Ca^+(g) + e^-$ $\Delta H_3^{\circ\,\prime} = 589.5$ kJ

            $Ca^+(g) \rightarrow Ca^{2+}(g) + e^-$ $\Delta H_3^{\circ\,\prime\prime} = 1145$ kJ

       (4) $2[Cl(g) + e^- \rightarrow Cl^-(g)]$ $\Delta H_4^\circ = 2(-349$ kJ$) = -698$ kJ

       (5) $Ca^{2+}(g) + 2Cl^-(g) \rightarrow CaCl_2(s)$ $\Delta H_5^\circ = ?$

            $Ca(s) + Cl_2(g) \rightarrow CaCl_2(s)$ $\Delta H_{overall}^\circ = -795$ kJ

Thus we write:

$$\Delta H_{overall}^\circ = \Delta H_1^\circ + \Delta H_2^\circ + \Delta H_3^{\circ\,\prime} + \Delta H_3^{\circ\,\prime\prime} + \Delta H_4^\circ + \Delta H_5^\circ$$

$$\Delta H_5^\circ = (-795 - 121 - 242.8 - 589.5 - 1145 + 698)\ \text{kJ} = \mathbf{-2195\ kJ}$$

The lattice energy is represented by the reverse of equation (5); therefore, the lattice energy is **+2195 kJ/mol**.

9.36   Classifying Chemical Bonds, Problem Type 1.

The amount of ionic character is based on the electronegativity difference between the two atoms. The larger the electronegativity difference, the greater the ionic character. Let's let ΔEN = electronegativity difference. The bonds arranged in order of increasing ionic character are:

     C–H ($\Delta EN = 0.4$) <   Br–H ($\Delta EN = 0.7$) <   F–H ($\Delta EN = 1.9$) <   Li–Cl ($\Delta EN = 2.0$)

     <   Na–Cl ($\Delta EN = 2.1$) <   K–F ($\Delta EN = 3.2$)

9.38   The order of increasing ionic character is:

Cl–Cl (zero difference in electronegativity) < Br–Cl (difference 0.2) < Si–C (difference 0.7)

< Cs–F (difference 3.3).

9.40   **(a)**   The two silicon atoms are the same. The bond is covalent.

(b)   The electronegativity difference between Cl and Si is 3.0 – 1.8 = 1.2. The bond is polar covalent.

(c)   The electronegativity difference between F and Ca is 4.0 – 1.0 = 3.0. The bond is ionic.

(d)   The electronegativity difference between N and H is 3.0 – 2.1 = 0.9. The bond is polar covalent.

9.44   Writing Lewis Structures, Problem Type 3.

(a)

Step 1: The outer-shell electron configuration of O is $2s^2 2p^4$. Also, we must add the negative charges to the number of valence electrons, Thus, there are

$$(2 \times 6) + 2 = 14 \text{ valence electrons}$$

Step 2: We draw a single covalent bond between each O, and then attempt to complete the octets for the O atoms.

$$\left[:\overset{(-1)}{\underset{..}{\ddot{O}}} - \overset{(-1)}{\underset{..}{\ddot{O}}}: \right]^{2-}$$

The octet rule is satisfied for both oxygen atoms. As a final check, we verify that there are 14 valence electrons in the Lewis structure of O_2^-.

Follow the same procedure as part (a) for parts (b), (c), and (d). The appropriate Lewis structures are:

(b) $\left[:\overset{(-1)}{C} \equiv \overset{(-1)}{C}: \right]^{2-}$ **(c)** $\left[:N \equiv \overset{(+1)}{O}: \right]^{+}$ **(d)** $\left[\begin{array}{c} H \\ | \\ H - \overset{(+1)}{N} - H \\ | \\ H \end{array} \right]^{+}$

9.46   **(a)**   Neither oxygen atom has a complete octet. The left-most hydrogen atom is forming two bonds (4 e⁻). Hydrogen can only be surrounded by at most two electrons.

(b)   The correct structure is:

$$\begin{array}{ccc} H & :\ddot{O}: \\ | & || \\ H - C - C - \ddot{O} - H \\ | \\ H \end{array}$$

Do the two structures have the same number of electrons? Is the octet rule satisfied for all atoms other than hydrogen, which should have a duet of electrons?

9.50 Drawing Resonance Structures, Problem Type 5.

The Lewis structure for ClO_3^- is shown below.

$$\overset{\displaystyle :\!\ddot{O}\!:}{\underset{(+1)}{\overset{(-1)\,..\quad\;\;\|\quad\;\;..\,(-1)}{:\ddot{O}\!-\!\ddot{Cl}\!-\!\ddot{O}:}}}$$

We can draw two more equivalent Lewis structures with the double bond between Cl and a different oxygen atom.

The resonance structures with formal charges are as follows:

$$\underset{(+1)}{\overset{(-1)}{\ddot{O}\!=\!\underset{..}{Cl}\!-\!\overset{..}{\ddot{O}:}}^{(-1)}} \quad\longleftrightarrow\quad \underset{(+1)}{\overset{(-1)\,..\;\overset{:\!\ddot{O}\!:}{\|}\;..\,(-1)}{:\ddot{O}\!-\!\ddot{Cl}\!-\!\ddot{O}:}} \quad\longleftrightarrow\quad \underset{(+1)}{\overset{..\,(-1)}{:\ddot{O}\!-\!\underset{..}{Cl}\!=\!\ddot{O}}}$$

9.52 The structures of the most important resonance forms are:

$$\underset{H-\overset{\overset{\displaystyle H}{|}}{\underset{(+1)}{C}}=\overset{(-1)}{\underset{..}{N}}=\ddot{N}:} \quad\longleftrightarrow\quad \underset{H-\overset{\overset{\displaystyle H}{|}}{\underset{..}{\underset{(-1)}{C}}}-\overset{(+1)}{N}\equiv N:}$$

9.54 Three reasonable resonance structures with the formal charges indicated are

$$\overset{(-1)\;..\;\;\;(+1)\;\;..}{:N\!=\!N\!=\!\ddot{O}:} \quad\longleftrightarrow\quad \overset{(+1)\;\;..\,(-1)}{:N\!\equiv\!N\!-\!\ddot{O}:} \quad\longleftrightarrow\quad \overset{(-2)\,..\;\;(+1)\;\;(+1)}{:\ddot{N}\!-\!N\!\equiv\!O:}$$

9.60 The incomplete octet, Problem Type 6A.

The octet rule for Be is *not* satisfied in this compound. The Lewis structure is:

$$:\!\ddot{Cl}\!-\!Be\!-\!\ddot{Cl}\!:$$

An octet of electrons on Be can only be formed by making two double bonds as shown below:

$$\overset{(+1)\;..\qquad(-2)\qquad..\,(+1)}{\ddot{Cl}\!=\!Be\!=\!\ddot{Cl}}$$

This places a high negative formal charge on Be and positive formal charges on the Cl atoms. This structure distributes the formal charges counter to the electronegativities of the elements. It is not a plausible Lewis structure.

9.62   The outer electron configuration of antimony is $5s^2 5p^3$. The Lewis structure is shown below. All five valence electrons are shared in the five covalent bonds. The octet rule is not obeyed. (The electrons on the chlorine atoms have been omitted for clarity.)

Can Sb have an expanded octet?

9.64   The reaction can be represented as:

$$:\ddot{C}l—Al—\ddot{C}l: \quad + \quad :\ddot{C}l:^{-} \quad \longrightarrow \quad \left[\begin{array}{c} :\ddot{C}l: \\ | \\ :\ddot{C}l—Al—\ddot{C}l: \\ | \\ :\ddot{C}l: \end{array} \right]^{-}$$

The new bond formed is called a coordinate covalent bond.

9.68   This problem is similar to Problem Type 7, Using Bond Energies to Estimate the Enthalpy of a Reaction.

There are two oxygen-to-oxygen bonds in ozone. We will represent these bonds as O–O. However, these bonds might not be true oxygen-to-oxygen single bonds.

$\Delta H° = \Sigma BE(\text{reactants}) - \Sigma BE(\text{products})$

$\Delta H° = BE(\text{O=O}) - 2BE(\text{O–O})$

In the problem, we are given $\Delta H°$ for the reaction, and we can look up the O=O bond energy in Table 9.4 of the text. Solving for the average bond energy in ozone,

$-2BE(\text{O–O}) = \Delta H° - BE(\text{O=O})$

$$BE(\text{O–O}) = \frac{BE(\text{O=O}) - \Delta H°}{2} = \frac{498.7 \text{ kJ / mol} + 107.2 \text{ kJ / mol}}{2} = \textbf{303.0 kJ/mol}$$

Considering the resonance structures for ozone, is it expected that the O–O bond energy in ozone is between the single O–O bond energy (142 kJ) and the double O=O bond energy (498.7 kJ)?

9.70   **(a)**

Bonds Broken	Number Broken	Bond Energy (kJ/mol)	Energy Change (kJ)
C – H	12	414	4968
C – C	2	347	694
O = O	7	498.7	3491

Bonds Formed	Number Formed	Bond Energy (kJ/mol)	Energy Change (kJ)
C = O	8	799	6392
O – H	12	460	5520

$$\Delta H° = \text{total energy input} - \text{total energy released}$$

$$= (4968 + 694 + 3491) - (6392 + 5520) = \textbf{-2759 kJ}$$

(b) $\Delta H° = 4\,\Delta H_f°\,(CO_2) + 6\,\Delta H_f°\,(H_2O) - [2\,\Delta H_f°\,(C_2H_6) + 7\,\Delta H_f°\,(O_2)]$

$\Delta H° = (4\ \text{mol})(-393.5\ \text{kJ/mol}) + (6\ \text{mol})(-241.8\ \text{kJ/mol}) - (2\ \text{mol})(-84.7\ \text{kJ/mol}) - (7\ \text{mol})(0) = \textbf{-2855 kJ}$

The answers for part (a) and (b) are different, because *average* bond energies are used for part (a).

9.72   Typically, ionic compounds are composed of a metal cation and a nonmetal anion. RbCl and KO_2 are ionic compounds.

Typically, covalent compounds are composed of two nonmetals. PF_5, BrF_3, and CI_4 are covalent compounds.

9.74   Recall that you can classify bonds as ionic or covalent based on electronegativity difference.

The melting points (°C) are shown in parentheses following the formulas.

Ionic: NaF (993) MgF_2 (1261) AlF_3 (1291)

Covalent: SiF_4 (−90.2) PF_5 (−83) SF_6 (−121) ClF_3 (−83)

Is there any correlation between ionic character and melting point?

9.76   KF is an ionic compound. It is a solid at room temperature made up of K^+ and F^- ions. It has a high melting point, and it is a strong electrolyte. Benzene, C_6H_6, is a covalent compound that exists as discrete molecules. It is a liquid at room temperature. It has a low melting point, is insoluble in water, and is a nonelectrolyte.

9.78   The resonance structures are:

$$\overset{(-1)}{:\!N}=\overset{(+1)}{N}=\overset{(-1)}{N\!:} \quad \longleftrightarrow \quad :N\!\overset{(+1)}{\equiv}\!N\overset{(-2)}{-\ddot{N}\!:} \quad \longleftrightarrow \quad \overset{(-2)}{:\!\ddot{N}}\overset{(+1)}{-N}\!\equiv\!N\!:$$

Which is the most plausible structure based on a formal charge argument?

9.80   **(a)**   An example of an aluminum species that satisfies the octet rule is the anion $AlCl_4^-$. The Lewis dot structure is drawn in Problem 9.64.

(b)   An example of an aluminum species containing an expanded octet is anion AlF_6^{3-}. (How many pairs of electrons surround the central atom?)

(c)   An aluminum species that has an incomplete octet is the compound $AlCl_3$. The dot structure is given in Problem 9.64.

9.82 CF_2 would be very unstable because carbon does not have an octet. (How many electrons does it have?)

LiO_2 would not be stable because the lattice energy between Li^+ and superoxide O_2^- would be too low to stabilize the solid.

$CsCl_2$ requires a Cs^{2+} cation. The second ionization energy is too large to be compensated by the increase in lattice energy.

PI_5 appears to be a reasonable species (compared to PF_5 in Example 9.10 of the text). However, the iodine atoms are too large to have five of them "fit" around a single P atom.

9.84   **(a)** false   **(b)** true   **(c)** false   **(d)** false

For question (c), what is an example of a second-period species that violates the octet rule?

9.86   The formation of CH_4 from its elements is:

$$C\,(s) \;+\; 2H_2\,(g) \;\longrightarrow\; CH_4\,(g)$$

The reaction could take place in two steps:

Step 1: $C\,(s) + 2\,H_2\,(g) \longrightarrow C\,(g) + 4\,H(g)$ $\Delta H^{\circ}_{rxn} = 716\ kJ + 872.8\ kJ = 1589\ kJ$

Step 2: $C\,(g) + 4\,H\,(g) \longrightarrow CH_4\,(g)$ $\Delta H^{\circ}_{rxn} \approx -4 \times$ (bond energy of C–H bond)

$$= -4 \times 414\ kJ = -1656\ kJ$$

Therefore, ΔH°_{f} (CH_4) would be approximately the sum of the enthalpy changes for the two steps. See Problem Type 4C, Chapter 6 (Hess's law).

$$\Delta H^{\circ}_{f}\,(CH_4) = \Delta H^{\circ}_{rxn}\,(1) + \Delta H^{\circ}_{rxn}\,(2)$$

$$\Delta H^{\circ}_{f}\,(CH_4) = 1589 - 1656\ kJ/mol = \textbf{-67 kJ/mol}$$

The actual value of ΔH°_{f} (CH_4) = $-74.85\ kJ/mol$

9.88   Only N_2 has a triple bond. Therefore, it has the shortest bond length.

9.90   To be isoelectronic, molecules must have the same number and arrangement of valence electrons. NH_4^+ and CH_4 are isoelectronic (8 valence electrons), as are CO and N_2 (10 valence electrons), as are $B_3N_3H_6$ and C_6H_6 (30 valence electrons). Draw Lewis structures to convince yourself that the electron arrangements are the same in each isoelectronic pair.

9.92   The reaction can be represented as:

$$H-\ddot{N}\overset{..}{}^{-} \;+\; H-\ddot{O}: \;\longrightarrow\; H-\ddot{N}-H \;+\; \overset{..}{}\ddot{O}-H$$
$$\underset{H}{|} \qquad\quad \underset{H}{|} \qquad\qquad\quad \underset{H}{|}$$

9.94   The central iodine atom in I_3^- has *ten* electrons surrounding it: two bonding pairs and three lone pairs. The central iodine has an expanded octet. Elements in the second period such as fluorine cannot have an expanded octet as would be required for F_3^-.

9.96   The skeletal structure is:

$$
\begin{array}{ccccc}
 & & \text{H} & & \\
\text{H} & \text{C} & \text{N} & \text{C} & \text{O} \\
 & & \text{H} & &
\end{array}
$$

The number of valence electron is: $(1 \times 3) + (2 \times 4) + 5 + 6 = 22$ valence electrons

We can draw two resonance structures for methyl isocyanate.

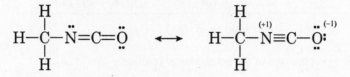

9.98   **(a)**   This is a very good resonance form; there are no formal charges and each atom satisfies the octet rule.

       (b)   This is a second choice after (a) because of the positive formal charge on the oxygen (high electronegativity).

       (c)   This is a poor choice for several reasons. The formal charges are placed counter to the electronegativities of C and O, the oxygen atom does not have an octet, and there is no bond between that oxygen and carbon!

       (d)   This is a mediocre choice because of the large formal charge and lack of an octet on carbon.

9.100   The nonbonding electron pairs around Cl and F are omitted for simplicity.

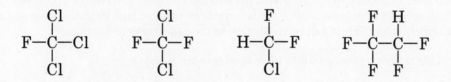

9.102   **(a)**   Using Equation (9.4) of the text,

            $\Delta H = \sum BE(\text{reactants}) - \sum BE(\text{products})$

            $\boldsymbol{\Delta H} = [(436.4 + 151.0) - 2(298.3)] = \boldsymbol{-9.2 \text{ kJ}}$

       (b)   Using Equation (6.8) of the text,

            $\Delta H^{\circ} = 2 \Delta H_{f}^{\circ} [\text{HI}(g)] - \{ \Delta H_{f}^{\circ} [\text{H}_2(g)] + \Delta H_{f}^{\circ} [\text{I}_2(g)] \}$

            $\Delta H^{\circ} = (2 \text{ mol})(25.9 \text{ kJ/mol}) - [(0) + (1 \text{ mol})(61.0 \text{ kJ/mol})] = \boldsymbol{-9.2 \text{ kJ}}$

9.104   The Lewis structures are:

(a) $:\overset{(-1)}{\text{C}}\equiv\overset{(+1)}{\text{O}}:$ **(b)** $:\text{N}\equiv\overset{(+1)}{\text{O}}:$ **(c)** $:\overset{(-1)}{\text{C}}\equiv\text{N}:$ **(d)** $:\text{N}\equiv\text{N}:$

9.106   True. Each noble gas atom already has completely filled ns and np subshells.

9.108   **(a)**   The bond energy of F_2^- is the energy required to break up F_2^- into an F atom and an F^- ion.

$$F_2^-\,(g) \longrightarrow F\,(g) + F^-\,(g)$$

We can arrange the equations given in the problem so that they add up to the above equation. See Problem Type 4C, Chapter 6 (Hess's law).

$$F_2^-\,(g) \longrightarrow F_2\,(g) + e^- \qquad \Delta H° = 290 \text{ kJ}$$
$$F_2\,(g) \longrightarrow 2F\,(g) \qquad \Delta H° = 156.9 \text{ kJ}$$
$$F\,(g) + e^- \longrightarrow F^-\,(g) \qquad \Delta H° = -333 \text{ kJ}$$
$$\overline{F_2^-\,(g) \longrightarrow F\,(g) + F^-\,(g)}$$

The bond energy of F_2^- is the sum of the enthalpies of reaction.

$$BE(F_2^-) = 290 \text{ kJ} + 156.9 \text{ kJ} + (-333 \text{ kJ}) = \textbf{114 kJ}$$

(b)   The bond in F_2^- is weaker (114 kJ) than the bond in F_2 (156.9 kJ), because the extra electron increases repulsion between the F atoms.

9.110   In **(a)** there is a lone pair on the C atom and the negative formal charge is on the less electronegative C atom.

9.112   **(a)**

$$:\!\overset{\bullet}{N}\!=\!\overset{..}{\underset{..}{O}} \quad \longleftrightarrow \quad \overset{(-)}{:}\!\overset{..}{N}\!=\!\overset{\bullet}{O}{}^{(+)}$$

The first structure is the most important. Both N and O have formal charges of zero. In the second structure, the more electronegative oxygen atom has a formal charge of +1. Having a positive formal charge on an highly electronegative atom is not favorable. In addition, both structures leave one atom with an incomplete octet. This cannot be avoided due to the odd number of electrons.

(b)   It is not possible to draw a structure with a triple bond between N and O.

$$:\!N\!\equiv\!\overset{\bullet}{O}\!:$$

Any structure drawn with a triple bond will lead to an expanded octet. Elements in the second row of the period table cannot exceed the octet rule.

9.114   The OCOO structure violates the octet rule (expanded octet). The structure shown below satisfies the octet rule with 22 valence electrons. However, CO_3^{2-} has 24 valence electrons. Adding two more electrons to the structure would cause at least one atom to exceed the octet rule.

$$\overset{..}{\underset{..}{O}}\!=\!C\!=\!\overset{..}{\underset{..}{O}}\!-\!\overset{..}{\underset{..}{O}}\!:$$

9.116

The arrows indicate coordinate covalent bonds.

9.118   There are four C–H bonds in CH_4, so the average bond energy of a C–H bond is:

$$\frac{1656 \text{ kJ / mol}}{4} = 414 \text{ kJ / mol}$$

The Lewis structure of propane is:

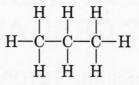

There are eight C–H bonds and two C–C bonds. We write:

$$8(\text{C–H}) + 2(\text{C–C}) = 4006 \text{ kJ/mol}$$

$$8(414 \text{ kJ/mol}) + 2(\text{C–C}) = 4006 \text{ kJ/mol}$$

$$2(\text{C–C}) = 694 \text{ kJ/mol}$$

So, the average bond energy of a C–C bond is: $\dfrac{694}{2}$ kJ / mol = **347 kJ/mol**

9.120

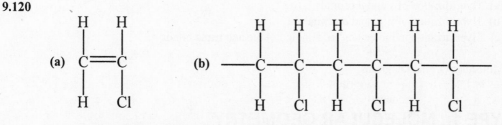

(c)   In the formation of poly(vinyl chloride) form vinyl chloride, for every C=C double bond broken, 2 C–C single bonds are formed. No other bonds are broken or formed. The energy changes for 1 mole of vinyl chloride reacted are:

total energy input (breaking C=C bonds) = 620 kJ

total energy released (forming C–C bonds) = 2 × 347 kJ = 694 kJ

$\Delta H° = 620 \text{ kJ} - 694 \text{ kJ} = -74 \text{ kJ}$

The negative sign shows that this is an exothermic reaction. To find the total heat released when 1.0×10^3 kg of vinyl chloride react, we proceed as follows:

$$\text{heat released} = 1.0 \times 10^6 \text{ g } C_2H_3Cl \times \frac{1 \text{ mol } C_2H_3Cl}{62.49 \text{ g } C_2H_3Cl} \times \frac{-74 \text{ kJ}}{1 \text{ mol } C_2H_3Cl} = \mathbf{-1.2 \times 10^6 \text{ kJ}}$$

9.122 $EN(O) = \dfrac{1314 + 141}{2} = 727.5$ $EN(F) = \dfrac{1680 + 328}{2} = 1004$ $EN(Cl) = \dfrac{1251 + 349}{2} = 800$

Using Mulliken's definition, the electronegativity of chlorine is greater than that of oxygen, and fluorine is still the most electronegative element. We can convert to the Pauling scale by dividing each of the above by 230 kJ/mol.

$EN(O) = \dfrac{727.5}{230} = \mathbf{3.16}$ $EN(F) = \dfrac{1004}{230} = \mathbf{4.37}$ $EN(Cl) = \dfrac{800}{230} = \mathbf{3.48}$

These values compare to the Pauling values for oxygen of 3.5, fluorine of 4.0, and chlorine of 3.0.

CHAPTER 10
CHEMICAL BONDING II: MOLECULAR GEOMETRY AND HYBRIDIZATION OF ATOMIC ORBITALS

PROBLEM-SOLVING STRATEGIES AND TUTORIAL SOLUTIONS

TYPES OF PROBLEMS

Problem Type 1: Molecular Geometry.
      (a) Molecules in which the central atom has *no* lone pairs.
      (b) Molecules in which the central atom has one or more lone pairs.
      (c) Geometry of molecules with more than one central atom.

Problem Type 2: Predicting Dipole Moments.

Problem Type 3: Hybridization of Atomic Orbitals.
      (a) Hybridization of *s* and *p* orbitals.
      (b) Hybridization of *s*, *p*, and *d* orbitals.
      (c) Hybridization in molecules containing double and triple bonds.

Problem Type 4: Molecular Orbital Diagrams.

PROBLEM TYPE 1: MOLECULAR GEOMETRY

The model that we are going to use to study molecular geometry is called the **valence-shell electron-pair repulsion (VSEPR) model**. It accounts for the geometric arrangement of electron pairs around a central atom in terms of the repulsion between electron pairs. The geometry that a molecule ultimately adopts *minimizes* electron-pair repulsion.

Guidelines for Applying the VSEPR Model

1. Write the Lewis structure of the molecule.

2. Only consider the electron pairs around the *central atom* (the atom that is bonded to more than one other atom). Count the number of electron pairs around the central atom (bonding pairs and lone pairs). For counting purposes, treat double and triple bonds as though they were single bonds. Refer to Table 10.1 of your text to predict the overall arrangement of the electron pairs.

3. Use Tables 10.1 and 10.2 of your text to predict the *geometry* of the molecule.

4. In predicting bond angles, note that a lone pair repels another lone pair or a bonding pair more strongly than a bonding pair repels another bonding pair. There is no easy way to predict bond angles accurately when the central atom possesses one or more lone pairs.

A. Molecules in which the central atom has *no* lone pairs

For simplicity, we will only consider molecules that contain atoms of two elements, A and B, where A is the central atom. These molecules have the general formula AB_x, where x is an integer 2, 3, ... In most cases, x is between 2 and 6.

Table 10.1 of your text shows five possible arrangements of electron pairs around the central atom A. Remember, these arrangements are adopted because electron-pair repulsions are minimized.

Below is a condensed version of Table 10.1 from your text.

Arrangement of electron pairs around a central atom (A) in a molecule, and geometry of molecules if the central atom has *no* lone pairs.

Number of electron pairs around central atom	Arrangement of electron pairs	Molecular geometry
2	Linear, 180°	Linear, AB_2
3	Trigonal planar, 120°	Trigonal planar, AB_3
4	Tetrahedral, 109.5°	Tetrahedral, AB_4
5	Trigonal bipyramidal, 120°, 90°	Trigonal bipyramidal, AB_5
6	Octahedral, 90°	Octahedral, AB_6

Note: Since there are *no* lone pairs around the central atom in all the examples shown above, the molecular geometry is *always* the same as the arrangement of electron pairs around the central atom.

EXAMPLE 10.1

Use the VSEPR model to predict the geometry of the following molecules and ions: (a) $HgCl_2$, (b) $SnCl_4$, (c) NO_3^-, (d) PF_5.

(a)

Step 1: Write the Lewis structure of the molecule (see Problem Type 3, Chapter 9).

$$:\overset{\cdot\cdot}{\underset{\cdot\cdot}{Cl}}-Hg-\overset{\cdot\cdot}{\underset{\cdot\cdot}{Cl}}:$$

Step 2: Count the number of electron pairs around the central atom. There are two electron pairs around Hg.

Step 3: Since there are two electron pairs around Hg, the electron-pair arrangement that minimizes repulsion is **linear**.

In addition, since there are no lone pairs around the central atom, the geometry is also **linear** (AB_2).

(b)

Step 1: Write the Lewis structure of the molecule.

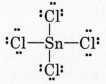

Step 2: Count the number of electron pairs around the central atom. There are four electron pairs around Sn.

Step 3: Since there are four electron pairs around Sn, the electron-pair arrangement that minimizes repulsion is **tetrahedral**.

In addition, since there are no lone pairs around the central atom, the geometry is also **tetrahedral** (AB_4).

(c)

Step 1: Write the Lewis structure of the molecule.

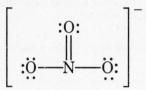

Step 2: Count the number of electron pairs around the central atom. There are three electron pairs around N. Remember, for VSEPR purposes, a double or triple bond counts the same as a single bond.

Step 3: Since there are three electron pairs around N, the electron-pair arrangement that minimizes repulsion is **trigonal planar**.

In addition, since there are no lone pairs around the central atom, the geometry is also **trigonal planar** (AB$_3$).

(d)

Step 1: Write the Lewis structure of the molecule.

Step 2: Count the number of electron pairs around the central atom. There are five electron pairs around P.

Since there are five electron pairs around P, the electron-pair arrangement that minimizes repulsion is **trigonal bipyramidal**.

In addition, since there are no lone pairs around the central atom, the geometry is also **trigonal bipyramidal** (AB$_5$).

PRACTICE EXERCISE

1.   Use the VSEPR model to predict the geometry of the following molecules: (a) CO_2, (b) CCl_4, (c) SO_3, (d) SF_6.

Text Problems: **10.8**, 10.10, 10.12

B. Molecules in which the central atom has one or more lone pairs

If lone pairs are present on the central atom, the overall arrangement of electron pairs is *not* the same as the geometry of the molecule.

To keep track of the total number of bonding pairs and lone pairs, we will designate molecules with lone pairs as AB_xE_y, where A is the central atom, B is a surrounding atom, and E is a lone pair on the central atom, A. Both x and y are integers: x = 2, 3, . . . , and y = 1, 2, ...

When working with molecules that have lone pairs on the central atom, remember to count all electron pairs on the central atom, both bonding pairs and lone pairs. The total number of electron pairs around the central atom determines the electron arrangement around the central atom. See Table 10.1. However, the molecular geometry will not be the same as this electron arrangement. Use Table 10.2 of your text to determine the molecular geometry, or a better option is to build a model. Gum drops and toothpicks make an effective and inexpensive model kit. And besides, you can eat your model when you are finished. (Not the toothpicks!)

Below is a condensed version of Table 10.2 of your text.

Geometry of simple molecules and ions in which the central atom has one or more lone pairs.

Class of molecule	Total # of electron pairs on central atom	Number of bonding pairs	Arrangement of electron pairs	Number of lone pairs	Molecular geometry
AB_2E	3	2	Trigonal planar	1	Bent
AB_3E	4	3	Tetrahedral	1	Trigonal pyramid
AB_2E_2	4	2	Tetrahedral	2	Bent
AB_4E	5	4	Trigonal bipyramidal	1	Distorted tetrahedron
AB_3E_2	5	3	Trigonal bipyramidal	2	T-shaped
AB_2E_3	5	2	Trigonal bipyramidal	3	Linear
AB_5E	6	5	Octahedral	1	Square pyramidal
AB_4E_2	6	4	Octahedral	2	Square planar

> **Note:** The arrangement of 3 electron-pairs is always trigonal planar; the arrangement of 4 electron-pairs is tetrahedral; the arrangement of 5 electron-pairs is trigonal bipyramidal; the arrangement of 6 electron-pairs is octahedral. However, if there are any lone pairs on the central atom, the molecular geometry will be different from the electron arrangement. **Build models!**

EXAMPLE 10.2

Use the VSEPR model to predict the geometry of the following molecules: (a) O_3, (b) XeF_2, (c) IF_5.

(a)

Step 1: Write the Lewis structure of the molecule (see Problem Type 3, Chapter 9).

$$\ddot{O}{=}\ddot{O}{-}\ddot{\underset{..}{O}}{:}$$

Step 2: Count the number of electron pairs around the central atom. There are three electron pairs around the central oxygen. Remember, for VSEPR purposes, a double or triple bond counts the same as a single bond.

Step 3: Since there are three electron pairs around the central O, the electron-pair arrangement that minimizes repulsion is **trigonal planar**.

However, there is one lone pair of electrons around the central atom. Consulting a model or Table 10.2 of your text, you should find that the geometry of the molecule is **bent** (AB_2E).

(b)

Step 1: Write the Lewis structure of the molecule.

$$:\ddot{\underset{..}{F}}{-}\ddot{\underset{..}{Xe}}{-}\ddot{\underset{..}{F}}{:}$$

Step 2: Count the number of electron pairs around the central atom. There are five electron pairs around xenon.

Step 3: Since there are five electron pairs around xenon, the electron-pair arrangement that minimizes repulsion is **trigonal bipyramidal**.

However, there are three lone pairs of electrons around the central atom. Consulting a model or Table 10.2 of the text, you should find that the geometry of the molecule is **linear** (AB_2E_3).

(c)

Step 1: Write the Lewis structure of the molecule.

Step 2: Count the number of electron pairs around the central atom. There are six electron pairs around iodine.

Step 3: Since there are six electron pairs around iodine, the electron-pair arrangement that minimizes repulsion is **octahedral**.

However, there is one lone pair of electrons around the central atom. Consulting a model or Table 10.2 of the text, you should find that the geometry of the molecule is **square pyramidal** (AB_5E).

PRACTICE EXERCISE

2.   Use the VSEPR model to predict the geometry of the following molecules: (a) SO_2, (b) XeF_4, (c) SF_4.

Text Problems: 10.10, 10.12

C. Geometry of molecules with more than one central atom

So far, we have discussed the geometry of molecules having only one central atom. (The term "central atom" means an atom that is not a terminal atom in a polyatomic molecule.) Many molecules will have more than one "central atom". In these cases, you have to apply the VSEPR method presented in parts A and B above to each of the central atoms.

EXAMPLE 10.3

Use the VSEPR model to predict the geometry of C_2H_6.

Step 1: Write the Lewis structure of the molecule. Hydrogens must be in terminal positions. The two carbons must be the "central atoms".

Step 2: Count the number of electron pairs around the "central atoms". There are four electron pairs around each carbon.

Step 3: Since there are four electron pairs around each carbon, the electron-pair arrangement around each carbon that minimizes repulsion is **tetrahedral**.

In addition, since there are no lone pairs around the central atoms, their geometries are also **tetrahedral**.

PRACTICE EXERCISE

3.   Use the VSEPR model to predict the geometry of C_2H_2.

PROBLEM TYPE 2: PREDICTING DIPOLE MOMENTS

To determine if a molecule has a dipole moment (a measure of electrical charge separation in a molecule) you must consider two factors.

1. Are the *bonds* in the molecule polar?

    A bond will be polar if there is a difference in electronegativity between the two atoms comprising the bond (see Section 9.5 of your text). For example, in an O–H bond, there is a shift in electron density from H to O because O is more electronegative than H. The shift in electron density is symbolized by placing a crossed arrow ($\longmapsto$) above the bond to indicate the direction of the shift in electron density. This is called a *bond* moment. For example:

$$\overset{\longleftarrow\!\!\!+}{\text{O – H}}$$

    The consequent charge separation can be represented as:

$$\overset{\delta^- \quad \delta^+}{\text{O – H}}$$

    where δ (delta) represents a partial charge.

2. Is the *molecule* polar?

    Diatomic molecules containing atoms of the same element do not have dipole moments and so are **nonpolar molecules**. There is no difference in electronegativity since the two elements are the same.

    However, most molecules will have polar bonds (bond moments). If a molecule has polar bonds, does this mean that it is a polar molecule? Not necessarily. The *bond moment* is a vector quantity, which means that it has both a magnitude and direction. The measured *dipole moment* of the molecule is equal to the vector sum of the bond moments. For example, in CO_2, the two bond moments are equal in magnitude and opposite in direction. The sum or resultant dipole moment will be *zero*. Hence, CO_2 is a **nonpolar molecule**.

$$\overset{\longleftarrow\!+ \quad +\!\longrightarrow}{\ddot{\text{O}}\!=\!\text{C}\!=\!\ddot{\text{O}}}$$

    In summary, to determine if a molecule is **polar** (i.e., does the molecule have a dipole moment), you must sum the individual bond moments to determine if there is a resultant dipole moment.

EXAMPLE 10.4

Predict whether each of the following molecules has a dipole moment: (a) CCl_4 and (b) $CHCl_3$.

(a)

Step 1: Write the Lewis structure for the molecule. The lone pairs on Cl have been omitted.

Step 2: Shown on the Lewis structure are the bond moments. Chlorine is more electronegative than C, so the arrows indicate the shift in electron density toward Cl. However, these polar bonds are arranged in a symmetric tetrahedral fashion about the central C atom. The sum or resultant dipole moment is *zero*. CCl_4 is a **nonpolar molecule**.

(b)

Step 1: Write the Lewis structure for the molecule. The lone pairs on Cl have been omitted.

Step 2: Shown on the Lewis structure are the bond moments. Chlorine is more electronegative than C, so the arrows indicate the shift in electron density toward Cl. The electronegativity difference between C and H is very small, so a C–H bond is essentially nonpolar. In this case, the three bond moments partially reinforce each other. Thus, $CHCl_3$ has a dipole moment and is therefore a **polar molecule**.

PRACTICE EXERCISE

4.   Predict whether each of the following molecules has a dipole moment: (a) CO, (b) BCl_3, and (c) XeF_4.

Text Problems: 10.18, 10.20, **10.22**

PROBLEM TYPE 3: HYBRIDIZATION OF ATOMIC ORBITALS

The **VSEPR** model is very powerful considering its simplicity; however, it does not explain chemical bond formation in any detail. In the 1930s, **valence bond (VB) theory** was introduced to account for chemical bond formation. **VB** theory describes covalent bonding as the overlapping of atomic orbitals. This means that the orbitals share a common region in space.

VB theory uses a concept called **hybridization** to explain covalent bonding. Hybridization is the mixing of atomic orbitals in an atom (usually a central atom) to generate a set of new atomic orbitals, called *hybrid orbitals*. Hybrid orbitals are atomic orbitals obtained when two or more nonequivalent orbitals of the same atom combine. The hybrid orbitals are used to form covalent bonds.

A. Hybridization of *s* and *p* orbitals

1. *sp* hybrid orbitals

Let's consider the central atom Be in the BeH_2 molecule. Be has a ground state electron configuration of $1s^2 2s^2$. Only valence electrons are involved in bonding, so an orbital diagram of the valence electrons is:

$$\underset{2s}{\underline{\uparrow\downarrow}} \qquad \underset{2p}{\underline{\quad}\ \underline{\quad}\ \underline{\quad}}$$

With this ground-state electron configuration, Be cannot form bonds with H because its valence electrons are paired in a 2s orbital.

To explain the bonding, first an electron is promoted from the 2s orbital to a 2p orbital.

$$\underset{2s}{\underline{\uparrow}} \qquad \underset{2p}{\underline{\uparrow}\ \underline{\quad}\ \underline{\quad}}$$

Now, we have two different orbitals that can bond to the two hydrogens, which would result in two nonequivalent Be–H bonds. However, experimental evidence shows that there are two equivalent Be–H bonds.

This is where hybridization comes in. By mixing the 2s orbital with one of the 2p orbitals, we can generate two equivalent sp hybrid orbitals.

$$\underset{\substack{sp \text{ hybrid} \\ \text{orbitals}}}{\uparrow \quad \uparrow} \qquad \underset{\substack{\text{empty } p \\ \text{orbitals}}}{\rule{1em}{0.4pt} \quad \rule{1em}{0.4pt}}$$

Figure 10.10 of your text shows the shape and orientation of the sp hybrid orbitals. These two hybrid orbitals lie along the same line so that the angle between them is 180° (linear arrangement). Each of the BeH bonds is formed by the overlap of a Be sp hybrid and a H 1s orbital. The resulting BeH_2 molecule has a linear geometry.

2. **sp^2 hybrid orbitals**

Following the same type of argument for sp hybrids, sp^2 hybrid orbitals are formed by mixing an s orbital with two p orbitals. Three equivalent sp^2 hybrid orbitals are formed.

$$\underset{s}{\uparrow} \qquad \underset{p}{\uparrow \quad \uparrow} \quad \rule{1em}{0.4pt}$$

Mixing an s with two p orbitals gives:

$$\underset{\substack{sp^2 \text{ hybrid} \\ \text{orbitals}}}{\uparrow \quad \uparrow \quad \uparrow} \qquad \underset{\substack{\text{empty } p \\ \text{orbital}}}{\rule{1em}{0.4pt}}$$

Figure 10.12 of your text shows the shape and orientation of the sp^2 hybrid orbitals. These three hybrid orbitals lie in a plane with an angle between any two hybrids of 120°. The three sp^2 hybrid orbitals are arranged in a trigonal planar fashion.

3. **sp^3 hybrid orbitals**

Again, following the same argument as above, sp^3 hybrid orbitals are formed by mixing an s orbital with three p orbitals. Four equivalent sp^3 hybrid orbitals are formed.

$$\underset{s}{\uparrow} \qquad \underset{p}{\uparrow \quad \uparrow \quad \uparrow}$$

Mixing an s with three p orbitals gives:

$$\underset{\substack{sp^3 \text{ hybrid} \\ \text{orbitals}}}{\uparrow \quad \uparrow \quad \uparrow \quad \uparrow}$$

Figure 10.7 of your text shows the shape and orientation of the sp^3 hybrid orbitals. These four equivalent hybrid orbitals are directed toward the four corners of a regular tetrahedron with 109.5° bond angles.

Summarizing,

Type of hybrid	No. of equivalent hybrid orbitals	Arrangement of hybrid orbitals	No. of empty p orbitals
sp	2	linear, 180°	2
sp^2	3	trigonal planar, 120°	1
sp^3	4	tetrahedral, 109.5°	0

EXAMPLE 10.5

Determine the hybridization of the central (underlined) atom in each of the following molecules: (a) $\underline{C}Cl_4$ and (b) $\underline{B}Cl_3$.

(a)

Step 1: Write the Lewis structure of the molecule. The lone pairs on the chlorines have been omitted.

Step 2: Count the number of electron pairs around the central atom. Since there are four electron pairs around C, the electron arrangement that minimizes electron-pair repulsion is **tetrahedral**.

We conclude that C is sp^3 **hybridized** because it has the electron arrangement of four sp^3 hybrid orbitals.

(b)

Step 1: Write the Lewis structure of the molecule. The lone pairs on the chlorines have been omitted.

Step 2: Count the number of electron pairs around the central atom. Since there are three electron pairs around B, the electron arrangement that minimizes electron-pair repulsion is **trigonal planar**.

We conclude that B is sp^2 **hybridized** because it has the electron arrangement of three sp^2 hybrid orbitals.

PRACTICE EXERCISE

5.   Determine the hybridization of the central (underlined) atom in each of the following molecules or ions:
      (a) $\underline{C}O_2$ and (b) $\underline{C}O_3^{2-}$.

Text Problems: **10.32**, 10.34, 10.36, **10.38**, **10.40**

B.   Hybridization of *s*, *p*, and *d* orbitals

1. sp^3d hybrid orbitals

We use the same approach that we used for hybridizing *s* and *p* orbitals, but now we are also mixing in a *d* orbital. The sp^3d hybrid orbitals are formed by mixing an *s* orbital, three *p* orbitals, and a *d* orbital. Five equivalent sp^3d hybrid orbitals are formed.

Mixing an s orbital, three p orbitals, and a d orbital gives:

↑ ↑ ↑ ↑ ↑   ___ ___ ___ ___
sp^3d hybrid   empty d orbitals
orbitals

Table 10.4 of your text shows the shape and orientation of the sp^3d hybrid orbitals. These five equivalent hybrid orbitals are directed toward the five corners of a trigonal bipyramid with bond angles of 120° and 90°.

2. **sp^3d^2 hybrid orbitals**

We use the same approach that we used above, but now we are mixing in two d orbitals. The sp^3d^2 hybrid orbitals are formed by mixing an s orbital, three p orbitals, and two d orbitals. Six equivalent sp^3d^2 hybrid orbitals are formed.

↑   ↑ ↑ ↑   ↑ ↑ ___ ___ ___
s p d

Mixing an s orbital, three p orbitals, and two d orbitals gives:

↑ ↑ ↑ ↑ ↑ ↑   ___ ___ ___
sp^3d^2 hybrid   empty d orbitals
orbitals

Table 10.4 of your text shows the shape and orientation of the sp^3d^2 hybrid orbitals. These six equivalent hybrid orbitals are directed toward the six corners of an octahedron with bond angles of 90°.

Summarizing,

Type of hybrid	No. of equivalent hybrid orbitals	Arrangement of hybrid orbitals
sp^3d	5	trigonal bipyramid, 120°, 90°
sp^3d^2	6	octahedral, 90°

EXAMPLE 10.6
Describe the hybridization of xenon in xenon tetrafluoride (XeF_4).

Step 1: Write the Lewis structure for XeF_4. The lone pairs of electrons on F have been omitted.

Step 2: Count the number of electron pairs around the central atom. Since there are six electron pairs around Xe, the electron arrangement that minimizes electron-pair repulsions is **octahedral**.

We conclude that Xe is **sp^3d^2 hybridized** because it has the electron arrangement of six sp^3d^2 hybrid orbitals.

> **Tip:** It is important to use the electron arrangement to determine the hybridization of the central atom and not the geometry. The two lone pairs on Xe are occupying two of the hybrid orbitals, so the lone pairs must be included to determine the correct hybridization.

PRACTICE EXERCISE

6.   Describe the hybridization of sulfur in sulfur tetrafluoride, SF_4.

| Text Problem: 10.40 |

C.   Hybridization in molecules containing double or triple bonds

We can determine the hybridization of molecules containing double or triple bonds in the same manner as molecules with single bonds. Furthermore, we would like to determine which orbitals overlap to form the double or triple bond.

In a double or triple bond, there are two types of covalent bonds formed. One involves end-to-end overlap of orbitals in which the electron density is concentrated between the nuclei of the bonding atoms. A bond of this type is called a **sigma bond (σ bond)**. The second type involves side-to-side overlap of orbitals in which electron density is concentrated above and below the plane of the nuclei of the bonding atoms. This type of bond is called a **pi bond (π bond)**. For the molecules we will be considering, a π bond is formed from the side-to-side overlap of two p orbitals.

EXAMPLE 10.7
Describe the bonding in carbon dioxide, CO_2.

Step 1: Write the Lewis structure for CO_2.

$$\ddot{O}\!=\!C\!=\!\ddot{O}$$

Step 2: Count the number of electron pairs around the central atom. Since there are two electron pairs around C, the electron arrangement that minimizes electron-pair repulsions is **linear**.

We conclude that C is *sp* **hybridized** because it has the electron arrangement of two *sp* hybrid orbitals.

Step 3: Count the number of electron pairs around each oxygen atom. Since there are three electron pairs around each O, the electron arrangement that minimizes repulsions is **trigonal planar**.

We conclude that each O is sp^2 **hybridized** because it has the electron arrangement of three sp^2 hybrid orbitals.

Step 4: Each of the *sp* orbitals of the C atom forms a sigma bond with an sp^2 hybrid on each of the O atoms. Carbon has two "pure" *p* orbitals that did not mix with the *s* orbital. Each of these *p* orbitals of the C atom forms a pi bond by overlapping in a side-to-side fashion with a *p* orbital on each of the oxygen atoms. Each double bond is composed of one σ bond and one π bond. Finally, the two lone pairs on each O atom are placed in its two remaining sp^2 orbitals.

Tip: A double bond is composed of one σ bond and one π bond. A triple bond is composed of one σ bond and two π bonds.

PRACTICE EXERCISE

7.   Describe the bonding in a nitrogen molecule, N_2.

| Text Problems: 10.36, **10.38**, 10.42 |

PROBLEM TYPE 4: MOLECULAR ORBITAL DIAGRAMS

When working molecular orbital problems, realize that this is another bonding model that is different from the models we have encountered. So far, we have looked at the Lewis model and valence bond theory. We will focus on molecular orbital diagrams for homonuclear diatomic molecules of second-period elements. Examples include N_2, O_2, and F_2.

A. Writing electron configurations

For homonuclear diatomic molecules of second-period elements, the types of molecular orbitals used in bonding will be similar. The order of filling molecular orbitals for Li_2, B_2, C_2, and N_2 is:

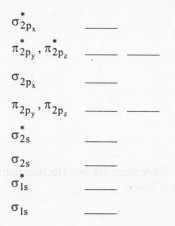

The order of filling molecular orbitals for O_2 and F_2 is:

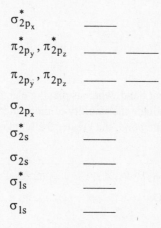

Note that for O_2 and F_2, the σ_{2p_x} is lower in energy than the π_{2p_y}, π_{2p_z} molecular orbitals.

EXAMPLE 10.8

Write the electron configuration for the ion, O_2^-.

Step 1: Count the number of electrons in the ion.

Each oxygen has 8 electrons, plus we need to add one electron for the negative charge. The total number of electrons in the ion is 17.

Step 2: Place the electrons in molecular orbitals following this convention.

1. Build up from the lowest energy molecular orbital to higher energy orbitals.
2. Each molecular orbital can hold a maximum of two electrons.
3. Follow Hund's rule: the most stable arrangement of electrons in molecular orbitals with equal energy is the one with the greatest number of parallel spins.
4. When electrons are paired in a molecular orbital, they must have opposite spin.

Placing 17 electrons following the above rules gives:

$\sigma_{2p_x}^*$	____	
$\pi_{2p_y}^*, \pi_{2p_z}^*$	↑↓	↑
π_{2p_y}, π_{2p_z}	↑↓	↑↓
σ_{2p_x}	↑↓	
σ_{2s}^*	↑↓	
σ_{2s}	↑↓	
σ_{1s}^*	↑↓	
σ_{1s}	↑↓	

The electron configuration is: $(\sigma_{1s})^2(\sigma_{1s}^*)^2(\sigma_{2s})^2(\sigma_{2s}^*)^2(\sigma_{2p_x})^2(\pi_{2p_y})^2(\pi_{2p_z})^2(\pi_{2p_y}^*)^2(\pi_{2p_z}^*)^1$

PRACTICE EXERCISE

8.   Write the electron configuration for C_2^+.

Text Problems: 10.50, 10.52, 10.54, 10.56, 10.58

B. Calculating bond order

Placing electrons in a "bonding" molecular orbital yields a stable covalent bond, whereas placing electrons in an "antibonding" molecular orbital results in an unstable bond. We can evaluate the stability of molecules or ions by calculating their **bond order**. The bond order indicates the strength of the bond; the greater the bond order, the stronger the bond. We can calculate bond order as follows:

$$\text{bond order} = \frac{1}{2}\left(\begin{array}{c}\text{number of electrons} \\ \text{in bonding MOs}\end{array} - \begin{array}{c}\text{number of electrons} \\ \text{in antibonding MOs}\end{array}\right)$$

EXAMPLE 10.9

Determine the bond order of O_2^-.

The bond order is:

$$\textbf{bond order} = \frac{1}{2}(10 - 7) = \textbf{1.5}$$

The bond order of O_2 is 2. This indicates that O_2 is more stable than O_2^-. Is this what you would expect?

PRACTICE EXERCISE

9.   Determine the bond order of C_2^+.

Text Problems: 10.50, 10.52, 10.54, 10.56

C. Determining the magnetic character of a molecule or ion

In Section 7.8 of the text, the terms "paramagnetic" and "diamagnetic" were discussed. A *paramagnetic* substance contains unpaired electrons and is *attracted* by an external magnetic field. Any substance with an *odd* number of electrons must be paramagnetic, because we need an even number of electrons for complete pairing. In a *diamagnetic* substance, all the electron spins are paired. Diamagnetic substances are slightly *repelled* by an external magnetic field.

Substances containing an even number of electrons may be either diamagnetic or paramagnetic. A molecular orbital diagram can be helpful in determining the magnetic character of a molecule or ion with an even number of electrons. O_2 is an example of a molecule with an even number of electrons (16) that is paramagnetic (see Table 10.5 of the text).

EXAMPLE 10.10

Determine the magnetic character of O_2^-.

Since O_2^- has an odd number of electrons (17), it is paramagnetic. We could also write the molecular orbital diagram for O_2^- to determine its magnetic character. Looking at the MO diagram for O_2^- in Example 10.8 above, we see that there is a single unpaired electron in the $\pi_{2p_z}^*$ molecular orbital. Hence, O_2^- is paramagnetic.

PRACTICE EXERCISE

10.   Determine the magnetic character of C_2^+.

Text Problems: 10.54, 10.56

ANSWERS TO PRACTICE EXERCISES

1. **(a)** linear   **(b)** tetrahedral   **(c)** trigonal planar   **(d)** octahedral

2. **(a)** bent   **(b)** square planar **(c)** distorted tetrahedron (seesaw)

3. The electron arrangement around each C that minimizes electron-pair repulsion is linear. Since there are no lone pairs around each C, the geometry is also **linear**.

4. **(a)** Yes, the molecule is polar.
    (b) No, the molecule is nonpolar.
    (c) No, the molecule is nonpolar.

5. **(a)** sp **(b)** sp^2

6. When you draw the Lewis structure for SF_4, you will find five electron pairs around the central atom, S (four bonding pairs and one lone pair). The electron arrangement that minimizes electron-pair repulsion is trigonal bipyramidal. You should conclude that S is dsp^3 hybridized because it has the electron arrangement of five dsp^3 hybrid orbitals.

7.

$$:N{\equiv}N:$$

The structure of N_2 is **linear**. We conclude that each N is *sp* **hybridized** because it has the electron arrangement of two *sp* hybrid orbitals. An *sp* orbital of one N atom overlaps with an *sp* orbital on the other N to form a sigma bond. The other *sp* orbital of each N contains the lone pair of electrons. Each nitrogen atom has two "pure" *p* orbitals that did not mix with the *s* orbital. The two *p* orbitals on one N can form two pi bonds by overlapping side-to-side with the two *p* orbitals on the other N atom. The triple bond is composed of one σ bond and two π bonds.

8. $(\sigma_{1s})^2(\sigma_{1s}^*)^2(\sigma_{2s})^2(\sigma_{2s}^*)^2(\pi_{2p_y})^2(\pi_{2p_z})^1$

9.   bond order = 1.5

10. C_2^+ is paramagnetic.

SOLUTIONS TO SELECTED TEXT PROBLEMS

10.8   Molecular Geometry, Problem Type 1.

In each case, you should use the following approach.

Step 1: Write the Lewis structure of the molecule. The lone pairs on the terminal atoms have been omitted.

Step 2: Count the number of electron pairs around the central atom.

Step 3: Build a model, or consult Table 10.1 of your text to predict the *geometry* of the molecule.

	Lewis structure	Electron pairs on central atom	Electron arrangement	Lone pairs	Geometry
(a)	Cl | Cl—Al—Cl	3	trigonal planar	0	trigonal planar, AB_3
(b)	Cl—Zn—Cl	2	linear	0	linear, AB_2
(c)	$\begin{bmatrix} & Cl & \\ & \| & \\ Cl— & Zn & —Cl \\ & \| & \\ & Cl & \end{bmatrix}^{2-}$	4	tetrahedral	0	tetrahedral, AB_4

10.10 (a) AB_4 tetrahedral   **(f)** AB_4 tetrahedral

      (b) AB_2E_2 bent   **(g)** AB_5 trigonal bipyramid

      (c) AB_3 trigonal planar   **(h)** AB_3E trigonal pyramid

      (d) AB_2E_3 linear   **(i)** AB_4 tetrahedral

      (e) AB_4E_2 square planar

10.12 Only molecules with four bonds to the central atom and no lone pairs are tetrahedral (AB_4).

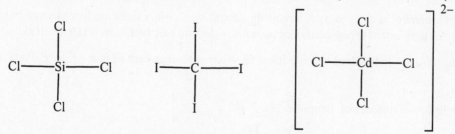

What are the Lewis structures and shapes for XeF_4 and SeF_4?

10.18   The electronegativity of the halogens decreases from F to I. Thus, the polarity of the H–X bond (where X denotes a halogen atom) also decreases from HF to HI. This difference in electronegativity accounts for the decrease in dipole moment.

10.20   Draw the Lewis structures. Both molecules are linear (AB_2). In CS_2, the two C–S bond moments are equal in magnitude and opposite in direction. The sum or resultant dipole moment will be *zero*. Hence, CS_2 is a **nonpolar molecule**. Even though OCS is linear, the C–O and C–S bond moments are not exactly equal, and there will be a small net dipole moment. Hence, OCS has a larger dipole moment than CS_2 (zero).

10.22   Predicting Dipole Moments, Problem Type 2.

Each vertex of the hexagonal structure of benzene represents the location of a C atom. Around the ring, there is no difference in electronegativity between C atoms, so the only bonds we need to consider are the polar C–Cl bonds.

The molecules shown in **(b)** and **(d)** are nonpolar. Due to the high symmetry of the molecules and the equal magnitude of the bond moments, the bond moments in each molecule cancel one another. The resultant dipole moment will be *zero*. For the molecules shown in **(a)** and **(c)**, the bond moments do not cancel and there will be net dipole moments. The dipole moment of the molecule in **(a)** is larger than that in **(c)**, because in **(a)** all the bond moments point in the same relative direction, reinforcing each other (see Lewis structure below). Therefore, the order of increasing dipole moments is:

$$(b) = (d) < (c) < (a).$$

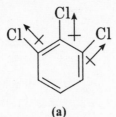

(a)

10.32   Hybridization of Atomic Orbitals, Problem Type 3.

(a)
Step 1: Write the Lewis structure of the molecule.

$$\text{H–Si–H}$$

with H above and H below the Si.

Step 2: Count the number of electron pairs around the central atom. Since there are four electron pairs around Si, the electron arrangement that minimizes electron-pair repulsion is **tetrahedral**.

We conclude that Si is sp^3 **hybridized** because it has the electron arrangement of four sp^3 hybrid orbitals.

(b)
Step 1: Write the Lewis structure of the molecule.

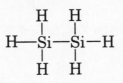

Step 2: Count the number of electron pairs around the "central atoms". Since there are four electron pairs around each Si, the electron arrangement that minimizes electron-pair repulsion for each Si is **tetrahedral**.

We conclude that each Si is sp^3 **hybridized** because it has the electron arrangement of four sp^3 hybrid orbitals.

10.34 Draw the Lewis structures. Before the reaction, boron is sp^2 hybridized (trigonal planar electron arrangement) in BF_3 and nitrogen is sp^3 hybridized (tetrahedral electron arrangement) in NH_3. After the reaction, boron and nitrogen are both sp^3 hybridized (tetrahedral electron arrangement).

10.36 **(a)** Each carbon has four bond pairs and no lone pairs and therefore has a tetrahedral electron pair arrangement. This implies sp^3 hybrid orbitals.

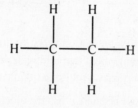

(b) The left-most carbon is tetrahedral and therefore has sp^3 hybrid orbitals. The two carbon atoms connected by the double bond are trigonal planar with sp^2 hybrid orbitals.

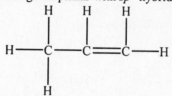

(c) Carbons 1 and 4 have sp^3 hybrid orbitals. Carbons 2 and 3 have sp hybrid orbitals.

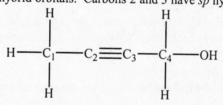

(d) The left-most carbon is tetrahedral (sp^3 hybrid orbitals). The carbon connected to oxygen is trigonal planar (why?) and has sp^2 hybrid orbitals.

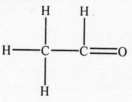

(e)   The left-most carbon is tetrahedral (sp^3 hybrid orbitals). The other carbon is trigonal planar with sp^2 hybridized orbitals.

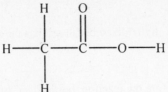

10.38   Hybridization of Atomic Orbitals, Problem Type 3.

Step 1: Write the Lewis structure of the molecule. Several resonance forms with formal charges are shown.

Step 2: Count the number of electron pairs around the central atom. Since there are two electron pairs around N, the electron arrangement that minimizes electron-pair repulsion is **linear** (AB_2). Remember, for VSEPR purposes a multiple bond counts the same as a single bond.

We conclude that N is *sp* **hybridized** because it has the electron arrangement of two *sp* hybrid orbitals.

10.40   Hybridization of Atomic Orbitals, Problem Type 3.

Step 1: Write the Lewis structure of the molecule.

$$
\begin{array}{c}
:\!\ddot{F}\!: \\
| \\
:\!\ddot{F}\!-\!P\!-\!\ddot{F}\!: \\
| \quad :\!\ddot{F}\!: \\
:\!\ddot{F}\!:
\end{array}
$$

Step 2: Count the number of electron pairs around the central atom. Since there are five electron pairs around P, the electron arrangement that minimizes electron-pair repulsion is **trigonal bipyramidal** (AB_5).

We conclude that P is sp^3d **hybridized** because it has the electron arrangement of five sp^3d hybrid orbitals.

10.42   A single bond is usually a sigma bond, a double bond is usually a sigma bond and a pi bond, and a triple bond is always a sigma bond and two pi bonds. Therefore, there are nine pi bonds and nine sigma bonds in the molecule.

10.48   In order for the two hydrogen atoms to combine to form a H_2 molecule, the electrons must have opposite spins. Furthermore, the combined energy of the two atoms must not be too great. Otherwise, the H_2 molecule will possess too much energy and will break apart into two hydrogen atoms.

10.50   The electron configurations are listed. Refer to Table 10.5 for the molecular orbital diagram.

Li_2: $(\sigma_{1s})^2(\sigma_{1s}^*)^2(\sigma_{2s})^2$ bond order = 1

Li_2^+: $(\sigma_{1s})^2(\sigma_{1s}^*)^2(\sigma_{2s})^1$ bond order = $\dfrac{1}{2}$

$$\text{Li}_2^- : \qquad (\sigma_{1s})^2(\sigma_{1s}^*)^2(\sigma_{2s})^2(\sigma_{2s}^*)^1 \qquad\qquad \text{bond order} = \frac{1}{2}$$

Order of increasing stability: $\text{Li}_2^- = \text{Li}_2^+ < \text{Li}_2$

In reality, Li_2^+ is more stable than Li_2^- because there is less electrostatic repulsion in Li_2^+.

10.52   See Table 10.5 of the text. Removing an electron from B_2 (bond order = 1) gives B_2^+, which has a bond order of (1/2). Therefore, B_2^+ has a weaker and longer bond than B_2.

10.54   In both the Lewis structure and the molecular orbital energy level diagram (Table 10.5), the oxygen molecule has a double bond (bond order = 2). The principal difference is that the molecular orbital treatment predicts that the molecule will have two unpaired electrons (paramagnetic). Experimentally this is found to be true.

10.56   We refer to Table 10.5.

O_2 has a bond order of 2 and is paramagnetic (two unpaired electrons).

O_2^+ has a bond order of 2.5 and is paramagnetic (one unpaired electron).

O_2^- has a bond order of 1.5 and is paramagnetic (one unpaired electron).

O_2^{2-} has a bond order of 1 and is diamagnetic.

Based on molecular orbital theory, the stability of these molecules increases as follows:

$$O_2^{2-} < O_2^- < O_2 < O_2^+$$

10.58   As discussed in the text (see Table 10.5), the single bond in B_2 is a pi bond (the electrons are in a pi *bonding* molecular orbital) and the double bond in C_2 is made up of two pi bonds (the electrons are in the pi *bonding* molecular orbitals).

10.62   The symbol on the left shows the pi bond delocalized over the entire molecule. The symbol on the right shows only one of the two resonance structures of benzene; it is an incomplete representation.

10.64   **(a)**   Two Lewis resonance forms (with lone electron pairs on fluorine omitted) are shown below. Formal charges different than zero are indicated.

(b)   There are no lone pairs on the nitrogen atom; it should have a trigonal planar electron pair arrangement and therefore use sp^2 hybrid orbitals.

(c)   The bonding consists of sigma bonds joining the nitrogen atom to the fluorine and oxygen atoms. In addition there is a pi molecular orbital delocalized over the entire molecule. Is nitryl fluoride isoelectronic with the carbonate ion?

10.66   The Lewis structures of ozone are:

$$\ddot{O}=\ddot{O}-\ddot{\underset{..}{O}}: \quad \longleftrightarrow \quad :\ddot{\underset{..}{O}}-\ddot{O}=\ddot{O}$$

The central oxygen atom is sp^2 hybridized (AB_2E). The unhybridized $2p_z$ orbital on the central oxygen overlaps with the $2p_z$ orbitals on the two end atoms.

10.68   Molecular Geometry, Problem Type 1.

Step 1: Write the Lewis structure of the molecule (see Problem Type 3, Chapter 9).

$$:\ddot{\underset{..}{Br}}-Hg-\ddot{\underset{..}{Br}}:$$

Step 2: Count the number of electron pairs around the central atom. There are two electron pairs around Hg.

Step 3: Since there are two electron pairs around Hg, the electron-pair arrangement that minimizes electron-pair repulsion is **linear**.

In addition, since there are no lone pairs around the central atom, the geometry is also **linear** (AB_2).

You could establish the geometry of $HgBr_2$ by measuring its dipole moment. If mercury(II) bromide were bent, it would have a measurable dipole moment. Experimentally, it has no dipole moment and therefore must be linear.

10.70   According to valence bond theory, a pi bond is formed through the side-to-side overlap of a pair of *p* orbitals. As atomic size increases, the distance between atoms is too large for *p* orbitals to overlap effectively in a side-to-side fashion. If two orbitals overlap poorly, that is, they share very little space in common, then the resulting bond will be very weak. This situation applies in the case of pi bonds between silicon atoms as well as between any other elements not found in the second period. It is usually far more energetically favorable for silicon, or any other heavy element, to form two single (sigma) bonds to two other atoms than to form a double bond (sigma + pi) to only one other atom.

10.72   The Lewis structures and VSEPR geometries of these species are shown below. The three nonbonding pairs of electrons on each fluorine atom have been omitted for simplicity.

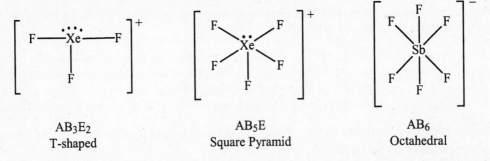

AB₃E₂   AB₅E   AB₆
T-shaped   Square Pyramid   Octahedral

10.74   To predict the bond angles for the molecules, you would have to draw the Lewis structure and determine the geometry using the VSEPR model. From the geometry, you can predict the bond angles.

(a) $BeCl_2$: AB_2 type, 180° (linear).

(b) BCl_3: AB_3 type, 120° (trigonal planar).

(c) CCl_4: AB_4 type, 109.5° (tetrahedral).

(d) CH_3Cl: AB_4 type, 109.5° (tetrahedral with a possible slight distortion resulting from the different sizes of the chlorine and hydrogen atoms).

(e) Hg_2Cl_2:   Each mercury atom is of the AB_2 type. The entire molecule is linear, 180° bond angles.

(f) $SnCl_2$: AB_2E type, roughly 120° (bent).

(g) H_2O_2:   The atom arrangement is HOOH. Each oxygen atom is of the AB_2E_2 type and the H–O–O angles will be roughly 109.5°.

(h) SnH_4: AB_4 type, 109.5° (tetrahedral).

10.76   Since arsenic and phosphorus are both in the same group of the periodic table, this problem is exactly like Problem 10.40. AsF_5 is an AB_5 type molecule, so the geometry is trigonal bipyramidal. We conclude that As is sp^3d **hybridized** because it has the electron arrangement of five sp^3d hybrid orbitals.

10.78   Only ICl_2^- and $CdBr_2$ will be linear. The rest are bent.

10.80   (a)   Hybridization of Atomic Orbitals, Problem Type 3.

Step 1: The geometry around each nitrogen is identical. To complete an octet of electrons around N, you must add a lone pair of electrons. Count the number of electron pairs around N. There are three electron pairs around each N.

Step 2: Since there are three electron pairs around N, the electron-pair arrangement that minimizes electron-pair repulsion is **trigonal planar**.

We conclude that each N is sp^2 **hybridized** because it has the electron arrangement of three sp^2 hybrid orbitals.

(b)   Predicting Dipole Moments, Problem Type 2.

An N–F bond is polar because F is more electronegative than N. The structure on the right has a dipole moment because the two N–F bond moments do not cancel each other out and so the molecule has a net dipole moment. On the other hand, the two N–F bond moments in the left-hand structure cancel. The sum or resultant dipole moment will be *zero*.

10.82   In 1,2-dichloroethane, the two C atoms are joined by a sigma bond. Rotation about a sigma bond does not destroy the bond, and the bond is therefore free (or relatively free) to rotate. Thus, all angles are permitted and the molecule is nonpolar because the C–Cl bond moments cancel each other because of the averaging effect brought about by rotation. In *cis*-dichloroethylene the two C–Cl bonds are locked in position. The π bond between the C atoms prevents rotation (in order to rotate, the π bond must be broken, using an energy source such as light or heat). Therefore, there is no rotation about the C=C in *cis*-dichloroethylene, and the molecule is polar.

10.84 O_3, CO, CO_2, NO_2, CH_4, and $CFCl_3$ are greenhouse gases.

10.86   The Lewis structure is:

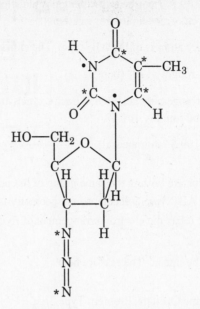

The carbon atoms and nitrogen atoms marked with an asterisk (C* and N*) are sp^2 hybridized; unmarked carbon atoms and nitrogen atoms marked with a dot (C and N•) are sp^3 hybridized; and the unmarked nitrogen atom is sp hybridized.

10.88   C has no d orbitals but Si does ($3d$). Thus, H_2O molecules can add to Si in hydrolysis (valence-shell expansion).

10.90   The carbons are in sp^2 hybridization states. The nitrogens are in the sp^3 hybridization state, except for the ring nitrogen double-bonded to a carbon that is sp^2 hybridized. The oxygen atom is sp^2 hybridized.

10.92   **(a)**   Use a conventional oven. A microwave oven would not cook the meat from the outside toward the center (it penetrates).

(b)   Polar molecules absorb microwaves and would interfere with the operation of radar.

(c)   Too much water vapor (polar molecules) absorbed the microwaves, interfering with the operation of radar.

10.94   The smaller size of F compared to Cl results in a shorter F–F bond than a Cl–Cl bond. The closer proximity of the lone pairs of electrons on the F atoms results in greater electron-electron repulsions that weaken the bond.

10.96 $1 \text{ D} = 3.336 \times 10^{-30} \text{ C·m}$

electronic charge (e) $= 1.6022 \times 10^{-19} \text{ C}$

$$\frac{\mu}{ed} \times 100\% = \frac{1.92 \text{ D} \times \dfrac{3.336 \times 10^{-30} \text{ C·m}}{1 \text{ D}}}{(1.6022 \times 10^{-19} \text{ C}) \times (91.7 \times 10^{-12} \text{ m})} \times 100\% = \textbf{43.6\%} \text{ ionic character}$$

CHAPTER 11
INTERMOLECULAR FORCES AND LIQUIDS AND SOLIDS

PROBLEM-SOLVING STRATEGIES AND TUTORIAL SOLUTIONS

TYPES OF PROBLEMS

Problem Type 1: Identifying Intermolecular Forces.

Problem Type 2: Identifying Hydrogen Bonds.

Problem Type 3: Counting the Number of Atoms in a Unit Cell.

Problem Type 4: Calculating Density from Crystal Structure and Atomic Radius.

PROBLEM TYPE 1: IDENTIFYING INTERMOLECULAR FORCES

Intermolecular forces are attractive forces between molecules. Intermolecular forces account for the existence of the condensed states of matter--liquids and solids. To understand the properties of condensed matter, you must understand the different types of intermolecular forces.

   (1) **Dipole-dipole forces:** These are attractive forces that act between polar molecules, that is, between molecules that possess dipole moments (see Section 11.2 of your text). The larger the dipole moments, the greater the attractive force.

   (2) **Ion-dipole forces:** These are attractive forces that occur between an ion (either a cation or an anion) and a polar molecule (see Figure 11.2 of your text). The strength of this interaction depends on the charge and size of the ion and on the magnitude of the dipole moment and the size of the molecule.

   (3) **Dispersion forces:** These are attractive forces that arise as a result of temporary dipoles induced in the atoms or molecules. This is the only type of intermolecular force that exists between nonpolar atoms or molecules. The likelihood of a dipole moment being induced depends on the polarizability of the atom or molecule. *Polarizability* is the ease with which the electron distribution in the atom or molecule can be distorted. Dispersion forces usually increase with molar mass.

For a complete discussion of intermolecular forces, see Section 11.2 of your text.

EXAMPLE 11.1
Indicate all the different types of intermolecular forces that exist in each of the following substances:
(a) CCl_4 (l) and (b) HBr (l).

A good place to start with this type of problem is to determine if the molecule is polar or nonpolar. See Problem Type 2, Chapter 10 if you need to review this topic.

(a) CCl_4 is nonpolar. The only type of intermolecular forces present in nonpolar molecules is *dispersion forces*.

(b)   HBr is a polar molecule. The types of intermolecular forces present are *dipole-dipole* and *dispersion forces*. There is no hydrogen bonding in HBr. The Br atom is too large and is not electronegative enough.

PRACTICE EXERCISE

1. Which of the following substances should have the strongest intermolecular attractive forces: N_2, Ar, F_2, or Cl_2?

2. The dipole moment (μ) in HCl is 1.03 D, and in HCN it is 2.99 D. Which substance should have the higher boiling point?

Text Problems: **11.8**, 11.10, 11.16, **11.18**

PROBLEM TYPE 2: IDENTIFYING HYDROGEN BONDS

The boiling points of NH_3, H_2O, and HF are much higher than expected if the boiling point is solely based on molar mass (see Figure 11.7 of your text). The high boiling points in these compounds are due to extensive *hydrogen bonding*. The hydrogen bond is a special type of dipole-dipole interaction between the hydrogen atom in a polar bond, such as N–H, O–H, or F–H, and an electronegative O, N, or F atom. This interaction is written

$$A-H\cdots B \qquad \text{or} \qquad A-H\cdots A$$

where A and B represent O, N, or F. A–H is one molecule or part of a molecule and B is a part of another molecule. The dotted line represents the hydrogen bond.

The average energy of a hydrogen bond is quite large for a dipole-dipole interaction (up to 40 kJ/mol). Thus, hydrogen bonds are a powerful force in determining the structures and properties of many compounds.

EXAMPLE 11.2
Predict whether hydrogen bonding intermolecular attractions are present in the following substances:
(a) CH_3OH (*l*) and (b) $CH_3CH_2OCH_2CH_3$ (*l*).

(a) CH_3OH (methanol) is polar and has a hydrogen atom bound to an oxygen atom. Hydrogen bonding intermolecular forces are present between CH_3OH molecules. Dipole-dipole and dispersion forces are also present.

(b) $CH_3CH_2OCH_2CH_3$ (diethyl ether) is polar and does contain both oxygen atoms and hydrogen atoms. However, all the hydrogen atoms are bonded to carbon, not oxygen. There is no hydrogen bonding in diethyl ether, because carbon is not electronegative enough. The intermolecular attractive forces present are dipole-dipole and dispersion forces.

PRACTICE EXERCISE

3. Which member of each pair has the stronger intermolecular forces of attraction:
   (a) H_2O or H_2S, (b) HCl or HF, and (c) NH_3 or PH_3?

Text Problems: 11.12, **11.14**, **11.18**, 11.20

PROBLEM TYPE 3: COUNTING THE NUMBER OF ATOMS IN A UNIT CELL

To solve this problem, you must realize that because every unit cell in a crystalline solid is adjacent to other unit cells, most of a cell's atoms are shared by neighboring cells. In cubic cells, each corner atom is shared by eight unit cells [see Figure 11.19(a) of your text]. A face-centered atom is shared by two unit cells [See Figure 11.19(b) of your text].

A center atom in a unit cell is solely contained by that unit cell and is not shared. An edge atom is shared by four unit cells. Table 11.1 summarizes this information.

TABLE 11.1

Type of atom	Amount of atom contained in unit cell
Corner	1/8
Edge	1/4
Face-centered	1/2
Center	1

EXAMPLE 11.4

If atoms of a solid occupy a face-centered cubic lattice, how many atoms are there per unit cell?

In a face-centered cubic unit cell, there are atoms at each of the eight corners, and there is one atom in each of the six faces.

(8 corner atoms)(1/8 atom per corner) + (6 face-centered atoms)(1/2 atom per face) = **4 atoms/unit cell**

PRACTICE EXERCISE

4.   Atoms of polonium (Po) occupy a simple cubic lattice. How many Po atoms are there per unit cell?

Text Problems: 11.38, **11.44**

PROBLEM TYPE 4: CALCULATING DENSITY FROM CRYSTAL STRUCTURE AND ATOMIC RADIUS

Step 1: To solve this type of problem, you must know how many atoms are contained in the different types of cubic unit cells. Table 11.2 summarizes the number of atoms per unit cell.

TABLE 11.2

Type of cubic unit cell	Number of atoms/unit cell
Simple cubic	1
Body-centered cubic	2
Face-centered cubic	4

Step 2: You can look up the relationship between edge length (a) and atomic radius (r). See Figure 11.22 of the text. Table 11.3 summarizes these relationships.

TABLE 11.3

Type of cubic unit cell	Relationship between a and r
Simple cubic	$a = 2r$
Body-centered cubic	$a = \dfrac{4}{\sqrt{3}}r$
Face-centered cubic	$a = \sqrt{8}\,r$

Step 3: Density = mass/volume. See Problem Type 1, Chapter 1. The volume of a cube is equal to the edge length cubed.

$$V = a^3$$

You will know how many atoms are in a unit cell (*Step 1* above). To calculate the mass, you need to convert from atoms/unit cell to grams/unit cell. A reasonable strategy would be

$$\frac{atoms}{unit\ cell} \rightarrow \frac{mol}{unit\ cell} \rightarrow \frac{grams}{unit\ cell}$$

Step 4: Substitute the mass and volume into the density equation to solve for the density.

EXAMPLE 11.3
Nickel crystallizes in a face-centered cubic lattice with an edge length of 352 pm. Calculate the density of nickel.

Step 1: A face-centered cubic unit cell contains four atoms per unit cell (see Table 11.2).

Step 2: Density = mass/volume. See Problem Type 1, Chapter 1. The volume of a cube is equal to the edge length cubed.

$$V = a^3$$

$$V = (325\ pm)^3 = 3.43 \times 10^7\ pm^3$$

You should convert pm^3 to cm^3 because the density of a solid is typically expressed in units of g/cm^3.

$$3.43 \times 10^7\ pm^3 \times \left(\frac{10^{-12}\ m}{1\ pm}\right)^3 \times \left(\frac{1\ cm}{10^{-2}\ m}\right)^3 = 3.43 \times 10^{-23}\ cm^3$$

There are four atoms/unit cell. To calculate the mass, you need to convert from atoms/unit cell to grams/unit cell. A reasonable strategy would be:

$$atoms/unit\ cell \rightarrow mol/unit\ cell \rightarrow grams/unit\ cell$$

$$\frac{4\ Ni\ atoms}{unit\ cell} \times \frac{1\ mol\ Ni}{6.022 \times 10^{23}\ Ni\ atoms} \times \frac{58.71\ g\ Ni}{1\ mol\ Ni} = \frac{3.900 \times 10^{-22}\ g\ Ni}{1\ unit\ cell}$$

Step 3: Substitute the mass and volume into the density equation to solve for the density.

$$d = \frac{m}{V} = \frac{3.900 \times 10^{-22}\ g\ Ni}{3.43 \times 10^{-23}\ cm^3} = \textbf{11.4 g/cm}^3$$

PRACTICE EXERCISE

5.   Potassium crystallizes in a body-centered cubic lattice and has a density of 0.856 g/cm^3 at 25°C.
      (a) How many atoms are there per unit cell?
      (b) What is the length of an edge of the unit cell?

Text Problems: **11.40**, 11.42

ANSWERS TO PRACTICE EXERCISES

1. Cl_2 2.   HCN   3.   **(a)** H_2O **(b)** HF   **(c)** NH_3

4.   1 atom/unit cell   5.   **(a)** 2 atoms/unit cell   **(b)** $a = 5.34 \times 10^{-8}$ cm = 534 pm

SOLUTIONS TO SELECTED TEXT PROBLEMS

11.8   Identifying Intermolecular Forces, Problem Type 1.

The three molecules are essentially nonpolar. There is little difference in electronegativity between carbon and hydrogen. Thus, the only type of intermolecular attraction in these molecules is dispersion forces. Other factors being equal, the molecule with the greater number of electrons will exert greater intermolecular attractions. By looking at the molecular formulas you can predict that the order of increasing boiling points will be $CH_4 < C_3H_8 < C_4H_{10}$.

On a very cold day, propane and butane would be liquids (boiling points −44.5°C and −0.5°C, respectively); only methane would still be a gas (boiling point −161.6°C).

11.10   **(a)**   Benzene (C_6H_6) molecules are nonpolar. Only dispersion forces will be present.

         (b)   Chloroform (CH_3Cl) molecules are polar (why?). Dispersion and dipole-dipole forces will be present.

         (c)   Phosphorus trifluoride (PF_3) molecules are polar. Dispersion and dipole-dipole forces will be present.

         (d)   Sodium chloride (NaCl) is an ionic compound. Ion-ion (and dispersion) forces will be present.

         (e)   Carbon disulfide (CS_2) molecules are nonpolar. Only dispersion forces will be present.

11.12   In this problem you must identify the species capable of hydrogen bonding among themselves, not with water. In order for a molecule to be capable of hydrogen bonding with another molecule like itself, it must have at least one hydrogen atom bonded to N, O, or F. Of the choices, only (e) CH_3COOH (acetic acid) shows this structural feature. The others cannot form hydrogen bonds among themselves.

11.14   Identifying Hydrogen Bonds, Problem Type 2.

1-butanol has the higher boiling point because the molecules can form hydrogen bonds with each other. Diethyl ether molecules do contain both oxygen atoms and hydrogen atoms. However, all the hydrogen atoms are bonded to carbon, not oxygen. There is no hydrogen bonding in diethyl ether, because carbon is not electronegative enough.

11.16   **(a)**   Xe: it has more electrons and therefore stronger dispersion forces.

         (b) CS_2:   it has more electrons (both molecules nonpolar) and therefore stronger dispersion forces.

         (c) Cl_2:   it has more electrons (both molecules nonpolar) and therefore stronger dispersion forces.

         (d)   LiF: it is an ionic compound, and the ion-ion attractions are much stronger than the dispersion forces between F_2 molecules.

         (e) NH_3:   it can form hydrogen bonds and PH_3 cannot.

11.18   Identifying Intermolecular Forces and Hydrogen Bonding, Problem Types 1 and 2.

         (a)   Water has O−H bonds. Therefore, water molecules can form hydrogen bonds. The attractive forces that must be overcome are hydrogen bonding and dispersion forces.

         (b)   Bromine (Br_2) molecules are nonpolar. Only dispersion forces must be overcome.

         (c)   Iodine (I_2) molecules are nonpolar. Only dispersion forces must be overcome.

(d)   In this case, the F–F bond must be broken. This is an *intra*molecular force between two F atoms, not an *inter*molecular force between F_2 molecules. The attractive forces of the covalent bond must be overcome.

11.20   The lower melting compound (shown below) can form hydrogen bonds only with itself (*intra*molecular hydrogen bonds), as shown in the figure. Such bonds do not contribute to *inter*molecular attraction and do not help raise the melting point of the compound. The other compound can form *inter*molecular hydrogen bonds; therefore, it will take a higher temperature to provide molecules of the liquid with enough kinetic energy to overcome these attractive forces to escape into the gas phase.

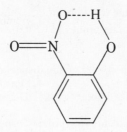

11.32   Ethylene glycol has two –OH groups, allowing it to exert strong intermolecular forces through hydrogen bonding. Its viscosity should fall between ethanol (1 OH group) and glycerol (3 OH groups).

11.38   A corner sphere is shared equally among eight unit cells, so only one-eighth of each corner sphere "belongs" to any one unit cell. A face-centered sphere is divided equally between the two unit cells sharing the face. A body-centered sphere belongs entirely to its own unit cell.

In a *simple cubic cell* there are eight corner spheres. One-eighth of each belongs to the individual cell giving a total of **one** whole sphere per cell. In a *body-centered cubic cell*, there are eight corner spheres and one body-center sphere giving a total of **two** spheres per unit cell (one from the corners and one from the body-center). In a *face-center* sphere, there are eight corner spheres and six face-centered spheres (six faces). The total number of spheres would be **four**: one from the corners and three from the faces.

11.40   Similar to Problem Type 4, Calculating the Density from Crystal Structure and Atomic Radius.

To solve this problem, we must first find the volume that a Ba atom occupies. Then, we need to calculate the volume one mole of Ba atoms occupies. Once we have these two pieces of information, we can multiply them together to end up with the number of Ba atoms per mol of Ba .

$$\frac{\text{number of Ba atoms}}{\text{cm}^3} \times \frac{\text{cm}^3}{1 \text{ mol Ba}} = \frac{\text{number of Ba atoms}}{1 \text{ mol Ba}}$$

Step 1:   The volume that contains two barium atoms is the volume of the body-centered cubic unit cell. Some of this volume is empty space because packing is only 68.0 percent efficient. But, this will not affect our calculation.

$$V = a^3$$

Let's also convert to cm^3.

$$V = (502 \text{ pm})^3 \times \left(\frac{10^{-12} \text{ m}}{1 \text{ pm}}\right)^3 \times \left(\frac{1 \text{ cm}}{10^{-2} \text{ m}}\right)^3 = \frac{1.27 \times 10^{-22} \text{ cm}^3}{2 \text{ Ba atoms}}$$

Step 2: The volume that contains one mole of barium atoms can be calculated from the density using the following strategy:

$$\frac{\text{volume}}{\text{mass of Ba}} \quad \rightarrow \quad \frac{\text{volume}}{\text{mol Ba}}$$

$$\frac{1 \text{ cm}^3}{3.50 \text{ g Ba}} \times \frac{137.34 \text{ g Ba}}{1 \text{ mol Ba}} = \frac{39.2 \text{ cm}^3}{1 \text{ mol Ba}}$$

Step 3: We can now calculate the number of barium atoms in one mole using the strategy presented in *Step 1.*

$$\frac{\text{number of Ba atoms}}{\text{cm}^3} \times \frac{\text{cm}^3}{1 \text{ mol Ba}} = \frac{\text{number of Ba atoms}}{1 \text{ mol Ba}}$$

$$\left(\frac{2 \text{ Ba atoms}}{1.27 \times 10^{-22} \text{ cm}^3} \right) \left(\frac{39.2 \text{ cm}^3}{1 \text{ mol Ba}} \right) = \mathbf{6.17 \times 10^{23} \text{ atoms / mol}}$$

This is close to Avogadro's number, 6.022×10^{23} particles/mol.

11.42   The mass of the unit cell is the mass in grams of two europium atoms.

$$m = \left(\frac{2 \text{ Eu atoms}}{1 \text{ unit cell}} \right) \left(\frac{1 \text{ mol Eu}}{6.022 \times 10^{23} \text{ Eu atoms}} \right) \left(\frac{152.0 \text{ g Eu}}{1 \text{ mol Eu}} \right) = 5.048 \times 10^{-22} \text{ g Eu / unit cell}$$

$$V = \left(\frac{5.048 \times 10^{-22} \text{ g}}{1 \text{ unit cell}} \right) \left(\frac{1 \text{ cm}^3}{5.26 \text{ g}} \right) = 9.60 \times 10^{-23} \text{ cm}^3 \text{ / unit cell}$$

The edge length (a) is:

$$a = V^{1/3} = (9.60 \times 10^{-23} \text{ cm}^3)^{1/3} = \mathbf{4.58 \times 10^{-8} \text{ cm} = 458 \text{ pm}}$$

11.44   Similar to Problem Type 4, Counting the Number of Atoms in a Unit Cell.

In a face-centered cubic unit cell, there are atoms at each of the eight corners, and there is one atom in each of the six faces. Only one-half of each face-centered atom and one-eighth of each corner atom belongs to the unit cell.

X atoms/unit cell = (8 corner atoms)(1/8 atom per corner) = 1 X atom/unit cell

Y atoms/unit cell = (6 face-centered atoms)(1/2 atom per face) = 3 Y atoms/unit cell

The unit cell is the smallest repeating unit in the crystal; therefore, the empirical formula is **XY₃**.

11.48   Rearranging the Bragg equation, we have:

$$\lambda = \frac{2d \sin \theta}{n} = \frac{2(282 \text{ pm})(\sin 23.0°)}{1} = 220 \text{ pm} = \mathbf{0.220 \text{ nm}}$$

11.56   In diamond, each carbon atom is covalently bonded to four other carbon atoms. Because these bonds are strong and uniform, diamond is a very hard substance. In graphite, the carbon atoms in each layer are linked by strong bonds, but the layers are bound by weak dispersion forces. As a result, graphite may be cleaved easily between layers and is not hard.

11.78   *Step 1:* Warming ice to the melting point.

$$q_1 = ms\Delta t = 866 \text{ g H}_2\text{O} \times 2.03 \text{ J/g}°\text{C} \times [0 - (-10)]°\text{C} = 17.6 \text{ kJ}$$

Step 2: Converting ice at the melting point to liquid water at 0°C (See Table 11.8 for the heat of fusion of water.)

$$q_2 = 866 \text{ g H}_2\text{O} \left(\frac{1 \text{ mol}}{18.02 \text{ g H}_2\text{O}} \right) \left(\frac{6.01 \text{ kJ}}{1 \text{ mol}} \right) = 289 \text{ kJ}$$

Step 3: Heating water from 0°C to 100°C.

$$q_3 = ms\Delta t = 866 \text{ g H}_2\text{O} \times 4.184 \text{ J/g}°\text{C} \times (100 - 0)°\text{C} = 362 \text{ kJ}$$

Step 4: Converting water at 100°C to steam at 100°C (See Table 11.6 for the heat of vaporization of water.)

$$q_4 = 866 \text{ g H}_2\text{O} \left(\frac{1 \text{ mol}}{18.02 \text{ g H}_2\text{O}} \right) \left(\frac{40.79 \text{ kJ}}{1 \text{ mol}} \right) = 1.96 \times 10^3 \text{ kJ}$$

Step 5: Heating steam from 100°C to 126°C.

$$q_5 = ms\Delta t = 866 \text{ g H}_2\text{O} \times 1.99 \text{ J/g}°\text{C} \times (126 - 100)°\text{C} = 44.8 \text{ kJ}$$

$$q_{\text{total}} = q_1 + q_2 + q_3 + q_4 + q_5 = \mathbf{2.67 \times 10^3 \text{ kJ}}$$

How would you set up and work this problem if you were computing the heat lost in cooling steam from 126°C to ice at −10°C?

11.80 $\Delta H_{\text{vap}} = \Delta H_{\text{sub}} - \Delta H_{\text{fus}} = 62.30 \text{ kJ/mol} - 15.27 \text{ kJ/mol} = \mathbf{47.03 \text{ kJ/mol}}$

11.82   Two phase changes occur in this process. First, the liquid is turned to solid (freezing), then the solid ice is turned to gas (sublimation).

11.84   When steam condenses to liquid water at 100°C, it releases a large amount of heat equal to the enthalpy of vaporization. Thus steam at 100°C exposes one to more heat than an equal amount of water at 100°C.

11.86   We can use a modified form of the Clausius-Clapeyron equation to solve this problem. See Equation (11.5) in the text.

$P_1 = 40.1 \text{ mmHg}$ $\mathbf{P_2 = ?}$
$T_1 = 7.6°\text{C} = 280.6 \text{ K}$ $T_2 = 60.6°\text{C} = 333.6 \text{ K}$

$$\ln \frac{P_1}{P_2} = \frac{\Delta H_{\text{vap}}}{R} \left(\frac{1}{T_2} - \frac{1}{T_1} \right)$$

$$\ln \frac{40.1}{P_2} = \frac{31000 \text{ J/mol}}{8.314 \text{ J/K} \cdot \text{mol}} \left(\frac{1}{333.6 \text{ K}} - \frac{1}{280.6 \text{ K}} \right)$$

$$\ln \frac{40.1}{P_2} = -2.11$$

Taking the antilog of both sides, we have:

$$\frac{40.1}{P_2} = 0.121$$

$$P_2 = \mathbf{331\ mmHg}$$

11.88 Using Equation (11.5):

$$\ln \frac{P_1}{P_2} = \frac{\Delta H_{vap}}{R}\left(\frac{1}{T_2} - \frac{1}{T_1}\right)$$

$$\ln\left(\frac{1}{2}\right) = \left(\frac{\Delta H_{vap}}{8.314\ J/K\cdot mol}\right)\left(\frac{1}{368\ K} - \frac{1}{358\ K}\right) = \Delta H_{vap}\left(\frac{-7.59 \times 10^{-5}}{8.314\ J/mol}\right)$$

$$\Delta H_{vap} = 7.59 \times 10^4\ J/mol = \mathbf{75.9\ kJ/mol}$$

11.92 Initially, the ice melts because of the increase in pressure. As the wire sinks into the ice, the water above the wire refreezes. Eventually the wire actually moves completely through the ice block without cutting it in half.

11.94 Region labels: The region containing point A is the solid region. The region containing point B is the liquid region. The region containing point C is the gas region.

(a) Raising the temperature at constant pressure beginning at A implies starting with solid ice and warming until melting occurs. If the warming continued, the liquid water would eventually boil and change to steam. Further warming would increase the temperature of the steam.

(b) At point C water is in the gas phase. Cooling without changing the pressure would eventually result in the formation of solid ice. Liquid water would never form.

(c) At B the water is in the liquid phase. Lowering the pressure without changing the temperature would eventually result in boiling and conversion to water in the gas phase.

11.96 **(a)** A low surface tension means the attraction between molecules making up the surface is weak. Water has a high surface tension; water bugs could not "walk" on the surface of a liquid with a low surface tension.

(b) A low critical temperature means a gas is very difficult to liquefy by cooling. This is the result of weak intermolecular attractions. Helium has the lowest known critical temperature (5.3 K).

(c) A low boiling point means weak intermolecular attractions. It takes little energy to separate the particles. All ionic compounds have extremely high boiling points.

(d) A low vapor pressure means it is difficult to remove molecules from the liquid phase because of high intermolecular attractions. Substances with low vapor pressures have high boiling points (why?).

Thus, only choice (d) indicates strong intermolecular forces in a liquid. The other choices indicate weak intermolecular forces in a liquid.

11.98 The properties of hardness, high melting point, poor conductivity, and so on, could place boron in either the ionic or covalent categories. However, boron atoms will not alternately form positive and negative ions to achieve an ionic crystal. The structure is covalent because the units are single boron atoms.

11.100 CCl_4. Generally, the larger the number of electrons and the more diffuse the electron cloud in an atom or a molecule, the greater its polarizability. Recall that polarizability is the ease with which the electron distribution in an atom or molecule can be distorted.

11.102 The vapor pressure of mercury (as well as all other substances) is 760 mmHg at its normal boiling point.

11.104 It has reached the critical point; the point of critical temperature (T_c) and critical pressure (P_c).

11.106 Crystalline SiO_2. Its regular structure results in a more efficient packing.

11.108 (a) **False**. Permanent dipoles are usually much stronger than temporary dipoles.

   (b) **False**. The hydrogen atom must be bonded to N, O, or F.

   (c) **True**.

   (d) **False**. The magnitude of the attraction depends on both the ion charge and the polarizability of the neutral atom or molecule.

11.110 Sublimation temperature is −78°C or 195 K at a pressure of 1 atm.

$$\ln\frac{P_1}{P_2} = \frac{\Delta H_{sub}}{R}\left(\frac{1}{T_2} - \frac{1}{T_1}\right)$$

$$\ln\frac{1}{P_2} = \frac{25.9 \times 10^3 \text{ J/mol}}{8.314 \text{ J/mol}\cdot\text{K}}\left(\frac{1}{150 \text{ K}} - \frac{1}{195 \text{ K}}\right)$$

$$\ln\frac{1}{P_2} = 4.79$$

Taking the antilog of both sides gives:

$$P_2 = 8.3 \times 10^{-3} \text{ atm}$$

11.112 (a) K_2S: Ionic forces are much stronger than the dipole-dipole forces in $(CH_3)_3N$.

   (b) Br_2: Both molecules are nonpolar; but Br_2 has more electrons. (The boiling point of Br_2 is 50°C and that of C_4H_{10} is −0.5°C.)

11.114 CH_4 is a tetrahedral, nonpolar molecule that can only exert weak dispersion type attractive forces. SO_2 is bent (why?) and possesses a dipole moment, which gives rise to stronger dipole-dipole attractions. Sulfur dioxide will have a larger value of "a" in the van der Waals equation (a is a measure of the strength of the interparticle attraction) and will behave less like an ideal gas than methane.

11.116 The standard enthalpy change for the formation of gaseous iodine from solid iodine is simply the difference between the standard enthalpies of formation of the products and the reactants in the equation:

$$I_2 (s) \rightarrow I_2 (g)$$

$$\Delta H_{vap} = \Delta H_f^\circ [I_2(g)] - \Delta H_f^\circ [I_2(s)] = 62.4 \text{ kJ/mol} - 0 \text{ kJ/mol} = \textbf{62.4 kJ/mol}$$

11.118 Smaller ions have more concentrated charges (charge densities) and are more effective in ion-dipole interaction. The greater the ion-dipole interaction, the larger is the heat of hydration.

11.120 **(a)**   For the process: $Br_2(l) \rightarrow Br_2(g)$

$\Delta H° = \Delta H_f[Br_2(g)] - \Delta H_f[Br_2(l)] = (1 \text{ mol})(30.7 \text{ kJ/mol}) - 0 = $ **30.7 kJ**

(b)   For the process: $Br_2(g) \rightarrow 2Br(g)$

$\Delta H° = $ **192.5 kJ** (from Table 9.4)

As expected, the bond energy represented in part (b) is much greater than the energy of vaporization represented in part (a). It requires more energy to break the bond than to vaporize the molecule.

11.122 **(a)**   Decreases   **(b)**   No change   **(c)**   No change

11.124 $CaCO_3(s) \rightarrow CaO(s) + CO_2(g)$

Three phases (two solid and one gas). $CaCO_3$ and CaO constitute two separate solid phases because they are separated by well-defined boundaries.

11.126 SiO_2 has an extensive three-dimensional structure. CO_2 exists as discrete molecules. It will take much more energy to break the strong network covalent bonds of SiO_2; therefore, SiO_2 has a much higher boiling point than CO_2.

11.128 The moles of water vapor can be calculate using the ideal gas equation.

$$n = \frac{PV}{RT} = \frac{\left(187.5 \text{ mmHg} \times \frac{1 \text{ atm}}{760 \text{ mmHg}}\right)(5.00 \text{ L})}{\left(0.0821 \frac{L \cdot atm}{mol \cdot K}\right)(338 \text{ K})} = 0.0445 \text{ mol}$$

mass of water vapor $= 0.0445 \text{ mol} \times 18.02 \text{ g/mol} = 0.802 \text{ g}$

Now, we can calculate the percentage of the 1.20 g sample of water that is vapor.

$$\textbf{\% of H}_2\textbf{O vaporized} = \frac{0.802 \text{ g}}{1.20 \text{ g}} \times 100\% = \textbf{66.8\%}$$

11.130 The packing efficiency is: $\dfrac{\text{volume of atoms in unit cell}}{\text{volume of unit cell}} \times 100\%$

An atom is assumed to be spherical, so the volume of an atom is $(4/3)\pi r^3$. The volume of a cubic unit cell is a^3 (a is the length of the cube edge). The packing efficiencies are calculated below:

(a)   Simple cubic cell: cell edge $(a) = 2r$

$$\text{Packing efficiency} = \frac{\left(\frac{4\pi r^3}{3}\right) \times 100\%}{(2r)^3} = \frac{4\pi r^3 \times 100\%}{24r^3} = \frac{\pi}{6} \times 100\% = \textbf{52.4\%}$$

(b)   Body-centered cubic cell: cell edge $= \dfrac{4r}{\sqrt{3}}$

$$\text{Packing efficiency } = \frac{2 \times \left(\dfrac{4\pi r^3}{3}\right) \times 100\%}{\left(\dfrac{4r}{\sqrt{3}}\right)^3} = \frac{2 \times \left(\dfrac{4\pi r^3}{3}\right) \times 100\%}{\left(\dfrac{64r^3}{3\sqrt{3}}\right)} = \frac{2\pi\sqrt{3}}{16} \times 100\%$$

$$= \mathbf{68.0\%}$$

Remember, there are two atoms per body-centered cubic unit cell.

(c)   Face-centered cubic cell: cell edge $= \sqrt{8}r$

$$\text{Packing efficiency } = \frac{4 \times \left(\dfrac{4\pi r^3}{3}\right) \times 100\%}{\left(\sqrt{8}\,r\right)^3} = \frac{\left(\dfrac{16\pi r^3}{3}\right) \times 100\%}{8r^3\sqrt{8}} = \frac{2\pi}{3\sqrt{8}} \times 100\% = \mathbf{74.0\%}$$

Remember, there are four atoms per face-centered cubic unit cell.

11.132 For a face-centered cubic unit cell, the length of an edge (a) is given by:

$$a = \sqrt{8}r$$

$$a = \sqrt{8}\,(191\,\text{pm}) = 5.40 \times 10^2\,\text{pm}$$

The volume of a cube equals the edge length cubed (a^3).

$$V = a^3 = (5.40 \times 10^2\,\text{pm})^3 \left(\frac{1 \times 10^{-12}\,\text{m}}{1\,\text{pm}}\right)^3 \left(\frac{1\,\text{cm}}{1 \times 10^{-2}\,\text{m}}\right)^3 = 1.57 \times 10^{-22}\,\text{cm}^3$$

Now that we have the volume of the unit cell, we need to calculate the mass of the unit cell in order to calculate the density of Ar. The number of atoms in one face centered cubic unit cell is four.

$$m = \frac{4\,\text{atoms}}{1\,\text{unit cell}} \times \frac{1\,\text{mol}}{6.022 \times 10^{23}\,\text{atoms}} \times \frac{39.95\,\text{g}}{1\,\text{mol}} = \frac{2.65 \times 10^{-22}\,\text{g}}{1\,\text{unit cell}}$$

$$d = \frac{m}{V} = \frac{2.65 \times 10^{-22}\,\text{g}}{1.57 \times 10^{-22}\,\text{cm}^3} = \mathbf{1.69\ g/cm^3}$$

11.134 (a)   Two triple points: Diamond/graphite/liquid and graphite/liquid/vapor.

(b)   Diamond.

(c)   Apply high pressure at high temperature.

11.136 The cane is made of many molecules held together by intermolecular forces. The forces are strong and the molecules are packed tightly. Thus, when the handle is raised, all the molecules are raised because they are held together.

CHAPTER 12
PHYSICAL PROPERTIES OF SOLUTIONS

PROBLEM-SOLVING STRATEGIES AND TUTORIAL SOLUTIONS

TYPES OF PROBLEMS

Problem Type 1: Predicting Solubility Based on Intermolecular Forces.

Problem Type 2: Types of Concentration Units.
   (a) Percent by mass.
   (b) Molarity.
   (c) Molality.

Problem Type 3: Converting between Concentration Units.
   (a) Converting molality to molarity.
   (b) Converting molarity to molality.
   (c) Converting percent by mass to molality.

Problem Type 4: Effect of Pressure on Solubility: Henry's Law.

Problem Type 5: Colligative Properties of Nonelectrolytes.
   (a) Vapor-pressure lowering, Raoult's law.
   (b) Boiling-point elevation.
   (c) Freezing-point depression.
   (d) Osmotic pressure.

Problem Type 6: Determining Molar Mass Using Colligative Properties.
   (a) Calculating molar mass from freezing-point depression.
   (b) Calculating molar mass from osmotic pressure.

Problem Type 7: Colligative Properties of Electrolytes.

PROBLEM TYPE 1: PREDICTING SOLUBILITY BASED ON INTERMOLECULAR FORCES

Two substances with intermolecular forces of similar type and magnitude are likely to be soluble in each other. The saying "*like dissolves like*" will help you predict the solubility of a substance in a solvent.

EXAMPLE 12.1
Which of the following would be a better solvent for molecular I_2 (s): CCl_4 or H_2O?

I_2 is a nonpolar molecule (see Problem Type 2, Chapter 10). Using the like-dissolves-like rule, I_2 will be more soluble in the nonpolar solvent, CCl_4, than in the polar solvent, H_2O.

PRACTICE EXERCISE

1. In which solvent will NaBr be more soluble, benzene (C_6H_6) or water?

Text Problems: 12.10, 12.12

PROBLEM TYPE 2: TYPES OF CONCENTRATION UNITS

We will focus on three of the most common units of concentration: percent by mass, molarity, and molality.

A. Percent by mass

The percent by mass is defined as:

$$\text{percent by mass of solute} = \frac{\text{mass of solute}}{\text{mass of solute} + \text{mass of solvent}} \times 100\% \qquad (12.1)$$

$$= \frac{\text{mass of solute}}{\text{mass of soln}} \times 100\%$$

The percent by mass has no units because it is a ratio of two similar quantities.

EXAMPLE 12.2

The dehydrated form of Epsom salt is magnesium sulfate. What is the percent $MgSO_4$ by mass in a solution made from 16.0 g $MgSO_4$ and 100 mL of H_2O at 25°C? The density of water at 25°C is 0.997 g/mL.

Step 1: Calculate the mass of 100 mL of water using the density of water as a conversion factor.

$$? \text{ g of water} = 100 \text{ mL } H_2O \times \frac{0.997 \text{ g } H_2O}{1 \text{ mL } H_2O} = 99.7 \text{ g } H_2O$$

Step 2: Substitute the mass of solute and the mass of solvent into Equation (12.1) to calculate the percent by mass of $MgSO_4$.

$$\text{percent by mass } MgSO_4 = \frac{\text{mass of solute}}{\text{mass of solute} + \text{mass of solvent}} \times 100\%$$

$$= \frac{16.0 \text{ g}}{16.0 \text{ g} + 99.7 \text{ g}} \times 100\% = \mathbf{13.8 \%}$$

PRACTICE EXERCISE

2.   An aqueous solution contains 167 g $CuSO_4$ in 820 mL of solution. The density of the solution is 1.195 g/mL. Calculate the percent $CuSO_4$ by mass in the solution.

Text Problem: 12.16

B. Molarity (*M*)

We have already defined molarity in Chapter 4 (see Problem Type 5, Chapter 4). Molarity is defined as

$$\text{molarity} = \frac{\text{moles of solute}}{\text{liters of soln}} \qquad (12.2)$$

Molarity has units of mol/L.

EXAMPLE 12.3

What is the molarity of the $MgSO_4$ solution made in Example 12.2? Assume that the density remains unchanged upon addition of $MgSO_4$ to the water.

Step 1: Calculate the number of moles $MgSO_4$ in 16.0 g. Use the molar mass of $MgSO_4$ as a conversion factor.

$$? \text{ mol } MgSO_4 = 16.0 \text{ g } MgSO_4 \times \frac{1 \text{ mol } MgSO_4.}{120.38 \text{ g } MgSO_4} = 0.133 \text{ mol } MgSO_4$$

Step 2: Calculate the volume of the solution. The total mass of the solution is:

$$16.0 \text{ g } MgSO_4 + \left(100 \text{ mL } H_2O \times \frac{0.997 \text{ g } H_2O}{1 \text{ mL } H_2O} \right) = 115.7 \text{ g}$$

Assuming that the density of the solution is the same as that of water, we can calculate the volume of the solution as follows:

$$\text{Volume of solution} = 115.7 \text{ g} \times \frac{1 \text{ mL}}{0.997 \text{ g}} = 116.0 \text{ mL soln}$$

Step 3: Calculate the molarity of the solution by substituting the mol of solute and volume of solution (in L) into Equation (12.2).

$$M = \frac{\text{moles of solute}}{\text{liters of soln}} = \frac{0.133 \text{ mol}}{0.116 \text{ L}} = \mathbf{1.15 \, M}$$

PRACTICE EXERCISE

3.   An aqueous solution contains 167 g $CuSO_4$ in 820 mL of solution. The density of the solution is 1.195 g/mL. Calculate the molarity of the solution.

C.   Molality (*m*)

Molality is defined as the number of moles of solute per mass of solvent (in kg).

$$\text{molality} = \frac{\text{moles of solute}}{\text{mass of solvent (kg)}} \qquad (12.3)$$

EXAMPLE 12.4

What is the molality of the $MgSO_4$ solution made in Example 12.2?

Step 1: Calculate the number of moles $MgSO_4$ in 16.0 g. Use the molar mass of $MgSO_4$ as a conversion factor.

$$? \text{ mol } MgSO_4 = 16.0 \text{ g } MgSO_4 \times \frac{1 \text{ mol } MgSO_4}{120.38 \text{ g } MgSO_4} = 0.133 \text{ mol } MgSO_4$$

Step 2: Calculate the mass (in kg) of 100 mL of H_2O. Use the density of water as a conversion factor.

$$? \text{ mass of } H_2O = 100 \text{ mL } H_2O \times \frac{0.997 \text{ g } H_2O}{1 \text{ mL } H_2O} \times \frac{1 \text{ kg}}{1000 \text{ g}} = 0.0997 \text{ kg } H_2O$$

Step 3: Calculate the molality by substituting the mol of solute and the mass of solvent (in kg) into Equation (12.3).

$$m = \frac{\text{moles of solute}}{\text{mass of solvent (kg)}} = \frac{0.133 \text{ mol}}{0.0997 \text{ kg}} = \mathbf{1.33 \, m}$$

PRACTICE EXERCISE

4.   An aqueous solution contains 167 g $CuSO_4$ in 820 mL of solution. The density of the solution is 1.195 g/mL. Calculate the molality of the solution.

Text Problems: 12.18, 12.20

PROBLEM TYPE 3: CONVERTING BETWEEN CONCENTRATION UNITS

There are advantages and disadvantages to each type of concentration unit. An advantage of molality and percent by mass is that the concentration is temperature independent. On the other hand, molarity changes with temperature, because solution volume typically increases with increasing temperature. However, the advantage of molarity is that it is generally easier to measure the volume of solution than to weigh the solvent or solution.

A. Converting molality to molarity

EXAMPLE 12.5
Calculate the molarity of a 2.44 m NaCl solution given that its density is 1.089 g/mL.

Step 1: The mass of the solution must be converted to volume of solution to convert from molality to molarity. To calculate the mass of the solution, you must calculate the mass of the solute and then add that to the mass of water (1 kg = 1000 g). Remember that 2.44 m means 2.44 mol of solute per 1 kg (1000 g) of solvent.

Use the molar mass of the solute as a conversion factor to convert from moles of solute to mass of solute.

$$? \text{ mass of solute} = 2.44 \text{ mol NaCl} \times \frac{58.44 \text{ g NaCl}}{1 \text{ mol NaCl}} = 143 \text{ g NaCl}$$

$$? \text{ mass of solution} = \text{mass of solute} + \text{mass of solvent}$$

$$= 143 \text{ g} + 1000 \text{ g} = 1143 \text{ g}$$

Step 2: From the mass of the solution, you can calculate the volume of solution using the solution density as a conversion factor.

$$? \text{ volume of solution} = 1143 \text{ g} \times \frac{1 \text{ mL}}{1.089 \text{ g}} = 1.050 \times 10^3 \text{ mL} = 1.050 \text{ L}$$

Step 3: The moles of solute are given in the molarity (2.44 mol), and the volume of solution was calculated in *Step 2*. Divide moles of solute by liters of solution to calculate the molarity of the solution.

$$M = \frac{\text{mol of solute}}{\text{L of soln}} = \frac{2.44 \text{ mol}}{1.050 \text{ L}} = \mathbf{2.32 \textit{ M}}$$

PRACTICE EXERCISE
5.   Concentrated hydrochloric acid is 15.7 m. Calculate the molarity of concentrated HCl given that its density is 1.18 g/mL.

Text Problem: 12.22

B. Converting molarity to molality

EXAMPLE 12.6
Calculate the molality of a 2.55 M NaCl solution given that its density is 1.089 g/mL.

Step 1: From the volume of solution, you can calculate the mass of the solution using the solution density as a conversion factor. Remember that 1.55 *M* means 1.55 mol of solute per 1 L (1000 mL) of solution.

$$? \text{ mass of solution } = 1000 \text{ mL soln } \times \frac{1.089 \text{ g}}{1 \text{ mL soln}} = 1089 \text{ g}$$

Step 2: To calculate the mass of the *solvent* (water), you need to subtract the mass due to the *solute* from the mass of solution calculated in *Step 1*. You can calculate the mass of the solute from the moles of solute using molar mass as a conversion factor.

$$? \text{ mass of solute } = 2.55 \text{ mol NaCl } \times \frac{58.44 \text{ g NaCl}}{1 \text{ mol NaCl}} = 149 \text{ g NaCl}$$

$$? \text{ mass of solvent } = \text{ mass of soln } - \text{ mass of solute}$$

$$= 1089 \text{ g } - 149 \text{ g } = 9.40 \times 10^2 \text{ g } = 0.940 \text{ kg solvent}$$

Step 3: The moles of solute are given in the molarity (2.55 mol). Divide moles of solute by mass of solvent in kg to calculate the molality of the solution.

$$m = \frac{\text{mol of solute}}{\text{kg of solvent}} = \frac{2.55 \text{ mol}}{0.940 \text{ kg}} = \textbf{2.71 } \boldsymbol{m}$$

PRACTICE EXERCISE

6. Concentrated sulfuric acid, H_2SO_4 is 18.0 *M*. Calculate the molality of concentrated H_2SO_4 given that its density is 1.83 g/mL.

C. Converting percent by mass to molality

EXAMPLE 12.7
Concentrated hydrochloric acid is 36.5 percent HCl by mass. Its density is 1.18 g/mL. Calculate the molality of concentrated HCl.

Step 1: Assume that the mass of the solution is 100.0 g. The percent by mass of the solute can then be converted directly to grams. Then, the mass of solute can be converted to moles of solute using molar mass as a conversion factor.

Assuming 100.0 g solution,

$$36.5 \text{ g HCl } \times \frac{1 \text{ mol HCl}}{36.46 \text{ g HCl}} = 1.00 \text{ mol HCl}$$

Step 2: The mass of the solvent (water) can then be calculated by subtracting the mass of solute from the mass of the solution (100.0 g). Remember, the mass of the solution is equal to mass of solute + mass of solvent.

$$? \text{ mass of solvent } = \text{ mass of solution } - \text{ mass of solute}$$

$$= 100.0 \text{ g } - 36.5 \text{ g } = 63.5 \text{ g } = 0.0635 \text{ kg solvent (water)}$$

Step 3: Moles of solute was calculated in *Step 1*, and mass of solvent was calculated in *Step 2*. Divide moles of solute by mass of solvent in kg to calculate the molality of the solution.

$$m = \frac{\text{mol of solute}}{\text{kg of solvent}} = \frac{1.00 \text{ mol}}{0.0635 \text{ kg}} = \textbf{15.7 } \boldsymbol{m}$$

PRACTICE EXERCISE

7.   What is the molality of a 3.0 percent hydrogen peroxide (H_2O_2) aqueous solution? The density of the solution is 1.0 g/mL.

> **Text Problem: 12.22**

PROBLEM TYPE 4: EFFECT OF PRESSURE ON SOLUBILITY: HENRY'S LAW

Solubility is defined as the maximum amount of a solute that will dissolve in a given quantity of solvent at a specific temperature. For all practical purposes, external pressure has no influence on the solubilities of liquids and solids, but it does greatly affect the solubility of gases. There is a quantitative relationship between gas solubility and pressure called **Henry's law**, which states that the solubility of a gas in a liquid is proportional to the pressure of the gas over the solution:

$$c = kP$$

where,

   c is the molar concentration (mol/L) of the dissolved gas.

   P is the pressure, in atmospheres, of the gas over the solution.

   k is a constant for a given gas that depends only on the temperature. k has units of mol/L·atm.

As the pressure of the gas over the solution increases, more gas molecules strike the surface of the liquid increasing the number of gas molecules that dissolve in the liquid. For further discussion, see Section 12.5 of your text.

EXAMPLE 12.8

What is the concentration of O_2 at 25°C in water that is saturated with *air* at an atmospheric pressure of 645 mmHg? The Henry's law constant (k) for oxygen is 3.5×10^{-4} mol/L·atm. Assume that the mole fraction of oxygen in air is 0.209.

Step 1: We want to calculate the molar concentration (c) of O_2 in water. Using Henry's law, $c_{O_2} = kP_{O_2}$. k is given in the problem. To calculate the molar concentration, we must first calculate the partial pressure of oxygen in air. The partial pressure of O_2 is found using Dalton's law of partial pressures (see Problem Type 3, Chapter 5).

$$P_{O_2} = X_{O_2} P_T = (0.209)(645 \text{ mmHg}) \times \frac{1 \text{ atm}}{760 \text{ mmHg}} = 0.177 \text{ atm}$$

Step 2: Substitute the partial pressure of O_2 and k into the Henry's law expression to solve for the molar concentration of O_2 in water.

$$c_{O_2} = kP_{O_2} = (3.5 \times 10^{-4} \text{ mol/L·atm})(0.177 \text{ atm}) = \mathbf{6.2 \times 10^{-5} \text{ mol/L}}$$

PRACTICE EXERCISE

8.   The Henry's law constant for CO is 9.73×10^{-4} mol/L·atm at 25°C. What is the concentration of dissolved CO in water if the partial pressure of CO in the air is 0.015 mmHg?

> **Text Problems: 12.36, 12.38**

PROBLEM TYPE 5: COLLIGATIVE PROPERTIES OF NONELECTROLYTES

Several important properties of solutions depend on the number of solute particles in solution and not on the nature of the solute particles. These properties are called **colligative properties**.

A. Vapor pressure lowering, Raoult's law

If a solute is nonvolatile, the vapor pressure of its solution is always less than that of the pure solvent. Raoult's law quantifies this relationship by stating that the partial pressure of a solvent over a solution, P_1, is given by the vapor pressure of the pure solvent, P_1°, times the mole fraction of the solvent in the solution, X_1:

$$P_1 = X_1 P_1^\circ \tag{12.4}$$

If the solution contains only one solute, $X_1 = 1 - X_2$, where X_2 is the mole fraction of the solute. Substituting for X_1 in Equation (12.4) gives:

$$P_1 = (1 - X_2) P_1^\circ$$
$$P_1 = P_1^\circ - X_2 P_1^\circ$$
$$P_1^\circ - P_1 = \Delta P = X_2 P_1^\circ \tag{12.5}$$

Equation (12.5) shows that a decrease in vapor pressure, ΔP, is directly proportional to the concentration of the solute in solution, X_2.

EXAMPLE 12.9

Calculate the vapor pressure of an aqueous solution at 30°C made from 1.00×10^2 g of sucrose ($C_{12}H_{22}O_{11}$) and 1.00×10^2 g of water. The vapor pressure of pure water at 30°C is 31.8 mmHg.

Step 1: Equation (12.4) above gives a relationship between the vapor pressure of the solvent over a solution and the mole fraction of the solvent. If you can calculate the mole fraction of the solvent, you can calculate the vapor pressure of the solution. Problem Type 3, Chapter 5 defines mole fraction.

$$\text{mol water} = 1.00 \times 10^2 \text{ g water} \times \frac{1 \text{ mol water}}{18.02 \text{ g water}} = 5.55 \text{ mol}$$

$$\text{mol sucrose} = 1.00 \times 10^2 \text{ g sucrose} \times \frac{1 \text{ mol sucrose}}{342.3 \text{ g sucrose}} = 0.292 \text{ mol}$$

$$X_{\text{water}} = \frac{\text{mol}_{\text{water}}}{\text{mol}_{\text{water}} + \text{mol}_{\text{sucrose}}} = \frac{5.55 \text{ mol}}{5.55 \text{ mol} + 0.292 \text{ mol}} = 0.950$$

Step 2: Substitute X_{water} and the vapor pressure of pure water into Equation (12.4) to solve for the vapor pressure of the solution.

$$P_{\text{soln}} = X_{\text{water}} P_{\text{water}}^\circ = (0.950)(31.8 \text{ mmHg}) = \textbf{30.2 mmHg}$$

> **Tip:** You could also have solved this problem using Equation (12.5). You could calculate the change in vapor pressure (ΔP) from the mole fraction of solute. Then, you could calculate the vapor pressure of the solution by subtracting ΔP from the vapor pressure of the pure solvent (water).

PRACTICE EXERCISE

9. At 25°C, the vapor pressure of pure water is 23.76 mmHg and that of an aqueous sucrose ($C_{12}H_{22}O_{11}$) is 23.28 mmHg. Calculate the molality of the solution.

Text Problems: **12.52**, 12.54, 12.56

B. Boiling-point elevation

The **boiling point** of a solution is the temperature at which its vapor pressure equals the external atmospheric pressure (see Section 11.8 of the text). We just saw in Part (a) that a nonvolatile solute always decreases the vapor pressure of the solution relative to the pure solvent. Consequently, the boiling point of the solution is *higher* than the pure solvent, because more energy in the form of heat must be added to raise the vapor pressure of the solution to the external atmospheric pressure. The change in boiling point is proportional to the concentration of solute.

$$\Delta T_b = K_b m \qquad\qquad (12.6)$$

where,

$\Delta T_b = T_b - T_b^\circ$ (where T_b is the boiling point of the solution and T_b° is the boiling point of the pure
solvent)

m is the molal concentration of the solute

K_b is the molal boiling-point elevation constant of the solvent with units of $°C/m$

Do you know why molality is used for the concentration instead of molarity? Because we are dealing with a system that is not kept at constant temperature. We cannot express the concentration in molarity because molarity changes with temperature.

EXAMPLE 12.10

What is the boiling point of an "antifreeze/coolant" solution made from a 50-50 mixture (by volume) of ethylene glycol, $C_2H_6O_2$ and water? Assume the density of water is 1.00 g/mL and the density of ethylene glycol is 1.11 g/mL.

Using Equation (12.6), you can calculate the change in boiling point, ΔT_b, by calculating the molality of the solution and then multiplying by K_b for water (see Table 12.2 of your text).

Step 1: For simplicity, assume that you have 100.0 mL of solution. Since the mixture is 50-50 by volume, there are 50.0 mL of ethylene glycol and 50.0 mL of water. To calculate the molality of the solution, you need the moles of solute (ethylene glycol) and the mass of solvent (water) in kg.

$$\text{mol of ethylene glycol} = 50.0 \text{ mL } C_2H_6O_2 \times \frac{1.11 \text{ g } C_2H_6O_2}{1 \text{ mL } C_2H_6O_2} \times \frac{1 \text{ mol } C_2H_6O_2}{62.07 \text{ g } C_2H_6O_2} = 0.894 \text{ mol}$$

$$\text{mass of water} = 50.0 \text{ mL } H_2O \times \frac{1.00 \text{ g } H_2O}{1 \text{ mL } H_2O} \times \frac{1 \text{ kg}}{1000 \text{ g}} = 0.0500 \text{ kg}$$

$$\text{molality} = \frac{\text{mol solute}}{\text{kg solvent}} = \frac{0.894 \text{ mol}}{0.0500 \text{ kg}} = 17.9 \text{ } m$$

Step 2: Substitute the molality of the solution and K_b into Equation (12.6) to solve for the change in boiling point. Then, add the change in boiling point to the normal boiling point of water (100.0°C) to calculate the boiling point of the solution.

$$\Delta T_b = K_b m = (0.52°C/m)(17.9 \text{ } m) = 9.3°C$$

$$\text{b.p. of soln} = 100.0°C + 9.3°C = \textbf{109.3°C}$$

PRACTICE EXERCISE

10. What is the boiling point of an aqueous solution of a nonvolatile solute that freezes at −3.0°C?

C. Freezing-point depression

The **freezing point** of a liquid (or the melting point of a solid) is the temperature at which the solid and liquid phases coexist in equilibrium. It might be easier to understand freezing-point depression by looking at the opposite of freezing, melting. To melt a solid, the intermolecular forces holding the solid molecules together must be overcome. Adding another solid substance to a pure solid disrupts the intermolecular forces of the formerly pure solid. Hence, it is easier to overcome the intermolecular forces and the mixture melts at a lower temperature than the pure solid. The melting point (or the freezing point) is depressed. The depression of freezing point can be represented by the following equation.

$$\Delta T_f = K_f m \qquad\qquad (12.7)$$

where,

$\Delta T_f = T_f^\circ - T_f$ (where T_f is the freezing point of the solution and T_f° is the freezing point of the pure solvent)

m is the molal concentration of the solute

K_f is the molal freezing-point depression constant with units of $°C/m$

EXAMPLE 12.11

How many grams of isopropyl alcohol, C_3H_7OH, should be added to 1.0 L of water to give a solution that will not freeze above $-16°C$?

We want to lower the freezing point of 1.0 L of water by 16°C. Using Equation (12.7), we can calculate the molality of the solution needed. Then, from the molality, we can calculate the grams of isopropyl alcohol needed.

Step 1: Rearrange Equation (12.7) to solve for the molality. You can look up K_f in Table 12.2 of your text.

$$m = \frac{\Delta T_f}{K_f} = \frac{16 \; °C}{1.86 \; °C / m} = 8.6 \; m$$

Step 2: 8.6 m means that the solution contains 8.6 mol of solute per 1 kg of solvent. Assume that the density of water is 1.0 g/mL; thus, 1 kg (1000 g) of water has a volume of 1 L. Convert 8.6 mol of isopropyl alcohol to grams of isopropyl alcohol using the molar mass as a conversion factor.

$$\textbf{? g isopropyl alcohol} = 8.6 \; mol \; C_3H_7OH \times \frac{60.09 \; g \; C_3H_7OH}{1 \; mol \; C_3H_7OH} = \textbf{5.2} \times \textbf{10}^2 \textbf{ g}$$

PRACTICE EXERCISE

11. Benzene melts at 5.50°C. When 2.11 g of naphthalene, $C_{10}H_8$, is added to 100 g of benzene, the solution freezes at 4.65°C. Calculate the freezing-point depression constant (K_f) for benzene.

| Text Problems: 12.58, 12.62 |

D. Osmotic pressure

Osmosis is the net movement of solvent molecules through a semipermeable membrane from a pure solvent or from a dilute solution to a more concentrated solution. The **osmotic pressure** (π) of a solution is the pressure required to stop osmosis. The osmotic pressure of a solution is given by:

$$\pi = MRT \qquad\qquad (12.8)$$

where,

M is the molarity of the solution

R is the gas constant (0.0821 L·atm/K·mol)

T is the absolute temperature in K

EXAMPLE 12.12

The average osmotic pressure of seawater is about 30.0 atm at 25°C. Calculate the molar concentration of an aqueous solution of urea (NH_2CONH_2) that is isotonic with seawater.

A solution of urea that is isotonic with seawater must have the same osmotic pressure, 30.0 atm. Solve Equation (12.8) algebraically for the molar concentration. Then, substitute π, R, and T (in K) into the equation to solve for the molar concentration of the urea solution.

$$M = \frac{\pi}{RT} = \frac{30.0 \text{ atm}}{298 \text{ K}} \times \frac{\text{mol} \cdot \text{K}}{0.0821 \text{ L} \cdot \text{atm}} = 1.23 \, M$$

PRACTICE EXERCISE

12. The walls of red blood cells are semipermeable membranes, and the solution of NaCl within those walls exerts an osmotic pressure of 7.82 atm at 37°C. What concentration of NaCl must a *surrounding* solution have so that this pressure is balanced and cell rupture (hemolysis) is prevented?

PROBLEM TYPE 6: DETERMINING MOLAR MASS USING COLLIGATIVE PROPERTIES

Any of the four colligative properties discussed in Problem Type 5 can be used to calculate the molar mass of the solute. However, in practice, only freezing-point depression and osmotic pressure are used because they show the most pronounced changes.

A. Calculating molar mass from freezing-point depression

You can solve this type of problem by first calculating the molality of the solute using Equation (12.7) and then calculating the molar mass from the molality.

EXAMPLE 12.13

Benzene has a normal freezing point of 5.51°C. The addition of 1.25 g of an unknown compound to 85.0 g of benzene produces a solution with a freezing point of 4.52°C. What is the molar mass of the unknown compound?

Step 1: Solve Equation (12.7) algebraically for molality (m), then substitute ΔT_f and K_f into the equation to calculate the molality.

$$\Delta T_f = 5.51°C - 4.52°C = 0.99°C$$

$$m = \frac{\Delta T_f}{K_f} = \frac{0.99 \,°C}{5.12 \,°C/m} = 0.19 \, m$$

Step 2: Multiplying the molality by the mass of solvent (in kg) gives moles of unknown solute. Then, dividing the mass of solute (in g) by the moles of solute, gives the molar mass of the unknown solute.

$$? \text{ mol of unknown solute} = \frac{0.19 \text{ mol solute}}{1 \text{ kg benzene}} \times 0.085 \text{ kg benzene}$$

$$= 0.016 \text{ mol solute}$$

$$\text{molar mass of unknown} = \frac{1.25 \text{ g}}{0.016 \text{ mol}} = 78 \text{ g/mol}$$

PRACTICE EXERCISE

13. When 48 g of glucose (a nonelectrolyte) is dissolved in 500 g of H_2O, the solution has a freezing point of −0.94°C. What is the molar mass of glucose?

14. In the course of research, a chemist isolates a new compound. An elemental analysis shows the following: C, 50.7 percent; H, 4.25 percent; O, 45.1 percent. If 5.01 g of the compound is dissolved in 100 g of water, a solution with a freezing point of −0.65°C is produced. What is the molecular formula of the compound?

| Text Problems: 12.60, 12.64 |

B. Calculating the molar mass from osmotic pressure

You can solve this type of problem by first calculating the molarity of the solution using Equation (12.8) and then calculating the molar mass from the molarity.

EXAMPLE 12.14

30.0 g of sucrose is dissolved in water making 1.00×10^2 mL of solution. The solution has an osmotic pressure of 20.8 atm at 16.0°C. What is the molar mass of sucrose?

Step 1: Solve Equation (12.8) algebraically for molarity (M), then substitute π, R, and T (in K) into the equation to calculate the molarity.

$$M = \frac{\pi}{RT} = \frac{20.8 \text{ atm}}{289 \text{ K}} \times \frac{\text{mol} \cdot \text{K}}{0.0821 \text{ L} \cdot \text{atm}} = 0.877 \ M$$

Step 2: Multiplying the molarity by the volume of solution (in L) gives moles of solute. Then, dividing the mass of solute (in g) by the moles of solute, gives the molar mass of the solute.

$$? \text{ mol of sucrose} = (0.877 \text{ mol/L})(0.100 \text{ L}) = 0.0877 \text{ mol sucrose}$$

$$\textbf{molar mass of sucrose} = \frac{30.0 \text{ g sucrose}}{0.0877 \text{ mol sucrose}} = \textbf{342 g/mol}$$

PRACTICE EXERCISE

15. Peruvian Indians use a dart poison from root extracts called curare. It is a nonelectrolyte. The osmotic pressure at 20.0°C of an aqueous solution containing 0.200 g of curare in 1.00×10^2 mL of solution is 56.2 mmHg. Calculate the molar mass of curare.

| Text Problems: 12.66, 12.68 |

PROBLEM TYPE 7: COLLIGATIVE PROPERTIES OF ELECTROLYTES

Electrolytes dissociate into ions in solution, so this requires us to take a slightly different approach than that used for the colligative properties of nonelectrolytes. Remember, it is the number of solute particles that determines the colligative properties of a solution. To account for the dissociation of an electrolyte into ions, the equations for colligative properties must be modified as follows:

$$\Delta T_b = iK_b m$$
$$\Delta T_f = iK_f m \tag{12.9}$$
$$\pi = iMRT$$

where,

i is the van't Hoff factor which is defined as:

$$i = \frac{\text{actual number of particles in soln after dissociation}}{\text{number of formula units initially dissolved in soln}}$$

Thus, i should be 1 for all nonelectrolytes. For strong electrolytes such as KCl and $BaSO_4$, i should be 2, and for strong electrolytes such as $CaCl_2$ and Na_2CO_3, i should be 3. In reality, the colligative properties of *electrolyte* solutions are usually smaller than anticipated. At higher concentrations, electrostatic forces come into play, drawing cations and anions together. The ion pairs that are formed decrease the number of solute particles in solution. Table 12.3 of your text lists the van't Hoff factor (i) for various solutes.

EXAMPLE 12.15
Calculate the value of i for an electrolyte that should dissociate into 2 ions, if a 1.0 m aqueous solution of the electrolyte freezes at $-3.28°C$. Why is the value of i less than 2?

Solve Equation (12.9) algebraically for i, then substitute the values of ΔT_f, K_f, and m into the equation to solve for i.

$\Delta T_f = 3.28°C$

$$i = \frac{\Delta T_f}{K_f m} = \frac{3.28\,°C}{(1.86\,°C/m)(1.0\,m)} = 1.76$$

The value if i is less than 2 because ion pairing reduces the number of particles in solution. Remember, colligative properties depend only on the number of solute particles in solution and *not* on the nature of the solute particles.

PRACTICE EXERCISE
16. Arrange the following aqueous solutions in order of increasing boiling points: 0.100 m ethanol, 0.050 m $Ca(NO_3)_2$, 0.100 m NaBr, 0.050 m HCl.

Text Problems: 12.74, 12.76, 12.78, **12.80**, 12.82

ANSWERS TO PRACTICE EXERCISES

1. Remember that "like dissolves like". Therefore, the ionic solid NaBr will be more soluble in the polar solvent, **H_2O**.

2. 17.0 percent $CuSO_4$

3. 1.28 M $CuSO_4$

4. 1.29 m $CuSO_4$

5. 11.8 M HCl

6. $m = 3 \times 10^2\ m$

7. 0.91 m H_2O_2

8. $1.9 \times 10^{-8}\ M$ CO

9. $m = 1.1\ m$

10. b.p. of soln = 100.84°C

11. K_f (benzene) = 5.15°C/m

12. 0.15 M NaCl

13. molar mass = 1.9×10^2 g/mol

14. $C_6H_6O_4$

15. molar mass of curare = 6.50×10^2 g/mol

16. ethanol $\cong$ HCl < $Ca(NO_3)_2$ < NaBr

SOLUTIONS TO SELECTED TEXT PROBLEMS

12.10   Predicting Solubility Based on Intermolecular Forces, Problem Type 1.

Recall that "like dissolves like". Strong hydrogen bonding (dipole-dipole attraction) is the principal intermolecular attraction in liquid ethanol, but in liquid cyclohexane the intermolecular forces are dispersion forces because cyclohexane is nonpolar. Cyclohexane cannot form hydrogen bonds with ethanol, and therefore cannot attract ethanol molecules strongly enough to form a solution.

12.12   The longer the C–C chain, the more the molecule "looks like" a hydrocarbon and the less important the –OH group becomes. Hence, as the C–C chain length increases, the molecule becomes less polar. Since "like dissolves like", as the molecules become more nonpolar, the solubility in polar water decreases. The –OH group of the alcohols can form strong hydrogen bonds with water molecules, but this property decreases as the chain length increases.

12.16   Percent by Mass Calculation, Problem Type 2.

(a)   The percent by mass is defined as

$$\text{percent by mass of solute} = \frac{\text{mass of solute}}{\text{mass of solute} + \text{mass of solvent}} \times 100\%$$

We are given percent by mass of solute and the mass of solute in the problem. Substituting those two quantities into the above equation, we can solve for the mass of solvent (water).

$$16.2\% = \frac{5.00 \text{ g urea}}{5.00 \text{ g urea} + \text{mass of water}} \times 100\%$$

$$(0.162)(\text{mass of water}) = 5.00 \text{ g} - (0.162)(5.00\text{g})$$

mass of water = 25.9 g

(b)   Similar to part (a),

$$1.5\% = \frac{26.2 \text{ g MgCl}_2}{26.2 \text{ g MgCl}_2 + \text{mass of water}} \times 100\%$$

mass of water $= 1.72 \times 10^3$ g

12.18 $\text{molality} = \dfrac{\text{moles of solute}}{\text{mass of solvent (kg)}}$

(a) $\text{mass of 1 L soln} = 1000 \text{ mL} \left(\dfrac{1.08 \text{ g}}{1 \text{ mL}} \right) = 1080 \text{ g}$

$$\text{mass of water} = 1080 \text{ g} - 2.50 \text{ mol NaCl} \left(\frac{58.44 \text{ g NaCl}}{1 \text{ mol NaCl}} \right) = 934 \text{ g} = 0.934 \text{ kg}$$

$$m = \frac{2.50 \text{ mol NaCl}}{0.934 \text{ kg H}_2\text{O}} = \textbf{2.68 } \boldsymbol{m}$$

(b) 100 g of the solution contains 48.2 g KBr and 51.8 g H_2O.

$$\text{mol of KBr} = 48.2 \text{ g KBr} \times \left(\frac{1 \text{ mol KBr}}{119.0 \text{ g KBr}} \right) = 0.405 \text{ mol KBr}$$

$$\text{mass of } H_2O \text{ (in kg)} = 51.8 \text{ g } H_2O \times \frac{1 \text{ kg } H_2O}{1000 \text{ g } H_2O} = 0.0518 \text{ kg } H_2O$$

$$m = \frac{0.405 \text{ mol KBr}}{0.0518 \text{ kg } H_2O} = \textbf{7.82 } \textbf{\textit{m}}$$

12.20 Let's assume that we have 1.0 L of a 0.010 M solution.

Assuming a solution density of 1.0 g/mL, the mass of 1.0 L of the solution is:

$$\text{Mass of 1.0 L} = 1.0 \text{ L} \left(\frac{1000 \text{ mL}}{1 \text{ L}} \right) \left(\frac{1.0 \text{ g}}{1 \text{ mL}} \right) = 1.0 \times 10^3 \text{ g} \quad .$$

The mass of 0.010 mole of urea is:

$$0.010 \text{ mol urea} \left(\frac{60.06 \text{ g urea}}{1 \text{ mol urea}} \right) = 0.60 \text{ g urea}$$

The mass of the solvent is:

$$\text{(solution mass)} - \text{(solute mass)} = (1.0 \times 10^3 \text{ g}) - (0.60 \text{ g}) = 1.0 \times 10^3 \text{ g} = 1.0 \text{ kg}$$

$$m = \frac{\text{moles solute}}{\text{mass solvent}} = \frac{0.010 \text{ mol}}{1.0 \text{ kg}} = \textbf{0.010 } \textbf{\textit{m}}$$

12.22 Converting between Concentration Units, Problem Type 3.

(a) Converting percent by mass to molality

Step 1: Assume that the mass of the solution is 100.0 g. The percent by mass of the solute can then be converted directly to grams. Then, the mass of solute can be converted to moles of solute using molar mass as a conversion factor. Assuming 100.0 g solution,

$$98.0 \text{ g } H_2SO_4 \times \frac{1 \text{ mol } H_2SO_4}{98.09 \text{ g } H_2SO_4} = 0.999 \text{ mol } H_2SO_4$$

Step 2: The mass of the solvent (water) can then be calculated by subtracting the mass of solute from the mass of the solution (100.0 g). Remember, the mass of the solution is equal to the mass of solute + mass of solvent.

$$? \text{ mass of solvent} = \text{mass of solution} - \text{mass of solute}$$

$$= 100.0 \text{ g} - 98.0 \text{ g} = 2.0 \text{ g} = 2.0 \times 10^{-3} \text{ kg solvent (water)}$$

Step 3: Moles of solute was calculated in *Step 1*, and mass of solvent was calculated in *Step 2*. Divide moles of solute by mass of solvent in kg to calculate the molality of the solution.

$$m = \frac{\text{mol of solute}}{\text{kg of solvent}} = \frac{0.999 \text{ mol}}{2.0 \times 10^{-3} \text{ kg}} = \textbf{5.0} \times \textbf{10}^2 \textbf{\textit{m}}$$

(b) Converting molality to molarity.

Step 1: The mass of the solution must be converted to volume of solution to convert from molality to molarity. We assumed 100.0 g of solution in part (a). Use the solution density as a conversion factor to convert to volume of solution.

$$? \text{ volume of solution } = 100.0 \text{ g} \times \frac{1 \text{ mL}}{1.83 \text{ g}} = 54.6 \text{ mL} = 0.0546 \text{ L}$$

Step 2: We calculated moles of solute in part (a), 0.999 mol H_2SO_4. Divide moles of solute by liters of solution to calculate the molarity of the solution.

$$M = \frac{\text{mol of solute}}{\text{L of soln}} = \frac{0.999 \text{ mol}}{0.0546 \text{ L}} = \textbf{18.3 } \boldsymbol{M}$$

12.24   Assume 100.0 g of solution.

(a)   The mass of ethanol in the solution is $0.100 \times 100.0 \text{ g} = 10.0 \text{ g}$. The mass of the water is $100.0 \text{ g} - 10.0 \text{ g} = 90.0 \text{ g} = 0.0900 \text{ kg}$. The amount of ethanol in moles is:

$$10.0 \text{ g ethanol}\left(\frac{1 \text{ mol}}{46.07 \text{ g}}\right) = 0.217 \text{ mol ethanol}$$

$$m = \frac{\text{mol solute}}{\text{kg solvent}} = \frac{0.217 \text{ mol}}{0.0900 \text{ kg}} = \textbf{2.41 } \boldsymbol{m}$$

(b)   The volume of the solution is:

$$100.0 \text{ g}\left(\frac{1 \text{ mL}}{0.984 \text{ g}}\right) = 102 \text{ mL} = 0.102 \text{ L}$$

The amount of ethanol in moles is 0.217 mol [part (a)].

$$M = \frac{\text{mol solute}}{\text{liters of soln}} = \frac{0.217 \text{ mol}}{0.102 \text{ L}} = \textbf{2.13 } \boldsymbol{M}$$

(c)   **Solution volume** $= 0.125 \text{ mol}\left(\dfrac{1 \text{ L}}{2.13 \text{ mol}}\right) = \textbf{0.0587 L = 58.7 mL}$

12.28   At 75°C, 155 g of KNO_3 dissolves in 100 g of water to form 255 g of solution. When cooled to 25°C, only 38.0 g of KNO_3 remain dissolved. This means that $(155 - 38.0) \text{ g} = 117 \text{ g of } KNO_3$ will crystallize.

The amount of KNO_3 formed when 100 g of saturated solution at 75°C is cooled to 25°C can be found by a simple unit conversion.

$$(100 \text{ g saturated soln}) \times \left(\frac{117 \text{ g } KNO_3 \text{ crystallized}}{255 \text{ g saturated soln}}\right) = \textbf{45.9 g } \boldsymbol{KNO_3}$$

12.36   According to Henry's law, the solubility of a gas in a liquid increases as the pressure increases ($c = kP$). The soft drink tastes flat at the bottom of the mine because the carbon dioxide pressure is greater and the dissolved gas is not released from the solution. As the miner goes up in the elevator, the atmospheric carbon dioxide pressure decreases and dissolved gas is released from his stomach.

12.38   Effect of Pressure on Solubility, Problem Type 4. To solve this problem, you need to compare the solubilities of N_2 in blood under normal pressure (0.80 atm) and under a greater pressure that a deep-sea diver might experience (4.0 atm).

Step 1: Calculate the Henry's law constant, k, using the concentration of N_2 in blood at 0.80 atm.

$$k = \frac{c}{P}$$

$$k = \frac{5.6 \times 10^{-4} \text{ mol / L}}{0.80 \text{ atm}} = 7.0 \times 10^{-4} \text{ mol/L·atm}$$

Step 2: Calculate the concentration of N_2 in blood at 4.0 atm using k calculated above.

$$c = kP$$

$$c = (7.0 \times 10^{-4} \text{ mol/L·atm})(4.0 \text{ atm}) = 2.8 \times 10^{-3} \text{ mol/L}$$

Step 3: From each of the concentrations of N_2 in blood, we can calculate the number of moles of N_2 dissolved by multiplying by the total blood volume of 5.0 L. Then, we can calculate the number of moles of N_2 released when the diver returns to the surface.

The number of moles of N_2 in 5.0 L of blood at 0.80 atm is:

$$(5.6 \times 10^{-4} \text{ mol/L}) \times 5.0 \text{ L} = 2.8 \times 10^{-3} \text{ mol}$$

The number of moles of N_2 in 5.0 L of blood at 4.0 atm is:

$$(2.8 \times 10^{-3} \text{ mol/L}) \times 5.0 \text{ L} = 1.4 \times 10^{-2} \text{ mol}$$

The amount of N_2 released in moles when the diver returns to the surface is:

$$1.4 \times 10^{-2} \text{ mol} - 2.8 \times 10^{-3} \text{ mol} = 1.1 \times 10^{-2} \text{ mol}$$

Step 4: We can now calculate the volume of N_2 released using the ideal gas equation. The total pressure pushing on the N_2 that is released is atmospheric pressure (1 atm).

The volume of N_2 released is:

$$V_{N_2} = \frac{nRT}{P}$$

$$V_{N_2} = \frac{(1.1 \times 10^{-2} \text{ mol})(273 + 37)\text{K}}{(1.0 \text{ atm})} \times \frac{0.0821 \text{ L·atm}}{\text{mol·K}} = \textbf{0.28 L}$$

12.52   Vapor-pressure lowering, Raoult's law, Problem Type 5A.

Step 1: Using Equation (12.5) from Problem Type 5A, we can calculate the mole fraction of sucrose that causes a 2.0 mmHg drop in vapor pressure.

$$\Delta P = X_2 P_1^\circ \qquad\qquad (12.5)$$

$$\Delta P = X_{sucrose} P_{water}^\circ$$

$$X_{sucrose} = \frac{\Delta P}{P_{water}^\circ} = \frac{2.0 \text{ mmHg}}{17.5 \text{ mmHg}} = 0.11$$

Step 2: From the definition of mole fraction, we can calculate the number of moles of sucrose needed to cause a 2.0 mmHg drop in vapor pressure.

$$X_{sucrose} = \frac{n_{sucrose}}{n_{water} + n_{sucrose}}$$

$$\text{moles of water} = 552 \text{ g} \times \left(\frac{1 \text{ mol}}{18.02 \text{ g}} \right) = 30.6 \text{ mol H}_2\text{O}$$

$$X_{sucrose} = 0.11 = \frac{n_{sucrose}}{30.6 + n_{sucrose}}$$

$$n_{sucrose} = 3.8 \text{ mol sucrose}$$

Step 3: Using the molar mass of sucrose as a conversion factor, we can calculate the mass of sucrose.

$$\textbf{mass of sucrose} = 3.8 \text{ mol sucrose} \times \left(\frac{342.3 \text{ g sucrose}}{1 \text{ mol sucrose}} \right) = \textbf{1.3} \times \textbf{10}^\textbf{3} \textbf{ g sucrose}$$

12.54   For any solution the sum of the mole fractions of the components is always 1.00, so the mole fraction of 1–propanol is 0.700. The partial pressures are:

$$\boldsymbol{P}_{ethanol} = X_{ethanol} \times P_{ethanol}^\circ = 0.300 \times 100 \text{ mmHg} = \textbf{30.0 mmHg}$$

$$\boldsymbol{P}_{1-propanol} = X_{1-propanol} \times P_{1-propanol}^\circ = 0.700 \times 37.6 \text{ mmHg} = \textbf{26.3 mmHg}$$

Is the vapor phase richer in one of the components than the solution? Which component? Should this always be true for ideal solutions?

12.56   This problem is very similar to Problem 12.52.

$$\Delta P = X_{urea} P_{water}^\circ$$

$$2.50 \text{ mmHg} = X_{urea}(31.8 \text{ mmHg})$$

$$X_{urea} = 0.0786$$

The number of moles of water is:

$$n_{water} = 450 \text{ g H}_2\text{O} \left(\frac{1 \text{ mol H}_2\text{O}}{18.02 \text{ g H}_2\text{O}} \right) = 25.0 \text{ mol H}_2\text{O}$$

$$X_{urea} = \frac{n_{urea}}{n_{water} + n_{urea}}$$

$$0.0786 = \frac{n_{urea}}{25.0 + n_{urea}}$$

$$n_{urea} = 2.13 \text{ mol}$$

$$\textbf{mass of urea} = 2.13 \text{ mol urea} \times \left(\frac{60.06 \text{ g urea}}{1 \text{ mol urea}} \right) = \textbf{128 g of urea}$$

12.58 $m = \dfrac{\Delta T_f}{K_f} = \dfrac{1.1°C}{1.86°C/m} = \textbf{0.59 } \boldsymbol{m}$

12.60   This is a combination of Problem Type 4B, Chapter 3, and Problem Type 6A, Chapter 12. First, we need to find the empirical formula from mass percent data. Then, we can determine the molar mass from the freezing-point depression. Finally, from the empirical formula and the molar mass, we can find the molecular formula.

METHOD 1:

Step 1: Assume you have exactly 100 g of substance. 100 g is a convenient amount, because all the percentages sum to 100 percent. In 100 g of the solid extracted from gum arabic there will be 40.0 g C, 6.7 g H, and 53.3 g O.

Step 2: Calculate the number of moles of each element in the compound. Remember, an *empirical formula* tells us which elements are present and the simplest whole-number ratio of their atoms. This ratio is also a mole ratio. Let n_C, n_H, and n_O be the number of moles of elements present. Use the molar masses of these elements as conversion factors to convert to moles.

$$n_C = 40.0 \text{ g C} \times \frac{1 \text{ mol C}}{12.01 \text{ g C}} = \textbf{3.33 mol C}$$

$$n_H = 6.7 \text{ g H} \times \frac{1 \text{ mol H}}{1.008 \text{ g H}} = \textbf{6.6 mol H}$$

$$n_O = 53.3 \text{ g O} \times \frac{1 \text{ mol O}}{16.00 \text{ g O}} = \textbf{3.33 mol O}$$

Thus, we arrive at the formula $C_{3.33}H_{6.6}O_{3.3}$, which gives the identity and the ratios of atoms present. However, chemical formulas are written with whole numbers.

Step 3: Try to convert to whole numbers by dividing all the subscripts by the smallest subscript.

C: $\dfrac{3.33}{3.33} = 1.00$ **H:** $\dfrac{6.6}{3.33} = 2.0$ **O:** $\dfrac{3.33}{3.33} = 1.00$

This gives us the empirical, **CH_2O.**

Step 4: Use the freezing point data to determine the molar mass. First, calculate the molality of the solution.

$$m = \frac{\Delta T_f}{K_f} = \frac{1.56°C}{8.00°C/m} = 0.195 \, m$$

Step 5: Multiplying the molality by the mass of solvent (in kg) gives moles of unknown solute. Then, dividing the mass of solute (in g) by the moles of solute, gives the molar mass of the unknown solute.

$$? \text{ mol of unknown solute} = \frac{0.195 \text{ mol solute}}{1 \text{ kg diphenyl}} \times 0.0278 \text{ kg diphenyl}$$

$$= 0.00542 \text{ mol solute}$$

$$\textbf{molar mass of unknown} = \frac{0.650 \text{ g}}{0.00542 \text{ mol}} = \textbf{1.20} \times \textbf{10}^2 \textbf{ g/mol}$$

Step 6: Calculate the empirical molar mass. Then, determine the number of CH_2O units present in the molecular formula by taking the ratio of molar mass to empirical mass.

$$\text{empirical molar mass} = 12.01 \text{ g} + 2(1.008 \text{ g}) + 16.00 \text{ g} = 30.03 \text{ g/mol}$$

The number of (CH_2O) units present in the molecular formula is:

$$\frac{\text{molar mass}}{\text{empirical molar mass}} = \frac{1.20 \times 10^2 \text{ g}}{30.03 \text{ g}} = 4.00$$

Thus, there are four CH_2O units in each molecule of the compound, so the molecular formula is $(CH_2O)_4$, or $\textbf{C}_4\textbf{H}_8\textbf{O}_4$.

METHOD 2:

Step 1: Use the freezing point data to determine the molar mass. First, calculate the molality of the solution.

$$m = \frac{\Delta T_f}{K_f} = \frac{1.56°\text{C}}{8.00°\text{C}/m} = 0.195 \text{ } m$$

Step 2: Multiplying the molality by the mass of solvent (in kg) gives moles of unknown solute. Then, dividing the mass of solute (in g) by the moles of solute, gives the molar mass of the unknown solute.

$$? \text{ mol of unknown solute} = \frac{0.195 \text{ mol solute}}{1 \text{ kg diphenyl}} \times 0.0278 \text{ kg diphenyl}$$

$$= 0.00542 \text{ mol solute}$$

$$\textbf{molar mass of unknown} = \frac{0.650 \text{ g}}{0.00542 \text{ mol}} = \textbf{1.20} \times \textbf{10}^2 \textbf{ g/mol}$$

Step 3: Multiply the mass % (converted to a decimal) of each element by the molar mass to convert to grams of each element. Then, use the molar mass to convert to moles of each element.

$$n_C = (0.400) \times (1.20 \times 10^2 \text{ g}) \times \frac{1 \text{ mol C}}{12.01 \text{ g C}} = \textbf{4.00 mol C}$$

$$n_H = (0.067) \times (1.20 \times 10^2 \text{ g}) \times \frac{1 \text{ mol H}}{1.008 \text{ g H}} = \textbf{7.98 mol H}$$

$$n_O = (0.533) \times (1.20 \times 10^2 \text{ g}) \times \frac{1 \text{ mol O}}{16.00 \text{ g O}} = \textbf{4.00 mol O}$$

Step 4: Since we used the molar mass to calculate the moles of each element present in the compound, this method directly gives the molecular formula. The formula is $\textbf{C}_4\textbf{H}_8\textbf{O}_4$.

12.62   We first find the number of moles of gas using the ideal gas equation.

$$n = \frac{PV}{RT} = \frac{(748/760) \text{ atm } (4.00 \text{ L})}{(27+273) \text{ K}} \times \frac{\text{mol} \cdot \text{K}}{0.0821 \text{ L} \cdot \text{atm}} = 0.160 \text{ mol}$$

$$\text{molality} = \frac{0.160 \text{ mol}}{0.0580 \text{ kg benzene}} = 2.76 \text{ } m$$

$$\Delta T_f = K_f m = (5.12°\text{C}/m)(2.76 \text{ } m) = 14.1°\text{C}$$

freezing point $= 5.5°\text{C} - 14.1°\text{C} = \textbf{-8.6°C}$

12.64   First, from the freezing point depression we can calculate the molality of the solution. See Table 12.2 of the text for the normal freezing point and K_f value for benzene.

$$\Delta T_f = (5.5 - 4.3)°\text{C} = 1.2°\text{C}$$

$$m = \frac{\Delta T_f}{K_f} = \frac{1.2°\text{C}}{5.12°\text{C}/m} = 0.23 \text{ } m$$

Multiplying the molality by the mass of solvent (in kg) gives moles of unknown solute. Then, dividing the mass of solute (in g) by the moles of solute, gives the molar mass of the unknown solute.

$$? \text{ mol of unknown solute} = \frac{0.23 \text{ mol solute}}{1 \text{ kg benzene}} \times 0.0250 \text{ kg benzene}$$

$$= 0.0058 \text{ mol solute}$$

$$\textbf{molar mass of unknown} = \frac{2.50 \text{ g}}{0.0058 \text{ mol}} = \textbf{4.3} \times \textbf{10}^2 \textbf{ g/mol}$$

The empirical molar mass of C_6H_5P is 108.1 g/mol. Therefore, the molecular formula is $(C_6H_5P)_4$ or $\textbf{C}_{24}\textbf{H}_{20}\textbf{P}_4$.

12.66   Calculating molar mass from osmotic pressure, Problem Type 6B.

Step 1: Solve Equation (12.8) from Problem Type 6B algebraically for molarity (M), then substitute π, R, and T (in K) into the equation to calculate the molarity.

$$\pi = MRT \tag{12.8}$$

$$M = \frac{\pi}{RT} = \frac{\left(5.20 \text{ mmHg} \times \dfrac{1 \text{ atm}}{760 \text{ mmHg}}\right)}{298 \text{ K}} \times \frac{\text{mol} \cdot \text{K}}{0.0821 \text{ L} \cdot \text{atm}} = 2.80 \times 10^{-4} \text{ } M$$

Step 2: Multiplying the molarity by the volume of solution (in L) gives moles of solute. Then, dividing the mass of solute (in g) by the moles of solute, gives the molar mass of the solute.

$$? \text{ mol of polymer} = (2.80 \times 10^{-4} \text{ mol/L})(0.170 \text{ L}) = 4.76 \times 10^{-5} \text{ mol polymer}$$

$$\textbf{molar mass of polymer} = \frac{0.8330 \text{ g polymer}}{4.76 \times 10^{-5} \text{ mol polymer}} = \textbf{1.75} \times \textbf{10}^4 \textbf{ g/mol}$$

12.68   We use the osmotic pressure data to determine the molarity.

$$M = \frac{\pi}{RT} = \frac{4.61 \text{ atm}}{(20+273) K} \times \frac{\text{mol} \cdot K}{0.0821 \text{ L} \cdot \text{atm}} = 0.192 \text{ mol} / \text{L}$$

Next we use the density and the solution mass to find the volume of the solution.

mass of soln $= 6.85 \text{ g} + 100.0 \text{ g} = 106.9 \text{ g soln}$

volume of soln $= 106.9 \text{ g soln} \times \frac{1 \text{ mL}}{1.024 \text{ g}} = 104.4 \text{ mL} = 0.1044 \text{ L}$

Multiplying the molarity by the volume (in L) gives moles of solute (carbohydrate).

mol of solute $= M \times \text{L} = (0.192 \text{ mol/L})(0.1044 \text{ L}) = 0.0200 \text{ mol solute}$

Finally, dividing mass of carbohydrate by moles of carbohydrate gives the molar mass of the carbohydrate.

$$\textbf{molar mass} = \frac{6.85 \text{ g carbohydrate}}{0.0200 \text{ mol carbohydrate}} = \textbf{343 g/mol}$$

12.74   Boiling point, vapor pressure, and osmotic pressure all depend on particle concentration. Therefore, these solutions also have the same boiling point, osmotic pressure, and vapor pressure.

12.76   The freezing point will be depressed most by the solution that contains the most solute particles. You should try to classify each solute as a strong electrolyte, a weak electrolyte, or a nonelectrolyte. All three solutions have the same concentration, so comparing the solutions is straightforward. HCl is a strong electrolyte, so under ideal conditions it will completely dissociate into two particles per molecule. The concentration of particles will be 1.00 *m*. Acetic acid is a weak electrolyte, so it will only dissociate to a small extent. The concentration of particles will be greater than 0.50 *m*, but less than 1.00 *m*. Glucose is a nonelectrolyte, so glucose molecules remain as glucose molecules in solution. The concentration of particles will be 0.50 *m*. For these solutions, the order in which the freezing points become *lower* is:

0.50 *m* glucose $>$ 0.50 *m* acetic acid $>$ 0.50 *m* HCl

In other words, the HCl solution will have the lowest freezing point (greatest freezing point depression).

12.78   Using Equation (12.5) from Problem Type 5A, we can find the mole fraction of the NaCl. We use subscript 1 for H_2O and subscript 2 for NaCl.

$$\Delta P = X_2 P_1^{\circ}$$

$$X_2 = \frac{\Delta P}{P_1^{\circ}}$$

$$X_2 = \frac{23.76 \text{ mmHg} - 22.98 \text{ mmHg}}{23.76 \text{ mmHg}} = 0.03283$$

Let's assume that we have 1000 g (1 kg) of water as the solvent, because the definition of molality is moles of solute per kg of solvent. We can find the number of moles of particles dissolved in the water using the definition of mole fraction.

$$X_2 = \frac{n_2}{n_1 + n_2}$$

$$n_1 = 1000 \text{ g H}_2\text{O} \left(\frac{1 \text{ mol H}_2\text{O}}{18.02 \text{ g H}_2\text{O}} \right) = 55.49 \text{ mol H}_2\text{O}$$

$$\frac{n_2}{55.49 + n_2} = 0.03283$$

$$n_2 = 1.884 \text{ mol}$$

Since NaCl dissociates to form two particles (ions), the number of moles of NaCl is half of the above result.

$$\text{Moles NaCl} = 1.884 \text{ mol particles} \times \frac{1 \text{ mol NaCl}}{2 \text{ mol particles}} = 0.9420 \text{ mol}$$

The molality of the solution is:

$$\frac{0.9420 \text{ mol}}{1.000 \text{ kg}} = \textbf{0.9420 } \boldsymbol{m}$$

12.80   Colligative Properties of Electrolytes, Problem Type 7.

Step 1: To calculate the osmotic pressure of "physiological saline", we must first calculate the molarity of the solution since $\pi = iMRT$.

To calculate molarity, let's assume that we have 1.000 L of solution (1.000×10^3 mL). We can use the solution density as a conversion factor to calculate the mass of 1.000×10^3 mL of solution.

$$1.000 \times 10^3 \text{ mL soln} \times \frac{1.005 \text{ g soln}}{1 \text{ mL soln}} = 1005 \text{ g of soln}$$

Since the solution is 0.86% by mass NaCl, the mass of NaCl in the solution is:

$$1005 \text{ g} \times \frac{0.86\%}{100\%} = 8.6 \text{ g NaCl}$$

The molarity of the solution is:

$$\left(\frac{8.6 \text{ g NaCl}}{1.000 \text{ L}} \right) \left(\frac{1 \text{ mol NaCl}}{58.44 \text{ g NaCl}} \right) = 0.15 \text{ } M$$

Step 2: Since NaCl is a strong electrolyte, let's assume that the van't Hoff factor is 2. Substituting i, M, R, and T into the equation for osmotic pressure gives:

$$\pi = iMRT = 2 \left(\frac{0.15 \text{ mol}}{L} \right) \times \left(\frac{0.0821 \text{ L} \cdot \text{atm}}{\text{mol} \cdot \text{K}} \right) \times (310 \text{ K}) = \textbf{7.6 atm}$$

12.82   From Table 12.3, $i = 1.3$

$$\pi = iMRT$$

$$\pi = (1.3) \times \left(\frac{0.0500 \text{ mol}}{L} \right) \times \left(\frac{0.0821 \text{ L} \cdot \text{atm}}{\text{mol} \cdot \text{K}} \right) \times (298 \text{ K})$$

$$\pi = \textbf{1.6 atm}$$

12.86   At constant temperature, the osmotic pressure of a solution is proportional to the molarity. When equal volumes of the two solutions are mixed, the molarity will just be the mean of the molarities of the two solutions (assuming additive volumes). Since the osmotic pressure is proportional to the molarity, the osmotic pressure of the solution will be the mean of the osmotic pressure of the two solutions.

$$\pi = \frac{2.4\ \text{atm} + 4.6\ \text{atm}}{2} = \textbf{3.5 atm}$$

12.88   **(a)**   You can use Equation (12.4) from Problem Type 5A to calculate the vapor pressure of each component.

$$P_1 = X_1 P_1^\circ \tag{12.4}$$

First, you must calculate the mole fraction of each component.

$$X_A = \frac{n_A}{n_A + n_B} = \frac{1.00\ \text{mol}}{1.00\ \text{mol} + 1.00\ \text{mol}} = 0.500$$

Similarly,

$$X_B = 0.500$$

Substitute the mole fraction calculated above and the vapor pressure of the pure solvent into Equation (12.4) to calculate the vapor pressure of each component of the solution.

$$P_A = X_A P_A^\circ = 0.500 \times 76\ \text{mmHg} = 38\ \text{mmHg}$$

$$P_B = X_B P_B^\circ = 0.500 \times 132\ \text{mmHg} = 66\ \text{mmHg}$$

The total vapor pressure is the sum of the vapor pressures of the two components.

$$P_{\textbf{Total}} = P_A + P_B = 38\ \text{mmHg} + 66\ \text{mmHg} = \textbf{104 mmHg}$$

(b)   This problem is solved similarly to part (a).

$$X_A = \frac{n_A}{n_A + n_B} = \frac{2.00\ \text{mol}}{2.00\ \text{mol} + 5.00\ \text{mol}} = 0.286$$

Similarly,

$$X_B = 0.714$$

$$P_A = X_A P_A^\circ = 0.286 \times 76\ \text{mmHg} = 22\ \text{mmHg}$$

$$P_B = X_B P_B^\circ = 0.714 \times 132\ \text{mmHg} = 94\ \text{mmHg}$$

$$P_{\textbf{Total}} = P_A + P_B = 22\ \text{mmHg} + 94\ \text{mmHg} = \textbf{116 mmHg}$$

12.90   From the osmotic pressure, you can calculate the molarity of the solution.

$$M = \frac{\pi}{RT} = \frac{\left(30.3\ \text{mmHg} \times \dfrac{1\ \text{atm}}{760\ \text{mmHg}}\right)}{308\ \text{K}} \times \frac{\text{mol} \cdot \text{K}}{0.0821\ \text{L} \cdot \text{atm}} = 1.58 \times 10^{-3}\ \text{mol/L}$$

Multiplying molarity by the volume of solution in liters gives the moles of solute.

$$(1.58 \times 10^{-3}\ \text{mol solute/L soln}) \times (0.262\ \text{L soln}) = 4.14 \times 10^{-4}\ \text{mol solute}$$

Divide the grams of solute by the moles of solute to calculate the molar mass.

$$\text{molar mass of solute} = \frac{1.22 \text{ g}}{4.14 \times 10^{-4} \text{ mol}} = 2.95 \times 10^3 \text{ g/mol}$$

12.92   Solve Equation (12.7) from Problem Type 6A algebraically for molality (m), then substitute ΔT_f and K_f into the equation to calculate the molality. You can find the normal freezing point for benzene and K_f for benzene in Table 12.2 of your text.

$\Delta T_f = 5.5°\text{C} - 3.9°\text{C} = 1.6°\text{C}$

$$m = \frac{\Delta T_f}{K_f} = \frac{1.6°\text{C}}{5.12°\text{C}/m} = 0.31 \ m$$

Multiplying the molality by the mass of solvent (in kg) gives moles of unknown solute. Then, dividing the mass of solute (in g) by the moles of solute, gives the molar mass of the unknown solute.

$$? \text{ mol of unknown solute} = \frac{0.31 \text{ mol solute}}{1 \text{ kg benzene}} \times 8.0 \times 10^{-3} \text{ kg benzene}$$

$$= 2.5 \times 10^{-3} \text{ mol solute}$$

$$\text{molar mass of unknown} = \frac{0.50 \text{ g}}{2.5 \times 10^{-3} \text{ mol}} = 2.0 \times 10^2 \text{ g/mol}$$

The molar mass for cocaine $C_{17}H_{21}NO_4 = 303$ g/mol, so the compound is not cocaine. We assume in our analysis that the compound is a pure, monomeric, nonelectrolyte.

12.94   The molality of the solution assuming $AlCl_3$ to be a nonelectrolyte is:

$$1.00 \text{ g } AlCl_3 \left(\frac{1 \text{ mol } AlCl_3}{133.3 \text{ g } AlCl_3} \right) \frac{1}{0.0500 \text{ kg}} = 0.150 \ m$$

The molality calculated with Equation (12.7) from the text is:

$$m = \frac{\Delta T_f}{K_f} = \frac{1.11°\text{C}}{1.86°\text{C}/m} = 0.597 \ m$$

The ratio $\dfrac{0.597 \ m}{0.150 \ m}$ is 4. Thus each $AlCl_3$ dissociates as follows:

$$AlCl_3(s) \rightarrow Al^{3+}(aq) + 3Cl^{-}(aq)$$

12.96   First, we tabulate the concentration of all of the ions. Notice that the chloride concentration comes from more than one source.

$MgCl_2$:	If $[MgCl_2] = 0.054 \ M$,	$[Mg^{2+}] = 0.054 \ M$	$[Cl^-] = 2 \times 0.054 \ M$
Na_2SO_4:	if $[Na_2SO_4] = 0.051 \ M$,	$[Na^+] = 2 \times 0.051 \ M$	$[SO_4^{2-}] = 0.051 \ M$
$CaCl_2$:	if $[CaCl_2] = 0.010 \ M$,	$[Ca^{2+}] = 0.010 \ M$	$[Cl^-] = 2 \times 0.010 \ M$
$NaHCO_3$:	if $[NaHCO_3] = 0.0020 \ M$	$[Na^+] = 0.0020 \ M$	$[HCO_3^-] = 0.0020 \ M$
KCl:	if $[KCl] = 0.0090 \ M$	$[K^+] = 0.0090 \ M$	$[Cl^-] = 0.0090 \ M$

The subtotal of chloride ion concentration is:

$$[Cl^-] = (2 \times 0.0540) + (2 \times 0.010) + (0.0090) = 0.137 \ M$$

Since the required $[Cl^-]$ is 2.60 M, the difference $(2.6 - 0.137 = 2.46 \ M)$ must come from NaCl.

The subtotal of sodium ion concentration is:

$$[Na^+] = (2 \times 0.051) + (0.0020) = 0.104 \ M$$

Since the required $[Na^+]$ is 2.56 M, the difference $(2.56 - 0.104 = 2.46 \ M)$ must come from NaCl.

Now, calculating the mass of the compounds required:

NaCl: $\qquad 2.46 \ mol \times \dfrac{58.44 \text{ g NaCl}}{1 \text{ mol NaCl}} = \mathbf{143.8 \ g}$

MgCl$_2$: $\qquad 0.054 \ mol \times \dfrac{95.21 \text{ g MgCl}_2}{1 \text{ mol MgCl}_2} = \mathbf{5.14 \ g}$

Na$_2$SO$_4$: $\qquad 0.051 \ mol \times \dfrac{142.1 \text{ g Na}_2\text{SO}_4}{1 \text{ mol Na}_2\text{SO}_4} = \mathbf{7.25 \ g}$

CaCl$_2$: $\qquad 0.010 \ mol \times \dfrac{111.0 \text{ g CaCl}_2}{1 \text{ mol CaCl}_2} = \mathbf{1.11 \ g}$

KCl: $\qquad 0.0090 \ mol \times \dfrac{74.55 \text{ g KCl}}{1 \text{ mol KCl}} = \mathbf{0.67 \ g}$

NaHCO$_3$: $\qquad 0.0020 \ mol \times \dfrac{84.01 \text{ g NaHCO}_3}{1 \text{ mol NaHCO}_3} = \mathbf{0.17 \ g}$

12.98 <u>Solution A:</u> Let molar mass be $\mathcal{M}$.

$$\Delta P = X_A P_A^\circ$$

$$(760 - 754.5) = X_A(760)$$

$$X_A = 7.237 \times 10^{-3}$$

$$n = \frac{\text{mass}}{\text{molar mass}}$$

$$X_A = \frac{n_A}{n_A + n_{water}} = \frac{5.00/\mathcal{M}}{5.00/\mathcal{M} + 100/18.02} = 7.237 \times 10^{-3}$$

$$\mathcal{M} = \mathbf{124 \ g/mol}$$

<u>Solution B:</u> Let molar mass be $\mathcal{M}$

$$\Delta P = X_B P_B^\circ$$

$$X_B = 7.237 \times 10^{-3}$$

$$n = \frac{\text{mass}}{\text{molar mass}}$$

$$X_B = \frac{n_B}{n_B + n_{\text{benzene}}} = \frac{2.31/\mathcal{M}}{2.31/\mathcal{M} + 100/78.11} = 7.237 \times 10^{-3}$$

$$\mathcal{M} = \textbf{248 g/mol}$$

The molar mass in benzene is about twice that in water. This suggests some sort of dimerization is occurring in a nonpolar solvent such as benzene.

12.100 As the chain becomes longer, the alcohols become more like hydrocarbons (nonpolar) in their properties. The alcohol with five carbons (*n*-pentanol) would be the best solvent for iodine (a) and *n*-pentane (c) (why?). Methanol (CH_3OH) is the most water like and is the best solvent for an ionic solid like KBr.

12.102 I_2 – H_2O: Dipole - induced dipole.

I_3^- – H_2O: Ion - dipole. Stronger interaction causes more I_2 to be converted to I_3^-.

12.104 **(a)** If the membrane is permeable to all the ions and to the water, the result will be the same as just removing the membrane. You will have two solutions of equal NaCl concentration.

(b) This part is tricky. The movement of one ion but not the other would result in one side of the apparatus acquiring a positive electric charge and the other side becoming equally negative. This has never been known to happen, so we must conclude that migrating ions always drag other ions of the opposite charge with them. In this hypothetical situation only water would move through the membrane from the dilute to the more concentrated side.

(c) This is the classic osmosis situation. Water would move through the membrane from the dilute to the concentrated side.

12.106 First, we calculate the number of moles of HCl in 100 g of solution.

$$n_{HCl} = 100 \text{ g soln} \times \frac{37.7 \text{ g HCl}}{100 \text{ g soln}} \times \frac{1 \text{ mol HCl}}{36.46 \text{ g HCl}} = 1.03 \text{ mol HCl}$$

Next, we calculate the volume of 100 g of solution.

$$V = 100 \text{ g} \times \frac{1 \text{ mL}}{1.19 \text{ g}} \times \frac{1 \text{ L}}{1000 \text{ mL}} = 0.0840 \text{ L}$$

Finally, the molarity of the solution is:

$$\frac{1.03 \text{ mol}}{0.0840 \text{ L}} = \textbf{12.3 } \textit{\textbf{M}}$$

12.108 Let the mass of NaCl be *x* g. Then, the mass of sucrose is $(10.2 - x)$g.

We know that the equation representing the osmotic pressure is:

$$\pi = MRT$$

π, R, and T are given. Using this equation and the definition of molarity, we can calculate the percentage of NaCl in the mixture.

$$\text{molarity} = \frac{\text{mol solute}}{\text{L soln}}$$

Remember that NaCl dissociates into two ions in solution; therefore, we multiply the moles of NaCl by two.

$$\text{mol solute} = 2\left(x \text{ g NaCl} \times \frac{1 \text{ mol NaCl}}{58.44 \text{ g NaCl}}\right) + \left((10.2 - x)\text{g sucrose} \times \frac{1 \text{ mol sucrose}}{342.3 \text{ g sucrose}}\right)$$

$$\text{mol solute} = 0.03422x + 0.02980 - 0.002921x$$

$$\text{mol solute} = 0.03130x + 0.02980$$

$$\text{Molarity of solution} = \frac{\text{mol solute}}{\text{L soln}} = \frac{0.03130x + 0.02980}{0.250 \text{ L}}$$

Substitute molarity into the equation for osmotic pressure to solve for x.

$$\pi = MRT$$

$$7.32 \text{ atm} = \left(\frac{(0.03130x + 0.02980)\,\text{mol}}{0.250 \text{ L}}\right)\left(0.0821 \frac{\text{L} \cdot \text{atm}}{\text{mol} \cdot \text{K}}\right)(296 \text{ K})$$

$$0.0753 = 0.03130x + 0.02980$$

$$x = 1.45 \text{ g} = \text{mass of NaCl}$$

$$\textbf{Mass \% NaCl} = \frac{1.45 \text{ g}}{10.2 \text{ g}} \times 100\% = \textbf{14.2\%}$$

12.110 **(a)**   Solubility decreases with increasing lattice energy.

(b)   Ionic compounds are more soluble in a polar solvent.

(c)   Solubility increases with enthalpy of hydration of the cation and anion.

12.112 $\text{molality} = \dfrac{98.0 \text{ g H}_2\text{SO}_4 \times \dfrac{1 \text{ mol H}_2\text{SO}_4}{98.09 \text{ g H}_2\text{SO}_4}}{2.0 \text{ g H}_2\text{O} \times \dfrac{1 \text{ kg H}_2\text{O}}{1000 \text{ g H}_2\text{O}}} = 5.0 \times 10^2 \; m$

We can calculate the density of sulfuric acid from the molarity.

$$\text{molarity} = 18\,M = \frac{18.0 \text{ mol H}_2\text{SO}_4}{1 \text{ L soln}}$$

The 18 mol of H_2SO_4 has a mass of:

$$18 \text{ mol H}_2\text{SO}_4 \times \frac{98.0 \text{ g H}_2\text{SO}_4}{1 \text{ mol H}_2\text{SO}_4} = 1.8 \times 10^3 \text{ g H}_2\text{SO}_4$$

$$1 \text{ L} = 1000 \text{ mL}$$

$$\textbf{density} = \frac{\text{mass H}_2\text{SO}_4}{\text{volume}} = \frac{1.8 \times 10^3 \text{ g}}{1000 \text{ mL}} = \textbf{1.80 g/mL}$$

12.114 $P_A = X_A P_A^{\circ}$

$P_{ethanol} = 0.62 \times 108 \text{ mmHg} = 67.0 \text{ mmHg}$

$P_{1\text{-propanol}} = 0.38 \times 40.0 \text{ mmHg} = 15.2 \text{ mmHg}$

In the vapor phase:

$$X_{ethanol} = \frac{67.0}{67.0 + 15.2} = \mathbf{0.815}$$

12.116 NH_3 can form hydrogen bonds with water; NCl_3 cannot. (Like dissolves like.)

12.118   We can calculate the molality of the solution from the freezing point depression.

$$\Delta T_f = K_f m$$
$$0.203 = 1.86 \, m$$
$$m = \frac{0.203}{1.86} = 0.109 \, m$$

The molality of the original solution was $0.106 \, m$. Some of the solution has ionized to H^+ and CH_3COO^-.

$$CH_3COOH \rightleftharpoons CH_3COO^- + H^+$$

	CH_3COOH	CH_3COO^-	H^+
Initial	$0.106 \, m$	0	0
Change	$-x$	$+x$	$+x$
Equil.	$0.106 \, m - x$	x	x

At equilibrium, the total concentration of species in solution is $0.109 \, m$.

$$(0.106 - x) + 2x = 0.109 \, m$$
$$x = 0.003 \, m$$

The percentage of acid that has undergone ionization is:

$$\frac{0.003 \, m}{0.106 \, m} \times 100\% = \mathbf{2.8\%}$$

12.120   First, we can calculate the molality of the solution from the freezing point depression.

$$\Delta T_f = (5.12)m$$
$$(5.5 - 3.5) = (5.12)m$$
$$m = 0.39$$

Next, from the definition of molality, we can calculate the moles of solute.

$$m = \frac{\text{mol solute}}{\text{kg solvent}}$$

$$0.39 \, m = \frac{\text{mol solute}}{80 \times 10^{-3} \text{ kg benzene}}$$

$$\text{mol solute} = 0.031 \text{ mol}$$

The molar mass ($\mathcal{M}$) of the solute is:

$$\frac{3.8 \text{ g}}{0.031 \text{ mol}} = 1.2 \times 10^2 \text{ g/mol}$$

The molar mass of CH_3COOH is 60.05 g/mol. Since the molar mass of the solute calculated from the freezing point depression is twice this value, the structure of the solute most likely is a dimer that is held together by hydrogen bonds.

$$CH_3C \underset{O\text{---}H\text{---}O}{\overset{O\text{---}H\text{---}O}{\diagup}} C\text{---}CH_3 \qquad \text{A dimer}$$

12.122 **(a)** $\Delta T_f = K_f m$

$2 = (1.86)(m)$

molality = **1.1 *m***

This concentration is too high and is *not* a reasonable physiological concentration.

(b) Although the protein is present in low concentrations, it can prevent the formation of ice crystals.

CHAPTER 13
CHEMICAL KINETICS

PROBLEM-SOLVING STRATEGIES AND TUTORIAL SOLUTIONS

TYPES OF PROBLEMS

Problem Type 1: Writing Rate Expressions.

Problem Type 2: Determining the Rate Law of a Reaction.

Problem Type 3: First-Order Reactions.
  (a) Analyzing a first-order reaction.
  (b) Determining the half-life of a first-order reaction.
  (c) Analyzing first-order kinetics.

Problem Type 4: Second-Order Reactions.
  (a) Analyzing a second-order reaction.
  (b) Determining the half-life of a second-order reaction.
  (c) Analyzing second-order kinetics.

Problem Type 5: The Arrhenius Equation.
  (a) Applying the Arrhenius equation.
  (b) Applying a modified form of the Arrhenius equation that relates the rate constants at two different temperatures.

Problem Type 6: Studying Reaction Mechanisms.

PROBLEM TYPE 1: WRITING RATE EXPRESSIONS

As a chemical reaction proceeds, the concentrations of reactants and products change with time. As the reaction

$$A + B \longrightarrow C$$

progresses, the concentration of C increases. The rate can be expressed as the change in concentration of C during the time interval Δt.

$$\text{rate} = \frac{\Delta[C]}{\Delta t}$$

For a specific reaction, we need to take into account the stoichiometry of the balanced equation. For example, let's express the rate of the following reaction in terms of the concentrations of the individual reactants and products.

$$2\,NO\,(g) + O_2\,(g) \longrightarrow 2\,NO_2\,(g)$$

Notice from the balanced equation that the concentration of NO will decrease twice as fast as that of O_2. We can write the rate as

$$\text{rate} = -\frac{1}{2}\frac{\Delta[NO]}{\Delta t} \quad \text{or} \quad \text{rate} = -\frac{\Delta[O_2]}{\Delta t} \quad \text{or} \quad \text{rate} = \frac{1}{2}\frac{\Delta[NO_2]}{\Delta t}$$

Division of each concentration by the coefficient from the balanced equation makes all of the above rates equal. Notice also that a negative sign is inserted before terms involving reactants. The $\Delta[NO]$ is a negative quantity because the concentration of NO *decreases* with time. Therefore, multiplying $\Delta[NO]$ by a negative sign makes the rate of reaction a positive quantity.

For a general reaction:

$$a\text{A} + b\text{B} \longrightarrow c\text{C} + d\text{D}$$

the rate of reaction can be expressed in terms of any reactant or product.

$$\text{rate} = -\frac{1}{a}\frac{\Delta[A]}{\Delta t} = -\frac{1}{b}\frac{\Delta[B]}{\Delta t} = \frac{1}{c}\frac{\Delta[C]}{\Delta t} = \frac{1}{d}\frac{\Delta[D]}{\Delta t}$$

EXAMPLE 13.1

Write expressions for the rate of the following reaction in terms of each of the reactants and products.

$$2\,N_2O_5\,(g) \longrightarrow 4\,NO_2\,(g) + O_2\,(g)$$

Recall that the rate is defined as the change in concentration of a reactant or product with time. Each "change in concentration" term is divided by the corresponding stoichiometric coefficient. Terms involving reactants are preceded by a minus sign.

$$\textbf{Rate} = -\frac{1}{2}\frac{\Delta[N_2O_5]}{\Delta t} = \frac{1}{4}\frac{\Delta[NO_2]}{\Delta t} = \frac{\Delta[O_2]}{\Delta t}$$

PRACTICE EXERCISE

1. Oxygen gas can be formed by the decomposition of nitrogen monoxide (nitric oxide).

$$2\,NO\,(g) \longrightarrow O_2\,(g) + N_2\,(g)$$

   If the rate of formation of O_2 is 0.054 M/s, what is the rate of change of the NO concentration?

> **Text Problems:** 13.6, **13.8**

PROBLEM TYPE 2: DETERMINING THE RATE LAW OF A REACTION

The **rate law** is an expression relating the rate of a reaction to the rate constant and the concentrations of reactants. For a general reaction of the type

$$a\text{A} + b\text{B} \longrightarrow c\text{C} + d\text{D}$$

the rate law takes the form

$$\text{rate} = k[A]^x[B]^y$$

The term k is the **rate constant**, a constant of proportionality between the reaction rate and the concentrations of the reactants. The sum of the powers to which all reactant concentrations appearing in the rate law are raised is called the overall **reaction order**. In the rate law expression shown above, the overall reaction order is given by x + y.

k, x, and y must be determined experimentally. One method to determine x and y is to keep the concentrations of all but one reactant fixed. Then, the rate of reaction is measured as a function of the concentration of the one reactant whose concentration is varied. Any variation in rate is due solely to the variation in this reactant's concentration.

The example below will show you how to determine the rate law and how to calculate the rate constant.

EXAMPLE 13.2

The following rate data were collected for the reaction

$$2\,NO + 2H_2 \longrightarrow N_2 + 2H_2O$$

(a) Determine the rate law.

(b) Calculate the rate constant.

Experiment	$[NO]_0$ (M)	$[H_2]_0$ (M)	$\Delta[N_2]/\Delta t$ (M/h)
1	0.60	0.15	0.076
2	0.60	0.30	0.15
3	0.60	0.60	0.30
4	1.20	0.60	1.21

(a)   To determine the rate law, we must determine the exponents in the equation

$$\text{rate} = k[NO]^x[H_2]^y$$

To determine the order of the reaction with respect to H_2, find two experiments in which the [NO] is held constant. Compare the data from experiments 1 and 2. When the concentration of H_2 is doubled, the reaction rate doubles. Thus, the reaction is *first-order* in H_2.

To determine the order with respect to NO, compare experiments 3 and 4. When the NO concentration is doubled, the reaction rate quadruples. Thus, the reaction is *second-order* in NO.

The rate law is:

$$\text{rate} = k[NO]^2[H_2]$$

(b)

Step 1: Rearrange the rate law from part (a), solving for k.

$$k = \frac{\text{rate}}{[NO]^2[H_2]}$$

Step 2: Substitute the data from any one of the experiments to calculate k. Using the data from Experiment 1,

$$k = \frac{0.076\ M/\text{hr}}{(0.60\ M)^2(0.15\ M)} = 1.4\ M^{-2}\text{hr}^{-1}$$

PRACTICE EXERCISE

2.   The following experimental data were obtained for the reaction

$$2A + B \longrightarrow \text{products}$$

What is the rate law for this reaction?

Experiment	$[A]_0$ (M)	$[B]_0$ (M)	Rate (M/s)
1	0.40	0.20	5.6×10^{-3}
2	0.80	0.20	5.5×10^{-3}
3	0.40	0.40	22.3×10^{-3}

Text Problems: 13.16, **13.18**, 13.20, **13.50 a,b**

PROBLEM TYPE 3: FIRST-ORDER REACTIONS

One of the most widely encountered kinetic forms is the first-order rate equation. Consider the reaction,

$$A \longrightarrow products$$

For a first-order reaction, the exponent of [A] in the rate law is 1. We can write:

$$rate = k[A]$$

We also know that

$$rate = \frac{-\Delta[A]}{\Delta t}$$

Combining the two equations gives:

$$k[A] = \frac{-\Delta[A]}{\Delta t}$$

Using calculus, we can integrate both sides of the above equation to give:

$$\ln \frac{[A]}{[A]_0} = -kt \qquad\qquad (13.1)$$

where ln is the natural logarithm, and $[A]_0$ and $[A]$ are the concentrations of A at times $t = 0$ and $t = t$, respectively.

A. Analyzing a first-order reaction

EXAMPLE 13.3
Methyl isocyanide undergoes a first-order isomerization to from methyl cyanide.

$$CH_3NC\ (g) \longrightarrow CH_3CN\ (g)$$

The reaction was studied at 199°C. The initial concentration of CH_3NC was 0.0258 mol/L and after 11.4 min, analysis showed the concentration of the product to be 1.30×10^{-3} mol/L.

(a)   What is the first-order rate constant?

(b)   How long will it take for 90.0 percent of the CH_3NC to react?

(a)   This problem illustrates the use of the first-order rate equation

$$\ln \frac{[CH_3NC]}{[CH_3NC]_0} = -kt$$

where k is the rate constant and $[CH_3NC]$ is the reactant concentration at time t.

Step 1: We can calculate $[CH_3NC]$ by realizing that the amount of product formed equals the amount of reactant lost due to the 1:1 mole ratio between reactant and product. Thus,

$$[CH_3NC] = [CH_3NC]_0 - 1.30 \times 10^{-3}\ M$$

$$= 0.0258\ M - 1.30 \times 10^{-3}\ M = 0.0245\ M$$

Step 2: Substitute the concentrations and the time into the first-order rate equation to calculate the rate constant k.

$$\ln \frac{[CH_3NC]}{[CH_3NC]_0} = -kt$$

$$\ln \frac{0.0245 \; M}{0.0258 \; M} = -k(11.4 \; \text{min})$$

$$k = -\frac{\ln 0.9496}{11.4 \; \text{min}} = 4.54 \times 10^{-3} \; \text{min}^{-1}$$

(b) If 90 percent of the initial CH_3NC is consumed, then 10 percent remains. We can write,

$$[CH_3NC] = 0.10[CH_3NC]_0$$

Step 1: Substitution into the first-order rate equation gives:

$$\ln \frac{0.10[CH_3NC]_0}{[CH_3NC]_0} = -kt$$

Step 2: Substitute in the rate constant k calculated in part (a) to solve for t.

$$\ln \frac{0.10[CH_3NC]_0}{[CH_3NC]_0} = -(4.54 \times 10^{-3} \; \text{min}^{-1})t$$

$$t = -\frac{\ln 0.10}{4.54 \times 10^{-3} \; \text{min}^{-1}} = 507 \; \text{min}$$

PRACTICE EXERCISE

3.   The hydrolysis of sucrose ($C_{12}H_{22}O_{11}$) yields the simple sugars glucose ($C_6H_{12}O_6$) and fructose ($C_6H_{12}O_6$), which happen to be isomers.

$$C_{12}H_{22}O_{11} + H_2O \longrightarrow C_6H_{12}O_6 + C_6H_{12}O_6$$

The reaction is first-order in glucose concentration.

$$\text{rate} = k[C_{12}H_{22}O_{11}]$$

At 27°C, the rate constant is $2.1 \times 10^{-6} \; \text{s}^{-1}$. Starting with a sucrose solution with a concentration of $0.10 \; M$ at 27°C, what would the concentration of sucrose be 24 hours later? (The solution is maintained at 27°C.)

Text Problem: 13.28

B. Determining the half-life of a first-order reaction

The **half-life** of a reaction, $t_{\frac{1}{2}}$, is the time required for the concentration of a reactant to decrease to half of its initial concentration. From Equation (13.1), we can write

$$\ln \frac{[A]_0}{[A]} = kt \qquad (13.2)$$

$$t = \frac{1}{k} \ln \frac{[A]_0}{[A]} \qquad (13.3)$$

From the definition of half-life, when $t = t_{\frac{1}{2}}$

$$[A] = \frac{[A]_0}{2}$$

Substituting into Equation (13.3) gives:

$$t_{\frac{1}{2}} = \frac{1}{k} \ln \frac{[A]_0}{\dfrac{[A]_0}{2}}$$

or

$$t_{\frac{1}{2}} = \frac{1}{k} \ln 2 = \frac{0.693}{k} \qquad (13.4)$$

Equation (13.4) shows that the half-life of a first-order reaction is *independent* of the initial concentration of the reactant. Thus, it would take the same time for the concentration of the reactant to decrease from 1.0 M to 0.50 M as it would to decrease from 0.10 M to 0.050 M.

The half-life can also be used to determine the rate constant of a first-order reaction.

EXAMPLE 13.4
Methyl isocyanide undergoes a first-order isomerization to form methyl cyanide.

$$CH_3NC\ (g) \longrightarrow CH_3CN\ (g)$$

The reaction was studied at 199°C. The initial concentration of CH_3NC was 0.0258 mol/L and after 11.4 min, analysis showed the concentration of the product to be 1.30×10^{-3} mol/L. Using the rate constant calculated in Example 13.3, calculate the half-life of methyl isocyanide.

The rate constant calculated in Example 13.3 is 4.54×10^{-3} min^{-1}. Substituting the rate constant into Equation (13.4) gives,

$$t_{\frac{1}{2}} = \frac{0.693}{k} = \frac{0.693}{4.54 \times 10^{-3} \text{ min}^{-1}} = \textbf{153 min}$$

PRACTICE EXERCISE
4.   The hydrolysis of sucrose ($C_{12}H_{22}O_{11}$) yields the simple sugars glucose ($C_6H_{12}O_6$) and fructose ($C_6H_{12}O_6$), which happen to be isomers.

$$C_{12}H_{22}O_{11} + H_2O \longrightarrow C_6H_{12}O_6 + C_6H_{12}O_6$$

The reaction is first-order in sucrose concentration.

$$rate = k[C_{12}H_{22}O_{11}]$$

At 27°C, the rate constant is 2.1×10^{-6} s^{-1}. What is the half-life of sucrose at 27°C?

> **Text Problem: 13.28**

C. Analyzing first-order kinetics
From Equation (13.1),

$$\ln \frac{[A]}{[A]_0} = -kt \qquad (13.1)$$

$$\ln[A] - \ln[A]_0 = -kt$$

or

$$\ln[A] = -kt + \ln[A]_0 \qquad (13.5)$$

Equation (13.5) has the form of a linear equation.

$$y = mx + b$$

where,

    m is the slope of the line
    b is the y-intercept

A plot of ln[A] versus t (y vs. x) gives a straight line with a slope of $-k$ (m). Thus, we can calculate the rate constant k from the slope of the plot.

EXAMPLE 13.5
At 500 K, butadiene gas converts to cyclobutene gas:

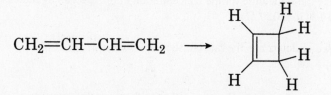

Given the following data, is the reaction first-order in butadiene concentration?

Time from start (s)	Concentration of butadiene (M)
195	0.0162
604	0.0147
1246	0.0129
2180	0.0110
4140	0.0084
8135	0.0057

If the reaction is first-order in butadiene, then a plot of ln[butadiene] versus t will be a straight line.

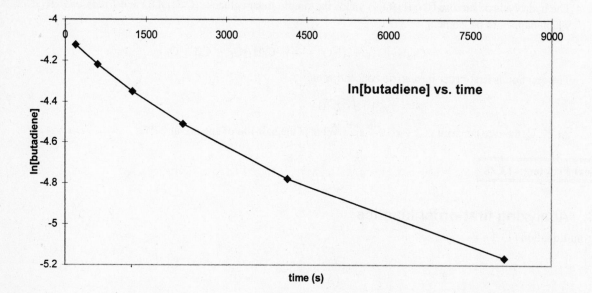

As you can see, a plot of ln[butadiene] versus t does *not* give a straight line. Hence, the reaction is *not* first-order in butadiene.

PRACTICE EXERCISE

5.   In a certain experiment, the rate of hydrogen peroxide decomposition,

$$2 H_2O_2 \rightarrow 2 H_2O + O_2$$

is followed by titration against a potassium permanganate solution. At regular intervals, an equal volume of H_2O_2 is withdrawn to give the following data:

Time (min)	0	10.0	20.0	30.0
Volume of KMnO₄ used (mL)	22.8	13.8	8.25	5.00

Confirm that the reaction is first-order in hydrogen peroxide and calculate the rate constant.

Hint: The concentration of hydrogen peroxide is directly proportional to the volume (in mL) of $KMnO_4$ used in each titration.

Text Problem: 13.22

PROBLEM TYPE 4: SECOND-ORDER REACTIONS

Second-order reactions are also encountered quite often. Consider the reaction,

$$A \longrightarrow \text{products}$$

For a second-order reaction, the exponent of [A] in the rate law is 2. We can write:

$$\text{rate} = k[A]^2$$

We also know that

$$\text{rate} = \frac{-\Delta[A]}{\Delta t}$$

Combining the two equations gives:

$$k[A]^2 = \frac{-\Delta[A]}{\Delta t}$$

Using calculus, we can integrate both sides of the above equation to give:

$$\frac{1}{[A]} = \frac{1}{[A]_0} + kt \tag{13.6}$$

where $[A]_0$ and $[A]$ are the concentrations of A at times $t = 0$ and $t = t$, respectively.

A. Analyzing a second-order reaction

EXAMPLE 13.6
At 500 K, butadiene gas converts to cyclobutene gas:

$$CH_2{=}CH{-}CH{=}CH_2 \longrightarrow$$

At 500 K, the rate constant for the reaction is 0.0143/M·s. If the initial concentration of butadiene is 0.272 M, calculate the concentration of butadiene after 30.0 min.

This problem illustrates the use of the second-order rate equation:

$$\frac{1}{[\text{butadiene}]} = \frac{1}{[\text{butadiene}]_0} + kt$$

where k is the rate constant and [butadiene] is the reactant concentration at time t.

We can calculate [butadiene] by substituting k, t, and the initial concentration of butadiene into the above equation.

$$\frac{1}{[\text{butadiene}]} = \frac{1}{0.272\ M} + (0.0143\ /\ M \cdot \text{s})(1.80 \times 10^3\ \text{s})$$

$$\frac{1}{[\text{butadiene}]} = 29.4\ M^{-1}$$

[butadiene] = 0.0340 M

PRACTICE EXERCISE

6.   For the reaction shown in Example 13.6 above, how long will it take for the concentration of butadiene to decrease from its initial concentration of 0.272 M to 0.100 M?

Text Problem: 13.30

B. Determining the half-life of a second-order reaction

The **half-life** of a reaction, $t_{\frac{1}{2}}$, is the time required for the concentration of a reactant to decrease to half of its initial concentration. From Equation (13.6), we can write:

$$\frac{1}{[\text{A}]} = \frac{1}{[\text{A}]_0} + kt \tag{13.6}$$

$$t = \left(\frac{1}{k}\right)\left(\frac{1}{[\text{A}]} - \frac{1}{[\text{A}]_0}\right) \tag{13.7}$$

From the definition of half-life, when $t = t_{\frac{1}{2}}$

$$[\text{A}] = \frac{[\text{A}]_0}{2}$$

Substituting into Equation (13.7) gives:

$$t_{\frac{1}{2}} = \left(\frac{1}{k}\right)\left(\frac{1}{\dfrac{[\text{A}]_0}{2}} - \frac{1}{[\text{A}]_0}\right) = \left(\frac{1}{k}\right)\left(\frac{2}{[\text{A}]_0} - \frac{1}{[\text{A}]_0}\right)$$

or

$$t_{\frac{1}{2}} = \frac{1}{k[\text{A}]_0} \tag{13.8}$$

Equation (13.8) shows that the half-life of a second-order reaction is *dependent* on the initial concentration of the reactant, unlike the half-life of a first-order reaction.

EXAMPLE 13.7
The following reaction follows second-order kinetics:

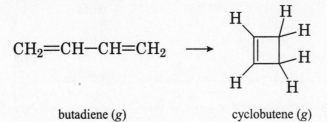

butadiene (*g*)   cyclobutene (*g*)

At 500 K, the rate constant is 0.0143/M·s. The initial concentration of butadiene is 0.272 M. What is the half-life for this reaction?

The rate constant is 0.0143 /M·s. Substituting the rate constant and the initial concentration into the Equation (13.8) gives:

$$t_{\frac{1}{2}} = \frac{1}{k[A]_0} = \frac{1}{(0.0143/M \cdot s)(0.272\ M)} = \textbf{257 s}$$

PRACTICE EXERCISE
7.   For the reaction shown in Example 13.7 above, the half-life of the reaction is determined to be 66.6 s at
     500 K. What is the initial concentration of butadiene? (*k* = 0.0143 /m·s at 500 K)

C. Analyzing second-order kinetics

Equation (13.6) has the form of a linear equation.

$$\frac{1}{[A]} = kt + \frac{1}{[A]_0} \tag{13.6}$$

$$y = mx + b$$

where,

   m is the slope of the line
   b is the *y*-intercept

A plot of $\frac{1}{[A]}$ versus *t* (*y* vs. *x*) gives a straight line with a slope of *k* (*m*). Thus, we can calculate the rate constant *k*

from the slope of the plot.

EXAMPLE 13.8
At 500 K, butadiene gas converts to cyclobutene gas:

$$CH_2{=}CH{-}CH{=}CH_2 \longrightarrow$$

Given the following data, is the reaction second-order in butadiene concentration?

Time from start (s)	Concentration of butadiene (M)
195	0.0162
604	0.0147
1246	0.0129
2180	0.0110
4140	0.0084
8135	0.0057

If the reaction is second-order in butadiene, then a plot of $\dfrac{1}{[\text{butadiene}]}$ versus t will be a straight line.

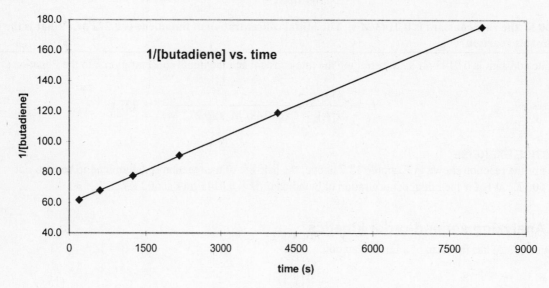

As you can see, a plot of $\dfrac{1}{[\text{butadiene}]}$ versus t *does* give a straight line. Hence, the reaction is second-order in butadiene.

PRACTICE EXERCISE

8. If a plot of $\dfrac{1}{[A]}$ versus time produces a straight line, which of the following must be *true*?

   a.   The reaction is first-order in A.
   b.   The reaction is second-order in A.
   c.   The reaction is first-order in two reactants.
   d.   The rate of the reaction does not depend on the concentration of A.
   e.   None of the above

Text Problem: 13.22

PROBLEM TYPE 5: THE ARRHENIUS EQUATION

The dependence of the rate constant of a reaction on temperature can be expressed by the following equation, called the **Arrhenius equation**.

$$k = Ae^{-E_a/RT}$$

where, E_a is the activation energy of the reaction (in kJ/mol)
        R is the gas constant (8.314 J/mol·K)
        T is the absolute temperature (in K)
        e is the base of the natural logarithm scale (see Appendix 4)

The quantity A represents the collision frequency and is called the *frequency factor*. It can be treated as a constant for a given reacting system over a fairly wide temperature range. The Arrhenius equation shows that the rate constant is directly proportional to A. Therefore, as the number of collisions increase, the rate increases.

The minus sign associated with the exponent E_a/RT indicates that the rate constant decreases with increasing activation energy and increases with increasing temperature.

The Arrhenius equation can be expressed in a more useful form by taking the natural logarithm of both sides.

$$\ln k = \ln\left[Ae^{-E_a/RT}\right]$$

$$\ln k = \ln A - \frac{E_a}{RT} \tag{13.9}$$

or

$$\ln k = \left(-\frac{E_a}{R}\right)\left(\frac{1}{T}\right) + \ln A \tag{13.10}$$

Equation (13.10) has the form of the linear equation

$$y = mx + b$$

where,

    m is the slope of the line
    b is the y-intercept

A plot of $\ln[k]$ versus $\frac{1}{T}$ (y vs. x) gives a straight line with a slope of $\frac{-E_a}{R}$ (m). Thus, we can calculate the activation energy (E_a) from the slope of the plot.

A. Applying the Arrhenius Equation

EXAMPLE 13.9
Variation of the rate constant with temperature for the reaction

$$\text{NO} + \text{O}_3 \rightarrow \text{NO}_2 + \text{O}_2$$

is given in the following table. Determine graphically the activation energy for the reaction.

Temperature (K)	$k\ (M^{-1}s^{-1})$
283	9.30×10^6
293	1.08×10^7
303	1.25×10^7
313	1.43×10^7
323	1.62×10^7

A plot of $\ln[k]$ versus $\dfrac{1}{T}$ should give a straight line with a slope of $\dfrac{-E_a}{R}$. Thus, we can calculate the activation energy (E_a) from the slope of the plot.

From the given data we obtain:

$1/T$ (K^{-1})	$\ln k$
3.53×10^{-3}	16.05
3.41×10^{-3}	16.20
3.30×10^{-3}	16.34
3.19×10^{-3}	16.48
3.10×10^{-3}	16.60

These data, when plotted, yield the graph shown below.

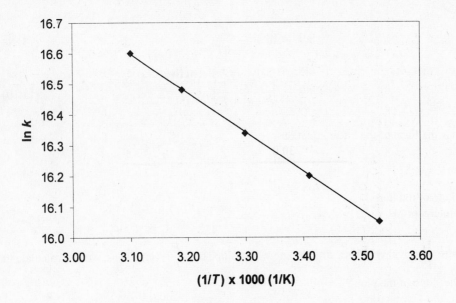

Calculating the slope from the first and last points gives:

$$\text{slope} = \frac{16.05 - 16.60}{(3.53 \times 10^{-3}) - (3.10 \times 10^{-3})} = -1.28 \times 10^3 \text{ K}$$

The slope is equal to $-E_a/R$ or

$$E_a = -\text{slope} \times R$$

$$E_a = -(-1.28 \times 10^3 \text{ K})(8.314 \text{ J/mol}\cdot\text{K}) = 1.06 \times 10^4 \text{ J/mol} = 10.6 \text{ kJ/mol}$$

B. Applying a modified form of the Arrhenius equation that relates the rate constants at two different temperatures

Starting with Equation (13.9), we can write,

$$\ln k_1 = \ln A - \frac{E_a}{RT_1}$$

and

$$\ln k_2 = \ln A - \frac{E_a}{RT_2}$$

Subtracting $\ln k_2$ from $\ln k_1$ gives:

$$\ln k_1 - \ln k_2 = -\frac{E_a}{RT_1} + \frac{E_a}{RT_2}$$

$$\ln \frac{k_1}{k_2} = \frac{E_a}{R}\left(\frac{1}{T_2} - \frac{1}{T_1}\right)$$

or

$$\ln \frac{k_1}{k_2} = \frac{E_a}{R}\left(\frac{T_1 - T_2}{T_1 T_2}\right) \qquad\qquad (13.11)$$

Using Equation (13.11), we can calculate the activation energy if we know the rate constants at two temperatures or find the rate constant at another temperature if the activation energy is known.

EXAMPLE 13.10
For the reaction

$$\textbf{NO + O}_3 \longrightarrow \textbf{NO}_2 + \textbf{O}_2$$

the following rate constants have been obtained. Calculate the activation energy for this reaction.

Temperature (°C)	k ($M^{-1}s^{-1}$)
10.0	9.3×10^6
30.0	1.25×10^7

We can calculate the activation energy using Equation (13.11).

$$\ln \frac{k_1}{k_2} = \frac{E_a}{R}\left(\frac{T_1 - T_2}{T_1 T_2}\right) \qquad\qquad (13.11)$$

Let,

$$k_1 = 9.3 \times 10^6 \ M^{-1}s^{-1} \qquad \text{at} \qquad T_1 = 273 + 10.0 = 283 \text{ K}$$
$$k_2 = 1.25 \times 10^7 \ M^{-1}s^{-1} \qquad \text{at} \qquad T_2 = 273 + 30.0 = 303 \text{ K}$$

Recall that $R = 8.314$ J/mol·K. Substitute the rate constants, the temperatures, and R into Equation (13.11) to solve for E_a.

$$\ln\left(\frac{9.3 \times 10^6 \text{ M}^{-1}\text{s}^{-1}}{1.25 \times 10^7 \text{ M}^{-1}\text{s}^{-1}}\right) = \frac{E_a}{8.314 \text{ J}/\text{mol} \cdot \text{K}}\left(\frac{283 \text{ K} - 303 \text{ K}}{(283 \text{ K})(303 \text{ K})}\right)$$

$$(\ln 0.744)(8.314 \text{ J/mol·K}) = E_a\left(\frac{-20 \text{ K}}{8.57 \times 10^4 \text{ K}^2}\right)$$

$$(-0.296)\left(8.314 \ \frac{\text{J}}{\text{mol·K}}\right) = E_a\left(-2.33 \times 10^{-4} \ \frac{1}{\text{K}}\right))$$

$$E_a = \textbf{1.06} \times \textbf{10}^4 \textbf{ J/mol} = \textbf{10.6 kJ/mol}$$

Compare the answer in this example with that obtained in Example 13.9.

PRACTICE EXERCISE

9.   At 300 K, the rate constant is $1.5 \times 10^{-5} \, M^{-1} \cdot s^{-1}$ for the following reaction:

$$2 \, NOCl \longrightarrow 2 \, NO \, + \, Cl_2$$

The activation energy is 90.2 kJ/mol. Calculate the value of the rate constant at 310 K.

Text Problems: **13.38**, 13.40, 13.42

PROBLEM TYPE 6: STUDYING REACTION MECHANISMS

For a complete discussion of reaction mechanisms, see Section 13.5 of your text. Experimental studies of reaction mechanisms begin with the collection of data (rate measurements). Next the data are analyzed to determine the rate constant and the order of reaction, so that the rate law can be written. Finally, a plausible mechanism is suggested for the reaction in terms of elementary steps. This sequence of steps is summarized in Figure 13.18 of the text.

The elementary steps of the proposed mechanism must satisfy two requirements:

- The sum of the elementary steps must give the overall balanced equation for the reaction.

- The **rate-determining step**, which is the slowest step in the sequence of steps leading to product formation, should predict the same rate law as is determined experimentally.

EXAMPLE 13.11
The rate law for the substitution of NH_3 for H_2O in the following reaction is first order in $Ni(H_2O)_6^{2+}$ and zero order in NH_3.

$$Ni(H_2O)_6^{2+} \, (aq) \, + \, NH_3 \, (aq) \, \longrightarrow \, Ni(H_2O)_5(NH_3)^{2+} \, (aq) \, + \, H_2O \, (l)$$

Show that the following mechanism is consistent with the experimental rate law.

$$Ni(H_2O)_6^{2+} \, (aq) \, \longrightarrow \, Ni(H_2O)_5^{2+} \, (aq) \, + \, H_2O \, (l) \qquad \text{(slow)}$$

$$Ni(H_2O)_5^{2+} \, (aq) \, + \, NH_3 \, (aq) \, \longrightarrow \, Ni(H_2O)_5(NH_3)^{2+} \, (aq) \qquad \text{(fast)}$$

First, does the sum of the elementary steps give the overall balanced equation. Yes, the sum of the steps gives:

$$Ni(H_2O)_6^{2+} \, (aq) \, + \, NH_3 \, (aq) \, \longrightarrow \, Ni(H_2O)_5(NH_3)^{2+} \, (aq) \, + \, H_2O \, (l)$$

Next, the reaction was experimentally determined to be first order in $Ni(H_2O)_6^{2+}$ and zero order in NH_3. We can write the rate law for the reaction.

$$\text{rate} \, = \, k[Ni(H_2O)_6^{2+}]$$

Does the rate law match the rate law of the proposed mechanism? We can write the rate law from the rate determining step of the proposed mechanism. Step 1 is the slow step, so we can write:

$$\text{rate} \, = \, k[Ni(H_2O)_6^{2+}]$$

which matches the experimentally determined rate law.

The elementary steps of the proposed mechanism satisfy the two requirements outlined above; therefore, it is a valid mechanism.

PRACTICE EXERCISE

10. The reaction of nitric oxide and chlorine,

$$2 \, NO \, (g) \, + \, Cl_2 \, (g) \, \rightarrow \, 2 \, NOCl \, (g)$$

has been proposed to proceed by the following two-step mechanism:

$$NO \, (g) \, + \, Cl_2 \, (g) \, \rightarrow \, NOCl_2 \, (g)$$

$$NO \, (g) \, + \, NOCl_2 \, (g) \, \rightarrow \, 2 \, NOCl \, (g)$$

What rate law is predicted if the first step of the proposed mechanism is the rate-determining step?

Text Problems: **13.50c**, 13.52, 13.60

ANSWERS TO PRACTICE EXERCISES

1. Rate of change of NO = −0.11 M/s

2. Rate = $k[B]^2$

3. [sucrose] = 0.083 M after 24 h

4. $t_{\frac{1}{2}}$ = 3.3 × 10^5 s

5. The slope of the straight line plot equals −k. You should find **k = 0.0504 min^{-1}**.

6. 442 s

7. 1.05 M

8. **(b)**

9. k = 4.8 × 10^{-5} $M^{-1}s^{-1}$

10. rate = $k[NO][Cl_2]$

SOLUTIONS TO SELECTED TEXT PROBLEMS

13.6   **(a)**   rate $= -\dfrac{1}{2}\dfrac{\Delta[H_2]}{\Delta t} = -\dfrac{\Delta[O_2]}{\Delta t} = \dfrac{1}{2}\dfrac{\Delta[H_2O]}{\Delta t}$

    (b)   rate $= -\dfrac{1}{4}\dfrac{\Delta[NH_3]}{\Delta t} = -\dfrac{1}{5}\dfrac{\Delta[O_2]}{\Delta t} = \dfrac{1}{4}\dfrac{\Delta[NO]}{\Delta t} = \dfrac{1}{6}\dfrac{\Delta[H_2O]}{\Delta t}$

13.8   Writing Rate Expressions, Problem Type 1.

The rate is defined as the change in concentration of a reactant or product with time. Each "change in concentration" term is divided by the corresponding stoichiometric coefficient. Terms involving reactants are proceeded by a minus sign.

$$\text{rate} = -\dfrac{\Delta[N_2]}{\Delta t} = -\dfrac{1}{3}\dfrac{\Delta[H_2]}{\Delta t} = \dfrac{1}{2}\dfrac{\Delta[NH_3]}{\Delta t}$$

(a)   If hydrogen is reacting at the rate of −0.074 M/s, the rate at which ammonia is being formed is

$$\dfrac{1}{2}\dfrac{\Delta[NH_3]}{\Delta t} = -\dfrac{1}{3}\dfrac{\Delta[H_2]}{\Delta t}$$

or

$$\dfrac{\Delta[NH_3]}{\Delta t} = -\dfrac{2}{3}\dfrac{\Delta[H_2]}{\Delta t}$$

$$\dfrac{\Delta[NH_3]}{\Delta t} = -\dfrac{2}{3}(-0.074\ M/\text{s}) = \mathbf{0.049\ \textit{M}/s}$$

(b)   The rate at which nitrogen is reacting must be:

$$\dfrac{\Delta[N_2]}{\Delta t} = \dfrac{1}{3}\dfrac{\Delta[H_2]}{\Delta t} = \dfrac{1}{3}(-0.074\ M/\text{s}) = \mathbf{-0.025\ \textit{M}/s}$$

Will the rate at which ammonia forms always be twice the rate of reaction of nitrogen, or is this true only at the instant described in this problem?

13.16   Assume the rate law has the form:

$$\text{rate} = k[F_2]^x[ClO_2]^y$$

To determine the order of the reaction with respect to F_2, find two experiments in which the $[ClO_2]$ is held constant. Compare the data from experiments 1 and 3. When the concentration of F_2 is doubled, the reaction rate doubles. Thus, the reaction is *first-order* in F_2.

To determine the order with respect to ClO_2, compare experiments 1 and 2. When the ClO_2 concentration is quadrupled, the reaction rate quadruples. Thus, the reaction is *first-order* in ClO_2.

The rate law is:

$$\text{rate} = k[F_2][ClO_2]$$

The value of k can be found using the data from any of the experiments. If we take the numbers from the second experiment we have:

$$k = \frac{\text{rate}}{[F_2][ClO_2]} = \frac{4.8 \times 10^{-3}\ M/s}{(0.10\ M)(0.040\ M)} = 1.2\ M^{-1}s^{-1}$$

Verify that the same value of k can be obtained from the other sets of data.

Since we now know the rate law and the value of the rate constant, we can calculate the rate at any concentration of reactants.

$$\textbf{rate} = k[F_2][ClO_2] = (1.2\ M^{-1}s^{-1})(0.010\ M)(0.020\ M) = \textbf{2.4} \times \textbf{10}^{-4}\ \textbf{\textit{M/s}}$$

13.18   Determining the Rate Law of a Reaction, Problem Type 2.

(a)   Compare the second and fifth set of data. Doubling [X] at constant [Y] increases the rate by a factor of 0.509/0.127 or 4. Therefore, the reaction is second-order in X. Next, compare the second and fourth set of data. Doubling [Y] at constant [X] increases the rate by a factor of 0.254/0.127 or 2 times. Therefore, the reaction is first order in Y. The rate law is:

$$\text{rate} = k[X]^2[Y]$$

The order of the reaction is $(2 + 1) = \textbf{3}$. The reaction is *3rd-order*.

(b)
Step 1: First, use any set of data to calculate the rate constant k. Using the first set of data,

$$k = \frac{\text{rate}}{[X]^2[Y]} = \frac{0.053\ M/s}{(0.10\ M)^2(0.50\ M)} = 10.6\ M^{-2}s^{-1}$$

Step 2: Substitute the concentrations of X and Y into the rate law to calculate the initial rate of disappearance of X.

$$\textbf{rate} = (10.6\ M^{-2}s^{-1})(0.30\ M)^2(0.40\ M) = \textbf{0.38}\ \textbf{\textit{M/s}}$$

13.20   **(a)**   For a reaction first-order in A,

$$\text{Rate} = k[A]$$

$$1.6 \times 10^{-2}\ M/s = k(0.35\ M)$$

$$k = \textbf{0.046}\ \textbf{s}^{-1}$$

(b)   For a reaction second-order in A,

$$\text{Rate} = k[A]^2$$

$$1.6 \times 10^{-2}\ M/s = k(0.35\ M)^2$$

$$k = \textbf{0.13}\ \textbf{\textit{M}}^{-1}\textbf{s}^{-1}$$

13.22   Let P_0 be the pressure of $ClCO_2CCl_3$ at $t = 0$, and let x be the decrease in pressure after time t. Note that from the coefficients in the balanced equation that the loss of 1 atmosphere of $ClCO_2CCl_3$ results in the formation of two atmospheres of $COCl_2$. We write:

$$ClCO_2CCl_3 \rightarrow 2COCl_2$$

Time	$[ClCO_2CCl_3]$	$[COCl_2]$
$t = 0$	P_0	0
$t = t$	$P_0 - x$	$2x$

Thus the change (increase) in pressure is x. We have:

t(s)	P (mmHg)	$\Delta P = x$	$P_{ClCO_2CCl_3}$	$\ln P_{ClCO_2CCl_3}$	$1/P_{ClCO_2CCl_3}$
0	15.76	0.00	15.76	2.757	0.0635
181	18.88	3.12	12.64	2.537	0.0791
513	22.79	7.03	8.73	2.167	0.115
1164	27.08	11.32	4.44	1.491	0.225

If the reaction is first order, then a plot of $\ln P_{ClCO_2CCl_3}$ vs. t would be linear. If the reaction is second order, a plot of $1/P_{ClCO_2CCl_3}$ vs. t would be linear. The two plots are shown below.

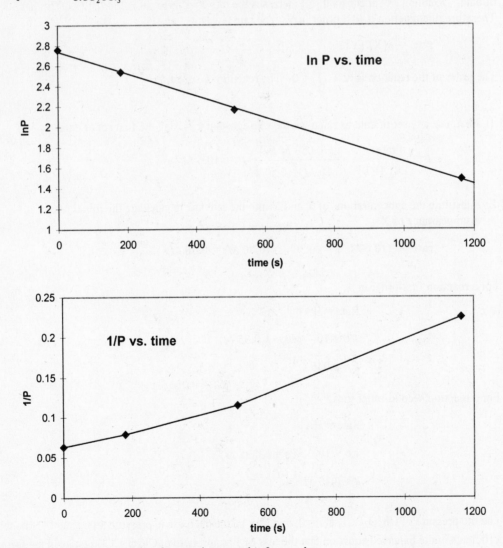

From the graphs we see that the reaction must be *first-order*.

13.28   Determining the half-life of a first-order reaction, Problem Type 3B.

(a)   For any first order reaction the rate constant can be found from the half–life using the following equation.

$$k = \frac{0.693}{t_{\frac{1}{2}}}$$

$$k = \frac{0.693}{t_{\frac{1}{2}}} = \frac{0.693}{35.0\text{ s}} = \textbf{0.0198 s}^{-1}$$

(b)   The time required for 95% of the phosphine to decompose can be found using the following equation.

$$\ln\frac{[A]}{[A]_0} = -kt$$

or

$$t = \frac{1}{k}\ln\frac{[A]_0}{[A]}$$

The value of [A] after 95% of the reactant has decomposed is $0.0500[A]_0$. Substitute the rate constant calculated in part (a) and $[A] = 0.0500[A]_0$ into the above equation to calculate the time required for 95% of the phosphine to decompose.

$$t = \frac{1}{k}\ln\frac{[A]_0}{[A]} = \left(\frac{1}{0.0198\text{ s}^{-1}}\right)\ln\frac{[A]_0}{0.0500[A]_0}$$

$$t = \frac{\ln 20.0}{0.0198\text{ s}^{-1}} = \textbf{151 s}$$

13.30 $\dfrac{1}{[A]} = \dfrac{1}{[A]_0} + kt$

$$\frac{1}{0.28} = \frac{1}{0.62} + 0.54t$$

$$t = \textbf{3.6 s}$$

13.38   Applying a modified form of the Arrhenius equation, Problem Type 5B.

The ratio of the rates at two different temperatures is equivalent to the ratio of the rate constants at the two temperatures.

Use the modified form of the Arrhenius equation shown below to calculate the energy of activation.

Let $T_1 = 250°C = 523$ K and $T_2 = 150°C = 423$ K.

$$\ln\frac{k_1}{k_2} = \frac{E_a}{R}\left(\frac{T_1 - T_2}{T_1 T_2}\right)$$

$$\ln(1.50 \times 10^3) = \frac{E_a}{8.314\text{ J}/\text{mol}\cdot\text{K}}\left(\frac{523\text{ K} - 423\text{ K}}{(523\text{ K})(423\text{ K})}\right)$$

$$7.31 = \frac{E_a}{8.314 \dfrac{J}{mol \cdot K}}\left(4.52 \times 10^{-4}\, \frac{1}{K}\right)$$

$$E_a = 1.35 \times 10^5 \text{ J/mol} = 135 \text{ kJ/mol}$$

13.40 Use the modified form of the Arrhenius equation to calculate the temperature at which the rate constant is $8.80 \times 10^{-4}\text{ s}^{-1}$.

$$\ln\frac{k_1}{k_2} = \frac{E_a}{R}\left(\frac{1}{T_2} - \frac{1}{T_1}\right)$$

$$\ln\left(\frac{4.60 \times 10^{-4}\text{ s}^{-1}}{8.80 \times 10^{-4}\text{ s}^{-1}}\right) = \frac{1.04 \times 10^5 \text{ J/mol}}{8.314 \text{ J/mol} \cdot K}\left(\frac{1}{T_2} - \frac{1}{623\text{ K}}\right)$$

$$(\ln 0.523) = 1.25 \times 10^4 \text{ K}\left(\frac{1}{T_2} - \frac{1}{623\text{ K}}\right)$$

$$-0.649 + 20.1 = \frac{1.25 \times 10^4 \text{ K}}{T_2}$$

$$19.5\, T_2 = 1.25 \times 10^4 \text{ K}$$

$$T_2 = 641 \text{ K} = 368°C$$

13.42 Since the ratio of rates is equal to the ratio of rate constants, we can write:

$$\ln\frac{rate_1}{rate_2} = \ln\frac{k_1}{k_2}$$

$$\ln\frac{k_1}{k_2} = \ln\left(\frac{2.0 \times 10^2}{39.6}\right) = \frac{E_a}{8.314 \text{ J/K} \cdot mol}\left(\frac{(300\text{ K} - 278\text{ K})}{(300\text{ K})(278\text{ K})}\right)$$

$$E_a = 5.10 \times 10^4 \text{ J/mol} = 51.0 \text{ kJ/mol}$$

13.50 **(a)** Determining the Rate Law of a Reaction, Problem Type 2.

Since the reaction rate doubles when the X_2 concentration is doubled, the reaction is first-order in X. The reaction rate triples when the concentration of Y is tripled, so the reaction is also first-order in Y. The concentration of Z has no effect on the rate, so the reaction is zero-order in Z.

The rate law is:

$$rate = k[X_2][Y]$$

(b) If a change in the concentration of Z has no effect on the rate, the concentration of Z is not a term in the rate law. This implies that Z does not participate in the rate-determining step of the reaction mechanism.

(c)   Studying Reaction Mechanisms, Problem Type 6.

The rate law shows that the slow step involves reaction of a molecule of X_2 with a molecule of Y. Since Z is not present in the rate law, it does not take part in the slow step and must appear in a fast step at a later time. (If the fast step involving Z happened before the rate-determining step, the rate law would involve Z in a more complex way.) A mechanism that is consistent with the rate law could be:

$$X_2 + Y \longrightarrow XY + X \qquad \text{(slow)}$$

$$X + Z \longrightarrow XZ \qquad \text{(fast)}$$

The rate law only tells us about the slow step. Other mechanisms with different subsequent fast steps are possible. Try to invent one.

13.52   The experimentally determined rate law is first order in H_2 and second order in NO. In Mechanism I the slow step is bimolecular and the rate law would be:

$$\text{rate} = k[H_2][NO]$$

Mechanism I can be discarded.

The rate-determining step in Mechanism II involves the simultaneous collision of two NO molecules with one H_2 molecule. The rate law would be:

$$\text{rate} = k[H_2][NO]^2$$

Mechanism II is a possibility.

In Mechanism III we assume the forward and reverse reactions in the first fast step are in dynamic equilibrium, so their rates are equal:

$$k_f[NO]^2 = k_r[N_2O_2]$$

The slow step is bimolecular and involves collision of a hydrogen molecule with a molecule of N_2O_2. The rate would be:

$$\text{rate} = k_2[H_2][N_2O_2]$$

If we solve the dynamic equilibrium equation of the first step for $[N_2O_2]$ and substitute into the above equation, we have the rate law:

$$\text{rate} = \frac{k_2 k_f}{k_r}[H_2][NO]^2 = k[H_2][NO]^2$$

Mechanism III is also a possibility. Can you suggest an experiment that might help to decide between the two mechanisms?

13.60   The rate-determining step involves the breakdown of ES to E and P. The rate law for this step is:

$$\text{rate} = k_2[ES]$$

In the first elementary step, the intermediate ES is in equilibrium with E and S. The equilibrium relationship is:

$$\frac{[ES]}{[E][S]} = \frac{k_1}{k_{-1}}$$

or

$$[ES] = \frac{k_1}{k_{-1}}[E][S]$$

Substitute [ES] into the rate law expression.

$$\textbf{rate} = k_2[ES] = \frac{k_1 k_2}{k_{-1}}\textbf{[E][S]}$$

13.62   Temperature, energy of activation, concentration of reactants, and a catalyst.

13.64   First, calculate the radius of the 10.0 cm^3 sphere.

$$V = \frac{4}{3}\pi r^3$$

$$10.0\ \text{cm}^3 = \frac{4}{3}\pi r^3$$

$$r = 1.34\ \text{cm}$$

The surface area of the sphere is:

$$\textbf{area} = 4\pi r^2 = 4\pi(1.34\ \text{cm})^2 = \textbf{22.6 cm}^2$$

Next, calculate the radius of the 1.25 cm^3 sphere.

$$V = \frac{4}{3}\pi r^3$$

$$1.25\ \text{cm}^3 = \frac{4}{3}\pi r^3$$

$$r = 0.668\ \text{cm}$$

The surface area of one sphere is:

$$\text{area} = 4\pi r^2 = 4\pi(0.668\ \text{cm})^2 = 5.61\ \text{cm}^2$$

$$\textbf{The total area of 8 spheres} = 5.61\ \text{cm}^2 \times 8 = \textbf{44.9 cm}^2$$

Obviously, the surface area of the eight spheres (44.9 cm^2) is greater than that of one larger sphere (22.6 cm^2). A greater surface area promotes the catalyzed reaction more effectively.

13.66   The overall rate law is of the general form: rate $= k[H_2]^x[NO]^y$

(a)   Comparing Experiment #1 and Experiment #2, we see that the concentration of NO is constant and the concentration of H_2 has decreased by one-half. The initial rate has also decreased by one-half. Therefore, the initial rate is directly proportional to the concentration of H_2; x $= 1$.

Comparing Experiment #1 and Experiment #3, we see that the concentration of H_2 is constant and the concentration of NO has decreased by one-half. The initial rate has decreased by one-fourth. Therefore, the initial rate is proportional to the squared concentration of NO; $y = 2$.

The overall rate law is: $\text{rate} = k[H_2][NO]^2$

(b) Using Experiment #1 to calculate the rate constant,

$$\text{rate} = k[H_2][NO]^2$$

$$k = \frac{\text{rate}}{[H_2][NO]^2}$$

$$k = \frac{2.4 \times 10^{-6} \ M/s}{(0.010 \ M)(0.025 \ M)^2} = 0.38 \ M^{-2}s^{-1}$$

(c) Consulting the rate law, we assume that the slow step in the reaction mechanism will probably involve one H_2 molecule and two NO molecules. Additionally the hint tells us that O atoms are an intermediate.

$$
\begin{array}{ll}
H_2 + 2NO \rightarrow N_2 + H_2O + O & \text{slow step} \\
\underline{O + H_2 \rightarrow H_2O} & \text{fast step} \\
2H_2 + 2NO \rightarrow N_2 + 2H_2O &
\end{array}
$$

13.68 If water is also the solvent in this reaction, it is present in vast excess over the other reactants and products. Throughout the course of the reaction, the concentration of the water will not change by a measurable amount. As a result, the reaction rate will not appear to depend on the concentration of water.

13.70 Since the reaction is first order in both A and B, then we can write the rate law expression:

$$\text{rate} = k[A][B]$$

Substituting in the values for the rate, [A], and [B]:

$$4.1 \times 10^{-4} \ M/s = k(1.6 \times 10^{-2})(2.4 \times 10^{-3})$$

$$k = 10.7 \ M^{-1}s^{-1}$$

Knowing that the overall reaction was second order, could you have predicted the units for k?

13.72 Recall that the pressure of a gas is directly proportional to the number of moles of gas. This comes from the ideal gas equation.

$$P = \frac{nRT}{V}$$

The balanced equation is:

$$2 \, N_2O \, (g) \longrightarrow 2 \, N_2 \, (g) + O_2 \, (g)$$

From the stoichiometry of the balanced equation, for every one mole of N_2O that decomposes, one mole of N_2 and 0.5 moles of O_2 will be formed. Let's assume that we had 2 moles of N_2O at $t = 0$. After one half-life there will be one mole of N_2O remaining and one mole of N_2 and 0.5 moles of O_2 will be formed. The total number of moles of gas after one half-life will be:

$$n_T = n_{N_2O} + n_{N_2} + n_{O_2} = 1 \text{ mol} + 1 \text{ mol} + 0.5 \text{ mol} = 2.5 \text{ moles}$$

At $t = 0$, there were 2 mol of gas. Now, at $t_{\frac{1}{2}}$, there are 2.5 mol of gas. Since the pressure of a gas is directly proportional to the number of moles of gas, we can write:

$$\frac{2.10 \text{ atm}}{2 \text{ mol gas } (t = 0)} \times 2.5 \text{ mol gas } \left(\text{at } t_{\frac{1}{2}} \right) = \textbf{2.63 atm after one half-life}$$

13.74   The rate expression for a third order reaction is:

$$\text{rate} = -\frac{\Delta[A]}{\Delta t} = k[A]^3$$

The units for the rate law are:

$$\frac{M}{s} = kM^3$$

$$k = M^{-2}s^{-1}$$

13.76   Both compounds, A and B, decompose by first-order kinetics. Therefore, we can write a first-order rate equation for A and also one for B.

$$[A] = [A]_0 e^{-k_A t} \qquad\qquad [B] = [B]_0 e^{-k_B t}$$

We can calculate each of the rate constants, k_A and k_B, from their respective half-lives.

$$k_A = \frac{0.693}{50.0 \text{ min}} = 0.0139 \text{ min}^{-1} \qquad\qquad k_B = \frac{0.693}{18.0 \text{ min}} = 0.0385 \text{ min}^{-1}$$

The initial concentration of A and B are equal. $[A]_0 = [B]_0$. Therefore, from the first-order rate equations, we can write:

$$\frac{[A]}{[B]} = 4 = \frac{[A]_0 e^{-k_A t}}{[B]_0 e^{-k_B t}} = \frac{e^{-k_A t}}{e^{-k_B t}} = e^{(k_B - k_A)t} = e^{(0.0385 - 0.0139)t}$$

$$4 = e^{0.0246t}$$

$$t = \textbf{56.4 min}$$

13.78   **(a)**   Changing the concentration of a reactant has no effect on k.

(b)   If a reaction is run in a solvent other than in the gas phase, then the reaction mechanism will probably change and will thus change k.

(c)   Doubling the pressure simply changes the concentration. No effect on k, as in (a).

(d)   The rate constant k changes with temperature.

(e)   A catalyst changes the reaction mechanism and therefore changes k.

13.80 Mathematically, the amount left after ten half–lives is:

$$\left(\frac{1}{2}\right)^{10} = 9.8 \times 10^{-4} = \textbf{0.098\%}$$

13.82 The net ionic equation is:

$$\text{Zn}\,(s) + 2\,\text{H}^+\,(aq) \longrightarrow \text{Zn}^{2+}\,(aq) + \text{H}_2\,(g)$$

(a) Changing from the same mass of granulated zinc to powdered zinc increases the rate because the surface area of the zinc (and thus its concentration) has increased.

(b) Decreasing the mass of zinc (in the same granulated form) will decrease the rate because the total surface area of zinc has decreased.

(c) The concentration of protons has decreased in changing from the strong acid (hydrochloric) to the weak acid (acetic); the rate will decrease.

(d) An increase in temperature will increase the rate constant k; therefore, the rate of reaction increases.

13.84 In terms of the initial concentration of A, the concentration of A remaining after 4.90 min is $0.645[\text{A}]_0$. Using the first-order rate equation,

$$\ln\frac{[\text{A}]}{[\text{A}]_0} = -kt$$

or

$$\ln\frac{0.645[\text{A}]_0}{[\text{A}]_0} = -kt$$

$$0.439 = k(4.90 \text{ min})$$

$$k = \frac{0.439}{4.90 \text{ min}} = \textbf{0.0896 min}^{-1}$$

13.86 The first-order rate equation can be arranged to take the form of a straight line.

$$\ln[\text{A}] = -kt + \ln[\text{A}]_0$$

If a reaction obeys first-order kinetics, a plot of $\ln[\text{A}]$ vs. t will be a straight line with a slope of $-k$.

The slope of a plot of $\ln[\text{N}_2\text{O}_5]$ vs. t is -6.18×10^{-4} min^{-1}. Thus,

$$k = 6.18 \times 10^{-4} \text{ min}^{-1}$$

The equation for the half-life of a first-order reaction is:

$$t_{\frac{1}{2}} = \frac{0.693}{k}$$

$$t_{\frac{1}{2}} = \frac{0.693}{6.18 \times 10^{-4} \text{ min}^{-1}} = \textbf{1.12} \times \textbf{10}^3 \textbf{ min}$$

13.88   **(a)**   In the two-step mechanism the rate-determining step is the collision of a hydrogen molecule with two iodine atoms. If visible light increases the concentration of iodine atoms, then the rate must increase. If the true rate-determining step were the collision of a hydrogen molecule with an iodine molecule (the one-step mechanism), then the visible light would have no effect (it might even slow the reaction by depleting the number of available iodine molecules).

   (b)   To split hydrogen molecules into atoms, one needs ultraviolet light of much higher energy.

13.90   **(a)**   We can write the rate law for an elementary step directly from the stoichiometry of the balanced reaction. In this rate-determining elementary step three molecules must collide simultaneously (one X and two Y's). This makes the reaction termolecular, and consequently the rate law must be third order: first order in X and second order in Y.

   The rate law is:
   $$\text{rate} = k[\text{X}][\text{Y}]^2$$

   (b)   The value of the rate constant can be found by solving algebraically for k.

   $$k = \frac{\text{rate}}{[\text{X}][\text{Y}]^2} = \frac{3.8 \times 10^{-3}\ M/s}{(0.26\ M)(0.88\ M)^2} = 1.9 \times 10^{-2}\ M^{-2} s^{-1}$$

   Could you write the rate law if the reaction shown were the overall balanced equation and not an elementary step?

13.92

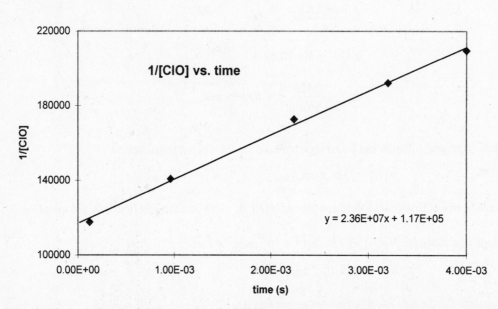

   Reaction is second-order because a plot of 1/[ClO] vs. time is a straight line. The slope of the line equals the rate constant, k.

   $$k = \text{Slope} = 2.4 \times 10^7\ /M\cdot s$$

13.94   During the first five minutes or so the engine is relatively cold, so the exhaust gases will not fully react with the components of the catalytic converter. Remember, for almost all reactions, the rate of reaction increases with temperature.

13.96   (a) E_a has a large value.

(b) $E_a \approx 0$. Orientation factor is not important.

13.98   First, solve for the rate constant, k, from the half-life of the decay.

$$t_{\frac{1}{2}} = 2.44 \times 10^5 \text{ yr} = \frac{0.693}{k}$$

$$k = \frac{0.693}{2.44 \times 10^5 \text{ yr}} = 2.84 \times 10^{-6} \text{ yr}^{-1}$$

Now, we can calculate the time for the plutonium to decay from 5.0×10^2 g to 1.0×10^2 g using the equation for a first-order reaction relating concentration and time.

$$\ln \frac{[A]}{[A]_0} = -kt$$

$$\ln \frac{1.0 \times 10^2}{5.0 \times 10^2} = -(2.84 \times 10^{-6} \text{ yr}^{-1})t$$

$$1.6 = (2.84 \times 10^{-6} \text{ yr}^{-1})t$$

$$t = \mathbf{5.6 \times 10^5 \text{ yr}}$$

13.100   (a)   Catalyst: Mn^{2+}; intermediate: Mn^{3+}

First step is rate-determining.

(b)   Without the catalyst, the reaction would be a termolecular one involving 3 cations! (Tl^+ and two Ce^{4+}). The reaction would be slow.

(c)   The catalyst is a homogeneous catalyst because it has the same phase (aqueous) as the reactants.

13.102   Initially, the number of moles of gas in terms of the volume is:

$$n = \frac{PV}{RT} = \frac{(0.350 \text{ atm})V}{\left(0.0821 \frac{\text{L} \cdot \text{atm}}{\text{mol} \cdot \text{K}}\right)(450 + 273)\text{K}} = 5.90 \times 10^{-3} \, V$$

We can calculate the concentration of dimethyl ether from the following equation.

$$\frac{[(CH_3)_2O]}{[(CH_3)_2O]_0} = e^{-kt}$$

Since, the volume is held constant, it will cancel out of the equation. The concentration of dimethyl ether after 8.0 minutes is:

$$[(CH_3)_2O] = \left(\frac{5.90 \times 10^{-3} \, V}{V}\right) e^{-\left(3.2 \times 10^{-4} \frac{1}{s}\right)\left(8.0 \text{ min} \times 60 \frac{s}{\text{min}}\right)}$$

$$[(CH_3)_2O] = 5.06 \times 10^{-3} \, M$$

After 8.0 min, the concentration of $(CH_3)_2O$ has decreased by $(5.90 \times 10^{-3} - 5.06 \times 10^{-3})$ M or 8.4×10^{-4} M. Since three moles of product form for each mole of dimethyl ether that reacts, the concentrations of the products are $(3 \times 8.4 \times 10^{-4}$ $M) = 2.5 \times 10^{-3}$ M.

The pressure of the system after 8.0 minutes is:

$$P = \frac{nRT}{V} = \left(\frac{n}{V}\right)RT = MRT$$

$$P = (5.06 \times 10^{-3} + 2.5 \times 10^{-3})M \times (0.0821 \text{ L·atm/mol·K}) \times (723 \text{ K})$$

$$\boldsymbol{P = 0.45 \text{ atm}}$$

13.104 **(a)** $\dfrac{\Delta[B]}{\Delta t} = k_1[A] - k_2[B]$

(b) If, $\dfrac{\Delta[B]}{\Delta t} = 0$

Then, from part (a) of this problem:

$k_1[A] = k_2[B]$

$[B] = \dfrac{k_1}{k_2}[A]$

13.106 **(a)** First, let's determine how the surface area of Mg affects the reaction rate by comparing reactions (1) and (4).

$$\left(\frac{140}{170}\right)^a = \frac{3.70}{2.07}$$

$$2^a = 1.8$$

$$a = 0.85 \approx 1$$

Also, comparing reactions (1) and (5):

$$\left(\frac{70}{35}\right)^a = \frac{2.07}{0.97}$$

$$a = 1.1 \approx 1$$

Next, let's determine how the $[H^+]$ affects the reaction rate by comparing reactions (1) and (2).

$$\left(\frac{1.0}{2.0}\right)^b = \frac{2.07}{0.0749}$$

$$b = 2$$

Thus, the rate law is:

rate = k[surface area Mg][H$^+$]^2

The rate constants calculated from each experiment are:

Rate constants: (1) 0.0296, (2) 0.0268, (3) 0.0255, (4) 0.0264, (5) 0.0277.

Average k = 0.0272 torr/mm^2·M^2·s

(b)   From the ideal gas equation, we can calculate the pressure of H_2 gas under ideal conditions.

$$\text{mol } H_2 = 0.0083 \text{ g Mg} \times \frac{1 \text{ mol Mg}}{24.31 \text{ g Mg}} \times \frac{1 \text{ mol } H_2}{1 \text{ mol Mg}} = 3.4 \times 10^{-4} \text{ mol } H_2$$

$$P = \frac{nRT}{V} = \frac{(3.4 \times 10^{-4} \text{ mol})\left(0.0821 \frac{L \cdot atm}{mol \cdot K}\right)(32 + 273)K}{0.0695 \text{ L}} = 0.123 \text{ atm} = 93.5 \text{ torr}$$

$$\textbf{percent error} = \frac{93.5 - 93.2}{93.5} \times 100\% = \textbf{0.32\%}$$

13.108

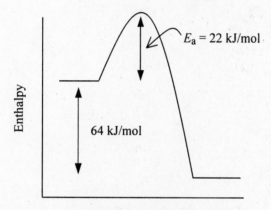

E_a for B → A is (22 + 64) kJ/mol = **86 kJ/mol**

13.110   Let $k_{cat} = k_{uncat}$

Then

$$Ae^{\frac{-E_a(\text{cat})}{RT_1}} = Ae^{\frac{-E_a(\text{uncat})}{RT_2}}$$

Since the frequency factor is the same, we can write:

$$e^{\frac{-E_a(\text{cat})}{RT_1}} = e^{\frac{-E_a(\text{uncat})}{RT_2}}$$

Taking the natural log (*ln*) of both sides of the equation gives:

$$\frac{-E_a(\text{cat})}{RT_1} = \frac{-E_a(\text{uncat})}{RT_2}$$

or,

$$\frac{E_a(\text{cat})}{T_1} = \frac{E_a(\text{uncat})}{T_2}$$

Substituting in the given values:

$$\frac{7.0 \text{ kJ} / \text{mol}}{293 \text{ K}} = \frac{42 \text{ kJ} / \text{mol}}{T_2}$$

$$T_2 = 1.8 \times 10^3 \text{ K}$$

This temperature is much too high to be practical.

CHAPTER 14
CHEMICAL EQUILIBRIUM

PROBLEM-SOLVING STRATEGIES AND TUTORIAL SOLUTIONS

TYPES OF PROBLEMS

Problem Type 1: Homogeneous Equilibria.
  - **(a)** Writing expressions for K_c and K_p.
  - **(b)** Calculating equilibrium partial pressures.
  - **(c)** Converting between K_c and K_p.

Problem Type 2: Heterogeneous Equilibria, Calculating K_p and K_c.

Problem Type 3: The Form of K and the Equilibrium Equation.

Problem Type 4: Using the Reaction Quotient (Q_c) to Predict the Direction of a Reaction.

Problem Type 5: Calculating Equilibrium Concentrations.

Problem Type 6: Factors that Affect Chemical Equilibrium.
  - **(a)** Changes in concentration.
  - **(b)** Changes in pressure and volume.
  - **(c)** Changes in temperature.
  - **(d)** The effect of a catalyst.

PROBLEM TYPE 1: HOMOGENEOUS EQUILIBRIA

Let's start by considering the following reversible reaction.

$$a\text{A} + b\text{B} \rightleftharpoons c\text{C} + d\text{D}$$

where *a, b, c,* and *d* are the stoichiometric coefficients for the reacting species A, B, C, and D. The equilibrium constant for the reaction at a particular temperature is:

$$K = \frac{[\text{C}]^c[\text{D}]^d}{[\text{A}]^a[\text{B}]^b}$$

This equation is the mathematical form of the **law of mass action**. It relates the concentrations of reactants and products at equilibrium in terms of a quantity called the **equilibrium constant (K)**.

The magnitude of the equilibrium constant, K, is important. If the equilibrium constant is much greater than one ($K \gg 1$), we say that the equilibrium lies to the right and favors the products. Conversely, if $K \ll 1$, the equilibrium lies to the left and favors the reactants.

A.   Writing expressions for K_c and K_p

The term **homogeneous equilibrium** applies to reactions in which all reacting species are in the same phase. An example of a homogeneous gas-phase equilibrium is the reaction between sulfur dioxide and oxygen to form sulfur trioxide.

$$2\text{ SO}_2\,(g) + \text{O}_2\,(g) \rightleftharpoons 2\text{ SO}_3\,(g)$$

The equilibrium constant as given in Equation (14.2) of your text is:

$$K_c = \frac{[SO_3]^2}{[SO_2]^2[O_2]}$$

The subscript c of K_c denotes that the concentrations of all species are expressed in mol/L.

The concentrations of reactants and products in gas-phase reactions can also be expressed in terms of their partial pressures. Starting with the ideal gas equation, we can write:

$$P = \left(\frac{n}{V}\right)RT$$

At constant temperature, the pressure (P) of a gas is directly related to the concentration of the gas in mol/L, $\left(\frac{n}{V}\right)$.

Thus, for the equilibrium shown above, we can write:

$$K_p = \frac{P_{SO_3}^2}{P_{SO_2}^2 P_{O_2}}$$

where P_{SO_3}, P_{SO_2}, and P_{O_2} are the equilibrium partial pressures (in atmospheres) of SO_3, SO_2, and O_2, respectively. The subscript p of K_p indicates the equilibrium concentrations are expressed in terms of pressure.

EXAMPLE 14.1
Write the equilibrium constant expression for the following reversible reaction:

$$4\,NH_3\,(g)\; +\; 5\,O_2\,(g)\; \rightleftharpoons\; 4\,NO\,(g)\; +\; 6\,H_2O\,(g)$$

Remember that the concentration of each component in the equilibrium constant expression is raised to a power equal to its coefficient in the balanced equation.

$$K_c = \frac{[NO]^4[H_2O]^6}{[NH_3]^4[O_2]^5}$$

You could also write the equilibrium constant expression in terms of the partial pressures of the gaseous components.

$$K_p = \frac{P_{NO}^4 P_{H_2O}^6}{P_{NH_3}^4 P_{O_2}^5}$$

PRACTICE EXERCISE
1.   Write the equilibrium constant expression for the following reaction.

$$2\,N_2O\,(g)\; +\; 3\,O_2\,(g)\; \rightleftharpoons\; 2\,N_2O_4\,(g)$$

Text Problem: See Review Question 14.8

B. Calculating equilibrium partial pressures

EXAMPLE 14.2
At 400°C, $K_p = 64$ for the following reaction:

$$H_2\,(g)\; +\; I_2\,(g)\; \rightleftharpoons\; 2\,HI\,(g)$$

At equilibrium, the partial pressures of H_2 and I_2 in a closed container are 0.20 atm and 0.50 atm respectively. What is the partial pressure of HI in the mixture?

Step 1: Write K_p in terms of the partial pressures of the reacting species.

$$K_p = \frac{P_{HI}^2}{P_{H_2} P_{I_2}}$$

Step 2: Solve the above equation algebraically for the unknown partial pressure. Then, substitute the known values into the equation to solve for the unknown.

$$P_{HI}^2 = K_p P_{H_2} P_{I_2}$$

$$P_{HI} = \sqrt{K_p P_{H_2} P_{I_2}}$$

$$P_{HI} = \sqrt{(64)(0.20)(0.50)} = \textbf{2.5 atm}$$

PRACTICE EXERCISE

2. Consider the following reaction at a temperature of 250°C.

$$PCl_5\,(g) \rightleftharpoons PCl_3\,(g) + Cl_2\,(g)$$

At equilibrium, the partial pressures of PCl_5, PCl_3, and Cl_2 are 0.0704 atm, 0.340 atm, and 0.218 atm, respectively. What is the value of K_p?

C. Converting K_c to K_p

In general, K_c is not equal to K_p, since the partial pressures of reactants and products are not equal to their concentrations expressed in mol/L. A simple relationship between K_p and K_c can be derived. See Section 14.2 of your text for the derivation. The relationship is:

$$K_p = K_c(RT)^{\Delta n}$$

where,

Δn = mol of gaseous product in balanced equation − mol of gaseous reactants in balanced eq.

If we express pressure in atmospheres, the gas constant R is given by 0.0821 L·atm/mol·K. We can write the relationship between K_p and K_c as

$$K_p = K_c(0.0821\ T)^{\Delta n} \tag{14.1}$$

In general, $K_p \neq K_c$ except in the special case when $\Delta n = 0$. In this case, Equation (14.1) can be written as:

$$K_p = K_c(0.0821\ T)^0$$

and

$$K_p = K_c$$

EXAMPLE 14.3

In the decomposition of carbon dioxide at 2000°C,

$$2\ CO_2\,(g) \rightleftharpoons 2\ CO\,(g) + O_2\,(g)$$

$K_p = 1.2 \times 10^{-4}$. **Calculate K_c for this reaction.**

Step 1: Rearrange Equation (14.1), which relates K_p and K_c, solving for K_c.

$$K_p = K_c(0.0821\ T)^{\Delta n}$$

$$K_c = \frac{K_p}{(0.0821\ T)^{\Delta n}}$$

Step 2: Substitute the given values into the above equation to solve for K_c. Temperature must be in units of Kelvin. **Note:** In the balanced equation, there are a total of three moles of gaseous products and two moles of gaseous reactants, so $\Delta n = 3 - 2$.

$$K_c = \frac{K_p}{(0.0821\,T)^{\Delta n}} = \frac{1.2 \times 10^{-4}}{[(0.0821)(2273\text{K})]^{(3-2)}} = \mathbf{6.4 \times 10^{-7}}$$

PRACTICE EXERCISE

3. At 400°C, $K_p = 64$ for the following reaction:

$$H_2\,(g) + I_2\,(g) \rightleftharpoons 2\,HI\,(g)$$

What is the value of K_c for this reaction?

> **Text Problems: 14.16**, 14.20

PROBLEM TYPE 2: HETEROGENEOUS EQUILIBRIA, CALCULATING K_p AND K_c

Whenever a reaction involves reactants and products that exist in different phases, it is called a *heterogeneous reaction*. If the reaction is carried out in a closed container, a **heterogeneous equilibrium** will result. For example, when steam is brought into contact with charcoal in a closed container, the following equilibrium is established:

$$C\,(s) + H_2O\,(g) \rightleftharpoons H_2\,(g) + CO\,(g)$$

At equilibrium, we might write the equilibrium constant as:

$$K_c' = \frac{[H_2][CO]}{[C][H_2O]} \tag{14.2}$$

However, the "concentration" of a solid, like its density, is an intensive property and thus does not depend on the amount of substance present. For this reason, the term [C] is a constant and can be combined with the equilibrium constant. We can simplify Equation (14.2) by writing

$$[C]\,K_c' = K_c = \frac{[H_2][CO]}{[H_2O]}$$

Keep in mind that the value of K_c does not depend on how much carbon is present, as long as some amount is present at equilibrium.

The argument presented above for solids also applies to pure liquids. Thus, if a liquid is a reactant or product, we can treat its concentration as a constant and omit it from the equilibrium constant expression.

EXAMPLE 14.4

What are the values of K_p and K_c at 1000°C for the reaction

$$CaCO_3\,(s) \rightleftharpoons CaO\,(s) + CO_2\,(g)$$

if the pressure of CO_2 in equilibrium with $CaCO_3$ and CaO is 3.87 atm?

Step 1: As stated above, if a solid or a liquid is a reactant or product, we can treat its concentration as a constant and omit it from the equilibrium constant expression. Thus, enough information is given to calculate K_p for this heterogeneous equilibrium.

$$K_p = P_{CO_2} = 3.87$$

Step 2: To calculate K_c, rearrange the equation relating K_p and K_c, solving for K_c. Then, substitute the given values into the equation to obtain the answer.

$$K_p = K_c(0.0821\ T)^{\Delta n}$$

$$K_c = \frac{K_p}{(0.0821\ T)^{\Delta n}}$$

$$\boldsymbol{K_c} = \frac{3.87}{[(0.0821)(1273K)]^{(1-0)}} = \boldsymbol{0.0370}$$

Text Problems: 14.20, **14.22**

PROBLEM TYPE 3: THE FORM OF *K* AND THE EQUILIBRIUM EQUATION

The equilibrium constant expression and its value depend on how an equation is balanced. Often an equation can be balanced with more than one set of coefficients. For example,

$$2\ SO_2\ (g)\ +\ O_2\ (g)\ \rightleftharpoons\ 2\ SO_3\ (g) \qquad K_c = \frac{[SO_3]^2}{[SO_2]^2[O_2]}$$

and

$$SO_2\ (g)\ +\ 1/2\ O_2\ (g)\ \rightleftharpoons\ SO_3\ (g) \qquad K_c^{'} = \frac{[SO_3]}{[SO_2][O_2]^{1/2}}$$

Is there a relationship between the equilibrium constants for the two reactions? Note that $K_c^{'}$ is the square root of the equilibrium constant for the first reaction, K_c.

$$K_c^{'} = \frac{[SO_3]}{[SO_2][O_2]^{1/2}} = \sqrt{\frac{[SO_3]^2}{[SO_2]^2[O_2]}} = \sqrt{K_c}$$

The general relationship is that you raise K_c to the power by which the equation was multiplied. To come up with the second balanced equation above, we had to multiply the first equation by 1/2. Thus,

$$K_c^{'} = (K_c)^{1/2} = \sqrt{K_c}$$

What if the reaction is written in the reverse direction?

$$2SO_3\ (g)\ \rightleftharpoons\ 2\ SO_2\ (g)\ +\ O_2\ (g) \qquad K_c^{''} = \frac{[SO_2]^2[O_2]}{[SO_3]^2}$$

Is there a relationship between $K_c^{''}$ and K_c? By inspection, you should find that $K_c^{''}$ is the reciprocal of K_c for the forward reaction.

$$K_c'' = \frac{[SO_2]^2[O_2]}{[SO_3]^2} = \frac{1}{K_c}$$

Remember, always use the K_c expression and value that are consistent with the way in which the balanced equation is written.

EXAMPLE 14.5

For the reaction, 2 HBr (g) $\rightleftharpoons$ H$_2$ (g) + Br$_2$ (g), $K_p = 1.4 \times 10^{-5}$ at 700 K. What are the values of K_p for the following reactions at the same temperature?

(a)   4 HBr (g) $\rightleftharpoons$ 2 H$_2$ (g) + 2 Br$_2$ (g)

(b)   H$_2$ (g) + Br$_2$ (g) $\rightleftharpoons$ 2 HBr (g)

(a)   In Equation (a), the original equation has been multiplied by *two*. The general relationship is that you raise K to the power by which the equation was multiplied by. Thus,

$$K_p' = (K_p)^2 = (1.4 \times 10^{-5})^2 = \mathbf{2.0 \times 10^{-10}}$$

(b)   Equation (b) is the reverse of the original equation. By inspection, you should find that K_p'' is the reciprocal of K_p.

$$K_p'' = \frac{1}{K_p} = \frac{1}{1.4 \times 10^{-5}} = \mathbf{7.1 \times 10^4}$$

PRACTICE EXERCISE

4.   For the reaction, 2 HBr (g) $\rightleftharpoons$ H$_2$ (g) + Br$_2$ (g), $K_p = 1.4 \times 10^{-5}$ at 700 K. What is the value of K_p for the following reaction at the same temperature?

$$\text{HBr } (g) \rightleftharpoons 1/2 \text{ H}_2 (g) + 1/2 \text{ Br}_2 (g)$$

Text Problems: 14.28, 14.30

PROBLEM TYPE 4: USING THE REACTION QUOTIENT (Q) TO PREDICT THE DIRECTION OF A REACTION

Equilibrium constants provide useful information about chemical reaction systems. For instance, equilibrium constants can be used to predict the direction in which a reaction will proceed to establish equilibrium.

The reaction quotient, Q_c, is a useful aid in predicting whether or not a reaction system is at equilibrium. Again, consider the reaction

$$2 \text{ SO}_2 (g) + \text{O}_2 (g) \rightleftharpoons 2 \text{ SO}_3 (g)$$

The reaction quotient is:

$$Q_c = \frac{[SO_3]_0^2}{[SO_2]_0^2[O_2]_0}$$

You should notice that Q_c has the same algebraic form as K_c. However, the concentrations are not necessarily equilibrium concentrations. We will call them initial concentrations, represented by a subscript 0 after the square brackets, [ ]$_0$. Substituting initial concentrations into the reaction quotient, gives a value for Q_c. In order to predict whether the system is at equilibrium, the magnitude of Q_c must be compared with that of K_c.

If $Q_c = K_c$,   The initial concentrations are equilibrium concentrations. The system is at equilibrium.

If $Q_c > K_c$,   The ratio of initial concentrations of products to reactants is too large. To reach equilibrium, products must be converted to reactants. The system proceeds from right to left (consuming products, forming reactants) to reach equilibrium.

If $Q_c < K_c$,   The ratio of initial concentrations of products to reactants is too small. To reach equilibrium, reactants must be converted to products. The system proceeds from left to right (consuming reactants, forming products) to reach equilibrium.

EXAMPLE 14.6

At a certain temperature, the reaction $CO\ (g)\ +\ Cl_2\ (g)\ \rightleftharpoons\ COCl_2\ (g)$, has an equilibrium constant

$K_c = 13.8$. Is the following mixture an equilibrium mixture? If not, in which direction (right or left) will the reaction proceed to reach equilibrium?

$$[CO]_0 = 2.5\ M, [Cl_2]_0 = 1.2\ M, \text{ and } [COCl_2]_0 = 5.0\ M$$

Step 1: Recall that for a system to be at equilibrium, $Q_c = K_c$. Substitute the given concentrations into the equation for the reaction quotient to calculate Q_c.

$$Q_c = \frac{[COCl_2]_0}{[CO]_0[Cl_2]_0} = \frac{5.0}{(2.5)(1.2)} = 1.7$$

Step 2: Compare Q_c to K_c. Since $Q_c < K_c$, the ratio of initial concentrations of products to reactants is too small. To reach equilibrium, reactants must be converted to products. The system proceeds from **left to right** (consuming reactants, forming products) to reach equilibrium.

PRACTICE EXERCISE

5.   Given the reaction,

$$N_2\ (g)\ +\ O_2\ (g)\ \rightleftharpoons\ 2\ NO\ (g) \qquad\qquad K_c = 2.5 \times 10^{-3} \text{ at } 2130°C$$

decide whether the following mixture is at equilibrium or if a net forward or reverse reaction will occur. [NO] = 0.0050 M, [O_2] = 0.25 M, and [N_2] = 0.020 M

Text Problem: 14.38

PROBLEM TYPE 5: CALCULATING EQUILIBRIUM CONCENTRATIONS

The expected concentrations at equilibrium can be calculated from a knowledge of the initial concentrations and the equilibrium constant. In these types of problems, it will be very helpful to recall that

equilibrium concentration = initial concentration $\pm$ the change due to reaction.

The next *two* examples illustrate this important type of calculation.

EXAMPLE 14.7

A 0.25 mol sample of N_2O_4 dissociates and comes to equilibrium in a 1.5 L flask at 100°C. The reaction is

$$N_2O_4 \, (g) \; \rightleftharpoons \; 2 \, NO_2 \, (g) \qquad\qquad K_c = 0.36 \text{ at } 100°C$$

What are the equilibrium concentrations of NO_2 and N_2O_4?

Step 1: Express the equilibrium concentrations of all species in terms of initial concentrations and a single unknown x, that represents the change in concentration. Let $(-x)$ be the depletion in concentration (mol/L) of N_2O_4.

From the stoichiometry of the reaction, it follows that the increase in concentration for NO_2 must be $2x$. Complete a table that lists the initial concentrations, the change in concentrations, and the equilibrium concentrations.

	$N_2O_4 \, (g)$	$\rightleftharpoons$	$2 \, NO_2 \, (g)$
Initial (*M*):	0.25 mol/1.5 L = 0.17		0
Change (*M*):	$-x$		$+2x$
Equilibrium (*M*):	$0.17 - x$		$2x$

Step 2: Write the equilibrium constant expression in terms of the equilibrium concentrations. Knowing the value of the equilibrium constant, solve for x.

$$K_c = \frac{[NO_2]^2}{[N_2O_4]}$$

$$0.36 = \frac{(2x)^2}{0.17 - x}$$

$$0.061 - 0.36x = 4x^2$$

$$4x^2 + 0.36\,x - 0.061 = 0$$

The above equation is a quadratic equation of the form $ax^2 + bx + c = 0$. The solution for a quadratic equation is :

$$x = \frac{-b \pm \sqrt{b^2 - 4ac}}{2a}$$

Here, we have a = 4, b = 0.36, and c = –0.061. Substituting into the above equation,

$$x = \frac{-0.36 \pm \sqrt{(0.36)^2 - 4(4)(-0.061)}}{2(4)}$$

$$x = \frac{-0.36 \pm 1.05}{8}$$

$$x = 0.086 \, M \quad \text{or} \quad x = -0.18 \, M$$

The second solution is physically impossible because you cannot have a negative concentration. The first solution is the correct answer.

> **Tip:** In solving a quadratic equation of this type, one answer is always physically impossible, so the choice of which value to use for x is easy to make.

Step 3: Having solved for *x*, calculate the equilibrium concentrations of all species.

$$[N_2O_4] = (0.17 - 0.086)M = \textbf{0.08 } \textbf{\textit{M}}$$

$$[NO_2] = 2(0.086 \text{ } M) = \textbf{0.17 } \textbf{\textit{M}}$$

EXAMPLE 14.8

A 1.00 L vessel initially contains 0.777 mol of SO₃ (*g*) at 1100 K. What is the value of *K*c for the following reaction if 0.520 mol of SO₃ remain at equilibrium?

$$\textbf{2 SO}_3 \textbf{ (}\textbf{\textit{g}}\textbf{)} \rightleftharpoons \textbf{2 SO}_2 \textbf{ (}\textbf{\textit{g}}\textbf{)} + \textbf{O}_2 \textbf{ (}\textbf{\textit{g}}\textbf{)}$$

If we can calculate the equilibrium concentrations for all species, we can calculate the equilibrium constant K_c.

Step 1: In this problem, we are given both the initial and equilibrium concentrations of SO₃. Recalling that

equilibrium concentration = initial concentration ± the change due to reaction

we can calculate the change in concentration of SO₃ due to reaction. Let's call this change, 2*x*, because of the coefficient of 2 for SO₃ in the balanced equation.

change in concentration of SO₃ = 2*x* = 0.777 *M* – 0.520 *M* = 0.257 *M*

x = 0.129 *M*

Complete a table that lists the initial concentrations, the change in concentrations, and the equilibrium concentrations.

	2 SO₃ (*g*) ⇌	2 SO₂ (*g*) +	O₂ (*g*)
Initial (*M*):	0.777	0	0
Change (*M*):	−2*x* = −0.257	+2*x*	+*x*
Equilibrium (*M*):	0.520	2*x*	*x*

So, at equilibrium,

$[SO_2] = 2x = 0.257 \text{ } M$

$[O_2] = x = 0.129 \text{ } M$

Tip: You probably could have come up with the equilibrium concentrations of SO₂ and O₂ without the use of a table. However, a table is a simple way to keep all the data organized.

Step 2: Substitute the equilibrium concentrations into the equilibrium constant expression to solve for K_c.

$$K_c = \frac{[SO_2]^2[O_2]}{[SO_3]^2}$$

$$K_c = \frac{(0.257)^2(0.129)}{(0.520)^2} = \textbf{0.0315}$$

PRACTICE EXERCISE

6.   For the reaction,

$$N_2\ (g)\ +\ O_2\ (g)\ \rightleftharpoons\ 2\ NO\ (g) \qquad\qquad K_p = 3.80 \times 10^{-4}\ \text{at } 2000°C$$

what equilibrium pressures of N_2, O_2, and NO will result when a 10.0 L reactor vessel is filled with 2.00 atm of N_2 and 0.400 atm of O_2 and the reaction is allowed to come to equilibrium?

7.   Initially a 1.0 L vessel contains 10.0 mol of NO and 6.0 mol of O_2 at a certain temperature.

$$2\ NO\ (g)\ +\ O_2\ (g)\ \rightleftharpoons\ 2\ NO_2\ (g)$$

At equilibrium, the vessel contains 8.8 mol of NO_2. Determine the value of K_c at this temperature.

Text Problems: **14.40**, **14.42**, 14.44, 14.46

PROBLEM TYPE 6: FACTORS THAT AFFECT CHEMICAL EQUILIBRIUM

What effect does a change in concentration, pressure, volume, or temperature have on a system at equilibrium? This question can be answered qualitatively by using **Le Chatelier's principle**. It states that when an external stress is applied to a system at equilibrium, the system adjusts in such a way that the stress is partially offset. The word "stress" here means a change in concentration, pressure, volume, or temperature that removes a system from the equilibrium state.

A. Changes in concentration

EXAMPLE 14.9
For the following reaction at equilibrium in a closed container,

$$2\ NaHCO_3\ (s)\ \rightleftharpoons\ Na_2CO_3\ (s)\ +\ H_2O\ (g)\ +\ CO_2\ (g)$$

state the effects (increase, decrease, or no change) of the following stresses on the number of moles of sodium carbonate, Na_2CO_3. Note that Na_2CO_3 is a solid (this is a heterogeneous equilibrium); its concentration will remain constant, but its amount can change.

(a)   Removing CO_2 (g).

(b)   Adding H_2O (g).

(c)   Adding $NaHCO_3$ (s).

Applying Le Chatelier's principle,

(a)   If CO_2 concentration is lowered, the system will react to offset the change. That is, a shift to the right will replace some of the removed CO_2. Moles of Na_2CO_3 will **increase**.

(b)   The system will respond to the stress of added H_2O by shifting to the left to remove some of the water. Moles of Na_2CO_3 will **decrease**.

(c)   The position of a heterogeneous equilibrium does not depend on the amounts of pure solids or liquids present. Remember that the concentrations of pure solids and liquids are constant and thus do not enter into the equilibrium constant expression. Hence, there is no shift in the equilibrium, and the amount of Na_2CO_3 is **unchanged**.

B. Changes in pressure and volume

The pressure of a system of gases in chemical equilibrium can be increased by decreasing the available volume. The ideal gas equation shows this inverse relationship between pressure and volume.

$$PV = nRT$$

A decrease in volume causes the concentrations of all *gas-phase* components to increase. Remember that pressure is directly proportional to the concentration of a gas.

$$P = \left(\frac{n}{V}\right)RT$$

The stress caused by the increased pressure will be partially offset by a net reaction that will lower the total pressure. In other words, the system will adjust by lowering the total concentration of gas molecules to reestablish equilibrium. Again, consider the following equilibrium:

$$2\,SO_2\,(g)\ +\ O_2\,(g)\ \rightleftharpoons\ 2\,SO_3\,(g)$$

When the molecules of the above gases are compressed into a smaller volume, the total pressure increases and hence the total concentration increases (this is a stress). A net forward reaction (shift to the right) will bring the system to a new state of equilibrium, in which the *total concentration* of all molecules will be lowered somewhat. This partially offsets the initial stress on the system. The total concentration is lowered somewhat because when 2 moles of SO_2 react with 1 mole of O_2 (a total of 3 moles), only 2 moles of SO_3 are produced.

In general, an increase in pressure (decrease in volume) favors the net reaction that decreases the total number of moles of gases. On the other hand, a decrease in pressure (increase in volume) favors the net reaction that increases the total number of moles of gases.

EXAMPLE 14.10
For the following reaction at equilibrium in a closed container,

$$2\,NaHCO_3\,(s)\ \rightleftharpoons\ Na_2CO_3\,(s)\ +\ H_2O\,(g)\ +\ CO_2\,(g)$$

state the effect (increase, decrease, or no change) of increasing the volume of the container on the number of moles of sodium carbonate, Na_2CO_3. Note that Na_2CO_3 is a solid (this is a heterogeneous equilibrium); its concentration will remain constant, but its amount can change.

Pressure and volume are inversely related. Thus, an increase in the volume of the container corresponds to a decrease in pressure. The system will shift to offset the decreased pressure by undergoing the reaction that increases the total number of moles of gases. Looking at the above reaction, there are zero moles of gas on the reactants' side, and two moles of gas on the products' side.

The reaction will shift to the right to establish equilibrium, and the amount of Na_2CO_3 will **increase**.

C. Changes in temperature

A change in concentration, pressure, or volume may alter the equilibrium position, but it does not change the value of the equilibrium constant. However, a change in temperature can alter the equilibrium constant.

To decide how a temperature stress will affect a system at equilibrium, you must look at whether the reaction is endothermic or exothermic. The following reaction is endothermic (positive ΔH).

$$PCl_5\,(g)\ \rightleftharpoons\ PCl_3\,(g)\ +\ Cl_2\,(g)\qquad\qquad \Delta H° = 92.9\ kJ$$

What happens if we heat the above system at equilibrium? We have added heat (a stress). The system will shift in the direction that will partially offset this stress by removing some heat. Since endothermic processes absorb heat from the surroundings (reducing the stress), heating favors dissociation of PCl_5 into PCl_3 and Cl_2 molecules (the forward reaction). Consequently, the equilibrium constant for the above reaction, given by

$$K_c = \frac{[PCl_3][Cl_2]}{[PCl_5]}$$

increases with temperature.

In general, a temperature *increase* favors an *endothermic* reaction, and a temperature *decrease* favors an *exothermic* reaction.

EXAMPLE 14.11
For the following reaction at equilibrium in a closed container,

$$2\,NaHCO_3\,(s) \rightleftharpoons Na_2CO_3\,(s) + H_2O\,(g) + CO_2\,(g) \qquad\qquad \Delta H = 128\ \text{kJ}$$

state the effect (increase, decrease, or no change) of decreasing the temperature on the number of moles of sodium carbonate, Na_2CO_3. Note that Na_2CO_3 is a solid (this is a heterogeneous equilibrium); its concentration will remain constant, but its amount can change.

Recall that a temperature decrease favors an exothermic reaction. The above equation, as written, is endothermic (positive ΔH). However, the reverse reaction is exothermic with $\Delta H = -128$ kJ.

$$Na_2CO_3 + H_2O + CO_2 \rightleftharpoons 2\,NaHCO_3 \qquad\qquad \Delta H = -128\ \text{kJ}$$

Thus, the reverse reaction is favored, and the amount of Na_2CO_3 will **decrease**.

PRACTICE EXERCISE
8.   For the decomposition of calcium carbonate

$$CaCO_3\,(s) \rightleftharpoons CaO\,(s) + CO_2\,(g) \qquad\qquad \Delta H° = 175\ \text{kJ}$$

how will the amount (not concentration) of $CaCO_3$ (s) change with the following stresses to the system at equilibrium?

(a) CO_2 (g) is removed.
(b) CaO (s) is added.
(c) The temperature is raised.
(d) The volume of the container is decreased.

Text Problems: 14.52, **14.54, 14.56**, 14.58, 14.60

D. The effect of a catalyst

A catalyst increases the rate at which a reaction occurs. For a reversible reaction, a catalyst affects the rate in the forward and reverse directions to the same extent. Therefore, the presence of a catalyst does not alter the equilibrium constant, nor does it shift the position of an equilibrium system.

ANSWERS TO PRACTICE EXERCISES

1. If concentration is expressed in mol/L, $K_c = \dfrac{[N_2O_4]^2}{[N_2O]^2[O_2]^3}$

    or, in terms of partial pressures, $K_p = \dfrac{P_{N_2O_4}^2}{P_{N_2O}^2 P_{O_2}^3}$

2. $K_p = 1.05$ 3. $K_p = K_c = 64$ 4. $K_p' = 3.7 \times 10^{-3}$

5. Since $Q_c > K_c$, the ratio of initial concentrations of products to reactants is too large. To reach equilibrium, products must be converted to reactants. The system proceeds from **right to left** (consuming products, forming reactants) to reach equilibrium.

6. $P_{NO_2} = 1.74 \times 10^{-2}$ atm

    $P_{N_2} = 1.99$ atm

    $P_{O_2} = 0.391$ atm

7. $K_c = 34$

8. **(a)** The amount of $CaCO_3$ will **decrease**.   **(b)** The amount of $CaCO_3$ will **not change**.
    (c) The amount of $CaCO_3$ will **decrease**.   **(d)** The amount of $CaCO_3$ will **increase**.

SOLUTIONS TO SELECTED TEXT PROBLEMS

14.14   The problem states that the system is at equilibrium, so we simply substitute the equilibrium concentrations into the equilibrium constant expression to calculate K_c.

Step 1: Calculate the concentrations of the components in units of mol/L. The molarities can be calculated by simply dividing the number of moles by the volume of the flask.

$$[H_2] = \frac{2.50 \text{ mol}}{12.0 \text{ L}} = 0.208 \text{ } M$$

$$[S_2] = \frac{1.35 \times 10^{-5} \text{ mol}}{12.0 \text{ L}} = 1.13 \times 10^{-6} \text{ } M$$

$$[H_2S] = \frac{8.70 \text{ mol}}{12.0 \text{ L}} = 0.725 \text{ } M$$

Step 2: Once the molarities are known, K_c can be found by substituting the molarities into the equilibrium constant expression.

$$K_c = \frac{[H_2S]^2}{[H_2]^2[S_2]} = \frac{(0.725)^2}{(0.208)^2(1.13 \times 10^{-6})} = \mathbf{1.08 \times 10^7}$$

If you forget to convert moles to moles/liter, will you get a different answer? Under what circumstances will the two answers be the same?

14.16   Converting between K_c and K_p, Problem Type 1C.

Step 1: Rearrange the equation relating K_p and K_c, solving for K_c.

$$K_p = K_c(0.0821 \text{ } T)^{\Delta n}$$

$$K_c = \frac{K_p}{(0.0821 T)^{\Delta n}}$$

Step 2: Substitute the given values into the above equation to solve for K_c. Temperature must be in units of Kelvin.

$$K_c = \frac{K_p}{(0.0821 \text{ } T)^{\Delta n}} = \frac{1.8 \times 10^{-5}}{[(0.0821)(623 \text{ K})]^{(3-2)}} = \mathbf{3.5 \times 10^{-7}}$$

14.18   The equilibrium constant expressions are:

(a) $K_c = \dfrac{[NH_3]^2}{[N_2][H_2]^3}$

(b) $K_c = \dfrac{[NH_3]}{[N_2]^{1/2}[H_2]^{3/2}}$

Substituting the given equilibrium concentration gives:

(a) $K_c = \dfrac{(0.25)^2}{(0.11)(1.91)^3} = 0.082$

(b) $K_c = \dfrac{(0.25)}{(0.11)^{1/2}(1.91)^{3/2}} = 0.29$

Is there a relationship between the K_c values from parts (a) and (b)?

14.20 Because pure solids do not enter into an equilibrium constant expression, we can calculate K_p directly from the pressure that is due solely to CO_2 (g).

$$K_p = P_{CO_2} = 0.105$$

Now, we can convert K_p to K_c using the following equation.

$$K_p = K_c(0.0821\ T)^{\Delta n}$$

$$K_c = \dfrac{K_p}{(0.0821T)^{\Delta n}}$$

$$K_c = \dfrac{0.105}{(0.0821 \times 623)^{(1-0)}} = 2.05 \times 10^{-3}$$

14.22 Heterogeneous Equilibria, Calculating K_p. Problem Type 2.

Step 1: If a solid or a liquid is a reactant or product, we can treat its concentration as a constant and omit it from the equilibrium constant expression. Therefore, the equilibrium constant expression for the reaction is

$$K_p = P_{NH_3}^2 P_{CO_2}$$

Step 2: The total pressure in the flask (0.363 atm) is a sum of the partial pressures of NH_3 and CO_2.

$$P_T = P_{NH_3} + P_{CO_2} = 0.363\ \text{atm}$$

Let the partial pressure of $CO_2 = x$. From the stoichiometry of the balanced equation, you should find that $P_{NH_3} = 2 P_{CO_2}$ Therefore, the partial pressure of $NH_3 = 2x$. Substituting into the equation for total pressure gives:

$$P_T = P_{NH_3} + P_{CO_2} = 2x + x = 3x$$

$$3x = 0.363\ \text{atm}$$

$$x = P_{CO_2} = 0.121\ \text{atm}$$

$$P_{NH_3} = 2x = 0.242\ \text{atm}$$

Step 3: Substitute the equilibrium pressures into the equilibrium constant expression to solve for K_p.

$$K_p = P_{NH_3}^2 P_{CO_2} = (0.242)^2(0.121) = 7.09 \times 10^{-3}$$

14.24   If the CO pressure at equilibrium is 0.497 atm, the balanced equation requires the chlorine pressure to have the same value. The initial pressure of phosgene gas can be found from the ideal gas equation.

$$P = \frac{nRT}{V} = \frac{(3.00 \times 10^{-2}\ \text{mol})(0.0821\ \text{L·atm / mol·K})(800\ \text{K})}{(1.50\ \text{L})} = 1.31\ \text{atm}$$

Since there is a 1:1 mole ratio between phosgene and CO, the partial pressure of CO formed equals the partial pressure of phosgene reacted. The phosgene pressure at equilibrium is:

$$P_{COCl_2} = (1.31 - 0.497)\ \text{atm} = 0.81\ \text{atm}$$

The value of K_p is then found by substitution.

$$K_p = \frac{P_{COCl_2}}{P_{CO}P_{Cl_2}} = \frac{0.81}{(0.497)^2} = \mathbf{3.3}$$

14.26   In this problem, you are asked to calculate K_c.

Step 1: Calculate the initial concentration of NOCl

$$[\text{NOCl}]_0 = \frac{2.50\ \text{mol}}{1.50\ \text{L}} = 1.67\ M$$

Step 2: If 28.0 percent of the NOCl has dissociated at equilibrium, 72.0 percent remains at equilibrium. The concentration of NOCl is 72.0 percent of its initial value.

$$[\text{NOCl}] = (0.720)(1.67\ M) = 1.20\ M$$

Since there is a 1:1 mole ratio between NOCl and NO, the concentration of NO at equilibrium is equal to the 28 percent of NOCl that dissociated.

$$[\text{NO}] = (0.280)(1.67\ M) = 0.468\ M$$

The mole ratio between NOCl and Cl_2 is 2:1. This means that the concentration of Cl_2 is equal to half of the 28 percent NOCl that dissociated.

$$[\text{Cl}_2] = \frac{1}{2}(0.280)(1.67\ M) = 0.234\ M$$

Step 3: The equilibrium constant K_c can be calculated by substituting the above concentrations into the equilibrium constant expression.

$$K_c = \frac{[\text{NO}]^2[\text{Cl}_2]}{[\text{NOCl}]^2} = \frac{(0.468)^2(0.234)}{(1.20)^2} = \mathbf{0.0356}$$

14.28 $K = K'K''$

$K = (6.5 \times 10^{-2})(6.1 \times 10^{-5})$

$K = \mathbf{4.0 \times 10^{-6}}$

14.30 For the equilibrium: $2SO_2(g) \rightleftharpoons 2S(s) + 2O_2(g)$

$$\left(\frac{1}{K_c{}'}\right)^2 = \left(\frac{1}{4.2 \times 10^{52}}\right)^2$$

For the equilibrium: $2SO_2(g) + O_2(g) \rightleftharpoons 2SO_3(g)$

$$K_c = K_c{}''\left(\frac{1}{K_c{}'}\right)^2 = \left(9.8 \times 10^{128}\right)\left(\frac{1}{4.2 \times 10^{52}}\right)^2 = \mathbf{5.6 \times 10^{23}}$$

14.34 At equilibrium, the value of K_c is equal to the ratio of the forward rate constant to the rate constant for the reverse reaction.

$$K_c = \frac{k_f}{k_r} = \frac{k_f}{5.1 \times 10^{-2}} = 12.6$$

$$k_f = (12.6)(5.1 \times 10^{-2}) = 0.64$$

The forward reaction is third order, so the units of k_f must be:

$$\text{rate} = k_f(A)^2(B)$$

$$k_f = \frac{\text{rate}}{(\text{concentration})^3} = \frac{M/s}{M^3} = 1/M^2 \cdot s$$

$$\mathbf{k_f = 0.64/M^2 \cdot s}$$

14.38 Using the Reaction Quotient (Q_c) to Predict the Direction of a Reaction, Problem Type 4.

Step 1: Recall that for a system to be at equilibrium, $Q_c = K_c$. Substitute the given concentrations into the equation for the reaction quotient to calculate Q_c.

$$Q_c = \frac{[NH_3]^2}{[N_2][H_2]^3} = \frac{[0.48]^2}{[0.60][0.76]^3} = 0.87$$

Step 2: Compare Q_c to K_c. Since $Q_c < K_c$ (0.87 < 1.2), the ratio of initial concentrations of products to reactants is too small. To reach equilibrium, reactants must be converted to products. The system proceeds from left to right (consuming reactants, forming products) to reach equilibrium.

Therefore, **[NH$_3$] will increase** and **[N$_2$] and [H$_2$] will decrease** at equilibrium.

14.40 Calculating Equilibrium Concentrations, Problem Type 5.

Step 1: Since the reaction started with only pure NO$_2$, the equilibrium concentration of NO must be twice the equilibrium concentration of O$_2$, due to the 2:1 mole ratio of the balanced equation. Therefore, the equilibrium partial pressure of **NO is 0.50 atm**.

Step 2: We can find the equilibrium NO_2 pressure by rearranging the equilibrium constant expression, then substituting in the known values.

$$K_p = \frac{P_{NO}^2 P_{O_2}}{P_{NO_2}^2}$$

$$P_{NO_2} = \sqrt{\frac{(P_{NO})^2 P_{O_2}}{K_p}} = \sqrt{\frac{(0.50)^2(0.25)}{158}} = \mathbf{0.020\ atm}$$

14.42   Calculating Equilibrium Concentrations, Problem Type 5.

Step 1: Complete a table that lists the initial concentrations, the change in concentrations, and the equilibrium concentrations. The initial concentration of I_2 is 0.0456 mol/2.30 L = 0.0198 *M*. Let *x* be the amount (in mol/L) of I_2 dissociated.

	$I_2\ (g)$	$\rightleftharpoons$	$2\ I\ (g)$
Initial (*M*):	0.0198		0.000
Change (*M*):	−*x*		+2*x*
Equilibrium (*M*):	(0.0198 − *x*)		2*x*

Step 2: Write the equilibrium constant expression in terms of the equilibrium concentrations. Knowing the value of the equilibrium constant, solve for *x*.

$$K_c = \frac{[I]^2}{[I_2]} = \frac{(2x)^2}{(0.0198 - x)} = 3.80 \times 10^{-5}$$

$$4x^2 + (3.80 \times 10^{-5})x - (7.52 \times 10^{-7}) = 0$$

The above equation is a quadratic equation of the form $ax^2 + bx + c = 0$. The solution for a quadratic equation is

$$x = \frac{-b \pm \sqrt{b^2 - 4ac}}{2a}$$

Here, we have a = 4, b = 3.80×10^{-5}, and c = -7.52×10^{-7}. Substituting into the above equation,

$$x = \frac{-3.80 \times 10^{-5} \pm \sqrt{(3.80 \times 10^{-5})^2 - 4(4)(-7.52 \times 10^{-7})}}{2(4)}$$

$$x = \frac{-3.80 \times 10^{-5} \pm 3.47 \times 10^{-3}}{8}$$

$$x = 4.29 \times 10^{-4}\ M \quad \text{or} \quad x = -4.39 \times 10^{-4}\ M$$

The second solution is physically impossible because you cannot have a negative concentration. The first solution is the correct answer.

Step 3: Having solved for *x*, calculate the equilibrium concentrations of all species.

$$[I] = 2x = (2)(4.29 \times 10^{-4}\ M) = \mathbf{8.58 \times 10^{-4}\ M}$$

$$[I_2] = (0.0198 - x) = (0.0198 - 4.29 \times 10^{-4})\ M = \mathbf{0.0194\ M}$$

> **Tip:** We could have simplified this problem by assuming that x was small compared to 0.0198. We could then assume that $0.0198 - x \approx 0.0198$. By making this assumption, we could have avoided solving a quadratic equation.

14.44 **(a)** The equilibrium constant, K_c, can be found by simple substitution.

$$K_c = \frac{[H_2O][CO]}{[CO_2][H_2]} = \frac{(0.040)(0.050)}{(0.086)(0.045)} = \mathbf{0.52}$$

(b) The magnitude of the reaction quotient Q_c for the system after the concentration of CO_2 becomes 0.50 mol/L, but before equilibrium is reestablished, is:

$$Q_c = \frac{(0.040)(0.050)}{(0.50)(0.045)} = 0.089$$

The value of Q_c is smaller than K_c; therefore, the system will shift to the right, increasing the concentrations of CO and H_2O and decreasing the concentrations of CO_2 and H_2. Let x be the depletion in the concentration of CO_2 at equilibrium. The stoichiometry of the balanced equation then requires that the decrease in the concentration of H_2 must also be x, and that the concentration increases of CO and H_2O be equal to x as well. The changes in the original concentrations are shown in the table.

	CO_2	+	H_2	$\rightleftharpoons$	CO	+	H_2O
Initial (*M*):	0.50		0.045		0.050		0.040
Change (*M*):	$-x$		$-x$		$+x$		$+x$
Equilibrium (*M*):	$(0.50 - x)$		$(0.045 - x)$		$(0.050 + x)$		$(0.040 + x)$

The equilibrium constant expression is:

$$K_c = \frac{[H_2O][CO]}{[CO_2][H_2]} = \frac{(0.040 + x)(0.050 + x)}{(0.50 - x)(0.045 - x)} = 0.52$$

$$0.52(x^2 - 0.545x + 0.0225) = x^2 + 0.090x + 0.0020$$

$$0.48x^2 + 0.373x - (9.7 \times 10^{-3}) = 0$$

The positive root of the equation is $x = 0.025$.

The equilibrium concentrations are:

$[CO_2]$ = $(0.50 - 0.025)\,M$ = **0.48 *M***
$[H_2]$ = $(0.045 - 0.025)\,M$ = **0.020 *M***
$[CO]$ = $(0.050 + 0.025)\,M$ = **0.075 *M***
$[H_2O]$ = $(0.040 + 0.025)\,M$ = **0.065 *M***

14.46 The initial concentrations are $[H_2]$ = 0.80 mol/5.0 L = 0.16 *M* and $[CO_2]$ = 0.80 mol/5.0 L = 0.16 *M*.

	H_2 (g)	+	CO_2 (g)	$\rightleftharpoons$	H_2O (g)	+	CO_2 (g)
Initial (*M*):	0.16		0.16		0.00		0.00
Change (*M*):	$-x$		$-x$		$+x$		$+x$
Equilibrium (*M*):	$(0.16 - x)$		$(0.16 - x)$		x		x

$$K_c = \frac{[H_2O][CO]}{[H_2][CO_2]} = 4.2 = \frac{x^2}{(0.16-x)^2}$$

Taking the square root of both sides, we obtain:

$$\frac{x}{0.16-x} = 2.0$$

$$x = 0.11 \ M$$

The equilibrium concentrations are:

$$[H_2] = [CO_2] = (0.16 - 0.11) \ M = \mathbf{0.05 \ M}$$

$$[H_2O] = [CO] = \mathbf{0.11 \ M}$$

14.52 (a) Removal of CO_2 (g) from the system would shift the position of equilibrium to the **right**.

(b) Addition of more solid Na_2CO_3 would have **no effect**. [Na_2CO_3] does not appear in the equilibrium constant expression.

(c) Removal of some of the solid $NaHCO_3$ would have **no effect**. Same reason as (b).

14.54 Factors that Affect Chemical Equilibrium, Problem Type 6.

(a) Changes in pressure ordinarily do not affect the concentrations of reacting species in condensed phases because liquids and solids are virtually incompressible. Pressure change should have **no effect** on this system.

(b) Same situation as (a).

(c) Only the product is in the gas phase. An increase in pressure should favor the reaction (forward reaction or reverse reaction) that decreases the total number of moles of gas. The equilibrium should shift to the **left**, that is, the amount of B should decrease and that of A should increase.

(d) In this equation there are equal moles of gaseous reactants and products. A shift in either direction will have no effect on the total number of moles of gas present. There will be **no change** when the pressure is increased.

(e) A shift in the direction of the reverse reaction (**left**) will have the result of decreasing the total number of moles of gas present.

14.56 Factors that Affect Chemical Equilibrium, Problem Type 6.

(a) A temperature *increase* favors an *endothermic* reaction. The reaction is endothermic in the forward direction, so the equilibrium will shift to the **right**.

(b) The system will shift in the direction to remove some of the added Cl_2. The equilibrium will shift to the **left**.

(c) The system will shift to replace some of the PCl_3 that was removed. The equilibrium will shift to the **right**.

(d) An increase in pressure should favor the reaction that decreases the total number of moles of gas. The equilibrium will shift to the **left**.

(e) Adding a catalyst will not affect the position of the equilibrium.

14.58   There will be no change in the pressures. A catalyst has no effect on the position of the equilibrium.

14.60   For this system, $K_p = [CO_2]$.

This means that to remain at equilibrium, the pressure of carbon dioxide must stay at a fixed value as long as the temperature remains the same.

(a)   If the volume is increased, the pressure of CO_2 will drop (Boyle's law, pressure and volume are inversely proportional). Some $CaCO_3$ will break down to form more CO_2 and CaO. (**Shift right**)

(b)   Assuming that the amount of added solid CaO is not so large that the volume of the system is altered significantly, there should be no change at all. If a huge amount of CaO were added, this would have the effect of reducing the volume of the container. What would happen then?

(c)   Assuming that the amount of $CaCO_3$ removed doesn't alter the container volume significantly, there should be no change. Removing a huge amount of $CaCO_3$ will have the effect of increasing the container volume. The result in that case will be the same as in part (a).

(d)   The pressure of CO_2 will be greater and will exceed the value of K_p. Some CO_2 will combine with CaO to form more $CaCO_3$. (**Shift left**)

(e)   Carbon dioxide combines with aqueous NaOH according to the equation

$$CO_2\ (g)\ +\ NaOH\ (aq)\ \rightarrow\ NaHCO_3\ (aq)$$

This will have the effect of reducing the CO_2 pressure and causing more $CaCO_3$ to break down to CO_2 and CaO. (**Shift right**)

(f)   Carbon dioxide does not react with hydrochloric acid, but $CaCO_3$ does.

$$CaCO_3\ (s)\ +\ 2\ HCl\ (aq)\ \rightarrow\ CaCl_2\ (aq)\ +\ CO_2\ (g)\ +\ H_2O\ (l)$$

The CO_2 produced by the action of the acid will combine with CaO as discussed in (d) above. (**Shift left**)

(g)   This is a decomposition reaction. Decomposition reactions are endothermic. Increasing the temperature will favor this reaction and produce more CO_2 and CaO. (**Shift right**)

14.62   **(a)**   Since the total pressure is 1.00 atm, the sum of the partial pressures of NO and Cl_2 is

1.00 atm − partial pressure of NOCl = 1.00 atm − 0.64 atm = 0.36 atm

The stoichiometry of the reaction requires that the partial pressure of NO be twice that of Cl_2. Hence, the partial pressure of NO is **0.24 atm** and the partial pressure of Cl_2 is **0.12 atm**.

(b)   The equilibrium constant K_p is found by substituting the partial pressures calculated in part (a) into the equilibrium constant expression.

$$K_p\ =\ \frac{P_{NO}^2 P_{Cl_2}}{P_{NOCl}^2}\ =\ \frac{(0.24)^2(0.12)}{(0.64)^2}\ =\ \mathbf{0.017}$$

14.64   The equilibrium expression for this system is given by:

$$K_p = P_{CO_2} P_{H_2O}$$

(a) In a closed vessel the decomposition will stop when the product of the partial pressures of CO_2 and H_2O equals K_p. Adding more sodium bicarbonate will have no effect.

(b) In an open vessel, CO_2 (g) and H_2O (g) will escape from the vessel, and the partial pressures of CO_2 and H_2O will never become large enough for their product to equal K_p. Therefore, equilibrium will never be established. Adding more sodium bicarbonate will result in the production of more CO_2 and H_2O.

14.66 (a) The equation that relates K_p and K_c is:

$$K_p = K_c(0.0821\ T)^{\Delta n}$$

For this reaction, $\Delta n = 3 - 2 = 1$

$$K_c = \frac{K_p}{(0.0821\ T)} = \frac{2 \times 10^{-42}}{(0.0821 \times 298)} = 8 \times 10^{-44}$$

(b) Because of a very large activation energy, the reaction of hydrogen with oxygen is infinitely slow without a catalyst or an initiator. The action of a single spark on a mixture of these gases results in the explosive formation of water.

14.68 (a) Calculate the value of K_p by substituting the equilibrium partial pressures into the equilibrium constant expression.

$$K_p = \frac{P_B}{P_A^2} = \frac{(0.60)}{(0.60)^2} = 1.7$$

(b) The total pressure is the sum of the partial pressures for the two gaseous components, A and B. We can write:

$$P_A + P_B = 1.5\ \text{atm}$$

and

$$P_B = 1.5 - P_A$$

Substituting into the expression for K_p gives:

$$K_p = \frac{(1.5 - P_A)}{P_A^2} = 1.7$$

$$1.7\,P_A^2 + P_A - 1.5 = 0$$

Solving the quadratic equation, we obtain:

$$P_A = 0.69\ \text{atm}$$

and by difference,

$$P_B = 0.81\ \text{atm}$$

14.70 Total number of moles of gas is:

$$0.020 + 0.040 + 0.96 = 1.02\ \text{mol of gas}$$

You can calculate the partial pressure of each gaseous component from the mole fraction and the total pressure (See Dalton's Law of Partial Pressures, Problem Type 3, Chapter 5).

$$P_{NO} = X_{NO}P_T = \frac{0.040}{1.02} \times 0.20 = 0.0078 \text{ atm}$$

$$P_{O_2} = X_{O_2}P_T = \frac{0.020}{1.02} \times 0.20 = 0.0039 \text{ atm}$$

$$P_{NO_2} = X_{NO_2}P_T = \frac{0.96}{1.02} \times 0.20 = 0.19 \text{ atm}$$

Calculate K_p by substituting the partial pressures into the equilibrium constant expression.

$$K_p = \frac{P_{NO_2}^2}{P_{NO}^2 P_{O_2}} = \frac{(0.19)^2}{(0.0078)^2(0.0039)} = \mathbf{1.5 \times 10^5}$$

14.72   Set up a table that contains the initial concentrations, the change in concentrations, and the equilibrium concentration. Assume that the vessel has a volume of 1 L.

The reaction is:

	H_2	$+$	Cl_2	$\rightleftharpoons$	$2\ HCl$
Initial (M):	0.47		0		3.59
Change (M):	$+x$		$+x$		$-2x$
Equilibrium (M):	$(0.47 + x)$		x		$(3.59 - 2x)$

Substitute the equilibrium concentrations into the equilibrium constant expression, then solve for x. Since $\Delta n = 0$, $K_c = K_p$.

$$K_c = \frac{[HCl]^2}{[H_2][Cl_2]} = \frac{(3.59 - 2x)^2}{(0.47 + x)x} = 193$$

Solving the quadratic equation,

$$x = 0.10$$

Having solved for x, calculate the equilibrium concentrations of all species.

$$[H_2] = 0.57\ M \qquad [Cl_2] = 0.10\ M \qquad [HCl] = 3.39\ M$$

Since we assumed that the vessel had a volume of 1 L, the above molarities also correspond to the number of moles of each component.

From the mole fraction of each component and the total pressure, we can calculate the partial pressure of each component (See Dalton's Law of Partial Pressures, Problem Type 3, Chapter 5).

Total number of moles $= 0.57 + 0.10 + 3.39 = 4.06$ mol

$$P_{H_2} = \frac{0.57}{4.06} \times 2.00 = \mathbf{0.28\ atm}$$

$$P_{Cl_2} = \frac{0.10}{4.06} \times 2.00 = \mathbf{0.049\ atm}$$

$$P_{HCl} = \frac{3.39}{4.06} \times 2.00 = \mathbf{1.67\ atm}$$

14.74   This is a difficult problem. Express the equilibrium number of moles in terms of the initial moles and the change in number of moles (x). Next, calculate the mole fraction of each component. Using the mole fraction, you should come up with a relationship between partial pressure and total pressure for each component. Substitute the partial pressures into the equilibrium constant expression to solve for the total pressure, P_T.

The reaction is:

	N_2	$+$	$3 H_2$	$\rightleftharpoons$	$2 NH_3$
Initial (mol):	1		3		0
Change (mol):	$-x$		$-3x$		$2x$
Equilibrium (mol):	$(1-x)$		$(3-3x)$		$2x$

$$\text{Mole fraction of } NH_3 = \frac{\text{mol of } NH_3}{\text{total number of moles}}$$

$$X_{NH_3} = \frac{2x}{(1-x)+(3-3x)+2x} = \frac{2x}{4-2x}$$

$$0.21 = \frac{2x}{4-2x}$$

$$x = 0.35 \text{ mol}$$

Substituting x into the following mole fraction equations, the mole fractions of N_2 and H_2 can be calculated.

$$X_{N_2} = \frac{1-x}{4-2x} = \frac{1-0.35}{4-2(0.35)} = 0.20$$

$$X_{H_2} = \frac{3-3x}{4-2x} = \frac{3-3(0.35)}{4-2(0.35)} = 0.59$$

The partial pressures of each component are equal to the mole fraction multiplied by the total pressure.

$$P_{NH_3} = 0.21P_T \qquad P_{N_2} = 0.20P_T \qquad P_{H_2} = 0.59P_T$$

Substitute the partial pressures above (in terms of P_T) into the equilibrium constant expression, and solve for P_T.

$$K_p = \frac{P_{NH_3}^2}{P_{H_2}^3 P_{N_2}}$$

$$4.31 \times 10^{-4} = \frac{(0.21)^2 P_T^2}{(0.59P_T)^3(0.20P_T)}$$

$$4.31 \times 10^{-4} = \frac{1.07}{P_T^2}$$

$$P_T = 5.0 \times 10^1 \text{ atm}$$

14.76   The initial number of moles of SO_2Cl_2 is:

$$\text{mol } SO_2Cl_2 = 6.75 \text{ g } SO_2Cl_2 \times \frac{1 \text{ mol } SO_2Cl_2}{135.0 \text{ g } SO_2Cl_2} = 0.0500 \text{ mol}$$

At equilibrium, there is 0.0345 mol of SO_2 (and therefore also 0.0345 mol of Cl_2). The amount of SO_2Cl_2 left = (0.0500 mol − 0.0345 mol) = (0.0155 mol). Thus, the equilibrium concentrations are:

$$[SO_2Cl_2] = 0.0155 \text{ mol}/2.00 \text{ L} = 0.00775 \, M$$

$$[SO_2] = [Cl_2] = 0.0345 \text{ mol}/2.00 \text{ L} = 0.0173 \, M$$

Substitute the equilibrium concentrations into the equilibrium constant expression to calculate K_c.

$$K_c = \frac{[SO_2][Cl_2]}{[SO_2Cl_2]} = \frac{(0.0173)(0.0173)}{(0.00775)} = \mathbf{3.86 \times 10^{-2}}$$

14.78 $I_2 (g) \rightleftharpoons 2\,I\,(g)$

Assuming 1 mole of I_2 is present originally and α moles reacts, at equilibrium: $[I_2] = 1 - \alpha$, $[I] = 2\alpha$. The total number of moles present in the system = $(1 - \alpha) + 2\alpha = 1 + \alpha$. From Problem 14.73(a) in the text, we know that K_p is equal to:

$$K_p = \frac{4\alpha^2}{1-\alpha^2} P \qquad (1)$$

If there were no dissociation, then the pressure would be:

$$P = \frac{nRT}{V} = \frac{\left(1.00 \text{ g} \times \frac{1 \text{ mol}}{253.8 \text{ g}}\right)\left(0.0821 \frac{\text{L}\cdot\text{atm}}{\text{mol}\cdot\text{K}}\right)(1473 \text{ K})}{0.500 \text{ L}} = 0.953 \text{ atm}$$

$$\frac{\text{observed pressure}}{\text{calculated pressure}} = \frac{1.51 \text{ atm}}{0.953 \text{ atm}} = \frac{1+\alpha}{1}$$

$$\alpha = 0.584$$

Substituting in equation (1) above:

$$K_p = \frac{4\alpha^2}{1-\alpha^2} P = \frac{4 \times (0.584)^2}{1-(0.584)^2} \times 1.51 = \mathbf{3.13}$$

14.80   According to the ideal gas law, pressure is directly proportional to the concentration of a gas in mol/L if the reaction is at constant volume and temperature. Therefore, pressure may be used as a concentration unit. The reaction is:

	N_2	+	$3\,H_2$	$\rightleftharpoons$	$2\,NH_3$
Initial (atm):	0.862		0.373		0
Change (atm):	$-x$		$-3x$		$+2x$
Equilibrium (atm):	$(0.862 - x)$		$(0.373 - 3x)$		$2x$

$$K_p = \frac{P_{NH_3}^2}{P_{NH_2}^3 P_{N_2}}$$

$$4.31 \times 10^{-4} = \frac{(2x)^2}{(0.373 - 3x)^3 (0.862 - x)}$$

At this point, we need to make two assumptions that $3x$ is very small compared to 0.373 and that x is very small compared to 0.862. Hence,

$$0.373 - 3x \approx 0.373$$

and

$$0.862 - x \approx 0.862$$

$$4.31 \times 10^{-4} \approx \frac{(2x)^2}{(0.373)^3(0.862)}$$

Solving for x.

$$x = 2.20 \times 10^{-3} \text{ atm}$$

The equilibrium pressures are:

$$P_{N_2} = (0.862 - 2.20 \times 10^{-3}) \text{ atm } = \textbf{0.860 atm}$$

$$P_{H_2} = (0.373 - 3 \times 2.20 \times 10^{-3}) \text{ atm } = \textbf{0.366 atm}$$

$$P_{NH_3} = 2(2.20 \times 10^{-3}) \text{ atm } = \textbf{4.40} \times \textbf{10}^{-3} \textbf{ atm}$$

Was the assumption valid that we made above? Typically, the assumption is considered valid if x is less than 5 percent of the number that we said it was very small compared to. Is this the case?

14.82   **(a)**   The equation is:

	fructose	$\rightleftharpoons$	glucose
Initial (M):	0.244		0
Change (M):	−0.131		(0.244 − 0.113)
Equilibrium (M):	0.113		0.131

Calculating the equilibrium constant,

$$K_c = \frac{[\text{glucose}]}{[\text{fructose}]} = \frac{0.131}{0.113} = \textbf{1.16}$$

   (b)   **The percent converted** $= \dfrac{\text{amount of fructose converted}}{\text{original amount of fructose}} \times 100\%$

$$= \frac{0.244 - 0.113}{0.244} \times 100\% = \textbf{53.7\%}$$

14.84   **(a)**   There is only one gas phase component, O_2. The equilibrium constant is simply

$$K_p = P_{O_2} = \textbf{0.49 atm}$$

   (b)   From the ideal gas equation, we can calculate the moles of O_2 produced by the decomposition of CuO.

$$n_{O_2} = \frac{PV}{RT} = \frac{0.49 \text{ atm} \times 2.0 \text{ L}}{0.0821 \text{ L} \cdot \text{atm} / \text{K} \cdot \text{mol} \times 1297 \text{ K}} = 9.2 \times 10^{-3} \text{ mol } O_2$$

From the balanced equation,

$$9.2 \times 10^{-3} \text{ mol } O_2 \times \frac{4 \text{ mol CuO}}{1 \text{ mol } O_2} = 3.7 \times 10^{-2} \text{ mol CuO decomposed}$$

$$\textbf{Fraction of CuO decomposed} = \frac{\text{amount of CuO lost}}{\text{original amount of CuO}}$$

$$= \frac{3.7 \times 10^{-2} \text{ mol}}{0.16 \text{ mol}} = \textbf{0.23}$$

(c) If a 1.0 mol sample were used, the pressure of oxygen would still be the same (0.49 atm) and it would be due to the same quantity of O_2. Remember, a pure solid does not affect the equilibrium position. The moles of CuO lost would still be 3.7×10^{-2} mol. Thus the fraction decomposed would be:

$$\frac{0.037}{1.0} = \textbf{0.037}$$

(d) If the number of moles of CuO were less than 3.7×10^{-2} mol, the equilibrium could not be established because the pressure of O_2 would be less than 0.49 atm. Therefore, the smallest number of moles of CuO needed to establish equilibrium must be slightly greater than 3.7×10^{-2} mol.

14.86 We first must find the initial concentrations of all the species in the system.

$$[H_2]_0 = \frac{0.714 \text{ mol}}{2.40 \text{ L}} = 0.298 \text{ } M$$

$$[I_2]_0 = \frac{0.984 \text{ mol}}{2.40 \text{ L}} = 0.410 \text{ } M$$

$$[HI]_0 = \frac{0.886 \text{ mol}}{2.40 \text{ L}} = 0.369 \text{ } M$$

Calculate the reaction quotient by substituting the initial concentrations into the appropriate equation.

$$Q_c = \frac{[HI]_0^2}{[H_2]_0[I_2]_0} = \frac{(0.369)^2}{(0.298)(0.410)} = 1.11$$

We find that Q_c is less than K_c. The equilibrium will shift to the right, decreasing the concentrations of H_2 and I_2 and increasing the concentration of HI.

We set up the usual table. Let x be the decrease in concentration of H_2 and I_2.

	H_2	$+$	I_2	$\rightleftharpoons$	$2 HI$
Initial (*M*):	0.298		0.410		0.369
Change (*M*):	$-x$		$-x$		$+2x$
Equilibrium (*M*):	$(0.298 - x)$		$(0.410 - x)$		$(0.369 + 2x)$

The equilibrium constant expression is:

$$K_c = \frac{[HI]^2}{[H_2][I_2]} = \frac{(0.369 + 2x)^2}{(0.298 - x)(0.410 - x)} = 54.3$$

This becomes the quadratic equation

$$50.3x^2 - 39.9x + 6.48 = 0$$

The smaller root is $x = 0.228\ M$. (The larger root is physically impossible.)

Having solved for x, calculate the equilibrium concentrations.

$$[H_2] = (0.298 - 0.228)\ M = \mathbf{0.070\ M}$$
$$[I_2] = (0.410 - 0.228)\ M = \mathbf{0.182\ M}$$
$$[HI] = 0.369 + 2(0.228)\ M = \mathbf{0.825\ M}$$

14.88   The gas cannot be **(a)** because the color became lighter with heating. Heating **(a)** to 150°C would produce some HBr, which is colorless and would lighten rather than darken the gas.

The gas cannot be **(b)** because Br_2 doesn't dissociate into Br atoms at 150°C, so the color shouldn't change.

The gas must be **(c)**. From 25°C to 150°C, heating causes N_2O_4 to dissociate into NO_2, thus darkening the color (NO_2 is a brown gas).

$$N_2O_4\ (g) \rightarrow 2\ NO_2\ (g)$$

Above 150°C, the NO_2 breaks up into colorless NO and O_2.

$$2\ NO_2\ (g) \rightarrow 2\ NO\ (g) + O_2\ (g)$$

An increase in pressure shifts the equilibrium back to the left, forming NO_2, thus darkening the gas again.

$$2\ NO\ (g) + O_2\ (g) \rightarrow 2\ NO_2\ (g)$$

14.90   Given the following: $K_c = \dfrac{[NH_3]^2}{[N_2][H_2]^3} = 1.2$

(a)   Temperature must have units of Kelvin.

$$K_p = K_c(0.0821\ T)^{\Delta n}$$
$$\mathbf{K_p} = (1.2)(0.0821 \times 648)^{(2-4)} = \mathbf{4.2 \times 10^{-4}}$$

(b)   Recalling that,

$$K_{\text{forward}} = \frac{1}{K_{\text{reverse}}}$$

Therefore,

$$\mathbf{K_c'} = \frac{1}{1.2} = \mathbf{0.83}$$

(c)   Since the equation

$$\frac{1}{2}\ N_2\ (g) + \frac{3}{2}\ H_2\ (g) \rightleftharpoons NH_3\ (g)$$

is equivalent to

$$\frac{1}{2}\ [N_2\ (g) + 3\ H_2\ (g) \rightleftharpoons 2\ NH_3\ (g)],$$

then, K_c' for the reaction:

$$\frac{1}{2}\ N_2\ (g) + \frac{3}{2}\ H_2\ (g) \rightleftharpoons NH_3\ (g)$$

equals $K_c^{1/2}$ for the reaction:

$$N_2\,(g)\ +\ 3\,H_2\,(g)\ \rightleftharpoons\ 2\,NH_3\,(g)$$

Thus,

$$K_c' = K_c^{1/2} = \sqrt{1.2} = \textbf{1.1}$$

(d) For K_p in part (b):

$$K_p = (0.83)(0.0821 \times 648)^{+2} = \textbf{2.3} \times \textbf{10}^3$$

and for K_p in part (c):

$$K_p = (1.1)(0.0821 \times 648)^{-1} = \textbf{0.021}$$

14.92 The vapor pressure of water is equivalent to saying the partial pressure of $H_2O\,(g)$.

$$K_p = P_{H_2O} = \textbf{0.0231}$$

$$K_c = \frac{K_p}{(0.0821\,T)^{\Delta n}} = \frac{0.0231}{(0.0821 \times 293)^1} = \textbf{9.60} \times \textbf{10}^{-4}$$

14.94 This problem involves two types of problems from Chapter 5. Calculations using density or molar mass, Problem Type 2C, and Dalton's Law of Partial Pressures, Problem Type 3.

We can calculate the average molar mass of the gaseous mixture from the density. From Problem Type 2C, Chapter 5,

$$\mathcal{M} = \frac{dRT}{P}$$

Let $\overline{\mathcal{M}}$ be the average molar mass of NO_2 and N_2O_4. The above equation becomes:

$$\overline{\mathcal{M}} = \frac{dRT}{P} = \frac{(2.3\,\text{g}/\text{L})(0.0821\,\text{L}\cdot\text{atm}/\text{K}\cdot\text{mol})(347\,\text{K})}{1.3\,\text{atm}}$$

$$\overline{\mathcal{M}} = 50.4\,\text{g/mol}$$

The average molar mass is equal to the sum of the molar masses of each component times the respective mole fractions. Setting this up, we can calculate the mole fraction of each component.

$$\overline{\mathcal{M}} = X_{NO_2}\mathcal{M}_{NO_2} + X_{N_2O_4}\mathcal{M}_{N_2O_4} = 50.4\,\text{g/mol}$$

$$X_{NO_2}(46.01\,\text{g/mol}) + (1 - X_{NO_2})(92.01\,\text{g/mol}) = 50.4\,\text{g/mol}$$

$$X_{NO_2} = 0.905$$

We can now calculate the partial pressure of NO_2 from the mole fraction and the total pressure (see Problem Type 3, Chapter 5).

$$P_{NO_2} = X_{NO_2} P_T$$

$$P_{NO_2} = (0.905)(1.3\,\text{atm}) = \textbf{1.2 atm}$$

We can calculate the partial pressure of N_2O_4 by difference.

$$P_{N_2O_4} = P_T - P_{NO_2}$$

$$P_{N_2O_4} = (1.3 - 1.18) \text{ atm} = \textbf{0.12 atm}$$

Finally, we can calculate K_p for the dissociation of N_2O_4.

$$K_p = \frac{P_{NO_2}^2}{P_{N_2O_4}} = \frac{(1.2)^2}{0.12} = \textbf{12}$$

14.96   **(a)**   shifts to right   **(b)**   shifts to right   **(c)**   no change   **(d)**   no change
             (e)   no change   **(f)**   shifts to left

14.98   The equilibrium is: $N_2O_4 (g) \rightleftharpoons 2 NO_2(g)$

$$K_p = \frac{(P_{NO_2})^2}{P_{N_2O_4}} = \frac{0.15^2}{0.20} = 0.113$$

Volume is doubled so pressure is halved. Let's calculate Q_p and compare it to K_p.

$$Q_p = \frac{\left(\dfrac{0.15}{2}\right)^2}{\left(\dfrac{0.20}{2}\right)} = 0.0563 < K_p$$

Equilibrium will shift to the right. Some N_2O_4 will react, and some NO_2 will be formed. Let x = amount of N_2O_4 reacted.

At equilibrium: $P_{N_2O_4} = \dfrac{0.20}{2} - x$ $P_{NO_2} = \dfrac{0.15}{2} + 2x$

$$K_p = 0.113 = \frac{\left(\dfrac{0.15}{2} + 2x\right)^2}{\dfrac{0.20}{2} - x}$$

$$0.113 = \frac{(0.075 + 2x)^2}{0.10 - x}$$

$$4x^2 + 0.413x - 5.67 \times 10^{-3} = 0$$

$$x = 0.0123$$

At equilibrium:

$$P_{NO_2} = 0.075 + 2(0.0123) = 0.0996 \approx \textbf{0.10 atm}$$

$$P_{N_2O_4} = 0.10 - 0.0123 = \textbf{0.088 atm}$$

Check:

$$K_p = \frac{(0.10)^2}{0.088} = 0.114 \qquad \text{close enough to } 0.113$$

14.100 **(a)** Molar mass of PCl_5 = 208.2 g/mol

$$P = \frac{nRT}{V} = \frac{\left(2.50\ g \times \dfrac{1\ mol}{208.2\ g}\right)\left(0.0821\dfrac{L \cdot atm}{mol \cdot K}\right)(523\ K)}{0.500\ L} = \textbf{1.03 atm}$$

(b)

	PCl_5	$\rightleftharpoons$	PCl_3	+	Cl_2
Initial (atm)	1.03		0		0
Change (atm)	$-x$		$+x$		$+x$
Equilibrium (atm)	$1.03 - x$		x		x

$$K_p = 1.05 = \frac{x^2}{1.03 - x}$$

$$x^2 + 1.05x - 1.08 = 0$$

$$x = 0.639$$

At equilibrium:

$$P_{PCl_5} = 1.03 - 0.639 = \textbf{0.39 atm}$$

(c) $P_T = (1.03 - x) + x + x = 1.03 + 0.639 = \textbf{1.67 atm}$

(d) $\dfrac{0.639\ atm}{1.03\ atm} = \textbf{0.620 or 62.0\%}$

14.102 **(a)** $K_p = P_{Hg} = 0.0020\ mmHg = 2.6 \times 10^{-6}\ atm = \mathbf{2.6 \times 10^{-6}}$ (equil. constants are expressed without units)

$$K_c = \frac{K_p}{(0.0821T)^{\Delta n}} = \frac{2.6 \times 10^{-6}}{(0.0821 \times 299)^1} = \mathbf{1.1 \times 10^{-7}}$$

(b) Volume of lab = (6.1 m)(5.3 m)(3.1 m) = 100 m^3

$[Hg] = K_c$

$$\textbf{Total mass of Hg vapor} = \frac{1.1 \times 10^{-7}\ mol}{1\ L} \times \frac{200.6\ g}{1\ mol} \times \frac{1\ L}{1\ dm^3} \times \left(\frac{1\ dm}{0.1\ m}\right)^3 \times 100\ m^3 = \textbf{2.2 g}$$

The concentration of mercury vapor in the room is:

$$\frac{2.2\ g}{100\ m^3} = 0.22\ g/m^3 = \textbf{22 mg/m}^3$$

Yes! This concentration exceeds the safety limit of 0.05 mg/m^3. Better clean up the spill!

CHAPTER 15
ACIDS AND BASES

PROBLEM-SOLVING STRATEGIES AND TUTORIAL SOLUTIONS

TYPES OF PROBLEMS

Problem Type 1: Identifying Conjugate Acid-Base Pairs.

Problem Type 2: The Ion-Product Constant (K_w), Calculating $[H^+]$ from $[OH^-]$.

Problem Type 3: pH Calculations.
      (a) Calculating pH from $[H^+]$.
      (b) Calculating $[H^+]$ from pH.

Problem Type 4: Calculating the pH of a Strong Acid and a Strong Base.

Problem Type 5: Weak Acids.
      (a) Ionization of a weak monoprotic acid.
      (b) Determining K_a from a pH measurement.

Problem Type 6: Calculating the pH of a Diprotic Acid.

Problem Type 7: Calculating the pH of a Weak Base.

Problem Type 8: Predicting the Acid-Base Properties of Salt Solutions.

PROBLEM TYPE 1: IDENTIFYING CONJUGATE ACID-BASE PAIRS

A **conjugate acid-base pair** can be defined as *an acid and its conjugate base or a base and its conjugate acid*. The conjugate base of a Brønsted acid is the species that remains when *one* proton has been removed from the acid. Conversely, a conjugate acid results from the addition of *one* proton to a Brønsted base.

EXAMPLE 15.1
Consider the reaction

$$HSO_4^-\ (aq)\ +\ HCO_3^-\ (aq)\ \rightleftharpoons\ SO_4^{2-}\ (aq)\ +\ H_2CO_3\ (aq)$$

(a) Identify the acids and bases for the forward and reverse reactions.
(b) Identify the conjugate acid-base pairs.

(a)   In the forward reaction, HSO_4^- is the proton donor, which makes it an acid. The proton acceptor, HCO_3^-, is a base. In the reverse reaction, the proton donor (acid) is H_2CO_3, and SO_4^{2-} is the proton acceptor (base).

(b) $HSO_4^-\ (aq)\ +\ HCO_3^-\ (aq)\ \rightleftharpoons\ SO_4^{2-}\ (aq)\ +\ H_2CO_3\ (aq)$

      acid$_1$ base$_2$ base$_1$ acid$_2$

The subscripts 1 and 2 designate the two conjugate acid-base pairs.

PRACTICE EXERCISE

1.   Identify the conjugate bases of the following acids:

   (a) CH_3COOH (b) H_2S (c) HSO_3^- (d) $HClO$

> **Text Problems:** 15.4, 15.6, 15.8

PROBLEM TYPE 2: THE ION-PRODUCT CONSTANT (K_w), CALCULATING [H⁺] FROM [OH⁻]

Water is a very weak electrolyte and therefore a poor conductor of electricity, but it does ionize to a small extent:

$$H_2O \, (l) \; \rightleftharpoons \; H^+ \, (aq) \; + \; OH^- \, (aq)$$

We can write the equilibrium constant for the autoionization of water as

$$K_c = \frac{[H^+][OH^-]}{[H_2O]}$$

Since a very small fraction of water molecules are ionized, the concentration of water, $[H_2O]$, remains virtually unchanged. Therefore, we assume that the concentration of water is constant, and we write:

$$K_c[H_2O] = K_w = [H^+][OH^-] \tag{15.1}$$

In pure water at 25°C, the concentrations of H^+ and OH^- ions are equal and found to be $[H^+] = [OH^-] = 1.0 \times 10^{-7} \, M$. Substituting these concentrations into Equation (15.1),

$$K_w = [H^+][OH^-] = (1 \times 10^{-7})(1 \times 10^{-7}) = 1 \times 10^{-14}$$

Thus, knowing K_w and one of the concentrations, either $[H^+]$ or $[OH^-]$, we can easily calculate the other concentration.

EXAMPLE 15.2

The OH^- ion concentration in a certain solution at 25°C is $5.0 \times 10^{-5} \, M$. What is the H^+ concentration?

The ion product of water is applicable to all aqueous solutions. At 25°C,

$$K_w = 1.0 \times 10^{-14} = [H^+][OH^-]$$

Rearrange the equation to solve for $[H^+]$.

$$[H^+] = \frac{1.0 \times 10^{-14}}{[OH^-]} = \frac{1.0 \times 10^{-14}}{5.0 \times 10^{-5}} = 2.0 \times 10^{-10} \, M$$

PRACTICE EXERCISE

2.   The H^+ concentration in a certain solution is $6.6 \times 10^{-4} \, M$. What is the OH^- concentration?

> **Text Problems:** 15.16, 15.18, **15.20c**

PROBLEM TYPE 3: pH CALCULATIONS

Because the concentrations of H^+ and OH^- ions in aqueous solutions are frequently very small numbers making them inconvenient to work with, the Danish biochemist Soren Sorensen in 1909 proposed a more practical measure called pH. The **pH** of a solution is defined as *the negative log of the hydrogen ion concentration* (in moles per liter).

$$pH = -\log[H^+]$$

The pOH of a solution is defined in a similar manner.

$$pOH = -\log[OH^-]$$

A useful relationship between pH and pOH is:

$$pH + pOH = 14.$$

See Section 15.3 of your text for a complete discussion.

A. Calculating pH from [H⁺]

EXAMPLE 15.3

The [H⁺] of a solution is 0.015 M. What is the pH of the solution?

The relationship between $[H^+]$ and pH is

$$pH = -\log[H^+]$$

Substitute the H^+ ion concentration into the above equation to calculate the pH of the solution.

$$pH = -\log[H^+] = -\log[0.015] = \mathbf{1.82}$$

PRACTICE EXERCISE

3.   The OH^- concentration of a certain ammonia solution is 7.2×10^{-4} *M*. What is the pOH and pH?

Text Problem: 15.18

B. Calculating [H⁺] from pH

EXAMPLE 15.4

What is the H⁺ concentration in a solution with a pOH of 3.9?

Step 1: Recall that pH + pOH = 14. Since pOH is given in the problem, we can calculate the pH.

$$pH = 14.0 - pOH$$

$$pH = 14.0 - 3.9 = 10.1$$

Step 2: The H^+ concentration can now be determined from the pH.

$$pH = -\log[H^+]$$

or,

$$-pH = \log[H^+]$$

Taking the antilog of both sides of the equation,

$$10^{-pH} = [H^+]$$

$$[H^+] = 10^{-10.1} = 8 \times 10^{-11} \, M$$

PRACTICE EXERCISE

4.   The pH of a solution is 4.45. What is the H^+ concentration?

Text Problem: 15.20a,b

PROBLEM TYPE 4: CALCULATING THE pH OF A STRONG ACID AND A STRONG BASE

Strong acids and strong bases are strong electrolytes that are assumed to ionize completely in water. For example, consider the strong acid, hydrochloric acid [HCl (aq)]. We assume that it ionizes completely in water.

$$HCl \, (aq) \rightarrow H^+ \, (aq) + Cl^- \, (aq)$$

Since strong acids ionize completely in water, we can easily calculate the $[H^+]$ in solution. For example, consider a 0.10 M HCl solution. Let's set up a table to determine the $[H^+]$ of the solution.

	HCl (aq) $\rightarrow$	H$^+$ (aq) +	Cl$^-$ (aq)
initial conc.	0.10 M	0	0
conc. after ionization	0	0.10 M	0.10 M

Since there is a one:one mole ratio between HCl and H^+, the H^+ concentration after ionization equals the initial concentration of HCl.

Note that the above reaction can be written more accurately as:

$$HCl \, (aq) + H_2O \, (l) \rightarrow H_3O^+ \, (aq) + Cl^- \, (aq)$$

HCl is an acid and donates a proton (H^+) to the weak base, H_2O. H^+ in aqueous solution is really shorthand notation for H_3O^+.

You need to know the six strong acids and the six strong bases. Half the battle in pH calculations is recognizing the type of species in solution (i.e., strong acid, strong base, weak acid, weak base, or salt).

The six strong acids	
$HClO_4$	perchloric acid
HI	hydroiodic acid
HBr	hydrobromic acid
HCl	hydrochloric acid
H_2SO_4	sulfuric acid
HNO_3	nitric acid

The six **strong bases** include the five alkali metal hydroxides (LiOH, NaOH, KOH, RbOH, and CsOH), plus barium hydroxide, Ba(OH)$_2$.

EXAMPLE 15.5

Calculate the pH of a 0.10 M Ba(OH)$_2$ solution.

Step 1: First, you must recognize that barium hydroxide is one of the six strong bases. Strong bases ionize completely in solution. Let's write the reaction and set up a table to calculate the OH$^-$ concentration in solution.

$$Ba(OH)_2\ (aq)\ \rightarrow\ Ba^{2+}\ (aq)\ +\ 2\ OH^-\ (aq)$$

initial conc.	0.10 M	0	0
conc. after ionization	0	0.10 M	2(0.10 M) = 0.20 M

Since there is a 2:1 mole ratio between OH$^-$ and Ba(OH)$_2$, the OH$^-$ concentration is double the Ba(OH)$_2$ concentration.

Step 2: Calculate the pOH from the OH$^-$ concentration.

$$pOH = -\log[OH^-]$$

$$= -\log[0.20\ M] = 0.70$$

Step 3: Use the relationship, pH + pOH = 14.0 to calculate the pH of the Ba(OH)$_2$ solution.

$$\textbf{pH} = 14.00 - 0.70 = \textbf{13.30}$$

PRACTICE EXERCISE

5.   What is the pH of a 0.025 M HCl solution?

Text Problems: 15.24, 15.26

PROBLEM TYPE 5: WEAK ACIDS

Most acids ionize only to a limited extent in water. Such acids are classified as **weak acids**. For example, HNO$_2$ (nitrous acid), is a weak acid. Its ionization in water is represented by

$$HNO_2\ (aq)\ +\ H_2O\ (l)\ \rightleftharpoons\ H_3O^+\ (aq)\ +\ NO_2^-\ (aq)$$

or simply

$$HNO_2\ (aq)\ \rightleftharpoons\ H^+\ (aq)\ +\ NO_2^-\ (aq)$$

The equilibrium constant for this acid ionization is called the **acid ionization constant, K_a**. It is given by:

$$K_a = \frac{[H^+][NO_2^-]}{[HNO_2]}$$

In general,

$$K_a = \frac{[H^+][A^-]}{[HA]}$$

where,

A$^-$ is the conjugate base of the weak acid, HA.

A. Ionization of a weak monoprotic acid

This problem is very similar to Problem Type 5, Chapter 14: Calculating Equilibrium Concentrations. Try to think of this problem as just another equilibrium calculation. The only difference is that in Chapter 14, we dealt mostly with gas phase reactions. Now, we are dealing with weak acids in aqueous solution. Some problems will ask you to calculate the equilibrium concentrations of all species, others will ask for the pH of the solution.

EXAMPLE 15.6

Calculate the pH of a 0.20 M acetic acid, CH_3COOH, solution.

First, recognize that acetic acid is a weak acid. It is not one of the six strong acids, so it must be a weak acid.

Step 1: Express the equilibrium concentrations of all species in terms of initial concentrations and a single unknown x, that represents the change in concentration. Let $(-x)$ be the depletion in concentration (mol/L) of CH_3COOH. From the stoichiometry of the reaction, it follows that the increase in concentration for both H^+ and CH_3COO^- must be x. Complete a table that lists the initial concentrations, the change in concentrations, and the equilibrium concentrations.

	$CH_3COOH\ (aq)$	$\rightleftharpoons$	$H^+\ (aq)\ +$	$CH_3COO^-\ (aq)$
Initial (M):	0.20		0	0
Change (M):	$-x$		$+x$	$+x$
Equilibrium (M):	$0.20 - x$		x	x

Step 2: Write the ionization constant expression in terms of the equilibrium concentrations. Knowing the value of the equilibrium constant (K_a), solve for x.

$$K_a = \frac{[H^+][CH_3COO^-]}{[CH_3COOH]}$$

You can look up the K_a value for acetic acid in Table 15.3 of your text.

$$1.8 \times 10^{-5} = \frac{(x)(x)}{(0.20 - x)}$$

At this point, we can make an assumption that x is very small compared to 0.20. Hence,

$$0.20 - x \approx 0.20$$

Oftentimes, assumptions such as these are valid if K is very small. A very small value of K means that a very small amount of reactants go to products. Hence, x is small. If we did not make this assumption, we would have to solve a quadratic equation.

$$1.8 \times 10^{-5} \approx \frac{(x)(x)}{0.20}$$

Solving for x.

$$x = 1.9 \times 10^{-3}\ M = [H^+]$$

Checking the validity of the assumption,

$$\frac{1.9 \times 10^{-3}}{0.20} \times 100\% = 0.95\% < 5\%$$

The assumption was valid.

Step 3: Having solved for the $[H^+]$, calculate the pH of the solution.

$$pH = -\log[H^+] = -\log(1.9 \times 10^{-3}) = \mathbf{2.72}$$

PRACTICE EXERCISE

6.   Calculate the pH of a 0.50 M nitrous acid (HNO_2) solution. The K_a value for HNO_2 is 4.5×10^{-4}.

Text Problems: **15.42**, 15.44, 15.46, 15.48

B.   Determining K_a from a pH measurement

This problem is the reverse of Example 15.6 above. From the pH of the solution, you can calculate the $[H^+]$. Then from the $[H^+]$, you can calculate the equilibrium concentrations of the other species in solution. Substitute these equilibrium concentrations into the ionization constant expression to solve for K_a.

EXAMPLE 15.7

The pH of a 0.10 M solution of a weak monoprotic acid is 5.15. What is the K_a of the acid?

Step 1: Calculate the $[H^+]$ from the pH.

$$pH = -\log[H^+]$$

or,

$$-pH = \log[H^+]$$

Taking the antilog of both sides of the equation,

$$10^{-pH} = [H^+]$$

$$[H^+] = 10^{-5.15} = 7.1 \times 10^{-6}\ M$$

Step 2: Complete a table that lists the initial concentrations, the change in concentrations, and the equilibrium concentrations.

	HA (aq)	$\rightleftharpoons$	H^+ (aq) +	A^- (aq)
Initial (M):	0.10		0	0
Change (M):	$-x$		$+x$	$+x$
Equilibrium (M):	$0.10 - x$		x	x

In this case, x is known.

$$x = [H^+] = [A^-] = 7.1 \times 10^{-6}\ M$$

$$[HA] = 0.10 - x = (0.10 - 7.1 \times 10^{-6})M = 0.10\ M$$

Step 3: Substitute the equilibrium concentrations into the ionization constant expression to solve for K_a.

$$K_a = \frac{[H^+][A^-]}{[HA]}$$

$$\mathbf{K_a} = \frac{(7.1 \times 10^{-6})^2}{(0.10)} = \mathbf{5.0 \times 10^{-10}}$$

PRACTICE EXERCISE

7.   The pH of a 0.50 M monoprotic weak acid is 2.24. What is the K_a of the acid?

Text Problem: 15.52

PROBLEM TYPE 6: CALCULATING THE pH OF A DIPROTIC ACID

Diprotic acids may yield more than one hydrogen ion per molecule. These acids ionize in a stepwise manner, that is, they lose one proton at a time. An ionization constant expression can be written for each ionization stage. Consequently, two equilibrium constant expressions must be used to calculate the concentrations of species in the acid solution. For the generic diprotic acid, H_2A, we can write:

$$H_2A\ (aq) \rightleftharpoons H^+\ (aq) + HA^-\ (aq) \qquad\qquad K_{a_1} = \frac{[H^+][HA^-]}{[H_2A]}$$

$$HA^-\ (aq) \rightleftharpoons H^+\ (aq) + A^{2-}\ (aq) \qquad\qquad K_{a_2} = \frac{[H^+][A^{2-}]}{[HA^-]}$$

Note that the conjugate base in the first ionization becomes the acid in the second ionization.

EXAMPLE 15.8

Calculate the concentrations of H_2A, HA^-, H^+, and A^{2-} in a 1.0 M H_2A solution. The first and second ionization constants for H_2A are 1.3×10^{-2} and 6.3×10^{-8}, respectively.

Step 1: Complete a table showing the concentrations for the first ionization stage

	$H_2A\ (aq)$	$\rightleftharpoons$ $H^+\ (aq)$	+ $HA^-\ (aq)$
Initial (M):	1.0	0	0
Change (M):	$-x$	$+x$	$+x$
Equilibrium (M):	$1.0 - x$	x	x

Step 2: Write the ionization constant expression for K_{a_1}. Then, solve for x.

$$K_{a_1} = \frac{[H^+][HA^-]}{[H_2A]}$$

$$1.3 \times 10^{-2} = \frac{(x)(x)}{(1.0 - x)}$$

Since K_{a_1} is quite large, we cannot make the assumption

$$1.0 - x \approx 1.0$$

Therefore, we must solve a quadratic equation.

$$x^2 + 0.013x - 0.013 = 0$$

The above equation is a quadratic equation of the form $ax^2 + bx + c = 0$. The solution for a quadratic equation is:

$$x = \frac{-b \pm \sqrt{b^2 - 4ac}}{2a}$$

Here, we have a = 1, b = 0.013, and c = −0.013. Substituting into the above equation,

$$x = \frac{-0.013 \pm \sqrt{(0.013)^2 - 4(1)(-0.013)}}{2(1)}$$

$$x = \frac{-0.013 \pm 0.23}{2}$$

$$x = 0.11 \, M \quad \text{or} \quad x = -0.12 \, M$$

The second solution is physically impossible because you cannot have a negative concentration. The first solution is the correct answer.

Step 3: Having solved for x, calculate the concentrations when the equilibrium for the first stage of ionization is reached.

Because $K_{a_1} \gg K_{a_2}$, we assume that essentially all the H^+ comes from the first ionization stage. Hence,

$$[H^+] = [HA^-] = x = 0.11 \, M$$

$$[H_2A] = 1.0 - x = 1.00 - 0.11 = 0.89 \, M$$

Step 4: Now, consider the second stage of ionization. Set up a table showing the concentrations for the second ionization stage. Let y be the change in concentration.

	$HA^- \, (aq)$	$\rightleftharpoons$	$H^+ \, (aq)$	$+ \quad A^{2-} \, (aq)$
Initial (*M*):	0.11		0.11	0
Change (*M*):	−y		+y	+y
Equilibrium (*M*):	0.11 − y		0.11 + y	y

Step 5: Write the ionization constant expression for K_{a_2}. Then, solve for y.

$$K_{a_2} = \frac{[H^+][A^{2-}]}{[HA^-]}$$

$$6.3 \times 10^{-8} = \frac{(0.11 + y)(y)}{(0.11 - y)}$$

Since K_{a_2} is very small, we can make an assumption that y is very small compared to 0.11.

Hence,

$$0.11 \pm y \approx 0.11$$

$$6.3 \times 10^{-8} = \frac{(0.11)(y)}{0.11}$$

Solving for y.

$$y = 6.3 \times 10^{-8} \, M = [A^{2-}]$$

Checking the validity of the assumption,

$$\frac{6.3 \times 10^{-8}}{0.11} \times 100\% = 5.7 \times 10^{-5}\% < 5\%$$

The assumption was valid.

PRACTICE EXERCISE

8.   The first and second ionization constants of H_2CO_3 are 4.2×10^{-7} and 4.8×10^{-11}, respectively. Calculate the concentrations of H^+, HCO_3^-, CO_3^{2-}, and unionized H_2CO_3 in a 0.080 M H_2CO_3 solution.

Text Problems: 15.60, 15.62

PROBLEM TYPE 7: CALCULATING THE pH OF A WEAK BASE

The procedure used to calculate the pH of a weak base solution is essentially the same as the one used for weak acids.

EXAMPLE 15.9

What is the pH of a 0.10 M C_5H_5N (pyridine) solution?

Step 1: Express the equilibrium concentrations of all species in terms of initial concentrations and a single unknown x, that represents the change in concentration. Let $(-x)$ be the depletion in concentration (mol/L) of C_5H_5N.

From the stoichiometry of the reaction, it follows that the increase in concentration for both OH^- and $C_5H_5NH^+$ must be x. Complete a table that lists the initial concentrations, the change in concentrations, and the equilibrium concentrations.

	C_5H_5N (aq) + H_2O (l) $\rightleftharpoons$	$C_5H_5NH^+$ (aq) +	OH^- (aq)
Initial (M):	0.10	0	0
Change (M):	$-x$	$+x$	$+x$
Equilibrium (M):	$0.10 - x$	x	x

Step 2: Write the ionization constant expression in terms of the equilibrium concentrations. Knowing the value of the equilibrium constant (K_b), solve for x.

$$K_b = \frac{[C_5H_5NH^+][OH^-]}{[C_5H_5N]}$$

You can look up the K_b value for pyridine in Table 15.4 of your text.

$$1.7 \times 10^{-9} = \frac{(x)(x)}{(0.10 - x)}$$

At this point, we can make an assumption that x is very small compared to 0.10. Hence,

$$0.10 - x \approx 0.10$$

Oftentimes, assumptions such as these are valid if K is very small. A very small value of K means that a very small amount of reactants go to products. Hence, x is small. If we did not make this assumption, we would have to solve a quadratic equation.

$$1.7 \times 10^{-9} \approx \frac{(x)(x)}{0.10}$$

Solving for x.

$$x = 1.3 \times 10^{-5} M = [OH^-]$$

Step 3: Having solved for the $[OH^-]$, calculate the pOH of the solution. Then use the relationship, pH + pOH = 14, to solve for the pH of the solution.

$$pOH = -\log[OH^-] = -\log(1.3 \times 10^{-5}) = 4.89$$

$$\textbf{pH} = 14 - pOH = 14.00 - 4.89 = \textbf{9.11}$$

PRACTICE EXERCISE

9.   What is the pH of a 2.0 M NH_3 solution?

Text Problem: 15.54

PROBLEM TYPE 8: PREDICTING THE ACID-BASE PROPERTIES OF SALT SOLUTIONS

Salts, when dissolved in water, can produce neutral, acidic, or basic solutions. We need to consider four possibilities.

1.   **Salts that produce neutral solutions**. It is generally true that salts containing an alkali metal ion or an alkaline earth metal ion (except Be^{2+}) and the conjugate base of a strong acid (for example, Cl^-, NO_3^-, and ClO_4^-) do not undergo hydrolysis, and thus their solutions are neutral.

2.   **Salts that produce basic solutions**. Salts that contain the conjugate base of a weak acid and an alkali metal or alkaline earth metal ion, produce basic solutions. As an example, consider potassium fluoride. The dissociation of KF in water is given by:

$$KF\ (s)\ \xrightarrow{\ H_2O\ }\ K^+\ (aq)\ +\ F^-\ (aq)$$

A hydrated alkali or alkaline earth metal ion has no acidic or basic properties. However, the conjugate base of a weak acid is a weak base (it has an affinity for H^+ ions). F^- is the conjugate base of the weak acid HF. The hydrolysis reaction is given by:

$$F^-\ (aq)\ +\ H_2O\ (l)\ \rightleftharpoons\ HF\ (aq)\ +\ OH^-\ (aq)$$

Because this reaction produces OH^- ions, the potassium fluoride solution will be basic.

3.   **Salts that produce acidic solutions**. Salts that contain the conjugate acid of a weak base and the conjugate base of a strong acid, produce acidic solutions. As an example, consider ammonium iodide. The dissociation of NH_4I in water is given by

$$NH_4I\ (s)\ \xrightarrow{\ H_2O\ }\ NH_4^+\ (aq)\ +\ I^-\ (aq)$$

The conjugate base of a strong acid, such as I^-, does not undergo hydrolysis. However, the conjugate acid of a weak base is a weak acid and does undergo hydrolysis. The reaction is

$$NH_4^+\ (aq)\ +\ H_2O\ (l)\ \rightleftharpoons\ NH_3\ (aq)\ +\ H_3O^+\ (aq)$$

Since this reaction produces H_3O^+ ions, the ammonium iodide solution will be acidic.

4. **Salts in which both the cation and anion hydrolyze.** In these cases, you must compare the base strength of the anion to the acid strength of the cation. We consider three situations.

- $K_b > K_a$. If K_b for the anion is greater than K_a for the cation, the solution is basic because the anion will hydrolyze to a greater extent than the cation. At equilibrium, there will be more OH^- ions than H^+ ions.

- $K_b < K_a$. Conversely, if K_b for the anion is smaller than K_a for the cation, the solution will be acidic because cation hydrolysis will be more extensive than anion hydrolysis. At equilibrium, there will be more H^+ ions than OH^- ions.

- $K_b \approx K_a$. If K_b is approximately equal to K_a, the solution will be close to neutral.

EXAMPLE 15.10
Is a 0.10 M solution of Na_2CO_3 acidic, basic, or neutral?

Na_2CO_3 is a salt. Salts that contain the conjugate base of a weak acid and an alkali metal or alkaline earth metal ion, produce basic solutions. The dissociation of Na_2CO_3 in water is given by

$$Na_2CO_3\ (s) \xrightarrow{\ H_2O\ } 2\,Na^+\ (aq)\ +\ CO_3^{2-}\ (aq)$$

A hydrated alkali or alkaline earth metal ion has no acidic or basic properties. However, the conjugate base of a weak acid is a weak base (it has an affinity for H^+ ions). CO_3^{2-} is the conjugate base of the weak acid HCO_3^-.

The hydrolysis reaction is given by

$$CO_3^{2-}\ (aq)\ +\ H_2O\ (l)\ \rightleftharpoons\ HCO_3^-\ (aq)\ +\ OH^-\ (aq)$$

Because this reaction produces OH^- ions, the sodium carbonate solution will be basic.

PRACTICE EXERCISE
10. Predict whether the following aqueous solutions will be acidic, basic, or neutral.

(a) KI
(b) NH_4Cl
(c) CH_3COOK

Text Problems: **15.76**, 15.78, 15.80

ANSWERS TO PRACTICE EXERCISES

1. (a) CH_3COO^- (b) HS^- (c) SO_3^{2-} (d) ClO^-

2. $[OH^-] = 1.5 \times 10^{-11}\ M$

3. pOH = 3.14
    pH = 10.86

4. $[H^+] = 3.5 \times 10^{-5}\ M$

5. pH = 1.60

6. pH = 1.82

7. $K_a = 6.69 \times 10^{-5}$

8. $[H^+] = [HCO_3^-] = 1.8 \times 10^{-4}\ M$ $[H_2CO_3] = 0.080\ M$ $[CO_3^{2-}] = 4.8 \times 10^{-11}\ M$

9. pH = 11.8

10. (a) neutral   (b) acidic   (c) basic

SOLUTIONS TO SELECTED TEXT PROBLEMS

15.4   Recall that the conjugate base of a Brønsted acid is the species that remains when *one* proton has been removed from the acid.

(a)   nitrite ion: NO_2^-

(b)   hydrogen sulfate ion (also called bisulfate ion): HSO_4^-

(c)   hydrogen sulfide ion (also called bisulfide ion): HS^-

(d)   cyanide ion: CN^-

(e)   formate ion: $HCOO^-$

15.6   The conjugate acid of any base is just the base with a proton added.

(a) H_2S (b) H_2CO_3 (c) HCO_3^- (d) H_3PO_4 (e) $H_2PO_4^-$

(f) HPO_4^{2-} (g) H_2SO_4 (h) HSO_4^- (i) HSO_3^-

15.8   The conjugate base of any acid is simply the acid minus one proton.

(a) CH_2ClCOO^- (b) IO_4^- (c) $H_2PO_4^-$ (d) HPO_4^{2-} (e) PO_4^{3-}

(f) HSO_4^- (g) SO_4^{2-} (h) IO_3^- (i) SO_3^{2-} (j) NH_3

(k) HS^- (l) S^{2-} (m) OCl^-

15.16 $[OH^-] = 0.62\ M$

$$[H^+] = \frac{1.0 \times 10^{-14}}{0.62} = 1.6 \times 10^{-14}\ M$$

15.18   (a) $Ba(OH)_2$ is ionic and fully ionized in water. The concentration of the hydroxide ion is $5.6 \times 10^{-4}\ M$ (Why? What is the concentration of Ba^{2+}?) We find the hydrogen ion concentration.

$$[H^+] = \frac{K_w}{[OH^-]} = \frac{1.0 \times 10^{-14}}{5.6 \times 10^{-4}} = 1.8 \times 10^{-11}\ M$$

The pH is then: **pH** $= -\log[H^+] = -\log(1.8 \times 10^{-11}) = $ **10.74**

(b)   Nitric acid is a strong acid, so the concentration of hydrogen ion is also $5.2 \times 10^{-4}\ M$. The pH is:

$$\mathbf{pH} = -\log[H^+] = -\log(5.2 \times 10^{-4}) = \mathbf{3.28}$$

15.20   For (a) and (b) we can calculate the H^+ concentration using the equation representing the definition of pH. Problem Type 3B.

(a)   pH $= -\log[H^+]$

$[H^+] = 10^{-5.20} = \mathbf{6.3 \times 10^{-6}\ M}$

(b) $pH = -\log[H^+]$

$[H^+] = 10^{-16.00} = \mathbf{1.0 \times 10^{-16}\ \textit{M}}$

(c)   For part (c), it is probably easiest to calculate the $[H^+]$ from the ion product of water, Problem Type 2. The ion product of water is applicable to all aqueous solutions. At 25°C,

$$K_w = 1.0 \times 10^{-14} = [H^+][OH^-]$$

Rearrange the equation to solve for $[H^+]$.

$$[H^+] = \frac{1.0 \times 10^{-14}}{[OH^-]} = \frac{1.0 \times 10^{-14}}{3.7 \times 10^{-9}} = \mathbf{2.7 \times 10^{-6}\ \textit{M}}$$

15.22   **(a)** acidic   **(b)** neutral   **(c)** basic

15.24 $5.50\ \cancel{mL}\left(\dfrac{1\ \cancel{L}}{1000\ \cancel{mL}}\right)\left(\dfrac{0.360\ mol}{1\ \cancel{L}}\right) = \mathbf{1.98 \times 10^{-3}\ mol\ KOH}$

KOH is a strong base and therefore ionizes completely. The OH^- concentration equals the KOH concentration, because there is a 1:1 mole ratio between KOH and OH^-.

$$[OH^-] = 0.360\ M$$

$$\mathbf{pOH = -\log[OH^-] = 0.444}$$

15.26   Molarity of the HCl solution is: $18.4\ \cancel{g\ HCl}\left(\dfrac{1\ mol\ HCl}{36.46\ \cancel{g\ HCl}}\right)\left(\dfrac{1}{662\ \cancel{mL}}\right)\left(\dfrac{1000\ \cancel{mL}}{1\ L}\right) = 0.762\ M$

$$\mathbf{pH = -\log(0.762) = 0.118}$$

15.32   **(a)** strong base   **(b)** weak base   **(c)** weak base   **(d)** weak base   **(e)** strong base

15.34   **(a)**   false, they are equal   **(b)**   true, find the value of $\log(1.00)$ on your calculator
          (c)   true   **(d)**   false, if the acid is strong, $[HA] = 0.00\ M$

15.36 Cl^- is the conjugate base of the strong acid, HCl. It is a negligibly weak base and has no affinity for protons. Therefore, the reaction will *not* proceed from left to right to any measurable extent.

Another way to think about this problem is to consider the possible products of the reaction.

$$CH_3COOH\ (aq) + Cl^-\ (aq) \rightarrow HCl\ (aq) + CH_3COO^-\ (aq)$$

The favored reaction is the one that proceeds from right to left. HCl is a strong acid and will ionize completely, donating all its protons to the base, CH_3COO^-.

15.42   Ionization of a weak monoprotic acid, Problem Type 5A.

Step 1: Calculate the concentration of acetic acid before ionization.

$$0.0560 \text{ g acetic acid} \times \frac{1 \text{ mol acetic acid}}{60.05 \text{ g acetic acid}} = 9.33 \times 10^{-4} \text{ mol acetic acid}$$

$$\frac{9.33 \times 10^{-4} \text{ mol}}{0.0500 \text{ L soln}} = 0.0187 \, M \text{ acetic acid}$$

Step 2: Next, recognize that acetic acid is a weak, monoprotic acid. It is not one of the six strong acids, so it must be a weak acid. Express the equilibrium concentrations of all species in terms of initial concentrations and a single unknown x, that represents the change in concentration. Let $(-x)$ be the depletion in concentration (mol/L) of CH_3COOH. From the stoichiometry of the reaction, it follows that the increase in concentration for both H^+ and CH_3COO^- must be x. Complete a table that lists the initial concentrations, the change in concentrations, and the equilibrium concentrations.

	$CH_3COOH \, (aq)$	$\rightleftharpoons$	$H^+ \, (aq)$	$+$	$CH_3COO^- \, (aq)$
Initial (M):	0.0187		0		0
Change (M):	$-x$		$+x$		$+x$
Equilibrium (M):	$0.0187 - x$		x		x

Step 3: Write the ionization constant expression in terms of the equilibrium concentrations. Knowing the value of the equilibrium constant (K_a), solve for x.

$$K_a = \frac{[H^+][CH_3COO^-]}{[CH_3COOH]}$$

You can look up the K_a value for acetic acid in Table 15.3 of your text.

$$1.8 \times 10^{-5} = \frac{(x)(x)}{(0.0187 - x)}$$

At this point, we can make an assumption that x is very small compared to 0.0187. Hence,

$$0.0187 - x \approx 0.0187$$

$$1.8 \times 10^{-5} = \frac{(x)(x)}{0.0187}$$

Solving for x.

$$x = 5.8 \times 10^{-4} \, M = [H^+] = [CH_3COO^-]$$

$$[CH_3COOH] = (0.0187 - 5.8 \times 10^{-4})M = 0.0181 \, M$$

Checking the validity of the assumption,

$$\frac{5.8 \times 10^{-4}}{0.0187} \times 100\% = 3.1\% < 5\%$$

The assumption is valid.

15.44   A pH of 3.26 corresponds to a $[H^+]$ of 5.5×10^{-4} M. Let the original concentration of formic acid be x so that:

	HCOOH (aq)	$\rightleftharpoons$	H^+ (aq)	+	$HCOO^-$ (aq)
Initial (M):	x		0		0
Change (M):	-5.5×10^{-4}		5.5×10^{-4}		5.5×10^{-4}
Equilibrium (M):	$x - 5.5 \times 10^{-4}$		5.5×10^{-4}		5.5×10^{-4}

Substitute K_a and the equilibrium concentrations into the ionization constant expression to solve for x.

$$\frac{[H^+][HCOO^-]}{[HCOOH]} = 1.7 \times 10^{-4}$$

$$\frac{(5.5 \times 10^{-4})^2}{(x - 5.5 \times 10^{-4})} = 1.7 \times 10^{-4}$$

$$x = [HCOOH] = 2.3 \times 10^{-3} \, M$$

15.46   Percent ionization is defined as:

$$\text{percent ionization} = \frac{\text{ionized acid concentration at equilibrium}}{\text{initial concentration of acid}} \times 100\%$$

For a monoprotic acid, HA, the concentration of acid that undergoes ionization is equal to the concentration of H^+ ions or the concentration of A^- ions at equilibrium. Thus, we can write:

$$\text{percent ionization} = \frac{[H^+]}{[HA]_0} \times 100\%$$

(a)   First, recognize that hydrofluoric acid is a weak acid. It is not one of the six strong acids, so it must be a weak acid.

Step 1: Express the equilibrium concentrations of all species in terms of initial concentrations and a single unknown x, that represents the change in concentration. Let $(-x)$ be the depletion in concentration (mol/L) of HF. From the stoichiometry of the reaction, it follows that the increase in concentration for both H^+ and F^- must be x. Complete a table that lists the initial concentrations, the change in concentrations, and the equilibrium concentrations.

	HF (aq)	$\rightleftharpoons$	H^+ (aq)	+	F^- (aq)
Initial (M):	0.60		0		0
Change (M):	$-x$		$+x$		$+x$
Equilibrium (M):	$0.60 - x$		x		x

Step 2: Write the ionization constant expression in terms of the equilibrium concentrations. Knowing the value of the equilibrium constant (K_a), solve for x.

$$K_a = \frac{[H^+][F^-]}{[HF]}$$

You can look up the K_a value for hydrofluoric acid in Table 15.3 of your text.

$$7.1 \times 10^{-4} = \frac{(x)(x)}{(0.60 - x)}$$

At this point, we can make an assumption that x is very small compared to 0.60. Hence,

$$0.60 - x \approx 0.60$$

Oftentimes, assumptions such as these are valid if K is very small. A very small value of K means that a very small amount of reactants go to products. Hence, x is small. If we did not make this assumption, we would have to solve a quadratic equation.

$$7.1 \times 10^{-4} = \frac{(x)(x)}{0.60}$$

Solving for x.

$$x = 0.021 \, M = [H^+]$$

Step 3: Having solved for the $[H^+]$, calculate the percent ionization.

$$\textbf{percent ionization} = \frac{[H^+]}{[HF]_0} \times 100\%$$

$$= \frac{0.021 \, M}{0.60 \, M} \times 100\% = \textbf{3.5\%}$$

(b) – (c) are worked in a similar manner to part (a). However, as the initial concentration of HF becomes smaller, the assumption that x is very small compared to this concentration will no longer be valid. You must solve a quadratic equation.

(b) $\quad K_a = \dfrac{[H^+][F^-]}{[HF]} = \dfrac{x^2}{(0.0046 - x)} = 7.1 \times 10^{-4}$

$$x^2 + (7.1 \times 10^{-4})x - (3.3 \times 10^{-6}) = 0$$

$$x = 1.5 \times 10^{-3} \, M$$

$$\textbf{Percent ionization} = \frac{1.5 \times 10^{-3} \, M}{0.0046 \, M} \times 100\% = \textbf{33\%}$$

(c) $\quad K_a = \dfrac{[H^+][F^-]}{[HF]} = \dfrac{x^2}{(0.00028 - x)} = 7.1 \times 10^{-4}$

$$x^2 + (7.1 \times 10^{-4})x - (2.0 \times 10^{-7}) = 0$$

$$x = 2.2 \times 10^{-4} \, M$$

$$\textbf{Percent ionization} = \frac{2.2 \times 10^{-4} \, M}{0.00028 \, M} \times 100\% = \textbf{79\%}$$

As the solution becomes more dilute, the percent ionization increases.

15.48 $\quad C_9H_8O_4 \, (aq) \rightleftharpoons H^+ \, (aq) + C_9H_7O_4^-$

(a) $\quad K_a = 3.0 \times 10^{-4} = \dfrac{[H^+][C_9H_7O_4^-]}{[C_9H_8O_4]} = \dfrac{x^2}{(0.20 - x)}$

Assuming $(0.20 - x) \approx 0.20$

$$x = [H^+] = 7.7 \times 10^{-3} M$$

Percent ionization $= \dfrac{[C_9H_7O_4^-]}{[C_9H_8O_4]} \times 100\%$

$$= \dfrac{7.7 \times 10^{-3} M}{0.20 M} \times 100\% = \mathbf{3.9\%}$$

(b) At pH 1.00 the concentration of hydrogen ion is 0.10 M (Why only two significant figures?) This will tend to suppress the ionization of the weak acid (LeChatelier's principle, Section 14.5). The extra hydrogen ion shifts the position of equilibrium in the direction of the un-ionized acid, and to two-significant-figure accuracy, we can safely ignore the contribution of the weak acid to the total hydrogen ion concentration. The percent ionization of the acid is then

$$\text{Percent ionization} = \dfrac{[C_9H_7O_4^-]}{[C_9H_8O_4]} \times 100\%$$

and,

$$K_a = \dfrac{[H^+][C_9H_7O_4^-]}{[C_9H_8O_4]} \qquad \text{or} \qquad \dfrac{K_a}{[H^+]} = \dfrac{[C_9H_7O_4^-]}{[C_9H_8O_4]}$$

Substituting into the equation for percent ionization gives

percent ionization $= \dfrac{K_a}{[H^+]} \times 100\%$

$$= \dfrac{3.0 \times 10^{-4}}{0.10} \times 100\% = \mathbf{0.30\%}$$

The high acidity of the gastric juices appears to enhance the rate of absorption of unionized aspirin molecules through the stomach lining. In some cases this can irritate these tissues and cause bleeding.

15.52 Similar to Problem Type 5B, Determining K_a from a pH measurement.

Step 1: Calculate the pOH from the pH. Then, calculate the OH^- concentration from the pOH.

$$pOH = 14.00 - pH = 14.00 - 10.66 = 3.34$$

$$pOH = -\log[OH^-]$$

or,

$$-pOH = \log[OH^-]$$

Taking the antilog of both sides of the equation,

$$10^{-pOH} = [OH^-]$$

$$[OH^-] = 10^{-3.34} = 4.6 \times 10^{-4} M$$

Step 2: Complete a table that lists the initial concentrations, the change in concentrations, and the equilibrium concentrations. Let B represent the weak base.

$$B \ (aq) + H_2O \ (l) \ \rightleftharpoons \ BH^+ \ (aq) + OH^- \ (aq)$$

Initial (M):	0.30	0	0
Change (M):	$-x$	$+x$	$+x$
Equilibrium (M):	$0.30 - x$	x	x

In this case, x is known.

$$x = [BH^+] = [OH^-] = 4.6 \times 10^{-4} \ M$$

$$[B] = 0.30 - x = (0.30 - 4.6 \times 10^{-4})M = 0.30 \ M$$

Step 3: Substitute the equilibrium concentrations into the ionization constant expression to solve for K_b.

$$K_b = \frac{[BH^+][OH^-]}{[B]}$$

$$K_b = \frac{(4.6 \times 10^{-4})^2}{(0.30)} = \mathbf{7.1 \times 10^{-7}}$$

15.54   The reaction is:

$$NH_3 \ (aq) + H_2O \ (l) \rightleftharpoons NH_4^+ \ (aq) + OH^- \ (aq)$$

Let $[OH^-] = [NH_4^+] = x$

At equilibrium we have:

$$\frac{[NH_4^+][OH^-]}{[NH_3]} = \frac{x^2}{(0.080-x)} \approx \frac{x^2}{0.080} = 1.8 \times 10^{-5}$$

$$x = 1.2 \times 10^{-3} \ M$$

$$\textbf{Percent NH}_3 \textbf{ present as NH}_4^+ = \frac{1.2 \times 10^{-3}}{0.080} \times 100\% = \mathbf{1.5\%}$$

15.60   The pH of a 0.040 M HCl solution (strong acid) is **1.40**. Follow the procedure for calculating the pH of a diprotic acid, Problem Type 6, to calculate the pH of the sulfuric acid solution.

Step 1: H_2SO_4 is a strong acid. The first ionization stage goes to completion. Therefore, the H^+ ion concentration and the HSO_4^- concentration from the first ionization stage equal the initial concentration of sulfuric acid, 0.040 M.

$$H_2SO_4 \ (aq) \rightarrow H^+ \ (aq) + HSO_4^- \ (aq)$$

Step 2: Now, consider the second stage of ionization. HSO_4^- is a weak acid. Set up a table showing the concentrations for the second ionization stage. Let x be the change in concentration.

$$HSO_4^- \ (aq) \ \rightleftharpoons \ H^+ \ (aq) + SO_4^{2-} \ (aq)$$

Initial (M):	0.040	0.040	0
Change (M):	$-x$	$+x$	$+x$
Equilibrium (M):	$0.040 - x$	$0.040 + x$	x

Step 3: Write the ionization constant expression for K_a. Then, solve for x.

$$K_a = \frac{[H^+][SO_4^{2-}]}{[HSO_4^-]}$$

$$1.3 \times 10^{-2} = \frac{(0.040 + x)(x)}{(0.040 - x)}$$

Since K_a is quite large, we cannot make the assumptions that

$$0.040 - x \approx 0.040 \quad \text{and} \quad 0.040 + x \approx 0.040$$

Therefore, we must solve a quadratic equation.

$$x^2 + 0.053x - 5.2 \times 10^{-4} = 0$$

$$x = \frac{-0.053 \pm \sqrt{(0.053)^2 - 4(1)(-5.2 \times 10^{-4})}}{2(1)}$$

$$x = \frac{-0.053 \pm 0.070}{2}$$

$$x = 8.5 \times 10^{-3}\,M \quad \text{or} \quad x = -0.062\,M$$

The second solution is physically impossible because you cannot have a negative concentration. The first solution is the correct answer.

Step 4: Having solved for x, we can calculate the H^+ concentration after equilibrium for the second stage of ionization is reached. Then, we can calculate the pH from the H^+ concentration.

$$[H^+] = 0.040\,M + x = (0.040 + 8.5 \times 10^{-3})M = 0.049\,M$$

$$\textbf{pH} = -\log(0.049) = \textbf{1.31}$$

Without doing any calculations, could you have known that the pH of the sulfuric acid would be lower (more acidic) than that of the hydrochloric acid?

15.62 For the first stage of ionization:

	$H_2CO_3\ (aq)$	$\rightleftharpoons$	$H^+\ (aq)$	$+$	$HCO_3^-\ (aq)$
Initial (*M*):	0.025		0.00		0.00
Change (*M*):	$-x$		$+x$		$+x$
Equilibrium (*M*):	$(0.025 - x)$		x		x

$$K_{a_1} = \frac{[H^+][HCO_3^-]}{[H_2CO_3]} = \frac{x^2}{(0.025 - x)} = 4.2 \times 10^{-7}$$

Assuming $(0.025 - x) \approx 0.025$,

$$x = 1.0 \times 10^{-4}\,M$$

For the second ionization,

$$HCO_3^-\ (aq) \rightleftharpoons H^+\ (aq) + CO_3^{2-}\ (aq)$$

$$K_{a_2} = \frac{[H^+][CO_3^{2-}]}{[HCO_3^-]} = 4.8 \times 10^{-11}$$

Since HCO_3^- is a very weak acid, there is little ionization at this stage. Therefore we have:

$$[H^+] = [HCO_3^-] = 1.0 \times 10^{-4} \ M \ \text{and} \ [CO_3^{2-}] = 4.8 \times 10^{-11} \ M$$

15.66 All the listed pairs are oxoacids that contain different central atoms whose elements are in the same group of the periodic table and have the same oxidation number. In this situation the acid with the most electronegative central atom will be the strongest.

(a) $H_2SO_4 > H_2SeO_4$.

(b) $H_3PO_4 > H_3AsO_4$

15.68 The conjugate bases are $C_6H_5O^-$ from phenol and CH_3O^- from methanol. The $C_6H_5O^-$ is stabilized by resonance:

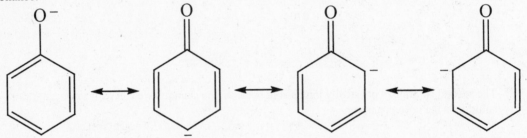

The CH_3O^- ion has no such resonance stabilization. A more stable conjugate base means an increase in the strength of the acid.

15.76 Predicting the Acid-Base Properties of Salt Solutions, Problem Type 8.

There is an inverse relationship between acid strength and conjugate base strength. As acid strength decreases, the proton accepting power of the conjugate base increases. In general the weaker the acid, the stronger the conjugate base. All three of the potassium salts ionize completely to form the conjugate base of the respective acid. The greater the pH, the stronger the conjugate base, and therefore, the weaker the acid.

The order of increasing acid strength is **HZ < HY < HX**.

15.78 The salt ammonium chloride completely ionizes upon dissolution, producing 0.42 M [NH_4^+] and 0.42 M [Cl^-] ions. NH_4^+ will undergo hydrolysis because it is a weak acid (NH_4^+ is the conjugate acid of the weak base, NH_3).

Step 1: Express the equilibrium concentrations of all species in terms of initial concentrations and a single unknown x, that represents the change in concentration. Let $(-x)$ be the depletion in concentration (mol/L) of NH_4^+. From the stoichiometry of the reaction, it follows that the increase in concentration for both H_3O^+ and NH_3 must be x. Complete a table that lists the initial concentrations, the change in concentrations, and the equilibrium concentrations.

	$NH_4^+ \ (aq) + H_2O \ (l)$	$\rightleftharpoons$	$NH_3 \ (aq) +$	$H_3O^+ \ (aq)$
Initial (M):	0.42		0.00	0.00
Change (M):	$-x$		$+x$	$+x$
Equilibrium (M):	$(0.42 - x)$		x	x

Step 2: You can calculate the K_a value for NH_4^+ from the K_b value of NH_3. The relationship is

$$K_a \times K_b = K_w$$

or

$$K_a = \frac{K_w}{K_b} = \frac{1.0 \times 10^{-14}}{1.8 \times 10^{-5}} = 5.6 \times 10^{-10}$$

Step 3: Write the ionization constant expression in terms of the equilibrium concentrations. Knowing the value of the equilibrium constant (K_a), solve for x.

$$K_a = \frac{[NH_3][H_3O^+]}{[NH_4^+]} = \frac{x^2}{0.42 - x} = 5.6 \times 10^{-10}$$

Assuming $(0.42 - x) \approx 0.42$,

$$x = [H^+] = 1.5 \times 10^{-5} \, M$$

$$\textbf{pH} = -\log(1.5 \times 10^{-5}) = \textbf{4.82}$$

Since NH_4Cl is the salt of a weak base (aqueous ammonia) and a strong acid (HCl), we expect the solution to be slightly acidic, which is confirmed by the calculation.

15.80   The acid and base reactions are:

   acid: $HPO_4^{2-} (aq) \rightleftharpoons H^+ (aq) + PO_4^{3-} (aq)$

   base: $HPO_4^{2-} (aq) + H_2O (l) \rightleftharpoons H_2PO_4^- (aq) + OH^- (aq)$

K_a for HPO_4^{2-} is 4.8×10^{-13}. Note that HPO_4^{2-} is the conjugate base of $H_2PO_4^-$, so K_b is 1.6×10^{-7}. Comparing the two K's, we conclude that the monohydrogen phosphate ion is a much stronger proton acceptor (base) than a proton donor (acid). The solution will be basic.

15.84   The most basic oxides occur with metal ions having the lowest positive charges (or lowest oxidation numbers).

   (a) $Al_2O_3 < BaO < K_2O$ **(b)** $CrO_3 < Cr_2O_3 < CrO$

15.86 $Al(OH)_3$ is an amphoteric hydroxide. The reaction is:

$$Al(OH)_3(s) + OH^-(aq) \rightarrow Al(OH)_4^-(aq)$$

This is a Lewis acid-base reaction. Can you identify the acid and base?

15.90 $AlCl_3$ is a Lewis acid with an incomplete octet of electrons and Cl^- is the Lewis base donating a pair of electrons.

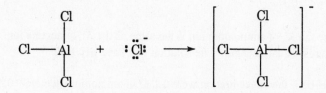

15.92   By definition Brønsted acids are proton donors, therefore such compounds must contain at least one hydrogen atom. In Problem 15.89, Lewis acids that do not contain hydrogen, and therefore are not Brønsted acids, are CO_2, SO_2, and BCl_3. Can you name others?

15.94   We first find the number of moles of CO_2 produced in the reaction:

$$0.350 \, g \, NaHCO_3 \left(\frac{1 \, mol \, NaHCO_3}{84.01 \, g \, NaHCO_3} \right) \left(\frac{1 \, mol \, CO_2}{1 \, mol \, NaHCO_3} \right) = 4.17 \times 10^{-3} \, mol \, CO_2$$

$$V_{CO_2} = \frac{n_{CO_2} RT}{P} = \frac{(4.17 \times 10^{-3} \, mol)(0.0821 \, L \cdot atm \, / \, K \cdot mol)(37.0 + 273) \, K}{(1.00 \, atm)} = \mathbf{0.106 \, L}$$

15.96   If we assume that the unknown monoprotic acid is a strong acid that is 100% ionized, then the $[H^+]$ concentration will be 0.0642 M.

$$pH = -\log (0.0642) = 1.19$$

Since the actual pH of the solution is higher, the acid must be a weak acid.

15.98   The reaction of a weak acid with a strong base is driven to completion by the formation of water. Irrespective of whether the strong base is reacting with a strong monoprotic acid or a weak monoprotic acid, the same number of moles of acid is required to react with a constant number of moles of base. Therefore the volume of base required to react with the same concentration of acid solutions (either both weak, both strong, or one strong and one weak) will be the same.

15.100 High oxidation state leads to covalent compounds and low oxidation state leads to ionic compounds. Therefore, CrO is ionic and basic and CrO_3 is covalent and acidic.

15.102 We can write two equilibria that add up to the equilibrium in the problem.

$CH_3COOH \, (aq) \rightleftharpoons H^+ \, (aq) + CH_3COO^- \, (aq)$ $\qquad K_a = \dfrac{[H^+][CH_3COO^-]}{[CH_3COOH]} = 1.8 \times 10^{-5}$

$H^+ \, (aq) + NO_2^- \, (aq) \rightleftharpoons HNO_2 \, (aq)$ $\qquad K_a' = \dfrac{1}{K_a(HNO_2)} = \dfrac{1}{4.5 \times 10^{-4}} = 2.2 \times 10^3$

$\qquad\qquad\qquad\qquad\qquad\qquad\qquad\qquad K_a' = \dfrac{[HNO_2]}{[H^+][NO_2^-]}$

$CH_3COOH \, (aq) + NO_2^- \, (aq) \rightleftharpoons CH_3COO^- \, (aq) + HNO_2 \, (aq)$ $\quad K = \dfrac{[CH_3COO^-][HNO_2]}{[CH_3COOH][NO_2^-]} = K_a \times K_a'$

The equilibrium constant for this sum is the product of the equilibrium constants of the component reactions.
$$K = K_a \times K_a' = (1.8 \times 10^{-5})(2.2 \times 10^3) = \mathbf{4.0 \times 10^{-2}}$$

15.104 In this specific case the K_a of ammonium ion is the same as the K_b of acetate ion. The two are of exactly (to two significant figures) equal strength. The solution will have **pH 7.00**.

What would the pH be if the concentration were 0.1 M in ammonium acetate? 0.4 M?

15.106 The fact that fluorine attracts electrons in a molecule more strongly than hydrogen should cause NF_3 to be a poor electron pair donor and a poor base. NH_3 is the stronger base.

15.108 The autoionization for deuterium-substituted water is: $D_2O \rightleftharpoons D^+ + OD^-$

$$[D^+][OD^-] = 1.35 \times 10^{-15} \qquad (1)$$

(a) The definition of pD is: $\quad \mathbf{pD} = -\log[D^+] = -\log\sqrt{1.35 \times 10^{-15}} = \mathbf{7.43}$

(b) To be acidic, the **pD** must be **< 7.43**.

(c) Taking $-\log$ of both sides of equation (1) above:

$$-\log[D^+] + -\log[OD^-] = -\log(1.35 \times 10^{-15})$$

$$\mathbf{pD + pOD = 14.87}$$

15.110 First we must calculate the molarity of the trifluoromethane sulfonic acid. (Molar mass = 150.1 g)

$$\text{Molarity} = 0.616\,\cancel{g} \times \frac{1\,\text{mol}}{150.1\,\cancel{g}} \times \frac{1}{0.250\,\text{L}} = 0.0164\,M$$

Since trifluoromethane sulfonic acid is a strong acid and is 100% ionized, the $[H^+]$ is 0.0165 *M*.

$$\mathbf{pH} = -\log(0.0164) = \mathbf{1.79}$$

15.112 The reactions are $\quad HF \rightleftharpoons H^+ + F^- \qquad (1)$

$$F^- + HF \rightleftharpoons HF_2^- \qquad (2)$$

Note that for equation (2), the equilibrium constant is relatively large with a value of 5.2. This means that the equilibrium lies to the right. Applying Le Chatelier's principle, as HF ionizes in the first step, the F^- that is produced is partially removed in the second step. More HF must ionize to compensate for the removal of the F^-, at the same time producing more H^+.

15.114 (a) We must consider both the complete ionization of the strong acid, and the partial ionization of water.

$$HA \longrightarrow H^+ + A^-$$
$$H_2O \rightleftharpoons H^+ + OH^-$$

From the above two equations, the $[H^+]$ in solution is:

$$[H^+] = [A^-] + [OH^-] \qquad (1)$$

We can also write:

$$[H^+][OH^-] = K_w$$

$$[OH^-] = \frac{K_w}{[H^+]}$$

Substituting into Equation (1):

$$[H^+] = [A^-] + \frac{K_w}{[H^+]}$$

$$[H^+]^2 = [A^-][H^+] + K_w$$

$$[H^+]^2 - [A^-][H^+] - K_w = 0$$

Solving a quadratic equation:

$$[H^+] = \frac{[A^-] \pm \sqrt{[A^-]^2 + 4K_w}}{2}$$

(b)   For the strong acid, HCl, with a concentration of $1.0 \times 10^{-7} \, M$, the $[Cl^-]$ will also be $1.0 \times 10^{-7} \, M$.

$$[H^+] = \frac{[Cl^-] \pm \sqrt{[Cl^-]^2 + 4K_w}}{2} = \frac{1 \times 10^{-7} \pm \sqrt{\left(1 \times 10^{-7}\right)^2 + 4\left(1 \times 10^{-14}\right)}}{2}$$

$$[H^+] = 1.6 \times 10^{-7} \, M \, (\text{or} -6.0 \times 10^{-8} \, M, \text{which is impossible})$$

$$\textbf{pH} = -\log[1.6 \times 10^{-7}] = \textbf{6.80}$$

15.116   The solution for the first step is standard:

	$H_3PO_4(aq)$	$\rightleftharpoons$	$H^+(aq)$	$+$	$H_2PO_4^-(aq)$
Initial (M):	0.100		0.000		0.000
Change (M):	$-x$		$+x$		$+x$
Equil. (M):	$(0.100 - x)$		x		x

$$K_{a_1} = \frac{[H^+][H_2PO_4^-]}{[H_3PO_4]} = \frac{x^2}{(0.100 - x)} = 7.5 \times 10^{-3}$$

In this case we probably cannot say that $(0.100 - x) \approx 0.100$ due to the magnitude of K_a. We obtain the quadratic equation:

$$x^2 + (7.5 \times 10^{-3})x - (7.5 \times 10^{-4}) = 0$$

The positive root is $x = 0.0239 \, M$. We have:

$$[H^+] = [H_2PO_4^-] = 0.0239 \, M$$

$$[H_3PO_4] = (0.100 - 0.0239) \, M = 0.076 \, M$$

For the second ionization:

	$H_2PO_4^-(aq)$	$\rightleftharpoons$	$H^+(aq)$	$+$	$HPO_4^{2-}(aq)$
Initial (M):	0.0239		0.0239		0.000
Change (M):	$-y$		$+y$		$+y$
Equil (M):	$(0.0239 - y)$		$0.0239 + y$		y

$$K_{a_2} = \frac{[H^+][HPO_4^{2-}]}{[H_2PO_4^-]} = \frac{(0.0239 + y)(y)}{(0.0239 - y)} = 6.2 \times 10^{-8}$$

Since K_{a_2} is small, we can assume $(0.0239 + y) \approx (0.0239 - y)$, so that $y = 6.2 \times 10^{-8}\ M$. Thus,

$$[H^+] = [H_2PO_4^-] = 0.0239\ M$$

$$[HPO_4^{2-}] = K_{a_2} = 6.2 \times 10^{-8}\ M$$

We set up the problem for the third ionization in the same manner.

	$HPO_4^{2-}(aq)$	$\rightleftharpoons$	$H^+(aq)$	$+$	$PO_4^{3-}(aq)$
Initial (M):	6.2×10^{-8}		0.0239		0
Change (M):	$-z$		$+z$		$+z$
Equil. (M):	$(6.2 \times 10^{-8} - z)$		$0.0239 + z$		z

$$K_{a_3} = \frac{[H^+][PO_4^{3-}]}{[HPO_4^{2-}]} = \frac{(0.0239 + z)(z)}{6.2 \times 10^{-8} - z} = 4.8 \times 10^{-13}$$

We can assume $(0.0239 + z) \approx 0.0239$ and $(6.2 \times 10^{-8}) - z \approx 6.2 \times 10^{-8}$

Therefore,

$$\frac{(0.0239)(z)}{6.2 \times 10^{-8}} = 4.8 \times 10^{-13}$$

$$z = 1.2 \times 10^{-18}\ M$$

The equilibrium concentrations are:

$$[H^+] = [H_2PO_4^-] = 0.0239\ M$$

$$[H_3PO_4] = 0.076\ M$$

$$[HPO_4^{2-}] = 6.2 \times 10^{-8}\ M$$

$$[PO_4^{3-}] = 1.2 \times 10^{-18}\ M$$

15.118 $0.100\ M\ Na_2CO_3 \rightarrow 0.200\ M\ Na^+ + 0.100\ M\ CO_3^{2-}$

<u>First stage:</u>

	$CO_3^{2-}(aq)$	$+$	$H_2O(l)$	$\rightleftharpoons$	$HCO_3^-(aq)$	$+$	$OH^-(aq)$
	0.100 M				x M		x M

$$K_1 = \frac{K_w}{K_2} = \frac{1.0 \times 10^{-14}}{4.8 \times 10^{-11}} = 2.1 \times 10^{-4}$$

$$K_1 = 2.1 \times 10^{-4} = \frac{[HCO_3^-][OH^-]}{[CO_3^{2-}]} = \frac{x^2}{0.100 - x}$$

$$x = 4.6 \times 10^{-3}\ M = [HCO_3^-] = [OH^-]$$

Second stage:

$$HCO_3^-(aq) \quad + \quad H_2O(l) \quad \rightleftharpoons \quad H_2CO_3(aq) \quad + \quad OH^-(aq)$$

$$(4.6 \times 10^{-3} - y) \, M \qquad\qquad\qquad y \, M \qquad\qquad (4.6 \times 10^{-3} + y) \, M$$

$$K_2 = \frac{[H_2CO_3][OH^-]}{[HCO_3^-]}$$

$$2.4 \times 10^{-8} = \frac{y\left(4.6 \times 10^{-3} + y\right)}{4.6 \times 10^{-3} - y}$$

$$y = 2.4 \times 10^{-8} \, M$$

At equilibrium:

$$[Na^+] = \mathbf{0.200 \, M}$$

$$[HCO_3^-] = (4.6 \times 10^{-3}) \, M - (2.4 \times 10^{-8}) \, M \approx \mathbf{4.6 \times 10^{-3} \, M}$$

$$[H_2CO_3] = \mathbf{2.4 \times 10^{-8} \, M}$$

$$[OH^-] = (4.6 \times 10^{-3}) \, M + (2.4 \times 10^{-8}] \, M \approx \mathbf{4.6 \times 10^{-3} \, M}$$

$$[H^+] = \frac{1.0 \times 10^{-14}}{4.6 \times 10^{-3}} = \mathbf{2.2 \times 10^{-12} \, M}$$

15.120 When NaCN is treated with HCl, the following reaction occurs.

$$NaCN + HCl \rightarrow NaCl + HCN$$

HCN is a very weak acid, and only partially ionizes in solution.

$$HCN \, (aq) \rightleftharpoons H^+(aq) + CN^-(aq)$$

The main species in solution is HCN which has a tendency to escape into the gas phase.

$$HCN(aq) \rightleftharpoons HCN(g)$$

Since the HCN(g) that is produced is a highly poisonous compound, it would be dangerous to treat NaCN with acids without proper ventilation.

15.122 $pH = 2.53 = -\log[H^+]$

$[H^+] = 2.95 \times 10^{-3} \, M$

The equation representing the ionization of formic acid is:

$$HCOOH(aq) \quad \rightleftharpoons \quad H^+(aq) \quad + \quad HCOO^-(aq)$$

$$y - (2.95 \times 10^{-3}) \qquad (2.95 \times 10^{-3}) \qquad (2.95 \times 10^{-3})$$

$$K_a = 1.7 \times 10^{-4} = \frac{[H^+][HCOO^-]}{[HCOOH]} = \frac{\left(2.95 \times 10^{-3}\right)^2}{\left(y - 2.95 \times 10^{-3}\right)}$$

$$y = 0.054 \, M$$

There are 0.054 moles of formic acid in 1000 mL of solution. The mass of formic acid in 100 mL is:

$$\frac{0.054 \text{ mol formic acid}}{1000 \text{ mL soln}} \times \frac{46.03 \text{ g formic acid}}{1 \text{ mol formic acid}} \times 100 \text{ mL} = \mathbf{0.25 \text{ g formic acid}}$$

15.124 The balanced equation is: $Mg + 2HCl \rightarrow MgCl_2 + H_2$

$$\text{mol of Mg} = \frac{\text{g Mg}}{\text{molar mass Mg}} = \frac{1.87 \text{ g Mg}}{24.31 \text{ g / mol}} = 0.0769 \text{ mol}$$

From the balanced equation:

mol of HCl required for reaction = 2 × mol Mg = 2 × 0.0769 mol = 0.154 mol HCl

The concentration of HCl:

pH = −0.544, thus $[H^+]$ = 3.50 M

mol HCl = M × Vol (L) = 3.50 M × 0.0800 L = 0.280 mol HCl

Moles of HCl left after reaction:

total mol HCl − mol HCl reacted = 0.280 mol − 0.154 mol = 0.126 mol HCl

Molarity of HCl left after reaction:

M = mol/L = 0.126 mol/0.080 L = 1.58 M **pH = −0.20**

15.126 The important equation is the hydrolysis of NO_2^-: $NO_2^- + H_2O \rightleftharpoons HNO_2 + OH^-$

(a) Addition of HCl will result in the reaction of the H^+ from the HCl with the OH^- that was present in the solution. The OH^- will effectively be removed and the equilibrium will shift to the right to compensate (more hydrolysis).

(b) Addition of NaOH is effectively addition of more OH^- which places stress on the right hand side of the equilibrium. The equilibrium will shift to the left (less hydrolysis) to compensate for the addition of OH^-.

(c) Addition of NaCl will have no effect.

(d) Recall that the percent ionization of a weak acid increases with dilution (see Figure 15.4). The same is true for weak bases. Thus dilution will cause more hydrolysis, shifting the equilibrium to the right.

15.128 In Chapter 11, we found that salts with their formal electrostatic intermolecular attractions had low vapor pressures and thus high boiling points. Ammonia and its derivatives (amines) are molecules with dipole-dipole attractions; as long as the nitrogen has one direct N–H bond, the molecule will have hydrogen bonding. Even so, these molecules will have much higher vapor pressures than ionic species. Thus, if we could convert the neutral ammonia-type molecules into salts, their vapor pressures, and thus associated odors, would decrease. Lemon juice contains acids which can react with neutral ammonia-type (amine) molecules to form ammonium salts.

$$NH_3 + H^+ \rightarrow NH_4^+$$

$$RNH_2 + H^+ \rightarrow RNH_3^+$$

15.130 $HCOOH \rightleftharpoons H^+ + HCOO^-$

$0.400 - x \qquad\quad x \qquad\quad x$

Total concentration of particles in solution: $(0.400 - x) + x + x = 0.400 + x$

Assuming the molarity of the solution is equal to the molality, we can write:

$$\Delta T_f = K_f m$$

$$0.758 = (1.86)(0.400 + x)$$

$$0.408 = 0.400 + x$$

$$x = 0.00800 = [H^+] = [HCOO^-]$$

$$K_a = \frac{[H^+][HCOO^-]}{[HCOOH]} = \frac{(0.00800)(0.00800)}{0.400 - 0.00800} = \mathbf{1.6 \times 10^{-4}}$$

15.132 $SO_2(g) + H_2O(l) \rightleftharpoons H^+(aq) + HSO_3^-(aq)$

Recall that 0.12 ppm SO_2 would mean 0.12 parts SO_2 per 1 million (10^6) parts of air by volume. The number of particles of SO_2 per volume will be directly related to the pressure.

$$P_{SO_2} = \frac{0.12 \text{ parts } SO_2}{10^6 \text{ parts air}} \text{ atm} = 1.2 \times 10^{-7} \text{ atm}$$

We can now calculate the $[H^+]$ from the equilibrium constant expression.

$$K = \frac{[H^+][HSO_3^-]}{P_{SO_2}}$$

$$1.3 \times 10^{-2} = \frac{x^2}{1.2 \times 10^{-7}}$$

$$x^2 = (1.3 \times 10^{-2})(1.2 \times 10^{-7})$$

$$x = 3.9 \times 10^{-5} M = [H^+]$$

$$\mathbf{pH} = -\log(3.9 \times 10^{-5}) = \mathbf{4.40}$$

15.134   In inhaling the smelling salt, some of the powder dissolves in the basic solution. The ammonium ions react with the base as follows:

$$NH_4^+(aq) + OH^-(aq) \rightarrow NH_3(aq) + H_2O$$

It is the pungent odor of ammonia that prevents a person from fainting.

CHAPTER 16
ACID-BASE EQUILIBRIA AND
SOLUBILITY EQUILIBRIA

PROBLEM-SOLVING STRATEGIES AND TUTORIAL SOLUTIONS

TYPES OF PROBLEMS

Problem Type 1: Buffers.
   (a) Identifying buffer systems.
   (b) Calculating the pH of a buffer system.
   (c) Preparing a buffer solution with a specific pH.

Problem Type 2: Titrations.
   (a) Strong acid–strong base titrations.
   (b) Weak acid–strong base titrations.
   (c) Strong acid–weak base titrations.

Problem Type 3: Choosing Suitable Acid-Base Indicators.

Problem Type 4: Solubility Equilibria.
   (a) Calculating K_{sp} from molar solubility.
   (b) Calculating solubility from K_{sp}.
   (c) Predicting a precipitation reaction.
   (d) The effect of a common ion on solubility.

Problem Type 5: Complex Ion Equilibria and Solubility.

PROBLEM TYPE 1: BUFFERS

A **buffer solution** is a solution of (1) a weak acid or a weak base and (2) its salt; both components must be present. The solution has the ability to resist change in pH upon the addition of small amounts of either acid or base. A buffer resists change in pH, because the weak acid component reacts with small amounts of added base. The weak base component of the buffer reacts with small amounts of added acid.

$$\text{HA} \ (aq) \quad + \quad \text{OH}^- \ (aq) \quad \longrightarrow \quad \text{A}^- \ (aq) + \text{H}_2\text{O} \ (l)$$
weak acid   added base
of buffer

$$\text{A}^- \ (aq) \quad + \quad \text{H}_3\text{O}^+ \ (aq) \quad \longrightarrow \quad \text{HA} \ (aq) + \text{H}_2\text{O} \ (l)$$
weak base   added acid
of buffer

A. Identifying buffer systems

To identify a buffer system, look for a weak acid and its conjugate base (usually the anion in a soluble salt) or a weak base and its conjugate acid (usually the cation in a soluble salt).

EXAMPLE 16.1

Which of the following solutions are buffer systems: (a) CH_3COOH/CH_3COONa, (b) KNO_3/HNO_3, and (c) NH_4Cl/NH_3?

(a) CH_3COOH (acetic acid) is a weak acid, and its conjugate base, CH_3COO^- (acetate ion, the anion of the salt CH_3COONa), is a weak base. Therefore, this is a buffer system.

(b) Because HNO_3 is a strong acid, its conjugate base, NO_3^-, is an extremely weak base. This means that NO_3^- will not combine with H^+ in solution to form HNO_3. Thus, the system cannot act as a buffer system.

(c) NH_3 (ammonia) is a weak base and its conjugate acid, NH_4^+ (ammonium ion, the cation of the salt NH_4Cl), is a weak acid. Therefore, this is a buffer system.

PRACTICE EXERCISE

1.   Which of the following solutions are buffer systems: (a) KCN/HCN, (b) $NaCl/HCl$, and (c) HNO_2/KNO_2?

Text Problem: 16.8

B. Calculating the pH of a buffer system

Calculating the pH of a buffer system is similar to calculating the pH of a weak acid solution (Problem Type 5A, Chapter 15) or the pH of a weak base solution (Problem Type 7, Chapter 15). The difference is that the initial concentration of the conjugate of the weak acid or weak base is *not zero*. The initial concentration of the conjugate comes from the salt component of the buffer.

For example, consider a solution that is 1.0 M in CH_3COOH (acetic acid) and 1.0 M in CH_3COONa (sodium acetate). If this were only a 1.0 M acetic acid solution, the initial concentration of the conjugate base (CH_3COO^-) would be *zero*. However, CH_3COONa is a soluble salt that ionizes completely.

$$CH_3COONa\,(aq) \longrightarrow Na^+\,(aq) + CH_3COO^-\,(aq)$$

Initial	1.0 M	0	0
After dissociation	0	1.0 M	1.0 M

The CH_3COO^- ion is present initially and enters into the weak acid equilibrium of acetic acid. Thus, if we look at the weak acid equilibrium for acetic acid, we have:

$$CH_3COOH\,(aq) \rightleftharpoons H^+\,(aq) + CH_3COO^-\,(aq)$$

Initial (M):	1.0	0	1.0
Change (M):	$-x$	$+x$	$+x$
Equilibrium (M):	$1.0 - x$	x	$1.0 + x$

You should notice that the only difference between this equilibrium, and a weak acid equilibrium (see Problem Type 5A, Chapter 15), is that the initial concentration of the conjugate base, CH_3COO^-, is *not zero*, as it would be for a weak acid equilibrium. Remember, a buffer contains both a weak acid and its conjugate base or a weak base and its conjugate acid. The above system is a buffer because it initially contains both the weak acid (CH_3COOH) and its conjugate base (CH_3COO^-). Example 16.2 illustrates how to calculate the pH of a buffer system.

EXAMPLE 16.2

(a) Calculate the pH of a buffer system containing 0.25 M HF and 0.50 M NaF.

(b) What is the pH of the buffer system after the addition of 0.060 mol of gaseous HCl to 1.00 L of the solution? Assume that the volume of the solution does not change upon addition of the HCl.

(a)

Step 1: Recognize that NaF is a soluble salt that will completely dissociate. From the dissociation of NaF, we can calculate the initial concentration of the weak base, F^-.

$$NaF\,(aq) \longrightarrow Na^+\,(aq) + F^-\,(aq)$$

Initial	0.50 M	0	0
After dissociation	0	0.50 M	0.50 M

Step 2: Next, consider the weak acid equilibrium. Express the equilibrium concentrations of all species in terms of initial concentrations and a single unknown, x, which represents the change in concentration. Let $(-x)$ be the depletion in concentration (mol/L) of HF. From the stoichiometry of the reaction, it follows that the increase in concentration for both H^+ and F^- must be x. Complete a table that lists the initial concentrations, the change in concentrations, and the equilibrium concentrations.

$$HF\,(aq) \rightleftharpoons H^+\,(aq) + F^-\,(aq)$$

	HF	H^+	F^-
Initial (M):	0.25	0	0.50
Change (M):	$-x$	$+x$	$+x$
Equilibrium (M):	$0.25 - x$	x	$0.50 + x$

Step 3: Write the ionization constant expression in terms of the equilibrium concentrations. Knowing the value of the equilibrium constant (K_a), solve for x.

$$K_a = \frac{[H^+][F^-]}{[HF]}$$

You can look up the K_a value for hydrofluoric acid in Table 15.3 of your text.

$$7.1 \times 10^{-4} = \frac{(x)(0.50 + x)}{(0.25 - x)}$$

At this point, we will make two assumptions.

$$0.25 - x \approx 0.25$$

and

$$0.50 + x \approx 0.50$$

Oftentimes, assumptions such as these are valid if K is very small. A very small value of K means that a very small amount of reactants go to products. Hence, x is small. If we did not make this assumption, we would have to solve a quadratic equation!

$$7.1 \times 10^{-4} \approx \frac{(x)(0.50)}{0.25}$$

Solving for x,

$$x = 3.6 \times 10^{-4}\,M = [H^+]$$

Checking the validity of the assumption,

$$\frac{3.6 \times 10^{-4}}{0.25} \times 100\% = 0.14\% < 5\%$$

The assumption is valid.

Step 3: Having solved for the $[H^+]$, calculate the pH of the solution.

$$\textbf{pH} = -\log[H^+] = -\log(3.6 \times 10^{-4}) = \textbf{3.44}$$

(b)   We have added 0.060 mol of HCl (strong acid), which will react completely with the weak base component of the buffer. This reaction will change the equilibrium concentrations of both the acid and base components of the equilibrium. Therefore, after the reaction between HCl and the weak base of the buffer, the equilibrium must be reestablished.

Step 1: Write the reaction that occurs between the strong acid, HCl, and the weak base component of the buffer, F^-. After addition of HCl, complete ionization of HCl occurs.

	$HCl\,(aq)$	$\longrightarrow$	$H^+\,(aq)$	$+$	$Cl^-\,(aq)$
Initial	0.060 mol		0		0
After ionization	0		0.060 mol		0.060 mol

Next, H^+ will react with the weak base component of the buffer, F^-. Since H^+ is the strongest acid that can exist in water, this reaction will be driven to completion.

$$H^+\,(aq) + F^-\,(aq) \rightarrow HF\,(aq)$$

Since we have 1.00 L of buffer solution, the number of moles of F^- in solution is:

$$\text{mol } F^- = \left(0.50\,\frac{\text{mol}}{\text{L}}\right) \times 1.00\,\text{L} = 0.50\,\text{mol}$$

Similarly, the number of moles of HF is 0.25 mol.

We can now calculate the moles of F^- and HF that remain after F^- reacts with the strong acid HCl.

	$H^+\,(aq)$	$+$	$F^-\,(aq)$	$\longrightarrow$	$HF\,(aq)$
Initial (mol)	0.060		0.50		0.25
Change (mol)	−0.060		−0.060		+0.060
After reaction (mol)	0		0.44		0.31

Step 2: After the reaction above, HF and F^- are no longer in equilibrium. Equilibrium must be reestablished. The concentrations of HF and F^- are:

$$[HF] = \frac{0.31\,\text{mol}}{1.00\,\text{L}} = 0.31\,M$$

$$[F^-] = \frac{0.44\,\text{mol}}{1.00\,\text{L}} = 0.44\,M$$

Reestablishing equilibrium between HF and F^-, we have:

	$HF\,(aq)$	$\rightleftharpoons$	$H^+\,(aq)$	$+$	$F^-\,(aq)$
Initial (*M*):	0.31		0		0.44
Change (*M*):	−x		+x		+x
Equilibrium (*M*):	0.31 − x		x		0.44 + x

Step 3: Write the ionization constant expression in terms of the equilibrium concentrations. Knowing the value of the equilibrium constant (K_a), solve for x.

$$K_a = \frac{[H^+][F^-]}{[HF]}$$

$$7.1 \times 10^{-4} = \frac{(x)(0.44 + x)}{(0.31 - x)}$$

$$7.1 \times 10^{-4} \approx \frac{(x)(0.44)}{0.31}$$

$$x = [H^+] = 5.0 \times 10^{-4}\, M$$

Step 4: Having solved for the $[H^+]$, calculate the pH of the solution.

$$\textbf{pH} = -\log[H^+] = -\log(5.0 \times 10^{-4}) = \textbf{3.30}$$

The pH of the solution dropped only 0.15 pH units upon addition of 0.060 mol of the strong acid, HCl. Thus, this buffer solution resisted change in pH.

Does it make sense that the pH of the solution decreased upon the addition of a strong acid to the buffer?

PRACTICE EXERCISE

2. A buffer solution is prepared by mixing 500 mL of 0.600 M CH_3COOH with 500 mL of a 1.00 M CH_3COONa solution. What is the pH of the solution?

3. (a) Calculate the pH of a buffer system that is 0.0600 M HNO_2 and 0.160 M $NaNO_2$.
    (b) What is the pH after 2.00 mL of 2.00 M NaOH are added to 1.00 L of this buffer?

Text Problems: 16.10, 16.12, 16.16

C. Preparing a buffer solution with a specific pH

To prepare a buffer solution with a specific pH, we need to consider the Henderson-Hasselbalch equation. For a derivation of this equation, see Section 16.2 of the text.

$$pH = pK_a + \log\frac{[conjugate\ base]}{[acid]}$$

where,

$$pK_a = -\log K_a$$

If the molar concentrations of the acid and its conjugate base are approximately equal, that is,

$$[acid] \approx [conjugate\ base]$$

then,

$$\log\frac{[conjugate\ base]}{[acid]} \approx 0$$

Substituting into the Henderson-Hasselbalch equation gives:

$$pH \approx pK_a$$

Thus, to prepare a buffer solution, we should choose a weak acid with a pK_a value close to the desired pH. This choice not only gives the desired pH value for the buffer system, but also ensures that we have comparable amounts of the weak acid and its conjugate base present. Having pH $\approx pK_a$ is a prerequisite for the buffer system to function effectively.

EXAMPLE 16.3
Which of the following mixtures is suitable for making a buffer solution with an optimum pH of about 9.2?

(a) CH_3COONa/CH_3COOH (b) NH_4Cl/NH_3 (c) HF/NaF

(d) $HNO_2/NaNO_2$ (e) NaCl/HCl

We can rule out choice (e) immediately, because it contains a strong acid and its conjugate base. This solution is not a buffer.

For the other choices we need to calculate the pK_a and compare it to the desired pH. The solutions with a pK_a values close to the desired pH will be effective buffers.

$$(a) \quad pK_a = 4.74$$
$$(b) \quad pK_a = 9.26$$
$$(c) \quad pK_a = 3.15$$
$$(d) \quad pK_a = 3.35$$

Thus, the only solution that would make an effective buffer at pH = 9.2 is choice (b), NH_4Cl/NH_3.

> **Tip:** We could have saved some time by recognizing that if the K_a value is less than 1×10^{-7}, the pK_a value will be greater than 7. The only choice that had a K_a value less than 1×10^{-7} was choice (b). Thus, it was the only choice that has a pK_a value greater than 7.

PRACTICE EXERCISE

4.   How would you prepare a liter of an HF/F$^-$ buffer at a pH of 2.85?

> **Text Problem: 16.18**

PROBLEM TYPE 2: TITRATIONS

A. Strong acid–strong base titrations

The reaction between a strong acid (HNO_3) and a strong base (KOH) can be written as:

$$HNO_3\ (aq) + KOH\ (aq) \longrightarrow KNO_3\ (aq) + H_2O\ (l)$$

or in terms of the net ionic equation

$$H^+\ (aq) + OH^-\ (aq) \longrightarrow H_2O\ (l)$$

At the equivalence point of a titration, the $[H^+] = [OH^-]$. Thus, the only species present in solution other than water at the equivalence point is $KNO_3\ (aq)$. Potassium nitrate is a salt that does not undergo hydrolysis (see Problem Type 8, Chapter 15); therefore, the pH at the equivalence point of a strong acid–strong base titration is 7.

A typical strong acid–strong base titration problem involves calculating the pH at various points in the titration. Example 16.4 below illustrates this type of problem.

EXAMPLE 16.4

A 20.0 mL sample of 0.0200 M HNO$_3$ is titrated with 0.0100 M KOH.

(a) What is the pH at the equivalence point?
(b) How many mL of KOH are required to reach the equivalence point?
(c) What is the pH before any KOH is added?
(d) What will be the pH after 10.0 mL of KOH are added?
(e) What will be the pH after 45.0 mL of KOH are added?

(a)   Since this is a strong acid–strong base titration, the pH at the equivalence point is 7.

(b)   We worked this type of problem in Chapter 4 (see Problem Type 8, Titrations).

Step 1: In order to have the correct mole ratio to solve the problem, you must start with a balanced chemical equation.

$$\text{HNO}_3\,(aq) \,+\, \text{KOH}\,(aq) \,\longrightarrow\, \text{KNO}_3\,(aq) \,+\, \text{H}_2\text{O}\,(l)$$

Step 2: From the molarity and volume of the HNO$_3$ solution, you can calculate moles of HNO$_3$. Then, using the mole ratio from the balanced equation above, you can calculate moles of KOH.

20.0 mL = 0.0200 L

$$\text{mol KOH} = \frac{0.0200 \text{ mol HNO}_3}{1 \text{ L of solution}} \times (0.0200 \text{ L solution}) \times \frac{1 \text{ mol KOH}}{1 \text{ mol HNO}_3} = 4.00 \times 10^{-4} \text{ mol KOH}$$

Step 3: Solve the molarity equation algebraically for liters of solution. Then, substitute in the moles of KOH and molarity of KOH to solve for volume of KOH.

$$M = \frac{\text{moles of solute}}{\text{liters of solution}}$$

$$\text{liters of solution} = \frac{\text{moles of solute}}{M}$$

$$\textbf{volume of KOH} = \frac{4.00 \times 10^{-4} \text{ mol KOH}}{0.0100 \text{ mol} / \text{L}} = \textbf{0.0400 L} = \textbf{40.0 mL}$$

(c)   We only have the strong acid, HNO$_3$, in solution before any KOH is added. Thus, we need to calculate the pH of a strong acid (Problem Type 4, Chapter 15).

Step 1: Strong acids ionize completely in solution. Let's write the reaction and set up a table to calculate the H$^+$ ion concentration in solution.

	HNO$_3$ (aq) $\longrightarrow$	H$^+$ (aq) +	NO$_3^-$ (aq)
initial conc.	0.0200 M	0	0
conc. after ionization	0	0.0200 M	0.0200 M

Since there is a 1:1 mole ratio between H$^+$ and HNO$_3$, the H$^+$ concentration equals the HNO$_3$ concentration.

Step 2: Calculate the pH from the H$^+$ concentration.

$$\textbf{pH} = -\log[\text{H}^+] = -\log[0.0200\ M] = \textbf{1.70}$$

(d)   Any strong base added will react completely with the strong acid in solution. The reaction is:

$$HNO_3\ (aq)\ +\ KOH\ (aq)\ \longrightarrow\ KNO_3\ (aq)\ +\ H_2O\ (l)$$

Step 1: Calculate the number of moles of HNO_3 in 20.0 mL, and calculate the number of moles of KOH in 10.0 mL.

$$mol\ HNO_3\ =\ 20.0\ mL\ soln\ \times\ \frac{1.00\ L}{1000\ mL}\ \times\ \frac{0.0200\ mol\ HNO_3}{1.00\ L\ soln}\ =\ 4.00\times10^{-4}\ mol$$

$$mol\ KOH\ =\ 10.0\ mL\ soln\ \times\ \frac{1.00\ L}{1000\ mL}\ \times\ \frac{0.0100\ mol\ KOH}{1.00\ L\ soln}\ =\ 1.00\times10^{-4}\ mol$$

Step 2: Set up a table showing the number of moles of HNO_3 and KOH before and after the reaction.

	$HNO_3\ (aq)$ +	$KOH\ (aq)$ $\longrightarrow$	$KNO_3\ (aq)$ +	$H_2O\ (l)$
initial mol	4.00×10^{-4}	1.00×10^{-4}	0	
mol after reaction	3.00×10^{-4}	0	1.00×10^{-4}	

Thus, 3.00×10^{-4} mol of the strong acid HNO_3 remain after the addition of 10.0 mL of KOH.

Step 3: The total volume of solution is now the sum of the volume of the acid solution and the volume of the base solution.

$$V_{soln}\ =\ V_{acid}\ +\ V_{base}\ =\ 20.0\ mL\ +\ 10.0\ mL\ =\ 30.0\ mL$$

We can now calculate the concentration of H^+ in solution. Since HNO_3 ionizes completely, the number of moles of H^+ in solution is 3.00×10^{-4} mol. Calculate the molarity by dividing the moles by liters of solution. 30.0 mL = 0.0300 L

$$[H^+]\ =\ \frac{3.00\times10^{-4}\ mol}{0.0300\ L}\ =\ 0.0100\ M$$

Step 4: Calculate the pH from the H^+ concentration.

$$\mathbf{pH}\ =\ -log[H^+]\ =\ -log[0.0100\ M]\ =\ \mathbf{2.00}$$

(e)   Start this part by following the same procedure as in part (d).

Step 1: Calculate the number of moles of KOH in 45.0 mL.

$$mol\ KOH\ =\ 45.0\ mL\ soln\ \times\ \frac{1.00\ L}{1000\ mL}\ \times\ \frac{0.0100\ mol\ KOH}{1.00\ L\ soln}\ =\ 4.50\times10^{-4}\ mol$$

Step 2: Set up a table showing the number of moles of HNO_3 and KOH before and after the reaction.

	$HNO_3\ (aq)$ +	$KOH\ (aq)$ $\longrightarrow$	$KNO_3\ (aq)$ +	$H_2O\ (l)$
initial mol	4.00×10^{-4}	4.50×10^{-4}	0	
mol after reaction	0	5.00×10^{-5}	4.00×10^{-4}	

We have passed the equivalence point of the titration. The only species of significance that remains in solution is the strong base, KOH. 5.00×10^{-5} mol of the strong base KOH remain after the addition of 45.0 mL of KOH.

Step 3: The total volume of solution is now the sum of the volume of the acid solution and the volume of the base solution.

$$V_{\text{soln}} = V_{\text{acid}} + V_{\text{base}} = 20.0 \text{ mL} + 45.0 \text{ mL} = 65.0 \text{ mL}$$

We can now calculate the concentration of OH^- in solution. Since KOH ionizes completely, the number of moles of OH^- in solution is 5.00×10^{-5} mol. Calculate the molarity by dividing the moles by liters of solution.

$$65.0 \text{ mL} = 0.0650 \text{ L}$$

$$[OH^-] = \frac{5.00 \times 10^{-5} \text{ mol}}{0.0650 \text{ L}} = 7.69 \times 10^{-4} \ M$$

Step 4: Calculate the pOH from the OH^- concentration.

$$\text{pOH} = -\log[OH^-] = -\log[7.69 \times 10^{-4} \ M] = 3.11$$

Step 5: Use the relationship, pH + pOH = 14.00, to calculate the pH of the solution.

$$\textbf{pH} = 14.00 - 3.11 = \textbf{10.89}$$

PRACTICE EXERCISE

5.   Consider the titration of 25.0 mL of 0.250 M KOH with 0.100 M HCl.

  (a)   What is the pH at the equivalence point?
  (b)   How many mL of HCl are required to reach the equivalence point?
  (c)   What is the pH before any HCl is added?
  (d)   What will be the pH after 15.0 mL of HCl are added?
  (e)   What will be the pH after 75.0 mL of HCl are added?

Text Problem: 16.22

B. Weak acid–strong base titrations

Consider the neutralization between nitrous acid (a weak acid) and sodium hydroxide (a strong base):

$$HNO_2 \ (aq) + NaOH \ (aq) \longrightarrow NaNO_2 \ (aq) + H_2O \ (l)$$

The net ionic equation is:

$$HNO_2 \ (aq) + OH^- \ (aq) \longrightarrow NO_2^- \ (aq) + H_2O \ (l)$$

At the equivalence point of a titration, the $[HNO_2] = [OH^-]$. Thus, the major species present in solution other than water at the equivalence point are $NO_2^- \ (aq)$ and $Na^+ \ (aq)$. NO_2^- is a weak base; it is the conjugate base of the weak acid, HNO_2 (see Problem Type 8, Chapter 15). The hydrolysis reaction is given by:

$$NO_2^- \ (aq) + H_2O \ (l) \rightleftharpoons HNO_2 \ (aq) + OH^- \ (aq)$$

Therefore, at the equivalence point, the pH will be *greater than 7* as a result of the excess OH^- formed.

A typical weak acid–strong base titration problem involves calculating the pH at various points in the titration. Example 16.5 below illustrates this type of problem.

EXAMPLE 16.5

Consider the titration of 50.0 mL of 0.100 M CH_3COOH with 0.100 M NaOH.

(a)   How many mL of NaOH are required to reach the equivalence point?
(b)   What is the pH before any NaOH is added?
(c)   What will be the pH after 25.0 mL of NaOH are added?
(d)   What is the pH at the equivalence point?
(e)   What will be the pH after 60.0 mL of KOH are added?

(a)   We worked this type of problem in Chapter 4 (see Problem Type 8, Titrations) and in Example 16.4 above.

Step 1: In order to have the correct mole ratio to solve the problem, you must start with a balanced chemical equation.

$$CH_3COOH\ (aq)\ +\ NaOH\ (aq)\ \longrightarrow\ CH_3COONa\ (aq)\ +\ H_2O\ (l)$$

Step 2: We can take a shortcut on this problem by recognizing that both CH_3COOH and NaOH have the same concentration, 0.100 M. Since the mole ratio between CH_3COOH and NaOH is 1:1, it will require the same volume of each component to neutralize the other.

Thus, the volume of NaOH required to reach the equivalence point is **50.0 mL**.

(b)   We only have the weak acid, CH_3COOH, in solution before any NaOH is added. Thus, we need to calculate the pH of a weak acid (Problem Type 5A, Chapter 15).

Step 1: Express the equilibrium concentrations of all species in terms of initial concentrations and a single unknown x, that represents the change in concentration. Let $(-x)$ be the depletion in concentration (mol/L) of CH_3COOH. From the stoichiometry of the reaction, it follows that the increase in concentration for both H^+ and CH_3COO^- must be x. Complete a table that lists the initial concentrations, the change in concentrations, and the equilibrium concentrations.

	$CH_3COOH\ (aq)$	$\rightleftharpoons$	$H^+\ (aq)\ +$	$CH_3COO^-\ (aq)$
Initial (*M*):	0.100		0	0
Change (*M*):	$-x$		$+x$	$+x$
Equilibrium (*M*):	$0.100 - x$		x	x

Step 2: Write the ionization constant expression in terms of the equilibrium concentrations. Knowing the value of the equilibrium constant (K_a), solve for x.

$$K_a = \frac{[H^+][CH_3COO^-]}{[CH_3COOH]}$$

You can look up the K_a value for acetic acid in Table 15.3 of your text.

$$1.8 \times 10^{-5} = \frac{(x)(x)}{(0.100 - x)}$$

At this point, we can make an assumption that x is very small compared to 0.100. Hence,

$$0.100 - x \approx 0.100$$

and,

$$1.8 \times 10^{-5} = \frac{(x)(x)}{0.100}$$

Solving for x.

$$x = 1.34 \times 10^{-3} \, M = [H^+]$$

Checking the validity of the assumption,

$$\frac{1.34 \times 10^{-3}}{0.100} \times 100\% = 1.34\% < 5\%$$

The assumption was valid.

Step 3: Having solved for the $[H^+]$, calculate the pH of the solution.

$$\mathbf{pH} = -\log[H^+] = -\log(1.34 \times 10^{-3}) = \mathbf{2.87}$$

(c) You should recognize that when 25.0 mL of NaOH are added, we are half-way to the equivalence point. For a weak acid-strong base titration or a weak base-strong acid titration, the pH at the half-way point equals pK_a. This relation can be derived from the Henderson-Hasselbalch equation. At the half-way point, the concentration of the acid equals the concentration of its conjugate base, because half of the acid (CH_3COOH) has been neutralized, forming an equal amount of its conjugate base (CH_3COO^-).

$$pH = pK_a + \log \frac{[\text{conjugate base}]}{[\text{acid}]}$$

At the halfway point,

$$[\text{acid}] = [\text{conjugate base}]$$

and,

$$\log \frac{[\text{conjugate base}]}{[\text{acid}]} = 0$$

Substituting into the Henderson-Hasselbalch equation gives:

$$pH = pK_a$$

$$\mathbf{pH} = pK_a = -\log(1.8 \times 10^{-5}) = \mathbf{4.74}$$

(d) At the equivalence point the strong base has completely neutralized the weak acid in solution. The reaction is:

$$CH_3COOH \, (aq) + NaOH \, (aq) \longrightarrow CH_3COONa \, (aq) + H_2O \, (l)$$

Step 1: The number of moles of acetic acid equals the number of moles of NaOH at the equivalence point. Calculate the number of moles of acetic acid and sodium hydroxide.

$$\text{mol } CH_3COOH = 50.0 \text{ mL soln} \times \frac{1.00 \text{ L}}{1000 \text{ mL}} \times \frac{0.100 \text{ mol } CH_3COOH}{1.00 \text{ L soln}} = 5.00 \times 10^{-3} \text{ mol}$$

$$\text{mol KOH} = \text{mol } CH_3COOH$$

Step 2: Set up a table showing the number of moles of CH_3COOH and NaOH before and after the reaction.

$$CH_3COOH\ (aq)\ +\ NaOH\ (aq)\ \longrightarrow\ CH_3COONa\ (aq)\ +\ H_2O\ (l)$$

initial mol	5.00×10^{-3}	5.00×10^{-3}	0
mol after reaction	0	0	5.00×10^{-3}

The only species in solution of significance at the equivalence point is the salt CH_3COONa. This salt contains the conjugate base (CH_3COO^-) of the weak acid, CH_3COOH. The conjugate base is a weak base. The concentration of CH_3COO^- is

$$[CH_3COO^-] = \frac{5.00 \times 10^{-3}\ mol\ CH_3COO^-}{0.100\ L\ soln} = 0.0500\ M$$

Remember, to calculate the total volume of solution you must add the volume of the acid to the volume of the base.

$$50.0\ mL\ acid\ +\ 50.0\ mL\ base\ =\ 100.0\ mL\ soln\ =\ 0.100\ L\ soln$$

At this point, this problem just becomes a weak base calculation (see Problem Type 7, Chapter 15).

Step 3: Express the equilibrium concentrations of all species in terms of initial concentrations and a single unknown x, that represents the change in concentration. Let $(-x)$ be the depletion in concentration (mol/L) of CH_3COO^-. From the stoichiometry of the reaction, it follows that the increase in concentration for both OH^- and CH_3COOH must be x. Complete a table that lists the initial concentrations, the change in concentrations, and the equilibrium concentrations.

$$CH_3COO^-\ (aq)\ +\ H_2O\ (l)\ \rightleftharpoons\ CH_3COOH\ (aq)\ +\ OH^-\ (aq)$$

	CH_3COO^-		CH_3COOH	OH^-
Initial (M):	0.0500		0	0
Change (M):	$-x$		$+x$	$+x$
Equilibrium (M):	$0.0500 - x$		x	x

Step 4: Write the ionization constant expression in terms of the equilibrium concentrations. Knowing the value of the equilibrium constant (K_b), solve for x.

We can calculate K_b from the K_a value of acetic acid.

$$K_b = \frac{K_w}{K_a} = \frac{1.0 \times 10^{-14}}{1.8 \times 10^{-5}} = 5.6 \times 10^{-10}$$

$$K_b = \frac{[CH_3COOH][OH^-]}{[CH_3COO^-]}$$

$$5.6 \times 10^{-10} = \frac{(x)(x)}{(0.0500 - x)}$$

$$5.6 \times 10^{-10} \approx \frac{(x)(x)}{0.0500}$$

Solving for x.

$$x = 5.3 \times 10^{-6}\ M = [OH^-]$$

Step 5: Having solved for the $[OH^-]$, calculate the pOH of the solution. Then use the relationship, $pH + pOH = 14$, to solve for the pH of the solution.

$$pOH = -\log[OH^-] = -\log(5.3 \times 10^{-6}) = 5.28$$

$$\textbf{pH} = 14 - pOH = 14.00 - 5.28 = \textbf{8.72}$$

Does it make sense that the pH at the equivalence point is *greater than* 7?

(e)   After 60.0 mL of NaOH have been added, we have passed the equivalence point. The reaction is:

$$CH_3COOH\ (aq) + NaOH\ (aq) \rightarrow CH_3COONa\ (aq) + H_2O\ (l)$$

Step 1: Calculate the number of moles of sodium hydroxide in 60.0 mL of solution.

$$\text{mol NaOH} = 60.0 \text{ mL soln} \times \frac{1.00 \text{ L}}{1000 \text{ mL}} \times \frac{0.100 \text{ mol NaOH}}{1.00 \text{ L soln}} = 6.00 \times 10^{-3} \text{ mol}$$

Step 2: Set up a table showing the number of moles of CH_3COOH and NaOH before and after the reaction.

	$CH_3COOH\ (aq)$	$+$	$NaOH\ (aq)$	$\rightarrow$	$CH_3COONa\ (aq)$	$+$	$H_2O\ (l)$
initial mol	5.00×10^{-3}		6.00×10^{-3}		0		
mol after reaction	0		1.00×10^{-3}		5.00×10^{-3}		

The only species in solution of significance past the equivalence point is the strong base NaOH. The salt does contain a weak base, but since NaOH is a strong base, we can assume that all the OH^- comes from NaOH. The concentration of OH^- is:

$$[OH^-] = \frac{1.00 \times 10^{-3} \text{ mol } OH^-}{0.110 \text{ L soln}} = 9.09 \times 10^{-3} \ M$$

Step 3: Having solved for the $[OH^-]$, calculate the pOH of the solution. Then use the relationship, $pH + pOH = 14$, to solve for the pH of the solution.

$$pOH = -\log[OH^-] = -\log(9.09 \times 10^{-3}) = 2.04$$

$$\textbf{pH} = 14.00 - pOH = 14.00 - 2.04 = \textbf{11.96}$$

PRACTICE EXERCISE

6.   Consider the titration of 20.0 mL of 0.200 M HNO_2 with 0.100 M NaOH.

   (a)   How many mL of NaOH are required to reach the equivalence point?
   (b)   What is the pH before any NaOH is added?
   (c)   What will be the pH after 20.0 mL of NaOH are added?
   (d)   What is the pH at the equivalence point?
   (e)   What will be the pH after 50.0 mL of KOH are added?

Text Problems: 16.24, 16.26, 16.28

C. Strong acid–weak base titrations

Consider the neutralization between ammonia (a weak base) and nitric acid (a strong acid):

$$NH_3\ (aq) + HNO_3\ (aq) \rightarrow NH_4NO_3\ (aq)$$

The net ionic equation is:

$$NH_3\ (aq)\ +\ H^+\ (aq)\ \rightarrow\ NH_4^+\ (aq)$$

At the equivalence point of a titration, the $[NH_3]$ = $[H^+]$. Thus, the major species present in solution other than water at the equivalence point are $NH_4^+\ (aq)$ and $NO_3^-\ (aq)$. NH_4^+ is a weak acid; it is the conjugate acid of the weak base, NH_3 (see Problem Type 8, Chapter 15). The hydrolysis reaction is given by:

$$NH_4^+\ (aq)\ +\ H_2O\ (l)\ \rightleftharpoons\ NH_3\ (aq)\ +\ H_3O^+\ (aq)$$

Therefore, at the equivalence point, the pH will be *less than 7* as a result of the excess H_3O^+ ions formed.

A weak base–strong acid titration problem is worked similarly to a weak acid–strong base titration problem (see Example 16.5).

PROBLEM TYPE 3: CHOOSING SUITABLE ACID-BASE INDICATORS

The criterion for choosing an appropriate indicator for a given titration is whether the pH range over which the indicator changes color corresponds with the steep portion of the titration curve. If the indicator does not change color during this portion of the curve, it will not accurately identify the equivalence point. Table 16.1 of your text lists some common acid-base indicators and the pH ranges in which they change color.

Typically, the pH range for an indicator is equal to the pK_a of the indicator, plus or minus one pH unit.

$$pH\ range\ =\ pK_a\ \pm\ 1$$

EXAMPLE 16.6
Will bromophenol blue be a good choice as an indicator for a weak acid–strong base titration?

A typical titration curve for a weak acid–strong base titration is shown in Figure 16.4 of your text. The pH at the equivalence point is greater than 7. Checking Table 16.1, bromophenol blue changes color over the pH range, 3.0–4.6. This indicator would change color before the steep portion of the titration curve for a typical weak acid–strong base titration. Thus, bromophenol blue would *not* be a good choice as an indicator for a weak acid–strong base titration.

PRACTICE EXERCISE
7. Would methyl red be a suitable indicator for a titration that has a pH of 5.0 at the equivalence point?

Text Problems: 16.32, 16.34

PROBLEM TYPE 4: SOLUBILITY EQUILIBRIA

Solubility equilibria typically involve salts of low solubility. For example, calcium phosphate, $[Ca_3(PO_4)_2]$, is practically insoluble in water. However, a very small amount of calcium phosphate will dissolve and dissociate completely into Ca^{2+} and PO_4^{3-} ions. An equilibrium will then be established between undissolved calcium phosphate and the ions in solution. Consider a saturated solution of calcium phosphate that is in contact with solid calcium phosphate. The solubility equilibrium can be represented as

$$Ca_3(PO_4)_2\ (s)\ \rightleftharpoons\ 3\ Ca^{2+}\ (aq)\ +\ 2\ PO_4^{3-}\ (aq)$$

We know from Chapter 14 that for heterogeneous equilibria, the concentration of a solid is a constant. Thus, we can write the equilibrium constant for the dissolution of $Ca_3(PO_4)_2$ as

$$K_{sp} = [Ca^{2+}]^3[PO_4^{3-}]^2$$

where K_{sp} is called the solubility product constant or simply the solubility product. In general, the **solubility product** of a compound is the product of the molar concentrations of the constituent ions, each raised to the power of its stoichiometric coefficient in the balanced equilibrium equation.

A.   Calculating K_{sp} from molar solubility

Molar solubility is the number of moles of solute in 1 L of a saturated solution (moles per liter). The concentrations of the ions in the solubility product expression are also molar concentrations. From the molar solubility, we can calculate the molar concentrations of the ions in solution from the stoichiometry of the balanced equilibrium equation.

EXAMPLE 16.7

If the solubility of $Fe(OH)_2$ in water is 7.7×10^{-6} mol/L at a certain temperature, what is its K_{sp} value at that temperature?

Step 1: Write the equilibrium reaction. Then, from the equilibrium equation, write the solubility product expression.

$$Fe(OH)_2 \ (s) \rightleftharpoons Fe^{2+} \ (aq) + 2 \ OH^- \ (aq)$$

$$K_{sp} = [Fe^{2+}][OH^-]^2$$

Step 2: The molar solubility is the amount of $Fe(OH)_2$ that dissolves. From the stoichiometry of the equilibrium equation, you should find that:

$$[Fe^{2+}] = [Fe(OH)_2] = 7.7 \times 10^{-6} \ M$$

and

$$[OH^-] = 2[Fe(OH)_2] = 1.5 \times 10^{-5} \ M$$

Step 3: Substitute the equilibrium concentrations of Fe^{2+} and OH^- into the solubility product expression to calculate K_{sp}.

$$K_{sp} = [Fe^{2+}][OH^-]^2 = (7.7 \times 10^{-6})(1.5 \times 10^{-5})^2 = \mathbf{1.7 \times 10^{-15}}$$

PRACTICE EXERCISE

8.   At a certain temperature, the solubility of barium chromate ($BaCrO_4$) is 1.8×10^{-5} mol/L. What is the K_{sp} value at this temperature?

Text Problems: **16.42**, 16.44, 16.48

B.   Calculating solubility from K_{sp}

As in other equilibrium calculations, the expected concentrations at equilibrium can be calculated from a knowledge of the initial concentrations and the equilibrium constant (see Problem Type 5, Chapter 14). In these types of problems, it will be very helpful to recall that

equilibrium concentration = initial concentration ± the change due to reaction.

Example 16.8 illustrates this important type of calculation.

EXAMPLE 16.8

What is the molar solubility of silver phosphate (Ag_3PO_4) in water? $K_{sp} = 1.8 \times 10^{-18}$.

Step 1: Write the equilibrium reaction. Then, from the equilibrium equation, write the solubility product expression.

$$Ag_3PO_4\ (s) \rightleftharpoons 3\ Ag^+\ (aq) + PO_4^{3-}\ (aq)$$

$$K_{sp} = [Ag^+]^3[PO_4^{3-}]$$

Step 2: A certain amount of silver phosphate will dissociate in solution. Let's represent this amount as $-s$. Since one unit of Ag_3PO_4 yields 3 Ag^+ ions and one PO_4^{3-} ion, at equilibrium $[Ag^+]$ is $3s$ and $[PO_4^{3-}]$ is s. We summarize the changes in concentration as follows:

	$Ag_3PO_4\ (s)$	$\rightleftharpoons$	$3\ Ag^+\ (aq)$	$+$	$PO_4^{3-}\ (aq)$
Initial (M):			0		0
Change (M):	$-s$		$+3s$		$+s$
Equilibrium (M):			$3s$		s

Recall that the concentration of a pure solid does not enter into an equilibrium constant expression. Therefore, the concentration of Ag_3PO_4 is not important.

Step 3: Substitute the value of K_{sp} and the concentrations of Ag^+ and PO_4^{3-} in terms of s into the solubility product expression to solve for s, the molar solubility.

$$K_{sp} = [Ag^+]^3[PO_4^{3-}]$$

$$1.8 \times 10^{-18} = (3s)^3(s)$$

$$1.8 \times 10^{-18} = 27s^4$$

$$s = \text{molar solubility} = \mathbf{1.6 \times 10^{-5}\ mol/L}$$

The molar solubility indicates that 1.6×10^{-5} mol of Ag_3PO_4 will dissolve in 1 L of an aqueous solution.

PRACTICE EXERCISE

9.   The K_{sp} value of silver carbonate (Ag_2CO_3) is 8.1×10^{-12} at 25°C. What is the solubility in g/L of silver carbonate in water at 25°C?

Text Problems: **16.46**, 16.50

C.   Predicting a precipitation reaction

To predict whether a precipitate will form, we must calculate the **ion product, Q**. We will follow the same procedure outlined in Section 14.4 of the text or in Problem Type 4, Chapter 14. The ion product represents the molar concentrations of the ions raised to the power of their stoichiometric coefficients. Thus, for an aqueous solution containing Pb^{2+} and S^{2-} ions at 25°C,

$$Q = [Pb^{2+}]_0[S^{2-}]_0$$

The subscript 0 indicates that these are initial concentrations and do not necessarily correspond to equilibrium concentrations.

There are three possible relationships between Q and K_{sp}.

- $Q = K_{sp}$ $[Pb^{2+}]_0[S^{2-}]_0 = 3.4 \times 10^{-28}$ Since $Q = K_{sp}$, the reaction is already at equilibrium. The solution is saturated.

- $Q < K_{sp}$ $[Pb^{2+}]_0[S^{2-}]_0 < 3.4 \times 10^{-28}$ With $Q < K_{sp}$, the solution is not saturated. More solid could dissolve to produce more ions in solution. No precipitate will form.

- $Q > K_{sp}$ $[Pb^{2+}]_0[S^{2-}]_0 > 3.4 \times 10^{-28}$ With $Q > K_{sp}$, the solution is supersaturated. There are too many ions in solution. Some ions will combine to form PbS, which will precipitate out until the product of the ion concentrations is equal to 3.4×10^{-28}.

EXAMPLE 16.9

Predict whether a precipitate of PbI$_2$ will form when 200 mL of 0.015 M Pb(NO$_3$)$_2$ and 300 mL of 0.050 M NaI are mixed together. K_{sp} (PbI$_2$) = 1.4×10^{-8}.

The question asks whether the following reaction will occur:

$$Pb^{2+} (aq) + 2\,I^- (aq) \longrightarrow PbI_2 (s)$$

To answer that question, we need to consider the following equilibrium:

$$PbI_2 (s) \rightleftharpoons Pb^{2+} (aq) + I^- (aq)$$

Step 1: Calculate the concentrations of Pb(NO$_3$)$_2$ and NaI after mixing the two solutions together. This is a dilution problem. See Problem Type 6, Chapter 4. The total solution volume is 500 mL.

$$M_{initial} V_{initial} = M_{final} V_{final}$$

or,

$$M_{final} = \frac{M_{initial} V_{initial}}{V_{final}}$$

$$[Pb(NO_3)_2] = \frac{(0.015\ M)(200\ mL)}{500\ mL} = 6.0 \times 10^{-3}\ M$$

$$[NaI] = \frac{(0.050\ M)(300\ mL)}{500\ mL} = 0.030\ M$$

Both of these salts are soluble and ionize completely. We can write the concentrations of Pb^{2+} and I$^-$ directly from the salt concentrations.

$$[Pb^{2+}] = 6.0 \times 10^{-3}\ M \quad \text{and} \quad [I^-] = 0.030\ M$$

Step 2: Substitute the concentrations of Pb^{2+} and I$^-$ into the ion product expression to calculate Q. Then, compare Q to K_{sp} to determine if a precipitate will form.

$$PbI_2 (s) \rightleftharpoons Pb^{2+} (aq) + I^- (aq)$$

$$Q = [Pb^{2+}]_0[I^-]_0^2$$

$$Q = (6.0 \times 10^{-3})(0.030)^2 = 5.4 \times 10^{-6}$$

Comparing Q to K_{sp}, we find the $Q > K_{sp}$. This means that the solution is supersaturated. There are too many ions in solution. Some ions will combine to form PbI_2, which will precipitate out until the product of the ion concentrations is equal to K_{sp}, 1.4×10^{-8}.

PRACTICE EXERCISE

10. Will a precipitate of MgF_2 form when 6.00×10^2 mL of a solution that is 2.0×10^{-4} M in $MgCl_2$ is added to 3.00×10^2 mL of a 1.1×10^{-2} M NaF solution? K_{sp} $(MgF_2) = 6.6 \times 10^{-9}$.

Text Problem: 16.66

D. The effect of a common ion on solubility

Consider the slightly soluble salt copper (I) iodide. We can write its equilibrium reaction in water as follows:

$$CuI\ (s) \rightleftharpoons Cu^+\ (aq) + I^-\ (aq)$$

Now, suppose that instead of dissolving CuI in water, we attempted to dissolve a certain quantity of CuI in a potassium iodide (KI) solution. Would the presence of KI affect the solubility of the copper(I) iodide?

Remember, that KI is a soluble salt, so it will ionize completely into K^+ and I^- ions.

$$KI\ (aq) \rightarrow K^+\ (aq) + I^-\ (aq)$$

The I^- ions from KI *will affect* the copper(I) iodide equilibrium. Refer to Problem Type 6, Chapter 14, which discusses LeChatelier's principle. LeChatelier's principle states that when an external stress is applied to a system at equilibrium, the system adjusts in such a way that the stress is partially offset. In this system, the stress is additional I^- ions in solution (from KI). The CuI equilibrium will shift to the left to remove some of the additional I^- ions, decreasing the solubility of CuI.

In summary, the effect of adding a **common ion**, in this case I^-, is to **decrease** the solubility of the salt (CuI) in solution.

Calculating the solubility of a salt when a common ion is present is very similar to calculating the solubility from K_{sp} (Problem Type 4B). The only difference is that the initial concentration of the common ion is *not* zero. The initial concentration of the common ion comes from the soluble salt that is present in solution. Example 16.10 below illustrates this type of problem.

EXAMPLE 16.10
What is the molar solubility of $PbCl_2$ in a 0.50 M NaCl solution? K_{sp} $(PbCl_2) = 2.4 \times 10^{-4}$.

This problem is worked in a similar manner to Example 16.8.

Step 1: Write the equilibrium reaction. Then, from the equilibrium equation, write the solubility product expression.

$$PbCl_2\ (s) \rightleftharpoons Pb^{2+}\ (aq) + 2\ Cl^-\ (aq)$$

$$K_{sp} = [Pb^{2+}][Cl^-]^2$$

Step 2: A certain amount of lead(II) chloride will dissociate in solution. Let's represent this amount as $-s$. Since one unit of $PbCl_2$ yields one Pb^{2+} ion, at equilibrium $[Pb^{2+}]$ is s. There are two sources of Cl^- ions. NaCl ionizes completely to give a Cl^- concentration of $0.50\ M$. Also, one unit of $PbCl_2$ yields two Cl^- ions, so at equilibrium this will add $2s$ to the Cl^- concentration. We summarize the changes in concentration as follows:

$$PbCl_2\ (s) \ \rightleftharpoons\ Pb^{2+}\ (aq)\ +\ 2\ Cl^-\ (aq)$$

Initial (M):		0	0.50
Change (M):	$-s$	$+s$	$+2s$
Equilibrium (M):		s	$0.50 + s$

Recall that the concentration of a pure solid does not enter into an equilibrium constant expression. Therefore, the concentration of $PbCl_2$ is not important.

Step 3: Substitute the value of K_{sp} and the concentrations of Pb^{2+} and Cl^- into the solubility product expression to solve for s, the molar solubility.

$$K_{sp}\ =\ [Pb^{2+}][Cl^-]^2$$

$$2.4 \times 10^{-4}\ =\ (s)(0.50 + 2s)^2$$

Let's assume that $0.50 \gg 2s$. This is a valid assumption because K_{sp} is small.

$$2.4 \times 10^{-4} \approx (s)(0.50)^2$$

$$s\ =\ \text{molar solubility}\ =\ \mathbf{9.6 \times 10^{-4}\ mol/L}$$

Checking the validity of the assumption,

$$\frac{2(9.6 \times 10^{-4})}{0.50} \times 100\% = 0.38\% < 5\%$$

The molar solubility of lead(II) chloride in pure water is $3.9 \times 10^{-2}\ M$. Does it make sense that the molar solubility decreased in $0.50\ M$ NaCl?

PRACTICE EXERCISE

11. What is the molar solubility of Ag_3PO_4 in $0.20\ M$ $AgNO_3$? K_{sp} (Ag_3PO_4) = 1.8×10^{-18}.

Text Problems: **16.56**, 16.58

PROBLEM TYPE 5: COMPLEX ION EQUILIBRIA AND SOLUBILITY

We can define a **complex ion** as an ion containing a central metal cation bonded to one or more molecules or ions. Transition metals have a particular tendency to form complex ions. For example, silver chloride (AgCl) is insoluble in water. However, when aqueous ammonia is added to AgCl in water, the silver chloride dissolves to form $Ag(NH_3)_2^+$ (aq) and Cl^- (aq). The equilibrium equation is:

$$Ag^+\ (aq)\ +\ 2\ NH_3\ (aq)\ \rightleftharpoons\ Ag(NH_3)_2^+(aq\ )$$

A measure of the tendency of a metal ion to form a particular complex ion is given by the **formation constant** K_f (also called the stability constant). The formation constant is simply the equilibrium constant for complex ion formation. The larger the K_f value, the more stable the complex ion. Table 16.4 of the text lists the formation constants of a number of complex ions.

The formation constant expression for the formation of $Ag(NH_3)_2^+$ can be written as:

$$K_f = \frac{[Ag(NH_3)_2^+]}{[Ag^+][NH_3]^2} = 1.5 \times 10^7$$

The very large value of K_f in this case indicates the great stability of the complex ion $[Ag(NH_3)_2^+]$ in solution. Thus, the insoluble silver chloride dissolves to form the stable complex ion, $Ag(NH_3)_2^+$.

A typical complex ion equilibrium problem involves using the formation constant to calculate the equilibrium concentrations of species in solution. See Example 16.11 below.

EXAMPLE 16.11

Calculate the concentration of free Ag^+ ions in a solution formed by adding 0.20 mol of $AgNO_3$ to 1.0 L of 1.0 M NaCN. Assume no volume change upon addition of $AgNO_3$.

In solution, Ag^+ ions will complex with CN^- ions. The concentration of Ag^+ will be determined by the following equilibrium:

$$Ag^+ (aq) + 2\,CN^- (aq) \rightleftharpoons Ag(CN)_2^- \qquad K_f = 1.0 \times 10^{21}$$

Since K_f is so large, this equilibrium lies far to the right. We can safely assume that all the Ag^+ reacts to form 0.20 mol of $Ag(CN)_2^-$. There is more than enough CN^- available.

Step 1: To find the concentration of free Ag^+ at equilibrium, use the formation constant expression.

$$K_f = \frac{[Ag(CN)_2^-]}{[Ag^+][CN^-]^2}$$

Rearranging,

$$[Ag^+] = \frac{[Ag(CN)_2^-]}{K_f[CN^-]^2}$$

We already know that the concentration of $Ag(CN)_2^-$ is 0.20 M. We can calculate the concentration of CN^- from the stoichiometry of the reaction.

$$[CN^-] = [CN^-]_0 - 2[Ag^+] = 1.0\,M - 2(0.20\,M) = 0.60\,M$$

Step 2: Substitute the equilibrium concentrations into the formation constant expression to calculate the equilibrium concentration of Ag^+.

$$[Ag^+] = \frac{[Ag(CN)_2^-]}{K_f[CN^-]^2} = \frac{0.20}{(1.0 \times 10^{21})(0.60)^2} = 5.6 \times 10^{-22}\,M$$

This concentration corresponds to only *three* Ag^+ ions per 10 mL of solution!

PRACTICE EXERCISE

12. If 0.0100 mol of $Cu(NO_3)_2$ are dissolved in 1.00 L of 1.00 M NH_3, what are the concentrations of Cu^{2+}, $Cu(NH_3)_4^{2+}$, and NH_3 at equilibrium? Assume no volume change upon addition of the $Cu(NO_3)_2$ to the ammonia solution.

Text Problems: **16.68**, 16.70, 16.72

ANSWERS TO PRACTICE EXERCISES

1. Both (a) and (c).   **2.** pH = 4.96   **3.** **(a)** pH = 3.77   **(b)** pH = 3.81

4. To prepare a pH = 2.85 buffer you would need to use a ratio of $[F^-]$ to $[HF]$ of 0.50, or 1 to 2.

One way to prepare this buffer would be to add 0.50 mol of NaF (21 g) to 1.0 L of 1.0 M HF (assuming no change in volume).

5. **(a)** pH = 7   **(b)** 62.5 mL of HCl   **(c)** pH = 13.40
     (d) pH = 13.08   **(e)** pH = 1.90

6. **(a)** 40.0 mL of NaOH   **(b)** pH = 2.02   **(c)** Halfway point, pH = pK_a = 3.35
     (d) pH = 8.09   **(e)** pH = 12.16

7. Yes, methyl red would be a suitable indicator.

8. $K_{sp} = 3.2 \times 10^{-10}$ **9.** solubility = 0.035 g/L   **10.** No, a precipitate will *not* form.

11. s = molar solubility = $2.3 \times 10^{-16}\, M$ **12.** $[NH_3] = 0.96\, M$
$[Cu(NH_3)_4^{2+}] = 0.0100\, M$
$[Cu^{2+}] = 2.4 \times 10^{-16}\, M$

SOLUTIONS TO SELECTED TEXT PROBLEMS

16.4   **(a)**   This is a weak base calculation.

$$K_b = 1.8 \times 10^{-5} = \frac{[NH_4^+][OH^-]}{[NH_3]} = \frac{(x)(x)}{0.20 - x}$$

$$x = 1.90 \times 10^{-3} \, M = [OH^-]$$

$$pOH = 2.72$$

$$\mathbf{pH = 11.28}$$

(b)   Table 15.4 gives the value of K_a for the ammonium ion. Using this and the given concentrations with the Henderson-Hasselbalch equation gives:

$$pH = pK_a + \log\frac{[\text{conjugate base}]}{\text{acid}} = -\log\left(5.6 \times 10^{-10}\right) + \log\frac{(0.20)}{(0.30)}$$

$$\mathbf{pH} = 9.25 - 0.18 = \mathbf{9.07}$$

Is there any difference in the Henderson-Hasselbalch equation in the cases of a weak acid and its conjugate base and a weak base and its conjugate acid?

16.8   Identifying buffer systems, Problem Type 1A.

(a)   HCl (hydrochloric acid) is a strong acid. A buffer is a solution containing both a weak acid and a weak base. Therefore, this is *not* a buffer system.

(b) NH_3 (ammonia) is a weak base, and its conjugate acid, NH_4^+ is a weak acid. Therefore, this is a buffer system.

(c)   This solution contains both a weak acid, $H_2PO_4^-$ and its conjugate base, HPO_4^{2-}. Therefore, this is a buffer system.

(d) HNO_2 (nitrous acid) is a weak acid, and its conjugate base, NO_2^- (nitrite ion, the anion of the salt KNO_2), is a weak base. Therefore, this is a buffer system.

(e) H_2SO_4 (sulfuric acid) is a strong acid. A buffer is a solution containing both a weak acid and a weak base. Therefore, this is *not* a buffer system.

(f)   HCOOH (formic acid) is a weak acid, and its conjugate base, $HCOO^-$ (formate ion, the anion of the salt HCOOK), is a weak base. Therefore, this is a buffer system.

16.10   Calculating the pH of a buffer system, Problem Type 1B.

(a)   This problem is greatly simplified because the concentration of the weak acid (acetic acid) is equal to the concentration of its conjugate base (acetate ion). Let's set up a table of initial concentrations, change in concentrations, and equilibrium concentrations.

	$CH_3COOH \ (aq)$	$\rightleftharpoons$	$H^+ \ (aq) \ +$	$CH_3COO^- \ (aq)$
Initial	$2.0 \, M$		0	$2.0 \, M$
Change	$-x$		$+x$	$+x$
Equilibrium	$2.0 \, M - x$		x	$2.0 \, M + x$

$$K_a = \frac{[H^+][CH_3COO^-]}{[CH_3COOH]}$$

$$K_a = \frac{[H^+](2.0 + x)}{(2.0 - x)} \approx \frac{[H^+](2.0)}{2.0}$$

$$K_a = [H^+]$$

Taking the −log of both sides,

$$pK_a = pH$$

Thus, for a buffer system in which the [weak acid] = [weak base],

$$pH = pK_a$$

$$\mathbf{pH} = -\log(1.8 \times 10^{-5}) = \mathbf{4.74}$$

(b) Similar to part (a),

$$\mathbf{pH} = pK_a = \mathbf{4.74}$$

Buffer (a) will be a more effective buffer because the concentrations of acid and base components are ten times higher than those in (b). Thus, buffer (a) can neutralize 10 times more added acid or base compared to buffer (b).

16.12 *Step 1:* Write the equilibrium that occurs between $H_2PO_4^-$ and HPO_4^{2-}. Set up a table relating the initial concentrations, the change in concentration to reach equilibrium, and the equilibrium concentrations.

	$H_2PO_4^-$ (aq)	$\rightleftharpoons$	H^+ (aq)	+	HPO_4^{2-} (aq)
Initial (*M*):	0.15		0		0.10
Change (*M*):	−x		+x		+x
Equilibrium (*M*):	0.15 − x		x		0.10 + x

Step 2: Write the ionization constant expression in terms of the equilibrium concentrations. Knowing the value of the equilibrium constant (K_a), solve for x.

$$K_a = \frac{[H^+][HPO_4^{2-}]}{[H_2PO_4^-]}$$

You can look up the K_a value for dihydrogen phosphate in Table 15.5 of your text.

$$6.2 \times 10^{-8} = \frac{(x)(0.10 + x)}{(0.15 - x)}$$

$$6.2 \times 10^{-8} \approx \frac{(x)(0.10)}{(0.15)}$$

$$x = [H^+] = 9.3 \times 10^{-8} \, M$$

Step 3: Having solved for the $[H^+]$, calculate the pH of the solution.

$$\mathbf{pH} = -\log[H^+] = -\log(9.3 \times 10^{-8}) = \mathbf{7.03}$$

16.14   We can use the Henderson-Hasselbalch equation to calculate the ratio $[HCO_3^-]/[H_2CO_3]$. The Henderson-Hasselbalch equation is:

$$pH = pK_a + \log\frac{[\text{conjugate base}]}{[\text{acid}]}$$

For the buffer system of interest, HCO_3^- is the conjugate base of the acid, H_2CO_3. We can write:

$$pH = 7.40 = -\log(4.2\times10^{-7}) + \log\frac{[HCO_3^-]}{[H_2CO_3]}$$

$$7.40 = 6.38 + \log\frac{[HCO_3^-]}{[H_2CO_3]}$$

The [conjugate base]/[acid] ratio is:

$$\log\frac{[HCO_3^-]}{[H_2CO_3]} = 7.40 - 6.38 = 1.02$$

$$\mathbf{\frac{[HCO_3^-]}{[H_2CO_3]} = 10^{1.02} = 1.0 \times 10^1}$$

The buffer should be more effective against an added acid because ten times more base is present compared to acid. Note that a pH of 7.40 is only a two significant figure number (Why?); the final result should only have two significant figures.

16.16   As calculated in Problem 16.10, the pH of this buffer system is equal to pK_a.

$$\mathbf{pH = pK_a = -\log(1.8 \times 10^{-5}) = 4.74}$$

(a)   The added NaOH will react completely with the acid component of the buffer, CH_3COOH. NaOH ionizes completely; therefore, 0.080 mol of OH^- are added to the buffer.

Step 1: The neutralization reaction is:

	$CH_3COOH\ (aq)$	$+\ OH^-\ (aq)$	$\longrightarrow$	$CH_3COO^-\ (aq)$	$+\ H_2O\ (l)$
Initial (mol)	1.00	0.080		1.00	
After reaction (mol)	0.92	0		1.08	

Step 2: Now, the acetic acid equilibrium is reestablished. Since the volume of the solution is 1.00 L, we can convert directly from moles to molar concentration.

	$CH_3COOH\ (aq)$	$\rightleftharpoons$	$H^+\ (aq)$	$+\ CH_3COO^-\ (aq)$
Initial (*M*)	0.92		0	1.08
Change (*M*)	$-x$		$+x$	$+x$
Equilibrium (*M*)	$0.92 - x$		x	$1.08 + x$

Write the K_a expression, then solve for x.

$$K_a = \frac{[H^+][CH_3COO^-]}{[CH_3COOH]}$$

$$1.8 \times 10^{-5} = \frac{(x)(1.08 + x)}{(0.92 - x)} \approx \frac{x(1.08)}{0.92}$$

$$x = [H^+] = 1.5 \times 10^{-5} \, M$$

Step 3: Having solved for the $[H^+]$, calculate the pH of the solution.

$$\textbf{pH} = -\log[H^+] = -\log(1.5 \times 10^{-5}) = \textbf{4.82}$$

The pH of the buffer increased from 4.74 to 4.82 upon addition of 0.080 mol of strong base.

(b)   The added acid will react completely with the base component of the buffer, CH_3COO^-. HCl ionizes completely; therefore, 0.12 mol of H^+ ion are added to the buffer

Step 1: The neutralization reaction is:

	$CH_3COO^- \, (aq)$	$+ \; H^+ \, (aq)$	$\longrightarrow$	$CH_3COOH \, (aq)$
Initial (mol)	1.00	0.12		1.00
After reaction (mol)	0.88	0		1.12

Step 2: Now, the acetic acid equilibrium is reestablished. Since the volume of the solution is 1.00 L, we can convert directly from moles to molar concentration.

	$CH_3COOH \, (aq)$	$\rightleftharpoons$	$H^+ \, (aq)$	$+ \; CH_3COO^- \, (aq)$
Initial (M)	1.12		0	0.88
Change (M)	$-x$		$+x$	$+x$
Equilibrium (M)	$1.12 - x$		x	$0.88 + x$

Write the K_a expression, then solve for x.

$$K_a = \frac{[H^+][CH_3COO^-]}{[CH_3COOH]}$$

$$1.8 \times 10^{-5} = \frac{(x)(0.88 + x)}{(1.12 - x)} \approx \frac{x(0.88)}{1.12}$$

$$x = [H^+] = 2.3 \times 10^{-5} \, M$$

Step 3: Having solved for the $[H^+]$, calculate the pH of the solution.

$$\textbf{pH} = -\log[H^+] = -\log(2.3 \times 10^{-5}) = \textbf{4.64}$$

The pH of the buffer decreased from 4.74 to 4.64 upon addition of 0.12 mol of strong acid.

16.18   Preparing a buffer solution with a specific pH, Problem Type 1C.

Recall that to prepare a solution of a desired pH, , we should choose a weak acid with a pK_a value close to the desired pH.

Calculating the pK_a for each acid:

$$\text{For HA,} \qquad \textbf{p}K_a = -\log(2.7 \times 10^{-3}) = \textbf{2.57}$$

$$\text{For HB,} \qquad \textbf{p}K_a = -\log(4.4 \times 10^{-6}) = \textbf{5.36}$$

$$\text{For HC,} \qquad \text{p}K_{\mathbf{a}} = -\log(2.6 \times 10^{-9}) = \mathbf{8.59}$$

The buffer solution with a $\text{p}K_a$ closest to the desired pH is HC. Thus, **HC** is the best choice to prepare a buffer solution with pH = 8.60.

16.22   The neutralization reaction is:

$$2 \text{ KOH } (aq) + \text{H}_2\text{A } (aq) \longrightarrow \text{K}_2\text{A } (aq) + 2 \text{ H}_2\text{O } (l)$$

The number of moles of H_2A reacted is:

$$11.1 \text{ mL KOH} \times \left(\frac{1.00 \text{ mol KOH}}{1000 \text{ mL}} \right) \left(\frac{1 \text{ mol H}_2\text{A}}{2 \text{ mol KOH}} \right) = 5.55 \times 10^{-3} \text{ mol H}_2\text{A}$$

We know that 0.500 g of the diprotic acid were reacted (1/10 of the 250 mL was tested). Divide the number of grams by the number of moles to calculate the molar mass.

$$\mathcal{M}\,(\text{H}_2\text{A}) = \frac{0.500 \text{ g H}_2\text{A}}{5.55 \times 10^{-3} \text{ mol H}_2\text{A}} = \mathbf{90.1 \text{ g/mol}}$$

16.24 $2\text{HCOOH} + \text{Ba(OH)}_2 \rightarrow (\text{HCOO})_2\text{Ba} + 2\text{H}_2\text{O}$

$$\text{Number of moles of HCOOH reacted} = \frac{0.883 \text{ mol}}{1 \text{ L}} \times (20.4 \times 10^{-3} \text{ L}) = 0.0180 \text{ mol HCOOH}$$

The mole ratio between Ba(OH)_2 and HCOOH is 1:2.

Therefore, the molarity of the Ba(OH)_2 solution is:

$$0.0180 \text{ mol HCOOH} \times \frac{1 \text{ mol Ba(OH)}_2}{2 \text{ mol HCOOH}} \times \frac{1}{19.3 \times 10^{-3} \text{ L}} = \mathbf{0.466 \; M}$$

16.26   The resulting solution is not a buffer system. There is excess NaOH and the neutralization is well past the equivalence point.

$$\text{Moles NaOH} = 0.500 \text{ L} \left(\frac{0.167 \text{ mol}}{1 \text{ L}} \right) = 0.0835 \text{ mol}$$

$$\text{Moles CH}_3\text{COOH} = 0.500 \text{ L} \left(\frac{0.100 \text{ mol}}{1 \text{ L}} \right) = 0.0500 \text{ mol}$$

After neutralization, the amount of NaOH remaining in $0.0835 - 0.0500 = 0.0335$ mol. The volume of the resulting solution is 1.00 L

$$[\text{OH}^-] = \frac{0.0335 \text{ mol}}{1.00 \text{ L}} = \mathbf{0.0335 \; M}$$

$$[\text{Na}^+] = \frac{0.0835 \text{ mol}}{1.00 \text{ L}} = \mathbf{0.0835 \; M}$$

$$[\text{H}^+] = \frac{1.0 \times 10^{-14}}{0.0335} = \mathbf{3.0 \times 10^{-13} \; M}$$

$$[CH_3COO^-] = \frac{0.0500 \text{ mol}}{1.00 \text{ L}} = 0.0500 \; M$$

$$[CH_3COOH] = \frac{[H^+][CH_3COOH^-]}{K_a} = \frac{(3.0 \times 10^{-13})(0.0500)}{1.8 \times 10^{-5}} = 8.3 \times 10^{-10} \; M$$

16.28 Concentration of HCOONa at the equivalence point is 0.050 M, since the solution volume doubles (the volume of NaOH will equal the volume of HCOOH since the molarities are equal).

$$HCOO^- + H_2O \;\rightleftharpoons\; HCOOH \;+\; OH^-$$

$$0.050 - x \qquad\qquad\qquad x \qquad\quad x$$

$$\frac{x^2}{0.050 - x} = 5.9 \times 10^{-11}$$

Assume $0.050 - x \approx 0.050$

$$x = 1.7 \times 10^{-6} \; M = [OH^-]$$

$$pOH = 5.77$$

$$\mathbf{pH = 8.23}$$

16.32 CO_2 in the air dissolves in the solution:

$$CO_2 + H_2O \;\rightleftharpoons\; H_2CO_3$$

The carbonic acid neutralizes the NaOH.

16.34 According to Section 16.5 of the text, when $[HIn] \approx [In^-]$ the indicator color is a mixture of the colors of HIn and In^-. In other words, the indicator color changes at this point. When $[HIn] \approx [In^-]$ we can write:

$$\frac{[In^-]}{[HIn]} = \frac{K_a}{[H^+]} = 1$$

$$[H^+] = K_a = 2.0 \times 10^{-6}$$

$$\mathbf{pH = 5.70}$$

16.42 Calculating K_{sp} from molar solubility, Problem Type 4A.

In each case, we first calculate the number of moles of compound dissolved in one liter of solution (the molar solubility).

(a) $\left(\dfrac{7.3 \times 10^{-2} \text{ g SrF}_2}{1 \text{ L soln}}\right) \times \left(\dfrac{1 \text{ mol SrF}_2}{125.6 \text{ g SrF}_2}\right) = 5.8 \times 10^{-4} \text{ mol/L}$

Step 1: Write the equilibrium reaction. Then, from the equilibrium equation, write the solubility product expression.

$$SrF_2 \,(s) \;\rightleftharpoons\; Sr^{2+} \,(aq) \;+\; 2\,F^- \,(aq)$$

$$K_{sp} = [Sr^{2+}][F^-]^2$$

Step 2: The molar solubility is the amount of SrF_2 that dissolves. From the stoichiometry of the equilibrium equation, you should find that

$$[Sr^{2+}] = [SrF_2] = 5.8 \times 10^{-4}\ M$$

and

$$[F^-] = 2[SrF_2] = 1.16 \times 10^{-3}\ M$$

Step 3: Substitute the equilibrium concentrations of Sr^{2+} and F^- into the solubility product expression to calculate K_{sp}.

$$K_{sp} = [Sr^{2+}][F^-]^2 = (5.8 \times 10^{-4})(1.16 \times 10^{-3})^2 = \mathbf{7.8 \times 10^{-10}}$$

(b) $\left(\dfrac{6.7 \times 10^{-3}\ g\ Ag_3PO_4}{1\ L\ soln} \right) \times \left(\dfrac{1\ mol\ Ag_3PO_4}{418.7\ g\ Ag_3PO_4} \right) = 1.6 \times 10^{-5}\ mol/L$

(b) is solved in a similar manner to (a)

The equilibrium equation is:

$$Ag_3PO_4\ (s) \rightleftharpoons 3\ Ag^+\ (aq) + PO_4^{3-}\ (aq)$$

$$K_{sp} = [Ag^+]^3[PO_4^{3-}] = [3 \times (1.6 \times 10^{-5})]^3(1.6 \times 10^{-5}) = \mathbf{1.8 \times 10^{-18}}$$

16.44   First, we can convert the solubility of MX in g/L to mol/L.

$$\left(\dfrac{4.63 \times 10^{-3}\ g\ MX}{1\ L\ soln} \right) \times \left(\dfrac{1\ mol\ MX}{346\ g\ MX} \right) = 1.34 \times 10^{-5}\ mol/L = s\ (molar\ solubility)$$

The equilibrium reaction is:

$$MX\ (s) \rightleftharpoons M^{n+}\ (aq) + X^{n-}\ (aq)$$

Since the mole ratio of MX to each of the ions is 1:1, the equilibrium concentrations of each of the ions can also be represented by s. Solving for K_{sp},

$$K_{sp} = [M^{n+}][X^{n-}] = s^2 = (1.34 \times 10^{-5})^2 = \mathbf{1.80 \times 10^{-10}}$$

16.46   Calculating solubility from K_{sp}, Problem Type 4B.

Step 1: Write the equilibrium reaction. Then, from the equilibrium equation, write the solubility product expression.

$$CaF_2\ (s) \rightleftharpoons Ca^{2+}\ (aq) + 2\ F^-\ (aq)$$

$$K_{sp} = [Ca^{2+}][F^-]^2$$

Step 2: A certain amount of calcium fluoride will dissociate in solution. Let's represent this amount as $-s$. Since one unit of CaF_2 yields one Ca^{2+} ion and two F^- ions, at equilibrium $[Ca^{2+}]$ is s and $[F^-]$ is $2s$. We summarize the changes in concentration as follows:

$$CaF_2 \ (s) \ \rightleftharpoons \ Ca^{2+} \ (aq) \ + \ 2 \ F^- \ (aq)$$

Initial (M):		0	0
Change (M):	$-s$	$+s$	$+2s$
Equilibrium (M):		s	$2s$

Recall, that the concentration of a pure solid does not enter into an equilibrium constant expression. Therefore, the concentration of CaF_2 is not important.

Step 3: Substitute the value of K_{sp} and the concentrations of Ca^{2+} and F^- in terms of s into the solubility product expression to solve for s, the molar solubility.

$$K_{sp} = [Ca^{2+}][F^-]^2$$

$$4.0 \times 10^{-11} = (s)(2s)^2$$

$$4.0 \times 10^{-11} = 4s^3$$

$$s = \text{molar solubility} = \mathbf{2.2 \times 10^{-4} \ mol/L}$$

The molar solubility indicates that 2.2×10^{-4} mol of CaF_2 will dissolve in 1 L of an aqueous solution.

16.48   First we can calculate the OH^- concentration from the pH.

$$pOH = 14.00 - pH$$

$$pOH = 14.00 - 9.68 = 4.32$$

$$[OH^-] = 10^{-pOH} = 10^{-4.32} = 4.8 \times 10^{-5} \ M$$

The equilibrium equation is:

$$MOH \ (s) \ \rightleftharpoons \ M^+ \ (aq) \ + \ OH^- \ (aq)$$

From the balanced equation we know that $[M^+] = [OH^-]$

$$\boldsymbol{K_{sp}} = [M^+][OH^-] = (4.8 \times 10^{-5})^2 = \mathbf{2.3 \times 10^{-9}}$$

16.50   The net ionic equation is:

$$Sr^{2+} \ (aq) \ + \ 2 \ F^- \ (aq) \ \longrightarrow \ SrF_2 \ (s)$$

Let's find the limiting reagent in the precipitation reaction.

$$\text{Moles } F^- = 75 \ \text{mL} \times \left(\frac{1 \ L}{1000 \ \text{mL}} \right) \left(\frac{0.060 \ \text{mol}}{1 \ L} \right) = 0.0045 \ \text{mol}$$

$$\text{Moles } Sr^{2+} = 25 \ \text{mL} \times \left(\frac{1 \ L}{1000 \ \text{mL}} \right) \left(\frac{0.15 \ \text{mol}}{1 \ L} \right) = 0.0038 \ \text{mol}$$

From the stoichiometry of the balanced equation, twice as many moles of F^- are required to react with Sr^{2+}. This would require 0.0076 mol of F^-, but we only have 0.0045 mol. Thus, F^- is the limiting reagent.

Let's assume that the above reaction goes to completion. Then, we will consider the equilibrium that is established when SrF_2 partially dissociates into ions.

	Sr^{2+} (aq)	+ 2 F$^-$ (aq)	$\longrightarrow$	SrF_2 (s)
Initial (mol)	0.0038	0.0045		0
Change (mol)	−0.00225	−0.0045		+0.00225
After reaction (mol)	0.00155	0		0.00225

Now, let's establish the equilibrium reaction. The total volume of the solution is 100 mL = 0.100 L. Divide the above moles by 0.100 L to convert to molar concentration.

	SrF_2 (s)	$\rightleftharpoons$	Sr^{2+} (aq)	+ 2 F$^-$ (aq)
Initial (M)	0.0225		0.0155	0
Change (M)	−x		+x	+2x
Equilibrium (M)	0.0225 − x		0.0155 + x	2x

Write the solubility product expression, then solve for x.

$$K_{sp} = [Sr^{2+}][F^-]^2$$

$$2.0 \times 10^{-10} = (0.0155 + x)(2x)^2 \approx (0.0155)(2x)^2$$

$$x = 5.7 \times 10^{-5} M$$

$$[F^-] = 2x = 1.1 \times 10^{-4} M$$

$$[Sr^{2+}] = 0.0155 + x = 0.016 M$$

Both sodium ions and nitrate ions are spectator ions and therefore do not enter into the precipitation reaction.

$$[NO_3^-] = \frac{2(0.0038) \text{ mol}}{0.10 \text{ L}} = 0.076 M$$

$$[Na^+] = \frac{0.0045 \text{ mol}}{0.10 \text{ L}} = 0.045 M$$

16.52   For $Fe(OH)_3$, $K_{sp} = 1.1 \times 10^{-36}$. When $[Fe^{3+}] = 0.010 M$, the $[OH^-]$ value is:

$$K_{sp} = [Fe^{3+}][OH^-]^3$$

or

$$[OH^-] = \left(\frac{K_{sp}}{[Fe^{3+}]} \right)^{\frac{1}{3}}$$

$$[OH^-] = \left(\frac{1.1 \times 10^{-36}}{0.010} \right)^{\frac{1}{3}} = 4.8 \times 10^{-12} M$$

This $[OH^-]$ corresponds to a pH of 2.68. In other words, $Fe(OH)_3$ will begin to precipitate from this solution at pH of 2.68.

For $Zn(OH)_2$, $K_{sp} = 1.8 \times 10^{-14}$. When $[Zn^{2+}] = 0.010\ M$, the $[OH^-]$ value is:

$$[OH^-] = \left(\frac{K_{sp}}{[Zn^{2+}]} \right)^{\frac{1}{2}}$$

$$[OH^-] = \left(\frac{1.8 \times 10^{-14}}{0.010} \right)^{\frac{1}{2}} = 1.3 \times 10^{-6}\ M$$

This corresponds to a pH of 8.11. In other words $Zn(OH)_2$ will begin to precipitate from the solution at pH = 8.11. These results show that $Fe(OH)_3$ will precipitate when the pH just exceeds 2.68 and that $Zn(OH)_2$ will precipitate when the pH just exceeds 8.11. Therefore, to selectively remove iron as $Fe(OH)_3$, the pH must be *greater than* **2.68** but *less than* **8.11**.

16.56   The effect of a common ion on solubility, Problem Type 4D.

(a)   Set up a table to find the equilibrium concentrations in pure water.

	$PbBr_2\ (s)$	$\rightleftharpoons$	$Pb^{2+}\ (aq)$	$+$	$2\ Br^-\ (aq)$
Initial (M)			0		0
Change (M)	$-s$		$+s$		$+2s$
Equilibrium (M)			s		$2s$

$$K_{sp} = [Pb^{2+}][Br^-]^2$$

$$8.9 \times 10^{-6} = (s)(2s)^2$$

$$s = \text{molar solubility} = \mathbf{0.013\ M}$$

(b)   Set up a table to find the equilibrium concentrations in 0.20 M KBr. KBr is a soluble salt that ionizes completely giving a initial concentration of $Br^- = 0.20\ M$.

	$PbBr_2\ (s)$	$\rightleftharpoons$	$Pb^{2+}\ (aq)$	$+$	$2\ Br^-\ (aq)$
Initial (M)			0		0.20
Change (M)	$-s$		$+s$		$+2s$
Equilibrium (M)			s		$0.20 + 2s$

$$K_{sp} = [Pb^{2+}][Br^-]^2$$

$$8.9 \times 10^{-6} = (s)(0.20 + 2s)^2$$

$$8.9 \times 10^{-6} \approx (s)(0.20)^2$$

$$s = \text{molar solubility} = \mathbf{2.2 \times 10^{-4}\ M}$$

Thus, the molar solubility of $PbBr_2$ is reduced from 0.013 M to $2.2 \times 10^{-4}\ M$ as a result of the common ion (Br^-) effect.

(c)   Set up a table to find the equilibrium concentrations in 0.20 M $Pb(NO_3)_2$. $Pb(NO_3)_2$ is a soluble salt that dissociates completely giving an initial concentration of $[Pb^{2+}] = 0.20\ M$.

	$PbBr_2\ (s)$	$\rightleftharpoons$	$Pb^{2+}\ (aq)$	$+$	$2\ Br^-\ (aq)$
Initial (M):			0.20		0
Change (M):	$-s$		$+s$		$+2s$
Equilibrium (M):			$0.20 + s$		$2s$

$$K_{sp} = [Pb^{2+}][Br^-]^2$$

$$8.9 \times 10^{-6} = (0.20 + s)(2s)^2$$

$$8.9 \times 10^{-6} \approx (0.20)(2s)^2$$

$$s = \text{molar solubility} = \mathbf{3.3 \times 10^{-3}\ M}$$

Thus, the molar solubility of $PbBr_2$ is reduced from $0.013\ M$ to $3.3 \times 10^{-3}\ M$ as a result of the common ion (Pb^{2+}) effect.

16.58   The equilibrium reaction is:

$$BaSO_4\ (s) \rightleftharpoons Ba^{2+}\ (aq) + SO_4^{2-}\ (aq)$$

For both parts of the problem:

$$K_{sp} = [Ba^{2+}][SO_4^{2-}] = 1.1 \times 10^{-10}$$

(a)   In pure water, let $[Ba^{2+}] = [SO_4^{2-}] = s$

$$K_{sp} = [Ba^{2+}][SO_4^{2-}]$$

$$1.1 \times 10^{-10} = s^2$$

$$s = \mathbf{1.0 \times 10^{-5}\ M}$$

The molar solubility of $BaSO_4$ in pure water is 1.0×10^{-5} mol/L.

(b)   Assuming the molar solubility of $BaSO_4$ to be s, then

$$[Ba^{2+}] = s\ M \quad \text{and} \quad [SO_4^{2-}] = (1.0 + s)\ M \approx 1.0\ M$$

$$K_{sp} = [Ba^{2+}][SO_4^{2-}]$$

$$1.1 \times 10^{-10} = (s)(1.0)$$

$$s = \mathbf{1.1 \times 10^{-10}\ M}$$

Due to the common ion effect, the molar solubility of $BaSO_4$ decreases to 1.1×10^{-10} mol/L in $1.0\ M\ SO_4^{2-}(aq)$ compared to 1.0×10^{-5} mol/L in pure water.

16.60   **(b)** $SO_4^{2-}\ (aq)$ is a weak base

(c) $OH^-\ (aq)$ is a strong base

(d) $C_2O_4^{2-}\ (aq)$ is a weak base

(e) $PO_4^{3-}\ (aq)$ is a weak base.

The solubilities of the above will increase in acidic solution. Only (a), which contains an extremely weak base (I^- is the conjugate base of the strong acid HI) is unaffected by the acid solution.

16.62   From Table 16.2, the value of K_{sp} for iron(II) is 1.6×10^{-14}.

(a)   At pH = 8.00, pOH = 14.00 − 8.00 = 6.00, and $[OH^-] = 1.0 \times 10^{-6}\ M$

$$[Fe^{2+}] = \frac{K_{sp}}{[OH^-]^2} = \frac{1.6 \times 10^{-14}}{(1.0 \times 10^{-6})^2} = 0.016\ M$$

The *molar solubility* of iron(II) hydroxide at pH = 8.00 is **0.016 M**

(b)   At pH = 10.00, pOH = 14.00 − 10.00 = 4.00, and $[OH^-] = 1.0 \times 10^{-4}\ M$

$$[Fe^{2+}] = \frac{K_{sp}}{[OH^-]^2} = \frac{1.6 \times 10^{-14}}{(1.0 \times 10^{-4})^2} = 1.6 \times 10^{-6}\ M$$

The *molar solubility* of iron(II) hydroxide at pH = 10.00 is **$1.6 \times 10^{-6}\ M$**.

16.64   We first determine the effect of the added ammonia. Let's calculate the concentration of NH_3. This is a dilution problem.

$$M_I V_I = M_F V_F$$
$$(0.60\ M)(2.00\ mL) = M_2(1002\ mL)$$
$$M_2 = 0.0012\ M\ NH_3$$

Ammonia is a weak base ($K_b = 1.8 \times 10^{-5}$).

	NH_3 + H_2O	$\rightleftharpoons$	NH_4^+ +	OH^-
initial (*M*):	0.0012		0	0
change (*M*):	−x		+x	+x
equil. (*M*):	0.0012 − x		x	x

$$K_b = \frac{[NH_4^+][OH^-]}{[NH_3]} = \frac{x^2}{(0.0012 - x)} = 1.8 \times 10^{-5}$$

Solving the resulting quadratic equation gives $x = 0.00014$, or $[OH^-] = 0.00014\ M$

This is a solution of iron(II) sulfate, which contains Fe^{2+} ions. These Fe^{2+} ions could combine with OH^- to precipitate $Fe(OH)_2$. Therefore, we must use K_{sp} for iron(II) hydroxide. We compute the value of Q_c for this solution.

$$Q = [Fe^{2+}]_0[OH^-]_0^2 = (1.0 \times 10^{-3})(0.00014)^2 = 2.0 \times 10^{-11}$$

Q is larger than K_{sp} [$Fe(OH)_2$] = 1.6×10^{-14}; therefore **a precipitate of $Fe(OH)_2$ will form**.

16.68   Complex Ion Equilibria and Solubility, Problem Type 5.

In solution, Cd^{2+} ions will complex with CN^- ions. The concentration of Cd^{2+} will be determined by the following equilibrium

$$Cd^{2+}\ (aq) + 4\ CN^-\ (aq) \rightleftharpoons Cd(CN)_4^{2-} \qquad K_f = 7.1 \times 10^{16}$$

Since K_f is so large, this equilibrium lies far to the right. We can safely assume that all the Cd^{2+} reacts.

Step 1: Calculate the initial concentration of Cd^{2+} ions.

$$[Cd^{2+}]_0 = 0.50 \text{ g} \times \frac{1 \text{ mol Cd(NO}_3)_2}{236.42 \text{ g Cd(NO}_3)_2} \times \frac{1 \text{ mol Cd}}{1 \text{ mol Cd(NO}_3)_2} \times \frac{1}{0.50 \text{ L}} = 4.2 \times 10^{-3} M$$

Step 2: If we assume that the above equilibrium goes to completion, we can write

$$Cd^{2+}(aq) + 4 CN^-(aq) \longrightarrow Cd(CN)_4^{2-}$$

Initial (M)	4.2×10^{-3}	0.50	0
After reaction (M)	0	0.48	4.2×10^{-3}

Step 3: To find the concentration of free Cd^{2+} at equilibrium, use the formation constant expression.

$$K_f = \frac{[Cd(CN)_4^{2-}]}{[Cd^{2+}][CN^-]^4}$$

Rearranging,

$$[Cd^{2+}] = \frac{[Cd(CN)_4^{2-}]}{K_f[CN^-]^4}$$

Substitute the equilibrium concentrations calculated above into the formation constant expression to calculate the equilibrium concentration of Cd^{2+}.

$$[Cd^{2+}] = \frac{[Cd(CN)_4^{2-}]}{K_f[CN^-]^4} = \frac{4.2 \times 10^{-3}}{(7.1 \times 10^{16})(0.48)^4} = 1.1 \times 10^{-18} M$$

$$[CN^-] = 0.48 \, M + (1.1 \times 10^{-18} \, M) = \mathbf{0.48 \, M}$$

$$[Cd(CN)_4^{2-}] = (4.2 \times 10^{-3} \, M) - (1.1 \times 10^{-18}) = \mathbf{4.2 \times 10^{-3} \, M}$$

16.70 Silver iodide is only slightly soluble. It dissociates to form a small amount of Ag^+ and I^- ions. The Ag^+ ions then complex with NH_3 in solution to form the complex ion $Ag(NH_3)_2^+$. The balanced equations are:

$$AgI(s) \rightleftharpoons Ag^+(aq) + I^-(aq) \qquad\qquad K_{sp} = [Ag^+][I^-] = 8.3 \times 10^{-17}$$

$$Ag^+(aq) + 2 NH_3(aq) \rightleftharpoons Ag(NH_3)_2^+(aq) \qquad K_f = \frac{[Ag(NH_3)_2^+]}{[Ag^+][NH_3]^2} = 1.5 \times 10^7$$

Overall: $AgI(s) + 2 NH_3(aq) \rightleftharpoons Ag(NH_3)_2^+(aq) + I^-(aq) \qquad K = K_{sp} \times K_f = 1.2 \times 10^{-9}$

If s is the molar solubility of AgI then,

$$AgI(s) + 2 NH_3(aq) \rightleftharpoons Ag(NH_3)_2^+(aq) + I^-(aq)$$

initial (M):		1.0	0.0	0.0
change (M):	$-s$	$-2s$	$+s$	$+s$
equilibrium (M)		$(1.0 - 2s)$	s	s

Because K_f is large, we can assume all of the silver ions exist as $Ag(NH_3)_2^+$. Thus,

$$[Ag(NH_3)_2^+] = [I^-] = s$$

We can write the equilibrium constant expression for the above reaction, then solve for s.

$$K = 1.2 \times 10^{-9} = \frac{(s)(s)}{(1.0 - 2s)^2} \approx \frac{(s)(s)}{(1.0)^2}$$

$$s = 3.5 \times 10^{-5}\ M$$

At equilibrium, 3.5×10^{-5} moles of AgI dissolves in 1 L of $1.0\ M$ NH_3 solution.

16.72 **(a)** The equations are as follows:

$$CuI_2\ (s) \rightleftharpoons Cu^{2+}\ (aq) + 2\ I^-\ (aq)$$

$$\mathbf{Cu^{2+}\ (aq) + 4\ NH_3\ (aq) \rightleftharpoons [Cu(NH_3)_4]^{2+}\ (aq)}$$

The ammonia combines with the Cu^{2+} ions formed in the first step to form the complex ion $[Cu(NH_3)_4]^{2+}$, effectively removing the Cu^{2+} ions, causing the first equilibrium to shift to the right (resulting in more CuI_2 dissolving).

(b) Similar to part (a):

$$AgBr\ (s) \rightleftharpoons Ag^+\ (aq) + Br^-\ (aq)$$

$$\mathbf{Ag^+\ (aq) + 2\ CN^-\ (aq) \rightleftharpoons [Ag(CN)_2]^-\ (aq)}$$

(c) Similar to parts (a) and (b).

$$HgCl_2\ (s) \rightleftharpoons Hg^{2+}\ (aq) + 2Cl^-\ (aq)$$

$$\mathbf{Hg^{2+}\ (aq) + 4Cl^-\ (aq) \rightleftharpoons [HgCl_4]^{2-}\ (aq)}$$

16.76 Since some $PbCl_2$ precipitates, the solution is saturated.   From Table 16.2, the value of K_{sp} for lead(II) chloride is 2.4×10^{-4}. The equilibrium is:

$$PbCl_2\ (aq) \rightleftharpoons Pb^{2+}\ (aq) + 2\ Cl^-\ (aq)$$

We can write the solubility product expression for the equilibrium.

$$K_{sp} = [Pb^{2+}][Cl^-]^2$$

K_{sp} and $[Cl^-]$ are known. Solving for the Pb^{2+} concentration,

$$[Pb^{2+}] = \frac{K_{sp}}{[Cl^-]^2} = \frac{2.4 \times 10^{-4}}{(0.15)^2} = 0.011\ M$$

16.78 Chloride ion will precipitate Ag^+ but not Cu^{2+}. So, dissolve some solid in H_2O and add HCl. If a precipitate forms, the salt was $AgNO_3$. A flame test will also work. Cu^{2+} gives a green flame test.

16.80 We can use the Henderson-Hasselbalch equation to solve for the pH when the indicator is 90% acid / 10% conjugate base and when the indicator is 10% acid / 90% conjugate base.

$$pH = pK_a + \log\frac{[\text{conjugate base}]}{[\text{acid}]}$$

Solving for the pH with 90% of the indicator in the HIn form:

$$pH = 3.46 + \log\frac{[10]}{[90]} = 3.46 - 0.95 = 2.51$$

Next, solving for the pH with 90% of the indicator in the In$^-$ form:

$$pH = 3.46 + \log\frac{[90]}{[10]} = 3.46 + 0.95 = 4.41$$

Thus the pH range varies from **2.51 to 4.41** as the [HIn] varies from 90% to 10%.

16.82 First, calculate the pH of the 2.00 M weak acid (HNO$_2$) solution before any NaOH is added.

$$K_a = \frac{[H^+][NO_2^-]}{[HNO_2]}$$

$$4.5 \times 10^{-4} = \frac{x^2}{2.00 - x}$$

$$x = [H^+] = 0.030\ M$$

$$pH = -\log(0.030) = 1.52$$

Since the pH after the addition is 1.5 pH units greater, the new pH = 1.52 + 1.50 = 3.02.

From this new pH, we can calculate the [H$^+$] in solution.

$$[H^+] = 10^{-pH} = 10^{-3.02} = 9.55 \times 10^{-4}\ M$$

When the NaOH is added, we dilute our original 2.00 M HNO$_2$ solution to:

$$M_I V_I = M_F V_F$$
$$(2.00\ M)(400\ mL) = M_F(600\ mL)$$
$$M_F = 1.33\ M$$

Since we have not reached the equivalence point, we have a buffer solution. The reaction between HNO$_2$ and NaOH is:

$$HNO_2\ (aq) + NaOH\ (aq) \longrightarrow NaNO_2\ (aq) + H_2O\ (l)$$

Since the mole ratio between HNO$_2$ and NaOH is 1:1, the decrease in [HNO$_2$] is the same as the decrease in [NaOH].

We can calculate the decrease in [HNO$_2$] by setting up the weak acid equilibrium. From the pH of the solution, we know that the [H$^+$] at equilibrium is $9.55 \times 10^{-4}\ M$.

	$HNO_2 (aq)$	$\rightleftharpoons$	$H^+ (aq)$	$+$	$NO_2^- (aq)$
Initial (M)	1.33		0		0
Change (M)	$-x$				$+x$
Equilibrium (M)	$1.33 - x$		9.55×10^{-4}		x

We can calculate x from the equilibrium constant expression.

$$K_a = \frac{[H^+][NO_2^-]}{[HNO_2]}$$

$$4.5 \times 10^{-4} = \frac{(9.55 \times 10^{-4})(x)}{1.33 - x}$$

$$x = 0.426 \, M$$

Thus, x is the decrease in $[HNO_2]$ which equals the concentration of added OH^-. However, this is the concentration of NaOH after it has been diluted to 600 mL. We need to correct for the dilution from 200 mL to 600 mL to calculate the concentration of the original NaOH solution.

$$M_I V_I = M_F V_F$$
$$M_I(200 \text{ mL}) = (0.426 \, M)(600 \text{ mL})$$
$$[\text{NaOH}] = M_I = \textbf{1.28} \, \boldsymbol{M}$$

16.84 The resulting solution is not a buffer system. There is excess NaOH and the neutralization is well past the equivalence point. We have a solution of sodium acetate and sodium hydroxide.

$$\text{Moles NaOH} = 0.500 \, \text{L} \left(\frac{0.167 \text{ mol}}{1 \, \text{L}} \right) = 0.0835 \text{ mol}$$

$$\text{Moles CH}_3\text{COOH} = 0.500 \, \text{L} \left(\frac{0.100 \text{ mol}}{1 \, \text{L}} \right) = 0.0500 \text{ mol}$$

The reaction between sodium hydroxide and acetic acid is:

	$NaOH (aq)$	$+$	$CH_3COOH (aq)$	$\longrightarrow$	$NaCH_3COO (aq)$	$+$	$H_2O (l)$
Initial (mol)	0.0835		0.0500		0		
After Reaction (mol)	0.0335		0		0.0500		

Since the total volume of the solution is 1.00 L, we can convert the number of moles directly to molarity.

$$[\text{OH}^-] = \textbf{0.0335} \, \boldsymbol{M}$$
$$[\text{Na}^+] = \textbf{0.0335} \, \boldsymbol{M} + \textbf{0.0500} \, \boldsymbol{M} = \textbf{0.0835} \, \boldsymbol{M}$$
$$[\text{CH}_3\text{COO}^-] = \textbf{0.0500} \, \boldsymbol{M}$$

We can calculate the $[H^+]$ from the $[OH^-]$.

$$[\text{H}^+] = \frac{K_w}{[\text{OH}^-]} = \frac{1.00 \times 10^{-14}}{0.0335} = \textbf{2.99} \times \textbf{10}^{-13} \, \boldsymbol{M}$$

Finally, we can calculate the acetic acid concentration from the K_a expression.

$$K_a = \frac{[H^+][CH_3COO^-]}{[CH_3COOH]}$$

or,

$$[CH_3COOH] = \frac{[H^+][CH_3COO^-]}{K_a}$$

$$[CH_3COOH] = \frac{(2.99 \times 10^{-13})(0.0500)}{1.8 \times 10^{-5}} = \mathbf{8.3 \times 10^{-10}} \, M$$

16.86   The number of moles of Ba^{2+} present in the original 50.0 mL of solution is:

$$50.0 \, \text{mL} \times \frac{1.00 \, \text{mol Ba}^{2+}}{1 \, \text{L soln}} \times \frac{1 \, \text{L}}{1000 \, \text{mL}} = 0.0500 \, \text{mol Ba}^{2+}$$

The number of moles of SO_4^{2-} present in the original 86.4 mL of solution, assuming complete dissociation, is:

$$86.4 \, \text{mL} \times \frac{0.494 \, \text{mol SO}_4^{2-}}{1 \, \text{L soln}} \times \frac{1 \, \text{L}}{1000 \, \text{mL}} = 0.0427 \, \text{mol SO}_4^{2-}$$

The complete ionic equation is:

	Ba^{2+} (aq) +	2 OH⁻ (aq) +	2 H⁺ (aq) +	SO_4^{2-} (aq) $\rightarrow$	$BaSO_4$ (s) + H_2O (l)
Initial (mol):	0.0500	0.100	0.0854	0.0427	0
Change (mol)	−0.0427	−2(0.0427)	−2(0.0427)	−0.0427	+0.0427
After rxn (mol)	0.0073	0.015	0	0	0.0427

Thus the mass of $BaSO_4$ formed is:

$$(0.0427 \, \text{mol BaSO}_4) \times \frac{233.4 \, \text{g BaSO}_4}{1 \, \text{mol BaSO}_4} = \mathbf{9.97 \, g \, BaSO_4}$$

The pH can be calculated from the excess OH⁻ in solution. First, calculate the molar concentration of OH⁻. The total volume of solution is 136.4 mL = 0.1364 L.

$$[OH^-] = \frac{0.015 \, \text{mol}}{0.1364 \, \text{L}} = 0.11 \, M$$

$$pOH = -\log(0.11) = 0.96$$

$$\mathbf{pH} = 14.00 - pOH = 14.00 - 0.96 = \mathbf{13.04}$$

16.88   First, we calculate the molar solubility of $CaCO_3$.

	$CaCO_3$ (s) $\rightleftharpoons$	Ca^{2+} (aq) +	CO_3^{2-} (aq)
initial (M):	?	0	0
change (M):	−s	+s	+s
equil. (M):	? − s	s	s

$$K_{sp} = [Ca^{2+}][CO_3^{2-}] = s^2 = 8.7 \times 10^{-9}$$

$$s = 9.3 \times 10^{-5} \, M = 9.3 \times 10^{-5} \, \text{mol/L}$$

The moles of $CaCO_3$ in the kettle is:

$$\frac{116 \text{ g}}{100.1 \text{ g}/\text{mol}} = 1.16 \text{ moles } CaCO_3$$

The volume of distilled water needed to dissolve 1.16 moles of $CaCO_3$ is:

$$1.16 \text{ mol } CaCO_3 \times \frac{1 \text{ L}}{9.3 \times 10^{-5} \text{ mol } CaCO_3} = 1.2 \times 10^4 \text{ L}$$

The number of times the kettle would have to be filled is:

$$\frac{1.2 \times 10^4 \text{ L}}{2.0 \text{ L per filling}} = \mathbf{6.0 \times 10^3 \text{ fillings}}$$

Note that the very important assumption is made that each time the kettle is filled, the calcium carbonate is allowed to reach equilibrium before the kettle is emptied.

16.90   First we find the molar solubility and then convert moles to grams. The solubility equilibrium for silver carbonate is:

$$Ag_2CO_3(s) \rightleftharpoons 2Ag^+(aq) + CO_3^{2-}(aq)$$

$$K_{sp} = [Ag^+]^2[CO_3^{2-}] = (2s)^2(s) = 4s^3 = 8.1 \times 10^{-12}$$

$$s = \left(\frac{8.1 \times 10^{-12}}{4}\right)^{1/3} = 1.3 \times 10^{-4} \text{ } M$$

Converting from mol/L to g/L:

$$\left(\frac{1.3 \times 10^{-4} \text{ mol}}{1 \text{ L soln}}\right)\left(\frac{275.8 \text{ g}}{1 \text{ mol}}\right) = \mathbf{0.036 \text{ g/L}}$$

16.92   (a)   To 2.50×10^{-3} mol HCl (that is, 0.0250 L of 0.100 M solution) is added 1.00×10^{-3} mol CH_3NH_2 (that is, 0.0100 L of 0.100 M solution). After the acid-base reaction, we have 1.50×10^{-3} mol of HCl remaining. Since HCl is a strong acid, the $[H^+]$ will come from the HCl. The total solution volume is 35.0 mL = 0.0350 L.

$$[H^+] = \frac{1.50 \times 10^{-3} \text{ mol}}{0.0350 \text{ L}} = 0.0429 \text{ } M$$

pH = 1.37

(b)   When a total of 25.0 mL of CH_3NH_2 is added, we reach the equivalence point. That is, 2.50×10^{-3} mol HCl reacts with 2.50×10^{-3} mol CH_3NH_2 to form 2.50×10^{-3} mol CH_3NH_3Cl. Since there is a total of 50.0 mL of solution, the concentration of $CH_3NH_3^+$ is:

$$[CH_3NH_3^+] = \frac{2.50 \times 10^{-3} \text{ mol}}{0.0500 \text{ L}} = 5.00 \times 10^{-2} \text{ } M$$

This is a problem involving the hydrolysis of the weak acid $CH_3NH_3^+$

$$K_a = 2.3 \times 10^{-11} = \frac{[CH_3NH_2][H^+]}{[CH_3NH_3^+]} = \frac{x^2}{\left(5.00 \times 10^{-2} - x\right)} \approx \frac{x^2}{5.00 \times 10^{-2}}$$

$$1.15 \times 10^{-12} = x^2$$

$$x = 1.07 \times 10^{-6} M = [H^+]$$

pH = 5.97

(c)   When a total of 35.0 mL of 0.100 M CH_3NH_2 (3.50×10^{-3} mol) is added to the 25 mL of 0.100 M HCl (2.50×10^{-3} mol), the acid-base reaction produces 2.50×10^{-3} mol CH_3NH_3Cl with 1.00×10^{-3} mol of CH_3NH_2 in excess. Using the Henderson-Hasselbalch equation:

$$pH = pK_a + \log\frac{[\text{conjugate base}]}{[\text{acid}]}$$

$$\textbf{pH} = -\log(2.3 \times 10^{-11}) + \log\frac{(1.00 \times 10^{-3})}{(2.50 \times 10^{-3})} = \textbf{10.24}$$

16.94   The precipitate is HgI_2.

$$Hg^{2+}(aq) + 2\,I^-(aq) \longrightarrow HgI_2(s)$$

With further addition of I^-, a soluble complex ion is formed and the precipitate redissolves.

$$HgI_2(s) + 2\,I^-(aq) \longrightarrow HgI_4^{2-}(aq)$$

16.96   We can use the Henderson-Hasselbalch equation to solve for the pH when the indicator is 95% acid / 5% conjugate base and when the indicator is 5% acid / 95% conjugate base.

$$pH = pK_a + \log\frac{[\text{conjugate base}]}{[\text{acid}]}$$

Solving for the pH with 95% of the indicator in the HIn form:

$$pH = 9.10 + \log\frac{[5]}{[95]} = 9.10 - 1.28 = 7.82$$

Next, solving for the pH with 95% of the indicator in the In^- form:

$$pH = 9.10 + \log\frac{[95]}{[5]} = 9.10 + 1.28 = 10.38$$

Thus the pH range varies from **7.82 to 10.38** as the [HIn] varies from 95% to 5%.

16.98 **(a)** We abbreviate the name of cacodylic acid to CacH. We set up the usual table.

$$CacH\ (aq)\ \rightleftharpoons\ Cac^-\ (aq)\ +\ H^+\ (aq)$$

	CacH (aq)	Cac⁻ (aq)	H⁺ (aq)
Initial(M):	0.10	0.00	0.00
Change(M)	$-x$	$+x$	$+x$
Equilibrium(M):	$(0.10 - x)$	x	x

$$K_a = \frac{[H^+][Cac^-]}{[CacH]} = \frac{x^2}{0.10 - x} = 6.4 \times 10^{-7}$$

We assume that $(0.10 - x) \approx 0.10$. Then,

$$x = 2.5 \times 10^{-4}\ M = [H^+]$$

$$\mathbf{pH} = -\log(2.5 \times 10^{-4}) = \mathbf{3.60}$$

(b) We set up a table for the hydrolysis of the anion:

$$Cac^-\ (aq)\ +\ H_2O\ (l)\ \rightleftharpoons\ CacH\ (aq)\ +\ OH^-\ (aq)$$

	Cac⁻ (aq)	H₂O (l)	CacH (aq)	OH⁻ (aq)
Initial (M):	0.15		0.00	0.00
Change(M):	$-x$		$+x$	$+x$
Equilibrium(M):	$(0.15 - x)$		x	x

The ionization constant, K_b, for Cac⁻ is:

$$K_b = \frac{K_w}{K_a} = \frac{1.0 \times 10^{-14}}{6.4 \times 10^{-7}} = 1.6 \times 10^{-8}$$

$$\frac{x^2}{0.15 - x} = 1.6 \times 10^{-8}$$

$$x = 4.9 \times 10^{-5}\ M$$

$$pOH = -\log(4.9 \times 10^{-5}) = 4.31$$

$$\mathbf{pH} = 14.00 - 4.31 = \mathbf{9.69}$$

(c) Number of moles of CacH from (a) is:

$$50.0\ \text{mL CacH} \left(\frac{0.10\ \text{mol CacH}}{1000\ \text{mL}} \right) = 5.0 \times 10^{-3}\ \text{mol CacH}$$

Number of moles of Cac⁻ from (b) is:

$$25.0\ \text{mL CacNa} \left(\frac{0.15\ \text{mol CacNa}}{1000\ \text{mL}} \right) = 3.8 \times 10^{-3}\ \text{mol CacNa}$$

At this point we have a buffer solution.

$$\mathbf{pH} = pK_a + \log \frac{[Cac^-]}{[CacH]} = -\log(6.4 \times 10^{-7}) + \log \frac{3.8 \times 10^{-3}}{5.0 \times 10^{-3}} = \mathbf{6.07}$$

16.100 (a) $MCO_3 + 2HCl \rightarrow MCl_2 + H_2O + CO_2$

$HCl + NaOH \rightarrow NaCl + H_2O$

(b) Moles of HCl reacted with MCO_3 = Total moles of HCl − Moles of excess HCl

$$\text{Total moles of HCl} = 20.00 \text{ mL} \times \frac{1 \text{ L}}{1000 \text{ mL}} \times \frac{0.0800 \text{ mol}}{1 \text{ L}} = 1.60 \times 10^{-3} \text{ mol HCl}$$

$$\text{Moles of excess HCl} = 5.64 \text{ mL} \times \frac{1 \text{ L}}{1000 \text{ mL}} \times \frac{0.1000 \text{ mol}}{1 \text{ L}} = 5.64 \times 10^{-4} \text{ mol HCl}$$

$$\text{Moles of HCl reacted with } MCO_3 = 1.60 \times 10^{-3} \text{ mol} - 5.64 \times 10^{-4} = 1.04 \times 10^{-3} \text{ mol HCl}$$

$$\text{Moles of } MCO_3 \text{ reacted} = 1.04 \times 10^{-3} \text{ mol HCl} \times \frac{1 \text{ mol } MCO_3}{2 \text{ mol HCl}} = 5.20 \times 10^{-4} \text{ mol } MCO_3$$

$$\textbf{Molar mass of } \mathbf{MCO_3} = \frac{0.1022 \text{ g}}{5.20 \times 10^{-4} \text{ mol}} = \mathbf{197 \text{ g/mol}}$$

Molar mass of CO_3 = 60.01

Molar mass of M = 197 g/mol − 60.01 g/mol = 137 g/mol

The metal, M, is **Ba**!

16.102 The number of moles of NaOH reacted is:

$$15.9 \text{ mL NaOH} \left(\frac{0.500 \text{ mol NaOH}}{1000 \text{ mL}} \right) = 7.95 \times 10^{-3} \text{ mol NaOH}$$

Since two moles of NaOH combine with one mole of oxalic acid, the number of moles of oxalic acid reacted is 3.98×10^{-3} mol. This is the number of moles of oxalic acid hydrate in 25.0 mL of solution. In 250 mL, the number of moles present is 3.98×10^{-2} mol. Thus the molar mass is:

$$\frac{5.00 \text{ g}}{3.98 \times 10^{-2} \text{ mol}} = 126 \text{ g/mol}$$

From the molecular formula we can write:

$$2(1.008) \text{ g} + 2(12.01) \text{ g} + 4(16.00) \text{ g} + x(18.02)\text{g} = 126 \text{ g}$$

Solving for x:

$$x = 2$$

16.104 (a) $\text{pH} = pK_a + \log \dfrac{[\text{conjugate base}]}{[\text{acid}]}$

$$8.00 = 9.10 + \log \frac{[\text{ionized}]}{[\text{un} - \text{ionized}]}$$

$$\frac{[\text{un} - \text{ionized}]}{[\text{ionized}]} = \mathbf{12.6} \qquad\qquad (1)$$

(b)   The total concentration of the indicator is:

$$(2 \text{ drops}) \times \left(\frac{0.050 \text{ mL phenolphthalein}}{1 \text{ drop}} \right) \times \left(\frac{1}{50 \text{ mL soln}} \right) \times (0.060 \text{ } M) = 1.2 \times 10^{-4} \text{ } M$$

Using equation (1) above and letting $y = $ [ionized]

$$\frac{1.2 \times 10^{-4} - y}{y} = 12.6$$

$$y = 8.8 \times 10^{-6} \text{ } M$$

16.106 **(a)**   Add sulfate. Na_2SO_4 is soluble, $BaSO_4$ is not.

       (b)   Add sulfide. K_2S is soluble, PbS is not

       (c)   Add iodide. ZnI_2 is soluble, HgI_2 is not.

16.108 The amphoteric oxides cannot be used to prepare buffer solutions because they are insoluble in water.

16.110 The ionized polyphenols have a dark color. In the presence of citric acid from lemon juice, the anions are converted to the lighter-colored acids.

16.112 Assuming the density of water to be 1.00 g/mL, 0.05 g Pb^{2+} per 10^6 g water is equivalent to 5×10^{-5} g Pb^{2+}/L

$$\frac{0.05 \text{ g } Pb^{2+}}{1 \times 10^6 \text{ g } H_2O} \times \frac{1 \text{ g } H_2O}{1 \text{ mL } H_2O} \times \frac{1000 \text{ mL } H_2O}{1 \text{ L } H_2O} = 5 \times 10^{-5} \text{ g } Pb^{2+} / L$$

$$PbSO_4 \rightleftharpoons Pb^{2+} + SO_4^{2-}$$

$$1.6 \times 10^{-8} = s^2$$

$$s = 1.3 \times 10^{-4} \text{ } M$$

The solubility of $PbSO_4$ in g/L is:

$$1.3 \times 10^{-4} \frac{\text{mol}}{L} \times \frac{303.3 \text{ g}}{1 \text{ mol}} = 4.0 \times 10^{-2} \text{ g/L}$$

Yes. The [Pb^{2+}] exceeds the safety limit of 5×10^{-5} g Pb^{2+}/L.

16.114 **(c)** has the highest [H^+]

$$F^- + SbF_5 \rightarrow SbF_6^-$$

Removal of F^- promotes further ionization of HF.

16.116 **(a)**   This is a common ion (CO_3^{2-}) problem.

       The dissociation of Na_2CO_3 is:

$$Na_2CO_3 \text{ } (s) \xrightarrow{\text{ } H_2O \text{ }} 2 \text{ } Na^+ \text{ } (aq) + CO_3^{2-} \text{ } (aq)$$
$$2(0.050 \text{ } M) \qquad 0.050 \text{ } M$$

Let s be the molar solubility of $CaCO_3$ in Na_2CO_3 solution. We summarize the changes as:

$$CaCO_3\ (s) \rightleftharpoons Ca^{2+}\ (aq) + CO_3^{2-}$$

	Ca^{2+}	CO_3^{2-}
Initial (M)	0.00	0.050
Change (M)	+s	+s
Equil. (M)	+s	0.050 + s

$$K_{sp} = [Ca^{2+}][CO_3^{2-}]$$

$$8.7 \times 10^{-9} = s(0.050 + s)$$

Since s is small, we can assume that $0.050 + s \approx 0.050$

$$8.7 \times 10^{-9} = 0.050s$$

$$s = 1.7 \times 10^{-7}\ M$$

Thus, the addition of washing soda to permanent hard water removes most of the Ca^{2+} ions as a result of the common ion effect.

(b) Mg^{2+} is not removed by this procedure, because $MgCO_3$ is fairly soluble ($K_{sp} = 4.0 \times 10^{-5}$).

(c) The K_{sp} for $Ca(OH)_2$ is 8.0×10^{-6}.

$$Ca(OH)_2 \rightleftharpoons Ca^{2+} + 2OH^-$$

At equil: s $2s$

$$K_{sp} = 8.0 \times 10^{-6} = [Ca^{2+}][OH^-]^2$$

$$4s^3 = 8.0 \times 10^{-6}$$

$$s = 0.0126\ M$$

$$[OH^-] = 2s = 0.0252\ M$$

$$pOH = -\log(0.0252) = 1.60$$

pH = 12.40

(d) The $[OH^-]$ calculated above is $0.0252\ M$. At this rather high concentration of OH^-, most of the Mg^{2+} will be removed as $Mg(OH)_2$. The small amount of Mg^{2+} remaining in solution is due to the following equilibrium:

$$Mg(OH)_2\ (s) \rightleftharpoons Mg^{2+}\ (aq) + 2\ OH^-\ (aq)$$

$$K_{sp} = [Mg^{2+}][OH^-]^2$$

$$1.2 \times 10^{-11} = [Mg^{2+}](0.0252)^2$$

$$[Mg^{2+}] = 1.9 \times 10^{-8}\ M$$

(e) Remove Ca^{2+} first because it is present in larger amounts.

CHAPTER 17
CHEMISTRY IN THE ATMOSPHERE

Almost all the types of problems in this chapter have been encountered in previous chapters. Therefore, we will not repeat the problem types here, but will refer you to problem types where appropriate.

SOLUTIONS TO SELECTED TEXT PROBLEMS

17.6   Using the information in Table 17.1 and Problem 17.5, 0.033 percent of the volume (and therefore the pressure) of dry air is due to CO_2. The partial pressure of CO_2 is:

$$P_{CO_2} = X_{CO_2} P_T = (3.3 \times 10^{-4}) \times (754 \text{ mmHg}) \times \frac{1 \text{ atm}}{760 \text{ mmHg}} = \textbf{3.3} \times \textbf{10}^{-4} \textbf{ atm}$$

17.8   From Problem 5.92, the total mass of air is 5.25×10^{18} kg. Table 17.1 lists the composition of air by volume. Under the same conditions of P and T, $V \propto n$ (Avogadro's law).

$$\text{Total number of moles of gases} = \frac{5.25 \times 10^{21} \text{ g}}{29.0 \text{ g/mol}} = 1.81 \times 10^{20} \text{ mol}$$

Mass of N_2 (78.03%):

$$0.7803 \times (1.81 \times 10^{20} \text{ mol}) \times 28.02 \text{ g/mol} = 3.96 \times 10^{21} \text{ g} = \textbf{3.96} \times \textbf{10}^{18} \textbf{ kg}$$

Mass of O_2 (20.99%):

$$0.2099 \times (1.81 \times 10^{20} \text{ mol}) \times 32.00 \text{ g/mol} = 1.22 \times 10^{21} \text{ g} = \textbf{1.22} \times \textbf{10}^{18} \textbf{ kg}$$

Mass of CO_2 (0.033%):

$$(3.3 \times 10^{-4}) \times (1.81 \times 10^{20} \text{ mol}) \times 44.01 \text{ g/mol} = 2.63 \times 10^{18} \text{ g} = \textbf{2.63} \times \textbf{10}^{15} \textbf{ kg}$$

17.12   See Problem Types 1 and 2, Chapter 7.

$$\nu = \frac{c}{\lambda} = \frac{3.00 \times 10^8 \text{ m/s}}{558 \times 10^{-9} \text{ m}} = 5.38 \times 10^{14} \text{ /s}$$

$$\Delta E = h\nu = (6.63 \times 10^{-34} \text{ J·s})(5.38 \times 10^{14} \text{ /s})$$

$$\Delta E = \textbf{3.57} \times \textbf{10}^{-19} \textbf{ J}$$

17.22   The quantity of ozone lost is:

$$0.06 \times (3.2 \times 10^{12} \text{ kg}) = 1.9 \times 10^{11} \text{ kg of } O_3$$

Assuming no further deterioration, the kilograms of O_3 that would have to be manufactured on a daily basis are:

$$\frac{1.9 \times 10^{11} \text{ kg O}_3}{100 \text{ yr}} \times \frac{1 \text{ yr}}{365 \text{ days}} = \textbf{5.2} \times \textbf{10}^{\textbf{6}} \textbf{ kg/day}$$

The standard enthalpy of formation (from Appendix 3) for ozone:

$$\frac{3}{2}O_2 \rightarrow O_3 \qquad \Delta H_f^\circ = 142.2 \text{ kJ/mol}$$

The *total* energy required is:

$$1.9 \times 10^{14} \text{ g of O}_3 \times \frac{1 \text{ mol O}_3}{48.00 \text{ g O}_3} \times \frac{142.2 \text{ kJ}}{1 \text{ mol O}_3} = \textbf{5.6} \times \textbf{10}^{\textbf{14}} \textbf{ kJ}$$

17.24   The energy of the photons of UV radiation in the troposphere is insufficient (that is, the wavelength is too long and the frequency is too small) to break the bonds in CFCs.

17.26   See Problem Type 2, Chapter 7. First, we need to calculate the energy needed to break one bond.

$$\frac{276 \times 10^3 \text{ J}}{1 \text{ mol}} \times \frac{1 \text{ mol}}{6.022 \times 10^{23} \text{ molecules}} = 4.58 \times 10^{-19} \text{ J/molecule}$$

The longest wavelength required to break this bond is:

$$\lambda = \frac{hc}{E} = \frac{(3.00 \times 10^8 \text{ m/s})(6.63 \times 10^{-34} \text{ J} \cdot \text{s})}{4.58 \times 10^{-19} \text{ J}} = 4.34 \times 10^{-7} \text{ m} = \textbf{434 nm}$$

434 nm is in the visible region of the electromagnetic spectrum; therefore, CF_3Br will be decomposed in **both** the troposphere and stratosphere.

17.28   The Lewis structure of HCFC–123 is:

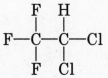

The Lewis structure for CF_3CFH_2 is:

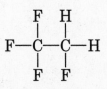

Lone pairs on the outer atoms have been omitted.

17.40   See Problem Type 7, Chapter 3. From the balanced equation, you should find a mole ratio of 1:1 between CaO and CO_2.

$$\text{mass CO}_2 = 1.7 \times 10^{13} \text{ g CaO} \times \frac{1 \text{ mol CaO}}{56.08 \text{ g CaO}} \times \frac{1 \text{ mol CO}_2}{1 \text{ mol CaO}} \times \frac{44.01 \text{ g}}{1 \text{ mol CO}_2}$$

$$= \textbf{1.3} \times \textbf{10}^{\textbf{13}} \textbf{ g CO}_2 = \textbf{1.3} \times \textbf{10}^{\textbf{10}} \textbf{ kg CO}_2$$

17.42   Ethane and propane are greenhouse gases. They would contribute to global warming.

17.50   Recall that ppm means the number of parts of substance per 1,000,000 parts. We can calculate the partial pressure of SO_2 in the troposphere.

$$P_{SO_2} = \frac{0.16 \text{ molecules of } SO_2}{10^6 \text{ parts of air}} \times 1 \text{ atm} = 1.6 \times 10^{-7} \text{ atm}$$

Next, we need to set up the equilibrium constant expression to calculate the concentration of H^+ in the rainwater. See Problem Type 5A, Chapter 15. From the concentration of H^+, we can calculate the pH.

$$SO_2 \quad + \quad H_2O \quad \rightleftharpoons \quad H^+ \quad + \quad HSO_3^-$$

equilibrium 1.6×10^{-7} atm x x

$$K = \frac{[H^+][HSO_3^-]}{P_{SO_2}} = 1.3 \times 10^{-2}$$

$$1.3 \times 10^{-2} = \frac{x^2}{1.6 \times 10^{-7}}$$

$$x^2 = 2.1 \times 10^{-9}$$

$$x = 4.6 \times 10^{-5} \, M = [H^+]$$

$$\mathbf{pH} = -\log(4.6 \times 10^{-5}) = \mathbf{4.34}$$

17.58   See Problem Type 2A, Chapter 5.

0.55 ppm by volume means:

$$\frac{V_{PAN}}{V_T} = \frac{0.55 \text{ L}}{1 \times 10^6 \text{ L}}$$

At STP, the number of moles of PAN in 1.0 L of air is:

$$n = \frac{PV}{RT} = \frac{(1 \text{ atm})\left(\dfrac{0.55 \text{ L}}{1 \times 10^6 \text{ L}} \times 1.0 \text{ L}\right)}{(0.0821 \text{ L} \cdot \text{atm} / \text{K} \cdot \text{mol})(273 \text{ K})} = 2.5 \times 10^{-8} \text{ mol}$$

See Problem Type 2, Chapter 13. Since the decomposition follows first-order kinetics, we can write:

$$\text{rate} = k[PAN]$$

$$\mathbf{rate} = (4.9 \times 10^{-4}/\text{s})\left(\frac{2.5 \times 10^{-8} \text{ mol}}{1.0 \text{ L}}\right) = \mathbf{1.2 \times 10^{-11} \, \textit{M/s}}$$

17.60   The Gobi desert lacks the primary pollutants (nitric oxide, carbon monoxide, hydrocarbons) to have photochemical smog. The primary pollutants are present both in New York City and in Boston. However, the sunlight that is required for the conversion of the primary pollutants to the secondary pollutants associated with smog is more likely in a July afternoon than one in January. Therefore, answer **(b)** is correct.

17.66   See Problem Type 7, Chapter 3.

$$CO_2 + Ca(OH)_2 \rightarrow CaCO_3 + H_2O$$

Moles of CO_2 reacted:

$$0.026 \text{ g CaCO}_3 \times \frac{1 \text{ mol CaCO}_3}{100.1 \text{ g CaCO}_3} \times \frac{1 \text{ mol CO}_2}{1 \text{ mol CaCO}_3} = 2.6 \times 10^{-4} \text{ mol CO}_2$$

See Problem Type 2A, Chapter 5. Total number of moles of air:

$$n = \frac{PV}{RT} = \frac{\left(747 \text{ mmHg} \times \frac{1 \text{ atm}}{760 \text{ mmHg}}\right)(5.0 \text{ L})}{(0.0821 \text{ L} \cdot \text{atm} / \text{mol} \cdot \text{K})(291 \text{ K})} = 0.21 \text{ mol air}$$

The percentage by volume of CO_2 in air is:

$$\frac{V_{CO_2}}{V_{air}} \times 100\% = \frac{n_{CO_2}}{n_{air}} \times 100\% = \frac{2.6 \times 10^{-4} \text{ mol}}{0.21 \text{ mol}} \times 100\% = \textbf{0.12\%}$$

17.68   Higher temperature has shifted the equilibrium to the right; therefore, the reaction is ***endothermic***.

17.70   The concentration of O_2 could be monitored. Formation of CO_2 must deplete O_2.

17.72   In Problem 17.6, we determined the partial pressure of CO_2 in dry air to be 3.3×10^{-4} atm. Using Henry's law, we can calculate the concentration of CO_2 in water. See Problem Type 4, Chapter 12.

$$c = kP$$

$$[CO_2] = (0.032 \text{ mol/L} \cdot \text{atm}) \times (3.3 \times 10^{-4} \text{ atm}) = 1.06 \times 10^{-5} \text{ mol/L}$$

We assume that all of the dissolved CO_2 is converted to H_2CO_3, thus giving us 1.06×10^{-5} mol/L of H_2CO_3. See Problem Type 5A, Chapter 15.

The equilibrium expression is:

	H_2CO_3	$\rightleftharpoons$	H^+	$+$	HCO_3^-
Initial (*M*):	1.06×10^{-5}		0		0
Change (*M*):	$-x$		$+x$		$+x$
Equilibrium (*M*):	$(1.06 \times 10^{-5}) - x$		x		x

$$K \text{ (from Table 15.5)} = 4.2 \times 10^{-7} = \frac{[H^+][HCO_3^-]}{[H_2CO_3]} = \frac{x^2}{1.06 \times 10^{-5} - x}$$

Solving the quadratic equation:

$$x = 1.9 \times 10^{-6} M = [H^+]$$

$$\textbf{pH} = -\log(1.9 \times 10^{-6}) = \textbf{5.72}$$

17.74 See Problem Types 4A and 5C, Chapter 6.

Consider reaction (1):

$$\Delta H° = \Delta H_f°(NO) + \Delta H_f°(O) - \Delta H_f°(NO_2)$$

$$\Delta H° = (1 \text{ mol})(90.4 \text{ kJ/mol}) + (1 \text{ mol})(249.4 \text{ kJ/mol}) - (1 \text{ mol})(33.85 \text{ kJ/mol})$$

$$\Delta H° = 306 \text{ kJ}$$

From Equation (6.15) of the text, $\Delta E° = \Delta H° - RT\Delta n$

$$\Delta E° = 306 \times 10^3 \text{ J} - (8.314 \text{ J/mol·K})(298 \text{ K})(1 \text{ mol})$$

$$\Delta E° = 304 \times 10^3 \text{ J}$$

See Problem Type 2, Chapter 7. This is the energy needed to dissociate 1 mole of NO_2. We need the energy required to dissociate *one molecule* of NO_2.

$$\frac{304 \times 10^3 \text{ J}}{1 \text{ mol } NO_2} \times \frac{1 \text{ mol } NO_2}{6.022 \times 10^{23} \text{ molecules } NO_2} = 5.05 \times 10^{-19} \text{ J/molecule}$$

The longest wavelength that can dissociate NO_2 is:

$$\lambda = \frac{hc}{E} = \frac{(6.63 \times 10^{-34} \text{ J·s})(3.00 \times 10^8 \text{ m/s})}{5.05 \times 10^{-19} \text{ J}} = 3.94 \times 10^{-7} \text{ m} = 394 \text{ nm}$$

17.76 This reaction has a high activation energy.

17.78 The size of tree rings can be related to CO_2 content, where the number of rings indicates the age of the tree. The amount of CO_2 in ice can be directly measured from portions of polar ice in different layers obtained by drilling. The "age" of CO_2 can be determined by radiocarbon dating and other methods.

17.80 See Problem Type 7, Chapter 9.

$$Cl_2 + O_2 \rightarrow 2ClO$$

$$\Delta H° = \Sigma BE(\text{reactants}) - \Sigma BE(\text{products})$$

$$\Delta H° = (1 \text{ mol})(242.7 \text{ kJ/mol}) + (1 \text{ mol})(498.7 \text{ kJ/mol}) - (2 \text{ mol})(206 \text{ kJ/mol})$$

$$\Delta H° = 329 \text{ kJ}$$

See Problem Type 4A, Chapter 6.

$$\Delta H° = 2\Delta H_f°(ClO) - 2\Delta H_f°(Cl_2) - 2\Delta H_f°(O_2)$$

$$329 \text{ kJ} = (2 \text{ mol})\Delta H_f°(ClO) - 0 - 0$$

$$\Delta H_f°(ClO) = \frac{329 \text{ kJ}}{2 \text{ mol}} = 165 \text{ kJ/mol}$$

CHAPTER 18
ENTROPY, FREE ENERGY,
AND EQUILIBRIUM

PROBLEM-SOLVING STRATEGIES AND TUTORIAL SOLUTIONS

TYPES OF PROBLEMS

Problem Type 1: Entropy.
 (a) Predicting entropy changes.
 (b) Calculating entropy changes of a system.

Problem Type 2: Free Energy.
 (a) Calculating standard free energy changes.
 (b) Calculating standard free energy change from enthalpy and entropy changes.
 (c) Entropy changes due to phase transitions.

Problem Type 3: Free Energy and Chemical Equilibrium.
 (a) Using the standard free energy change to calculate the equilibrium constant.
 (b) Using the free energy change to predict the direction of a reaction.

PROBLEM TYPE 1: ENTROPY

Entropy (S) is a direct measure of the randomness or disorder of a system. The greater the disorder of a system, the greater its entropy. Conversely, the more ordered a system, the lower its entropy.

A. Predicting entropy changes

For any substance, the particles in the solid are more ordered than those in the liquid state, which in turn are more ordered than those in the gaseous state. See Figure 18.2 of your text. Thus, for any substance, the entropy for the same molar amount always increases in the following order.

$$S_{solid} < S_{liquid} < S_{gas}$$

There are a number of other factors that you need to consider to predict entropy changes.

- Heating increases the entropy of a system because it increases the random motion of atoms and molecules.

- When a solid dissolves in water, the highly ordered structure of the solid and part of the order of the water are destroyed. Consequently, the solution possesses greater disorder than the pure solute and pure solvent.

- If a reaction produces more gas molecules than it consumes, the entropy of the system increases. If the total number of gas molecules diminishes during a reaction, the entropy of the system decreases. The dependence of the entropy of reaction on the number of gas molecules is due to the fact that gas molecules possess much greater entropy than either solid or liquid molecules.

- If there is no net change in the total number of gas molecules in a reaction, then the entropy of the system may increase or decrease, but it will be a relatively small change.

EXAMPLE 18.1

Predict whether the entropy increases, decreases, or remains essentially unchanged for the following reactions.

(a) H_2O_2 (*l*) $\longrightarrow$ H_2O (*l*) + 1/2 O_2 (*g*)

(b) H^+ (*aq*) + OH^- (*aq*) $\longrightarrow$ H_2O (*l*)

(c) $Ca(OH)_2$ (*s*) + CO_2 (*g*) $\longrightarrow$ $CaCO_3$ (*s*) + H_2O (*g*)

(a) The number of moles of gaseous compounds in the products is greater than in the reactant. Entropy increases during this reaction.

(b) Two reactants combine into one product in this reaction. Order is increased and so entropy decreases.

(c) The number of moles of gas phase products is the same as the reactants. The entropy remains essentially unchanged.

PRACTICE EXERCISE

1. Predict whether the entropy increases, decreases, or remains essentially unchanged for the following reactions:

(a) CaO (*s*) + CO_2 (*g*) $\longrightarrow$ $CaCO_3$ (*s*)

(b) $CuSO_4$ (*s*) $\longrightarrow$ Cu^{2+} (*aq*) + SO_4^{2-} (*aq*)

(c) $2 HCl$ (*g*) + Br_2 (*l*) $\longrightarrow$ $2 HBr$ (*g*) + Cl_2 (*g*)

Text Problems: 18.10, 18.14

B. Calculating entropy changes of a system

The universe is made up of the system and the surroundings. The entropy change in the universe (ΔS_{univ}) for any process is the *sum* of the entropy changes in the system (ΔS_{sys}) and in the surroundings (ΔS_{surr}).

$$\Delta S_{univ} = \Delta S_{sys} + \Delta S_{surr}$$

For a spontaneous process, the entropy of the universe increases.

$$\Delta S_{univ} = \Delta S_{sys} + \Delta S_{surr} > 0$$

In chemistry, we typically focus on the entropy of the system. Let's suppose that the system is represented by the following reaction:

$$a\text{A} + b\text{B} \longrightarrow c\text{C} + d\text{D}$$

As is the case for the enthalpy of a reaction [see Equation (6.7) in the text], the standard entropy change $\Delta S°$ is given by:

$$\Delta S_{rxn}° = [cS°(\text{C}) + dS°(\text{D})] - [aS°(\text{A}) + bS°(\text{B})]$$

or, using Σ to represent summation and *m* and *n* for the stoichiometric coefficients in the reaction,

$$\Delta S_{rxn}° = \Sigma nS°(\text{products}) - \Sigma mS°(\text{reactants})$$

Using the standard entropy values ($S°$) listed in Appendix 3, we can calculate $\Delta S_{rxn}°$, which corresponds to ΔS_{sys}.

EXAMPLE 18.2

Use standard entropy values to calculate the standard entropy change (ΔS°_{rxn}) for the reaction

$$H_2 (g) + 1/2\ O_2 (g) \longrightarrow H_2O (l)$$

The standard entropy change is given by:

$$\Delta S^\circ_{rxn} = S^\circ[H_2O\ (l)] - \{S^\circ[H_2\ (g)] + (1/2)S^\circ[O_2\ (g)]\}$$

You can look up the standard entropy values in Appendix 3 of the text.

$$\Delta S^\circ_{rxn} = (1\ \text{mol})\left(\frac{69.9\ J}{K \cdot mol}\right) - \left[(1\ \text{mol})\left(\frac{131.0\ J}{mol \cdot K}\right) + \left(\frac{1}{2}\ \text{mol}\right)\left(\frac{205.0\ J}{mol \cdot K}\right)\right]$$

$$\Delta S^\circ_{rxn} = 69.9\ J/K - 233.5\ J/K = -163.6\ J/K$$

> **Tip:** This reaction is known to be spontaneous, so you would think that ΔS should be positive. Remember, that the value of ΔS°_{rxn} applies only to the system; ΔS_{univ} will have a positive value.

PRACTICE EXERCISE

2.   Hydrate lime, $Ca(OH)_2$, can be reformed into quicklime, CaO, by heating:

$$Ca(OH)_2\ (s) \xrightarrow{\ \text{heat}\ } CaO\ (s) + H_2O\ (g)$$

Use standard entropy values to calculate the standard entropy change (ΔS°_{rxn}) for this reaction.

> **Text Problem: 18.12**

PROBLEM TYPE 2: FREE ENERGY

The second law of thermodynamics tells us that a spontaneous reaction increases the entropy of the universe; that is, $\Delta S_{univ} > 0$. To calculate ΔS_{univ}, we must calculate both ΔS_{surr} and ΔS_{sys}. However, it can be difficult to calculate ΔS_{surr}, and typically we are only concerned with what happens in a particular system.

Therefore, considering only the system, we use another thermodynamic function to determine if a reaction will occur spontaneously. This function, called **Gibbs free energy (*G*)**, or simply free energy is given by the following equation:

$$G = H - TS$$

where,
> *H* is the enthalpy
> *S* is the entropy
> *T* is the temperature (in K)

The change in free energy (ΔG) of a system for a reaction at constant temperature is:

$$\Delta G = \Delta H - T\Delta S$$

The sign of ΔG will allow us to predict whether a reaction is spontaneous at equilibrium. At constant temperature and pressure, we can summarize the following conditions in terms of ΔG.

- $\Delta G < 0$ A spontaneous process in the forward direction.

- $\Delta G > 0$ A nonspontaneous reaction as written. However, the reaction is spontaneous in the reverse direction.

- $\Delta G = 0$ The system is at equilibrium. There is no net change in the system.

A. Calculating standard free energy changes

Let's again suppose that the system is represented by the following reaction:

$$a\text{A} + b\text{B} \longrightarrow c\text{C} + d\text{D}$$

As is the case for the entropy of a reaction the standard free energy change $\Delta G°$ is given by:

$$\Delta G°_{rxn} = [c\,\Delta G°_f\,(\text{C}) + d\,\Delta G°_f\,(\text{D})] - [a\,\Delta G°_f\,(\text{A}) + b\,\Delta G°_f\,(\text{B})]$$

or,

$$\Delta G°_{rxn} = \Sigma n\,\Delta G°_f\,(\text{products}) - \Sigma m\,\Delta G°_f\,(\text{reactants})$$

where m and n are stoichiometric coefficients. The term $\Delta G°_f$ is the **standard free energy of formation** of a compound. It is the free energy change that occurs when 1 mole of the compound is synthesized from its elements in their standard states (see Table 18.2 of your text for conventions used for standard states). By definition, the standard free energy of formation of any element in its stable form is *zero*.

For example,

$$\Delta G°_f\,[\text{O}_2(g)] = 0 \quad \text{and} \quad \Delta G°_f\,[\text{Na}(s)] = 0$$

Other standard free energies of formation can be found in Appendix 3 of the text.

EXAMPLE 18.3

Calculate $\Delta G°_{rxn}$ at 25°C for the following reaction given that $\Delta G°_f\,(\text{Fe}_2\text{O}_3) = -741.0$ kJ/mol.

$$2\text{ Al }(s) + \text{Fe}_2\text{O}_3\,(s) \longrightarrow \text{Al}_2\text{O}_3\,(s) + 2\text{ Fe }(s)$$

The standard free energy change is given by:

$$\Delta G°_{rxn} = \{\Delta G°_f\,[\text{Al}_2\text{O}_3\,(s)] + 2\,\Delta G°_f\,[\text{Fe }(s)]\} - \{2\,\Delta G°_f\,[\text{Al }(s)] + \Delta G°_f\,[\text{Fe}_2\text{O}_3\,(s)]\}$$

You can look up the standard free energy of formation values in Appendix 3 of the text.

$$\Delta G°_{rxn} = [(1\text{ mol})(-1576.41\text{ kJ/mol}) + 0] - [0 + (1\text{ mol})(-741.0\text{ kJ/mol})]$$

$$\Delta G°_{rxn} = -1576.41\text{ kJ} + 741.0\text{ kJ} = \textbf{-835.4 kJ}$$

> **Tip:** Remember that the standard free energy of formation of any element in its stable form is *zero*. Therefore, the standard free energy of formation values for both Al (s) and Fe (s) are *zero*.

PRACTICE EXERCISE

3. Using Appendix 3 of the text, calculate $\Delta G°_{rxn}$ values for the following reactions:

    (a) $3\text{ CaO }(s) + 2\text{ Al }(s) \longrightarrow 3\text{ Ca }(s) + \text{Al}_2\text{O}_3\,(s)$

    (b) $\text{ZnO }(s) \longrightarrow \text{Zn }(s) + 1/2\text{ O}_2\,(g)$

Text Problems: 18.18, 18.32

B.   Calculating standard free energy change from enthalpy and entropy changes

We can determine the sign of ΔG if we know the signs of both ΔH and ΔS. A negative ΔH (an exothermic reaction) and a positive ΔS (increase in disorder), give a negative ΔG. In addition, temperature may influence the direction of a spontaneous reaction. There are four possible outcomes for the relationship

$$\Delta G = \Delta H - T\Delta S$$

- If both ΔH and ΔS are positive, ΔG will be negative only when $T\Delta S$ is greater in magnitude than ΔH. This condition is met when T is large (high temperature).

- If ΔH is positive and ΔS is negative, ΔG will always be positive, regardless of temperature. The reaction is nonspontaneous.

- If ΔH is negative and ΔS is positive, ΔG will always be negative, regardless of temperature. The reaction is spontaneous.

- If ΔH is negative and ΔS is negative, ΔG will be negative only when ΔH is greater in magnitude than $T\Delta S$. This condition is met when T is small (low temperature).

We can also calculate the value of ΔG if we know the values of ΔH, ΔS, and the temperature. Typically, $\Delta H°$ is calculated from standard enthalpies of formation.

$$\Delta H° = \Sigma n \, \Delta H_f° \text{ (products)} - \Sigma m \, \Delta H_f° \text{ (reactants)}$$

Also, $\Delta S°$ is calculated from standard entropy values (see Problem Type 1B).

$$\Delta S° = \Sigma n S°(\text{products}) - \Sigma m S°(\text{reactants})$$

Then, for reactions carried out under standard-state conditions, we can substitute $\Delta H°$, $\Delta S°$, and the temperature into the following equation to calculate $\Delta G°$.

$$\Delta G° = \Delta H° - T\Delta S°$$

EXAMPLE 18.4
Calculate $\Delta G°$ for the following reaction at 298 K:

$$2 \, H_2 \, (g) + CO \, (g) \rightleftharpoons CH_3OH \, (g)$$

given that $\Delta H° = -90.7$ kJ and $\Delta S° = -221.5$ J/K for this process.

The standard free energy change is given by:

$$\Delta G° = \Delta H° - T\Delta S°$$

Step 1: Let's convert $\Delta S°$ to units of kJ/K, so that we have consistent units.

$$-221.5 \frac{J}{K} \times \frac{1 \text{ kJ}}{1000 \text{ J}} = -0.2215 \text{ kJ/K}$$

Step 2: Substitute $\Delta H°$, $\Delta S°$, and the temperature into the above equation to calculate $\Delta G°$.

$$\Delta G° = \Delta H° - T\Delta S°$$

$$\Delta G° = -90.7 \text{ kJ} - [(298 \text{ K})(-0.2215 \text{ kJ/K})]$$

$$\Delta G° = -90.7 \text{ kJ} + 66.0 \text{ kJ} = \mathbf{-24.7 \text{ kJ}}$$

PRACTICE EXERCISE

4.   The synthesis of O_2 (g) is often carried out in general chemistry laboratories by the decomposition of $KClO_3$,

$$KClO_3 \ (s) \longrightarrow KCl \ (s) \ + \ 3/2 \ O_2 \ (g)$$

for which $\Delta H° = -44.7$ kJ and $\Delta S° = +59.1$ J/K. Is this reaction spontaneous at 25°C under standard conditions?

Text Problem: 18.20

C. Entropy changes due to phase transitions

At the transition temperature, the melting or boiling point, a system is at equilibrium and $\Delta G = 0$. Thus, we can write,

$$\Delta G \ = \ \Delta H - T\Delta S$$

or

$$0 \ = \ \Delta H - T\Delta S$$

and

$$\Delta S \ = \ \frac{\Delta H}{T} \qquad\qquad (18.1)$$

ΔS is the entropy change due to the phase transition.

EXAMPLE 18.5

The heat of fusion of water, ΔH_{fus}, at 0°C is 6.02 kJ/mol. What is ΔS_{fus} for 1 mol of H_2O at the melting point?

The entropy change due to the phase transition (the melting of water), can be calculated using Equation (18.1) above. Recall that the temperature must be in units of Kelvin (0°C = 273 K).

$$\Delta S \ = \ \frac{\Delta H}{T}$$

$$\Delta S \ = \ \frac{6.02 \ \frac{kJ}{mol}}{273 \ K} \ = \ \textbf{+0.0221 kJ/mol·K} \ = \ \textbf{+22.1 J/mol·K}$$

The increase in entropy upon melting the solid corresponds to the increase in molecular disorder in the liquid state compared to the solid state.

PRACTICE EXERCISE

5.   The enthalpy of vaporization of mercury is 58.5 kJ/mol and the normal boiling point is 630 K. What is the entropy of vaporization of mercury?

Text Problem: 18.60

PROBLEM TYPE 3: FREE ENERGY AND CHEMICAL EQUILIBRIUM

A. Using the standard free energy change to calculate the equilibrium constant

There is a relationship between the free energy change (ΔG) and the standard free energy change.

$$\Delta G = \Delta G° + RT\ln Q \qquad (18.2)$$

where,

R is the gas constant (8.314 J/mol·K)
T is the absolute temperature of the reaction (in K)
Q is the reaction quotient (see Problem Type 4, Chapter 14)

At equilibrium, by definition, $\Delta G = 0$ and $Q = K$, where K is the equilibrium constant. Thus,

$$0 = \Delta G° + RT\ln K$$

or

$$\Delta G° = -RT\ln K \qquad (18.3)$$

K can be either K_p, used for gases, or K_c used for reactions in solution. Note that the larger the value of K, the more negative the value of $\Delta G°$. This should make sense because a large value of K means that the equilibrium lies far to the right (toward products), and a negative value for $\Delta G°$ means that the reaction is spontaneous in the forward direction. See Table 18.4 of the text for a summary of the relation between K and $\Delta G°$.

> **Cautionary Note:** To be consistent with units in using Equation (18.2) and Equation (18.3), we need to express both $\Delta G°$ and ΔG in units of kJ/mol or J/mol. Here "per mol" means "per mol as the reaction is written".

EXAMPLE 18.6

The standard free energy change for the reaction

$$1/2\ N_2\ (g) + 3/2\ H_2\ (g) \rightleftharpoons NH_3\ (g)$$

is $\Delta G° = 26.9$ kJ/mol at 700 K. Calculate the equilibrium constant at this temperature.

The equilibrium constant is related to the standard free energy change by the Equation (18.3).

$$\Delta G° = -RT\ln K$$

Step 1: R has units of J/mol·K, so we must convert $\Delta G°$ from units of kJ/mol to J/mol.

$$26.9\ \frac{kJ}{mol} \times \frac{1000\ J}{1\ kJ} = 2.69 \times 10^4\ J/mol$$

Step 2: Substitute $\Delta G°$, R, and T into Equation (18.3) to calculate the equilibrium constant, K_p. We are calculating K_p in this problem since this a gas-phase reaction.

$$\Delta G° = -RT\ln K_p$$

$$2.69 \times 10^4\ J/mol = -(8.314\ J/mol\cdot K)(700.\ K)\ln K_p$$

$$-4.62 = \ln K_p$$

Taking the antilog of both sides gives:

$$e^{-4.62} = K_p$$

$$\boldsymbol{K_p = 9.9 \times 10^{-3}}$$

PRACTICE EXERCISE

6.   In Chapter 14, we saw that for the reaction

$$H_2\ (g) + I_2\ (g) \longrightarrow 2\ HI\ (g)$$

the equilibrium constant at 400°C is $K_p = 64$. Calculate the value of $\Delta G°$ at this temperature.

Text Problems: **18.24**, 18.26, **18.28**, 18.30

B. Using the free energy change to predict the direction of a reaction

To predict the direction of a reaction, you can calculate ΔG from Equation (18.2).

$$\Delta G = \Delta G^\circ + RT\ln Q \qquad\qquad (18.2)$$

To calculate the free energy change (ΔG), you must be given or you must calculate ΔG° and you must calculate the reaction quotient Q. Substitute the values of ΔG° and Q into Equation (18.2) to solve for ΔG. Recall the meaning of the sign of ΔG:

- $\Delta G < 0$ A spontaneous process in the forward direction.

- $\Delta G > 0$ A nonspontaneous reaction as written. However, the reaction is spontaneous in the reverse direction.

- $\Delta G = 0$ The system is at equilibrium. There is no net change in the system.

EXAMPLE 18.7
Using the reaction and data given in Example 18.6, calculate ΔG at 700 K if the reaction mixture consists of 30.0 atm of H_2, 20.0 atm of N_2, and 0.500 atm of NH_3.

Under nonstandard conditions, ΔG is related to the reaction quotient Q by Equation (18.2).

$$\Delta G = \Delta G^\circ + RT\ln Q_p$$

We are using Q_p in the equation because this is a gas-phase reaction.

Step 1: ΔG° is given in Example 18.6. We must calculate Q_p.

$$Q_p = \frac{P_{NH_3}}{P_{N_2}^{1/2} \cdot P_{H_2}^{3/2}} = \frac{0.50}{(20.0)^{1/2}(30.0)^{3/2}} = 6.80 \times 10^{-4}$$

Step 2: Substitute $\Delta G^\circ = 2.69 \times 10^4$ J/mol and Q_p into Equation (18.2) to calculate ΔG.

$$\Delta G = \Delta G^\circ + RT\ln Q_p$$

$$\Delta G = 2.69 \times 10^4 \text{ J/mol} + (8.314 \text{ J/mol·K})(700 \text{ K})\ln(6.80 \times 10^{-4})$$

$$\Delta G = 2.69 \times 10^4 \text{ J/mol} - 4.24 \times 10^4 \text{ J/mol} = \mathbf{-1.55 \times 10^4} \text{ J/mol} = \mathbf{-15.5 \text{ kJ/mol}}$$

By making the partial pressures of N_2 and H_2 high and that of NH_3 low, the reaction is spontaneous in the forward direction. This condition corresponds to $Q_p < K_p$, and so the reaction proceeds in the forward direction until $Q_p = K_p$ (see Problem Type 4, Chapter 14).

PRACTICE EXERCISE

7. Calculate ΔG for the following reaction at 25°C when the pressure of CO_2 is 0.0010 atm, given that $\Delta H^\circ = 177.8$ kJ and $\Delta S^\circ = 160.5$ J/K.

$$CaCO_3 \, (s) \longrightarrow CaO \, (s) + CO_2 \, (g)$$

Text Problem: 18.28

ANSWERS TO PRACTICE EXERCISES

1. (a) entropy decreases   (b) entropy increases   (c) entropy increases

2. $\Delta S^\circ_{rxn} = +145.1$ J/K   3. (a) $\Delta G^\circ_{rxn} = 236$ kJ   (b) $\Delta G^\circ_{rxn} = 318.2$ kJ

4. $\Delta G^\circ = -62.3$ kJ. Since $\Delta G^\circ < 0$, the reaction is spontaneous.   5. $\Delta S_{vap} = 92.9$ J/mol·K

6. $\Delta G^\circ = -23.2$ kJ   7. $\Delta G = 113$ kJ

SOLUTIONS TO SELECTED TEXT PROBLEMS

18.6 The probability (P) of finding all the molecules in the same flask becomes progressively smaller as the number of molecules increases. An equation that relates the probability to the number of molecules is given in the text.

$$P = \left(\frac{1}{2}\right)^N$$

where,

 N is the total number of molecules present.

Using the above equation, we find:

  (a) $P = 0.02$ **(b)** $P = 9 \times 10^{-19}$ **(c)** $P = 2 \times 10^{-181}$

18.10 In order of increasing entropy per mole at 25°C:

  (c) < (d) < (e) < (a) < (b)

  (c) Na (s): ordered, crystalline material.
  (d) NaCl (s): ordered crystalline material, but with more particles per mole than Na(s).
  (e) H_2: a diatomic gas, hence of higher entropy than a solid.
  (a) Ne (g): a monatomic gas of higher molar mass than H_2.
  (b) SO_2 (g): a polyatomic gas of higher molar mass than Ne.

18.12 Calculating entropy changes of a system, Problem Type 1B. The standard entropy change for a reaction can be calculated using the following equation.

$$\Delta S^\circ_{rxn} = \Sigma n S^\circ(\text{products}) - \Sigma m S^\circ(\text{reactants})$$

  (a) $\Delta S^\circ_{rxn} = S^\circ(\text{Cu}) + S^\circ(H_2O) - [S^\circ(H_2) + S^\circ(\text{CuO})]$

   $= (1 \text{ mol}) (33.3 \text{ J/K·mol}) + (1 \text{ mol})(188.7 \text{ J/K·mol}) - [(1 \text{ mol})(131.0 \text{ J/K·mol}) + (1 \text{ mol})(43.5 \text{ J/K·mol})]$

   $= \textbf{47.5 J/K}$

  (b) $\Delta S^\circ_{rxn} = S^\circ(Al_2O_3) + 3S^\circ(\text{Zn}) - [2S^\circ(\text{Al}) + 3S^\circ(\text{ZnO})]$

   $= (1 \text{ mol})(50.99 \text{ J/K·mol}) + (3 \text{ mol})(41.6 \text{ J/K·mol}) - [(2 \text{ mol})(28.3 \text{ J/K·mol}) + (3 \text{ mol})(43.9 \text{ J/K·mol})]$

   $= \textbf{–12.5 J/K}$

  (c) $\Delta S^\circ_{rxn} = S^\circ(CO_2) + 2S^\circ(H_2O) - [S^\circ(CH_4) + 2S^\circ(O_2)]$

   $= (1 \text{ mol})(213.6 \text{ J/K·mol}) + (2 \text{ mol})(69.9 \text{ J/K·mol}) - [(1 \text{ mol})(186.2 \text{ J/K·mol}) + (2 \text{ mol})(205.0 \text{ J/K·mol})]$

   $= \textbf{–242.8 J/K}$

Why was the entropy value for water different in parts (a) and (c)?

18.14 **(a)** $\Delta S < 0$; gas reacting with a liquid to form a solid (decrease in number of moles of gas).

(b) $\Delta S > 0$; solid decomposing to give a liquid and a gas.

(c) $\Delta S > 0$; increase in number of moles of gas.

(d) $\Delta S < 0$; gas reacting with a solid to form a solid (decrease in number of moles of gas).

18.18 Calculating standard free energy changes, Problem Type 2A. The standard free energy change for a reaction can be calculated using the following equation.

$$\Delta G^\circ_{rxn} = \Sigma n \, \Delta G^\circ_f \, (\text{products}) - \Sigma m \, \Delta G^\circ_f \, (\text{reactants})$$

(a) $\Delta G^\circ_{rxn} = 2 \, \Delta G^\circ_f \, (MgO) - [2 \, \Delta G^\circ_f \, (Mg) + \Delta G^\circ_f \, (O_2)]$

$\Delta G^\circ_{rxn} = (2 \text{ mol})(-569.6 \text{ kJ/mol}) - [(2 \text{ mol})(0) + (1 \text{ mol})(0)] = \mathbf{-1139 \text{ kJ}}$

(b) $\Delta G^\circ_{rxn} = 2 \, \Delta G^\circ_f \, (SO_3) - [2 \, \Delta G^\circ_f \, (SO_2) + \Delta G^\circ_f \, (O_2)]$

$\Delta G^\circ_{rxn} = (2 \text{ mol})(-370.4 \text{ kJ/mol}) - [(2 \text{ mol})(-300.4 \text{ kJ/mol}) + (1 \text{ mol})(0)] = \mathbf{-140.0 \text{ kJ}}$

(c) $\Delta G^\circ_{rxn} = 4 \, \Delta G^\circ_f \, [CO_2(g)] + 6 \, \Delta G^\circ_f \, [H_2O(l)] - \{2 \, \Delta G^\circ_f \, [C_2H_6(g)] + 7 \, \Delta G^\circ_f \, [O_2(g)]\}$

$\Delta G^\circ_{rxn} = (4 \text{ mol})(-394.4 \text{ kJ/mol}) + (6 \text{ mol})(-237.2 \text{ kJ/mol}) - [(2 \text{ mol})(-32.89 \text{ kJ/mol}) + (7 \text{ mol})(0)]$

$\Delta G^\circ_{rxn} = \mathbf{-2935.0 \text{ kJ}}$

18.20 **Reaction A:** Calculate ΔG from ΔH and ΔS.

$$\Delta G = \Delta H - T\Delta S = -126{,}000 \text{ J} - (298 \text{ K} \times 84 \text{ J/K}) = \mathbf{-151{,}000 \text{ J}}$$

The free energy change is negative so the reaction is spontaneous at 298 K. Since ΔH is negative and ΔS is positive, **the reaction is spontaneous at all temperatures**.

Reaction B: Calculate ΔG.

$$\Delta G = \Delta H - T\Delta S = -11{,}700 \text{ J} - [(298 \text{ K}) \times (-105 \text{ J/K})] = \mathbf{+19{,}600 \text{ J}}$$

The free energy change is positive at 298 K which means the reaction is not spontaneous at that temperature. The positive sign of ΔG results from the large negative value of ΔS. At lower temperatures, the $-T\Delta S$ term will be smaller thus allowing the free energy change to be negative.

ΔG will equal zero when $\Delta H = T\Delta S$.

Rearranging,

$$T = \frac{\Delta H}{\Delta S} = \frac{-11700 \text{ J}}{-105 \text{ J / K}} = \mathbf{111 \text{ K}}$$

At temperatures **below 111 K**, ΔG will be negative and the reaction will be spontaneous.

18.24   Similar to using the standard free energy change to calculate the equilibrium constant, Problem Type 3A.

The equilibrium constant is related to the standard free energy change by the following equation.

$$\Delta G° = -RT \ln K$$

Substitute K_w, R, and T into the above equation to calculate the standard free energy change, $\Delta G°$. The temperature at which $K_w = 1.0 \times 10^{-14}$ is 25°C = 298 K.

$$\Delta G° = -RT \ln K_w$$

$$\Delta G° = -(8.314 \text{ J/mol·K})(298 \text{ K})\ln(1.0 \times 10^{-14}) = \mathbf{8.0 \times 10^4 \text{ J/mol}} = \mathbf{8.0 \times 10^1 \text{ kJ/mol}}$$

18.26   Use standard free energies of formation from Appendix 3 to find the standard free energy difference.

$$\Delta G°_{rxn} = 2 \Delta G°_f [H_2(g)] + \Delta G°_f [O_2(g)] - 2 \Delta G°_f [H_2O(g)]$$

$$\Delta G°_{rxn} = (2 \text{ mol})(0) + (1 \text{ mol})(0) - (2 \text{ mol})(-228.6 \text{ kJ/mol})$$

$$\Delta G°_{rxn} = 457.2 \text{ kJ} = 4.572 \times 10^5 \text{ J}$$

We can calculate K_p using the following equation.

$$\Delta G° = -RT \ln K_p$$

$$4.572 \times 10^5 \text{ J/mol} = -(8.314 \text{ J/mol·K})(298 \text{ K})\ln K_p$$

$$-185 = \ln K_p$$

Taking the antilog of both sides,

$$e^{-185} = K_p$$

$$\mathbf{K_p = 4.5 \times 10^{-81}}$$

18.28 **(a)** Similar to using the standard free energy change to calculate the equilibrium constant, Problem Type 3A.

The equilibrium constant is related to the standard free energy change by the following equation.

$$\Delta G° = -RT \ln K$$

Substitute K_p, R, and T into the above equation to the standard free energy change, $\Delta G°$.

$$\Delta G° = -RT \ln K_p$$

$$\Delta G° = -(8.314 \text{ J/mol·K})(2000 \text{ K})\ln(4.40) = \mathbf{-2.46 \times 10^4 \text{ J/mol}} = \mathbf{-24.6 \text{ kJ/mol}}$$

(b) Similar to using the free energy change to predict the direction of a reaction, Problem Type 3B.

Under nonstandard conditions, ΔG is related to the reaction quotient Q by the following equation.

$$\Delta G = \Delta G° + RT \ln Q_p$$

We are using Q_p in the equation because this is a gas-phase reaction.

Step 1: $\Delta G°$ was calculated in part (a). We must calculate Q_p.

$$Q_p = \frac{P_{H_2O} \cdot P_{CO}}{P_{H_2} \cdot P_{CO_2}} = \frac{(0.66)(1.20)}{(0.25)(0.78)} = 4.1$$

Step 2: Substitute $\Delta G° = -2.46 \times 10^4$ J/mol and Q_p into the following equation to calculate ΔG.

$$\Delta G = \Delta G° + RT\ln Q_p$$

$$\Delta G = -2.46 \times 10^4 \text{ J/mol} + (8.314 \text{ J/mol·K})(2000 \text{ K})\ln(4.1)$$

$$\Delta G = -2.46 \times 10^4 \text{ J/mol} + 2.35 \times 10^4 \text{ J/mol}$$

$$\mathbf{\Delta G = -1.10 \times 10^3 \text{ J/mol} = -1.10 \text{ kJ/mol}}$$

18.30   We use the given K_p to find the standard free energy change.

$$\Delta G° = -RT\ln K$$

$$\Delta G° = -(8.314 \text{ J/K·mol})(298 \text{ K})\ln(5.62 \times 10^{35}) = 2.04 \times 10^5 \text{ J/mol} = -204 \text{ kJ/mol}$$

The standard free energy of formation of one mole of $COCl_2$ can now be found using the standard free energy of reaction calculated above and the standard free energies of formation of CO (g) and Cl_2 (g).

$$\Delta G°_{rxn} = \Sigma n \Delta G°_f \text{ (products)} - \Sigma m \Delta G°_f \text{ (reactants)}$$

$$\Delta G°_{rxn} = \Delta G°_f [COCl_2 (g)] - \{ \Delta G°_f [CO (g)] + \Delta G°_f [Cl_2 (g)]\}$$

$$-204 \text{ kJ} = (1\text{mol}) \Delta G°_f [COCl_2 (g)] - [(1 \text{ mol})(-137.3 \text{ kJ/mol}) + (1 \text{ mol})(0)]$$

$$\mathbf{\Delta G°_f [COCl_2 (g)] = -341 \text{ kJ/mol}}$$

18.32   The standard free energy change is given by:

$$\Delta G°_{rxn} = \Delta G°_f \text{ (graphite)} - \Delta G°_f \text{ (diamond)}$$

You can look up the standard free energy of formation values in Appendix 3 of the text.

$$\mathbf{\Delta G°_{rxn} = (1 \text{ mol})(0) - (1 \text{ mol})(2.87 \text{ kJ/mol}) = -2.87 \text{ kJ}}$$

Thus, the reaction is spontaneous under standard state conditions at 25°C. However, the rate of the diamond to graphite conversion is very slow so that it will take millions of years before the process is complete.

18.36   The equation for the coupled reaction is:

$$\text{glucose} + \text{ATP} \rightarrow \text{glucose 6-phosphate} + \text{ADP}$$

$$\Delta G° = 13.4 \text{ kJ} - 31 \text{ kJ} = -18 \text{ kJ}$$

As an estimate:

$$\ln K = \frac{-\Delta G°}{RT}$$

$$\ln K = \frac{-(-18 \times 10^3 \text{ J/mol})}{(8.314 \text{ J/K} \cdot \text{mol})(298 \text{ K})} = 7.3$$

$$\boldsymbol{K = 1 \times 10^3}$$

18.38   In each part of this problem we can use the following equation to calculate ΔG.

$$\Delta G = \Delta G° + RT \ln Q$$

or,

$$\Delta G = \Delta G° + RT \ln [H^+][OH^-]$$

(a)   In this case, the given concentrations are equilibrium concentrations at 25°C. Since the reaction is at equilibrium, $\Delta G = 0$. This is advantageous, because it allows us to calculate $\Delta G°$. Also recall that at equilibrium, $Q = K$. We can write:

$$\Delta G° = -RT \ln K_w$$

$$\boldsymbol{\Delta G°} = -(8.314 \text{ J/K} \cdot \text{mol})(298 \text{ K}) \ln(1.0 \times 10^{-14}) = \boldsymbol{8.0 \times 10^4 \text{ J/mol}}$$

(b) $\Delta G = \Delta G° + RT \ln Q = \Delta G° + RT \ln [H^+][OH^-]$

$$\boldsymbol{\Delta G} = 8.0 \times 10^4 \text{ J/mol} + (8.314 \text{ J/K} \cdot \text{mol})(298 \text{ K}) \ln (1.0 \times 10^{-3})(1.0 \times 10^{-4}) = \boldsymbol{4.0 \times 10^4 \text{ J/mol}}$$

(c) $\Delta G = \Delta G° + RT \ln Q = \Delta G° + RT \ln [H^+][OH^-]$

$$\boldsymbol{\Delta G} = 8.0 \times 10^4 \text{ J/mol} + (8.314 \text{ J/K} \cdot \text{mol})(298 \text{ K}) \ln (1.0 \times 10^{-12})(2.0 \times 10^{-8}) = \boldsymbol{-3.2 \times 10^4 \text{ J/mol}}$$

(d) $\Delta G = \Delta G° + RT \ln Q = \Delta G° + RT \ln [H^+][OH^-]$

$$\boldsymbol{\Delta G} = 8.0 \times 10^4 \text{ J/mol} + (8.314 \text{ J/K} \cdot \text{mol})(298 \text{ K}) \ln (3.5)(4.8 \times 10^{-4}) = \boldsymbol{6.4 \times 10^4 \text{ J/mol}}$$

18.40   One possible explanation is simply that no reaction is possible, namely that there is an unfavorable free energy difference between products and reactants ($\Delta G > 0$).

A second possibility is that the potential for spontaneous change is there ($\Delta G < 0$), but that the reaction is extremely slow (very large activation energy).

A remote third choice is that the student accidentally prepared a mixture in which the components were already at their equilibrium concentrations.

Which of the above situations would be altered by the addition of a catalyst?

18.42   For a solid to liquid phase transition (melting) the entropy always increases ($\Delta S > 0$) and the reaction is always endothermic ($\Delta H > 0$).

(a)   Melting is always spontaneous above the melting point, so $\Delta G < 0$.

(b)   At the melting point (−77.7°C), solid and liquid are in equilibrium, so $\Delta G = 0$.

(c)   Melting is not spontaneous below the melting point, so $\Delta G > 0$.

18.44   If the process is *spontaneous* as well as *endothermic*, the signs of ΔG and ΔH must be negative and positive, respectively. Since $\Delta G = \Delta H - T\Delta S$, the sign of $\boldsymbol{\Delta S}$ **must be positive** ($\Delta S > 0$) for ΔG to be negative.

18.46 **(a)** Using the relationship:

$$\frac{\Delta H_{vap}}{T_{b.p.}} = \Delta S_{vap} \approx 90 \text{ J/K·mol}$$

benzene $\Delta S_{vap} = 87.8 \text{ J/K·mol}$

hexane $\Delta S_{vap} = 90.1 \text{ J/K·mol}$

mercury $\Delta S_{vap} = 93.7 \text{ J/K·mol}$

toluene $\Delta S_{vap} = 91.8 \text{ J/K·mol}$

Most liquids have ΔS_{vap} approximately equal to a constant value because the order of the molecules in the liquid state is similar. The order of most gases is totally random; thus, ΔS for liquid $\rightarrow$ vapor should be similar for most liquids.

(b) Using the data in Table 11.6, we find:

ethanol $\Delta S_{vap} = 111.9 \text{ J/K·mol}$

water $\Delta S_{vap} = 109.4 \text{ J/K·mol}$

Both water and ethanol have a larger ΔS_{vap} because the liquid molecules are more ordered due to hydrogen bonding.

18.48 **(a)** $2CO + 2NO \rightarrow 2CO_2 + N_2$

(b) The oxidizing agent is NO; the reducing agent is CO.

(c) $\Delta G^\circ = 2 \Delta G_f^\circ (CO_2) + \Delta G_f^\circ (N_2) - 2 \Delta G_f^\circ (NO) - 2 \Delta G_f^\circ (NO)$

$\Delta G^\circ = (2 \text{ mol})(-394.4 \text{ kJ/mol}) + 0 - (2 \text{ mol})(-137.3 \text{ kJ/mol}) - (2 \text{ mol})(86.7 \text{ kJ/mol}) = -687.6 \text{ kJ}$

$\Delta G^\circ = -RT \ln K_p$

$$\ln K_p = \frac{6.876 \times 10^5 \text{ J/mol}}{(8.314 \text{ J/K·mol})(298 \text{ K})} = 278$$

$K_p = 1 \times 10^{121}$

(d) $Q_p = \dfrac{P_{N_2} P_{CO_2}^2}{P_{CO}^2 P_{NO}^2} = \dfrac{(0.80)(0.030)^2}{(5.0 \times 10^{-5})^2 (5.0 \times 10^{-7})^2} = 1.2 \times 10^{18}$

Since $Q_p \ll K_p$, the reaction will proceed from **left to right**.

(e) $\Delta H^\circ = 2 \Delta H_f^\circ (CO_2) + \Delta H_f^\circ (N_2) - 2 \Delta H_f^\circ (NO) - 2 \Delta H_f^\circ (NO)$

$\Delta H^\circ = (2 \text{ mol})(-393.5 \text{ kJ/mol}) + 0 - (2 \text{ mol})(-110.5 \text{ kJ/mol}) - (2 \text{ mol})(90.4 \text{ kJ/mol}) = -746.8 \text{ kJ}$

Since ΔH° is negative, raising the temperature will decrease K_p, thereby increasing the amount of reactants and decreasing the amount of products. **No**, the formation of N_2 and CO_2 is not favored by raising the temperature.

18.50   The equilibrium reaction is:

$$AgCl\,(s) \;\rightleftharpoons\; Ag^+\,(aq) \;+\; Cl^-\,(aq)$$

$$K_{sp} \;=\; [Ag^+][Cl^-] \;=\; 1.6 \times 10^{-10}$$

We can calculate the standard enthalpy of reaction from the standard enthalpies of formation in Appendix 3.

$$\Delta H^\circ \;=\; \Delta H_f^\circ\,(Ag^+) + \Delta H_f^\circ\,(Cl^-) - \Delta H_f^\circ\,(AgCl)$$

$$\Delta H^\circ \;=\; (1\ \text{mol})(105.9\ \text{kJ/mol}) + (1\ \text{mol})(-167.2\ \text{kJ/mol}) - (1\ \text{mol})(-127.0\ \text{kJ/mol}) \;=\; 65.7\ \text{kJ}$$

From Problem 18.49(a):

$$\ln \frac{K_2}{K_1} \;=\; \frac{\Delta H^\circ}{R}\left(\frac{T_2 - T_1}{T_1 T_2}\right)$$

$$K_1 = 1.6 \times 10^{-10} \qquad\qquad T_1 = 298\ \text{K}$$

$$K_2 = ? \qquad\qquad\qquad\quad T_2 = 333\ \text{K}$$

$$\ln\frac{K_2}{1.6\times10^{-10}} \;=\; \frac{6.57\times10^4\ \text{J}}{8.314\ \text{J/K}\cdot\text{mol}}\left(\frac{333\ \text{K} - 298\ \text{K}}{333\ \text{K} \times 298\ \text{K}}\right)$$

$$\ln\frac{K_2}{1.6\times10^{-10}} \;=\; 2.79$$

$$\frac{K_2}{1.6\times10^{-10}} \;=\; e^{2.79}$$

$$\boldsymbol{K_2 \;=\; 2.6 \times 10^{-9}}$$

The increase in K indicates that the solubility increases with temperature.

18.52   Assuming that both ΔH° and ΔS° are temperature independent, we can calculate both ΔH° and ΔS°.

$$\Delta H^\circ \;=\; [\,\Delta H_f^\circ\,(CO) + \Delta H_f^\circ\,(H_2)\,] - [\,\Delta H_f^\circ\,(H_2O) + \Delta H_f^\circ\,(C)\,]$$

$$\Delta H^\circ \;=\; [(1\ \text{mol})(-110.5\ \text{kJ/mol}) + (1\ \text{mol})(0\ \text{kJ/mol})] - [(1\ \text{mol})(-241.8\ \text{kJ/mol}) + (1\ \text{mol})(\,0\ \text{kJ/mol})]$$

$$\Delta H^\circ \;=\; 131.3\ \text{kJ}$$

$$\Delta S^\circ \;=\; [S^\circ(CO) + S^\circ(H_2)] - [S^\circ(H_2O) + S^\circ(C)]$$

$$\Delta S^\circ \;=\; [(1\ \text{mol})(197.9\ \text{J/K}\cdot\text{mol}) + (1\ \text{mol})(131.0\ \text{J/K}\cdot\text{mol})] - [(1\ \text{mol})(188.7\ \text{J/K}\cdot\text{mol}) + (1\ \text{mol})(5.69\ \text{J/K}\cdot\text{mol})]$$

$$\Delta S^\circ \;=\; 134.5\ \text{J/K}$$

It is obvious from the given conditions that the reaction must take place at a fairly high temperature (in order to have red–hot coke). Setting $\Delta G^\circ = 0$

$$0 = \Delta H^{\circ} - T\Delta S^{\circ}$$

$$T = \frac{\Delta H^{\circ}}{\Delta S^{\circ}} = \frac{(131.3\,\cancel{kJ})\left(\dfrac{1000\,J}{1\,\cancel{kJ}}\right)}{134.5\,J/K} = 976\ K = 703^{\circ}C$$

The temperature must be greater than 703°C for the reaction to be spontaneous.

18.54   For a reaction to be spontaneous at constant temperature and pressure, $\Delta G < 0$. The process of crystallization proceeds with more order (less disorder), so $\Delta S < 0$. We also know that

$$\Delta G = \Delta H - T\Delta S$$

Since ΔG must be negative, and since the entropy term will be positive ($-T\Delta S$, where ΔS is negative), then ΔH must be negative ($\Delta H < 0$). The reaction will be exothermic.

18.56   For the reaction to be spontaneous, ΔG must be negative.

$$\Delta G = \Delta H - T\Delta S$$

Given that $\Delta H = 19\ kJ = 19,000\ J$, then

$$\Delta G = 19,000\ J - (273\ K + 72\ K)(\Delta S)$$

Solving the equation with the value of $\Delta G = 0$

$$0 = 19,000\ J - (273\ K + 72\ K)(\Delta S)$$

$$\Delta S = 55\ J/K$$

This value of ΔS which we solved for is the value needed to produce a ΔG value of zero. The *minimum* value of ΔS that will produce a spontaneous reaction will be any value of entropy *greater than* 55 J/K.

18.58   The second law states that the entropy of the universe must increase in a spontaneous process. But the entropy of the universe is the sum of two terms: the entropy of the system plus the entropy of the surroundings. One of the entropies can decrease, but not both. In this case, the decrease in system entropy is offset by an increase in the entropy of the surroundings. The reaction in question is exothermic, and the heat released raises the temperature (and the entropy) of the surroundings.

Could this process be spontaneous if the reaction were endothermic?

18.60   Entropy changes due to phase transitions, Problem Type 2C.

Step 1: The entropy change due to the phase transition (the vaporization of ethanol), can be calculated using the following equation. Recall that the temperature must be in units of Kelvin (78.3°C = 351 K).

$$\Delta S_{vap} = \frac{\Delta H_{vap}}{T_{b.p.}}$$

$$\Delta S_{vap} = \frac{39.3\ \dfrac{kJ}{mol}}{351\ K} = +0.112\ kJ/mol \cdot K = +112\ J/mol \cdot K$$

Step 2: The problem asks for the change in entropy for the vaporization of 0.50 moles of ethanol. The ΔS calculated above is for 1 mole of ethanol.

$$\Delta S \text{ for 0.50 mol} = (112 \text{ J/mol·K})(0.50 \text{ mol}) = \textbf{56 J/K}$$

18.62 For the given reaction we can calculate the standard free energy change from the standard free energies of formation (see Appendix 3). Then, we can calculate the equilibrium constant, K_p, from the standard free energy change.

$$\Delta G° = \Delta G_f° [\text{Ni(CO)}_4] - [4 \Delta G_f° (\text{CO}) + \Delta G_f° (\text{Ni})]$$

$$\Delta G° = (1 \text{ mol})(-587.4 \text{ kJ/mol}) - [(4 \text{ mol})(-137.3 \text{ kJ/mol}) + (1 \text{ mol})(0)] = -38.2 \text{ kJ} = -3.82 \times 10^4 \text{ J}$$

Substitute $\Delta G°$, R, and T (in K) into the following equation to solve for K_p.

$$\Delta G° = -RT \ln K_p$$

$$\ln K_p = \frac{-\Delta G°}{RT} = \frac{-(-3.82 \times 10^4 \text{ J/mol})}{(8.314 \text{ J/K·mol})(353 \text{ K})}$$

$$K_p = \textbf{4.5} \times \textbf{10}^5$$

18.64 The equilibrium constant is related to the standard free energy change by the following equation:

$$\Delta G° = -RT \ln K_p$$

$$2.12 \times 10^5 \text{ J/mol} = -(8.314 \text{ J/mol·K})(298 \text{ K}) \ln K_p$$

$$-85.6 = \ln K_p$$

$$K_p = 6.7 \times 10^{-38}$$

We can write the equilibrium constant expression for the reaction.

$$K_p = \sqrt{P_{O_2}}$$

$$P_{O_2} = (K_p)^2$$

$$P_{O_2} = (6.7 \times 10^{-38})^2 = \textbf{4.5} \times \textbf{10}^{-75} \textbf{ atm}$$

This pressure is far too small to measure.

18.66 Both (a) and (b) apply to a reaction with a negative $\Delta G°$ value. Statement (c) is not always true. An endothermic reaction that has a positive $\Delta S°$ (increase in entropy) will have a negative $\Delta G°$ value at high temperatures.

18.68 We write the two equations as follows. The standard free energy change for the overall reaction will be the sum of the two steps.

$$\text{CuO } (s) \rightleftharpoons \text{Cu } (s) + \frac{1}{2} \text{ O}_2 (g) \qquad\qquad \Delta G° = 127.2 \text{ kJ}$$

$$\text{C (graphite)} + \frac{1}{2} \text{ O}_2 (g) \rightleftharpoons \text{CO } (g) \qquad\qquad \Delta G° = -137.3 \text{ kJ}$$

$$\overline{\text{CuO} + \text{C (graphite)} \rightleftharpoons \text{Cu } (s) + \text{CO } (g) \qquad\qquad \Delta G° = -10.1 \text{ kJ}}$$

We can now calculate the equilibrium constant from the standard free energy change, $\Delta G°$.

$$\ln K = \frac{-\Delta G°}{RT} = \frac{-(-10.1 \times 10^3 \text{ J/mol})}{(8.314 \text{ J/K} \cdot \text{mol})(673 \text{ K})}$$

$$\ln K = 1.81$$

$$\boldsymbol{K = 6.1}$$

18.70   1.   Crystal structure has disorder.

2.   There is impurity present in the crystal.

18.72   **(a)** $\Delta G° = \Delta G_f°(\text{H}_2) + \Delta G_f°(\text{Fe}^{2+}) - \Delta G_f°(\text{Fe}) - 2\,\Delta G_f°(\text{H}^+)$

$\Delta G° = (1 \text{ mol})(0) + (1 \text{ mol})(-84.9 \text{ kJ/mol}) - (1 \text{ mol})(0) - (2 \text{ mol})(0)$

$\Delta G° = -84.9 \text{ kJ}$

$\Delta G° = -RT\ln K$

$-84.9 \times 10^3 \text{ J/mol} = -(8.314 \text{ J/mol} \cdot \text{K})(298 \text{ K})\ln K$

$\boldsymbol{K = 7.6 \times 10^{14}}$

(b) $\Delta G° = (1 \text{ mol})[\Delta G_f°(\text{H}_2)] + (1 \text{ mol})[\Delta G_f°(\text{Cu}^{2+})] - \Delta G_f°(\text{Cu}) - (2 \text{ mol})[\Delta G_f°(\text{H}^+)]$

$\Delta G° = 64.98 \text{ kJ}$

$\Delta G° = -RT\ln K$

$64.98 \times 10^3 \text{ J/mol} = -(8.314 \text{ J/mol} \cdot \text{K})(298 \text{ K})\ln K$

$\boldsymbol{K = 4.1 \times 10^{-12}}$

The activity series is correct. The very large value of K for reaction (a) indicates that *products* are highly favored; whereas, the very small value of K for reaction (b) indicates that *reactants* are highly favored.

18.74   **(a)**   It is a "reverse" disproportionation redox reaction.

(b) $\Delta G° = (2 \text{ mol})(-228.6 \text{ kJ/mol}) - (2 \text{ mol})(-33.0 \text{ kJ/mol}) - (1 \text{ mol})(-300.4 \text{ kJ/mol})$

$\Delta G° = -90.8 \text{ kJ}$

$-90.8 \times 10^3 \text{ J/mol} = -(8.314 \text{ J/mol} \cdot \text{K})(298 \text{ K})\ln K$

$\boldsymbol{K = 8.2 \times 10^{15}}$

Because of the large value of K, this method is efficient for removing SO_2.

(c) $\Delta H° = (2 \text{ mol})(-241.8 \text{ kJ/mol}) + (3 \text{ mol})(0) - (2 \text{ mol})(-20.15 \text{ kJ/mol}) - (1 \text{ mol})(-296.1 \text{ kJ/mol})$

$\Delta H° = -147.2 \text{ kJ}$

$\Delta S° = (2 \text{ mol})(188.7 \text{ J/K} \cdot \text{mol}) + (3 \text{ mol})(31.88 \text{ J/K} \cdot \text{mol}) - (2 \text{ mol})(205.64 \text{ J/K} \cdot \text{mol}) - (1 \text{ mol})(248.5 \text{ J/K} \cdot \text{mol})$

$\Delta S° = -186.7 \text{ J/K} \cdot \text{mol}$

$$\Delta G° = \Delta H° - T\Delta S°$$

Due to the negative entropy change, $\Delta S°$, the free energy change, $\Delta G°$, will become positive at higher temperatures. Therefore, the reaction will be **less effective** at high temperatures.

18.76 $2O_3 \rightleftharpoons 3O_2$

$$\Delta G° = 3\,\Delta G_f°\,(O_2) - 2\,\Delta G_f°\,(O_3) = -(2 \text{ mol})(163.4 \text{ kJ/mol})$$

$$\Delta G° = -326.8 \text{ kJ}$$

$$-326.8 \times 10^3 \text{ J/mol} = -(8.314 \text{ J/mol·K})(243 \text{ K})\ln K_p$$

$$K_p = 1.8 \times 10^{70}$$

Due to the large magnitude of K, you would expect this reaction to be spontaneous in the forward direction. However, this reaction has a **large activation energy**, so the rate of reaction is extremely slow.

18.78

$Cu_2S \rightarrow 2Cu + S$	$\Delta G° = 86.1 \text{ kJ}$
$S + O_2 \rightarrow SO_2$	$\Delta G° = -300.4 \text{ kJ}$
$Cu_2S + O_2 \rightarrow 2Cu + SO_2$	$\Delta G° = -214.3 \text{ kJ}$

Since $\Delta G°$ is a large negative quantity, the coupled reaction is feasible for extracting sulfur.

18.80   First, we need to calculate $\Delta H°$ and $\Delta S°$ for the reaction in order to calculate $\Delta G°$.

$$\Delta H° = -41.2 \text{ kJ} \qquad\qquad \Delta S° = -42.0 \text{ J/K}$$

Next, we calculate $\Delta G°$ at 300°C or 573 K, assuming that $\Delta H°$ and $\Delta S°$ are temperature independent.

$$\Delta G° = \Delta H° - T\Delta S°$$

$$\Delta G° = -41.2 \times 10^3 \text{ J} - (573 \text{ K})(-42.0 \text{ J/K})$$

$$\Delta G° = -1.71 \times 10^4 \text{ J}$$

Having solved for $\Delta G°$, we can calculate K_p.

$$\Delta G° = -RT\ln K_p$$

$$-1.71 \times 10^4 \text{ J/mol} = -(8.314 \text{ J/K·mol})(573 \text{ K})\ln K_p$$

$$\ln K_p = 3.59$$

$$K_p = 36$$

Due to the negative entropy change calculated above, we expect that $\Delta G°$ will become positive at some temperature higher than 300°C. We need to find the temperature at which $\Delta G°$ becomes zero. This is the temperature at which reactants and products are equally favored ($K_p = 1$).

$$\Delta G° = \Delta H° - T\Delta S°$$

$$0 = \Delta H° - T\Delta S°$$

$$T = \frac{\Delta H^\circ}{\Delta S^\circ} = \frac{-41.2 \times 10^3 \text{ J}}{-42.0 \text{ J/K}}$$

$$T = 981 \text{ K} = 708°C$$

This calculation shows that at 708°C, $\Delta G^\circ = 0$ and the equilibrium constant $K_p = 1$. Above 708°C, ΔG° is positive and K_p will be smaller than 1, meaning that reactants will be favored over products. Note that the temperature 708°C is only an estimate, as we have assumed that both ΔH° and ΔS° are independent of temperature.

Using a more efficient catalyst will **not** increase K_p at a given temperature, because the catalyst will speed up both the forward and reverse reactions. The value of K_p will stay the same.

18.82   butane $\rightarrow$ isobutane

$$\Delta G^\circ = (1 \text{ mol}) \Delta G_f^\circ \text{ (isobutane)} - (1 \text{ mol}) \Delta G_f^\circ \text{ (butane)}$$

$$\Delta G^\circ = (1 \text{ mol})(-18.0 \text{ kJ/mol}) - (1 \text{ mol})(-15.9 \text{ kJ/mol})$$

$$\Delta G^\circ = -2.1 \text{ kJ}$$

For a mixture at equilibrium at 25°C:

$$\Delta G^\circ = -RT\ln K_p$$

$$-2.1 \times 10^3 \text{ J/mol} = -(8.314 \text{ J/mol·K})(298 \text{ K})\ln K_p$$

$$K_p = 2.3$$

$$K_p = \frac{P_{\text{isobutane}}}{P_{\text{butane}}} \; \alpha \; \frac{\text{mol isobutane}}{\text{mol butane}}$$

$$2.3 = \frac{\text{mol isobutane}}{\text{mol butane}}$$

This shows that there are 2.3 times as many moles of isobutane as moles of butane. Or, we can say for every one mole of butane, there are 2.3 moles of isobutane.

$$\textbf{mol \% isobutane} = \frac{2.3 \text{ mol}}{2.3 \text{ mol} + 1.0 \text{ mol}} \times 100\% = \textbf{70\%}$$

By difference, the mole % of butane is **30%**.

Yes, this result supports the notion that straight-chain hydrocarbons like butane are less stable than branched-chain hydrocarbons like isobutane.

18.84   We can calculate K_p from ΔG°.

$$\Delta G^\circ = (1 \text{ mol})(-394.4 \text{ kJ/mol}) + 0 - (1 \text{ mol})(-137.3 \text{ kJ/mol}) - (1 \text{ mol})(-255.2 \text{ kJ/mol})$$

$$\Delta G^\circ = -1.9 \text{ kJ}$$

$$-1.9 \times 10^3 \text{ J/mol} = -(8.314 \text{ J/mol·K})(1173 \text{ K})\ln K_p$$

$$K_p = 1.2$$

Now, from K_p, we can calculate the mole fractions of CO and CO_2.

$$K_p = \frac{P_{CO_2}}{P_{CO}} = 1.2 \qquad P_{CO_2} = 1.2\,P_{CO}$$

$$X_{CO} = \frac{P_{CO}}{P_{CO} + P_{CO_2}} = \frac{P_{CO}}{P_{CO} + 1.2\,P_{CO}} = \frac{1}{2.2} = \textbf{0.45}$$

$$X_{CO_2} = 1 - 0.45 = \textbf{0.55}$$

CHAPTER 19
ELECTROCHEMISTRY

PROBLEM-SOLVING STRATEGIES AND TUTORIAL SOLUTIONS

TYPES OF PROBLEMS

Problem Type 1: Balancing Redox Equations.

Problem Type 2: Standard Reduction Potentials.
    (a) Comparing strengths of oxidizing agents.
    (b) Calculating the standard emf ($E°$) of a galvanic cell.

Problem Type 3: Spontaneity of Redox Reactions.
    (a) Predicting whether a redox reaction is spontaneous.
    (b) Calculating $\Delta G°$ and K from $E°$.

Problem Type 4: The Nernst Equation.
    (a) Using the Nernst equation to predict the spontaneity of a redox reaction.
    (b) Using the Nernst equation to calculate concentration.

Problem Type 5: Electrolysis.
    (a) Predicting the products of electrolysis.
    (b) Calculating the quantity of products in electrolysis.

PROBLEM TYPE 1: BALANCING REDOX EQUATIONS

We will use the **ion-electron method** to balance redox reactions. In this approach, the overall reaction is divided into two half-reactions, one for oxidation and one for reduction. The two equations are balanced separately and then added together to give the overall balanced equation. Example 19.1 demonstrates how to balance a redox reaction by the ion-electron method.

EXAMPLE 19.1
Balance the following redox reaction in an acidic solution.

$$Sn + NO_3^- \longrightarrow Sn^{2+} + NO$$

Step 1: Write the unbalanced equation for the reaction in ionic form.

    The equation given in the problem is already in ionic form.

Step 2: Separate the equation into two half-reactions.

$$\overset{0}{Sn} \xrightarrow{\text{oxidation}} Sn^{2+}$$

$$\overset{+5}{N}O_3^- \xrightarrow{\text{reduction}} \overset{+2}{N}O$$

Step 3: Balance the atoms other than O and H in each half-reaction separately.

The oxidation half-reaction is already balanced for Sn atoms, and the reduction half-reaction is balanced for nitrogen atoms.

Step 4: For reactions in an acidic medium, add H_2O to balance the O atoms and H^+ to balance the H atoms.

Since the reaction takes place in an acidic medium, we add *two* H_2O molecules to the right-hand side of the reduction half-reaction to balance the O atoms.

$$NO_3^- \longrightarrow NO + 2\,H_2O$$

To balance the H atoms, we add $4\,H^+$ to the left-hand side.

$$4\,H^+ + NO_3^- \longrightarrow NO + 2\,H_2O$$

Step 5: Add electrons to one side of each half-reaction to balance the charge.

For the oxidation half-reaction we write:

$$Sn \longrightarrow Sn^{2+} + 2e^-$$

We added two electrons to the right-hand side so that the charge is *zero* on each side of the half-reaction.

For the reduction half-reaction, there are three net positive charges on the left-hand side and zero charge on the right-hand side. Therefore, we add three electrons on the left-hand side to balance the charge.

$$4\,H^+ + NO_3^- + 3e^- \longrightarrow NO + 2\,H_2O$$

These are the balanced half-reactions.

Step 6: In a redox reaction, the number of electrons gained must equal the number of electrons lost. If necessary, we must equalize the number of electrons in the two half-reactions by multiplying one or both half-reactions by appropriate coefficients.

We see that two electrons are lost in the oxidation half-reaction and three electrons are gained in the reduction half-reaction. We need to multiply the oxidation half-reaction by 3 and the reduction half-reaction by 2. Thus, six electrons are lost and six electrons are gained.

$$3\,Sn \longrightarrow 3\,Sn^{2+} + 6e^-$$

$$8\,H^+ + 2\,NO_3^- + 6e^- \longrightarrow 2\,NO + 4\,H_2O$$

Step 7: Add the two half-reactions and balance the final equation by inspection. The electrons on both sides of the equation must cancel.

$$8\,H^+ + 2\,NO_3^- + 3\,Sn + 6e^- \longrightarrow 2\,NO + 3\,Sn^{2+} + 4\,H_2O + 6e^-$$

The electrons on both sides cancel, and we are left with the balanced net ionic equation:

$$8\,H^+ + 2\,NO_3^- + 3\,Sn \longrightarrow 2\,NO + 3\,Sn^{2+} + 4\,H_2O$$

Step 8: Verify that the equation contains the same types and numbers of atoms and the same charges on both sides of the equation.

The equation is "atomically" balanced. There are 8 H, 2 N, 6 O, and 3 Sn atoms on each side of the equation. The equation is also "electrically" balanced. The net charge is +6 on each side of the equation.

PRACTICE EXERCISE

1.   The redox reaction between permanganate ion and iron(II) ions in acidic solution can be used to analyze iron ore for its iron content. Balance this redox reaction.

$$MnO_4^- \ (aq) \ + \ Fe^{2+} \ (aq) \ \longrightarrow \ Fe^{3+} \ (aq) \ + \ Mn^{2+} \ (aq)$$

Text Problem: 19.2

PROBLEM TYPE 2: STANDARD REDUCTION POTENTIALS

A. Comparing strengths of oxidizing agents

For a reduction reaction at an electrode when all solutes are 1 M and all gases are at 1 atm, the voltage is called the **standard reduction potential**. Table 19.1 lists the standard reduction potentials for a number of half-reactions. The more positive the value of $E°$, the greater the tendency for the substance to be reduced, and therefore the stronger its tendency to act as an oxidizing agent. Looking at Table 19.1, we see that F_2 is the strongest oxidizing agent and Li^+ the weakest.

EXAMPLE 19.2
Arrange the following species in order of increasing strength as oxidizing agents under standard-state conditions: Zn^{2+}, MnO_4^- (in acid solution), and Ag^+.

Consulting Table 19.1 of the text, we write the half-reactions in the order of decreasing standard reduction potentials, $E°$.

$$MnO_4^- \ (aq) \ + \ 8 \ H^+ \ (aq) \ + \ 5e^- \ \longrightarrow \ Mn^{2+} \ (aq) \ + \ 4 \ H_2O \ (l) \qquad E° = +1.51 \ V$$

$$Ag^+ \ (aq) \ + \ e^- \ \longrightarrow \ Ag \ (s) \qquad E° = +0.80 \ V$$

$$Zn^{2+} \ (aq) \ + \ 2e^- \ \longrightarrow \ Zn \ (s) \qquad E° = -0.76 \ V$$

Since the reduction of MnO_4^- has the highest reduction potential and Zn^{2+} has the lowest reduction potential, the order of increasing strength of oxidizing agents is:

$$Zn^{2+} \ < \ Ag^+ \ < \ MnO_4^-$$

The large positive reduction potential for MnO_4^- indicates the strong tendency for permanganate ion to be reduced, and therefore the stronger its tendency to act as an oxidizing agent.

PRACTICE EXERCISE

2.   Arrange the following species in order of increasing strength as oxidizing agents: Ce^{4+}, O_2, H_2O_2, and SO_4^{2-}.

Text Problems: 19.14, **19.18**

B. Calculating the standard emf ($E°$) of an electrochemical cell

In an electrochemical cell, electrons flow from one electrode to the other. This indicates that there is a voltage difference between the two electrodes. This voltage difference is called the **electromotive force**, or **emf (E)**, and it can be measured by connecting a voltmeter to both electrodes. The electromotive force is also called the cell voltage or cell potential. It is usually measured in volts. The electrode at which reduction occurs is called the *cathode*, and the electrode at which oxidation occurs is called the *anode*.

If all solutes have a concentration of 1 M and all gases have a pressure of 1 atm (standard conditions), the voltage difference between the two electrodes of the cell is called the **standard emf ($E°_{cell}$)**. The *standard emf* which is composed of a contribution from the *anode* and a contribution from the *cathode*, is given by:

$$E°_{cell} = E°_{cathode} - E°_{anode}$$

where $E°_{cathode}$ and $E°_{anode}$ are the standard reduction potentials of the cathode and anode, respectively.

Under standard-state conditions for reactants and products, the redox reaction is *spontaneous* in the forward direction if the standard emf of the cell is *positive*. If the standard cell emf is negative, the reaction is spontaneous in the opposite direction.

We can calculate the standard emf of the cell using a table of standard reduction potentials (Table 19.1). As an example, consider the galvanic cell represented by the following reaction:

$$Cu\ (s) + 2\ Ag^+\ (aq) \longrightarrow Cu^{2+}\ (aq) + Ag\ (s)$$

Let's break this reaction down into its two half-reactions.

$$Cu\ (s) \xrightarrow{\text{oxidation (anode)}} Cu^{2+}\ (aq) + 2e^-$$

$$2\ Ag^+\ (aq) + 2e^- \xrightarrow{\text{reduction (cathode)}} 2\ Ag\ (s)$$

We can calculate the standard cell emf by subtracting $E°_{anode}$ from $E°_{cathode}$.

$$E°_{cell} = E°_{cathode} - E°_{anode} = E°_{Ag^+/Ag} - E°_{Cu^{2+}/Cu}$$

$$E°_{cell} = 0.80\ V - 0.34\ V = 0.46\ V$$

EXAMPLE 19.3

Calculate the standard cell potential for an electrochemical cell in which the following reaction takes place:

$$Cl_2\ (g) + 2\ Br^-\ (aq) \longrightarrow Br_2\ (l) + 2\ Cl^-\ (aq)$$

Step 1: Let's separate the reaction into half-reactions.

$$Cl_2\ (g) + 2e^- \xrightarrow{\text{reduction (cathode)}} 2\ Cl^-\ (aq)$$

$$2\ Br^-\ (aq) \xrightarrow{\text{oxidation (anode)}} Br_2\ (l) + 2e^-$$

Step 2: We can look up standard reduction potentials in Table 19.1 of the text.

$$E°_{cathode} = E°_{Cl_2/Cl^-} = +1.36\ V$$

$$E°_{anode} = E°_{Br_2/Br^-} = +1.07\ V$$

Step 3: Finally, the standard cell emf is given by

$$E^{\circ}_{cell} = E^{\circ}_{cathode} - E^{\circ}_{anode} = E^{\circ}_{Cl_2/Cl^-} - E^{\circ}_{Br_2/Br^-}$$

$$E^{\circ}_{cell} = +1.36 \text{ V} - 1.07 \text{ V} = \textbf{+0.29 V}$$

PRACTICE EXERCISE

3.   Consider a uranium-bromine galvanic cell in which U is oxidized and Br_2 is reduced. The reduction half-reactions are:

$$U^{3+}(aq) + 3e^- \longrightarrow U(s) \qquad E^{\circ}_{U^{3+}/U} = ?$$

$$Br_2(l) + 2e^- \longrightarrow 2 Br^-(aq) \qquad E^{\circ}_{Br_2/Br^-} = 1.07 \text{ V}$$

If the standard cell emf is 2.91 V, what is the standard reduction potential for uranium?

Text Problem: 19.12

PROBLEM TYPE 3: SPONTANEITY OF REDOX REACTIONS

A. Predicting whether a redox reaction is spontaneous

There is a relationship between free energy change and cell emf.

$$\Delta G = -nFE_{cell} \qquad (19.1)$$

where,

      n is the number of moles of electrons transferred during the redox reaction

      F is the Faraday constant, which is the electrical charge contained in 1 mole of electrons

$$1 F = 96,500 \text{ C/mol} = 96,500 \text{ J/V·mol}$$

For a derivation of this relationship, see Section 19.4 of the text.

Both *n* and *F* are positive quantities, and we know from Chapter 18 that ΔG is *negative* for a spontaneous process. Therefore, E_{cell} is *positive* for a spontaneous process. Table 19.2 of the text summarizes the relationship between ΔG and E_{cell}.

For reactions in which reactants and products are in their standard states, Equation (19.1) becomes:

$$\Delta G^{\circ} = -nFE^{\circ}_{cell} \qquad (19.2)$$

EXAMPLE 19.4
Predict whether a spontaneous reaction will occur when the following reactants and products are in their standard states:

$$Fe^{2+} + Cr_2O_7^{2-} \longrightarrow Fe^{3+} + Cr^{3+}$$

Let's calculate the standard cell emf for this reaction. From the sign of E°_{cell}, we can determine if the reaction is spontaneous.

Separate the reaction into half-reactions to calculate the standard cell emf.

$$Fe^{2+}(aq) \xrightarrow{\text{oxidation (anode)}} Fe^{3+}(aq) + e^- \qquad E^{\circ}_{anode} = +0.77 \text{ V}$$

At this point, we could balance the reduction half-reaction and then come up with the overall balanced equation. But, we are not asked to do that. We can save time by looking at Table 19.1 and finding a reaction that contains both $Cr_2O_7^{2-}$ and Cr^{3+}.

$$Cr_2O_7^{2-}\ (aq)\ +\ 14\ H^+\ (aq)\ +\ 6e^- \xrightarrow{\text{reduction (cathode)}} 2\ Cr^{3+}\ (aq)\ +\ 7\ H_2O\ (l) \qquad E^\circ_{\text{cathode}}\ =\ +1.33\ V$$

We have not come up with a balanced equation, but we do not need a balanced equation to calculate E°_{cell}.

$$E^\circ_{\text{cell}}\ =\ E^\circ_{\text{cathode}}\ -\ E^\circ_{\text{anode}}$$

$$E^\circ_{\text{cell}}\ =\ 1.33\ V\ -\ 0.77\ V\ =\ \textbf{+0.56 V}$$

Since E°_{cell} is positive, the reaction is spontaneous.

PRACTICE EXERCISE

4.   Predict whether a spontaneous reaction will occur when the following reactants and products are in their standard states.

(a) $2\ Fe^{3+}\ (aq)\ +\ 2\ I^-\ (aq)\ \longrightarrow\ 2\ Fe^{2+}\ (aq)\ +\ I_2\ (s)$

(b) $Cu\ (s)\ +\ 2\ H^+\ (aq)\ \longrightarrow\ Cu^{2+}\ (aq)\ +\ H_2\ (g)$

Text Problem: 19.16

B. Calculating ΔG° and K from E°

Equation (19.2) shows that the relationship between standard free energy change and standard emf is:

$$\Delta G^\circ\ =\ -nFE^\circ_{\text{cell}} \tag{19.2}$$

Also recall that in Section 18.5 of the text, we saw that the standard free energy change for a reaction is related to its equilibrium constant K by the following equation:

$$\Delta G^\circ\ =\ -RT \ln K$$

Therefore, from the two equations we obtain:

$$-nFE^\circ_{\text{cell}}\ =\ -RT \ln K$$

Solving for E°_{cell}, we obtain:

$$E^\circ_{\text{cell}}\ =\ \frac{RT}{nF}\ \ln K \tag{19.3}$$

At 298 K, we can simplify Equation (19.3) by substituting for R and F.

$$E^\circ_{\text{cell}}\ =\ \frac{(8.314\ \text{J/mol} \cdot \text{K})(298\ \text{K})}{n(96,500\ \text{J/V} \cdot \text{mol})} \ln K$$

$$E^\circ_{\text{cell}}\ =\ \frac{0.0257\ \text{V}}{n} \ln K \tag{19.4}$$

We now have relationships between $\Delta G°$ and $E°_{cell}$, between $\Delta G°$ and K, and between $E°_{cell}$ and K. Figure 19.5 of the text summarizes the relationships among $\Delta G°$, K, and $E°_{cell}$.

EXAMPLE 19.5
Calculate $\Delta G°$ and the equilibrium constant at 25°C for the reaction

$$2\ Br^-\ (aq)\ +\ I_2\ (s)\ \longrightarrow\ Br_2\ (l)\ +\ 2\ I^-\ (aq)$$

First, we can calculate the standard cell emf from half-reaction potentials. Then, the standard Gibbs free energy change is:

$$\Delta G° = -nFE°_{cell} \tag{19.2}$$

The equilibrium constant at 25°C is related to $E°_{cell}$ by:

$$E°_{cell} = \frac{0.0257\ V}{n} \ln K \tag{19.4}$$

Step 1: Separate the reaction into half-reactions to calculate the standard cell emf.

$$I_2\ (s)\ +\ 2e^- \xrightarrow{\ \text{reduction (cathode)}\ } 2\ I^-\ (aq) \qquad E°_{cathode} = +0.53\ V$$

$$\underline{2\ Br^-\ (aq) \xrightarrow{\ \text{oxidation (anode)}\ } Br_2\ (l)\ +\ 2e^- \qquad E°_{anode} = +1.07\ V}$$

$$2\ Br^-\ (aq)\ +\ I_2\ (s)\ \longrightarrow\ Br_2\ (l)\ +\ 2\ I^-\ (aq) \qquad E°_{cell} = E°_{cathode} - E°_{anode} = -0.54\ V$$

Step 2: Substitute $E°_{cell}$ into Equation (19.2) to calculate the standard free energy change. In the balanced equation above, 2 moles of electrons are transferred. Therefore, $n = 2$.

$$\Delta G° = -nF\,E°_{cell}$$

$$\Delta G° = -(2)(96,500\ J/V \cdot mol)(-0.54\ V)$$

$$\Delta G° = 1.04 \times 10^5\ J/mol = 104\ kJ/mol$$

Both the positive value of $\Delta G°$ and the negative value for $E°_{cell}$ indicate that the reaction is not spontaneous under standard-state conditions.

Step 3: Rearrange Equation (19.4) to solve for the equilibrium constant, K.

$$\ln K = \frac{nE°}{0.0257\ V}$$

$$\ln K = \frac{(2)(-0.54\ V)}{0.0257\ V} = -42.0$$

Taking the anti-ln of both sides of the equation,

$$K = e^{-42.0} = 6 \times 10^{-19}$$

You should notice that the magnitude of K relates to what we have learned from $\Delta G°$ and $E°_{cell}$. A very small value for the equilibrium constant indicates that reactants are highly favored at equilibrium. This agrees with the signs of $\Delta G°$ and $E°_{cell}$, which indicates that the reaction is not spontaneous in the forward direction under standard-state conditions.

PRACTICE EXERCISE

5.   Calculate the equilibrium constant for the following redox reaction at 25°C.

$$2\,Fe^{2+}\,(aq)\,+\,Ni^{2+}\,(aq)\,\longrightarrow\,2\,Fe^{3+}\,(aq)\,+\,Ni\,(s)$$

Text Problems: **19.22**, 19.24, **19.26**, 19.38

PROBLEM TYPE 4: THE NERNST EQUATION

Thus far, we have only focused on redox reactions in which the reactants and products are in their standard states. However, standard-state conditions are often difficult and sometimes impossible to maintain. Therefore, we need a relationship between cell emf and the concentrations of reactants and products under *nonstandard-state* conditions.

In Chapter 18, we encountered a relationship between the free energy change (ΔG) and the standard free energy change ($\Delta G°$). See Problem Type 3, Chapter 18.

$$\Delta G\,=\,\Delta G°\,+\,RT\ln Q$$

where,
     Q is the reaction quotient.

We also know that:

$$\Delta G\,=\,-nFE_{cell}$$

and

$$\Delta G°\,=\,-nFE_{cell}°$$

Substituting for ΔG and $\Delta G°$ in the first equation, we find:

$$-nFE_{cell}\,=\,-nFE_{cell}°\,+\,RT\ln Q$$

Dividing both sides of the equation by $-nF$ gives:

$$E\,=\,E°\,-\,\frac{RT}{nF}\ln Q \qquad\qquad (19.5)$$

Equation (19.5) is known as the **Nernst equation**. At 298 K, Equation (19.5) can be simplified by substituting R, T, and F into the equation.

$$E\,=\,E°\,-\,\frac{0.0257\,V}{n}\ln Q \qquad\qquad (19.6)$$

At equilibrium, there is no net transfer of electrons, so $E = 0$ and $Q = K$, where K is the equilibrium constant of the redox reaction. Substituting into Equation (19.6) gives

$$E°\,=\,\frac{0.0257\,V}{n}\ln K$$

This is Equation (19.4) derived in Problem Type 3B above.

A. Using the Nernst equation to predict the spontaneity of a redox reaction

To predict the spontaneity of a redox reaction, we can use the Nernst equation to calculate the cell emf, E. If the cell emf is *positive*, the redox reaction is *spontaneous*. Conversely, if E is *negative*, the reaction is *not spontaneous* in the direction written.

EXAMPLE 19.6

Calculate the cell emf and predict whether the following reaction is spontaneous at 25°C.

$$Zn \; (s) \; + \; 2 \, H^+ \; (1 \times 10^{-4} \, M) \; \longrightarrow \; Zn^{2+} \; (1.5 \, M) \; + \; H_2 \; (1 \; atm)$$

The cell emf can be calculated using the Nernst equation.

$$E \; = \; E° \; - \; \frac{0.0257 \; V}{n} \ln Q$$

or

$$E \; = \; E° \; - \; \frac{0.0257 \; V}{n} \ln \frac{[Zn^{2+}]P_{H_2}}{[H^+]^2}$$

Step 1: Calculate the standard cell emf, $E°$, from standard reduction potentials (Table 19.1). Separate the reaction into its half-reactions.

$$2 \, H^+ \, (aq) \; + \; 2e^- \; \xrightarrow{\text{reduction (cathode)}} \; H_2 \, (g) \qquad E°_{\text{cathode}} \; = \; 0.00 \; V$$

$$Zn \, (s) \; \xrightarrow{\text{oxidation (anode)}} \; Zn^{2+} \, (aq) \; + \; 2e^- \qquad E°_{\text{anode}} \; = \; -0.76 \; V$$

$$Zn \, (s) \; + \; 2 \, H^+ \, (aq) \; \longrightarrow \; Zn^{2+} \, (aq) \; + \; H_2 \, (g) \qquad E°_{\text{cell}} = E°_{\text{cathode}} - E°_{\text{anode}} \; = \; +0.76 \; V$$

Step 2: Calculate the cell emf, E, from the Nernst equation. Two moles of electrons were transferred during the reaction, so $n = 2$.

$$E \; = \; E° \; - \; \frac{0.0257 \; V}{n} \ln \frac{[Zn^{2+}]P_{H_2}}{[H^+]^2}$$

$$E \; = \; (0.76 \; V) \; - \; \frac{0.0257 \; V}{2} \ln \frac{(1.5)(1)}{(1 \times 10^{-4})^2}$$

$$= \; 0.76 \; V - 0.24 \; V$$

$$= \; \mathbf{0.52 \; V}$$

The reaction is spontaneous because the cell emf, E, is positive. The low concentration of H^+ ($1 \times 10^{-4} \, M$) compared to its standard-state value ($1 \, M$) means that the driving force for the reaction will be less than in the standard state. This is confirmed because $E < E°$. We can also think about this in terms of Le Chatelier's principle. When H^+ ions are removed, the equilibrium will shift to the left to replace some of the H^+ ions that are removed. This will decrease the driving force for the forward reaction.

PRACTICE EXERCISE

6.   Calculate the cell emf for the following reaction:

$$2 \, Ag^+ \; (0.10 \, M) \; + \; H_2 \; (1 \; atm) \; \longrightarrow \; 2 \, Ag \; (s) \; + \; 2 \, H^+ \; (pH = 8)$$

Text Problems: 19.30, 19.32, 19.34

B. Using the Nernst equation to calculate concentration

Recall that the reaction quotient Q equals the concentrations of the products raised to the power of their stoichiometric coefficients divided by the concentrations of the reactants raised to the power of their stoichiometric coefficients.

$$Q \; = \; \frac{[\text{products}]^x}{[\text{reactants}]^y}$$

Since Q is in the Nernst equation, if the cell emf is measured and the standard cell emf is calculated, we can determine the concentration of one of the components if the concentrations of the other components are known. See Example 19.7 below.

EXAMPLE 19.7

An electrochemical cell is constructed from a silver half-cell and a copper half-cell. The copper half-cell contains 0.10 M Cu(NO$_3$)$_2$ and the concentration of silver ions in the other half-cell is unknown. If the Ag electrode is the cathode and the cell emf is measured and found to be 0.10 V at 25°C, what is the Ag$^+$ ion concentration?

We are given the cell emf, $E = 0.10$ V, in the problem. If we can write the reaction occurring in the cell, we can calculate the standard cell emf, $E°$. We are also given the Cu^{2+} concentration in the problem. Given all this information, we can calculate the Ag$^+$ ion concentration using the Nernst equation.

Step 1: Write the half-reactions to calculate the standard cell emf, $E°$. We are told that the Ag electrode is the cathode (reduction occurs at the cathode). We can write:

$$Ag^+ (aq) + e^- \xrightarrow{\text{reduction (cathode)}} Ag (s) \qquad E°_{cathode} = +0.80 \text{ V}$$

If Ag$^+$ is reduced, Cu must be oxidized.

$$Cu (s) \xrightarrow{\text{oxidation (anode)}} Cu^{2+} (aq) + 2e^- \qquad E°_{anode} = +0.34 \text{ V}$$

$$E°_{cell} = E°_{cathode} - E°_{anode}$$

$$E°_{cell} = 0.80 \text{ V} - 0.34 \text{ V} = \textbf{+0.46 V}$$

We must multiply the Ag half-reaction by two so that the number of electrons gained equals the number of electrons lost. The balanced reaction is:

$$2 Ag^+ (aq) + Cu (s) \longrightarrow 2 Ag (s) + Cu^{2+} (aq)$$

Step 2: Substitute E, $E°$, and [Cu^{2+}] into the Nernst equation to calculate the [Ag$^+$]. Also, $n = 2$ since two moles of electrons are transferred during the reaction.

$$E = E° - \frac{0.0257 \text{ V}}{n} \ln Q$$

or

$$E = E° - \frac{0.0257 \text{ V}}{n} \ln \frac{[Cu^{2+}]}{[Ag^+]^2}$$

$$0.10 \text{ V} = 0.46 \text{ V} - \frac{0.0257 \text{ V}}{2} \ln \frac{[0.10]}{[Ag^+]^2}$$

$$\frac{2(0.10 \text{ V} - 0.46 \text{ V})}{-0.0257 \text{ V}} = \ln \frac{0.10}{[Ag^+]^2}$$

$$28.0 = \ln \frac{0.10}{[Ag^+]^2}$$

Taking the antilog of both sides of the equation,

$$e^{28.0} = \frac{0.10}{[Ag^+]^2}$$

$$1.45 \times 10^{12} = \frac{0.10}{[Ag^+]^2}$$

$$[Ag^+] = \sqrt{\frac{0.10}{1.45 \times 10^{12}}}$$

$$[Ag^+] = 2.6 \times 10^{-7} \, M$$

PRACTICE EXERCISE

7.   When the concentration of Zn^{2+} is 0.15 M, the measured voltage of a Zn-Cu galvanic cell is 0.40 V. What is the Cu^{2+} ion concentration?

Text Problem: 19.110

PROBLEM TYPE 5: ELECTROLYSIS

Electrolysis is the process in which electrical energy is used to cause a nonspontaneous chemical reaction to occur. The same principles underlie electrolysis and the processes that take place in galvanic cells.

A.   Predicting the products of electrolysis

EXAMPLE 19.8

Predict the products of the electrolysis of an aqueous $MgCl_2$ solution.

This aqueous solution contains several species that could be oxidized or reduced. Two species that could be reduced are the metal ion (Mg^{2+}) and H_2O. The reduction half-reactions that might occur at the cathode are

$$(1) \quad Mg^{2+} \, (aq) + 2e^- \xrightarrow{\text{reduction (cathode)}} Mg \, (s) \qquad E^{\circ}_{\text{cathode}} = -2.37 \text{ V}$$

$$(2) \quad 2 \, H_2O \, (l) + 2e^- \xrightarrow{\text{reduction (cathode)}} H_2 \, (g) + 2 \, OH^- \, (aq) \qquad E^{\circ}_{\text{cathode}} = -0.83 \text{ V}$$

$$(3) \quad 2 \, H^+ \, (aq) + 2e^- \xrightarrow{\text{reduction (cathode)}} H_2 \, (g) \qquad E^{\circ}_{\text{cathode}} = 0.00 \text{ V}$$

We can rule out Reaction (1) immediately. The very negative standard reduction potential indicates that Mg^{2+} has essentially no tendency to undergo reduction. Reaction (3) is preferred over Reaction (2) under standard-state conditions. However, at a pH of 7 (as is the case for a $MgCl_2$ solution), they are equally probable. We generally use Reaction (2) to describe the cathode reaction because the concentration of H^+ ions is too low (about $1 \times 10^{-7} \, M$) to make (3) a reasonable choice.

The oxidation reactions that might occur at the anode are:

$$(4) \quad 2 \, Cl^- \, (aq) \xrightarrow{\text{oxidation (anode)}} Cl_2 \, (g) + 2e^- \qquad E^{\circ}_{\text{anode}} = +1.36 \text{ V}$$

$$(5) \quad 2 \, H_2O \, (l) \xrightarrow{\text{oxidation (anode)}} O_2 \, (g) + 4 \, H^+ \, (aq) + 2e^- \qquad E^{\circ}_{\text{anode}} = +1.23 \text{ V}$$

The standard reduction potentials of (4) and (5) are not very different, but the values do suggest that H_2O should be preferentially oxidized at the anode. However, we find by experiment that the gas liberated at the anode is Cl_2, not O_2. The large overvoltage for O_2 formation prevents its production when the Cl^- ion is there to compete (see Section 19.8 of the text).

The overall reaction is:

$$2\ H_2O\ (l)\ +\ 2e^-\ \longrightarrow\ H_2\ (g)\ +\ 2\ OH^-\ (aq)$$

$$\underline{2\ Cl^-\ (aq)\ \longrightarrow\ Cl_2\ (g)\ +\ 2e^-}$$

$$2\ H_2O\ (l)\ +\ 2\ Cl^-\ (aq)\ \longrightarrow\ H_2\ (g)\ +\ 2\ OH^-\ (aq)\ +\ Cl_2\ (g)$$

B. Calculating the quantity of products in electrolysis

During electrolysis, the mass of product formed (or reactant consumed) at an electrode is proportional to both the amount of electricity transferred at the electrode and the molar mass of the substance in question.

The steps involved in calculating the quantities of substances produced in electrolysis are shown below.

Current (amperes) and time	→	Charge in coulombs	→	Number of mol of electrons	→	Moles of substance reduced or oxidized	→	Grams of substance reduced or oxidized

To convert to coulombs, recall that

$$1\ C\ =\ 1\ A\ \times\ 1\ s$$

To convert from coulombs to moles of electrons, use the conversion factor

$$1\ mol\ e^-\ =\ 96{,}500\ C$$

Finally, we need to consider the number of moles of electrons transferred during the reaction. For example, to reduce Al^{3+} ions to Al metal, 3 moles of electrons are needed to reduce 1 mole of Al^{3+} ions.

$$Al^{3+}\ +\ 3e^-\ \longrightarrow\ Al$$

EXAMPLE 19.9

How many moles of electrons are transferred in an electrolytic cell when a current of 12 amps flows for 16 hours?

First, we can calculate the number of coulombs passing through the cell in 16 hours. Recall that

$$1\ C\ =\ 1\ A{\cdot}s$$

Then, we can convert from coulombs to moles of electrons using the following conversion factor.

$$1\ mol\ e^-\ =\ 96{,}500\ C$$

Step 1: Convert from amps to coulombs.

$$12\ A\ \times\ \frac{1\ \frac{C}{s}}{1\ A}\ \times\ \frac{3600\ s}{1\ hr}\ \times\ 16\ hr\ =\ 6.9\times10^5\ C$$

Step 2: Convert from coulombs to moles of electrons.

$$? \text{ mol e}^- = 6.9 \times 10^5 \, \cancel{C} \times \frac{1 \text{ mol e}^-}{96,500 \, \cancel{C}} = 7.2 \text{ mol e}^-$$

EXAMPLE 19.10

How many grams of copper metal would be deposited from a solution of $CuSO_4$ by the passage of 3.0 A of electrical current through an electrolytic cell for 2.0 h?

Step 1: Start by writing the half-reaction for the reduction of Cu^{2+} to Cu metal.

$$Cu^{2+} (aq) + 2e^- \longrightarrow Cu (s)$$

This half-reaction tells us that 2 mol of e^- are required to produce 1 mol of Cu (*s*).

Step 2: We can calculate the mass of Cu produced using the following strategy.

$$\text{current} \times \text{time} \rightarrow \text{coulombs} \rightarrow [\text{mol of e}^-] \rightarrow \text{mol Cu} \rightarrow \text{g Cu}$$

This is a large number of steps, so let's break it down into two parts. The moles of electrons are the link between the current measurements and the molar amount of Cu formed.

The number of moles of electrons is:

$$3.0 \, \cancel{A} \times \frac{1 \, \frac{C}{\cancel{s}}}{1 \, \cancel{A}} \times \frac{3600 \, \cancel{s}}{1 \, \cancel{hr}} \times 2.0 \, \cancel{hr} = 2.2 \times 10^4 \, \cancel{C} \times \frac{1 \text{ mol e}^-}{96,500 \, \cancel{C}} = 0.23 \text{ mol e}^-$$

Step 3: From the number of moles of electrons, we can calculate the moles of Cu produced. We can then convert to grams of Cu.

$$? \text{ g Cu} = 0.23 \, \cancel{\text{mol e}^-} \times \frac{1 \, \cancel{\text{mol Cu}}}{2 \, \cancel{\text{mol e}^-}} \times \frac{63.55 \text{ g Cu}}{1 \, \cancel{\text{mol Cu}}} = 7.3 \text{ g Cu}$$

Note: This predicted amount of Cu is based on a process that is 100 percent efficient. Any side reactions or any oxidation or reduction of impurities will cause the actual yield to be less than the theoretical yield.

EXAMPLE 19.11

How long will it take to electrodeposit (plate out) 1.0 g of Ni from a $NiSO_4$ solution using a current of 2.5 A?

Step 1: First, find the number of moles of electrons required to electrodeposit 1.0 g of Ni from solution. The half-reaction for the reduction of Ni^{2+} is:

$$Ni^{2+} (aq) + 2e^- \longrightarrow Ni (s)$$

2 moles of electrons are required to reduce 1 mol of Ni^{2+} ions to Ni metal. But, we are electrodepositing less than 1 mol of Ni (*s*). We need to complete the following conversions:

$$\text{g Ni} \rightarrow \text{mol Ni} \rightarrow \text{mol of e}^-$$

$$? \text{ mol of e}^- = 1.0 \, \cancel{\text{g Ni}} \times \frac{1 \, \cancel{\text{mol Ni}}}{58.7 \, \cancel{\text{g Ni}}} \times \frac{2 \text{ mol e}^-}{1 \, \cancel{\text{mol Ni}}} = 0.034 \text{ mol e}^-$$

Step 2: Determine how long it will take for 0.034 moles of electrons to flow through the cell when the current is 2.5 C/s. We need to complete the following conversions:

$$\text{mol of } e^- \rightarrow \text{coulombs} \rightarrow \text{seconds}$$

$$\textbf{? seconds} = 0.034 \text{ mol of } e^- \times \frac{96,500 \text{ C}}{1 \text{ mol } e^-} \times \frac{1 \text{ s}}{2.5 \text{ C}} = \textbf{1.3} \times \textbf{10}^3 \textbf{ s} \text{ (22 min)}$$

PRACTICE EXERCISES

8. How many moles of electrons are transferred in an electrolytic cell when a current of 2.0 amps flows for 6.0 hours?

9. How many grams of cobalt can be electroplated by passing a constant current of 5.2 A through a solution of $CoCl_3$ for 60.0 min?

10. How long will it take to produce 54 kg of Al metal by the reduction of Al^{3+} in an electrolytic cell using a current of 5.0×10^2 amps?

Text Problems: **19.46**, 19.48, **19.52**, 19.54, **19.56**, 19.58, 19.60

ANSWERS TO PRACTICE EXERCISES

1. $5 Fe^{2+} (aq) + MnO_4^- (aq) + 8 H^+ (aq) \longrightarrow 5 Fe^{3+} (aq) + Mn^{2+} (aq) + 4 H_2O (l)$

2. $SO_4^{2-} > O_2 > Ce^{4+} > H_2O_2$

3. $E^\circ_{U^{3+}/U} = -1.84$ V

4. (a) $E^\circ_{cell} = +0.24$ V. The positive value of the standard cell emf indicates that the reaction is spontaneous under standard-state conditions.

    (b) $E^\circ_{cell} = -0.34$ V. A negative standard cell emf indicates that this reaction is not spontaneous under standard-state conditions. Copper will not dissolve in 1 M HCl.

5. $K = 3.0 \times 10^{-35}$

6. $E^\circ = 1.21$ V

7. $[Cu^{2+}] = 3.1 \times 10^{-25} M$

8. Number of mol e^- transferred = 0.45 mol e^-

9. Grams of Co electroplated = 3.81 g Co

10. Time = 322 hr

SOLUTIONS TO SELECTED TEXT PROBLEMS

19.2   Balancing Redox Equations, Problem Type 1.

(a)

Step 1: Separate the equation into two half-reactions.

$$Mn^{2+} \xrightarrow{\text{oxidation}} MnO_2$$

$$H_2O_2 \xrightarrow{\text{reduction}} H_2O$$

Step 2: Balance the atoms other than O and H in each half-reaction separately.

The oxidation half-reaction is already balanced for Mn atoms.

Step 3: Assume that the reaction is taking place in an acidic medium. Add H_2O to balance the O atoms and H^+ to balance the H atoms.

For the oxidation half-reaction, add 2 H_2O to the left-hand side of the equation to balance the O atoms.

$$Mn^{2+} + 2\,H_2O \longrightarrow MnO_2$$

To balance the H atoms, we add 4 H^+ to the right-hand side.

$$Mn^{2+} + 2\,H_2O \longrightarrow MnO_2 + 4\,H^+$$

For the reduction half-reaction, we add one H_2O to the right-hand side of the equation to balance the O atoms.

$$H_2O_2 \longrightarrow 2\,H_2O$$

To balance the H atoms, we add 2 H^+ to the left-hand side.

$$H_2O_2 + 2\,H^+ \longrightarrow 2\,H_2O$$

Step 4: Since this reaction is in basic solution, we add one OH^- to both sides for each H^+ and combine pairs of H^+ and OH^- on the same side of the arrow to form H_2O.

$$Mn^{2+} + 2\,H_2O + 4\,OH^- \longrightarrow MnO_2 + 4\,H_2O$$

$$H_2O_2 + 2\,H_2O \longrightarrow 2\,H_2O + 2\,OH^-$$

We remove extra H_2O.

$$Mn^{2+} + 4\,OH^- \longrightarrow MnO_2 + 2\,H_2O$$

$$H_2O_2 \longrightarrow 2\,OH^-$$

Step 5: Add electrons to one side of each half-reaction to balance the charge.

$$Mn^{2+} + 4\,OH^- \longrightarrow MnO_2 + 2\,H_2O + 2e^-$$

$$H_2O_2 + 2e^- \longrightarrow 2\,OH^-$$

Step 6: We can now add the reactions because the number of electrons gained (2 e$^-$) equals the number of electrons lost (2 e$^-$). Canceling extra OH$^-$ and electrons gives the completed balanced equation.

$$Mn^{2+} + H_2O_2 + 2\,OH^- \longrightarrow MnO_2 + 2\,H_2O$$

Check to see that the equation is balanced by verifying that the equation has the same types and numbers of atoms and the same charges on both sides of the equation.

(b) This problem can be solved by the same methods used in part (a).

$$2\,Bi(OH)_3 + 3\,SnO_2^{2-} \longrightarrow 2\,Bi + 3\,H_2O + 3\,SnO_3^{2-}$$

(c)

Step 1: Separate the equation into two half-reactions.

$$C_2O_4^{2-} \xrightarrow{\text{oxidation}} CO_2$$
$$Cr_2O_7^{2-} \xrightarrow{\text{reduction}} Cr^{3+}$$

Step 2: Balance the atoms other than O and H in each half-reaction separately.

In the oxidation half-reaction, we need to balance the C atoms.

$$C_2O_4^{2-} \longrightarrow 2\,CO_2$$

In the reduction half-reaction, we need to balance the Cr atoms.

$$Cr_2O_7^{2-} \longrightarrow 2\,Cr^{3+}$$

Step 3: Add H_2O to balance the O atoms and H$^+$ to balance the H atoms.

In the oxidation half-reaction, the O atoms are already balanced.

$$Cr_2O_7^{2-} + 14\,H^+ \longrightarrow 2\,Cr^{3+} + 7\,H_2O$$

Step 4: Add electrons to one side of each half-reaction to balance the charge.

$$C_2O_4^{2-} \longrightarrow 2\,CO_2 + 2e^-$$
$$Cr_2O_7^{2-} + 14\,H^+ + 6e^- \longrightarrow 2\,Cr^{3+} + 7\,H_2O$$

Step 5: The number of electrons gained must equal the number of electrons lost. Multiply the oxidation half-reaction by 3, so that 6 e$^-$ are transferred.

$$3\,C_2O_4^{2-} \longrightarrow 6\,CO_2 + 6e^-$$
$$Cr_2O_7^{2-} + 14\,H^+ + 6e^- \longrightarrow 2\,Cr^{3+} + 7\,H_2O$$

Step 6: Add the two half-reactions and balance the final equation by inspection. The electrons on both sides of the equation must cancel.

$$Cr_2O_7^{2-} + 14\,H^+ + 3\,C_2O_4^{2-} \longrightarrow 2\,Cr^{3+} + 6\,CO_2 + 7\,H_2O$$

(d) This problem can be solved by the same methods used in part (c).

$$2\,Cl^- + 2\,ClO_3^- + 4\,H^+ \longrightarrow Cl_2 + 2\,ClO_2 + 2\,H_2O$$

19.12 Calculating the standard emf of a galvanic cell, Problem Type 2B.

Step 1: Let's separate the reaction into half-reactions.

$$Al\,(s) \xrightarrow{\text{oxidation (anode)}} Al^{3+}\,(aq) + 3e^-$$

$$Ag^+\,(aq) + e^- \xrightarrow{\text{reduction (cathode)}} Ag(s)$$

Step 2: We can look up standard reduction potentials in Table 19.1 of the text.

$$E^\circ_{\text{anode}} = E^\circ_{Al^{3+}/Al} = -1.66\ V$$

$$E^\circ_{\text{cathode}} = E^\circ_{Ag^+/Ag} = +0.80\ V$$

Step 3: Finally, we can calculate the standard cell emf by subtracting E°_{anode} from E°_{cathode}.

$$E^\circ_{\text{cell}} = E^\circ_{\text{cathode}} - E^\circ_{\text{anode}} = E^\circ_{Ag^+/Ag} - E^\circ_{Al^{3+}/Al}$$

$$E^\circ_{\text{cell}} = 0.80\ V - (-1.66\ V) = +2.46\ V$$

The cell reaction under standard conditions is:

$$Al\,(s) + 3\,Ag^+\,(aq) \longrightarrow 3\,Ag\,(s) + Al^{3+}\,(aq)$$

19.14 The half–reaction for oxidation is:

$$2\,H_2O\,(l) \xrightarrow{\text{oxidation (anode)}} O_2\,(g) + 4\,H^+\,(aq) + 4e^- \qquad E^\circ_{\text{anode}} = +1.23\ V$$

The species that can oxidize water to molecular oxygen must have an E°_{red} more positive than $+1.23$ V. From Table 19.1 we see that only **Cl_2 (g)** and **MnO_4^- (aq)** in acid solution can oxidize water to oxygen.

19.16 Predicting whether a redox reaction is spontaneous, Problem Type 3A.

E°_{cell} is *positive* for a spontaneous reaction. In each case, we can calculate the standard cell emf from the potentials for the two half-reactions.

$$E^\circ_{\text{cell}} = E^\circ_{\text{cathode}} - E^\circ_{\text{anode}}$$

(a) $E^\circ = -0.40\ V - (-2.87\ V) = \mathbf{2.47\ V}$. The reaction is spontaneous.

(b) $E^\circ = -0.14\ V - 1.07\ V = \mathbf{-1.21\ V}$. The reaction is not spontaneous.

(c) $E^\circ = -0.25\ V - 0.80\ V = \mathbf{-1.05\ V}$. The reaction is not spontaneous.

(d) $E^\circ = 0.77\ V - 0.15\ V = \mathbf{0.62\ V}$. The reaction is spontaneous.

19.18   Similar to Problem Type 2A, Comparing the strengths of oxidizing agents.

The greater the tendency for the substance to be oxidized, the stronger its tendency to act as a reducing agent. The species that has a stronger tendency to be oxidized will have a smaller reduction potential.

(a) Li   **(b)** H_2 **(c)** Fe^{2+} **(d)** Br^-

19.22   Calculating $E°$ from K. Similar to Problem Type 3B.

The equation that relates K and the standard cell emf is:

$$E°_{cell} = \frac{0.0257 \text{ V}}{n} \ln K$$

Substitute the equilibrium constant and the mol of e^- transferred ($n = 2$) during the redox reaction into the above equation to calculate $E°$.

$$E° = \frac{(0.0257 \text{ V}) \ln K}{n} = \frac{(0.0257 \text{ V}) \ln(2.69 \times 10^{12})}{2} = \textbf{0.368 V}$$

19.24   **(a)**   We break the equation into two half–reactions:

$$Mg\,(s) \xrightarrow{\text{oxidation (anode)}} Mg^{2+}\,(aq) + 2e^- \qquad E°_{anode} = -2.37 \text{ V}$$
$$Pb^{2+}\,(aq) + 2e^- \xrightarrow{\text{reduction (cathode)}} Pb\,(s) \qquad E°_{cathode} = -0.13 \text{ V}$$

The standard emf is given by

$$E°_{cell} = E°_{cathode} - E°_{anode} = -0.13 \text{ V} - (-2.37 \text{ V}) = 2.24 \text{ V}$$

We can calculate $\Delta G°$ from the standard emf.

$$\Delta G° = -nFE°_{cell}$$
$$= -(2)(96500 \text{ J/V·mol})(2.24 \text{ V}) = \textbf{-432 kJ/mol}$$

Next, we can calculate K using Equation (19.4) derived in Problem Type 3B.

$$E°_{cell} = \frac{0.0257 \text{ V}}{n} \ln K \qquad\qquad (19.4)$$

or

$$\ln K = \frac{nE°_{cell}}{0.0257 \text{ V}}$$

and

$$K = e^{nE°/0.0257}$$
$$= e^{(2)(2.24)/(0.0257)} = \textbf{5} \times \textbf{10}^{75}$$

Tip: You could also calculate K_c from the standard free energy change, $\Delta G°$, using the equation: $\Delta G° = -RT \ln K_c$.

(b)   We break the equation into two half–reactions:

$$Br_2\,(l) + 2e^- \xrightarrow{\text{reduction (cathode)}} 2\,Br^-\,(aq) \qquad E^\circ_{\text{cathode}} = 1.07\text{ V}$$

$$2\,I^-\,(aq) \xrightarrow{\text{oxidation (anode)}} I_2\,(s) + 2e^- \qquad E^\circ_{\text{anode}} = 0.53\text{ V}$$

The standard emf is

$$E^\circ_{\text{cell}} = E^\circ_{\text{cathode}} - E^\circ_{\text{anode}} = 1.07\text{ V} - 0.53\text{ V} = 0.54\text{ V}$$

We can calculate ΔG° from the standard emf.

$$\Delta G^\circ = -nFE^\circ_{\text{cell}}$$
$$= -(2)(96500\text{ J/V·mol})(0.54\text{ V}) = \mathbf{-104\text{ kJ/mol}}$$

Next, we can calculate K using Equation (19.4) derived in Problem Type 3B.

$$K = e^{nE^\circ/0.0257}$$
$$= e^{(2)(0.54)/(0.0257)} = \mathbf{2 \times 10^{18}}$$

(c)   This is worked in an analogous manner to parts (a) and (b).

$$E^\circ_{\text{cell}} = E^\circ_{\text{cathode}} - E^\circ_{\text{anode}}$$
$$= 1.23\text{ V} - 0.77\text{ V} = +0.46\text{ V}$$

$$\Delta G^\circ = -nFE^\circ_{\text{cell}}$$
$$= -(4)(96500\text{ J/V·mol})(0.46\text{ V}) = \mathbf{-178\text{ kJ/mol}}$$

$$K = e^{nE^\circ/0.0257}$$
$$= e^{(4)(0.46)/(0.0257)} = \mathbf{1 \times 10^{31}}$$

(d)   This is worked in an analogous manner to parts (a), (b), and (c).

$$E^\circ_{\text{cell}} = E^\circ_{\text{cathode}} - E^\circ_{\text{anode}}$$
$$= 0.53\text{ V} - (-1.66\text{ V}) = 2.19\text{ V}$$

$$\Delta G^\circ = -nFE^\circ_{\text{cell}}$$
$$= -(6)(96500\text{ J/V·mol})(2.19\text{ V}) = \mathbf{-1.27 \times 10^3\text{ kJ/mol}}$$

$$K = e^{nE^\circ/0.0257}$$
$$= e^{(6)(2.19)/(0.0257)} = \mathbf{8 \times 10^{211}}$$

19.26 Calculating $\Delta G°$ and K from $E°$, Problem Type 3B.

Step 1: Separate the reaction into half-reactions to calculate the standard cell emf.

$$Cu^+ (aq) \xrightarrow{\text{oxidation (anode)}} Cu^{2+} (aq) + e^- \qquad E°_{anode} = 0.15 \text{ V}$$

$$Cu^+ (aq) + e^- \xrightarrow{\text{reduction (cathode)}} Cu (s) \qquad E°_{cathode} = 0.52 \text{ V}$$

$$2\,Cu^+ (aq) \longrightarrow Cu^{2+} (aq) + Cu (s)$$

$$E°_{cell} = E°_{cathode} - E°_{anode} = 0.52 \text{ V} - 0.15 \text{ V} = \mathbf{0.37 \text{ V}}$$

Step 2: The equation that relates $E°_{cell}$ and the standard free energy change is shown below. Substitute $E°_{cell}$ into the equation to calculate $\Delta G°$. For this redox reaction, $n = 1$.

$$\Delta G° = -nFE°_{cell}$$
$$= -(1)(96500 \text{ J/V·mol})(0.37 \text{ V}) = \mathbf{-36 \text{ kJ/mol}}$$

Step 3: Next, we can calculate K using Equation (19.4) derived in Problem Type 3b.

$$E°_{cell} = \frac{0.0257 \text{ V}}{n} \ln K \tag{19.4}$$

or

$$\ln K = \frac{nE°_{cell}}{0.0257 \text{ V}}$$

and

$$K = e^{nE°/0.0257}$$
$$= e^{(1)(0.37)/(0.0257)} = e^{14.4} = \mathbf{2 \times 10^6}$$

Is this a spontaneous reaction? Can copper(I) salts exist in aqueous solution?

19.30 Using the Nernst equation, Problem Type 4.

(a)
Step 1: Use standard potentials for the two half-reactions to calculate the standard cell emf.

$$E°_{cell} = E°_{cathode} - E°_{anode} = -0.14 \text{ V} - (-2.37 \text{ V}) = \mathbf{2.23 \text{ V}}$$

Step 2: Use the Nernst equation to compute the emf at the given nonstandard concentrations.

$$E = E° - \frac{0.0257 \text{ V}}{n} \ln Q$$
$$= 2.23 \text{ V} - \frac{0.0257 \text{ V}}{2} \ln \frac{0.045}{0.035} = \mathbf{2.23 \text{ V}}$$

Step 3: We can find the free energy difference at the given concentrations using the following equation:

$$\Delta G = -nFE_{cell}$$
$$= -(2)(96500 \text{ J/V·mol})(2.23 \text{ V}) = \mathbf{-430 \text{ kJ/mol}}$$

(b)

Step 1: Use standard potentials for the two half-reactions to calculate the standard cell emf.

$$E_{cell}^o = E_{cathode}^\circ - E_{anode}^\circ = -0.74 \text{ V} - (-0.76 \text{ V}) = \textbf{0.02 V}$$

Step 2: Use the Nernst equation to compute the emf at the given nonstandard concentrations.

$$E = E^\circ - \frac{0.0257 \text{ V}}{n} \ln Q$$

$$= 0.02 \text{ V} - \frac{0.0257 \text{ V}}{6} \ln \frac{(0.0085)^3}{(0.010)^2} = \textbf{0.04 V}$$

Step 3: We can find the free energy difference at the given concentrations using the following equation:

$$\Delta G = -nFE_{cell}$$

$$= -(6)(96500 \text{ J/V·mol})(0.04 \text{ V}) = \textbf{-23 kJ/mol}$$

19.32   Let's write the two half-reactions to calculate the standard cell emf. (Oxidation occurs at the Pb electrode.)

$$\text{Pb } (s) \xrightarrow{\text{oxidation (anode)}} \text{Pb}^{2+} (aq) + 2e^- \qquad E_{anode}^\circ = -0.13 \text{ V}$$

$$\underline{2 \text{ H}^+ (aq) + 2e^- \xrightarrow{\text{reduction (cathode)}} \text{H}_2 (g) \qquad E_{cathode}^\circ = 0.00 \text{ V}}$$

$$2 \text{ H}^+ (aq) + \text{Pb } (s) \longrightarrow \text{H}_2 (g) + \text{Pb}^{2+} (aq)$$

$$E_{cell}^\circ = E_{cathode}^\circ - E_{anode}^\circ = 0.00 \text{ V} - (-0.13 \text{ V}) = 0.13 \text{ V}$$

Using the Nernst equation, we can calculate the cell emf, E.

$$E = E^\circ - \frac{0.0257 \text{ V}}{n} \ln \frac{[\text{Pb}^{2+}]P_{\text{H}_2}}{[\text{H}^+]^2}$$

$$= 0.13 \text{ V} - \frac{0.0257 \text{ V}}{2} \ln \frac{(0.10)(1.0)}{(0.050)^2} = \textbf{0.083 V}$$

19.34   All concentration cells have the same standard emf: *zero* volts.

$$\text{Mg}^{2+} (aq) + 2e^- \xrightarrow{\text{reduction (cathode)}} \text{Mg } (s) \qquad E_{cathode}^\circ = -2.37 \text{ V}$$

$$\text{Mg } (s) \xrightarrow{\text{oxidation (anode)}} \text{Mg}^{2+} (aq) + 2e^- \qquad E_{anode}^\circ = -2.37 \text{ V}$$

$$E_{cell}^\circ = E_{cathode}^\circ - E_{anode}^\circ = -2.37 \text{ V} - (-2.37 \text{ V}) = 0.00 \text{ V}$$

We use the Nernst equation to compute the emf. There are two moles of electrons transferred from the reducing agent to the oxidizing agent in this reaction, so $n = 2$.

$$E = E^\circ - \frac{0.0257 \text{ V}}{n} \ln Q = E^\circ - \frac{0.0257 \text{ V}}{n} \ln \frac{[\text{Mg}^{2+}]_{ox}}{[\text{Mg}^{2+}]_{red}}$$

$$E = 0 \text{ V} - \frac{0.0257 \text{ V}}{2} \ln \frac{0.24}{0.53} = \textbf{0.010 V}$$

What is the direction of spontaneous change in all concentration cells?

19.38   We can calculate the standard free energy change, $\Delta G°$, from the standard free energies of formation, $\Delta G_f°$ (see Problem Type 2A , Chapter 18). Then, we can calculate the standard cell emf, $E_{cell}°$, from $\Delta G°$.

The overall reaction is:

$$C_3H_8\ (g)\ +\ 5\ O_2\ (g)\ \longrightarrow\ 3\ CO_2\ (g)\ +\ 4\ H_2O\ (l)$$

$$\Delta G_{rxn}°\ =\ 3\ \Delta G_f°\ [CO_2(g)]\ +\ 4\ \Delta G_f°\ [H_2O(l)]\ -\ \{\ \Delta G_f°\ [C_3H_8(g)]\ +\ 5\ \Delta G_f°\ [O_2(g)]\}$$

$$\Delta G_{rxn}°\ =\ (3\ mol)(-394.4\ kJ/mol)\ +\ (4\ mol)(-237.2\ kJ/mol)\ -\ [(1\ mol)(-23.5\ kJ/mol)\ +\ (5\ mol)(0)]$$

$$=\ -2108.5\ kJ$$

We can now calculate the standard emf using the following equation:

$$\Delta G°\ =\ -nFE_{cell}°$$

or

$$E_{cell}°\ =\ \frac{-\Delta G°}{nF}$$

Check the half-reactions on p. 788 of the text to determine that 20 moles of electrons are transferred during this redox reaction.

$$E_{cell}°\ =\ \frac{-(-2108.5\times10^3\,J\,/\,mol)}{(20)(96500\,J\,/\,V\cdot mol)}\ =\ \textbf{1.09 V}$$

Does this suggest that, in theory, it should be possible to construct a galvanic cell (battery) based on any conceivable spontaneous reaction?

19.46   Calculating the quantity of products in electrolysis, Problem Type 5B.

(a)   The only ions present in molten $BaCl_2$ are Ba^{2+} and Cl^-. The electrode reactions are:

anode: $2\ Cl^-\ (aq)\ \longrightarrow\ Cl_2\ (g)\ +\ 2e^-$

cathode: $Ba^{2+}\ (aq)\ +\ 2e^-\ \longrightarrow\ Ba\ (s)$

This cathode half-reaction tells us that 2 mol of e^- are required to produce 1 mol of Ba (s).

(b)
Step 1: We can calculate the mass of Ba produced using the following strategy.

current $\times$ time $\rightarrow$ coulombs $\rightarrow$ [mol e^-] $\rightarrow$ mol Ba $\rightarrow$ g Ba

This is a large number of steps, so let's break it down into two parts. The moles of electrons are the link between the current measurements and the molar amount of Ba formed.

The number of moles of electrons is:

$$0.50\ \cancel{A}\ \times\ \frac{1\ \frac{C}{\cancel{s}}}{1\ \cancel{A}}\ \times\ \frac{60\ \cancel{s}}{1\ \cancel{min}}\ \times\ 30\ \cancel{min}\ =\ 9.0\times10^2\ \cancel{C}\ \times\ \frac{1\ mol\ e^-}{96{,}500\ \cancel{C}}\ =\ 9.3\times10^{-3}\ mol\ e^-$$

Step 3: From the number of moles of electrons, we can calculate the moles of Ba produced. We can then convert to grams of Ba.

$$? \text{ g Ba} = 9.3 \times 10^{-3} \text{ mol e}^- \times \frac{1 \text{ mol Ba}}{2 \text{ mol e}^-} \times \frac{137.3 \text{ g Ba}}{1 \text{ mol Ba}} = \textbf{0.64 g Ba}$$

19.48   The costs for producing various metals can be calculated by setting up the ratios of electrons needed to reduce each type of ion. You also need to take into account the atomic mass of each metal (i.e., What mass of metal ions is reduced by the moles of electrons in the balanced equation?). The reductions are:

$$Mg^{2+} + 2e^- \longrightarrow Mg$$
$$Al^{3+} + 3e^- \longrightarrow Al$$
$$Na^+ + e^- \longrightarrow Na$$
$$Ca^{2+} + 2e^- \longrightarrow Ca$$

(a)   For aluminum :

$$\left(\frac{\$155}{1 \text{ ton}}\right)\left(\frac{3 \text{ mol e}^- / \text{mol Al}}{2 \text{ mol e}^- / \text{mol Mg}}\right)\left(\frac{24.31 \text{ g Mg} / \text{mol Mg}}{26.98 \text{ g Al} / \text{mol Al}}\right) \times 10.0 \text{ tons} = \textbf{\$2.09} \times \textbf{10}^3$$

(b)   For sodium:

$$\left(\frac{\$155}{1 \text{ ton}}\right)\left(\frac{1 \text{ mol e}^- / \text{mol Na}}{2 \text{ mol e}^- / \text{mol Mg}}\right)\left(\frac{24.31 \text{ g Mg} / \text{mol Mg}}{22.99 \text{ g Na} / \text{mol Na}}\right) \times 30.0 \text{ tons} = \textbf{\$2.46} \times \textbf{10}^3$$

(c)   For calcium:

$$\left(\frac{\$155}{1 \text{ ton}}\right)\left(\frac{2 \text{ mol e}^- / \text{mol Ca}}{2 \text{ mol e}^- / \text{mol Mg}}\right)\left(\frac{24.31 \text{ g Mg} / \text{mol Mg}}{40.08 \text{ g Ca} / \text{mol Ca}}\right) \times 50.0 \text{ tons} = \textbf{\$4.70} \times \textbf{10}^3$$

19.50   **(a)**   The half–reaction is:

$$2 H_2O\,(l) \longrightarrow O_2\,(g) + 4 H^+\,(aq) + 4e^-$$

First, we can calculate the number of moles of oxygen produced using the ideal gas equation (see Problem Type 2A, Chapter 5).

$$n_{O_2} = \frac{PV}{RT}$$

$$n_{O_2} = \frac{(1.0 \text{ atm})(0.84 \text{ L})}{(0.0821 \text{ L} \cdot \text{atm/mol} \cdot \text{K})(298 \text{ K})} = 0.034 \text{ mol O}_2$$

Since 4 moles of electrons are needed for every 1 mole of oxygen, we will need 4 *F* of electrical charge to produce 1 mole of oxygen.

$$? F = 0.034 \text{ mol O}_2 \times \frac{4 F}{1 \text{ mol O}_2} = \textbf{0.14} \textbf{\textit{F}}$$

(b) The half–reaction is:

$$2\,Cl^-\,(aq) \longrightarrow Cl_2\,(g) + 2e^-$$

The number of moles of chlorine produced is:

$$n_{Cl_2} = \frac{PV}{RT}$$

$$n_{Cl_2} = \frac{\left(\dfrac{750\ mmHg}{760\ mmHg/atm}\right)(1.50\ L)}{(0.0821\ L\cdot atm/mol\cdot K)(298\ K)} = 0.0605\ mol\ Cl_2$$

Since 2 moles of electrons are needed for every 1 mole of chlorine gas, we will need 2 F of electrical charge to produce 1 mole of chlorine gas.

$$?\,F = 0.0605\ mol\ Cl_2 \times \frac{2\,F}{1\ mol\ Cl_2} = \mathbf{0.121\ F}$$

(c) The half–reaction is:

$$Sn^{2+}\,(aq) + 2e^- \longrightarrow Sn\,(s)$$

The number of moles of Sn (s) produced is

$$?\ mol\ Sn = 6.0\ g\ Sn \times \frac{1\ mol\ Sn}{118.7\ g\ Sn} = 0.051\ mol\ Sn$$

Since 2 moles of electrons are needed for every 1 mole of Sn, we will need 2 F of electrical charge to reduce 1 mole of Sn^{2+} ions to Sn metal.

$$?\,F = 0.051\ mol\ Sn \times \frac{2\,F}{1\ mol\ Sn} = \mathbf{0.10\ F}$$

19.52   Electrolysis, Problem Type 5.

(a)   The half–reaction is:

$$\mathbf{Ag^+\,(aq) + e^- \longrightarrow Ag\,(s)}$$

(b)   Since this reaction is taking place in an aqueous solution, the probable oxidation is the oxidation of water. (Neither Ag^+ nor NO_3^- can be further oxidized.)

$$\mathbf{2\,H_2O\,(l) \longrightarrow O_2\,(g) + 4\,H^+\,(aq) + 4e^-}$$

(c)   The half-reaction tells us that 1 mole of electrons is needed to reduce 1 mol of Ag^+ to Ag metal. We can set up the following strategy to calculate the quantity of electricity (in C) needed to deposit 0.67 g of Ag.

$$grams\ Ag \rightarrow mol\ Ag \rightarrow mol\ e^- \rightarrow coulombs$$

$$0.67\ g\ Ag\left(\frac{1\ mol\ Ag}{107.9\ g\ Ag}\right)\left(\frac{1\ mol\ e^-}{1\ mol\ Ag}\right)\left(\frac{96500\ C}{1\ mol\ e^-}\right) = \mathbf{6.0\times10^2\ C}$$

19.54   **(a)**   First find the amount of charge needed to produce 2.00 g of silver according to the half–reaction:

$$Ag^+ (aq) + e^- \longrightarrow Ag (s)$$

$$2.00 \text{ g Ag} \left(\frac{1 \text{ mol Ag}}{107.9 \text{ g Ag}} \right) \left(\frac{1 \text{ mol e}^-}{1 \text{ mol Ag}} \right) \left(\frac{96500 \text{ C}}{1 \text{ mol e}^-} \right) = 1.79 \times 10^3 \text{ C}$$

The half–reaction for the reduction of copper(II) is:

$$Cu^{2+} (aq) + 2e^- \longrightarrow Cu (s)$$

From the amount of charge calculated above, we can calculate the mass of copper deposited in the second cell.

$$1.79 \times 10^3 \text{ C} \left(\frac{1 \text{ mol e}^-}{96500 \text{ C}} \right) \left(\frac{1 \text{ mol Cu}}{2 \text{ mol e}^-} \right) \left(\frac{63.55 \text{ g Cu}}{1 \text{ mol Cu}} \right) = \textbf{0.589 g Cu}$$

(b)   We can calculate the current flowing through the cells using the following strategy.

Coulombs → Coulombs/hr → Coulombs/second

Recall that 1 C = 1 A·s

The current flowing through the cells is:

$$1.79 \times 10^3 \text{ A·s} \left(\frac{1}{3.75 \text{ hr}} \right) \left(\frac{1 \text{ hr}}{3600 \text{ s}} \right) = \textbf{0.133 A}$$

19.56   Electrolysis, Problem Type 5.

Step 1: Balance the half–reaction.

$$Cr_2O_7^{2-} (aq) + 14 H^+ (aq) + 12e^- \longrightarrow 2 Cr (s) + 7 H_2O (l)$$

Step 2: Calculate the quantity of chromium metal by calculating the volume and converting this to mass using the given density.

Volume Cr   = thickness × surface area

$$= 1.0 \times 10^{-2} \text{ mm} \left(\frac{1 \text{ m}}{1000 \text{ mm}} \right) \times 0.25 \text{ m}^2 = 2.5 \times 10^{-6} \text{ m}^3$$

Converting to cm³,

$$2.5 \times 10^{-6} \text{ m}^3 \times \left(\frac{100 \text{ cm}}{1 \text{ m}} \right)^3 = 2.5 \text{ cm}^3$$

Next, calculate the mass of Cr.

Mass = density × volume

$$\text{Mass Cr} = 2.5 \text{ cm}^3 \left(\frac{7.19 \text{ g}}{1 \text{ cm}^3} \right) = 18 \text{ g Cr}$$

Step 3: Find the number of moles of electrons required to electrodeposit 18 g of Cr from solution. The half-reaction is:

$$Cr_2O_7^{2-} (aq) + 14 H^+ (aq) + 12e^- \longrightarrow 2 Cr (s) + 7 H_2O (l)$$

Six moles of electrons are required to reduce 1 mol of Cr metal. But, we are electrodepositing less than 1 mol of Cr (s). We need to complete the following conversions:

$$g\ Cr \rightarrow mol\ Cr \rightarrow mol\ e^-$$

$$?\ faradays = 18\ g\ Cr \times \frac{1\ mol\ Cr}{52.00\ g\ Cr} \times \frac{6\ mol\ e^-}{1\ mol\ Cr} = 2.1\ mol\ e^-$$

Step 4: Determine how long it will take for 2.1 moles of electrons to flow through the cell when the current is 25.0 C/s. We need to complete the following conversions:

$$mol\ e^- \rightarrow coulombs \rightarrow seconds \rightarrow hours$$

$$\textbf{?\ hr} = 2.1\ mol\ e^- \times \frac{96,500\ C}{1\ mol\ e^-} \times \frac{1\ s}{25.0\ C} \times \frac{1\ hr}{3600\ s} = \textbf{2.3\ hr}$$

Would any time be saved by connecting several bumpers together in a series?

19.58   Based on the half-reaction, we know that one faraday will produce half a mole of copper.

$$Cu^{2+} (aq) + 2e^- \longrightarrow Cu (s)$$

First, let's calculate the charge (in C) needed to deposit 0.300 g of Cu.

$$(3.00\ A \times 304\ s)\left(\frac{1\ C}{1\ A \cdot s}\right) = 912\ C$$

We know that one faraday will produce half a mole of copper, but we don't have a half a mole of copper. We have:

$$0.300\ g\ Cu \times \frac{1\ mol\ Cu}{63.55\ g\ Cu} = 4.72 \times 10^{-3}\ mol$$

We calculated the number of coulombs (912 C) needed to produce 4.72×10^{-3} mol of Cu. How many coulombs will it take to produce 0.500 moles of Cu? This will be Faraday's constant.

$$\frac{912\ C}{4.72 \times 10^{-3}\ mol\ Cu} \times 0.500\ mol\ Cu = \textbf{9.66} \times \textbf{10}^4\ \textbf{C} = \textbf{1}\ \textbf{F}$$

19.60   First we can calculate the number of moles of hydrogen produced using the ideal gas equation.

$$n_{H_2} = \frac{PV}{RT}$$

$$n_{H_2} = \frac{(782/760)\ atm\ (0.845\ L)}{(0.0821\ L \cdot atm / K \cdot mol)(298\ K)} = 0.0355\ mol$$

The number of faradays passed through the solution is:

$$0.0355\ mol\ H_2 \left(\frac{2\ F}{1\ mol\ H_2}\right) = \textbf{0.0710}\ \textbf{F}$$

19.62   If you have difficulty balancing redox equations, see Problem Type 1. The balanced equation is:

$$Cr_2O_7^{2-} + 6\,Fe^{2+} + 14\,H^+ \longrightarrow 2\,Cr^{3+} + 6\,Fe^{3+} + 7\,H_2O$$

The remainder of this problem is a solution stoichiometry problem. See Problem Types 5 and 9, Chapter 4.

The number of moles of potassium dichromate in 26.0 mL of the solution is:

$$26.0\ \text{mL}\left(\frac{1\ \text{L}}{1000\ \text{mL}}\right)\left(\frac{0.0250\ \text{mol}}{1\ \text{L}}\right) = 6.50 \times 10^{-4}\ \text{mol}\ K_2Cr_2O_7$$

From the balanced equation it can be seen that 1 mole of dichromate is stoichiometrically equivalent to 6 moles of iron(II). The number of moles of iron(II) oxidized is therefore

$$6.50 \times 10^{-4}\ \text{mol}\ Cr_2O_7^{2-}\left(\frac{6\ \text{mol}\ Fe^{2+}}{1\ \text{mol}\ Cr_2O_7^{2-}}\right) = 3.90 \times 10^{-3}\ \text{mol}\ Fe^{2+}$$

Finally, the molar concentration of Fe^{2+} is:

$$3.90 \times 10^{-3}\ \text{mol}\left(\frac{1}{25.0\ \text{mL}}\right)\left(\frac{1000\ \text{mL}}{1\ \text{L}}\right) = 0.156\ \text{mol}/\text{L} = \textbf{0.156}\ \textbf{\textit{M}}\ \textbf{Fe}^{\textbf{2+}}$$

19.64   The balanced equation is:

$$MnO_4^- + 5\,Fe^{2+} + 8\,H^+ \longrightarrow Mn^{2+} + 5\,Fe^{3+} + 4\,H_2O$$

First, let's calculate the number of moles of potassium permanganate in 23.30 mL of solution.

$$23.30\ \text{mL}\left(\frac{1\ \text{L}}{1000\ \text{mL}}\right)\left(\frac{0.0194\ \text{mol}}{1\ \text{L}}\right) = 4.52 \times 10^{-4}\ \text{mol}\ KMnO_4$$

From the balanced equation it can be seen that 1 mole of permanganate is stoichiometrically equivalent to 5 moles of iron(II). The number of moles of iron(II) oxidized is therefore

$$4.52 \times 10^{-4}\ \text{mol}\ MnO_4^-\left(\frac{5\ \text{mol}\ Fe^{2+}}{1\ \text{mol}\ MnO_4^-}\right) = 2.26 \times 10^{-3}\ \text{mol}\ Fe^{2+}$$

The mass of Fe^{2+} oxidized is:

$$\text{mass}\ Fe^{2+} = 2.26 \times 10^{-3}\ \text{mol}\ Fe^{2+} \times \frac{55.85\ \text{g}\ Fe^{2+}}{1\ \text{mol}\ Fe^{2+}} = 0.126\ \text{g}\ Fe^{2+}$$

Finally, the mass percent of iron in the ore can be calculated.

$$\text{mass}\ \%\ Fe = \frac{\text{mass of iron}}{\text{total mass of sample}} \times 100\%$$

$$\textbf{\%Fe} = \frac{0.126\ \text{g}}{0.2792\ \text{g}} \times 100\% = \textbf{45.1\%}$$

19.66 **(a)** The half–reactions are:

(i) $MnO_4^- (aq) + 8 H^+ (aq) + 5e^- \longrightarrow Mn^{2+} (aq) + 4 H_2O (l)$

(ii) $C_2O_4^{2-} (aq) \longrightarrow 2 CO_2 (g) + 2e^-$

We combine the half–reactions to cancel electrons, that is, [2 × equation (i)] + [5 × equation (ii)]

$$2 MnO_4^- (aq) + 16 H^+ (aq) + 5 C_2O_4^{2-} (aq) \longrightarrow 2 Mn^{2+} (aq) + 10 CO_2 (g) + 8 H_2O (l)$$

(b) We can calculate the moles of $KMnO_4$ from the molarity and volume of solution.

$$24.0 \text{ mL } KMnO_4 \left(\frac{1 L}{1000 \text{ mL}} \right)\left(\frac{0.0100 \text{ mol } KMnO_4}{1 L} \right) = 2.40 \times 10^{-4} \text{ mol } KMnO_4$$

We can calculate the mass of oxalic acid from the stoichiometry of the balanced equation. The mole ratio between oxalate ion and permanganate ion is 5:2.

$$2.40 \times 10^{-4} \text{ mol } KMnO_4 \left(\frac{5 \text{ mol } H_2C_2O_4}{2 \text{ mol } KMnO_4} \right)\left(\frac{90.04 \text{ g } H_2C_2O_4}{1 \text{ mol } H_2C_2O_4} \right) = 0.0540 \text{ g } H_2C_2O_4$$

Finally, the percent by mass of oxalic acid in the sample is:

$$\textbf{\% oxalic acid} = \frac{0.0540 \text{ g}}{1.00 \text{ g}} \times 100\% = \textbf{5.40 \%}$$

19.68 The balanced equation is:

$$2 MnO_4^- + 5 C_2O_4^{2-} + 16 H^+ \longrightarrow 2 Mn^{2+} + 10 CO_2 + 8 H_2O$$

Therefore, 2 mol MnO_4^- reacts with 5 mol $C_2O_4^{2-}$

$$\text{Moles of } MnO_4^- \text{ reacted} = (24.2 \text{ mL})\left(\frac{1 L}{1000 \text{ mL}} \right)\left(\frac{9.56 \times 10^{-4} \text{ mol } MnO_4^-}{1 L} \right)$$

$$= 2.31 \times 10^{-5} \text{ mol } MnO_4^-$$

Recognize that the mole ratio of Ca^{2+} to $C_2O_4^{2-}$ is 1:1 in CaC_2O_4. The mass of Ca^{2+} in 10.0 mL is:

$$(2.31 \times 10^{-5} \text{ mol } MnO_4^-) \times \left(\frac{5 \text{ mol } Ca^{2+}}{2 \text{ mol } MnO_4^-} \right)\left(\frac{40.08 \text{ g } Ca^{2+}}{1 \text{ mol } Ca^{2+}} \right) = 2.31 \times 10^{-3} \text{ g } Ca^{2+}$$

Finally, converting to mg/mL, we have:

$$\left(\frac{2.31 \times 10^{-3} \text{ g } Ca^{2+}}{10.0 \text{ mL}} \right)\left(\frac{1000 \text{ mg}}{1 g} \right) = \textbf{0.231 mg } Ca^{2+} \textbf{/ mL blood}$$

19.70   **(a)**   The half–reactions are:

$$2\,H^+\,(aq) + 2e^- \longrightarrow H_2\,(g) \qquad\qquad E^\circ_{anode} = 0.00\,V$$

$$Ag^+\,(aq) + e^- \longrightarrow Ag\,(s) \qquad\qquad E^\circ_{cathode} = 0.80\,V$$

$$E^0_{cell} = E^\circ_{cathode} - E^\circ_{anode} = 0.80\,V - 0.00\,V = \mathbf{0.80\,V}$$

(b)   The spontaneous cell reaction under standard-state conditions is:

$$\mathbf{2\,Ag^+\,(aq) + H_2\,(g) \longrightarrow 2\,Ag\,(s) + 2\,H^+\,(aq)}$$

(c)   Using the Nernst equation we can calculate the cell potential under nonstandard-state conditions.

$$E = E^\circ - \frac{0.0257\,V}{n}\ln\frac{[H^+]^2}{[Ag^+]^2\,P_{H_2}}$$

(i) The potential is:

$$E = 0.80\,V - \frac{0.0257\,V}{2}\ln\frac{(1.0\times10^{-2})^2}{(1.0)^2(1.0)} = \mathbf{0.92\,V}$$

(ii) The potential is:

$$E = 0.80\,V - \frac{0.0257\,V}{2}\ln\frac{(1.0\times10^{-5})^2}{(1.0)^2(1.0)} = \mathbf{1.10\,V}$$

(d)   From the results in part (c), we deduce that this cell is a pH meter; its potential is a sensitive function of the hydrogen ion concentration. Each 1 unit increase in pH causes a voltage increase of 0.060 V.

19.72   The overvoltage of oxygen is not large enough to prevent its formation at the anode. Applying the diagonal rule, we see that water is oxidized before fluoride ion.

$$F_2\,(g) + 2e^- \longrightarrow 2\,F^-\,(aq) \qquad\qquad E^\circ = 2.87\,V$$

$$O_2\,(g) + 4\,H^+\,(aq) + 4e^- \longrightarrow 2\,H_2O\,(l) \qquad E^\circ = 1.23\,V$$

The very positive standard reduction potential indicates that F^- has essentially no tendency to undergo oxidation.

This fact was one of the major obstacles preventing the discovery of fluorine for many years. HF was usually chosen as the substance for electrolysis, but two problems interfered with the experiment. First, any water in the HF was oxidized before the fluoride ion. Second, pure HF without any water in it is a nonconductor of electricity (HF is a weak acid!). The problem was finally solved by dissolving KF in liquid HF to give a conducting solution.

19.74   We can calculate the amount of charge that 4.0 g of MnO_2 can produce.

$$4.0\,g\,MnO_2\left(\frac{1\,mol}{86.94\,g}\right)\left(\frac{2\,mol\,e^-}{2\,mol\,MnO_2}\right)\left(\frac{96500\,C}{1\,mol\,e^-}\right) = 4.44\times10^3\,C$$

Since a current of one ampere represents a flow of one coulomb per second, we can find the time it takes for this amount of charge to pass.

$$0.0050\ A\ =\ 0.0050\ C/s$$

$$4.44\times10^3\ \cancel{C}\left(\frac{1\ \cancel{s}}{0.0050\ \cancel{C}}\right)\left(\frac{1\ hr}{3600\ \cancel{s}}\right) = \mathbf{2.5\times10^2\ hr}$$

19.76   Since this is a concentration cell, the standard emf is zero. (Why?) Using the Nernst equation, we can write equations to calculate the cell voltage for the two cells.

$$(1)\qquad E_{cell} = -\frac{RT}{nF}\ln Q = -\frac{RT}{2F}\ln\frac{[Hg_2^{2+}]soln\ A}{[Hg_2^{2+}]soln\ B}$$

$$(2)\qquad E_{cell} = -\frac{RT}{nF}\ln Q = -\frac{RT}{1F}\ln\frac{[Hg^{+}]soln\ A}{[Hg^{+}]soln\ B}$$

In the first case, two electrons are transferred per mercury ion ($n = 2$), while in the second only one is transferred ($n = 1$). Note that the concentration ratio will be 1:10 in both cases. The voltages calculated at 18°C are:

$$(1)\qquad E_{cell} = \frac{-(8.314\ \cancel{J}/K\cdot\cancel{mol})(291\ \cancel{K})}{2(96500\ \cancel{J}\cdot V^{-1}\cancel{mol}^{-1})}\ln 10^{-1} = 0.0289\ V$$

$$(2)\qquad E_{cell} = \frac{-(8.314\ J/K\cdot mol)(291\ K)}{1(96500\ J\cdot V^{-1}mol^{-1})}\ln 10^{-1} = 0.0577\ V$$

Since the calculated cell potential for cell (1) agrees with the measured cell emf, we conclude that the mercury(I) ion exists as $\mathbf{Hg_2^{2+}}$ in solution.

19.78   We begin by treating this like an ordinary stoichiometry problem (see Chapter 3).

Step 1: Calculate the number of moles of Mg and Ag^{+}.

The number of moles of magnesium is:

$$1.56\ \cancel{g}\ Mg\left(\frac{1\ mol\ Mg}{24.31\ \cancel{g}\ Mg}\right) = 0.0642\ mol\ Mg$$

The number of moles of silver ion in the solution is:

$$\left(\frac{0.100\ mol\ Ag^{+}}{1\ \cancel{L}}\right)0.1000\ \cancel{L} = 0.0100\ mol\ Ag^{+}$$

Step 2: Calculate the mass of Mg remaining by determining how much Mg reacts with Ag^{+}.

The balanced equation for the reaction is:

$$2\ Ag^{+}\ (aq)\ +\ Mg\ (s)\ \longrightarrow\ 2\ Ag\ (s)\ +\ Mg^{2+}\ (aq)$$

Since you need twice as much Ag^{+} compared to Mg for complete reaction, Ag^{+} is the limiting reagent. The amount of Mg consumed is:

$$0.0100\ \cancel{mol}\ Ag^{+}\left(\frac{1\ mol\ Mg}{2\ \cancel{mol}\ Ag^{+}}\right) = 0.00500\ mol\ Mg$$

The amount of magnesium remaining is:

$$(0.0642 - 0.00500) \text{ mol Mg} \times 24.31 \text{ g/mol} = \textbf{1.44 g Mg}$$

Step 3: Assuming complete reaction, calculate the concentration of Mg^{2+} ions produced.

Since the mole ratio between Mg and Mg^{2+} is 1:1, the mol of Mg^{2+} formed will equal the mol of Mg reacted. The concentration of Mg^{2+} is:

$$[Mg^{2+}]_0 = \frac{0.00500 \text{ mol}}{0.100 \text{ L}} = 0.0500 \ M$$

Step 4: We can calculate the equilibrium constant for the reaction from the standard cell emf.

$$E^\circ_{cell} = E^\circ_{cathode} - E^\circ_{anode} = 0.80 \text{ V} - (-2.37 \text{ V}) = 3.17 \text{ V}$$

We can then compute the equilibrium constant.

$$K = e^{nE^\circ cell/0.0257}$$
$$= e^{(2)(3.17)/(0.0257)} = 1 \times 10^{107}$$

Step 5: To find equilibrium concentrations of Mg^{2+} and Ag^+, we have to solve an equilibrium problem (see Problem Type 5, Chapter 14).

Let x be the small amount of Mg^{2+} that reacts to achieve equilibrium. The concentration of Ag^+ will be $2x$ at equilibrium. Assume that essentially all Ag^+ has been reduced so that the initial concentration of Ag^+ is zero.

	$2 \text{ Ag}^+ (aq) + \text{Mg} (s)$	$\rightleftharpoons$	$2 \text{ Ag}(s) + \text{Mg}^{2+}(aq)$
Initial (*M*):	0.0000		0.0500
Change (*M*):	+2x		−x
Equilibrium (*M*):	2x		(0.0500 − x)

$$K = \frac{[Mg^{2+}]}{[Ag^+]^2}$$

$$1 \times 10^{107} = \frac{(0.0500 - x)}{(2x)^2}$$

We can assume $0.0500 - x \approx 0.0500$.

$$1 \times 10^{107} \approx \frac{0.0500}{(2x)^2}$$

$$(2x)^2 = \frac{0.0500}{1 \times 10^{107}} = 0.0500 \times 10^{-107}$$

$$(2x)^2 = 5.00 \times 10^{-109} = 50.0 \times 10^{-110}$$

$$2x = 7 \times 10^{-55} \ M$$

$$[Ag^+] = 2x = 7 \times 10^{-55} \ M$$

$$[Mg^{2+}] = 0.0500 - x = \textbf{0.0500} \ M$$

19.80 **(a)** Since this is an acidic solution, the gas must be hydrogen gas from the reduction of hydrogen ion. The two electrode reactions and the overall cell reaction are:

$$\text{anode:} \qquad Cu\,(s) \longrightarrow Cu^{2+}\,(aq) + 2e^-$$

$$\text{cathode:} \qquad 2\,H^+\,(aq) + 2e^- \longrightarrow H_2\,(g)$$

$$Cu\,(s) + 2\,H^+\,(aq) \longrightarrow Cu^{2+}\,(aq) + H_2\,(g)$$

Since 0.584 g of copper was consumed, the amount of hydrogen gas produced is:

$$0.584 \text{ g Cu} \left(\frac{1 \text{ mol Cu}}{63.55 \text{ g Cu}} \right) \left(\frac{1 \text{ mol } H_2}{1 \text{ mol Cu}} \right) = 9.20 \times 10^{-3} \text{ mol } H_2$$

At STP, 1 mole of an ideal gas occupies a volume of 22.41 L. Thus, the volume of H_2 at STP is:

$$V_{H_2} = (9.20 \times 10^{-3} \text{ mol } H_2)(22.41 \text{ L/mol}) = \textbf{0.206 L}$$

(b) From the current and the time, we can calculate the amount of charge:

$$(1.18 \text{ A}) \left(\frac{1 \text{ C/s}}{1 \text{ A}} \right) (1.52 \times 10^3 \text{ s}) = 1.79 \times 10^3 \text{ C}$$

Since we know the charge of an electron, we can compute the number of electrons.

$$(1.79 \times 10^3 \text{ C}) \left(\frac{1 \text{ } e^-}{1.6022 \times 10^{-19} \text{ C}} \right) = 1.12 \times 10^{22} \text{ } e^-$$

Using the amount of copper consumed in the reaction and the fact that 2 mol of e^- are produced for every 1 mole of copper consumed, we can calculate Avogadro's number.

$$\frac{1.12 \times 10^{22} \text{ } e^-}{9.20 \times 10^{-3} \text{ mol Cu}} \times \left(\frac{1 \text{ mol Cu}}{2 \text{ mol } e^-} \right) = \textbf{6.09} \times \textbf{10}^{\textbf{23}} \text{ } \textbf{e}^-\textbf{/mol}$$

In practice, Avogadro's number can be determined by electrochemical experiments like this. The charge of the electron can be found independently by Millikan's experiment.

19.82 **(a)** We can calculate $\Delta G°$ from standard free energies of formation.

$$\begin{aligned} \Delta G° &= 2\,\Delta G_f°\,(N_2) + 6\,\Delta G_f°\,(H_2O) - [4\,\Delta G_f°\,(NH_3) + 3\,\Delta G_f°\,(O_2)] \\ &= 0 + (6 \text{ mol})(-237.2 \text{ kJ/mol}) - [(4 \text{ mol})(-16.6 \text{ kJ/mol}) + 0] \\ &= \textbf{-1356.8 kJ} \end{aligned}$$

(b) The half–reactions are:

$$4\,NH_3\,(g) \longrightarrow 2\,N_2\,(g) + 12\,H^+\,(aq) + 12e^-$$

$$3\,O_2\,(g) + 12\,H^+\,(aq) + 12e^- \longrightarrow 6\,H_2O\,(l)$$

The overall reaction is a 12–electron process. We can calculate the standard cell emf from the standard free energy change, $\Delta G°$.

$$\Delta G° = -nFE°_{cell}$$

$$E°_{cell} = \frac{-\Delta G°}{nF} = \frac{-(-1356.8 \times 1000 \text{ J/mol})}{(12)(96500 \text{ J/V} \cdot \text{mol})} = \textbf{1.17 V}$$

19.84   Electrolysis, Problem Type 5.

The reduction of Ag^+ to Ag metal is:

$$Ag^+ (aq) + e^- \longrightarrow Ag$$

We can calculate both the moles of Ag deposited and the moles of Au deposited.

$$? \text{ mol Ag} = 2.64 \text{ g Ag} \times \frac{1 \text{ mol Ag}}{107.9 \text{ g Ag}} = 2.45 \times 10^{-2} \text{ mol Ag}$$

$$? \text{ mol Au} = 1.61 \text{ g Au} \times \frac{1 \text{ mol Au}}{197.0 \text{ g Au}} = 8.17 \times 10^{-3} \text{ mol Au}$$

We do not know the oxidation state of Au ions, so we will represent the ions as Au^{n+}. If we divide the mol of Ag by the mol of Au, we can determine the ratio of Ag^+ reduced compared to Au^{n+} reduced.

$$\frac{2.45 \times 10^{-2} \text{ mol Ag}}{8.17 \times 10^{-3} \text{ mol Au}} = 3$$

That is, the same number of electrons that reduced the Ag^+ ions to Ag reduced only one–third the number of moles of the Au^{n+} ions to Au. Thus, each Au^{n+} required three electrons per ion for every one electron for Ag^+. The oxidation state for the gold ion is +3; the ion is Au^{3+}.

$$Au^{3+} (aq) + 3e^- \longrightarrow Au$$

19.86   We reverse the first half–reaction and add it to the second to come up with the overall balanced equation

$Hg_2^{2+} \longrightarrow 2 Hg^{2+} + 2e^-$	$E°_{anode} = +0.92$ V
$Hg_2^{2+} + 2e^- \longrightarrow 2 Hg$	$E°_{cathode} = +0.85$ V
$2 Hg_2^{2+} \longrightarrow 2 Hg^{2+} + 2 Hg$	$E°_{cell} = 0.85$ V $- 0.92$ V $= -0.07$ V

Since the standard cell potential is an intensive property,

$$Hg_2^{2+} (aq) \longrightarrow Hg^{2+} (aq) + Hg (l) \qquad E°_{cell} = -0.07 \text{ V}$$

We calculate $\Delta G°$ from $E°$.

$$\Delta G° = -nFE° = -(1)(96500 \text{ J/V} \cdot \text{mol})(-0.07 \text{ V}) = \textbf{6.8 kJ/mol}$$

The corresponding equilibrium constant is:

$$K = \frac{[Hg^{2+}]}{[Hg_2^{2+}]}$$

We calculate K from $\Delta G°$.

$$\Delta G° = -RT \ln K$$

$$\ln K = \frac{-6.8 \times 10^3 \text{ J/mol}}{(8.314 \text{ J/K} \cdot \text{mol})(298 \text{ K})}$$

$$K = 0.064$$

19.88   The reactions for the electrolysis of NaCl (aq) are:

Anode:	$2 \text{ Cl}^- (aq) \longrightarrow \text{ Cl}_2 (g) + 2e^-$
Cathode:	$2 \text{ H}_2\text{O} (l) + 2e^- \longrightarrow \text{ H}_2 (g) + 2 \text{ OH}^- (aq)$
Overall:	$2\text{H}_2\text{O}(l) + 2\text{Cl}^- (aq) \rightarrow \text{H}_2(g) + \text{Cl}_2(g) + 2\text{OH}^- (aq)$

From the pH of the solution, we can calculate the OH^- concentration. From the $[\text{OH}^-]$, we can calculate the moles of OH^- produced. Then, from the moles of OH^- we can calculate the average current used.

$$\text{pH} = 12.24$$

$$\text{pOH} = 14.00 - 12.24 = 1.76$$

$$[\text{OH}^-] = 1.74 \times 10^{-2} \, M$$

The moles of OH^- produced are:

$$(1.74 \times 10^{-2} \text{ mol/L}) \times (0.300 \text{ L}) = 5.22 \times 10^{-3} \text{ mol OH}^-$$

From the balanced equation, it takes 1 mole of e^- to produce 1 mole of OH^- ions.

$$5.22 \times 10^{-3} \text{ mol OH}^- \times \frac{1 \text{ mol } e^-}{1 \text{ mol OH}^-} \times \frac{96500 \text{ C}}{1 \text{ mol } e^-} = 504 \text{ C}$$

Recall that $1 \text{ C} = 1 \text{ A} \cdot 1\text{s}$

$$504 \text{ C} \times \frac{1 \text{ A} \cdot \text{s}}{1 \text{ C}} \times \frac{1 \text{ min}}{60 \text{ s}} \times \frac{1}{6.00 \text{ min}} = \mathbf{1.4 \text{ A}}$$

19.90   The reaction is:

$$\text{Pt}^{n+} + ne^- \longrightarrow \text{Pt}$$

Thus, we can calculate the charge of the platinum ions by realizing that n mol of e^- are required per mol of Pt formed.

The moles of Pt formed are:

$$9.09 \text{ g Pt} \times \frac{1 \text{ mol Pt}}{195.1 \text{ g Pt}} = 0.0466 \text{ mol Pt}$$

Next, calculate the charge passed in C.

$$C = 2.00 \, \cancel{hr} \times \frac{3600 \, \cancel{s}}{1 \, \cancel{hr}} \times \frac{2.50 \, C}{1 \, \cancel{s}} = 1.80 \times 10^4 \, C$$

Convert to moles of electrons.

$$? \, mol \, e^- = (1.80 \times 10^4 \, \cancel{C}) \times \frac{1 \, mol \, e^-}{96500 \, \cancel{C}} = 0.187 \, mol \, e^-$$

We now know the number of moles of electrons ($0.187 \, mol \, e^-$) needed to produce $0.0466 \, mol$ of Pt metal. We can calculate the number of moles of electrons needed to produce 1 mole of Pt metal.

$$\frac{0.187 \, mol \, e^-}{0.0466 \, mol \, Pt} = 4.01 \, mol \, e^-/mol \, Pt$$

Since we need 4 moles of electrons to reduce 1 mole of Pt ions, the charge on the Pt ions must be **+4**.

19.92   The half–reaction for the oxidation of water to oxygen is:

$$2 \, H_2O \, (l) \xrightarrow{\text{oxidation (anode)}} O_2 \, (g) + 4 \, H^+ \, (aq) + 4e^-$$

Knowing that one mole of any gas at STP occupies a volume of 22.4 L, we find the number of moles of oxygen.

$$(4.26 \, \cancel{L} \, O_2)\left(\frac{1 \, mol}{22.4 \, \cancel{L}}\right) = 0.190 \, mol \, O_2$$

Since four electrons are required to form one oxygen molecule, the number of electrons must be:

$$(0.190 \, \cancel{mol} \, O_2)\left(\frac{4 \, \cancel{mol} \, e^-}{1 \, \cancel{mol} \, O_2}\right)\left(\frac{6.022 \times 10^{23} \, e^-}{1 \, \cancel{mol}}\right) = 4.58 \times 10^{23} \, e^-$$

The amount of charge passing through the solution is:

$$(6.00 \, \cancel{A})\left(\frac{1 \, C/\cancel{s}}{1 \, \cancel{A}}\right)\left(\frac{3600 \, \cancel{s}}{1 \, \cancel{hr}}\right)(3.40 \, \cancel{hr}) = 7.34 \times 10^4 \, C$$

We find the electron charge by dividing the amount of charge by the number of electrons.

$$\frac{7.34 \times 10^4 \, C}{4.58 \times 10^{23} \, e^-} = \mathbf{1.60 \times 10^{-19} \, C/e^-}$$

In actual fact, this sort of calculation can be used to find Avogadro's number, not the electron charge. The latter can be measured independently, and one can use this charge together with electrolytic data like the above to calculate the number of objects in one mole. See also Problem 19.80.

19.94   Cells of higher voltage require very reactive oxidizing and reducing agents, which are difficult to handle. (From Table 19.1 we see that 5.92 V is the theoretical limit of a cell made up of Li^+/Li and F_2/F^- electrodes under standard-state conditions.) Batteries made up of several cells in series are easier to use.

19.96   The half-reactions are:

$$Zn(s) + 4OH^-(aq) \rightarrow Zn(OH)_4^{2-}(aq) + 2e^- \qquad E^\circ_{anode} = -1.36 \text{ V}$$

$$Zn^{2+}(aq) + 2e^- \rightarrow Zn(s) \qquad\qquad\qquad E^\circ_{cathode} = -0.76 \text{ V}$$

$$\overline{Zn^{2+}(aq) + 4OH^-(aq) \rightarrow Zn(OH)_4^{2-}(aq) \qquad E^\circ_{cell} \quad -0.76 \text{ V} - (-1.36 \text{ V}) = 0.60 \text{ V}}$$

$$K_f = e^{nE^\circ/0.0257} = e^{(2)(0.60)/(0.0257)} = 2 \times 10^{20}$$

19.98   **(a)**   Since electrons flow from X to SHE, E° for X must be negative. Thus E° for Y must be positive.

(b)

$$Y^{2+} + 2e^- \rightarrow Y \qquad\qquad E^\circ_{cathode} = 0.34 \text{ V}$$

$$X \rightarrow X^{2+} + 2e^- \qquad\qquad E^\circ_{anode} = -0.25 \text{ V}$$

$$\overline{X + Y^{2+} \rightarrow X^{2+} + Y \qquad\qquad E^0_{cell} = 0.34 \text{ V} - (-0.25 \text{ V}) = \mathbf{0.59 \text{ V}}}$$

19.100   **(a)**   Gold does not tarnish in air because the reduction potential for oxygen is insufficient to result in the oxidation of gold.

$$O_2 + 4H^+ + 4e^- \rightarrow 2H_2O \qquad E^\circ_{cathode} = 1.23 \text{ V}$$

That is, $E^\circ_{cell} = E^\circ_{cathode} - E^\circ_{anode} < 0$, for either oxidation by O_2 to Au^+ or Au^{3+}.

$$E^\circ_{cell} = 1.23 \text{ V} - 1.50 \text{ V} < 0$$

or

$$E^\circ_{cell} = 1.23 \text{ V} - 1.69 \text{ V} < 0$$

(b)

$$3(Au^+ + e^- \rightarrow Au) \qquad E^\circ_{cathode} = 1.69 \text{ V}$$

$$Au \rightarrow Au^{3+} + 3e^- \qquad\quad E^\circ_{anode} = 1.50 \text{ V}$$

$$\overline{3Au^+ \rightarrow 2Au + Au^{3+} \qquad E^\circ_{cell} = 1.69 \text{ V} - 1.50 \text{ V} = 0.19 \text{ V}}$$

Calculating ΔG,

$$\Delta G^\circ = -nFE^\circ = = -(3)(96,500 \text{ J/V·mol})(0.19 \text{ V}) = \mathbf{-55.0 \text{ kJ/mol}}$$

For spontaneous electrochemical equations, ΔG° must be negative. Thus, **the disproportionation occurs spontaneously**.

(c)   Since the most stable oxidation state for gold is Au^{3+}, the predicted reaction is:

$$2Au + 3F_2 \rightarrow 2AuF_3$$

19.102   The balanced equation is: $5Fe^{2+} + MnO_4^- + 8H^+ \longrightarrow Mn^{2+} + 5Fe^{3+} + 4 H_2O$

Calculate the amount of iron(II) in the original solution using the mole ratio from the balanced equation.

$$23.0 \text{ mL} \left(\frac{1 \text{ L}}{1000 \text{ mL}} \right) \left(\frac{0.0200 \text{ mol KMnO}_4}{1 \text{ L}} \right) \left(\frac{5 \text{ mol Fe}^{2+}}{1 \text{ mol KMnO}_4} \right) = 0.00230 \text{ mol Fe}^{2+}$$

The concentration of iron(II) must be:

$$[Fe^{2+}] = \left(\frac{0.00230 \text{ mol}}{0.0250 \text{ L}} \right) = 0.0920 \ M$$

The total iron concentration can be found by simple proportion because the same sample volume (25.0 mL) and the same $KMnO_4$ solution were used.

$$[Fe]_{total} = \frac{40.0 \text{ mL } KMnO_4}{23.0 \text{ mL } KMnO_4} \times 0.0920 \ M = 0.160 \ M$$

$$[Fe^{3+}] = [Fe]_{total} - [Fe^{2+}] = 0.0680 \ M$$

Why are the two titrations with permanganate necessary in this problem?

19.104 (a) $1 A \cdot hr = 1 A \times 3600 s = 3600 \ C$

(b)   Anode: $Pb + SO_4^{2-} \rightarrow PbSO_4 + 2e^-$

Two moles of electrons are produced by 1 mole of Pb. Recall that the charge of 1 mol e^- is 96,500 C. We can set up the following conversions to calculate the capacity of the battery.

mol Pb $\rightarrow$ mol e^- $\rightarrow$ coulombs $\rightarrow$ ampere hour

$$406 \text{ g Pb} \times \frac{1 \text{ mol Pb}}{207.2 \text{ g Pb}} \times \frac{2 \text{ mol } e^-}{1 \text{ mol Pb}} \times \frac{96500 \text{ C}}{1 \text{ mol } e^-} = 3.74 \times 10^5 \text{ C} \times \frac{1 \text{ hr}}{3600 \text{ s}} = 104 \ A \cdot hr$$

This ampere·hour cannot be fully realized because the concentration of H_2SO_4 keeps decreasing.

(c) $E^o_{cell} = 1.70 \text{ V} - (-0.31 \text{ V}) = 2.01 \text{ V}$ (From Table 19.1)

$\Delta G^\circ = -nFE^\circ$

$\Delta G^\circ = -(2)(96500 \text{ J/V} \cdot \text{mol})(2.01 \text{ V}) = -3.88 \times 10^5 \text{ J/mol}$

Spontaneous as expected.

19.106 (a)   unchanged   **(b)**   unchanged   **(c)**   squared   **(d)**   doubled   **(e)**   doubled

19.108 $F_2 (g) + 2H^+ (aq) + 2e^- \rightarrow 2HF (g)$

$$E = E^\circ - \frac{RT}{2F} \ln \frac{P_{HF}^2}{P_{F_2}[H^+]^2}$$

With increasing $[H^+]$, E will be larger. F_2 will become a **stronger oxidizing agent**.

19.110 $Pb \rightarrow Pb^{2+} + 2e^- \qquad\qquad E^\circ_{anode} = -0.13 \text{ V}$

$\underline{2H^+ + 2e^- \rightarrow H_2 \qquad\qquad E^\circ_{cathode} = 0.00 \text{ V}}$

$Pb + 2H^+ \rightarrow Pb^{2+} + H_2 \qquad E^\circ_{cell} = 0.00 \text{ V} - (-0.13 \text{ V}) = 0.13 \text{ V}$

$$pH = 1.60$$

$$[H^+] = 10^{-1.60} = 0.025 \, M$$

$$E = E° - \frac{RT}{nF} \ln \frac{[Pb^{2+}]P_{H_2}}{[H^+]^2}$$

$$0 = 0.13 - \frac{0.0257 \text{ V}}{2} \ln \frac{(0.035)P_{H_2}}{0.025^2}$$

$$\frac{0.26}{0.0257} = \ln \frac{0.035 \, P_{H_2}}{0.025^2}$$

$$P_{H_2} = \mathbf{4.4 \times 10^2 \text{ atm}}$$

19.112 (a) The half-reactions are:

Anode: $Zn \rightarrow Zn^{2+} + 2e^-$
Cathode: $(1/2)O_2 + 2e^- \rightarrow O^{2-}$

Overall: $Zn + (1/2)O_2 \rightarrow ZnO$

To calculate the standard emf, we first need to calculate $\Delta G°$ for the reaction. From Appendix 3 we write:

$$\Delta G° = (1 \text{ mol}) \Delta G_f° (ZnO) - [(1 \text{ mol}) \Delta G_f° (Zn) + (\frac{1}{2} \text{ mol}) \Delta G_f° (O_2)]$$

$$\Delta G° = -318.2 \text{ kJ} - [0 + 0]$$
$$\Delta G° = -318.2 \text{ kJ}$$

$$\Delta G° = -nFE°$$
$$-318.2 \times 10^3 \text{ J/mol} = -(2)(96,500 \text{ J/V·mol})E°$$
$$E° = \mathbf{1.65 \text{ V}}$$

(b) We use the following equation:

$$E = E° - \frac{RT}{nF} \ln Q$$

$$E = 1.65 \text{ V} - \frac{0.0257 \text{ V}}{2} \ln \frac{1}{P_{O_2}}$$

$$E = 1.65 \text{ V} - \frac{0.0257 \text{ V}}{2} \ln \frac{1}{0.21}$$

$$E = 1.65 \text{ V} - 0.020 \text{ V}$$

$$E = \mathbf{1.63 \text{ V}}$$

(c) Since the free energy change represents the maximum work that can be extracted from the overall reaction, the maximum amount of energy that can be obtained from this reaction is the free energy change. To calculate the energy density, we multiply the free energy change by the number of moles of Zn present in 1 kg of Zn.

$$\textbf{energy density} = \frac{318.2 \text{ kJ}}{1 \text{ mol Zn}} \times \frac{1 \text{ mol Zn}}{65.39 \text{ g Zn}} \times \frac{1000 \text{ g Zn}}{1 \text{ kg Zn}} = \textbf{4.87} \times \textbf{10}^3 \textbf{ kJ/kg Zn}$$

(d)   One ampere is 1 C/s. The charge drawn every second is given by nF.

$$\text{charge} = nF$$

$$2.1 \times 10^5 \text{ C} = n(96,500 \text{ C/mol e}^-)$$

$$n = 2.2 \text{ mol e}^-$$

From the overall balanced reaction, we see that 4 moles of electrons will reduce 1 mole of O_2; therefore, the number of moles of O_2 reduced by 2.2 moles of electrons is:

$$\text{mol } O_2 = 2.2 \text{ mol e}^- \times \frac{1 \text{ mol } O_2}{4 \text{ mol e}^-} = 0.55 \text{ mol } O_2$$

The volume of oxygen at 1.0 atm partial pressure can be obtained by using the ideal gas equation.

$$V_{O_2} = \frac{nRT}{P} = \frac{(0.55 \text{ mol})(0.0821 \text{ L}\cdot\text{atm} / \text{mol}\cdot\text{K})(298 \text{ K})}{(1.0 \text{ atm})} = 13 \text{ L}$$

Since air is 21 percent oxygen by volume, the volume of air required every second is:

$$V_{\text{air}} = 13 \text{ L } O_2 \times \frac{100\% \text{ air}}{21\% \text{ } O_2} = \textbf{62 L of air}$$

19.114   We can calculate $\Delta G^\circ_{\text{rxn}}$ using the following equation.

$$\Delta G^\circ_{\text{rxn}} = \Sigma n \, \Delta G^\circ_{\text{f}} \text{ (products)} - \Sigma m \, \Delta G^\circ_{\text{f}} \text{ (reactants)}$$

$$\Delta G^\circ_{\text{rxn}} = 0 + 0 - [(1 \text{ mol})(-293.8 \text{ kJ} / \text{mol}) + 0] = 293.8 \text{ kJ}$$

Next, we can calculate E° using the equation

$$\Delta G^\circ = nFE^\circ$$

We use a more accurate value for Faraday's constant.

$$293.8 \times 10^3 \text{ J/mol} = (1)(96485.3 \text{ J/V}\cdot\text{mol})E^\circ$$

$$\boldsymbol{E^\circ = 3.05 \text{ V}}$$

CHAPTER 20
METALLURGY AND THE CHEMISTRY
OF METALS

This chapter is of a very descriptive nature. As such, there are essentially no defined problem types. Please read Chapter 20 of the text carefully before answering the end-of-chapter problems.

SOLUTIONS TO SELECTED TEXT PROBLEMS

20.12 The cathode reaction is: $Cu^{2+}(aq) + 2e^- \longrightarrow Cu(s)$

First, let's calculate the number of moles of electrons needed to reduce 5.0 kg of Cu.

$$5.00 \text{ kg Cu} \times \frac{1000 \text{ g}}{1 \text{ kg}} \times \frac{1 \text{ mol Cu}}{63.55 \text{ g Cu}} \times \frac{2 \text{ mol e}^-}{1 \text{ mol Cu}} = 1.57 \times 10^2 \text{ mol e}^-$$

Next, let's determine how long it will take for 1.57×10^2 moles of electrons to flow through the cell when the current is 37.8 C/s.

$$1.57 \times 10^2 \text{ mol e}^- \times \frac{96,500 \text{ C}}{1 \text{ mol e}^-} \times \frac{1 \text{ s}}{37.8 \text{ C}} \times \frac{1 \text{ hr}}{3600 \text{ s}} = \textbf{111 hr}$$

20.14 The sulfide ore is first roasted in air:

$$2ZnS(s) + 3O_2(g) \longrightarrow 2ZnO(s) + 2SO_2(g)$$

The zinc oxide is then mixed with coke and limestone in a blast furnace where the following reductions occur:

$$ZnO(s) + C(s) \longrightarrow Zn(g) + CO(g)$$
$$ZnO(s) + CO(g) \longrightarrow Zn(g) + CO_2(g)$$

The zinc vapor formed distills from the furnace into an appropriate receiver.

20.16 **(a)** We first find the mass of ore containing 2.0×10^8 kg of copper.

$$2.0 \times 10^8 \text{ kg Cu} \left(\frac{100 \text{ \% ore}}{0.80 \text{ \% Cu}} \right) = 2.5 \times 10^{10} \text{ kg ore}$$

We can then compute the volume from the density of the ore.

$$2.5 \times 10^{10} \text{ kg} \left(\frac{1000 \text{ g}}{1 \text{ kg}} \right) \left(\frac{1 \text{ cm}^3}{2.8 \text{ g}} \right) = \textbf{8.9} \times \textbf{10}^{12} \textbf{ cm}^3$$

(b)   From the formula of chalcopyrite it is clear that two moles of sulfur dioxide will be formed per mole of copper. The mass of sulfur dioxide formed will be:

$$2.0 \times 10^8 \text{ kg Cu} \left(\frac{1 \text{ mol Cu}}{0.06355 \text{ kg Cu}} \right) \left(\frac{2 \text{ mol SO}_2}{1 \text{ mol Cu}} \right) \left(\frac{0.06407 \text{ kg SO}_2}{1 \text{ mol SO}_2} \right) = \mathbf{4.0 \times 10^8 \text{ kg SO}_2}$$

20.18   Iron can be produced by reduction with coke in a blast furnace; whereas, aluminum is usually produced electrolytically, which is a much more expensive process.

20.28   **(a)** $2\text{Na}(s) + 2\text{H}_2\text{O}(l) \longrightarrow 2\text{NaOH}(aq) + \text{H}_2(g)$

(b) $2\text{NaOH}(aq) + \text{CO}_2(g) \longrightarrow \text{Na}_2\text{CO}_3(aq) + \text{H}_2\text{O}(l)$

(c) $\text{Na}_2\text{CO}_3(s) + 2\text{HCl}(aq) \longrightarrow 2\text{NaCl}(aq) + \text{CO}_2(g) + \text{H}_2\text{O}(l)$

(d) $\text{NaHCO}_3(aq) + \text{HCl}(aq) \longrightarrow \text{NaCl}(aq) + \text{CO}_2(g) + \text{H}_2\text{O}(l)$

(e) $2\text{NaHCO}_3(s) \longrightarrow \text{Na}_2\text{CO}_3(s) + \text{CO}_2(g) + \text{H}_2\text{O}(g)$

(f) $\text{Na}_2\text{CO}_3(s) \longrightarrow$ no reaction. Unlike $\text{CaCO}_3(s)$, $\text{Na}_2\text{CO}_3(s)$ is not decomposed by moderate heating.

20.30   The balanced equation is: $\text{Na}_2\text{CO}_3(s) + 2\text{HCl}(aq) \longrightarrow 2\text{NaCl}(aq) + \text{CO}_2(g) + \text{H}_2\text{O}(l)$

$$\text{mol CO}_2 \text{ produced } = 25.0 \text{ g Na}_2\text{CO}_3 \left(\frac{1 \text{ mol Na}_2\text{CO}_3}{106.0 \text{ g Na}_2\text{CO}_3} \right) \left(\frac{1 \text{ mol CO}_2}{1 \text{ mol Na}_2\text{CO}_3} \right) = 0.236 \text{ mol CO}_2$$

$$V_{\text{CO}_2} = \frac{nRT}{P} = \frac{(0.236 \text{ mol})(0.0821 \text{ L} \cdot \text{atm}/\text{K} \cdot \text{mol})(283 \text{ K})}{(746/760) \text{ atm}} = \mathbf{5.59 \text{ L}}$$

20.34   First magnesium is treated with concentrated nitric acid (redox reaction) to obtain magnesium nitrate.

$$3\text{Mg}(s) + 8\text{HNO}_3(aq) \longrightarrow 3\text{Mg(NO}_3)_2(aq) + 4\text{H}_2\text{O}(l) + 2\text{NO}(g)$$

The magnesium nitrate is recovered from solution by evaporation, dried, and heated in air to obtain magnesium oxide:

$$2\text{Mg(NO}_3)_2(s) \longrightarrow 2\text{MgO}(s) + 4\text{NO}_2(g) + \text{O}_2(g)$$

20.36   The electron configuration of magnesium is $[\text{Ne}]3s^2$. The $3s$ electrons are outside the neon core (shielded), so they have relatively low ionization energies. Removing the third electron means separating an electron from the neon (closed shell) core, which requires a great deal more energy.

20.38   Even though helium and the Group 2A metals have ns^2 outer electron configurations, helium has a closed shell noble gas configuration and the Group 2A metals do not. The electrons in He are much closer to and more strongly attracted by the nucleus. Hence, the electrons in He are not easily removed. Helium is inert.

20.40   **(a)**   quicklime: $\text{CaO}(s)$ **(b)**   slaked lime: $\text{Ca(OH)}_2(s)$

(c)   limewater: an aqueous suspension of Ca(OH)_2

20.44   The reduction reaction is: $Al^{3+}(aq) + 3e^- \rightarrow Al(s)$

First, we can calculate the amount of charge needed to deposit 664 g of Al.

$$664 \text{ g Al} \left(\frac{1 \text{ mol Al}}{26.98 \text{ g Al}} \right) \left(\frac{3 \text{ mol } e^-}{1 \text{ mol Al}} \right) \left(\frac{96,500 \text{ C}}{1 \text{ mol } e^-} \right) = 7.12 \times 10^6 \text{ C}$$

Since a current of one ampere represents a flow of one coulomb per second, we can find the time it takes to pass this amount of charge.

32.6 A = 32.6 C/s

$$7.12 \times 10^6 \text{ C} \times \frac{1 \text{ s}}{32.6 \text{ C}} \times \frac{1 \text{ hr}}{3600 \text{ s}} = \textbf{60.7 hr}$$

20.46   **(a)**   The relationship between cell voltage and free energy difference is:

$$\Delta G = -nFE$$

In the given reaction $n = 6$. We write:

$$E = \frac{-\Delta G}{nF} = \frac{-594 \times 10^3 \text{ J/mol}}{(6)(96500 \text{ J/V} \cdot \text{mol})} = \textbf{-1.03 V}$$

The balanced equation shows *two* moles of aluminum. Is this the voltage required to produce *one* mole of aluminum? If we divide everything in the equation by two, we obtain:

$$\frac{1}{2} Al_2O_3(s) + \frac{3}{2} C(s) \rightarrow Al(l) + \frac{3}{2} CO(g)$$

For the new equation $n = 3$ and ΔG is $\frac{1}{2} \times 594$ kJ/mol = 297 kJ/mol. We write:

$$E = \frac{-\Delta G}{nF} = \frac{-297 \times 10^3 \text{ J/mol}}{(3)(96500 \text{ J/V} \cdot \text{mol})} = \textbf{-1.03 V}$$

The minimum voltage required to produce one mole or one thousand moles of aluminum is the same; the amount of *current* will be different in each case. Why is the voltage *negative*?

(b)   First we convert 1.00 kg of Al to moles.

$$1.00 \text{ kg Al} \left(\frac{1 \text{ mol Al}}{0.02698 \text{ kg Al}} \right) = 37.1 \text{ mol Al}$$

The reaction in part (a) shows us that three moles of electrons are required to produce one mole of aluminum. The voltage is three times the minimum calculated above (namely, –3.09 V or –3.09 J/C). We can find the electrical energy by using the same equation with the either voltage.

$$\Delta G = -nFE = -(37.1 \text{ mol Al}) \left(\frac{3 \text{ mol } e^-}{1 \text{ mol Al}} \times \frac{96500 \text{ C}}{1 \text{ mol } e^-} \right) \left(\frac{-3.09 \text{ J}}{1 \text{ C}} \right) = \textbf{3.32} \times \textbf{10}^7 \textbf{ J} = \textbf{3.32} \times \textbf{10}^4 \textbf{ kJ}$$

This equation can be used because electrical work can be calculated by multiplying the voltage by the amount of charge transported through the circuit (joules = volts × coulombs). The nF term in Equation (19.3) of the text used above represents the amount of charge.

What is the significance of the positive sign of the free energy change? Would the manufacturing of aluminum be a different process if the free energy difference were negative?

20.48 $4Al(NO_3)_3(s) \longrightarrow 2Al_2O_3(s) + 12NO_2(g) + 3O_2(g)$

20.50   The "bridge" bonds in Al_2Cl_6 break at high temperature: $Al_2Cl_6(g) \rightleftharpoons 2AlCl_3(g)$.

This increases the number of molecules in the gas phase and causes the pressure to be higher than expected for pure Al_2Cl_6.

If you know the equilibrium constants for the above reaction at higher temperatures, could you calculate the expected pressure of the $AlCl_3$–Al_2Cl_6 mixture?

20.52   VSEPR analysis shows $AlCl_3$ to be an AB_3–type molecule (no lone pairs on the central atom). The geometry should be trigonal planar, and the aluminum atom should therefore be sp^2 **hybridized**. In Al_2Cl_6, each aluminum atom is surrounded by 4 bonding pairs of electrons (AB_4–type molecule), and therefore each aluminum atom is sp^3 **hybridized**.

20.54   The formulas of the metal oxide and sulfide are MO and MS (why?). The balanced equation must therefore be:

$$2MS(s) + 3O_2(g) \rightarrow 2MO(s) + 2SO_2(g)$$

The number of moles of MO and MS are equal. We let x be the molar mass of metal. The number of moles of metal oxide is:

$$0.972 \text{ g} \left(\frac{1 \text{ mol}}{(x + 16.00) \text{ g}} \right)$$

The number of moles of metal sulfide is:

$$1.164 \text{ g} \left(\frac{1 \text{ mol}}{(x + 32.07) \text{ g}} \right)$$

The moles of metal oxide equal the moles of metal sulfide.

$$\frac{0.972}{(x + 16.00)} = \frac{1.164}{(x + 32.07)}$$

We solve for x.

$$0.972(x + 32.07) = 1.164(x + 16.00)$$

$$x = \textbf{65.4 g/mol}$$

20.56   Copper(II) ion is more easily reduced than either water or hydrogen ion (How can you tell? See Section 19.3 of the text.) Copper metal is more easily oxidized than water. Water should not be affected by the copper purification process.

20.58   Using Equation 18.8 from the text:

(a) $\Delta G_{rxn}^{\circ} = 4\,\Delta G_f^{\circ}\,(\text{Fe}) + 3\,\Delta G_f^{\circ}\,(\text{O}_2) - 2\,\Delta G_f^{\circ}\,(\text{Fe}_2\text{O}_3)$

$\Delta G_{rxn}^{0} = (4\text{ mol})(0) + (3\text{ mol})(0) - (2\text{mol})(-741.0\text{ kJ/mol}) = \textbf{1482 kJ}$

(b) $\Delta G_{rxn}^{\circ} = 4\,\Delta G_f^{\circ}\,(\text{Al}) + 3\,\Delta G_f^{\circ}\,(\text{O}_2(g)) - 2\,\Delta G_f^{\circ}\,(\text{Al}_2\text{O}_3)$

$\Delta G_{rxn}^{0} = (4\text{ mol})(0) + (3\text{ mol})(0) - (2\text{ mol})(-1576.4\text{ kJ/mol}) = \textbf{3152.8 kJ}$

20.60   At high temperature, magnesium metal reacts with nitrogen gas to form magnesium nitride.

$$3\text{Mg}(s) + \text{N}_2(g) \longrightarrow \text{Mg}_3\text{N}_2(s)$$

Can you think of any gas other than a noble gas that could provide an inert atmosphere for processes involving magnesium at high temperature?

20.62   (a)   In water the aluminum(III) ion causes an increase in the concentration of hydrogen ion (lower pH). This results from the effect of the small diameter and high charge (3+) of the aluminum ion on surrounding water molecules. The aluminum ion draws electrons in the O–H bonds to itself, thus allowing easy formation of H^{+} ions.

(b) $Al(OH)_3$ is an amphoteric hydroxide. It will dissolve in strong base with the formation of a complex ion.

$$Al(OH)_3(s) + OH^-(aq) \longrightarrow Al(OH)_4^-(aq)$$

The concentration of OH^- in aqueous ammonia is too low for this reaction to occur.

20.64   Calcium oxide is a base. The reaction is a neutralization.

$$\text{CaO}(s) + 2\text{HCl}(aq) \longrightarrow \text{CaCl}_2(aq) + \text{H}_2\text{O}(l)$$

20.66   Metals have closely spaced energy levels and (referring to Figure 20.10) a very small energy gap between filled and empty levels. Consequently, many electronic transitions can take place with absorption and subsequent emission continually occurring. Some of these transitions fall in the visible region of the spectrum and give rise to the flickering appearance.

20.68   NaF is used in toothpaste to fight tooth decay.

Li_2CO_3 is used to treat mental illness.

$Mg(OH)_2$ is an antacid.

$CaCO_3$ is an antacid.

$BaSO_4$ is used to enhance X ray images of the digestive system.

$Al(OH)_2NaCO_3$ is an antacid.

20.70   Both Li and Mg form oxides (Li_2O and MgO). Other Group 1A metals (Na, K, etc.) also form peroxides and superoxides. In Group 1A, only Li forms nitride (Li_3N), like Mg (Mg_3N_2).

Li resembles Mg in that its carbonate, fluoride, and phosphate have low solubilities.

CHAPTER 21
NONMETALLIC ELEMENTS AND THEIR COMPOUNDS

This chapter is of a very descriptive nature. As such, there are essentially no defined problem types. Please read Chapter 21 of the text carefully before answering the end-of-chapter problems.

SOLUTIONS TO SELECTED TEXT PROBLEMS

21.12 **(a)** Hydrogen reacts with alkali metals to form ionic hydrides:

$$2Na(l) + H_2(g) \rightarrow 2NaH(s)$$

The oxidation number of hydrogen drops from 0 to -1 in this reaction.

(b) Hydrogen reacts with oxygen (combustion) to form water:

$$2H_2(g) + O_2(g) \rightarrow 2H_2O(l)$$

The oxidation number of hydrogen increases from 0 to $+1$ in this reaction.

21.14 Hydrogen forms an interstitial hydride with palladium, which behaves almost like a solution of hydrogen atoms in the metal. At elevated temperatures hydrogen atoms can pass through solid palladium; other substances cannot.

21.16 The number of moles of deuterium gas is:

$$n = \frac{PV}{RT} = \frac{0.90 \text{ atm} \times 2.0 \text{ L}}{0.0821 \text{ L} \cdot \text{atm} / \text{K} \cdot \text{mol} \times 298 \text{K}} = 0.074 \text{ mol}$$

If the abundance of deuterium is 0.015 percent, the number of moles of water must be:

$$0.074 \text{ mol D}_2 \times \frac{100\% \text{ H}_2\text{O}}{0.015\% \text{ D}_2} = 4.9 \times 10^2 \text{ mol H}_2\text{O}$$

At a recovery of 80 percent the amount of water needed is:

$$4.9 \times 10^2 \text{ mol H}_2\text{O} \times \frac{100\%}{80\%} \times \frac{0.018 \text{ kg H}_2\text{O}}{1.0 \text{ mol H}_2\text{O}} = \mathbf{11 \text{ kg H}_2\text{O}}$$

21.18 **(a)** $H_2 + Cl_2 \rightarrow 2HCl$

(b) $3H_2 + N_2 \rightarrow 2NH_3$

(c) $2Li + H_2 \rightarrow 2LiH$

$LiH + H_2O \rightarrow LiOH + H_2$

21.26   The Lewis structure is:

$$\left[\, :C\!\equiv\!C: \,\right]^{2-}$$

21.28   **(a)**   The reaction is: $2NaHCO_3(s) \rightarrow Na_2CO_3(s) + H_2O(g) + CO_2(g)$

Is this an endo- or an exothermic process?

(b)   The hint is generous. The reaction is:

$$Ca(OH)_2(aq) + CO_2(g) \rightarrow CaCO_3(s) + H_2O(l)$$

The visual proof is the formation of a white precipitate of $CaCO_3$. Why would a water solution of NaOH be unsuitable to qualitatively test for carbon dioxide?

21.30   Heat causes bicarbonates to decompose according to the reaction:

$$2HCO_3^- \rightarrow CO_3^{2-} + H_2O + CO_2$$

Generation of carbonate ion causes precipitation of the insoluble $MgCO_3$.

Do you think there is much chance of finding natural mineral deposits of calcium or magnesium bicarbonates?

21.32   The wet sodium hydroxide is first converted to sodium carbonate:

$$2NaOH(aq) + CO_2(g) \rightarrow Na_2CO_3(aq) + H_2O(l)$$

and then to sodium hydrogen carbonate: $Na_2CO_3(aq) + H_2O(l) + CO_2(g) \rightarrow 2NaHCO_3(aq)$

Eventually, the sodium hydrogen carbonate precipitates (the water solvent evaporates since $NaHCO_3$ is not hygroscopic). Thus, most of the white solid is $NaHCO_3$ plus some Na_2CO_3.

21.34   Carbon monoxide and molecular nitrogen are isoelectronic. Both have 14 electrons. What other diatomic molecules discussed in these problems are isoelectronic with CO?

21.40   **(a)** $2NaNO_3(s) \rightarrow 2NaNO_2(s) + O_2(g)$

(b) $NaNO_3(s) + C(s) \rightarrow NaNO_2(s) + CO(g)$

21.42   The balanced equation is: $2NH_3(g) + CO_2(g) \rightarrow (NH_2)_2CO(s) + H_2O(l)$

If pressure increases, the position of equilibrium will shift in the direction with the smallest number of molecules in the gas phase, that is, to the right. Therefore, the reaction is best run at high pressure.

Write the expression for Q_p for this reaction. Does increasing pressure cause Q_p to increase or decrease? Is this consistent with the above prediction?

21.44   The density of a gas depends on temperature, pressure, and the molar mass of the substance. When two gases are at the same pressure and temperature, the ratio of their densities should be the same as the ratio of their molar masses. The molar mass of ammonium chloride is 53.5 g/mol, and the ratio of this to the molar mass of molecular hydrogen (2.02 g/mol) is 26.8. The experimental value of 14.5 is roughly half this amount. Such results usually indicate breakup or dissociation into smaller molecules in the gas phase (note the temperature). The measured molar mass is the average of all the molecules in equilibrium.

$$NH_4Cl(g) \rightleftharpoons NH_3(g) + HCl(g)$$

Knowing that ammonium chloride is a stable substance at 298 K, is the above reaction exo- or endothermic?

21.46   The highest oxidation state possible for a Group 5A element is +5. This is the oxidation state of nitrogen in nitric acid (HNO_3).

21.48   Nitric acid is a strong oxidizing agent in addition to being a strong acid (see Table 19.1, $E_{red}^{\circ} = +0.96V$). The primary action of a good reducing agent like zinc is reduction of nitrate ion to ammonium ion.

$$4Zn(s) + NO_3^-(aq) + 10H^+(aq) \rightarrow 4Zn^{2+}(aq) + NH_4^+(aq) + 3H_2O(l)$$

21.50   One of the best Lewis structures for nitrous oxide is:

$$\overset{(-)}{\underset{\cdot\cdot}{\overset{\cdot\cdot}{N}}}=\overset{(+)}{N}=\overset{\cdot\cdot}{\underset{\cdot\cdot}{O}}$$

There are no lone pairs on the central nitrogen, making this an AB_2 VSEPR case. All such molecules are linear. Other resonance forms are:

$$:N\equiv\overset{(+)}{N}-\overset{\cdot\cdot}{\underset{\cdot\cdot}{\overset{(-)}{O}}}: \qquad \overset{(2-)}{\underset{\cdot\cdot}{:N}}-\overset{(+)}{N}\equiv\overset{(+)}{O}:$$

Are all the resonance forms consistent with a linear geometry?

21.52 $\Delta H^{\circ} = 4\,\Delta H_f^{\circ}\,[NO(g)] + 6\,\Delta H_f^{\circ}\,[H_2O(l)] - \{4\,\Delta H_f^{\circ}\,[NH_3(g)] + 5\,\Delta H_f^{\circ}\,[O_2(g)]\}$

$\Delta H^{\circ} = (4\text{ mol})(90.4\text{ kJ/mol}) + (6\text{ mol})(-285.8\text{ kJ/mol}) - (4\text{ mol})(-46.3\text{ kJ/mol}) = \textbf{-1168 kJ}$

21.54 $\Delta T_b = K_b m = 0.409°C$

$$\text{molality} = \frac{0.409°C}{2.34°C/m} = 0.175\ m$$

The number of grams of white phosphorus in 1 kg of solvent is:

$$\left(\frac{1.645\text{ g phosphorus}}{755\text{ g CS}_2}\right)\left(\frac{1000\text{ g}}{1\text{ kg}}\right) = 22.8\text{ g phosphorus/kg CS}_2$$

The molar mass of white phosphorus is:

$$\frac{22.8\text{ g phosphorus / kg CS}_2}{0.175\text{ mol phosphorus / kg CS}_2} = \textbf{125 g/mol}$$

Let the molecular formula of white phosphorus be P_n so that:

$$n \times 30.97\text{ g/mol} = 125\text{ g/mol}$$

$$n = 4$$

The molecular formula of white phosphorus is $\textbf{P}_4$.

21.56   The balanced equation is:

$$P_4O_{10}(s) + 4HNO_3(aq) \rightarrow 2N_2O_5(g) + 4HPO_3(l)$$

The theoretical yield of N_2O_5 is :

$$79.4 \text{ g } P_4O_{10} \left(\frac{1 \text{ mol } P_4O_{10}}{283.9 \text{ g } P_4O_{10}} \right) \left(\frac{2 \text{ mol } N_2O_5}{1 \text{ mol } P_4O_{10}} \right) \left(\frac{108.0 \text{ g } N_2O_5}{1 \text{ mol } N_2O_5} \right) = \textbf{60.4 g } N_2O_5$$

21.58 PH_4^+ is similar to NH_4^+. The hybridization of phosphorus in PH_4^+ is sp^3.

21.66 $\Delta G° = \Delta G_f°(NO_2) + \Delta G_f°(O_2) - \Delta G_f°(NO) - \Delta G_f°(O_3)$

$\Delta G° = (1 \text{ mol})(51.8 \text{ kJ/mol}) - (1 \text{ mol})(86.7 \text{ kJ/mol}) - (1 \text{ mol})(163.4 \text{ kJ/mol}) = \textbf{−198.3 kJ}$

$$\ln K_p = \frac{-\Delta G°}{RT} = \frac{198.3 \times 10^3 \text{ J/mol}}{(8.314 \text{ J/K} \cdot \text{mol})(298 \text{ K})}$$

$$K_p = \textbf{6} \times \textbf{10}^{\textbf{34}}$$

Since there is no change in the number of moles of gases, K_c is *equal* to K_p.

21.68   Following the rules given in Section 4.4 of the text, we assign hydrogen an oxidation number of +1 and **fluorine** an oxidation number of **−1**. Since HFO is a neutral molecule, the oxidation number of **oxygen** is **zero**. Can you think of other compounds in which oxygen has this oxidation number?

21.70   First, let's calculate the moles of sulfur in 48 million tons of sulfuric acid.

$$48 \times 10^6 \text{ ton } H_2SO_4 \times \frac{2000 \text{ lb}}{1 \text{ ton}} \times \frac{453.6 \text{ g}}{1 \text{ lb}} \times \frac{1 \text{ mol } H_2SO_4}{98.09 \text{ g } H_2SO_4} \times \frac{1 \text{ mol S}}{1 \text{ mol } H_2SO_4} = \textbf{4.4} \times \textbf{10}^{\textbf{11}} \textbf{ mol S}$$

Converting to grams of sulfur:

$$4.4 \times 10^{11} \text{ mol S} \times \frac{32.07 \text{ g S}}{1 \text{ mol S}} = \textbf{1.4} \times \textbf{10}^{\textbf{13}} \textbf{ g S}$$

21.72   There are actually several steps involved in removing sulfur dioxide from industrial emissions with calcium carbonate. First calcium carbonate is heated to form carbon dioxide and calcium oxide.

$$CaCO_3(s) \rightleftharpoons CaO(s) + CO_2(g)$$

The CaO combines with sulfur dioxide to form calcium sulfite.

$$CaO(s) + SO_2(g) \rightarrow CaSO_3(s)$$

Alternatively, calcium sulfate forms if enough oxygen is present.

$$2CaSO_3(s) + O_2(g) \rightarrow 2CaSO_4(s)$$

The amount of calcium carbonate (limestone) needed in this problem is:

$$50.6 \text{ g SO}_2 \left(\frac{1 \text{ mol SO}_2}{64.07 \text{ g SO}_2} \right) \left(\frac{1 \text{ mol CaCO}_3}{1 \text{ mol SO}_2} \right) \left(\frac{100.1 \text{ g CaCO}_3}{1 \text{ mol CaCO}_3} \right) = \textbf{79.1 g CaCO}_3$$

The calcium oxide–sulfur dioxide reaction is an example of a Lewis acid-base reaction (see Section 15.12 of the text) between oxide ion and sulfur dioxide. Can you draw Lewis structures showing this process? Which substance is the Lewis acid and which is the Lewis base?

21.74   The usual explanation for the fact that no chemist has yet succeeded in making SCl_6, SBr_6 or SI_6 is based on the idea of excessive crowding of the six chlorine, bromine, or iodine atoms around the sulfur. Others suggest that sulfur in the +6 oxidation state would oxidize chlorine, bromine, or iodine in the −1 oxidation state to the free elements. In any case, none of these substances has been made as of the date of this writing.

It is of interest to point out that thirty years ago all textbooks confidently stated that compounds like ClF_5 could not be prepared.

Note that PCl_6^- is a known species. How different are the sizes of S and P?

21.76   First we convert gallons of water to grams of water.

$$2.0 \times 10^2 \text{ gal} \left(\frac{3.785 \text{ L}}{1 \text{ gal}} \right) \left(\frac{1000 \text{ mL}}{1 \text{ L}} \right) \left(\frac{1.00 \text{ g H}_2\text{O}}{1 \text{ mL}} \right) = 7.6 \times 10^5 \text{ g H}_2\text{O}$$

An H_2S concentration of 22 ppm indicates that in 1 million grams of water, there will be 22 g of H_2S. First, let's calculate the number of moles of H_2S in 7.6×10^5 g of H_2O:

$$7.6 \times 10^5 \text{ g H}_2\text{O} \left(\frac{22 \text{ g H}_2\text{S}}{1.0 \times 10^6 \text{ g H}_2\text{O}} \right) \left(\frac{1 \text{ mol H}_2\text{S}}{34.09 \text{ g H}_2\text{S}} \right) = 0.49 \text{ mol H}_2\text{S}$$

The mass of chlorine required to react with 0.49 mol of H_2S is:

$$0.49 \text{ mol H}_2\text{S} \left(\frac{1 \text{ mol Cl}_2}{1 \text{ mol H}_2\text{S}} \right) \left(\frac{70.90 \text{ g Cl}_2}{1 \text{ mol Cl}_2} \right) = \textbf{35 g Cl}_2$$

21.78   A check of Table 19.1 shows that sodium ion cannot be reduced by any of the substances mentioned in this problem; it is a "spectator ion". We focus on the substances that are actually undergoing oxidation or reduction and write half-reactions for each.

$$2I^-(aq) \rightarrow I_2(s)$$

$$H_2SO_4(aq) \rightarrow H_2S(g)$$

Balancing the oxygen, hydrogen, and charge gives:

$$2I^-(aq) \rightarrow I_2(s) + 2e^-$$

$$H_2SO_4(aq) + 8H^+(aq) + 8e^- \rightarrow H_2S(g) + 4H_2O(l)$$

Multiplying the iodine half-reaction by four and combining gives the balanced redox equation.

$$H_2SO_4(aq) + 8I^-(aq) + 8H^+(aq) \rightarrow H_2S(g) + 4I_2(s) + 4H_2O(l)$$

The hydrogen ions come from extra sulfuric acid. We add one sodium ion for each iodide ion to obtain the final equation.

$$\mathbf{9H_2SO_4(aq) + 8NaI(aq) \rightarrow H_2S(g) + 4I_2(s) + 4H_2O(l) + 8NaHSO_4(aq)}$$

21.82 Sulfuric acid is added to solid sodium chloride, not aqueous sodium chloride. Hydrogen chloride is a gas at room temperature and can escape from the reacting mixture.

$$\mathbf{H_2SO_4(l) + NaCl(s) \rightarrow HCl(g) + NaHSO_4(s)}$$

The reaction is driven to the right by the continuous loss of $HCl(g)$ (Le Chatelier's principle).

What happens when sulfuric acid is added to a water solution of NaCl? Could you tell the difference between this solution and the one formed by adding hydrochloric acid to aqueous sodium sulfate?

21.84 The reaction is: $2Br^-(aq) + Cl_2(g) \rightarrow 2Cl^-(aq) + Br_2(l)$

The number of moles of chlorine needed is:

$$167 \, \text{g Br}^- \left(\frac{1 \, \text{mol Br}}{79.90 \, \text{g Br}} \right) \left(\frac{1 \, \text{mol Cl}_2}{2 \, \text{mol Br}^-} \right) = 1.05 \, \text{mol Cl}_2(g)$$

Use the ideal gas equation to calculate the volume of Cl_2 needed.

$$V_{Cl_2} = \frac{nRT}{P} = \frac{(1.05 \, \text{mol})(0.0821 \, \text{L} \cdot \text{atm}/\text{K} \cdot \text{mol})(293 \, \text{K})}{(1 \, \text{atm})} = \mathbf{25.3 \, L}$$

21.86 As with iodide salts, a redox reaction occurs between sulfuric acid and sodium bromide.

$$2H_2SO_4(aq) + 2NaBr \, (aq) \rightarrow SO_2(g) + Br_2(l) + 2H_2O(l) + Na_2SO_4(aq)$$

21.88 The balanced equation is:

$$Cl_2(g) + 2Br^-(aq) \rightarrow 2Cl^-(aq) + Br_2(g)$$

The number of moles of bromine is the same as the number of moles of chlorine, so this problem is essentially a gas law exercise in which P and T are changed for some given amount of gas.

$$V_2 = \frac{P_1V_1}{T_1} \times \frac{T_2}{P_2} = \frac{760 \, \text{mmHg} \times 2.00 \, \text{L}}{288 \, \text{K}} \times \frac{373 \, \text{K}}{700 \, \text{mmHg}} = \mathbf{2.81 \, L}$$

21.90 The balanced equation is:

$$\mathbf{I_2O_5(s) + 5CO(g) \rightarrow I_2(s) + 5CO_2(g)}$$

The oxidation number of iodine changes from +5 to 0 and the oxidation number of carbon changes from +2 to +4. **Iodine** is **reduced; carbon** is **oxidized**.

21.92 **(a)** $SiCl_4$ **(b)** F^- **(c)** F   **(d)** CO_2

21.94 There is no change in oxidation number; it is zero for both compounds.

21.96 **(a)** $2Na + 2D_2O \rightarrow 2NaOD + D_2$ **(d)** $CaC_2 + 2D_2O \rightarrow C_2D_2 + Ca(OD)_2$

      (b) $2D_2O \xrightarrow{\text{electrolysis}} 2D_2 + O_2$ **(e)** $Be_2C + 4D_2O \rightarrow 2Be(OD)_2 + CD_4$

            $D_2 + Cl_2 \rightarrow 2DCl$

      (c) $Mg_3N_2 + 6D_2O \rightarrow 3Mg(OD)_2 + 2ND_3$ **(f)** $SO_3 + D_2O \rightarrow D_2SO_4$

21.98 **(a)** At elevated pressures, water boils above 100°C.

      (b) Water is sent down the outermost pipe so that it is able to melt a larger area of sulfur.

      (c) Sulfur deposits are structurally weak. There will be a danger of the sulfur mine collapsing.

21.100 The oxidation is probably initiated by breaking a C–H bond (the rate-determining step). The C–D bond breaks at a slower rate than the C–H bond; therefore, replacing H by D decreases the rate of oxidation.

CHAPTER 22
TRANSITION METAL CHEMISTRY AND COORDINATION COMPOUNDS

PROBLEM-SOLVING STRATEGIES AND TUTORIAL SOLUTIONS

TYPES OF PROBLEMS

Problem Type 1: Assigning Oxidation Numbers to the Metal Atom in Coordination Compounds.

Problem Type 2: Naming Coordination Compounds.

Problem Type 3: Writing Formulas for Coordination Compounds.

Problem Type 4: Predicting the Number of Unpaired Spins in a Coordination Compound.

PROBLEM TYPE 1: ASSIGNING OXIDATION NUMBERS TO THE METAL ATOM IN COORDINATION COMPOUNDS

Transition metals exhibit variable oxidation states in their compounds. The charge on the central metal atom and its surrounding ligands sum to zero in a neutral coordination compound. In a complex ion, the charges on the central metal atom and the surrounding ligands sum to the net charge of the ion.

EXAMPLE 22.1
Specify the oxidation number of the central metal atom in each of the following compounds:
(a) $[Co(NH_3)_6]Cl_3$ **and (b)** $Co(CN)_6^{3-}$

(a) NH_3 is a neutral species. Since each chloride ion carries a –1 charge, and there are three Cl^- ions, the oxidation number of Co must be +3.

(b) Each cyanide ion has a charge of –1. The sum of the oxidation number of Co and the –6 charge for the six cyanide ions is –3. Therefore, the oxidation number of Co must be +3.

PRACTICE EXERCISE
1. Specify the oxidation number of the central metal atom in each of the following compounds:
   (a) $[Pt(NH_3)_3Cl_3]NO_3$ and (b) $Ni(CO)_4$.

Text Problems: 22.12, **22.14**

PROBLEM TYPE 2: NAMING COORDINATION COMPOUNDS

The complete set of rules for naming coordination compounds is given in Section 22.3 of your text. Presented below is an abridged version of those rules.

1. The cation is named before the anion, as is the case for other ionic compounds.

2. Within a complex ion the ligands are named first, in alphabetical order, and the metal ion is named last.

3.   The names of anionic ligands end with the letter o, whereas a neutral ligand is usually called by the name of the molecule. Exceptions are listed in Table 22.4 of the text.

4.   Greek prefixes are used to indicate the number of ligands of a particular kind present. If the ligand itself contains a Greek prefix, the prefixes *bis* (2), *tris* (3), and *tetrakis* (4) are used to indicate the number of ligands present.

5.   The oxidation number of the metal is written in Roman numerals following the name of the metal.

6.   If the complex is an anion, its name ends in "–ate".

EXAMPLE 22.2

Name the following coordination compounds and complex ions: (a) $[Ni(NH_3)_6]^{2+}$, (b) $K_2[Cu(CN)_4]$,
(c) $[Pt(NH_3)_4Cl_2]Cl_2$.

(a) NH_3 is a neutral species; therefore, the Ni must have a +2 charge. Ammonia as a ligand in a coordination compound is called ammine. The complex ion is called **hexaamminenickel(II) ion**.

(b)   Potassium is an alkali metal; it always has a +1 charge in ionic compounds. Since the two potassium ions have a total charge of +2, the charge on the complex ion $[Cu(CN)_4]$ must be –2. Cyanide ion has a –1 charge; therefore, Cu must have a +2 charge. Potassium is the cation and is named first. The compound is called **potassium tetracyanocuprate(II)**. We use an "–ate" ending because the complex is an anion.

(c)   The complex ion has a +2 charge balanced by the total charge of –2 for the two chloride ions. Focusing on the complex ion, NH_3 is a neutral species and the two chloride ligands each have a –1 charge. Since the charge of the complex ion is +2, platinum must have a +4 charge. This compound is called **tetraamminedichloroplatinum(IV) chloride**.

PRACTICE EXERCISE

2.   Name the following coordination compounds and complex ions:

   (a) $[Cr(OH)_4]^-$ (b) $[Pt(NH_3)_3Cl_3]NO_3$.

Text Problem: 22.16

PROBLEM TYPE 3: WRITING FORMULAS FOR COORDINATION COMPOUNDS

Follow the nomenclature rules given in Problem Type 2 above. Remember that the cation is named before the anion.

EXAMPLE 22.3

Write the formulas for the following compounds or complex ions:
(a) tetrahydroxoaluminate(III) ion, (b) potassium hexachloropalladate(IV), (c) diaquodicyanocopper(II)

(a)   Tetrahydroxo refers to four hydroxide ligands. The "–ate" of aluminate indicates that the complex is an anion. The Roman numeral (III) indicates a +3 charge on aluminum. Since each hydroxide ligand has a –1 charge, the charge on the ion is –1. The formula for the complex ion is $[Al(OH_4)]^-$.

(b)   The complex anion, hexachloropalladate(IV), has a –2 charge. Each of the six chloride ligands has a –1 charge (total of –6) and palladium has a +4 charge. To balance the –2 charge of the anion, there must be two potassium +1 ions. The formula for the compound is $K_2[PdCl_6]$.

(c)   Diaquo refers to two water ligands, and dicyano refers to two cyanide ligands. Each cyanide ligand has a –1 charge (total of –2), which balances the +2 charge on copper. The compound is electrically neutral. The formula for the compound is $[Cu(H_2O)_2(CN)_2]$.

PRACTICE EXERCISE

3.   Write the formulas for the following compounds:

   (a) tris(ethylenediamine)nickel(II) sulfate

   (b) tetraamminediaquocobalt(III) chloride

   (c) potassium hexacyanoferrate(II)

Text Problem: 22.18

PROBLEM TYPE 4: PREDICTING THE NUMBER OF UNPAIRED SPINS IN A COORDINATION COMPOUND

First, we must write the electron configuration for the transition metal of the coordination compound. Remember that the ns shell fills before the $(n-1)d$ shell. When the number of d electrons in the transition metal is known, we must decide how to place them in the d orbitals.

Crystal Field Theory tells us that in an octahedral complex, the five d orbitals are *not* equivalent in energy. The d_{xy}, d_{xz}, and d_{yz} are degenerate and at a lower energy than the degenerate set, d_{z^2} and $d_{x^2-y^2}$. The energy difference between these two sets of d orbitals is called the **crystal field splitting** (Δ).

Due to the crystal field splitting, for metals with electron configurations of d^4, d^5, d^6, or d^7, there are two ways to place the electrons in the five d orbitals. Let's consider a d^4 case. According to Hund's rule (see Section 7.8 of the text), maximum stability is reached if the electrons are placed in four separate orbitals with parallel spin. But, this arrangement can be achieved only at a cost; one of the four electrons must be energetically promoted to the higher energy d_{z^2} or $d_{x^2-y^2}$ orbital. This energy investment is not needed if all four electrons enter the d_{xy}, d_{xz}, and d_{yz} orbitals. However, in this electron arrangement, we must pair up two of the electrons. This pairing also takes energy. The two possible electron configurations are shown below.

high spin complex

$$\overline{\quad} \qquad \overline{\quad}$$
$$d_{z^2} \qquad d_{x^2-y^2}$$

low spin complex

$$\underline{\uparrow\downarrow} \qquad \underline{\uparrow} \qquad \underline{\uparrow}$$
$$d_{xy} \qquad d_{xz} \qquad d_{yz}$$

Whether a complex is high spin or low spin depends on the ligands that are bonded to the metal center. If a ligand is a weak-field ligand, the crystal field splitting (Δ) is small. Therefore, only a small energy expenditure is needed to keep all spins parallel, resulting in *high spin complexes*. On the other hand, the crystal field splitting is larger when a strong-field ligand is bound to the metal center. In these cases, it is energetically more favorable to pair up the electrons, resulting in *low spin complexes*.

In summary, the actual arrangement of the *d*-electrons is determined by the amount of stability gained by having maximum parallel spins versus the investment in energy required to promote electrons to higher *d* orbitals.

EXAMPLE 22.4

Predict the number of unpaired spins in the $[Co(CN)_6]^{3-}$ ion.

The electron configuration of Co^{3+} is $[Ar]3d^6$. Since CN^- is a strong-field ligand, we expect $[Co(CN)_6]^{3-}$ to be a low spin complex. All six electrons will be placed in the lower energy *d* orbitals (d_{xy}, d_{xz}, and d_{yz}), and there will be no unpaired spins.

$$\overline{\quad} \qquad \overline{\quad}$$
$$d_{z^2} \qquad d_{x^2-y^2}$$

$$\underline{\uparrow\downarrow} \qquad \underline{\uparrow\downarrow} \qquad \underline{\uparrow\downarrow}$$
$$d_{xy} \qquad d_{xz} \qquad d_{yz}$$

PRACTICE EXERCISE

4. Predict the number of unpaired spins in the $[Fe(H_2O)_6]^{3+}$ ion. Water is a weak-field ligand.

Text Problem: 22.60

ANSWERS TO PRACTICE EXERCISES

1. (a) Pt, +4   (b) Ni, 0

2. (a) tetrahydroxochromate(III) ion   (b) triamminetrichloroplatinum(IV) nitrate

3. (a) $[Ni(NH_2CH_2CH_2NH_2)_3]SO_4$ (b) $[Co(NH_3)_4(H_2O)_2]Cl_3$ (c) $K_4[Fe(CN)_6]$

4. five unpaired spins

SOLUTIONS TO SELECTED TEXT PROBLEMS

22.12 **(a)** The oxidation number of Cr is **+3**.

(b) The coordination number of Cr is **6**.

(c) **Oxalate ion** ($C_2O_4{}^{2-}$) is a bidentate ligand.

22.14 Assigning Oxidation Numbers to the Metal Atom in Coordination Compounds, Problem Type 1.

(a) Since **sodium** is always +1 and the oxygens are −2, **Mo** must have an oxidation number of **+6**.

(b) **Magnesium** is +2 and oxygen −2; therefore **W** is **+6**.

(c) CO ligands are neutral species, so the iron atom bears no net charge. The oxidation number of **Fe** is **0**.

22.16 Naming Coordination Compounds, Problem Type 2.

(a) Ethylenediamine is a neutral ligand, and each chloride has a −1 charge. Therefore, cobalt has a oxidation number of +3. The correct name for the ion is *cis*−**dichlorobis(ethylenediammine)cobalt(III)**. The prefix *bis* means two; we use this instead of *di* because *di* already appears in the name ethylenediamine.

(b) There are four chlorides each with a −1 charge; therefore, Pt has a +4 charge. The correct name for the compound is **pentaamminechloroplatinum(IV) chloride**.

(c) There are three chlorides each with a −1 charge; therefore, Co has a +3 charge. The correct name for the compound is **pentaamminechlorocobalt(III) chloride**.

22.18 Writing Formulas for Coordination Compounds, Problem Type 3.

(a) There are two ethylenediamine ligands and two chloride ligands. The correct formula is **[Cr(en)$_2$Cl$_2$]$^+$**.

(b) There are five carbonyl (CO) ligands. The correct formula is **Fe(CO)$_5$**.

(c) There are four cyanide ligands each with a −1 charge. Therefore, the complex ion has a −2 charge, and two K$^+$ ions are needed to balance the −2 charge of the anion. The correct formula is **K$_2$[Cu(CN)$_4$]**.

(d) There are four NH$_3$ ligands and two H$_2$O ligands. Two chloride ions are needed to balance the +2 charge of the complex ion. The correct formula is **[Co(NH$_3$)$_4$(H$_2$O)Cl]Cl$_2$**.

22.24 **(a)** In general for any MA$_2$B$_4$ octahedral molecule, only **two** geometric isomers are possible. The only real distinction is whether the two A−ligands are cis or trans. In Figure 22.11, (a) and (c) are the same compound (Cl atoms cis in both), and (b) and (d) are identical (Cl atoms trans in both).

(b) A model or a careful drawing is very helpful to understand the MA$_3$B$_3$ octahedral structure. There are only **two** possible geometric isomers. The first has all A's (and all B's) cis; this is called the facial isomer. The second has two A's (and two B's) at opposite ends of the molecule (trans). Try to make or draw other possibilities. What happens?

22.26 **(a)**   There are cis and trans geometric isomers (See Problem 22.24). No optical isomers.

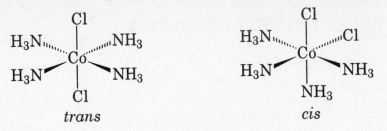

trans   *cis*

  (b)   There are two optical isomers. See Figure 22.7. The three bidentate en ligands are represented by the curved lines.

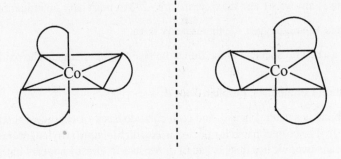

mirror

22.34 When a substance appears to be yellow, it is absorbing light from the blue–violet, high energy end of the visible spectrum. Often this absorption is just the tail of a strong absorption in the ultraviolet. Substances that appear green or blue to the eye are absorbing light from the lower energy red or orange part of the spectrum.

Cyanide ion is a very strong field ligand. It causes a larger crystal field splitting than water, resulting in the absorption of higher energy (shorter wavelength) radiation when a d electron is excited to a higher energy d orbital.

22.36 **(a)**   Wavelengths of 470 nm fall between blue and blue–green, corresponding to an observed color in the **orange** part of the spectrum.

  (b)   We convert wavelength to photon energy using the Planck relationship.

$$\Delta E = \frac{hc}{\lambda} = \frac{(6.63 \times 10^{-34}\ \text{J} \cdot \text{s})(3.00 \times 10^{8}\ \text{m/s})}{470 \times 10^{-9}\ \text{m}} = 4.23 \times 10^{-19}\ \text{J}$$

$$\left(\frac{4.23 \times 10^{-19}\ \text{J}}{1\ \text{photon}}\right)\left(\frac{6.022 \times 10^{23}\ \text{photons}}{1\ \text{mol}}\right)\left(\frac{1\ \text{kJ}}{1000\ \text{J}}\right) = \textbf{255 kJ / mol}$$

22.38 Calculating molar mass from freezing-point depression, Problem Type 6A, Chapter 12.

Step 1: The equation for freezing-point depression is

$$\Delta T_f = K_f m$$

Solve this equation algebraically for molality (m), then substitute ΔT_f and K_f into the equation to calculate the molality.

$$m = \frac{\Delta T_f}{K_f} = \frac{0.56°C}{1.86°C/m} = 0.30\ m$$

Step 2: Multiplying the molality by the mass of solvent (in kg) gives moles of unknown solute. Then, dividing the mass of solute (in g) by the moles of solute, gives the molar mass of the unknown solute.

$$?\ \text{mol of unknown solute} = \frac{0.30\ \text{mol solute}}{1\ \text{kg water}} \times 0.0250\ \text{kg water} = 0.0075\ \text{mol solute}$$

$$\text{molar mass of unknown} = \frac{0.875\ \text{g}}{0.0075\ \text{mol}} = 117\ \text{g/mol}$$

The molar mass of $Co(NH_3)_4Cl_3$ is 233.4 g/mol, which is twice the computed molar mass. This implies dissociation into two ions in solution. The formula must be:

$[Co(NH_3)_4Cl_2]Cl$

which contains the complex ion $[Co(NH_3)_4Cl_2]^+$ and a chloride ion, Cl^-. Refer to Problem 22.26 (a) for a diagram of the structure of the complex ion.

22.42 Use a radioactive label such as $^{14}CN^-$ (in NaCN). Add NaCN to a solution of $K_3Fe(CN)_6$. Isolate some of the $K_3Fe(CN)_6$ and check its radioactivity. If the complex shows radioactivity, then it must mean that the CN^- ion has participated in the exchange reaction.

22.44 The white precipitate is copper(II) cyanide.

$$Cu^{2+}(aq) + 2CN^-(aq) \rightarrow Cu(CN)_2(s)$$

This forms a soluble complex with excess cyanide.

$$Cu(CN)_2(s) + 2CN^-(aq) \rightarrow Cu(CN)_4^{2-}(aq)$$

Copper(II) sulfide is normally a very insoluble substance. In the presence of excess cyanide ion, the concentration of the copper(II) ion is so low that CuS precipitation cannot occur. In other words, the cyanide complex of copper has a very large formation constant.

22.46 The formation constant expression is:

$$K_f = \frac{[Fe(H_2O)_5NCS^{2+}]}{[Fe(H_2O)_6^{3+}][SCN^-]}$$

Notice that the original volumes of the Fe(III) and SCN^- solutions were both 1.0 mL and that the final volume is 10.0 mL. This represents a tenfold dilution, and the concentrations of Fe(III) and SCN^- become 0.020 M and 1.0×10^{-4} M, respectively. We make a table.

	$Fe(H_2O)_6^{3+}$ +	SCN^- $\rightleftharpoons$	$Fe(H_2O)_5NCS^{2+}$ +	H_2O
Initial (M):	0.020	1.0×10^{-4}	0	
Change (M):	-7.3×10^{-5}	-7.3×10^{-5}	$+7.3 \times 10^{-5}$	
Equilibrium (M):	0.020	2.7×10^{-5}	7.3×10^{-5}	

$$K_f = \frac{7.3 \times 10^{-5}}{(0.020)(2.7 \times 10^{-5})} = \mathbf{1.4 \times 10^2}$$

22.48 Mn^{3+} is $3d^4$ and Cr^{3+} is $3d^5$. Therefore, $\mathbf{Mn^{3+}}$ has a greater tendency to accept an electron and is a stronger oxidizing agent. The $3d^5$ electron configuration of Cr^{3+} is a stable configuration.

22.50   Ti is +3 and Fe is +3.

22.52   A 100.00 g sample of hemoglobin contains 0.34 g of iron. In moles this is:

$$0.34 \text{ g Fe} \left(\frac{1 \text{ mol}}{55.85 \text{ g}} \right) = 6.1 \times 10^{-3} \text{ mol Fe}$$

The amount of hemoglobin that contains one mole of iron must be:

$$\frac{100.00 \text{ g hemoglobin}}{6.1 \times 10^{-3} \text{ mol Fe}} = 1.6 \times 10^4 \text{ g hemoglobin/mol Fe}$$

We compare this to the actual molar mass of hemoglobin:

$$\frac{6.5 \times 10^4 \text{ g hemoglobin}}{1 \text{ mol hemoglobin}} \left(\frac{1 \text{ mol Fe}}{1.6 \times 10^4 \text{ g hemoglobin}} \right) = 4 \text{ mol Fe} / 1 \text{ mol hemoglobin}$$

The discrepancy between our minimum value and the actual value can be explained by realizing that one hemoglobin molecule contains **four** iron atoms.

22.54   **(a)** $[Cr(H_2O)_6]Cl_3$,   **(b)** $[Cr(H_2O)_5Cl]Cl_2 \cdot H_2O$,   **(c)** $[Cr(H_2O)_4Cl_2]Cl \cdot 2H_2O$

The compounds can be identified by a conductance experiment. Compare the conductances of equal molar solutions of the three compounds with equal molar solutions of NaCl, $MgCl_2$, and $FeCl_3$. The solution that has similar conductance to the NaCl solution contains (c); the solution with the conductance similar to $MgCl_2$ contains (b); and the solution with conductance similar to $FeCl_3$ contains (a).

22.56
$$Zn\,(s) \rightarrow Zn^{2+}(aq) + 2e^- \qquad\qquad E^\circ_{anode} = -0.76 \text{ V}$$
$$2[Cu^{2+}(aq) + e^- \rightarrow Cu^+(aq)] \qquad\qquad E^\circ_{cathode} = 0.15 \text{ V}$$

$$\overline{Zn(s) + 2Cu^{2+}(aq) \rightarrow Zn^{2+}(aq) + 2Cu^+(aq) \quad E^\circ_{cell} = E^\circ_{cathode} - E^\circ_{anode} = 0.15 \text{ V} - (-0.76 \text{ V}) = 0.91 \text{ V}}$$

$$\Delta G^\circ = -nFE^\circ = -(2)(96500 \text{ J/V·mol})(0.91 \text{ V}) = -1.8 \times 10^5 \text{ J/mol} = -1.8 \times 10^2 \text{ kJ/mol}$$

$$\ln K = \frac{-\Delta G^\circ}{RT} = \frac{1.8 \times 10^5 \text{ J/mol}}{(8.314 \text{ J/K·mol})(298 \text{ K})}$$

$$K = 4 \times 10^{31}$$

22.58   Iron is much more abundant than cobalt.

22.60   Oxyhemoglobin absorbs higher energy light than deoxyhemoglobin. Oxyhemoglobin is diamagnetic (low spin), while deoxyhemoglobin is paramagnetic (high spin). These differences occur because oxygen (O_2) is a strong–field ligand. The crystal field splitting diagrams are:

$$\uparrow \qquad\qquad \uparrow$$
$$d_{x^2-y^2} \qquad\quad d_{z^2}$$

$$\overline{\;d_{x^2-y^2}\;} \qquad \overline{\;d_{z^2}\;}$$

$$\uparrow\downarrow \qquad \uparrow \qquad \uparrow$$
$$d_{xy} \qquad d_{xz} \qquad d_{yz}$$

$$\uparrow\downarrow \qquad \uparrow\downarrow \qquad \uparrow\downarrow$$
$$d_{xy} \qquad d_{xz} \qquad d_{yz}$$

          deoxyhemoglobin   oxyhemoglobin

22.62   Complexes are expected to be colored when the highest occupied orbitals have between one and nine d electrons. Such complexes can therefore have $d \rightarrow d$ transitions (that are usually in the visible part of the electromagnetic radiation spectrum). The ions V^{5+}, Ca^{2+}, and Sc^{3+} have d^0 electron configurations and Cu^+, Zn^{2+}, and Pb^{2+} have d^{10} electron configurations: these complexes are colorless. The other complexes have outer electron configurations of d^1 to d^9 and are therefore colored.

22.64   Dipole moment measurement. Only the *cis* isomer has a dipole moment.

22.66   EDTA sequesters metal ions (like Ca^{2+} and Mg^{2+}) which are essential for growth and function, thereby depriving the bacteria to grow and multiply.

CHAPTER 23
NUCLEAR CHEMISTRY

PROBLEM-SOLVING STRATEGIES AND TUTORIAL SOLUTIONS

TYPES OF PROBLEMS
Problem Type 1: Balancing Nuclear Equations.

Problem Type 2: Nuclear Stability

Problem Type 3: Calculating Nuclear Binding Energy.

Problem Type 4: Kinetics of Radioactive Decay.

Problem Type 5: Balancing Nuclear Transmutation Equations.

PROBLEM TYPE 1: BALANCING NUCLEAR EQUATIONS

Writing a nuclear equation differs slightly from writing equations for chemical reactions. In addition to writing the symbols for various chemical elements, we must also explicitly indicate protons, neutrons, and electrons. In fact, we must show the numbers of protons and neutrons present in *every* species in the reaction.

You must know the symbols for elementary particles in order to balance a nuclear equation.

^1_1p or ^1_1H	^1_0n	$^{\,0}_{-1}e$ or $^{\,0}_{-1}\beta$	$^{\,0}_{+1}e$ or $^{\,0}_{+1}\beta$	^4_2He or $^4_2\alpha$
proton	neutron	electron	positron	α particle

In balancing any nuclear equation, we observe the following rules:

- The total number of protons plus neutrons in the products and in the reactants must be the same (conservation of mass number).

- The total number of nuclear charges in the products and in the reactants must be the same (conservation of atomic number).

If the atomic numbers and mass numbers of all the species but one in a nuclear equation are known, the unknown species can be identified by applying the above rules.

EXAMPLE 23.1
Identify X in the following nuclear reactions:

$$\text{(a)} \quad ^{14}_7N + ^1_0n \longrightarrow ^{14}_6C + X$$

$$\text{(b)} \quad ^{226}Ra \longrightarrow ^4_2\alpha + X$$

(a) According to rule 1, the sum of the mass numbers must be conserved. Thus, the unknown product will have a mass number of 1. According to rule 2, the sum of the atomic numbers must be conserved. Thus, the nuclear charge of the unknown product must be 1. The particle is a proton.

$$^{14}_7N + ^1_0n \longrightarrow ^{14}_6C + ^1_1\mathbf{p}$$

(b) Note that the atomic number of radium is missing. Look at a periodic table to find that Ra is element number 88. Balancing the mass numbers first, we find that the unknown product must have a mass of 222. Balancing the nuclear charges, we find that the nuclear charge of the unknown must be 86. Element number 86 is radon (Rn).

$$^{226}_{88}\text{Ra} \longrightarrow {}^{4}_{2}\alpha + {}^{222}_{86}\textbf{Rn}$$

PRACTICE EXERCISE

1.   Identify X in the following nuclear reactions:

   (a) $^{239}_{94}\text{Pu} \longrightarrow {}^{235}_{92}\text{U} + \text{X}$

   (b) $^{90}_{38}\text{Sr} \longrightarrow {}^{90}_{37}\text{Rb} + \text{X}$

Text Problems: 23.6, 23.28

PROBLEM TYPE 2: NUCLEAR STABILITY

The principal factor for determining the stability of a nucleus is the *neutron-to-proton ratio (n/p)*. For stable elements of low atomic number, the n/p ratio is close to 1. As the atomic number increases, the n/p ratios of stable nuclei become greater than 1. The deviation in the n/p ratio at higher atomic numbers arises because a larger number of neutrons is needed to stabilize the nucleus by counteracting the strong repulsion among the large number of protons. Figure 23.1 of the text shows a plot of the number of neutrons versus the number of protons in various isotopes. The stable nuclei are located in an area of the graph called the *belt of stability*. Most radioactive nuclei lie outside the belt.

The following rules are useful in predicting nuclear stability.

- Nuclei that contain 2, 8, 20, 50, 82, or 126 protons or neutrons are generally more stable than nuclei that do not possess these numbers. These numbers are called *magic numbers*.

- Nuclei with even numbers of both protons and neutrons are generally more stable than those with odd numbers of these particles (see Table 23.2 of the text).

- All isotopes of the elements starting with polonium (Po, Z = 84) are radioactive. All isotopes of technetium (Tc, Z = 43) and promethium (Pm, Z = 61) are also radioactive.

EXAMPLE 23.2
Rank the following nuclei in order of increasing nuclear stability:
$^{40}_{20}\text{Ca}$ $^{39}_{20}\text{Ca}$ $^{11}_{5}\text{B}$

Boron-11 has both an odd number of protons and an odd number of neutrons; therefore, it should be the least stable of the three nuclei. Calcium-39 has a magic number of protons (20), but an odd number of neutrons. Calcium-40 has both a magic number of protons and neutrons (20), so it should be more stable than calcium-39. The order of increasing nuclear stability is

$$^{11}_{5}\text{B} \quad < \quad {}^{39}_{20}\text{Ca} \quad < \quad {}^{40}_{20}\text{Ca}$$

PRACTICE EXERCISE

2.   Rank the following isotopes in order of increasing nuclear stability:
   $^{39}_{20}\text{Ca}$ $^{222}_{86}\text{Rn}$ $^{98}_{43}\text{Tc}$

Text Problems: 23.14, 23.16

PROBLEM TYPE 3: CALCULATING NUCLEAR BINDING ENERGY

A quantitative measure of nuclear stability is the **nuclear binding energy**, which is the energy required to break up a nucleus into its component protons and neutrons. The concept of nuclear binding energy evolved from studies showing that the masses of nuclei are always less than the sum of the masses of the **nucleons** (the protons and neutrons in a nucleus). The difference between the mass of an atom and the sum of the masses of its protons, neutrons, and electrons is called the **mass defect**. According to Einstein's mass-energy equivalence relationship

$$E = mc^2$$

where,

> E is energy
> m is mass
> c is the velocity of light

the loss in mass shows up as energy (heat) given off to the surroundings. We can calculate the amount of energy released by writing:

$$\Delta E = (\Delta m)c^2$$

where,

> ΔE = energy of products – energy of reactants

> Δm = mass of products – mass of reactants

See Example 23.3 below for a detailed calculation.

EXAMPLE 23.3
Calculate the nuclear binding energy of the light isotope of helium, ^{3}He. The atomic mass of $^3_2$He is 3.01603 amu.

The binding energy is the energy required for the process

$$^3_2\text{He} \longrightarrow 2\,^1_1\text{p} + \,^1_0\text{n}$$

Step 1: Calculate the mass difference (Δm) between the products and reactants.

$$\Delta m = [2(\text{proton mass}) + (\text{neutron mass})] - \,^3_2\text{He atomic mass}$$

$$= [2(1.007825 \text{ amu}) + 1.008665 \text{ amu}] - 3.01603 \text{ amu}$$

$$= 8.29 \times 10^{-3} \text{ amu}$$

Step 2: Use Einstein's equation to calculate the energy change for the process, ΔE.

$$\Delta E = (\Delta m)c^2$$

$$= 8.29 \times 10^{-3} \text{ amu} \times (3.00 \times 10^8 \text{ m/s})^2$$

$$= 7.46 \times 10^{14} \frac{\text{amu} \cdot \text{m}^2}{\text{s}^2}$$

Let's convert to useful units (J/He atom).

$$7.46 \times 10^{14} \frac{\text{amu} \cdot \text{m}^2}{\text{s}^2} \times \frac{1.00 \text{ g}}{6.022 \times 10^{23} \text{ amu}} \times \frac{1 \text{ kg}}{1000 \text{ g}} \times \frac{1 \text{ J}}{\frac{1 \text{ kg} \cdot \text{m}^2}{\text{s}^2}} = \mathbf{1.24 \times 10^{-12} \text{ J/atom}}$$

This is the nuclear binding energy. It's the energy required to break up one helium-3 nucleus into 2 protons and 1 neutron.

Step 3: When comparing the stability of any two nuclei we must account for the fact that they have different numbers of nucleons. For this reason, it is more meaningful to use the *nuclear binding energy per nucleon*, defined as

$$\text{nuclear binding energy per nucleon} = \frac{\text{nuclear binding energy}}{\text{number of nucleons}}$$

For the helium-3 nucleus,

$$\text{nuclear binding energy per nucleon} = \frac{1.24 \times 10^{-12} \text{ J/He atom}}{3 \text{ nucleons/He atom}}$$

$$= \mathbf{4.13 \times 10^{-13} \text{ J/nucleon}}$$

PRACTICE EXERCISE

3.   (a)   Calculate the binding energy and the binding energy per nucleon of $^{27}_{13}\text{Al}$. The atomic mass of $^{27}_{13}\text{Al}$ is 26.98154 amu.

     (b)   Compare the result from part (a) to the binding energy of $^{28}_{14}\text{Si}$, which has an even number of protons and neutrons. The atomic mass of $^{28}_{14}\text{Si}$ is 27.976928 amu.

Text Problem: 23.20

PROBLEM TYPE 4: KINETICS OF RADIOACTIVE DECAY

All radioactive decays obey first-order kinetics. To review first-order reactions, see Sections 13.2 and 13.3 of your text. The decay rate at any time t is given by

$$\text{rate of decay at time } t = \lambda N$$

where,

λ is the first-order rate constant
N is the number of radioactive nuclei present at time t

The relationship between the number of radioactive nuclei present at time zero (N_0) and the number of nuclei remaining at a later time t (N_t) is given by:

$$\ln \frac{N_0}{N_t} = \lambda t \qquad (23.1)$$

The corresponding half-life for a first-order reaction is given by:

$$t_{\frac{1}{2}} = \frac{0.693}{\lambda} \qquad (23.2)$$

Unlike ordinary chemical reactions, the rate constants for nuclear reactions are unaffected by changes in environmental conditions such as temperature and pressure.

EXAMPLE 23.4

Cobalt-60 is used in radiation therapy. It has a half-life of 5.26 years.

(a)   Calculate the rate constant for radioactive decay.

(b)   What fraction of a certain sample will remain after 12 years?

(a)   The rate constant for the decay can be calculated using Equation (23.2) for the half-life of a first-order reaction.

$$t_{\frac{1}{2}} = \frac{0.693}{\lambda} \hspace{3cm} (23.2)$$

Rearrange the equation to solve for the rate constant (λ).

$$\lambda = \frac{0.693}{t_{\frac{1}{2}}} = \frac{0.693}{5.26 \text{ yr}} = \mathbf{0.132 \text{ yr}^{-1}}$$

(b)   The fraction of a certain sample that will remain after 12 years is

$$\frac{N_t}{N_0}$$

where $t = 12$ yr.

$\dfrac{N_t}{N_0}$ can be calculated from Equation (23.1) above. We can rearrange the equation to solve for $\dfrac{N_t}{N_0}$.

$$\ln \frac{N_0}{N_t} = \lambda t$$

$$\frac{N_0}{N_t} = e^{\lambda t}$$

$$\frac{N_t}{N_0} = \frac{1}{e^{\lambda t}}$$

$$\frac{N_t}{N_0} = \frac{1}{e^{(0.132 \text{ yr}^{-1})(12 \text{ yr})}} = \mathbf{0.205}$$

In other words, 20.5 percent of the original sample will remain after 12 years.

PRACTICE EXERCISE

4.   Estimate the age of a bottle of wine that has a tritium $\left({}_{1}^{3}\text{H} \right)$ content 75.0 percent of the tritium content of environmental water obtained from the area where the grapes were grown. $t_{\frac{1}{2}} = 12.3$ yr.

| Text Problems: 23.24, 23.26, 23.30 |

PROBLEM TYPE 5: BALANCING NUCLEAR TRANSMUTATION EQUATIONS

Nuclei can undergo change as a result of bombardment by neutrons, protons, or other nuclei. This process is called **nuclear transmutation**. Unlike radioactive decay, nuclear transmutation is *not* a spontaneous process; consequently, nuclear transmutation reactions have more than one reactant. As an example, consider the synthesis of neptunium (Np), which was the first transuranium element to be synthesized by scientists.

First, uranium-238 is bombarded with neutrons to produce uranium-239.

$$^{238}_{92}U + ^{1}_{0}n \longrightarrow ^{239}_{92}U$$

This is a nuclear transmutation reaction. Uranium-239 is unstable and decays spontaneously to neptunium-239 by emitting a β particle.

$$^{239}_{92}U \longrightarrow ^{239}_{93}Np + ^{0}_{-1}\beta$$

To balance a nuclear transmutation reaction, follow the same rules used to balance nuclear equations, Problem Type 1.

- The total number of protons plus neutrons in the products and in the reactants must be the same (conservation of mass number).

- The total number of nuclear charges in the products and in the reactants must be the same (conservation of atomic number).

EXAMPLE 23.5

Write and balance the following reactions. When aluminum-27 is bombarded with α particles, phosphorus-30 and one other particle are produced. Phosphorus-30 has a low *n/p* ratio and decays spontaneously by positron emission.

The first reaction is a nuclear transmutation. You are given both reactants and one of the two products in the problem. To balance the equation, remember that both mass number and atomic number must be conserved.

$$^{27}_{13}Al + ^{4}_{2}\alpha \longrightarrow ^{30}_{15}P + X$$

To balance the mass number, the missing particle (X) must have a mass number of 1. To balance the atomic number, X must have an atomic number of 0. X must be a neutron.

$$^{27}_{13}Al + ^{4}_{2}\alpha \longrightarrow ^{30}_{15}P + ^{1}_{0}n$$

The second reaction is spontaneous so there is only one reactant, phosphorus-30. The problem indicates that phosphorus-30 decays by positron emission. Let's write down what we know so far.

$$^{30}_{15}P \longrightarrow X + ^{0}_{+1}\beta$$

To balance the mass number, the missing element (X) must have a mass number of 30. To balance the atomic number, X must have an atomic number of 14. X must be silicon-30.

$$^{30}_{15}P \longrightarrow ^{30}_{14}Si + ^{0}_{+1}\beta$$

PRACTICE EXERCISE

5. Write and balance the following reactions. When chlorine-37 is bombarded with neutrons, only one product is produced. The product is unstable and spontaneously decays by beta emission.

Text Problems: 23.34, 23.36

ANSWERS TO PRACTICE EXERCISES

1. (a) $X = {}^{4}_{2}He$

    (b) $X = {}^{0}_{+1}\beta$

2. Technetium-98 should be unstable because it has an odd number of both protons and neutrons. In fact, all isotopes of technetium are radioactive. Radon-222 will perhaps be slightly more stable than technetium-98, because it has an even number of both protons and neutrons. Remember, however, that all isotopes of the elements starting with polonium (Po, Z = 84) are radioactive. Calcium-39 has 20 protons (a "magic number") and should be the most stable of the three isotopes. The correct order of increasing nuclear stability is

    $$ {}^{98}_{43}Tc \quad < \quad {}^{222}_{86}Rn \quad < \quad {}^{39}_{20}Ca $$

3. (a) Binding energy $= 3.6 \times 10^{-11}$ J/Al atom $= 1.3 \times 10^{-12}$ J/nucleon

    (b) Binding energy $= 3.8 \times 10^{-11}$ J/Si atom $= 1.4 \times 10^{-12}$ J/nucleon

    When comparing the stability of any two nuclei, it is best to compare the binding energy in units of J/nucleon. Silicon has the greater binding energy per nucleon and hence is more stable than Al. You should have expected this result, because Si has an even number of protons and neutrons; whereas, Al has an odd number of protons and an even number of neutrons. Nuclei with even numbers of both protons and neutrons are generally more stable than those with odd numbers of these particles.

4. You can calculate the rate constant (λ) from the half-life.

    $$ \lambda = 0.0563 \text{ yr}^{-1} $$

    Next, $\dfrac{N_t}{N_0} = 0.750$ or $\dfrac{N_0}{N_t} = 1.33$. The age (t) of the bottle of wine can then be calculated from the equation,

    $$ \ln \frac{N_0}{N_t} = \lambda t $$
    $$ t = 5.07 \text{ yr} $$

5. $ {}^{1}_{0}n + {}^{37}_{17}Cl \longrightarrow {}^{38}_{17}Cl $

    $ {}^{38}_{17}Cl \longrightarrow {}^{38}_{18}Ar + {}^{0}_{-1}\beta $

SOLUTIONS TO SELECTED TEXT PROBLEMS

23.6   Balancing Nuclear Equations, Problem Type 1.

(a)   According to rule 1, the sum of the mass numbers must be conserved. Thus, the unknown product will have a mass number of 0. According to rule 2, the atomic number must be conserved. Thus, the nuclear charge of the unknown product must be –1. The particle is a β particle.

$$^{135}_{53}\text{I} \longrightarrow ^{135}_{54}\text{Xe} + ^{0}_{-1}\beta$$

(b)   Balancing the mass numbers first, we find that the unknown product must have a mass of 40. Balancing the nuclear charges, we find that the atomic number of the unknown must be 20. Element number 20 is calcium (Ca).

$$^{40}_{19}\text{K} \longrightarrow ^{0}_{-1}\beta + ^{40}_{20}\textbf{Ca}$$

(c)   Balancing the mass numbers, we find that the unknown product must have a mass of 4. Balancing the nuclear charges, we find that the nuclear charge of the unknown must be 2. The unknown particle is an alpha (α) particle.

$$^{59}_{27}\text{Co} + ^{1}_{0}\text{n} \longrightarrow ^{56}_{25}\text{Mn} + ^{4}_{2}\alpha$$

(d)   Balancing the mass numbers, we find that the unknown products must have a combined mass of 2. Balancing the nuclear charges, we find that the combined nuclear charge of the two unknown particles must be 0. The unknown particles are neutrons.

$$^{235}_{92}\text{U} + ^{1}_{0}\text{n} \longrightarrow ^{99}_{40}\text{Sr} + ^{135}_{52}\text{Te} + 2^{1}_{0}\textbf{n}$$

23.14   Nuclear Stability, Problem Type 2.

(a)   **Lithium-9** should be less stable. The neutron-to-proton ratio is too high. For small atoms, the *n/p* ratio will be close to 1:1.

(b)   **Sodium-25** is less stable. Its neutron-to-proton ratio is probably too high.

(c)   **Scandium-48** is less stable because of odd numbers of protons and neutrons. We would not expect calcium-48 to be stable even though it has a magic number of protons. Its *n/p* ratio is too high.

23.16   (a)   **Neon-17** should be radioactive. It falls below the belt of stability (low *n/p* ratio).

(b)   **Calcium-45** should be radioactive. It falls above the belt of stability (high *n/p* ratio).

(c)   All **technetium** isotopes are radioactive.

(d)   **Mercury-195** should be radioactive. Mercury-196 has an even number of both neutrons and protons.

(e)   All **curium** isotopes are unstable.

23.18   We can use the equation, $\Delta E = \Delta mc^2$, to solve the problem. Recall the following conversion factor:

$$1 \text{ J} = \frac{1 \text{ kg} \cdot \text{m}^2}{\text{s}^2}$$

The energy loss in one second is:

$$\Delta m = \frac{\Delta E}{c^2} = \frac{5 \times 10^{26} \frac{\text{kg} \cdot \text{m}^2}{\text{s}^2}}{(3.00 \times 10^8 \frac{\text{m}}{\text{s}})^2} = 6 \times 10^9 \text{ kg}$$

Therefore the rate of mass loss is **6×10^9 kg/s**.

23.20   Calculating Nuclear Binding Energy, Problem Type 3.

(a)   The binding energy is the energy required for the process

$$_2^4\text{He} \rightarrow 2\,_1^1\text{p} + 2\,_0^1\text{n}$$

Step 1: Calculate the mass difference (Δm) between the products and reactants.

$$\Delta m = [2(\text{proton mass}) + 2(\text{neutron mass})] - \,_2^4\text{He atomic mass}$$

$$= [2(1.007825 \text{ amu}) + 2(1.008665 \text{ amu})] - 4.0026 \text{ amu}$$

$$= 0.0304 \text{ amu}$$

Step 2: Use Einstein's equation to calculate the energy change for the process, ΔE.

$$\Delta E = (\Delta m)c^2$$

$$= 0.0304 \text{ amu} \times (3.00 \times 10^8 \text{ m/s})^2$$

$$= 2.74 \times 10^{15} \frac{\text{amu} \cdot \text{m}^2}{\text{s}^2}$$

Let's convert to useful units (J/He atom).

$$2.74 \times 10^{15} \frac{\text{amu} \cdot \text{m}^2}{\text{s}^2} \times \frac{1.00 \text{ g}}{6.022 \times 10^{23} \text{ amu}} \times \frac{1 \text{ kg}}{1000 \text{ g}} \times \frac{1 \text{ J}}{\frac{1 \text{ kg} \cdot \text{m}^2}{\text{s}^2}}$$

$$= \mathbf{4.55 \times 10^{-12} \text{ J}}$$

This is the nuclear binding energy. It's the energy required to break up one helium-4 nucleus into 2 protons and 2 neutrons.

Step 3: When comparing the stability of any two nuclei we must account for the fact that they have different numbers of nucleons. For this reason, it is more meaningful to use the *nuclear binding energy per nucleon*, defined as

$$\text{nuclear binding energy per nucleon} = \frac{\text{nuclear binding energy}}{\text{number of nucleons}}$$

For the helium-4 nucleus,

$$\text{nuclear binding energy per nucleon} = \frac{4.55 \times 10^{-12} \text{ J / He atom}}{4 \text{ nucleons / He atom}}$$

$$= \mathbf{1.14 \times 10^{-12} \text{ J/nucleon}}$$

(b)   The binding energy is the energy required for the process

$$^{184}_{74}\text{W} \rightarrow 74\,^{1}_{1}\text{p} + 110\,^{1}_{0}\text{n}$$

Step 1: Calculate the mass difference (Δm) between the products and reactants.

$$\Delta m = [74(\text{proton mass}) + 110(\text{neutron mass})] - {}^{184}_{74}\text{W atomic mass}$$

$$= [74(1.007825 \text{ amu}) + 110(1.008665 \text{ amu})] - 183.9510 \text{ amu}$$

$$= 1.5812 \text{ amu}$$

Step 2: Use Einstein's equation to calculate the energy change for the process, ΔE.

$$\Delta E = (\Delta m)c^2$$

$$= 1.5812 \text{ amu} \times (3.00 \times 10^8 \text{ m/s})^2$$

$$= 1.42 \times 10^{17} \frac{\text{amu} \cdot \text{m}^2}{\text{s}^2}$$

Let's convert to useful units (J/He atom).

$$1.42 \times 10^{17} \frac{\text{amu} \cdot \text{m}^2}{\text{s}^2} \times \frac{1.00 \text{ g}}{6.022 \times 10^{23} \text{ amu}} \times \frac{1 \text{ kg}}{1000 \text{ g}} \times \frac{1 \text{ J}}{1 \frac{\text{kg} \cdot \text{m}^2}{\text{s}^2}}$$

$$= \mathbf{2.36 \times 10^{-10} \text{ J}}$$

This is the nuclear binding energy. It's the energy required to break up one tungsten-184 nucleus into 74 protons and 110 neutrons.

Step 3: When comparing the stability of any two nuclei we must account for the fact that they have different numbers of nucleons. For this reason, it is more meaningful to use the *nuclear binding energy per nucleon*, defined as

$$\text{nuclear binding energy per nucleon} = \frac{\text{nuclear binding energy}}{\text{number of nucleons}}$$

For the tungsten-184 nucleus,

$$\text{nuclear binding energy per nucleon} = \frac{2.36 \times 10^{-10} \text{ J / W atom}}{184 \text{ nucleons / W atom}}$$

$$= \mathbf{1.28 \times 10^{-12} \text{ J/nucleon}}$$

23.24   Kinetics of Radioactive Decay, Problem Type 4.

Step 1: We can use the following equation to calculate the rate constant λ for each point.

$$\ln \frac{N_0}{N_t} = \lambda t$$

From day 0 to day 1, we have

$$\ln \frac{500}{389} = \lambda(1 \text{ d})$$

$$\lambda = 0.251 \text{ d}^{-1}$$

Following the same procedure for the other days,

t (d)	mass (g)	λ (d^{-1})
0	500	
1	389	0.251
2	303	0.250
3	236	0.250
4	184	0.250
5	143	0.250
6	112	0.249

The average value of λ is **0.250 d^{-1}**.

Step 2: Using the average value of λ, calculate the half-life.

$$t_{\frac{1}{2}} = \frac{0.693}{\lambda} = \frac{0.693}{0.250 \text{ d}^{-1}} = \textbf{2.77 d}$$

23.26   Since all radioactive decay processes have first–order rate laws, the decay rate is proportional to the amount of radioisotope at any time. The half-life is given by the following equation:

$$t_{\frac{1}{2}} = \frac{0.693}{\lambda} \qquad (1)$$

There is also an equation that relates the number of nuclei at time zero (N_0) and time t (N_t).

$$\ln \frac{N_0}{N_t} = \lambda t$$

We can use this equation to solve for the rate constant, λ. Then, we can substitute λ into Equation (1) to calculate the half-life.

The time interval is:

(2:15 p.m., 12/17/92) – (1:00 p.m., 12/3/92) = 14 d + 1 hr + 15 min = 20,235 min

$$\ln \left(\frac{9.8 \times 10^5 \text{ dis}/\text{min}}{2.6 \times 10^4 \text{ dis}/\text{min}} \right) = \lambda(20{,}235 \text{ min})$$

$$\lambda = 1.8 \times 10^{-4} \text{ min}^{-1}$$

Substitute λ into equation (1) to calculate the half-life.

$$t_{\frac{1}{2}} = \frac{0.693}{\lambda} = \frac{0.693}{1.8 \times 10^{-4} \text{ min}^{-1}} = \textbf{3.9} \times \textbf{10}^3 \textbf{ min or 2.7 d}$$

23.28 The equation for the overall process is:

$$^{232}_{90}\text{Th} \longrightarrow 6\,^{4}_{2}\text{He} + 4\,^{0}_{-1}\beta + \text{X}$$

The final product isotope must be $^{208}_{82}\text{Pb}$.

23.30 Let's consider the decay of A first.

$$\lambda = \frac{0.693}{t_{\frac{1}{2}}} = \frac{0.693}{4.50 \text{ s}} = 0.154 \text{ s}^{-1}$$

Let's convert λ to units of day^{-1}.

$$0.154 \, \frac{1}{s} \times \frac{3600 \, s}{1 \, hr} \times \frac{24 \, hr}{1 \, d} = 1.33 \times 10^{4} \text{ d}^{-1}$$

Next, use the first-order rate equation to calculate the amount of A left after 30 days.

$$\ln\frac{[A]_0}{[A]} = \lambda t$$

Let x be the amount of A left after 30 days.

$$\ln\frac{100}{x} = (1.33 \times 10^{4} \text{ d}^{-1})(30 \text{ d}) = 3.99 \times 10^{5}$$

$$x \approx 0$$

Thus, **no A remains**.

For B: As calculated above, all of A is converted to B in less than 30 days. In fact, essentially all of A is gone in less than 1 day! This means that at the beginning of the 30 day period, there is 1.00 mol of B present. The half life of B is 15 days, so that after two half-lives (30 days), there should be **0.25 mole of B** left.

For C: As in the case of A, the half-life of C is also very short. Therefore, at the end of the 30–day period, **no C is left**.

For D: D is not radioactive. 0.75 mol of B reacted in 30 days; therefore, due to a 1:1 mole ratio between B and D, there should be **0.75 mole of D** present after 30 days.

23.34 (a) $^{80}_{34}\text{Se} + \,^{2}_{1}\text{d} \longrightarrow \,^{81}_{34}\text{Se} + \,^{1}_{1}\text{p}$

(b) $^{9}_{4}\text{Be} + \,^{2}_{1}\text{d} \longrightarrow \,^{9}_{3}\text{Li} + 2\,^{1}_{1}\text{p}$

(c) $^{10}_{5}\text{B} + \,^{1}_{0}\text{n} \longrightarrow \,^{7}_{3}\text{Li} + \,^{4}_{2}\alpha$

23.36 Upon bombardment with neutrons, mercury–198 is first converted to mercury–199, which then emits a proton. The reaction is:

$$^{198}_{80}\text{Hg} + \,^{1}_{0}\text{n} \longrightarrow \,^{199}_{80}\text{Hg} \longrightarrow \,^{198}_{79}\text{Au} + \,^{1}_{1}\text{p}$$

23.48   The fact that the radioisotope appears only in the I_2 shows that the IO_3^- is formed only from the IO_4^-. Does this result rule out the possibility that I_2 could be formed from IO_4^- as well? Can you suggest an experiment to answer the question?

23.50   Add iron-59 to the person's diet, and allow a few days for the iron–59 isotope to be incorporated into the person's body. Isolate red blood cells from a blood sample and monitor radioactivity from the hemoglobin molecules present in the red blood cells.

23.52   Apparently there is a sort of Pauli exclusion principle for nucleons as well as for electrons. When neutrons pair with neutrons and when protons pair with protons, their spins cancel. Even–even nuclei are the only ones with no net spin.

23.54   **(a)**   One millicurie represents 3.70×10^7 disintegrations/s. The rate of decay of the isotope is given by the rate law: rate = λN, where N is the number of atoms in the sample. We find the value of λ in units of s^{-1}:

$$\lambda = \frac{0.693}{t_{\frac{1}{2}}} = \left(\frac{0.693}{2.20 \times 10^6 \text{ yr}}\right)\left(\frac{1 \text{ yr}}{365 \text{ d}}\right)\left(\frac{1 \text{ d}}{24 \text{ hr}}\right)\left(\frac{1 \text{ hr}}{3600 \text{ s}}\right) = 9.99 \times 10^{-15} \text{ s}^{-1}$$

The number of atoms (N) in a 0.500 g sample of neptunium–237 is:

$$0.500 \text{ g} \times \left(\frac{1 \text{ mol}}{237.0 \text{ g}}\right)\left(\frac{6.022 \times 10^{23} \text{ atoms}}{1 \text{ mol}}\right) = 1.27 \times 10^{21} \text{ atoms}$$

$$\text{rate of decay} = \lambda N$$
$$= (9.99 \times 10^{-15} \text{ s}^{-1})(1.27 \times 10^{21} \text{ atoms}) = 1.27 \times 10^7 \text{ atoms/s}$$

We can also say that:

$$\text{rate of decay} = 1.27 \times 10^7 \text{ disintegrations/s}$$

The activity in millicuries is:

$$1.27 \times 10^7 \text{ disintegrations/s} \times \left(\frac{1 \text{ millicurie}}{3.70 \times 10^7 \text{ disintegrations/s}}\right) = \textbf{0.343 millicuries}$$

(b)   The decay equation is:

$$^{237}_{93}\text{Np} \longrightarrow {}^{4}_{2}\alpha + {}^{233}_{91}\text{Pa}$$

23.56   We use the same procedure as in Problem 23.20.

	Isotope	Atomic Mass (amu)	Nuclear Binding Energy (J/nucleon)
(a)	^{10}B	10.0129	1.040×10^{-12}
(b)	^{11}B	11.00931	1.111×10^{-12}
(c)	^{14}N	14.00307	1.199×10^{-12}
(d)	^{56}Fe	55.9349	1.410×10^{-12}

23.58 When an isotope is above the belt of stability, the neutron/proton ratio is too high. The only mechanism to correct this situation is beta emission; the process turns a neutron into a proton. Direct neutron emission does not occur.

$$^{18}_{7}\text{N} \longrightarrow \ ^{18}_{8}\text{O} + \ ^{0}_{-1}\beta$$

Oxygen–18 is a stable isotope.

23.60 The age of the fossil can be determined by radioactively dating the age of the deposit that contains the fossil.

23.62 (a) $^{209}_{83}\text{Bi} + \ ^{4}_{2}\alpha \longrightarrow \ ^{211}_{85}\text{At} + 2\,^{1}_{0}\text{n}$

(b) $^{209}_{83}\text{Bi}(\alpha, 2\text{n})^{211}_{85}\text{At}$

23.64 Because of the relative masses, the force of gravity on the sun is much greater than it is on Earth. Thus the nuclear particles on the sun are already held much closer together than the equivalent nuclear particles on the earth. Less energy (lower temperature) is required on the sun to force fusion collisions between the nuclear particles.

23.66 Kinetics of Radioactive Decay, Problem Type 4.

Step 1: The half-life of carbon-14 is 5730 years. From the half-life, we can calculate the rate constant, λ.

$$\lambda = \frac{0.693}{t_{\frac{1}{2}}} = \frac{0.693}{5730 \text{ yr}} = 1.21 \times 10^{-4} \text{ yr}^{-1}$$

Step 2: The age of the object can now be calculated using the following equation.

$$\ln\frac{N_0}{N_t} = \lambda t$$

N = the number of radioactive nuclei. In the problem, we are given disintegrations per second per gram. The number of disintegrations is directly proportional to the number of radioactive nuclei. We can write,

$$t = \frac{1}{\lambda}\ln\frac{\text{decay rate of fresh sample}}{\text{decay rate of old sample}}$$

$$t = \frac{1}{1.21 \times 10^{-4} \text{ yr}^{-1}}\ln\frac{0.260 \text{ cps}/\text{g C}}{0.186 \text{ cps}/\text{g C}}$$

$$t = 2.77 \times 10^{3} \text{ yr}$$

23.68 (a) The balanced equation is:

$$^{40}_{19}\text{K} \longrightarrow \ ^{40}_{18}\text{Ar} + \ ^{0}_{+1}\beta$$

(b) First, calculate the rate constant λ.

$$\lambda = \frac{0.693}{t_{\frac{1}{2}}} = \frac{0.693}{1.2 \times 10^{9} \text{ yr}} = 5.8 \times 10^{-10} \text{ yr}^{-1}$$

Then, calculate the age of the rock by substituting λ into the following equation. ($N_t = 0.18N_0$)

$$t = \frac{1}{\lambda} \ln \frac{N_0}{N_t} = \left(\frac{1}{5.8 \times 10^{-10} \text{ yr}^{-1}} \right) \ln \left(\frac{1.00}{0.18} \right) = \mathbf{3.0 \times 10^9 \text{ yr}}$$

23.70   **(a)**   In the ^{90}Sr decay, the mass defect is:

$$\Delta m = (\text{mass } ^{90}\text{Y} + \text{mass e}^-) - \text{mass } ^{90}\text{Sr}$$

$$= [(89.907152 \text{ amu} + 5.4857 \times 10^{-4} \text{ amu}) - 89.907738 \text{ amu}] \times \frac{1 \text{ g}}{6.022 \times 10^{23} \text{ amu}}$$

$$= -6.216 \times 10^{-29} \text{ g} = -6.216 \times 10^{-32} \text{ kg}$$

The energy change is given by:

$$\Delta E = (\Delta m)c^2$$

$$= (-6.126 \times 10^{-32} \text{ kg})(3.00 \times 10^8 \text{ m/s})^2$$

$$= -5.59 \times 10^{-15} \text{ kg m}^2/\text{s}^2 = \mathbf{-5.59 \times 10^{-15} \text{ J}}$$

Similarly, for the ^{90}Y decay, we have

$$\Delta m = (\text{mass } ^{90}\text{Zr} + \text{mass e}^-) - \text{mass } ^{90}\text{Y}$$

$$= [(89.904703 \text{ amu} + 5.4857 \times 10^{-4} \text{ amu}) - 89.907152 \text{ amu}] \times \frac{1 \text{ kg}}{6.022 \times 10^{26} \text{ amu}}$$

$$= -3.156 \times 10^{-30} \text{ kg}$$

and the energy change is:

$$\Delta E = (-3.156 \times 10^{-30} \text{ kg})(3.00 \times 10^8 \text{ m/s})^2$$

$$= \mathbf{-2.84 \times 10^{-13} \text{ J}}$$

Thus the total amount of energy released is:

$$-5.59 \times 10^{-15} \text{ J} - 2.84 \times 10^{-13} \text{ J} = -2.90 \times 10^{-13} \text{ J}.$$

(b)   This calculation requires that we know the rate constant for the decay. From the half–life, we can calculate λ.

$$\lambda = \frac{0.693}{t_{\frac{1}{2}}} = \frac{0.693}{28.1 \text{ yr}} = 0.0247 \text{ yr}^{-1}$$

To calculate the number of moles of ^{90}Sr decaying in a year, we apply the following equation:

$$\ln \frac{N_0}{N_t} = \lambda t$$

$$\ln \frac{1.00}{x} = (0.0247 \text{ yr}^{-1})(1.00 \text{ yr})$$

where x is the number of moles of ^{90}Sr nuclei left over. Solving, we obtain:

$$x = 0.976 \text{ mol } ^{90}\text{Sr}$$

Thus the number of moles of nuclei which decay in a year is

$$(1.00 - 0.976) \text{ mol } = \textbf{0.024 mol}$$

This is a reasonable number since it takes 28.1 years for 0.5 mole of ^{90}Sr to decay.

(c)   Since the half–life of ^{90}Y is much shorter than that of ^{90}Sr, we can safely assume that *all* the ^{90}Y formed from ^{90}Sr will be converted to ^{90}Zr. The energy changes calculated in part (a) refer to the decay of individual nuclei. In 0.024 mole, the number of nuclei that have decayed is:

$$0.024 \text{ mol} \times 6.022 \times 10^{23} \text{ nuclei/mol} = 1.4 \times 10^{22} \text{ nuclei}$$

Realize that there are two decay processes occurring, so we need to add the energy released for each process calculated in part (a). Thus, the heat released from 1 mole of ^{90}Sr waste in a year is given by:

$$\textbf{heat released} = (-2.90 \times 10^{-13} \text{ J/nucleus})(1.4 \times 10^{22} \text{ nuclei}) = \textbf{--4.06} \times \textbf{10}^9 \textbf{ J} = \textbf{--4.06} \times \textbf{10}^6 \textbf{ kJ}$$

This amount is roughly equivalent to the heat generated by burning 50 tons of coal! Although the heat is released slowly during the course of a year, effective ways must be devised to prevent heat damage to the storage containers and subsequent leakage of radioactive material to the surroundings.

23.72   First, let's calculate the number of disintegrations/s to which 7.4 mC corresponds.

$$7.4 \text{ mC} \times \frac{1 \text{ Ci}}{1000 \text{ mC}} \times \frac{3.7 \times 10^{10} \text{ disintegrations/s}}{1 \text{ Ci}} = 2.7 \times 10^8 \text{ disintegrations/s}$$

This is the rate of decay. We can now calculate the number of iodine-131 atoms to which this radioactivity corresponds. First, we calculate the half-life in seconds:

$$t_{\frac{1}{2}} = 8.1 \text{ d} \times \frac{24 \text{ hr}}{1 \text{ d}} \times \frac{3600 \text{ s}}{1 \text{ hr}} = 7.0 \times 10^5 \text{ s}$$

$$\lambda = \frac{0.693}{t_{\frac{1}{2}}} \qquad \text{Therefore, } \lambda = \frac{0.693}{7.0 \times 10^5 \text{ s}} = 9.9 \times 10^{-7} \text{ s}^{-1}$$

$$\text{rate} = \lambda N$$

$$2.7 \times 10^8 \text{ disintegrations/s} = (9.9 \times 10^{-7} \text{ s}^{-1})N$$

$$N = \textbf{2.7} \times \textbf{10}^{14} \textbf{ iodine-131 atoms}$$

23.74   One curie represents 3.70×10^{10} disintegrations/s. The rate of decay of the isotope is given by the rate law: rate = λN, where N is the number of atoms in the sample and λ is the first-order rate constant. We find the value of λ in units of s^{-1}:

$$\lambda = \frac{0.693}{t_{\frac{1}{2}}} = \frac{0.693}{1.6 \times 10^3 \text{ yr}} = 4.3 \times 10^{-4} \text{ yr}^{-1}$$

$$\frac{4.3 \times 10^{-4}}{1 \text{ yr}} \times \frac{1 \text{ yr}}{365 \text{ d}} \times \frac{1 \text{ d}}{24 \text{ hr}} \times \frac{1 \text{ hr}}{3600 \text{ s}} = 1.4 \times 10^{-11} \text{ s}^{-1}$$

Now, we can calculate N, the number of Ra atoms in the sample.

$$\text{rate} = \lambda N$$

$$3.7 \times 10^{10} \text{ disintegrations/s} = (1.4 \times 10^{-11} \text{ s}^{-1})N$$

$$N = 2.6 \times 10^{21} \text{ Ra atoms}$$

By definition, 1 curie corresponds to exactly 3.7×10^{10} nuclear disintegrations per second which is the decay rate equivalent to that of *1 g of radium*. Thus, the mass of 2.6×10^{21} Ra atoms is 1 g.

$$\frac{2.6 \times 10^{21} \text{ Ra atoms}}{1.0 \text{ g Ra}} \times \frac{226.03 \text{ g Ra}}{1 \text{ mol Ra}} = \mathbf{5.9 \times 10^{23} \text{ atoms/mol}} = N_A$$

23.76 All except gravitational have a nuclear origin.

23.78 U–238, $t_{\frac{1}{2}} = 4.5 \times 10^9$ yr and Th–232, $t_{\frac{1}{2}} = 1.4 \times 10^{10}$ yr.

They are still present because of their long half lives.

23.80 $E = \dfrac{hc}{\lambda}$

$$\lambda = \frac{hc}{E} = \frac{(3.00 \times 10^8 \text{ m/s})(6.63 \times 10^{-34} \text{ J·s})}{2.4 \times 10^{-13} \text{ J}} = 8.3 \times 10^{-13} \text{ m} = \mathbf{8.3 \times 10^{-4} \text{ nm}}$$

This wavelength is clearly in the γ-ray region of the electromagnetic spectrum.

23.82 Only ^3H has a suitable half-life. The other half-lives are either too long or too short to accurately determine the time span of 6 years.

23.84 Obviously, a small scale chain reaction took place. Copper played the crucial role of reflecting neutrons from the splitting uranium-235 atoms back into the uranium sphere to trigger the chain reaction. Note that a sphere has the most appropriate geometry for such a chain reaction. In fact, during the implosion process prior to an atomic explosion, fragments of uranium-235 are pressed roughly into a sphere for the chain reaction to occur (see Section 23.5 of the text).

CHAPTER 24
ORGANIC CHEMISTRY

PROBLEM-SOLVING STRATEGIES AND TUTORIAL SOLUTIONS

TYPES OF PROBLEMS

Problem Type 1: Determining the Number of Structural Isomers.

Problem Type 2: Organic Nomenclature.
    (a) Alkanes.
    (b) Alkenes.
    (c) Alkynes.
    (d) Aromatic Compounds.

Problem Type 3: Addition Reactions.

Problem Type 4: Distinguishing between Structural and Geometric Isomers.

Problem Type 5: Functional Groups.

Problem Type 6: Chirality.

PROBLEM TYPE 1: DETERMINING THE NUMBER OF STRUCTURAL ISOMERS

Structural isomers are molecules that have the same molecular formula but a different order of linking the atoms. For small hydrocarbon molecules (eight or fewer carbon atoms), it is relatively easy to determine the number of structural isomers by trial and error.

EXAMPLE 24.1
How many structural isomers can be identified for hexane, C_6H_{14}?

Step 1: Write down the straight-chain structure.

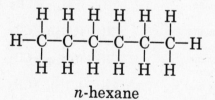

n-hexane

Step 2: By necessity, the other structures must have branched chains. First, try single methyl substituents.

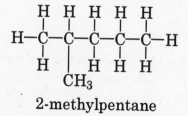

2-methylpentane

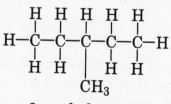

3-methylpentane

Then, try structures that have two methyl groups.

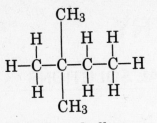

2,2-dimethylbutane   2,3-dimethylbutane

Hexane has five structural isomers, in which the numbers of carbon and hydrogen atoms remain unchanged despite the differences in structure.

PRACTICE EXERCISE

1.   How many structural isomers are there of $C_4H_{10}O$? **Hint:** Consider both alcohols and ethers.

Text Problems: **24.12**, **24.14**

PROBLEM TYPE 2: ORGANIC NOMENCLATURE

The first step in learning the nomenclature of hydrocarbons is to know the names of the first ten straight-chain alkanes. Except for the first four members, the number of carbon atoms in each alkane is identified by the Greek prefix. See Table 24.1 below.

TABLE 24.1
The First Ten Straight-Chain Alkanes

Name of hydrocarbon	Molecular formula
Methane	CH_4
Ethane	$CH_3 - CH_3$
Propane	$CH_3 - CH_2 - CH_3$
Butane	$CH_3 - (CH_2)_2 - CH_3$
Pentane	$CH_3 - (CH_2)_3 - CH_3$
Hexane	$CH_3 - (CH_2)_4 - CH_3$
Heptane	$CH_3 - (CH_2)_5 - CH_3$
Octane	$CH_3 - (CH_2)_6 - CH_3$
Nonane	$CH_3 - (CH_2)_7 - CH_3$
Decane	$CH_3 - (CH_2)_8 - CH_3$

Next, you need to learn the names of substituents (other than hydrogen) attached to hydrocarbon chains or aromatic compounds. For example, when a hydrogen atom is removed from methane, a $-CH_3$ fragment is left, which is called a *methyl* group. Similarly, removing a hydrogen atom from an ethane molecule gives an *ethyl* group, $-CH_2CH_3$.

These groups or substituents that are derived from alkanes are called *alkyl* groups. See Table 24.2 of your text for other common alkyl groups.

A. Alkanes

Alkanes have the general formula C_nH_{2n+2}, where $n = 1, 2, \ldots$. Alkanes contain only single covalent bonds. The bonds are said to be saturated because no more hydrogen atoms can be added to the carbon atoms. Thus, alkanes are also called *saturated hydrocarbons*.

There are other rules to follow when naming hydrocarbons.

1. Find the longest carbon chain in the molecule. The parent name of the compound is based on the longest carbon chain. For example, if the longest carbon chain contains five C atoms, the parent name of the compound is *pentane*. See Table 24.1 above.

2. Next, you must specify the name and location of any groups attached to the longest carbon chain. See Table 24.2 of your text for naming *alkyl* groups. To specify the location of the group or groups, start numbering the longest chain from the end that is closer to the carbon atom bearing the substituent group.

3. If there is more than one of a particular group attached to the longest carbon chain, you must specify the number of groups with a prefix. The prefixes are di– (2 groups), tri– (3 groups), tetra– (4 groups), and so on.

4. There can be many different types of substituents other than alkyl groups. Table 24.3 of your text lists the names of some common functional groups.

EXAMPLE 24.2
Give the correct name for the following structure:

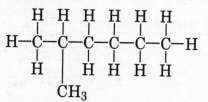

Step 1: Find the longest carbon chain. The longest chain has *six* carbons. Therefore, the parent name of the compound is *hexane*.

Step 2: Specify the name and location of any groups attached to the longest carbon chain. You should see that there is a methyl group attached to the second carbon from the left of the chain. You should number the carbon chain from the end that is closer to the carbon atom bearing the substituent group. If you start numbering from the left, the methyl group is on the second carbon in the chain. However, if you start numbering from the right, the methyl group is on the fifth carbon in the chain. Numbering from the left is correct.

The correct name for this compound is **2-methylhexane**.

Tip: You should always put a dash (–) between a number and a "word" when naming an organic compound.

EXAMPLE 24.3

Give the correct name for the following structure:

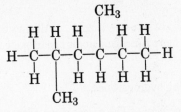

Step 1: You should find that the longest carbon chain has six carbons. The parent name is *hexane*.

Step 2: Numbering from the left, there are methyl groups on the second and fourth carbons. Since there are two methyl groups, we must specify this by using the prefix di–.

The correct name for the compound is **2,4-dimethylhexane**.

Let's examine the name of the compound if we had numbered the longest chain from right to left. The methyl groups would be on the third and fifth carbons. Hence, the name would be 3,5-dimethylhexane, which is incorrect. The correct name should always have the lowest numbering scheme as possible.

> **Tip:** Always use commas to separate numbers when naming organic compounds.

EXAMPLE 24.4

Give the correct name for each of the following structures:

(a)

Step 1: You should find that the longest carbon chain in the molecule is *three* carbons. The parent name is *propane*.

Step 2: Referring to Table 24.3 of your text, you should find that a –Cl group is called a chloro group. There are chloro groups on the 1 and 2 carbons, numbering from right to left. (Why not number from left to right?)

Hence, the correct name for the compound is **1,2-dichloropropane**.

(b)

Step 1: There is only one carbon in the molecule, so the parent name is *methane*.

Step 2: Referring to Table 24.3 of your text, you should find that a –NO2 group is called a nitro group. The correct name for the compound is **nitromethane**.

Why isn't the correct name 1-nitromethane?

PRACTICE EXERCISE

2.   Give the systematic name for each of the following structural formulas:

(a) $CH_3(CH_2)_7CH_3$

(b) $CH_3CH_2CH_2CHCHCH_2CH_2CH_3$
with substituent $CH_3\text{·}CHCH_3$ above and CH_2CH_3 below

(c) $BrCH_2CH_2CHBrCH_2Br$

Text Problem: 24.26

B. Alkenes

Alkenes are molecules that contain at least one carbon-carbon double bond. Alkenes have the general formula C_nH_{2n}, where $n = 2, 3, \ldots$.

To name alkenes, you follow the same rules as outlined for alkanes; except, you must specify the position(s) of the carbon-carbon double bond(s), and the name of alkenes ends with an "–ene" suffix.

EXAMPLE 24.5
Give the correct name for the following structure:

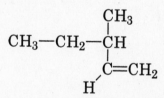

Step 1: The longest carbon chain in the molecule is five carbons. The parent name is pent*ene*, because the molecule has a double bond.

Step 2: For this molecule, we number from right to left, because we want the double bond to be at the lowest number possible. In this case, the double bond starts on the first carbon if we number from right to left. There is also a methyl substituent on the third carbon.

The correct name for this compound is **3-methyl-1-pentene**. The number 1 before pentene specifies that the double bond starts on the first carbon.

PRACTICE EXERCISE

3.   Draw the structural formula for 4-methyl-2-hexene.

Text Problems: 24.26, 24.28

C. Alkynes

Alkynes contain at least one carbon-carbon triple bond. They have the general formula C_nH_{2n-2}, where $n = 2, 3, \ldots$.

To name alkynes, you follow the same rules as outlined above for alkanes; except, you must specify the position(s) of the carbon-carbon triple bond(s), and the name of alkynes ends with an "–yne" suffix.

EXAMPLE 24.6
Give the correct name for the following structure:

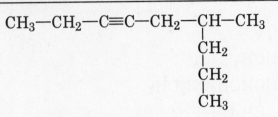

Step 1: The longest carbon chain in the molecule contains nine carbons. The parent name is non*yne* because there is a triple bond.

Step 2: You should number the carbon chain from left to right, placing the triple bond on the third carbon. If you numbered from right to left, the triple bond would be on the sixth carbon. You should also notice that there is a methyl substituent on the sixth carbon.

The correct name for this molecule is **6-methyl-3-nonyne**.

PRACTICE EXERCISE

4.   Give an acceptable name for each of the following structures:

(a) $(CH_3)_3CC \equiv CCH_2CH_3$

(b)

Text Problems: 24.26, 24.28

D. Aromatic Compounds

Benzene is the parent compound of this large family of organic substances. In naming aromatic compounds, we will consider both mono- and di-substituted benzene rings.

The naming of mono-substituted benzene rings (benzenes in which one H atom has been replaced by another atom or groups of atoms) is straightforward. You simply name the substituent followed by the name "benzene".

EXAMPLE 24.7
Give the correct name for each of the following structures:

CH₃   Br

(a)   (b)

(a)   The substituent is a methyl group, so the correct name is **methylbenzene**. However, there are older names that are still in common use for many compounds. Methylbenzene is usually called **toluene**.

(b)   The substituent is a bromo group, so the correct name is **bromobenzene**.

For di-substituted benzenes, we must indicate the location of the second group relative to the first. For example, three different dichlorobenzenes are possible.

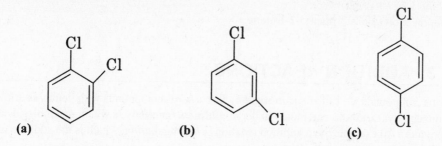

(a)   (b)   (c)

The systematic way to name these molecules is to number the carbon atoms of the benzene ring as follows:

Thus, (a) would be named 1,2-dichlorobenzene, (b) 1,3-dichlorobenzene, and (c) 1,4-dichlorobenzene. However, the prefixes *o-* (*ortho-*), *m-* (*meta*), and *p-* (*para-*) are used more often to denote the relative positions of the two substituted groups. *Ortho-* designates 1,2 substituents, *meta-* designates 1,3 substituents, and *para-* designates 1,4 substituents. Thus, (a) is named *o*-dichlorobenzene, (b) *m*-dichlorobenzene, and (c) *p*-dichlorobenzene.

In compounds with two different substituted groups, the positions of the substituents can be specified with numbers or with the *o-*, *m-*, or *p-* prefixes.

PRACTICE EXERCISE

5.   Draw structural formulas for:

    (a)   ethylbenzene   (b)   *m*-bromochlorobenzene   (c)   *p*-nitrotoluene

Text Problems: 24.26, **24.32**

Finally in some molecules, benzene is named as a substituent. A benzene molecule minus a hydrogen atom (C_6H_5) is called a *phenyl* group.

EXAMPLE 24.8
Give the correct name for the following structure:

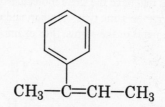

$$CH_3-C=CH-CH_3$$

For this molecule, it would be difficult to name the carbon chain as a substituent on the benzene ring. Therefore, we name the benzene ring as a substituent on the carbon chain.

Step 1: The longest carbon chain contains four carbons. The parent name is but*ene* because the molecule contains a double bond.

Step 2: You should number the carbon chain from left to right so that both the double bond and the phenyl group are on the second carbon.

The correct name for the molecule is **2-phenyl-2-butene**.

PROBLEM TYPE 3: ADDITION REACTIONS

Alkenes, alkynes, and aromatic compounds are called *unsaturated hydrocarbons*, compounds with double or triple carbon-carbon bonds. Unsaturated hydrocarbons commonly undergo **addition reactions** in which one molecule adds to another to form a single product. An example of an addition reaction is *hydrogenation*, which is the addition of molecular hydrogen to compounds containing C=C and C≡C bonds.

$$CH_2=CH_2 + H_2 \longrightarrow CH_3-CH_3$$

$$CH\equiv CH + 2\,H_2 \longrightarrow CH_3-CH_3$$

Alkanes are called *saturated hydrocarbons* because no more hydrogen can be added to the carbon atoms. The carbon atoms in a saturated hydrocarbon are already bonded to the maximum number of H atoms.

$$CH_3-CH_3 + H_2 \longrightarrow \text{ no reaction}$$

Other addition reactions involve an addition of HX or X_2 to the multiple bond, where X represents a halogen (Cl, Br, or I). Examples of these addition reactions are:

$$CH_2=CH_2 + HCl \longrightarrow CH_3-CH_2Cl$$

$$CH_2=CH_2 + Cl_2 \longrightarrow CH_2Cl-CH_2Cl$$

PROBLEM TYPE 4: DISTINGUISHING BETWEEN STRUCTURAL AND GEOMETRIC ISOMERS

Recall that **structural isomers** are molecules that have the same molecular formula but a different order of linking the atoms. **Geometric isomers** are molecules with the same type, number, and order of attachment of atoms and the same chemical bonds, but different spatial arrangements of the atoms. For example, 1,2-dichloroethene exists as two geometric isomers.

cis-1,2-dichloroethene   *trans*-1,2-dichloroethene

Note that in the *cis* isomer, the two Cl atoms (and the two H atoms) are adjacent to each other, whereas in the *trans* isomer, the Cl atoms are on the opposite side of the C=C bond.

EXAMPLE 24.9

Draw Lewis structures for all compounds, not including cyclic compounds, with the molecular formula C_5H_{10} and determine which are geometric isomers.

Alkenes have the general formula C_nH_{2n}. Thus the structures with molecular formula C_5H_{10} should be alkenes.

To draw the correct Lewis structures, follow the procedure outlined in Chapter 9. We can draw six Lewis structures as follows:

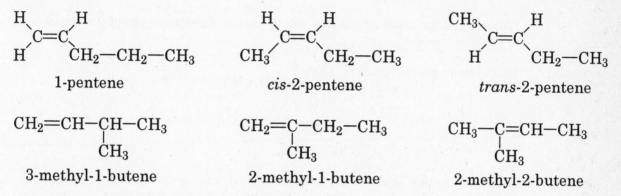

The geometric isomers are *cis*- and *trans*-2-pentene. 1-pentene, 3-methyl-1-butene, 2-methyl-1-butene, and 2-methyl-2-butene are structural isomers to the geometric isomers because the atoms are attached in a different order.

Why don't 1-pentene, 3-methyl-1-butene, 2-methyl-1-butene, and 2-methyl-2-butene have geometric isomers?

PRACTICE EXERCISE

6.   Draw the structural formula for 3-bromo-2,5-dimethyl-*trans*-3-hexene.

Text Problem: 24.24

PROBLEM TYPE 5: FUNCTIONAL GROUPS

We will discuss the most common organic functional groups, with emphasis on classes of compounds in which the functional groups include oxygen or nitrogen.

1.   **Alcohols.**   All alcohols contain the hydroxyl group, –OH. Figure 24.8 of your text shows some common alcohols. Alcohols typically undergo esterification (formation of an ester) reactions with carboxylic acids. Oxidation to aldehydes, ketones, and carboxylic acids are also common reactions.

2.   **Ethers.**   Ethers contain the R–O–R' linkage, where R and R' are either an alkyl group or a group derived from an aromatic hydrocarbon.

3.   **Aldehydes and Ketones.**   These compounds contain the carbonyl functional group.

$$\underset{\diagup \quad \diagdown}{\overset{\displaystyle O}{\overset{\displaystyle \|}{C}}}$$

The difference between aldehydes and ketones is that in aldehydes at least one hydrogen atom is bonded to the carbon atom of the carbonyl group. In ketones, no hydrogen atoms are bonded to the carbonyl carbon atom. Common reactions include reduction to yield alcohols. Oxidation of aldehydes yields carboxylic acids.

4.   **Carboxylic Acids.** These compounds contain the carboxyl group, $-COOH$.

Common reactions include esterification with alcohols and reaction with phosphorus pentachloride to yield acid chlorides.

5.   **Esters.**   Esters have the general formula R'COOR.

R' can be H, an alkyl, or an aromatic hydrocarbon group, and R is an alkyl or an aromatic hydrocarbon group. A common reaction is hydrolysis to yield acids and alcohols.

6.   **Amines.**   Amines are organic bases. They have the general formula R_3N, where one of the R groups must be an alkyl group or an aromatic hydrocarbon group. A common reaction is formation of ammonium salts with acids.

Table 24.4 of your text summarizes important functional groups. For more detailed discussions of reactions, see Section 24.4 of your text.

EXAMPLE 24.12
Identify the functional groups in the following molecules: (a) $C_5H_{11}OH$, (b) CH_3CHO, (c) $C_3H_7OCH_3$, (d) $CH_3COC_2H_5$, (e) CH_3COOCH_3.

Learning to recognize functional groups requires memorization of their structural formulas. Table 24.4 of your text shows a number of the important functional groups.

(a) $C_5H_{11}OH$ contains a *hydroxyl* group. It is an *alcohol*.

(b) CH_3CHO is a way to represent

on a single line of type. C=O is a carbonyl group. Since there is a hydrogen atom bonded to the carbonyl carbon, CH_3CHO is an *aldehyde*.

(c) $C_3H_7OC_2H_5$ contains a C–O–C group and is therefore an *ether*.

(d) $CH_3COC_2H_5$ contains a carbonyl group. Since a hydrogen atom is not bonded to the carbonyl carbon, this molecule is a *ketone*.

(e) CH_3COOCH_3 is a condensed structural formula for an *ester*.

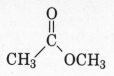

PRACTICE EXERCISE

7.   Indicate the functional groups by name that are in the following molecules.

$$CH_3CH{=}CHCHCH_2NH_2$$
$$\overset{|}{OH}$$

(a)

(b) $$HO{-}\overset{\overset{O}{\|}}{C}{-}CH_2{-}\overset{\overset{O}{\|}}{C}{-}CH_2CH_2\overset{\overset{O}{\|}}{C}H$$

Text Problems: **24.36**, 24.38, 24.40, 24.42

PROBLEM TYPE 6: CHIRALITY

Compounds that come as mirror-image pairs can be compared with left-handed and right-handed gloves and are thus referred to as **chiral**, or handed, molecules. While every molecule can have a mirror image, the *chiral* mirror-image pairs are *nonsuperimposable*. Conversely, achiral (nonchiral) pairs are superimposable. See Figure 24.3 of your text for an illustration of superimposable compared to nonsuperimposable.

Observations show that most simple chiral molecules contain at least one *asymmetric* carbon atom; that is, a carbon atom bonded to four different atoms or groups of atoms.

> **Tip:** Consider your hands when thinking about chirality. If you view your left hand in a mirror, the mirror image of your left hand is a right hand. However, your left and right hands are nonsuperimposable. To verify this, try putting a "left-handed" glove on your right hand.

EXAMPLE 24.13
Classify the following objects as chiral or achiral:

(a)   shoe   **(b)   screw**   **(c)   fork**   **(d)   coffee cup**

A shoe and a screw are chiral. The mirror image of a right shoe is a left shoe. A left shoe is not superimposable on a right shoe. A mirror image of a screw with clockwise threads will have counterclockwise threads. The screws would not be superimposable.

A fork and a coffee cup are achiral. Their mirror images are superimposable.

PRACTICE EXERCISE

8. Are the following molecules chiral?

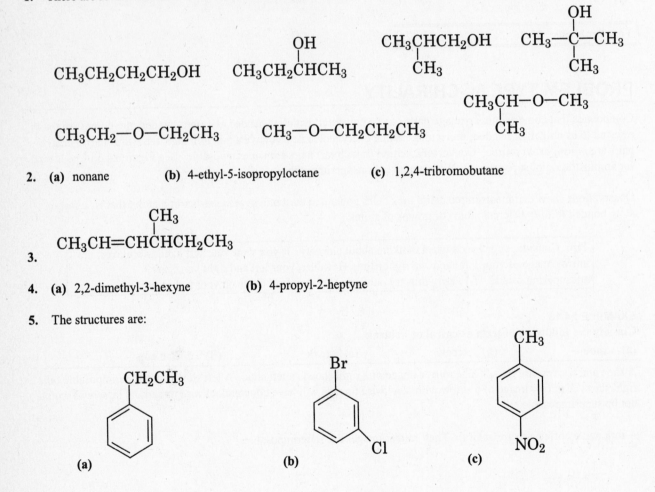

(a) 1,1-dichloroethane

(b) 1-bromo-1-chloroethane

Text Problem: 24.56

ANSWERS TO PRACTICE EXERCISES

1. There are seven structural isomers—four alcohols and three ethers. The structures are shown below.

2. **(a)** nonane   **(b)** 4-ethyl-5-isopropyloctane   **(c)** 1,2,4-tribromobutane

3.

4. **(a)** 2,2-dimethyl-3-hexyne   **(b)** 4-propyl-2-heptyne

5. The structures are:

(a)   (b)   (c)

6. The structure of 3-bromo-2,5-dimethyl-*trans*-3-hexene is:

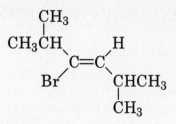

7. (a) carbon-carbon double bond, hydroxyl (–OH), and amine (R–NH$_2$).

   (b) carboxyl (–COOH), carbonyl (ketone), and carbonyl (aldehyde).

8. (a) There are only three different groups bonded to the second carbon atom. This molecule is *achiral*.

   (b) Replacing one of the chloro groups with a bromo group places four different groups on the second carbon atom. This molecule is *chiral*.

SOLUTIONS TO SELECTED TEXT PROBLEMS.

24.12   Determining the Number of Structural Isomers, Problem Type 1.

We are starting with *n*-pentane, so we do not need to worry about any branched chain structures. In the chlorination reaction, a Cl atom replaces one H atom. There are three different carbons on which the Cl atom can be placed. Hence, *three* structural isomers of chloropentane can be derived from *n*–pentane:

$CH_3CH_2CH_2CH_2CH_2Cl$ $CH_3CH_2CH_2CHClCH_3$ $CH_3CH_2CHClCH_2CH_3$

24.14   Both alkenes and cycloalkanes have the general formula C_nH_{2n}. Let's start with C_3H_6. It could be an alkene or a cycloalkane.

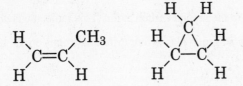

Now, let's replace one H with a Br atom to form C_3H_5Br. *Four* isomers are possible.

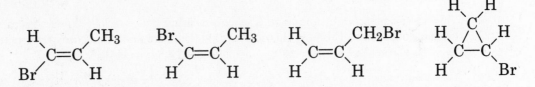

There is only one isomer for the cycloalkane. Note that all three carbons are equivalent in this structure.

24.16   **(a)**   This compound could be an **alkene** or a **cycloalkane**; both have the general formula, C_nH_{2n}.

(b)   This could be an **alkyne** with general formula, C_nH_{2n-2}. It could also be a hydrocarbon with two double bonds (a diene). It could be a cyclic hydrocarbon with one double bond (a cycloalkene).

(c)   This must be an **alkane**; the formula is of the C_nH_{2n+2} type.

(d)   This compound could be an **alkene** or a **cycloalkane**; both have the general formula, C_nH_{2n}.

(e)   This compound could be an **alkyne** with one triple bond, or it could be a cyclic alkene (unlikely because of ring strain).

24.18   If cyclobutadiene were square or rectangular, the C–C–C angles must be 90°. If the molecule is diamond-shaped, two of the C–C–C angles must be less than 90°. Both of these situations result in a great deal of distortion and strain in the molecule. Cyclobutadiene is very unstable for these and other reasons.

24.20   One compound is an alkane; the other is an alkene. Alkenes characteristically undergo addition reactions with hydrogen, with halogens (Cl_2, Br_2, I_2) and with hydrogen halides (HCl, HBr, HI). Alkanes do not react with these substances under ordinary conditions.

24.22   In this problem you are asked to calculate the standard enthalpy of reaction. This type of problem was covered in Chapter 6, Problem Type 4A.

$$\Delta H^\circ_{rxn} = \Sigma n \, \Delta H^\circ_f \text{ (products)} - \Sigma m \, \Delta H^\circ_f \text{ (reactants)}$$

$$\Delta H^\circ_{rxn} = \Delta H^\circ_f \, (C_6H_6) - 3 \, \Delta H^\circ_f \, (C_2H_2)$$

You can look up ΔH°_f values in Appendix 3 of your textbook.

$$\Delta H^\circ_{rxn} = (1 \text{ mol})(49.04 \text{ kJ/mol}) - (3 \text{ mol})(226.6 \text{ kJ/mol}) = \mathbf{-630.8 \text{ kJ}}$$

24.24   In this problem you must distinguish between *cis* and *trans* isomers. Recall that *cis* means that two particular atoms (or groups of atoms) are adjacent to each other, and *trans* means that the atoms (or groups of atoms) are on opposite sides in the structural formula.

In (a), the Cl atoms are adjacent to each other. This is the *cis* isomer. In (b), the Cl atoms are on opposite sides of the structure. This is the *trans* isomer.

The names are: **(a)** *cis*-**1,2-dichlorocyclopropane**; and **(b)** *trans*-**1,2-dichlorocyclopropane**.

Are any other dichlorocyclopropane isomers possible?

24.26   **(a)**   This is a branched hydrocarbon. The name is based on the longest carbon chain. The name is **2–methylpentane**.

   (b)   This is also a branched hydrocarbon. The longest chain includes the C_2H_5 group; the name is based on hexane, not pentane. This is an old trick. Carbon chains are flexible and don't have to lie in a straight line. The name is **2,3,4–trimethylhexane**. Why not 3,4,5–trimethylhexane?

   (c)   How many carbons in the longest chain? It doesn't have to be straight! The name is **3–ethylhexane**.

   (d)   An alkene with two double bonds is called a diene. The name is **3–methyl–1,4–pentadiene**.

   (e)   The name is **2–pentyne**.

   (f)   The name is **3–phenyl–1–pentene**.

24.28   The hydrogen atoms have been omitted from the skeletal structure for simplicity.

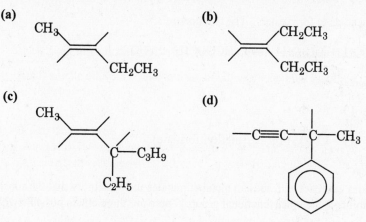

24.32 Organic Nomenclature, Problem Type 2.

When a benzene ring has more than *two* substituents, you must specify the location of the substituents with numbers. Remember to number the ring so that you end up with the lowest numbering scheme as possible, giving preference to alphabetical order.

(a) Since a chloro group comes alphabetically before a methyl group, let's start by numbering the top carbon of the ring as 1. If we number clockwise, this places the second chloro group on carbon 3 and a methyl group on carbon 4.

This compound is **1,3–dichloro–4–methylbenzene**.

(b) If we start numbering counterclockwise from the bottom carbon of the ring, the name is 2–ethyl–1,4–dinitrobenzene. Numbering clockwise from the top carbon gives 3–ethyl–1,4–dinitrobenzene.

Numbering as low as possible, the correct name is **2–ethyl–1,4–dinitrobenzene**.

(c) Again, keeping the numbers as low as possible, the correct name for this compound is **1,2,4,5–tetramethylbenzene**. You should number clockwise from the top carbon of the ring.

24.36 Functional Groups, Problem Type 5.

(a) $H_3C–O–CH_2–CH_3$ contains a C–O–C group and is therefore an **ether**.

(b) This molecule contains an RNH_2 group and is therefore an **amine**.

(c) This molecule is an **aldehyde**. It contains a carbonyl group in which one of the atoms bonded to the carbonyl carbon is a hydrogen atom.

(d) This molecule also contains a carbonyl group. However, in this case there are no hydrogen atoms bonded to the carbonyl carbon. This molecule is a **ketone**.

(e) This molecule contains a carboxyl group. It is a **carboxylic acid**.

(f) This molecule contains a hydroxyl group (–OH). It is an **alcohol**.

(g) This molecule has both an RNH_2 group and a carboxyl group. It is therefore both an *amine* and a *carboxylic acid*, commonly called an **amino acid**.

24.38 Alcohols react with carboxylic acids to form esters. The reaction is:

$$HCOOH + CH_3OH \longrightarrow HCOOCH_3 + H_2O$$

The structure of the product is:

$$\overset{\displaystyle O}{\underset{\displaystyle }{H-\overset{\|}{C}-O-CH_3}} \quad \text{(methyl formate)}$$

24.40 The fact that the compound does not react with sodium metal eliminates the possibility that the substance is an alcohol. The only other possibility is the ether functional group. There are three ethers possible with this molecular formula:

$$CH_3–CH_2–O–CH_2–CH_3 \qquad CH_3–CH_2–CH_2–O–CH_3 \qquad (CH_3)_2CH–O–CH_3$$

Light–induced reaction with chlorine results in substitution of a chlorine atom for a hydrogen atom (the other product is HCl). For the first ether there are only two possible chloro derivatives:

$ClCH_2-CH_2-O-CH_2-CH_3$ $CH_3-CHCl-O-CH_2-CH_3$

For the second there are four possible chloro derivatives. Three are shown below. Can you draw the fourth?

$CH_3-CHCl-CH_2-O-CH_3$ $CH_3-CH_2-CHCl-O-CH_3$ $CH_2Cl-CH_2-CH_2-O-CH_3$

For the third there are three possible chloro derivatives:

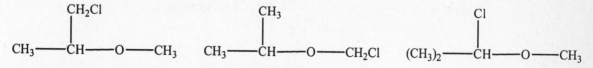

The **(CH$_3$)$_2$CH–O–CH$_3$** choice is the original compound.

24.42   **(a)**   ketone   **(b)**   ester   **(c)**   ether

24.44   See Problem Type 4C, Chapter 6.

If we rearrange the equations given and multiply times the necessary factors, we have:

$$2\, CO_2\,(g) + 2\, H_2O\,(l) \longrightarrow C_2H_4\,(g) + 3\, O_2\,(g) \qquad \Delta H^\circ = 1411\ kJ$$
$$C_2H_2\,(g) + \frac{5}{2}\, O_2\,(g) \longrightarrow 2\, CO_2\,(g) + H_2O\,(l) \qquad \Delta H^\circ = -1299.5\ kJ$$
$$H_2\,(g) + \frac{1}{2}\, O_2\,(g) \longrightarrow H_2O\,(l) \qquad \Delta H^\circ = -285.8\ kJ$$

$$\overline{C_2H_2\,(g) + H_2\,(g) \longrightarrow C_2H_4\,(g) \qquad \Delta H^\circ = \mathbf{-174\ kJ}}$$

The heat of hydrogenation for acetylene is **–174 kJ**.

24.46   To form a hydrogen bond *with water* a molecule must have at least one H–F, H–O, or H–N bond, *or* must contain an O, N, or F atom. The following can form hydrogen bonds with water:

(a)   carboxylic acids   **(c)**   ethers   **(d)**   aldehydes   **(f)**   amines

24.48   **(a)**   rubbing alcohol;   **(b)**   vinegar;   **(c)**   moth balls;   **(d)**   organic synthesis

(e)   organic synthesis;   **(f)**   antifreeze;   **(g)**   fuel (natural gas);   **(h)**   synthetic polymers.

24.50   **(a)**   2–butyne has **three** C–C sigma bonds.

(b)   Anthracene is:

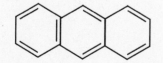

There are **sixteen** C–C sigma bonds.

(c)

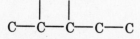

There are **six** C–C sigma bonds.

24.52   This problem relates back to Problem Type 5, Chapter 3—The Experimental Determination of Empirical Formulas.

(a)   The easiest way to calculate the mg of C in CO_2 is by mass ratio. There are 12.01 g of C in 44.01 g CO_2 or 12.01 mg C in 44.01 mg CO_2.

$$? \text{ mg C } = 57.94 \text{ mg } CO_2 \times \left(\frac{12.01 \text{ mg C}}{44.01 \text{ mg } CO_2} \right) = \textbf{15.81 mg C}$$

Similarly,

$$? \text{ mg H } = 11.85 \text{ mg } H_2O \times \left(\frac{2.016 \text{ mg H}}{18.02 \text{ mg } H_2O} \right) = \textbf{1.33 mg H}$$

The mg of oxygen can be found by difference.

$$? \text{ mg O } = 20.63 \text{ mg Y} - 15.81 \text{ mg C} - 1.33 \text{ mg H} = \textbf{3.49 mg O}$$

(b)   *Step 1:* Calculate the number of moles of each element present in the sample. Use molar mass as a conversion factor.

$$? \text{ mol C } = 15.81 \text{ mg C} \times \frac{1 \text{ g}}{1000 \text{ mg}} \times \frac{1 \text{ mol C}}{12.01 \text{ g C}} = 1.316 \times 10^{-3} \text{ mol C}$$

Similarly,

$$? \text{ mol H } = 1.33 \text{ mg H} \times \frac{1 \text{ g}}{1000 \text{ mg}} \times \frac{1 \text{ mol H}}{1.008 \text{ g H}} = 1.319 \times 10^{-3} \text{ mol H}$$

$$? \text{ mol O } = 3.49 \text{ mg O} \times \frac{1 \text{ g}}{1000 \text{ mg}} \times \frac{1 \text{ mol O}}{16.00 \text{ g O}} = 2.18 \times 10^{-4} \text{ mol O}$$

Thus, we arrive at the formula $C_{1.316 \times 10^{-3}} H_{1.319 \times 10^{-3}} O_{2.18 \times 10^{-4}}$, which gives the identity and the ratios of atoms present. However, chemical formulas are written with whole numbers.

Step 2: Try to convert to whole numbers by dividing all the subscripts by the smallest subscript.

$$\text{C: } \frac{1.316 \times 10^{-3}}{2.18 \times 10^{-4}} = 6.04 \approx 6 \qquad \text{H: } \frac{1.319 \times 10^{-3}}{2.18 \times 10^{-4}} = 6.05 \approx 6 \qquad \text{O: } \frac{2.18 \times 10^{-4}}{2.18 \times 10^{-4}} = 1.00$$

This gives us the empirical formula, **C_6H_6O**.

(c)   The presence of six carbons and a corresponding number of hydrogens suggests a benzene derivative. A plausible structure is shown below.

24.54   Since the masses are given in milligrams, it is more convenient to work with millimoles (mmol). The number of mmoles of carbon is:

$$\text{C:} \quad 9.708 \text{ mg CO}_2 \left(\frac{1 \text{ mmol CO}_2}{44.01 \text{ mg CO}_2} \right)\left(\frac{1 \text{ mmol C}}{1 \text{ mmol CO}_2} \right) = 0.2206 \text{ mmol C}$$

$$\text{H:} \quad 3.969 \text{ mg H}_2\text{O} \left(\frac{1 \text{ mmol H}_2\text{O}}{18.02 \text{ mg H}_2\text{O}} \right)\left(\frac{2 \text{ mmol H}}{1 \text{ mmol H}_2\text{O}} \right) = 0.441 \text{ mmol H}$$

The mass of oxygen is found by difference:

$$3.795 \text{ mg compound} - (2.649 \text{ mg C} + 0.445 \text{ mg H}) = 0.701 \text{ mg O}$$

$$\text{O:} \quad 0.701 \text{ mg} \left(\frac{1 \text{ mmol O}}{16.00 \text{ mg O}} \right) = 0.0438 \text{ mmol O}$$

This gives the formula is $C_{0.2206}H_{0.441}O_{0.0438}$. Dividing by the smallest number of moles gives the empirical formula, **$C_5H_{10}O$**.

The molar mass is found using the ideal gas equation:

$$\mathcal{M} = \frac{mRT}{PV} = \frac{(0.205 \text{ g})(0.0821 \text{ L} \cdot \text{atm} / \text{K} \cdot \text{mol})(473 \text{ K})}{(1.00 \text{ atm})(0.0898 \text{ L})} = \textbf{88.7 g/mol}$$

The formula mass of $C_5H_{10}O$ is 86.13 g, so this is also the molecular formula. Three possible structures are:

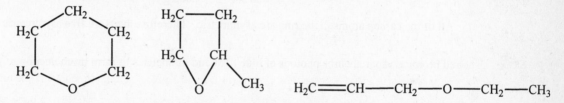

24.56   A carbon atom is asymmetric if it is bonded to four different atoms or groups. In the given structures the asymmetric carbons are marked with an asterisk (*).

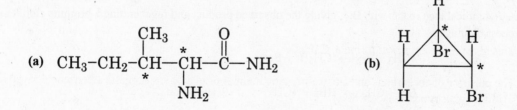

24.58   Acetone is a ketone with the formula, CH_3COCH_3. We must write the structure of an aldehyde that has the same number and types of atoms (C_3H_6O). Removing the aldehyde functional group (–CHO) from the formula leaves C_2H_5. This is the formula of an ethyl group. The aldehyde that is a structural isomer of acetone is:

$$\underset{}{CH_3CH_2}\overset{\displaystyle O}{\overset{\displaystyle \|}{C}}-H$$

24.60   **(a)**   alcohol   **(b)**   ether   **(c)**   aldehyde   **(d)**   carboxylic acid   **(e)**   amine

24.62   In Chapter 11, we found that salts with their electrostatic intermolecular attractions had low vapor pressures and thus high boiling points. Ammonia and its derivatives (amines) are molecules with dipole–dipole attractions. If the nitrogen has one direct N–H bond, the molecule will have hydrogen bonding. Even so, these molecules will have much weaker intermolecular attractions than ionic species and hence higher vapor pressures. Thus, if we could convert the neutral ammonia–type molecules into salts, their vapor pressures, and thus associated odors, would decrease. Lemon juice contains acids which can react with ammonia–type (amine) molecules to form ammonium salts.

$$NH_3 + H^+ \longrightarrow NH_4^+ \qquad\qquad RNH_2 + H^+ \longrightarrow RNH_3^+$$

24.64   Marsh gas (methane, CH_4); grain alcohol (ethanol, C_2H_5OH); wood alcohol (methanol, CH_3OH); rubbing alcohol (isopropyl alcohol, $(CH_3)_2CHOH$); antifreeze (ethylene glycol, CH_2OHCH_2OH); mothballs (naphthalene, $C_{10}H_8$); vinegar (acetic acid, CH_3COOH).

24.66   The asymmetric carbons are shown by asterisks:

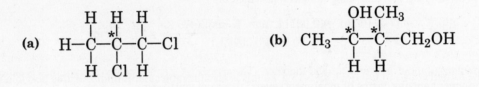

(c)   All of the carbon atoms in the ring are asymmetric. Therefore there are **five** asymmetric carbon atoms.

24.68   The red bromine vapor absorbs photons of blue light and dissociates to form bromine atoms.

$$Br_2 \rightarrow 2Br\bullet$$

The bromine atoms collide with methane molecules and abstract hydrogen atoms.

$$Br\bullet + CH_4 \rightarrow HBr + \bullet CH_3$$

The methyl radical then reacts with Br_2, giving the observed product and regenerating a bromine atom to start the process over again:

$$\bullet CH_3 + Br_2 \rightarrow CH_3Br + Br\bullet$$

$$Br\bullet + CH_4 \rightarrow HBr + \bullet CH_3 \qquad\qquad \text{and so on...}$$

CHAPTER 25
SYNTHETIC AND NATURAL
ORGANIC POLYMERS

PROBLEM-SOLVING STRATEGIES AND TUTORIAL SOLUTIONS

TYPES OF PROBLEMS

Problem Type 1: Synthetic Organic Polymers.
    (a) Addition reactions.
    (b) Condensation reactions.

Problem Type 2: Proteins.

Problem Type 3: Nucleic Acids.

PROBLEM TYPE 1: SYNTHETIC ORGANIC POLYMERS

The word **polymer** means "many parts". A **polymer** is a compound with an unusually high molecular mass, consisting of a large number of molecular units linked together. The small unit that is repeated many times is called a **monomer**. A typical polymer contains chains of monomers several thousand units long.

A. Addition reactions

Addition polymers are made by adding monomer to monomer until a long chain is produced. Ethylene (ethene) and its derivatives are excellent monomers for forming addition polymers. In Chapter 24, we saw that addition reactions occur with unsaturated compounds containing C=C and C≡C bonds. In an addition reaction, the polymerization process is initiated by a radical or an ion. When ethylene is heated to 250°C under high pressure (1000–3000 atm) in the presence of a little oxygen or benzoyl peroxide (the initiator), addition polymers with masses of about 30,000 amu are obtained. This reaction is represented by

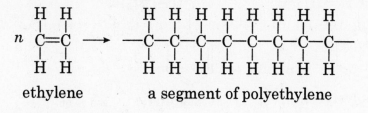

ethylene   a segment of polyethylene

The general equation for addition polymerization is:

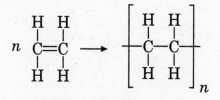

monomer   repeating unit

Substitution for one or more hydrogen atoms in ethylene with Cl atoms, acetate, CN, or F provides a wide selection of monomers from which to make addition polymers with various properties. For instance, substitution of a Cl atom for a H atom in ethylene gives the monomer called vinyl chloride, $CH_2=CHCl$. Polymerization of vinyl chloride yields the polymer polyvinyl chloride (PVC).

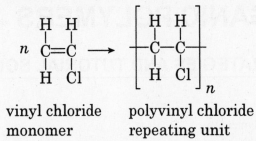

$$\begin{matrix} & \text{vinyl chloride} & \text{polyvinyl chloride} \\ & \text{monomer} & \text{repeating unit} \end{matrix}$$

Polymers that are made from one type of monomer such as polyvinyl chloride are called **homopolymers**. Table 25.1 of your text gives the names, structures, and uses of a number of monomers and addition polymers.

For more detailed information on the reaction mechanism of addition polymerization, see Section 25.2 of your text.

EXAMPLE 25.1
Write the formulas of the monomers used to prepare the following polymers:
(a) Teflon, (b) Polystyrene, and (c) PVC.

Refer to Table 25.1 of your text.

(a)   Teflon is an addition polymer with the formula

$$\left(\!\!-CF_2\!\!-\!\!CF_2\!\!-\!\right)_n$$

It is prepared from the monomer tetrafluoroethylene ($CF_2=CF_2$).

(b)   The monomer used to prepare polystyrene is styrene.

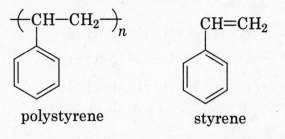

$$\begin{matrix} \text{polystyrene} & \qquad & \text{styrene} \end{matrix}$$

(c)   PVC (polyvinylchloride) is prepared by the successive addition of vinyl chloride molecules ($CH_2=CHCl$).

PRACTICE EXERCISE

1. Draw the structures of the monomers from which the following polymers are formed:

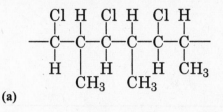

(a)

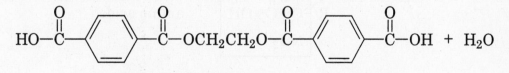

(b)

Text Problems: 25.8, 25.10, 25.12a

B. Condensation reactions

Copolymers are polymers that contain two or more different types of monomers. Polyesters, such as the well-known Dacron, are copolymers. In Dacron, one monomer is an alcohol and the other is a carboxylic acid, which can be joined by an *esterification* reaction. The alcohol and the acid both must contain two functional groups. The monomers of Dacron are the dicarboxylic acid called terephthalic acid and the dialcohol called ethylene glycol.

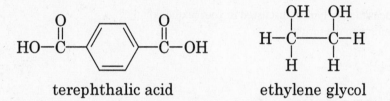

terephthalic acid   ethylene glycol

Condensation reactions differ from addition reactions in that the former always result in the formation of a small molecule such as water. When terephthalic acid and ethylene glycol react to form an ester, the first products are:

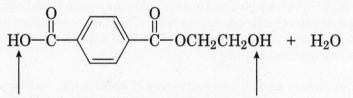

sites for further condensation reactions

When this product reacts with another molecule of the diacid, the polymer chain grows longer.

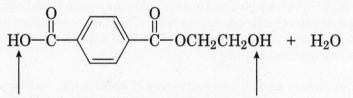

segment of a condensation polymer chain

The general formula for the polyester, Dacron, is:

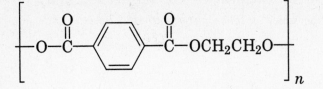

EXAMPLE 25.2
Draw structures for the monomers used to make the following polyester:

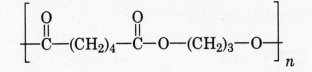

Answer

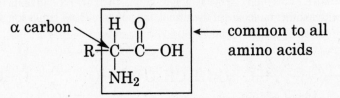

PRACTICE EXERCISE
2.   List two examples of condensation polymers discussed in your textbook.

┌─────────────────────────────┐
│ **Text Problem:** 25.12b   │
└─────────────────────────────┘

PROBLEM TYPE 2: PROTEINS

Proteins are *polymers of amino acids*. Proteins are truly giant molecules having molecular masses that range from about 10,000 to several million amu. Proteins play many roles in living organisms, where they function as catalysts (enzymes), transport molecules (hemoglobin), contractile fibers (muscle), protective agents (blood clots), hormones (chemical messengers), and structural members (feathers, horns, and nails). The word *protein* comes from the Greek work *proteios*, meaning "first". From the partial list of protein functions above, it is easy to see why proteins occupy "first place" among biomolecules in their importance to life.

Even though each protein is unique, all proteins are built from a set of only **20 amino acids**. An amino acid consists of an amino group, a carboxylic acid group, a hydrogen atom, and a distinctive R group, all bonded to the same carbon atom.

$$\alpha \text{ carbon} \longrightarrow \quad \begin{array}{c} H \quad O \\ | \quad || \\ R-C-C-OH \\ | \\ NH_2 \end{array} \quad \longleftarrow \begin{array}{l} \text{common to all} \\ \text{amino acids} \end{array}$$

All amino acids in proteins have a common structural feature that is the attachment of the amino group ($-NH_2$) to the carbon atom adjacent to the carboxylic acid group. This carbon is called the α carbon. The difference in amino acids is due to different R groups. Twenty different R groups are found in proteins from natural sources. In fact, all proteins in all species, from bacteria to humans, are constructed from the same set of 20 amino acids. The structural formulas of the 20 amino acids essential to living organisms are shown in Table 25.2 of the text.

In proteins, the amino acid units are joined together to form a *polypeptide* chain. The carboxyl group of one amino acid is joined to the amino group of another amino acid by the formation of a peptide bond.

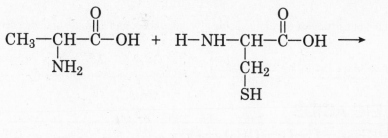

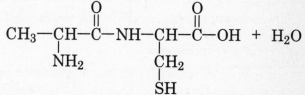

This type of reaction is another example of a condensation reaction. The new C–N covalent bond is called a *peptide* or an *amide* bond. The amide functional group present in all proteins is:

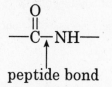

peptide bond

The molecule above, in which two amino acids are joined, is called a *dipeptide*. Peptides are structures intermediate in size between amino acids and proteins. The term *polypeptide* refers to long molecular chains containing many amino acid units. An amino acid unit in a polypeptide chain is called a *residue*.

Proteins are so complex that four levels of structural features have been identified. The structure of proteins is extremely important in determining how efficiently and effectively a protein will function. We will discuss the first two levels of protein structure.

1. **The Primary Structure**. Each protein has a unique amino acid sequence of its polypeptide chain. It is the amino acid sequence that distinguishes one protein from another. Proteins also differ in the numbers and types of amino acids, but the main difference is the sequence of amino acid residues.

2. **The Secondary Structure**. This refers to the spatial relationship of amino acid units that are close to one another in sequence. The configuration that appears in many proteins is the α helix, shown in Figure 25.11 of the textbook. In this configuration, the polypeptide is coiled much like the arrangement of stairs in a spiral staircase. Figure 25.11 also shows that the α helix is stabilized by the presence of intermolecular hydrogen bonds (dashed lines). The C=O group of each amino acid is hydrogen bonded to the NH group of the amino acid that is located four amino acids ahead in the sequence. The α helix is the main structural feature of the oxygen-transport protein, hemoglobin, and of many other proteins. The structure of hemoglobin is shown in Figure 25.13 of your text.

The β-pleated sheet is another common secondary structure. In this structure, a polypeptide chain interacts strongly with adjacent chains by forming many hydrogen bonds. See Figure 25.12 of your text.

EXAMPLE 25.3

What are the five chemical elements found in proteins?

Proteins are made up of the same elements as amino acids. Therefore, proteins contain C, H, O, and N. Two amino acids, methionine and cysteine, also contain sulfur, S.

PRACTICE EXERCISE

3.   Sketch a portion of the polypeptide chain consisting of the amino acids glycine, valine, and alanine, in that order.

Text Problems: 25.20, 25.22

PROBLEM TYPE 3: NUCLEIC ACIDS

The chemical composition of the cell nucleus was first studied in the 1860s by Friedrich Miescher. He found the major components to be protein and a new material not previously isolated. This material was found to be acidic, and so it was referred to as **nucleic acid**.

Nucleic acids are polymers that store genetic information and control protein synthesis. There are two types of nucleic acids, deoxyribonucleic acid (DNA) and ribonucleic acid (RNA). *DNA* carries all the genetic information necessary to carry on reproduction and to sustain an organism throughout its lifetime. *RNA* transcribes these instructions and controls the synthesis of proteins needed to implement them.

DNA molecules are among the largest known with molar masses up to 10 billion g/mol. Hydrolysis of nucleic acids show that they are composed of only four types of building blocks: a phosphate group, a sugar, purines, and pyrimidines.

Phosphate

One of two sugars is present, ribose in RNA and deoxyribose in DNA.

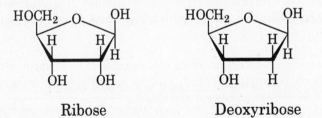

Ribose   Deoxyribose

Two purines, adenine and guanine, are found in both DNA and RNA.

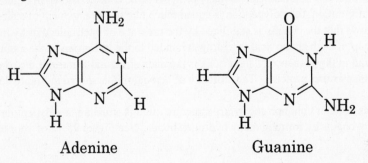

Adenine   Guanine

DNA also contains the pyrimidines thymine and cytosine. RNA also contains thymine and another pyrimidine, uracil.

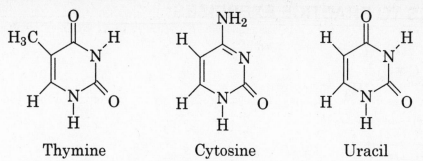

The purines and pyrimidines are collectively called **bases**. The table below summarizes the building blocks of DNA and RNA.

Building Blocks of DNA and RNA

	DNA	RNA
Sugar	Phosphate (P)	Phosphate (P)
	Deoxyribose (D)	Ribose (R)
Purines	Adenine (A)	Adenine (A)
	Guanine (G)	Guanine (G)
Pyrimidines	Thymine (T)	Thymine (T)
	Cytosine (C)	Uracil (U)

In the 1940s, Edwin Chargaff studied the composition of DNA. His analysis of the base composition of DNA showed that the amount of adenine always equaled that of thymine and that the amount of guanine equaled that of cytosine.

In 1953, Watson and Crick proposed a double-helical structure for DNA. The repeating unit in each strand is called a **nucleotide**, which consists of a base-deoxyribose-phosphate linkage (see Figure 25.18 of the text). The two strands are *not* identical, rather they are complementary. Adenine in one strand is always paired with thymine in the other strand. Guanine is always paired with cytosine. The base pairing is consistent with Chargaff's rules. Base pairing and the resulting association of the two strands are the result of *hydrogen bonding*. Hydrogen atoms in a base in one strand are attracted to unshared electron pairs on oxygen and nitrogen atoms of the base attached to the other strand (see Figure 25.19 of the text).

RNA, on the other hand, does not follow the base-pairing rules. X ray data and other evidence ruled out a double-helical structure for RNA. RNA is single-stranded.

EXAMPLE 25.4

What types of forces cause base pairing in the double-stranded helical DNA molecule?

The purine bases (adenine and guanine) in one strand of DNA form hydrogen bonds to the pyrimidine bases (thymine and cytosine) in the other DNA strand. Hydrogen atoms covalently bonded to nitrogen carry a partial positive charge. These H atoms are attracted to lone-electron pairs on oxygen and nitrogen atoms of another base. Since two complementary bases are attached to different strands, the hydrogen bonds hold the strands together.

PRACTICE EXERCISE

4. If the base sequence in one strand of DNA is A, T, G, C, T, what is the base sequence in the complementary strand?

Text Problem: 25.28

ANSWERS TO PRACTICE EXERCISES

1. (a)

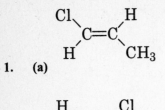

(b)

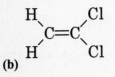

2. Nylon 66 (first prepared in 1931) and polyester.

3.

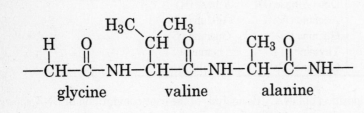

glycine   valine   alanine

4. Recall that according to Chargaff's rules, adenine (A) is always paired with thymine (T) and guanine (G) is always paired with cytosine (C). The complementary base sequence is T, A, C, G, A.

SOLUTIONS TO SELECTED TEXT PROBLEMS

25.8   The repeating structural unit of the polymer is:

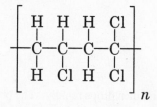

Does each carbon atom still obey the octet rule?

25.10   Polystyrene is formed by an addition polymerization reaction with the monomer, styrene, which is a phenyl–substituted ethylene. The structures of styrene and polystyrene are shown in Table 25.1 of your

25.12   The structures are shown.

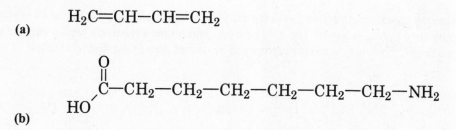

(a)

(b)

25.20   The main backbone of a polypeptide chain is made up of the α carbon atoms and the amide group repeating alternately along the chain.

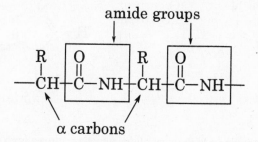

For each R group shown above, substitute the distinctive side groups of the two amino acids. Their are two possible dipeptides depending on how the two amino acids are connected, either glycine–lysine or lysine–glycine. The structures of the dipeptides are:

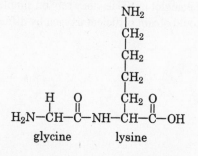

and

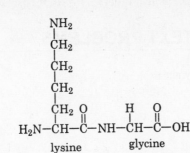

lysine   glycine

22   The rate increases in an expected manner from 10°C to 30°C and then drops rapidly. The probable reason for this is the loss of catalytic activity of the enzyme because of denaturation at high temperature.

Nucleic acids play an essential role in protein synthesis. Compared to proteins, which are made of up to 20 different amino acids, the composition of nucleic acids is considerably simpler. A DNA or RNA molecule contains only four types of building blocks: purines, pyrimidines, furanose sugars, and phosphate groups. Nucleic acids have simpler, uniform structures because they are primarily used for protein synthesis, whereas proteins have many uses.

The sample that has the higher percentage of C–G base pairs has a higher melting point because C–G base pairs are held together by three hydrogen bonds. The A–T base pair interaction is relatively weaker because it has only two hydrogen bonds. Hydrogen bonds are represented by dashed lines in the structures below.

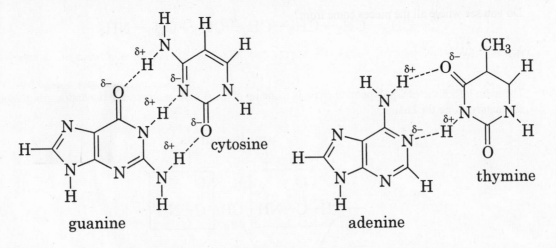

guanine   cytosine   adenine   thymine

25.32   Leg muscles are active having a high metabolism, which requires a high concentration of myoglobin. The high iron content from myoglobin makes the meat look dark after decomposition due to heating. The breast meat is "white" because of a low myoglobin content.

25.34   Insects have blood that contains no hemoglobin. Thus, they rely on simple diffusion to supply oxygen. It is unlikely that a human–sized insect could obtain sufficient oxygen by diffusion alone to sustain its metabolic requirements.

25.36 Calculating percent composition of a compound, Problem Type 4A, Chapter 3.

$$\% \text{ Fe } = \frac{\text{mass of Fe}}{\text{mass of compound (hemoglobin)}} \times 100\%$$

$$0.34\% = \frac{55.85 \text{ g}}{\text{mass of hemoglobin}} \times 100\%$$

minimum mass of hemoglobin = 1.6×10^4 g

Hemoglobin must contain **four Fe atoms per molecule** for the actual molar mass to be four times the minimum value calculated.

25.38 The type of intermolecular attractions that occur are mostly attractions between nonpolar groups. This type of intermolecular attraction is called a **dispersion force**.

25.40 This is as much a puzzle as it is a chemistry problem. The puzzle involves breaking up a nine-link chain in various ways and trying to deduce the original chain sequence from the various pieces. Examine the pieces and look for patterns. Remember that depending on how the chain is cut, the same link (amino acid) can show up in more than one fragment.

Since there are only seven different amino acids represented in the fragments, at least one must appear more than once. The nonapeptide is:

Gly–Ala–Phe–Glu–His–Gly–Ala–Leu–Val

Do you see where all the pieces come from?

25.42 No, the milk would *not* be fit to drink. Enzymes only act on one of two optical isomers of a compound.

Focus on
Object-Oriented Programming
with C++

Seventh Edition

Richard L. Stegman

Focus on Object-Oriented Programming with C++

Seventh Edition

Copyright © 2018, Richard L. Stegman

1111-10-00010010-0111-11010101-00000000

ISBN-13: 978-1979943697

ISBN-10: 1979943699

Table of Contents

Chapter 1. Data Abstraction & Abstract Data Types

Data Abstraction
Abstract Data Types

Data Abstraction

Data Abstraction

Data abstraction is the programming technique of inventing new data types that are well suited to an application, thereby making it easier to program. It is a powerful, general-purpose technique that can result in:

- Shorter programs
- More readable programs
- More flexible programs

A Motivation

Assume we are programming in C:

`first` contains a first name, such as Haley

`last` contains a last name, such as Laurel

We want to make `fullname` contain a first and last name separated by a period, such as `haley.laurel`

Release 1

```
char *first, *last;
char fullname[32];

strcpy(fullname, first);
strcat(fullname, ".");
strcat(fullname, last);
```

See if you can figure out why Release 1 crashes before reading further.

Why Release 1 Crashes

A user crashed Release 1 by using a `fullname` longer than 32 characters.

Release 2

```
char *first, *last;
char fullname[100];

if (strlen(first) + strlen(last) + 1 > 100)
    error();
strcpy(fullname, first);
strcat(fullname, ".") ;
strcat(fullname, last);
```

See if you can figure out why Release 2 crashes before reading further.

Why Release 2 Crashes
A user caused Release 2 to abort by using a `fullname` longer than 100 characters.

Release 3
We try doing it dynamically:

```
char *first, *last;
char *fullname;

fullname = malloc(strlen(first)+strlen(last)+1);
strcpy(fullname, first);
strcat(fullname, ".");
strcat(fullname, last);
```

See if you can figure out why Release 3 crashes before reading further.

Why Release 3 Crashes
Release 3 crashed at random times in random places, apparently unrelated to our subroutine.

An extensive debugging effort finally catches it: we are allocating one byte too little memory. Look closely at the `malloc()` declaration.

Release 4

```
char *first, *last;
char *fullname;

fullname = malloc(strlen(first)+strlen(last)+2);
strcpy(fullname, first);
strcat(fullname, ".");
strcat(fullname, last);
```

Release 4 worked fine in system tests. In the field, though, it was run on a machine with a very small memory and crashed when malloc() failed, returning NULL.

Release 5

```
char *first, *last;
char *fullname;

fullname = malloc(strlen(first) + strlen(last) + 2);
if (fullname == NULL)
    error();
strcpy(fullname, first);
strcat(fullname, ".");
strcat(fullname, last) ;
```

Release 5 no longer crashes. However, it gradually eats up memory when it is running for long enough.

Another huge debugging effort finds the problem, fixed in Release 6.

Release 6

```
char *first, *last;
char *fullname;

fullname = malloc(strlen(first)+ strlen(last) + 2);
if (fullname == NULL)
    error () ;
strcpy(fullname, first) ;
strcat(fullname, ".");
strcat(fullname, last) ;
free(fullname);
```

C++ Code using the String ADT

The problem with the C code is that the user has to worry about the implementation of character strings. In C++, given the abstract data type String, it's hard to avoid getting it right in Release 1:

```
#include "string.h"

String first, last;
String fullname;

fullname = first + "." + last;
```

Abstract Data Types

Fundamental Data Types
A data type provides a description of its variables that tells the computer such things as:

- how much memory to allocate for a variable
- how to interpret the data in that memory
- what range of values can be assigned to a variable
- what operations are permissible on that data

We do not have to write a description for the `int` or `float` data type. The author of the compiler did that for us and built it into the compiler. Data types that are built in to a compiler in this way are known as *fundamental data types*.

Abstract Data Types
Some programming languages have features that allow us to effectively extend the language by adding our own data types. A programmer-defined data type is called an *abstract-data type* (ADT) to distinguish it from a fundamental data type.

The term abstract refers to the way in which a programmer abstracts some concepts from a mass of programming detail and unifies these concepts to create a new data type.

ADT's in C++
In C++ we can implement our own abstract data types by declaring C++ classes and we can operate on them as if they had been built into the C++ language. This language feature of C++ allows programmers to become, in effect, language designers.

Being able to design and implement our own data types gives us lots of power that we can use to cope with programming large, complicated applications.

Procedural Programming in C

Here is an example of a C program that adds ints:

```
int main(void) {
    int a = 100;
    int b = 200;
    int c;

    c = a + b + 300;
    printf("%d\n", c);
}
```

Big Numbers

Now suppose we wish to perform a similar calculation, but this time `a` and `b` are big numbers, like the US national debt expressed in dollars. Such numbers are too big to be `int`s on most computers:

```
int a = 12345678901234;
```

would hopefully give us an error message and fail to compile.

Big integers have many practical applications, such as cryptography, symbolic algebra, and number theory where it can be necessary to perform arithmetic on numbers with hundreds or even thousands of digits.

We therefore need to construct a convenient way to handle such numbers.

Solution 1 - C Program Using Procedure Library

We can take advantage of a big integer C library called `BigInt` for performing very large integer arithmetic. The library defines a struct called `BigInt` for representing big integers and defines a variety of functions for operating on `BigInt`-type variables.

BigInt Library Function Prototypes

```
BigInt *itob(int n);
```
creates a new `BigInt` initialized to the value of the (small) integer n, and returns a pointer to this newly created `BigInt`

```
void badd(BigInt *a, BigInt *b, BigInt *c);
```
adds the `BigInt`s pointed to by a and b and returns their sum in the `BigInt` pointed to by c

```
void bmult(BigInt *a, BigInt *b, BigInt *c);
```
multiplies the `BigInt`s pointed to by a and b and returns their product in the `BigInt` pointed to by c

```
void bout(BigInt *a);
```
prints the `BigInt` pointed to by a

```
void bfree(BigInt *a);
```
frees the `BigInt` pointed to by a

Convert Big Integer Constant to BigInt
To use the `BigInt` library to solve our problem, we must write a function to convert an integer constant into a `BigInt` since the library does not provide a way to do this.

The function prototype:

```
BigInt *dtob(char *d);
```

takes a big integer written as a character string constant, and converts it into a `BigInt` one digit at a time.

Implementation of dtob() Function

```c
#include <BigInt.h>

BigInt *dtob(char *d) {
    BigInt *t, *ten;

    t = itob(0);
    ten = itob(10);
    while(*d) {
        BigInt *u;
        u = itob(*d++ - '0');
        bmult(t, ten, t);
        badd(t, u, t);
        bfree(u);
    }
    bfree(ten);
    return t;
}
```

C Program Using BigInt Library

```c
#include <BigInt.h>

BigInt *dtob(char *d);

int main(void) {
    BigInt *a, *b, *c, *t;

    a = dtob("12345678901234");
    b = dtob("24680135792356");
    c = itob(0);
    t = itob(300);
    badd(a, b, c);
    badd(c, t, c);
    bout(c);
    printf("\n");
    bfree(a);
    bfree(b);
    bfree(c);
    bfree(t);
}
```

Analysis of C Programs

Although the problem of adding big integers is conceptually similar to that of adding small integers, the two C programs are very different. In the 2nd example, we become so involved in the details of programming with `BigInts` that we lose sight of the essence of the problem we're trying to solve, making the program difficult to write and understand.

Procedural Programming Details

What details does the procedural programming model force us to deal with in the 2nd program?

- Initialize `BigInts` in separate statements, such as `c = itob(0);`. Failing to do this in `main()` is a frequent source of error.

- Handle temporary variables, such as the variable `t` in `main()`.

- Convert small integers to `BigInts`, as in `t = itob(300);` in `main()`.

- Use functions with distinctive names, like `badd()` and `bmult()`, instead of the traditional arithmetic operators.

- Manage storage, as in the calls to `bfree()` in `main()`.

C++ is not Magic

Code to perform operations on big integers and to handle the details of:

- initialization
- temporary variables
- type conversions
- storage management
- etc.

must still be written.

The important difference is that in C++, we can package this code as an ADT such that the programmer who codes the ADT, the provider, handles the complexities, while those who use the ADT, the clients, find it as convenient to use as a built-in data type.

C++ Code Using Data Abstraction

Using C++ and making the big integer software an ADT called `BigInt` instead of using a procedure library has a dramatic effect on writing a program to solve our problem:

```
#include "BigInt.h"

int main() {
    BigInt a = "12345678901234";
    BigInt b = "24680135792356";
    BigInt c;

    c = a + b + 300;
    c.print();
}
```

Notice how similar the 1st C program is to this C++ program which performs a similar calculation using `BigInt`s instead of `int`s. Obviously, the C++ version is more compact and easier to understand than its C counterpart in the 2nd C program.

Analyze C++ Program

```
#include "BigInt.h"

int main() {
    BigInt a = "12345678901234";
    BigInt b = "24680135792356";
    BigInt c;

    c = a + b + 300;
    c.print();
}
```

The first 3 statements in `main()` declare three type `BigInt` variables `a, b, c`. The compiler needs to know:

- how to create them
- how much space to allocate for them
- how to initialize them

The 1st and 2nd statements are similar. They initialize the `BigInt` variables `a` and `b` with big integer constants written as character strings containing only digits so the C++ compiler must be able to convert character strings into `BigInt`s. The 4th statement is most complicated. It adds `a` and `b` and the integer constant `300` and stores the result in `c`. The C++ compiler:

> - needs to be able to create a temporary `BigInt` variable to hold the sum of `a` and `b`
>
> - must then convert the `int` constant `300` into a `BigInt` and add this to the temporary variable
>
> - must assign this temporary `BigInt` variable to `c`

The 5th statement is a call to a special kind of function available in C++ known as a member function that sends the contents of `BigInt` variable `c` to standard output.

Even though the body of `main()` contains no more statements, the compiler is not finished yet. It must also destroy the `BigInt` variables `a`, `b`, and `c` and any `BigInt` temporary variables it may have created along the way before leaving a function such as `main()`.

What the C++ Compiler Must Do With `BigInts`

- create new instances of `BigInt` variables

- convert character strings and integers to `BigInt`s

- initialize the value of one `BigInt` with that of another `BigInt`

- assign the value of one `BigInt` to another

- add two `BigInt`s together

- print `BigInt`s

- destroy `BigInt`s when they are no longer needed

Specifications & Implementations

Where does the C++ compiler obtain this know-how? From the `BigInt.h` file which is included in the first line of the program. This file contains the specification of the `BigInt` ADT.

Specification - contains the information that programs that use an ADT, the client, need to have to be successfully compiled `(bigint.h)`.

Implementation - another file that contains most of the details of how the ADT works `(BigInt.cpp)`.

In our example, `BigInt.c` is compiled separately and the object code produced from it is linked with the program that uses the ADT. The specification and implementation of an ADT are combined with the source code of a client program to produce an executable program.

Separate Compilation

We separate the code for an ADT into a specification part and an implementation part to hide the implementation details from the client. We can then change the implementation details and be confident that the client programs will continue to work correctly after they are relinked with the modified object code.

This is useful when a team of programmers work on a large software project. Once they agree on the specification for the ADTs they need, each team member can implement one or more of them independently of the rest of the team. A well designed ADT also hides its complexity in its implementation, making it as easy as possible for clients to use.

ADT's and Programming Languages

Most other programming languages currently in widespread use make it difficult to practice data abstraction well. This is because data abstraction requires special language features not available in those languages.

Encapsulation

While it is possible to hide implementation details in almost any programming language, C++ explicitly supports it by providing mechanisms to prevent clients from accessing data and operations that are supposed to be hidden.

The provider of a C++ ADT can specify private data and private operations, so that if a client program attempts to use them, the C++ compiler issues an error message. This enforcement of the separation between specification and implementation is called *encapsulation*.

Encapsulation is an extremely important feature because without it we must rely on clients to cooperate by observing the separation between the specification and the implementation. Experience tells us that this is not practical, particularly on large software projects involving more than a few programmers. Somebody inevitably forgets the rules or succumbs to the temptation to use an implementation-dependent optimization, and the guarantee of being able to change an implementation without causing client problems to malfunction is gone.

Overview of C++ Support for Data Abstraction

C++ provides the class declaration for defining ADTs. A C++ class is similar to a C struct, but it allows you to associate both variables and functions with the name of a data type.

The variables and functions that belong to a class are called *members*. You can make a member either public or private. Any program can use the public members of a class. But only other member functions of the same class can use its private members.

You can practice encapsulation in C++ by declaring classes with all member variables private, so that the clients can access or modify them only as permitted by the class's public member functions.

Function Name Overloading

Unlike C, C++ requires you to specify the number any type of all of a functions arguments. You can then use the same function name for more than one function as long as the number and/or type of the function arguments are distinctive. This is called *function name overloading*. Function name overloading eliminates the need to use different function names when you wish to perform the same operation on different data types.

Operator Overloading

Operator overloading enables you to give new meanings to most of the C++ operators when you use them on instances of ADTs. Note that we used operator overloading to define the meaning of + when applied to `BigInts` in the C++ example.

Constructors

C++ classes may have two special kinds of member functions: constructors and destructors. Constructors create new instances of ADTs (classes) and are responsible for initializing them correctly. Since C++ automatically uses a constructor to initialize any instances of ADTs required by client programs, this feature eliminates this common programming error.

C++ can also use constructors to automatically perform type conversions. For example, if we define a constructor for the `BigInt` ADT that has an argument of type `int`, C++ will use a constructor automatically whenever it needs to convert a variable of type `int` to a `BigInt`.

Destructors

Destructors destroy instances of ADTs. We can use destructors together with constructors to hide the details of managing storage for ADTs from client programs.

Chapter 2. A First Look at C++

What is C++?
First C++ Program
Operator Precedence
C++ I/O Tutorial
Capturing Program Output

What Is C++?

C++ Executive Summary
- C++ is the work of Bjarne Stroustrup of AT&T Bell Labs.

- C++ is a mostly upward compatible extension of C that provides:

- C++ is A Better C
    - safer
    - less reliance on preprocessor and libraries
    - more flexibility

- Support for Data Abstraction
    - new data types

- Support for Object-Oriented Programming
    - supports classes
    - supports inheritance
    - supports polymorphism

- C++ supports these paradigms; it does not require their use

Introduction
- C++ is a better C

- C++ supports data abstraction

- C++ supports object-oriented programming

C++ Is a Better C - What's the Same?
- C++ contains a collection of features that make it more convenient to program in the traditional C paradigm.

- C++ is (almost) a superset of C.

- All operators and rules are the same in C++.

- A legal C program is a legal C++ program except when the C program:

  - uses C++ keywords as identifiers
  - depends on certain C type conversion rules that are different in C++

- Control structures (`for`, `while`, `do`, `if`, `if else`, `switch`, etc.) all work as before and cannot be extended

- Operators (`+`, `-`, `+=`, `++`, `--`, etc.) all work as before with the same precedence. However, operators can now be extended to work on new types (structs / classes).

- Still have global declarations and library declarations in `.h` files and implementation code in `.cpp` files, but levels of abstraction now tend to be separated into different source files,

  i.e., might have `TimeDate.h` declaration and `TimeDate.cpp` for implementation of functions, then `Postmark.h` and `Postmark.cpp` files that use `TimeDate` class

- `main()` as the program start point

- Still use { and } to delimit blocks

- Same primitive data types

- Same storage classes and scope rules

- Same memory layout and addressing scheme

- Functions still can't be nested

- Existing preprocessor features continue to work, but many uses of the preprocessor now done more elegantly in language itself

- Some dangerous / useful C features still present:

- no array bounds checking (but it's possible to define a safe array type or class)

C++ Is a Better C - What's New?

- Additional comment style available (//)

- Few more operators (scope resolution operator ::, new, delete)

- Function prototyping with checking of argument type and number of arguments is the most important "better C" feature.

      - They find lots of bugs at compile time that otherwise must be found at runtime

- Default initialization of function parameters

- Overloading of function names (allows name reuse)

- Overloading of operators

- Declarations as statements

- Inline functions allow for optimizations

- Structures are type names

- const declarations

- References allow call by reference to be specified by the function called and does not require addresses to be explicitly passed to the called function

- Type-safe linkage ensure correct calling syntax

- Improved approach to IO - Streams library is an IO data type that is used in place of a set of library routines

- Classes

- Single Inheritance

- Multiple Inheritance

- Templates (Standard Template Library)

- Exception Handling

- Less reliance on preprocessor and libraries

- Storage management is handled by `new` and `delete`. These operators understand the language and therefore can be more accurate than `malloc` and `free`.

C++ Supports Data Abstraction
Create abstract data types as "built-in" as fundamental data types

Operator overloading

User-defined objects

C++ Supports Object-Oriented Programming
Encapsulation
- classes
- objects
- scope control
- privacy

Inheritance
- sharing object class resources

Polymorphism
- semantic convenience
- virtual functions

C++/OOP Paradigm Shift
The key difference between C & C++ is that in C++ you try to work at a higher perceptual level. When people start writing a program in C, they get to the bits and bytes very fast. In a programming language like C++, and OOP in general, you start by saying:

Now what are the fundamental concepts in my application?

How do they relate to each other?

Can we express the relationships between those fundamental concepts directly?

In other words, the design phase becomes a natural part of programming.

- can write programs faster
- can get higher quality code
- can write code that's easier to maintain

Note that C++ supports data abstraction and OOP but does not require their use.

C++ Closing Comments

C++ provides a gradual migration path from C.

Biggest problem with the transition from C to C++ is that people don't know C!

You can still twiddle bits if you want to.

Run-time efficiency is very important... C++ provides the speed of C.

Experts can provide reusable packages to the rest of us that are:
- modular
- extensible by others

Primary C++ Benefits
- ability to mix paradigms
- experience base of C programmers
- convenience of higher level of abstraction
- programmers like it

Should be called ++C

First C++ Program

Note: If you copy and paste the code below into your C++ compiler, it will not compile properly until you change the slanted double quotes (from my Word document) into normal double quotes.

```cpp
// hello.cpp

#include <iostream>

using namespace std;

int main() {
    int age;

    cout << "Please enter your age: ";
    cin >> age;
    cout << "Hello, world! I am " << age << " today!" << endl;
}
```

Input / Output

The `iostream` header file automatically defines variables (objects) called `cout` and `cin`.

`cout` is a predefined object that accepts all data bound for the standard output stream (monitor).

`cin` is also a predefined object which corresponds to the standard input stream (keyboard).

To send data to the standard input stream, we use the << operator (which is the bitwise shift operator in C). This operator is overloaded in C++ to give it a new meaning when the operator is used with an object of a particular type. When used with `iostream` objects, << means *send to*.

Note that `cout` and the << operator know how to treat an integer and a string differently. If a string is sent, it is treated as text. If an integer is sent, it is treated as a number. This may seem obvious, but it's another example of operator overloading, a key feature of C++.

C programmers will remember that such functions as `printf()` need to be told about:

- both the variable to be displayed

- and the variable type

This older C syntax make things less intuitive.

Note that the send to operator $(<<)$ can be cascaded. This allows a series of values to be sent to standard output.

The $>>$ operator is the extraction operator or *get from* operator. It takes the value from the stream object on its left and places it in the variable on its right.

`iostream` Manipulators

Manipulators are operators used with the send to $(<<)$ operator to modify the way data is displayed. The two most common manipulators are:

- `endl`
- `setw`

`endl` generates a linefeed by inserting a linefeed into the output stream. It is equivalent to `'\n'`, but also clears the buffer.

`setw` causes the item following to be displayed in a field of the specified width. The default output field for a variable is just wide enough to hold the value. The `setw` manipulator acts on the number (or string) that follows it in the output stream. The item is displayed right-justified in the field.

For example:

```
cout << setw(8) << num << setw(12) << num << endl;
```

Additional manipulators:

```
cout << "The octal value is " << oct << num;
cout << "The decimal value is " << dec << num;
cout << "The hex value is " << hex << num;
```

The I/O manipulators cause the `num` values to be formatted differently than without any manipulators. The declarations for the manipulators are not in the usual `iostream` header file, but in a separate header file called `iomanip`:

```
#include <iomanip>
```

Why a New I/O Library?

Consider this C program:

```c
#include <stdio.h>

int main(void) {
    char a = 'a';
    char *b = "a string";
    double c = 3.14;

    printf(" a = %d\n b = %d\n c = %d\n", a, b, c);
}
```

When compiled and executed, the C program produces the following output:

```
a = 97
b = 72688
c = -171818680
```

Why? Because the type and number of the arguments are determined by the format character string and is left to the programmer to get it right. Here's the same program written in C++:

```cpp
#include <iostream>

using namespace std;

int main() {
    char a = 'a';
    char *b = "a string";
    double c = 3.14;

    cout << "a = " << a << endl;
    cout << "b = " << b << endl;
    cout << "c = " << c << endl;
}
```

When compiled and executed, the C++ program produces:

```
a = a
b = a string
c = 3.14
```

The burden of type resolution is lifted from the programmer, and therefore, the likelihood of error is reduced. The compiler chooses the correct instance of the output operator based on the types of the arguments, or flags the call as an error at compile time.

In C++, there actually is a different output operator defined for each type:

```
class ostream : public virtual ios {
    public:
        ostream& operator<<(int);
        ostream& operator<<(long);
        ostream& operator<<(float);
        ostream& operator<<(char);
        // ...
};
```

Don't worry about the syntax for now. But keep in mind that the programmer is shielded from these multiple instances through the mechanism of operator overloading:

```
cout << i << "     " << f << '\n';
```

The iostream Library is Extensible

To output a user-defined type in C, the programmer must provide a uniquely named user-defined print function, and the user of the type must either look-up or remember the name and interface to the function. For example:

```c
#include <stdio.h>

struct date {
    int month;
    int day;
    int year;
};

void printDate(struct date *pd) {
    printf("%d/%d/%d", pd->month, pd->day, pd->year);
}

void main(void) {
    struct date d = { 9, 18, 2012 };

    printf("Today's date is ");
    printDate(&d);
    printf("\n");
}
```

In C++, the designer of a user-defined type provides an instance of the output operator for the type. Clients of the type need know nothing about how its output is handled. For example:

```cpp
#include <iostream>

using namespace std;

struct date {
    int month;
    int day;
    int year;
};

ostream& operator<<(ostream& os, date &d) {
    os << d.month << "/" << d.day << "/", d.year;
    return os;
}

int main() {
    struct date d = { 9, 18, 99 };

    cout << "Today's date is " << d << endl;
}
```

Again, there's no need to worry about the syntax for now.

String Variables and Constants

Consider the following C++ program:

```cpp
#include <iostream>

using namespace std;

const int MAX = 50;

int main() {
    char buf[MAX];

    cout << "Enter a string: ";
    cin >> buf;
    cout << "You entered: " << buf;
}
```

What happens if the user enters a string that is longer than the array used to hold it? Simple. We overwrite memory. So how do we avoid buffer overflow? We can tell the >> operator to limit the number of characters it places in the array:

```cpp
#include <iostream>
#include <iomanip>

using namespace std;

const int MAX = 50;

int main() {
    char buf[MAX];

    cout << "Enter a string: ";
    cin >> setw(MAX) >> buf;
    cout << "You entered: " << buf;
}
```

The program uses the setw() manipulator to specify the maximum number of characters the input buffer can accept. The user may type more characters, but the >> operator won't insert them into the array. Actually, one character fewer than the number specified is inserted, so there is room in the buffer for the terminating null character.

Note: The extraction operator (>>) considers a space to be a terminating character. Thus a string is delimited by a space and anything typed after a space is thrown away.

Reading Strings with Embedded Blanks
To read text containing blanks, we can use another function, `cin.get()`, which invokes the member function `get()` of the `cin` stream class.

```cpp
#include <iostream>

using namespace std;

const int MAX = 50;

int main() {
    char buf[MAX];

    cout << "Enter a string: ";
    cin.get(buf, MAX);
    cout << "You entered: " << buf;
}
```

Reading Past a Carriage Return
How can we read strings with multiple lines? We can use the `cin::get()` function with a third argument which specifies the character that tells the function to halt. The function defaults to newline, `'\n'`, but we can call the function with some other character for this argument:

```cpp
#include <iostream>

using namespace std;

const int MAX = 50;

int main() {
    char buf[MAX];

    cout << "Enter a string: ";
    cin.get(buf, MAX, '$');
    cout << "You entered: " << buf;
}
```

Now you can type as many lines of input as you want. The function will continue to accept characters until you enter the terminating character (or until you exceed the size of the array). Remember, you must still press Enter after typing the `'$'` character.

Comments

All text between // and the end of the line is considered a comment. The comment automatically terminates at the end of the line.

```
// This is a comment

int num;     // This is another comment
```

Note that this is one of the exceptions to the rule that the compiler ignores white space. In this case, the carriage return \n line feed white space character defines the end of the comment.

For multi-line comments we can still use:

```
/*
int x;
int y;
int z;
*/
```

Comments may also be nested within /*   */. This allows you to comment out comments:

```
//    /*
int x;
int y;
//    */
```

or you could say:

```
/*   //   */    int x;
```

C++ Declarations

In standard C, a declaration must come at the beginning of a block before the first executable statement. This is not the case with C++. In C++, a declaration is a statement and can appear anywhere a statement can appear. Its scope extends to the end of the block in which it's defined.

```
void foo() {
    int x, y;

    y = faa();
    int z = y;
    cout << z;
}
```

Within a for statement, you can have declarations within the initializer portion:

```
for (int i = 0; i < num; ++i)
    a[i] = 0;
```

This is nice for two reasons:

• It makes programs more compact.

• It allows the variable in question to be declared at the point where it is initialized.

Note that an initialization cannot appear inside an expression:

```
• int x = 3 + (int y = 7);              // declaration of y is illegal

• for (int i = 0; int j > 5; ++i, ++j)  // declaration of j is illegal
```

Initializations also cannot appear within a function call:

```
• int x = foo(int i = 5);        // declaration of i is illegal

for (int i = 0; i < num; ++i)
    a[i] = 0;
```

Defining and initializing variables where they are used can make the source listing easier to understand, since you don't need to refer to the beginning of a block. However, use with discretion. Variables that are used in many places in a function are probably better defined at the start of the function.

Data Types

Traditional C data types are also available in C++:

```
char (the signedness of a char is machine dependent)
unsigned char
short
unsigned short
int
unsigned int
long
unsigned long
float
double
long double
```

C++ adds the logical boolean data type:

```
bool flag = true;
bool test = false;
```

The syntax for C casting looks like this:

```
(long)num
```

while the syntax for C++ casting looks like this:

```
long(num)
```

This looks a lot like a function call although it isn't really.

Scope Resolution Operator

The scope resolution operator `(::)` is a new C++ operator and has the highest precedence of any operator. It's used for differentiating a local variable from a global variable. For example:

```
int a = 1                   // external

void foo() {
    int a = 2;              // OK, but hides external a
    int b = a;              // b initialized to 2
    int c = ::a;            // c initialized to 1 (accesses external a
                            // this is unavailable in C)
        {                   // opens block
            int a = 3;      // hides a local to foo()
            int d = ::a;    // is d initialized to 3 or 2 or 1??
                            // answer is 1
                            // the outer local a is inaccessible
                            // doesn't go one scope up, but to the top

        }
```

So the unary form of the scope resolution operator is used to uncover or access a name that is at external scope and has been hidden by local or class scope. The scope resolution operator also has a binary form that is used to disambiguate names that are reused within classes. Its use is vital with inheritance.

```
Class Widgets {
    public:
        void foo();
};

class Gizmos {
    public:
        void foo();
};

void Widgets::foo() {
    ........
}

void Gizmos::foo() {
    ........
}
```

Don't worry yet about the syntax, but we use the scope resolution operator here to identify which `foo()` function we're giving definition to. One way to view the scope resolution operator is to view it as providing a path to an identifier. More later!

Structures

In C, structure names are not type names. The word `struct` must be used whenever the structure is referred to:

```
struct Employee {
    int salary;
    struct Employee *reportsTo;
};

struct Employee Haley;

struct Employee* updateReportsTo(struct Employee*, struct Employee*);
```

In C++, structure names are type names, just as if they were typedef'd. The struct keyword is superfluous:

```
struct Employee {
    int salary;
    Employee *reportsTo;
};

Employee Haley;

Employee* updateReportsTo(Employee*, Employee*);
```

Operator Precedence

In the following table of operator precedence, C++ operators are divided into 13 categories. The #1 category has the highest precedence, category #2 (Unary operators) takes second precedence, and so on to the Comma operator, which has lowest precedence. The operators within each category have equal precedence. The Unary (category #2), Conditional (category#11), and Assignment (category #12) operators associate right-to-left; all other operators associate left-to-right.

#	Category	Operator	What it is (or does)
1.	Highest		
		( )	Function call
		[ ]	Array subscript
		->	C++ indirect component selector
		: :	C++ scope access/resolution
		.	C++ direct component selector
2.	Unary		
		!	Logical negation (NOT)
		~	Bitwise (1's) complement
		+	Unary plus
		–	Unary minus
		++	Preincrement or postincrement
		– –	Predecrement or postdecrement
		&	Address
		*	Indirection
		sizeof	returns size of operand, in bytes
		new	dynamically allocates C++ storage
		delete	dynamically deallocates C++ storage
3.	Member access		
		.*	C++ dereference
		->*	C++ dereference

4. Multiplicative

*	Multiply
/	Divide
%	Remainder (modulus)

5. Additive

+	Binary plus
−	Binary minus

6. Shift

<<	Shift left
>>	Shift right

7. Relational

<	Less than
<=	Less than or equal to
>	Greater than
>=	Greater than or equal to

8. Equality

==	Equal to
!=	Not equal to

9. Bitwise

&	Bitwise AND
|	Bitwise OR
^	Bitwise XOR

10. Logical

&&	Logical AND
||	Logical OR

11. Conditional

?:	Conditional

12. Assignment

=	Simple assignment
*=	Assign product
/=	Assign quotient
%=	Assign remainder (modulus)
+=	Assign sum

$-=$		Assign difference
$\&=$		Assign bitwise AND
$\wedge=$		Assign bitwise XOR
$\mid=$		Assign bitwise OR
$<<=$		Assign left shift
$>>=$		Assign right shift

13.   Comma

,   Evaluate

All of the operators in this table can be overloaded except the following:

.	C++ direct component selector
.*	C++ dereference
::	C++ scope access/resolution
?:	Conditional

First C++ Program

This exercise is intended to make sure you can get your C++ compiler to work and to demonstrate a few simple features of the C++ I/O system. When you want to send output to the standard output device in a C++ program, you can use one of the output functions from the standard C library. For example, the line:

```
printf("String = %s\n", buf);
```

displays the value of the string `buf` on the standard output device. However, since we are writing C++ code, we use `cout` from the `ostream` class to send output to the standard output device. For example, the line:

```
cout << "String = " << buf << '\n';
```

displays the same string `buf` on the standard output device. Type in, compile and execute the source code for the following C++ program:

```cpp
// test.cpp - simple i/o test program

#include <iostream>

using namespace std;

int main() {
    char buf[80];

    cout << "Type your first name: ";
    cin >> buf;
    cout << "Congratulations, " << buf
        << ", you have successfully compiled a C++ program.\n";
    return 0;
}
```

In this program, `cin` and `cout` are input and output objects, and the `<<` and `>>` operators, which normally do bit-shifts in C, have been overloaded to mean *get from* (insertion) and *put to* (extraction). All of this is handled in the standard C++ I/O library, `iostream`. Note that C++ program files have the suffix `.cpp`.

Now, comment out the preprocessor directive:

```
// #include <iostream>
```

and compile the C++ program. Note the error messages that result.

Output Streams

While you can use either `printf()` or `cout`, using both of them in the same program can lead to problems. To see the problems that could arise, type in, compile and execute the following C++ program:

```
// bad.cpp - uses both printf() and cout to send output to the
// standard output device

#include <iostream>
#include <stdio.h>

using namespace std;

int main() {
    // turn on buffered output for cout
    cout.unsetf(ios::unitbuf);

    printf("Print this first.\n");
    cout << "Print this second.\n";
    printf("Print this third.\n");
    cout << "Print this last.\n";
    return 0;
}
```

Looking closely at `bad.cpp` you can see we've alternately used `printf()` and `cout` to display four strings. At first glance, the code in `bad.cpp` appears to have no bugs. However, some older C++ compilers will produce the following output:

```
Print this first.
Print this third.
Print this second.
Print this last.
```

while newer C++ compilers will produce the following output:

```
Print this first.
Print this second.
```

```
Print this third.
Print this last.
```

Note: For the sake of learning some important elements of C++ output streams, we're going to focus our attention of the output from the older C++ compilers.

Obviously, `bad.cpp` doesn't produce the output we'd expect. Instead of executing its lines in order, the program appears to call the `printf()` lines first, then the `cout` lines. Actually, `bad.cpp` does execute its lines of code properly. The problem is that `bad.cpp` uses *buffered* output for `cout`. When `cout` (or any `ostream` object) uses buffered output, `cout` doesn't display what you send until its output buffer is full or until the buffer is flushed. The two lines in `bad.cpp` that use `cout` don't fill the output buffer, and therefore the program doesn't output those strings until it flushes the `cout` stream when it terminates. Knowing this, we can follow through `bad.cpp`'s code and see what's happening.

First, `bad.cpp` calls `printf()` to display the string `Print this first.`, which it immediately displays. Next, `bad.cpp` inserts the string `Print this second.` into `cout`. Since that string doesn't fill `cout`'s buffer, it simply stores the string in memory to display later. Next, `bad.cpp` calls `printf()` with the string `Print this third.` which it immediately displays even though `Print this Second.` is still in `cout`'s buffer. Finally, `bad.cpp` inserts the string `Print this last.` into `cout`, which `cout` also stores in memory for later output. When `bad.cpp` terminates, it calls the destructor for `cout`, which automatically flushes any non-empty buffers. At this point, the program displays the two strings it inserted into `cout`.

Fortunately, there is a way you can force `cout` to flush the contents of its buffer every time you insert some output. (Note that the ability to flush a buffer's output is a very important task in working with output streams.) The `ostream` class defines the `flush()` member function, which you can call whenever you'd like. The `ostream` class also defines the `flush` manipulator, which you can insert along with other data in the output stream. The following program `ok1.cpp` uses the `flush` manipulator to force the display of the strings it inserts into `cout`.

Type in, compile and execute the following C++ program:

```cpp
// ok1.cpp - uses flush with cout

#include <iostream>
#include <stdio.h>

using namespace std;

int main() {
    // turn on buffered output for cout
    cout.unsetf(ios::unitbuf);

    printf("Print this first.\n");
    cout << "Print this second.\n" << flush;
    printf("Print this third.\n");
    cout << "Print this last.\n" << flush;
    return 0;
}
```

As you can see, you simply insert the `flush` manipulator whenever you want `cout` to flush the contents of the buffer. When you compile and run `ok1.cpp`, you get the output you'd expect on all compilers:

```
Print this first.
Print this second.
Print this third.
Print this last.
```

Instead of inserting the `flush` manipulator after each output string, we could have added a new line that calls the `ostream::flush()` member function. For example, the line:

```
cout.flush();
```

flushes `cout`'s buffer directly.

Type in, compile and execute the following C++ program:

```cpp
// ok2.cpp - uses flush member function with cout

#include <iostream>
#include <stdio.h>

using namespace std;

int main() {
    // turn on buffered output for cout
    cout.unsetf(ios::unitbuf);

    printf("Print this first.\n");
    cout << "Print this second.\n";
    cout.flush();
    printf("Print this third.\n");
    cout << "Print this last.\n";
    cout.flush();
    return 0;
}
```

For our example, we display individual strings separately on different lines. To do this, we must append ' \n ' (newline) characters to each string. Since we're also mixing calls to printf() in our program, we must also insert the flush manipulator. To save programming time, we can use the endl manipulator which inserts a newline character and calls the flush manipulator. The following program, better.cpp, uses the endl manipulator in place of adding newline characters and flushing the cout buffer.

Type in, compile and execute the following C++ program:

```
// better.cpp - uses endl with cout

#include <iostream>
#include <stdio.h>

using namespace std;

int main() {

    printf("Print this first.\n");
    cout << "Print this second." << endl;
    printf("Print this third.\n");
    cout << "Print this last." << endl;
    return 0;
}
```

Aside from using either the `flush` or `endl` manipulator, there is an even better way to mix calls to `printf()` with inserting output to `cout`. Since `cout`'s buffer causes the problem, we can simply instruct `cout` to flush its buffer after every insertion. This is accomplished by calling the `ostream` member function `setf()` and passing the `ios::unitbuf` flag.

For example, the line: `cout.unsetf(ios::unitbuf);`

turns on `cout`'s buffering of output and the line:

```
cout.setf(ios::unitbuf);
```

turns off `cout`'s buffering of output. Thus, we can fix the out-of-order output in our first example by adding the above call. The following program, `best.cpp`, uses *unbuffered* `cout` output and displays the strings in the proper order.

Type in, compile and execute the following C++ program:

```cpp
// best.cpp - uses setf and ios::unitbuf to always flush cout after
insertion

#include <iostream>
#include <stdio.h>

using namespace std;

int main() {
    // turn off buffered output for cout
    cout.setf(ios::unitbuf);

    printf("Print this first.\n");
    cout << "Print this second.\n";
    printf("Print this third.\n");
    cout << "Print this last.\n";
    return 0;

}
```

Of course, as with any solution, there is a trade-off. The `ostream` class uses buffering by default because buffered output is generally faster than non-buffered output. When you turn off `cout`'s buffering, your programs will take longer to execute because they stop to flush their buffers more often. However, since `printf()` flushes its output with each call, you must flush `cout` each time you mix calls to both `printf()` and `cout`.

That said, there is really never a need to include `printf()` statements in a C++ program.

Output Formats

Let's take a look at output formats. Type in, compile and execute the following C++ program that calculates an employee's salary:

```cpp
// employ1.cpp

#include <iostream>

using namespace std;

int main() {
    double hours, rate, salary;

    cout << "Enter the hours worked: ";
    cin >> hours;
    cout << "Enter the pay rate: ";
    cin >> rate;
    if (hours <= 40.0)
        salary = hours * rate;
    else
        salary = (40 * rate) + (hours - 40.0) * rate * 1.5;
    cout << "Hours worked = " << hours  << endl;
    cout << " Hourly rate = " << rate   << endl;
    cout << "      Salary = " << salary << endl;
    return 0;
}
```

The output produced by the preceding program:

```
Enter the hours worked: 52.5
Enter the pay rate: 12.75
Hours worked = 52.5
Hourly rate = 12.75
Salary = 749.062
```

is not really satisfactory because the `float` values are displayed with a precision that is not suitable for monetary values. This can be remedied by inserting format manipulators from the header file `<iomanip>` as items in the output list. These format manipulators, when inserted into an output list, specify the appearance (or format) of the output. A few manipulators are shown below:

`setw(width)`
Displays the next value in a field with the specified width (1 is the default). Affects only the next item to be displayed.

`setprecision(precision)`
Displays values with the specified precision (6 is a common default)

`setiosflags(flaglist)`

Sets the formatting flags in `flaglist`, where `flaglist` is a sequence of one or more 1-bit flags, separated with the l symbol: `flag1 | flag2 | flag3`

Here are a few of the formatting flags available:

• `ios::showpoint` - displays decimal point and trailing zeros, even in a number with no fractional part

• `ios::fixed` - displays real values in fixed-point form, avoiding exponential format

• `ios::scientific` - displays real value in floating-point form

• `ios::left` - displays values left-justified within a field

• `ios::right` - displays values right justified within a field

Note that at this point we do not need to know what the `ios` class is, or the reasons for the exact syntax used. To illustrate the use of format manipulators, type in, compile and execute the following program:

```cpp
// io1.cpp

#include <iostream>

using namespace std;

int main() {
    float x, y;

    x = 8.0 / 3.0;
    y = 9.0 / 3.0;
    cout << "(" << x << ")" << endl;
    cout << "(" << y << ")" << endl;
    return 0;
}
```

While the default format depends on the particular C++ implementation, the output displayed by these statements might appear as:

```
(2.666667)
(3)
```

Note the following:

• The default precision is 6, so that (up to) six positions are provided to hold the digits after the decimal point.

• The value of x is displayed in a field whose (default) width is 1, and C++ automatically widens fields that are too small to the minimum width required to display the value.

• By default, the flag `ios::showpoint` is not set, so that neither the decimal point nor the fractional values of y are displayed.

Using the `setiosflags()` manipulator to set the `ios::showpoint` flag will cause the decimal point and trailing zeros of y to be displayed, and using it to set the `ios::fixed` flag will ensure that values are displayed in fixed-point (rather than floating-point) form. Type in, compile, and execute the following program:

```cpp
// io2.cpp

#include <iostream>
#include <iomanip>

using namespace std;

int main() {
    float x, y;

    x = 8.0 / 3.0;
    y = 9.0 / 3.0;
    cout << setiosflags(ios::showpoint | ios::fixed);
    cout << "(" << x << ")" << endl;
    cout << "(" << y << ")" << endl;
}
```

It's output will appear as follows:

```
(2.666667)
(3.000000)
```

Suppose we wish to alter the precision to display only three decimal places. We can insert the `setprecision()` stream manipulator or we can use the `precision()` member function. A call to either sets the precision for all subsequent output operations until the next precision setting call.

Type in, compile and execute each of the two programs below:

```cpp
// io3.cpp

#include <iostream>
#include <iomanip>

using namespace std;

int main() {
    float x, y;

    x = 8.0 / 3.0;
    y = 9.0 / 3.0;
    cout << setiosflags(ios::showpoint | ios::fixed)
         << setprecision(3);
    cout << "(" << x << ")" << endl;
    cout << "(" << y << ")" << endl;
    return 0;
}
```

```cpp
// io4.cpp

#include <iostream>
#include <iomanip>

using namespace std;

int main() {
    float x, y;

    x = 8.0 / 3.0;
    y = 9.0 / 3.0;
    cout.precision(3);
    cout << setiosflags(ios::showpoint | ios::fixed);
    cout << "(" << x << ")" << endl;
    cout << "(" << y << ")" << endl;
    return 0;
}
```

Each of these programs will alter the output as follows:

```
(2.667)
(3.000)
```

Similarly, we can use either the `setw()` manipulator or the `width()` member function to change the width of the field in which x appears.

Type in, compile and execute each of the two programs below.

```cpp
// io5.cpp

#include <iostream>
#include <iomanip>

using namespace std;

int main() {
    float x, y;

    x = 8.0 / 3.0;
    y = 9.0 / 3.0;
    cout << setiosflags(ios::showpoint | ios::fixed) <<
setprecision(3);
    cout << "(" << setw(10) << x << ")" << endl;
    cout << "(" << y << ")" << endl;
    return 0;
}
```

```cpp
// io6.cpp

#include <iostream>
#include <iomanip>

using namespace std;

int main() {
    float x, y;

    x = 8.0 / 3.0;
    y = 9.0 / 3.0;
    cout << setiosflags(ios::showpoint | ios::fixed) <<
setprecision(3);
    cout << "(";
    cout.width(10);
    cout << x << ")" << endl;
    cout << "(" << y << ")" << endl;
    return 0;
}
```

Each of these programs will alter the output as follows:

```
(      2.667)
(3.000)
```

Note that if the values to be output are smaller than the field widths, fill characters are inserted as padding. A value wider than the designated width, however, will not be truncated. It will take what it needs. The width setting applies only for the next insertion or extraction, afterwards the width is implicitly set to zero.

We noted earlier that the output produced by the `employ1.cpp` program is not really satisfactory. Rewrite the program as `employ2.cpp` using format manipulators to display the results in a more acceptable format. Don't forget to include `iomanip`.

Fundamental Data Types

Write a program, `size.cpp`, to print out the size of the fundamental data types (`char, int, short int, long int, float, double, long double`). Also print the size of the pointer to each of these data types. Use the `sizeof` operator to find the size, and use the `cout` object to print them.

Capturing Program Output

When you submit your computer lab assignments, it will be necessary to include a file that contains the output generated by your program. While you might be tempted to simply copy and paste the output that appears on the screen into a file, often output generated from your programs cannot all fit on the screen and usually gets scrolled off the screen. Please follow the instructions below to capture the output of your program into a file called `csis.txt`:

• Include the header file for the `fstream` library in all `.cpp` source code that generates output to the file:

```
#include <fstream>
```

• Include the following global declaration (after `using namespace std;`) to your `.cpp` source code that contains the `main()` function:

```
ofstream csis;
```

• Include the following global declaration (after `using namespace std;`) in all `.cpp` source code, other than the file that contains `main()`, that generates output:

```
extern ofstream csis;
```

• Add the following statement to `main()` after your variable declarations. It will open the `csis.txt` file to capture your output.

```
csis.open("csis.txt");
```

• For every output statement in your program, you should include an appropriate statement to output the same data to the `csis.txt` file:

```
cout << "Hello" << endl;
csis << "Hello" << endl;
```

• Finally, the last statement in `main()` before the `return` statement should be:

```
csis.close();
```

This flushes the output buffer and closes the `csis.txt` file before the program ends execution.

• Here's an example of a C++ program that captures its output into the `csis.txt` file:

```
// testB.h

void foo();


// testB.cpp

#include <iostream>
#include <fstream>
#include "testB.h"

using namespace std;

extern ofstream csis;

void foo() {
      cout << "Haley" << endl;;
      csis << "Haley" << endl;
}
```

```
// testA.h

#include "testB.h"

void main();


// testA.cpp

#include <iostream>
#include <fstream>
#include "testA.h"

using namespace std;

ofstream csis;


void main() {
    int a = 3, b = 4, c = 5;

    csis.open("csis.txt");

    cout << "All you need is love." << endl;
    csis << "All you need is love." << endl;

    cout << "a = " << a <<    "   b = " << b
         <<    "   c = " << c << endl;
    csis << "a = " << a <<    "   b = " << b
         <<    "   c = " << c << endl;
    cout << 'A';
    csis << 'A';

    cout << endl;
    csis << endl;

    foo();

    cout << "Samantha" << endl;
    csis << "Samantha" << endl;

    csis.close();
}
```

```
// output from csis.txt file

All you need is love.
a = 3    b = 4    c = 5
A
Haley
Samantha
```

Please follow these instructions if you are using Xcode:

We must now tell Xcode to place our `csis.txt` output file in the project folder.

Select Product | Scheme | Edit Scheme

Select the Options tab and place a checkmark next to Working Directory.

Under Use Custom Working Directory, navigate to your project folder and click OK.

Your program's output file, `csis.txt`, will now appear in the project's folder after the program executes.

Chapter 3. Data Types and Data Protection

Data Types and Data Protection

Types = Structs + Functions
We're going to take a look at how C++ implements new data types and makes use of data protection.

C Structs
Suppose we want to build a system consisting of several elevators. In C, we typically attack the problem by creating a `struct` to hold data about each elevator:

```
struct elevator {
    int floorSelected;
    int floorNumber;
};
```

A structure is created which holds all the information we want to remember about each elevator. To create a variable, we use the `struct` keyword along with the structure name:

```
struct elevator e;
```

or, we can use the `typedef` identifier and say:

```
typedef struct elevator Elevator;
Elevator e;
```

The second example acts more like a built-in data type.

Pointers to C Structs as Function Arguments
When we create a function to act upon the `struct`, we must pass as an argument to the function a pointer to the `struct`.

The function below initializes the fields of the `struct`:

```
void initialize (Elevator *e) {
    e->floorSelected = 0;
    e->floorNumber = 0;
}
```

We create another function to select the floor:

```
void selectFloor(Elevator *e, int floor) {
    e->floorSelected = floor;
}
```

We can see that everywhere we want to use the `Elevator` structure, we must dereference a pointer. We are forced to use pointers rather than passing the `Elevator struct` by value because we are modifying variables outside the scope of the function. Passing the `Elevator struct` by value would simply make a local variable!

C++ Structs

Let's see how we might solve the same problem in C++.

We know that a type has characteristics and behavior. C++ implements a type as an extended `struct` that contains function declarations as well as data declarations. Collectively, these are called members:

- function members
- data members

Data members represent the state (characteristics) of the variable, and function members represent the operations (behavior) that can be performed on the data members.

```
#include <iostream>

using namespace std;

#define FLOORS 10

struct Elevator {
    int floorSelected;
    int floorNumber;
    void initialize();
    void selectFloor(int floor);
    void go();
};
```

Member Function Declarations

In C++, all member functions must be declared with full prototypes, i.e., all the argument types must be declared before the function can be called.

- Functions are declared inside a `struct`.

- The scope of these member functions is the scope of the `struct`.

- These are not global names.

Scope Resolution Operator

When defining member functions, we must tell the compiler that the function is associated with a particular type using the scope resolution operator (`::`).

```
void Elevator::initialize() {
    floorSelected = 0;
    floorNumber = 0;
};
```

Since the member function is in the scope of a `struct`, it has access to the other members of the `struct`. Therefore, no pointers are necessary to access and modify the data members of the `struct`!

```
int main() {
    Elevator e;

    e.initialize();
    e.selectFloor();
    e.go();
}
```

Defining a variable of a user-defined type is just like defining a variable of a built-in type. However, we did not have to create a `typedef` as we did in C.

When calling a member function, we must specify an object. Calling a member function or sending a message to an object is just like selecting a data member of a `struct`. Since the user doesn't have to pass the address of the variable into a function, member function calls are easier and cleaner to use.

Data Protection

So we've seen how a type is represented in C++ by extending a `struct` to include functions as well as data. But if a program still has direct access to all the members in a `struct`, the data is unprotected, and programmers can get into it and mess things up.

There are three reasons why we need to protect the data:

1. Sometimes modifying one data element may require modification of another to maintain the integrity of the object. If you can deny programmers access to certain data elements, that forces them to use member functions which you can design to preserve object integrity.

2. In a `struct`, where all members may be modified, it can be difficult for programmers to know which interface they're supposed to use. By denying access to some of the members within a `struct`, they could only use members that are part of the public interface.

3. It helps you isolate bugs. Say a programmer makes an incorrect modification to a data member and suddenly you have an error somewhere in your system. If all the members are public, that error could be coming from anywhere. But if some of those members are private, you know the problem must be coming from functions that have permission to access data.

Access Specifiers

So how does C++ address these security issues? C++ adds three new keywords to provide data protection, called access specifiers:

`public:` means that members declared after this point are accessible by everyone. Public members are like members of C structs. C++ structs default to public access.

`private:` means that all members declared after this point are only accessible to other members of the object. Other programmers cannot access private members so they are safe from accidental changes and are not part of the public interface to the type. C++ classes default to private access.

`protected:` members, like private members, are not accessible by other programmers and can only be modified by other members of the type. However, member functions in derived types, a subject we'll cover later, can access protected members of the base type.

Data Protection Example

Let's see how we can use access specifiers to increase the integrity of the `Elevator` type from before:

```cpp
struct Elevator {
    private:
        int floorSelected;
        int floorNumber;
    public:
        void initialize();
        void selectFloor(int floor);
        void go();
};
```

Now, the only way to modify the private data elements is through the public interface. For example,

```cpp
int main() {
    Elevator e;

    e.floorSelected = 12;
}
```

is now an invalid statement. The compiler will generate an error message. The data member `floorSelected` is now a private data member and can only be accessed within the scope of the `Elevator struct`.

Restricted Access to Member Functions

Member functions can also have restricted access as shown below:

```
struct Coordinates {
    private:
        int c1, c2, c3;
        void translate(int, int, int);
    public:
        void move (int, int, int);
};
```

Here we can assume that the member function `move()` calls the `translate()` member function, since programmers using the objects cannot.

So we've seen why we need to protect the data and functions within types and how the keywords public, private and protected can help us do that. This will lead us into a discussion of the class.

❖ Data Types and Data Protection Homework

1. Draw a hierarchy to represent types of `Food`.

a) Put `Food` at the root of the hierarchy, with `Prepared` and `Basic` branching from it.

b) From `Prepared`, branch to a number of different cuisines by nationality (`Chinese` and `American`, for example) and derive subtypes from those styles (such as `Szechwan` and `Mandarin` from `Chinese`, and `Cajun` and `Amish` from `American`).

c) From `Basic`, branch the *four basic food groups* you probably learned about in elementary school, and then derive subgroups from each of these groups.

2. Draw a hierarchy to represent types of electronic equipment.

a) Put `electronic equipment` at the root of the hierarchy, with `business` and `entertainment` branching from it.

b) `Business` includes such things as `cash registers`, `copiers`, etc.

c) `Entertainment` includes such things as `television`, `video games`, and `music systems`.

d) How do you classify things that could belong to both groups, like telephones and computers?

Note: For questions 3-8, <u>you will need to submit your modified code and resulting output for each part of the question</u>. It's okay to cut and paste your output into a document for this assignment. However, cut and paste will never be allowed for any of our programming assignments. We'll talk about capturing program output shortly. <u>Be sure to properly label your output.</u>

3. Enter `elevator.cpp` and perform the following operations:

a) Examine the source code, compile, and execute the program. Submit the output.

b) Modify `elevator.cpp` to add a new member function called `getFloor()`, which returns the floor the elevator is currently on. Then call this function from within `main()` and have `main()` output the returned value. Compile and execute the program. Submit the code and the output.

c) Create an array `E` of three elevators inside `main()` and initialize each elevator in the array. Send the first elevator to the 8th floor, the second elevator to the 3rd floor, and the third elevator to the 5th floor. Compile and execute the program generating appropriate output. Submit the code and the output.

d) Pointers to struct variables work the same way they do in C. Create a pointer `eptr` to an elevator and initialize it to the second elevator. Use this pointer to send the elevator to the 4th floor. Compile and execute the program. Submit the code and the output.

4. Member functions do not take up space inside each variable of the new type. The `struct` is only as big as the data members it contains. Prove this to yourself by writing a program `strsize.cpp` which creates a `struct` with both data and function members, and an identical `struct` with data members only. Print the `sizeof` each `struct` variable and compare the two values. Submit the code and the output.

5. An implementation in C, `elevator.c`, appears on the following pages. When using the `Elevator` struct and its associated functions, the end user is responsible for calling the function to initialize the `struct`, and passing the address of the `struct (initialize(&B))`. However, there's no guarantee users will call the initialization function. They may not understand the need for the function, or they may feel confident and do it themselves `(A.floorNumber = 8)`. This example is fairly innocuous, but suppose the `struct` contained pointers? An uninitialized pointer almost always spells disaster. We will see later how this problem will be solved by enforcing initialization with a constructor. Enter, compile, and execute the program. Submit the output.

```
#include <iostream>
#define FLOORS 10

using namespace std;

struct elevator {
    int floorSelected;
    int  floorNumber;
    void initialize();
    void selectFloor(int floor);
    void go();
};

void elevator::initialize() {
    floorSelected = 0;
    floorNumber = 0;
}

void elevator::selectFloor(int floor) {
    if (floor < 0 || floor >= FLOORS)
        return;
    floorSelected = floor;
}

void elevator::go() {
    if (floorNumber < floorSelected)
        while (++floorNumber < floorSelected)
            cout << "floor: " << floorNumber << " going up!" << endl;
    else if (floorNumber > floorSelected)
        while (--floorNumber > floorSelected)
            cout << "floor: " << floorNumber << " going down!" <<
            endl;
    cout << "floor: " << floorNumber << " stopping." << endl;
}

int main() {
    elevator A;
    A.initialize();
    A.selectFloor(7);
    A.go();
}
```

```
// elevator.c

#include <stdio.h>
#define FLOORS 10

typedef struct elev {
int floorSelected;
int floorNumber;
} elevator;

/* Functions to work with elevator variables. We have no choice but to
use pointers. */

void initialize(elevator * E) {
E->floorSelected = 0;
E->floorNumber = 0;
}

void selectFloor(elevator * E, int floor) {
if(floor < 0 || floor >= FLOORS)
return;
E->floorSelected = floor;
}

void go(elevator * E) {
if (E->floorNumber < E->floorSelected)
while (++E->floorNumber < E->floorSelected)
printf("Floor: %d; going up!\n", E->floorNumber);
else if (E->floorNumber > E->floorSelected)
while (--E->floorNumber > E->floorSelected)
printf("Floor: %d; going down!\n", E->floorNumber);
printf("Floor: %d; stopping.\n", E->floorNumber);
}

int main() {
elevator A,B;
initialize(&B);              /* User forced to work with addresses */  /*  */
selectFloor(&B, 7);          /*                     and pointers.  */  /*  */
A.floorNumber = 8;           /* User chooses to access elements    */  /*  */
A.floorSelected = 1;         /* directly, inviting errors          */  /*  */
go(&A);
go(&B);
}
```

```
// array1.cpp
// note: this program will compile, but it won't link without the
// set() definition

#include <iostream>
#define SIZE 10

using namespace std;

struct intArray {
  public:
    int a[SIZE];
    void set(int index, int value);
    void print();
};

void intArray::print() {
    for(int i = 0; i < SIZE; i++)
        cout << "a[" << i << "] = " << a[i] << endl;
}

int main() {
    intArray num;

    for(int i = 0; i < SIZE; i++)
        num.set(i,i);
    num.print();
}
```

```cpp
// farray1.cpp - friend function
// note: this program will compile, but it won't link without
// the set() definition

#include <iostream>
#define SIZE 10

using namespace std;

struct intArray {
  private:
      int a[SIZE];
  public:
      void set(int index, int value);
      friend void print(intArray);
};

void print(intArray x) {
    for(int i = 0; i < SIZE; i++)
        cout << "a[" << i << "] = " << x.a[i] << endl;
}

int main() {
    intArray num;

    for(int i = 0; i < SIZE; i++)
        num.set(i,i);
    print(num);
}
```

```cpp
// employee.cpp - entire structs can be friends

struct employee {
  private:
      int salary;
      int benefitLevel;
  public:
      void initialize(int startingPay);
      void raise(float percent);
      friend struct accountant;
};

int main() {
}
```

Chapter 4. Classes

C++ Classes
Preprocessor Commands
Distance Objects
Classes Lab (Poker)

C Structs

Conventional C programming techniques often make programmers work harder than necessary. For instance, consider how you might represent the date and time, a typical task that most programs need to do.

In C, you could declare a `struct` like:

```
struct Time {
    int month;
    int day;
    int year;
    int hour;
    int min;
    int sec;
};
```

The members of the `Time struct` store the component values of a single date and time.

Given this `struct`, you can declare a `Time` variable such as:

```
struct Time appt;
```

and you can assign values to `appt`'s members with statements such as:

```
appt.month = 9;
appt.day = 23;
appt.year = 2012;
appt.hour = 8;
appt.min = 30;
appt.sec = 0;
```

Using C Structs

A program that uses lots of `Time` structs probably needs functions to:

- display times and dates
- update struct members
- compare two dates

In C, you could write a display function like this:

```
void display(struct Time *tp) {
    printf("Date: %02d / %02d / %04d\n", tp->month, tp->day, tp->year);
    printf("Time: %02d : %02d : %02d\n", tp->hour, tp->min, tp->sec);
}
```

To display a date and time, a statement can pass to `display()` the address of a `Time struct`, such as:

```
display(&appt);
```

So far, so good!!

Problems with C Structs

Consider some problems that could arise with the code.

After designing a `struct` like `Time` and writing a zillion `date()` and `time()` functions, you discover a superior way to store the date and time that would greatly improve your program's runtime (either faster execution time or better memory utilization).

For example, you might use a `long int` that represents the number of seconds from a fixed date. This change also makes it possible to use date and time library functions that recognize the date and time stored in this way.

Too bad you didn't think of this great idea earlier! To change the program's representation at this late stage, you have to:

- modify the `Time` struct, deleting the current members and adding a single new `long int` data member... or you might not even decide to use a `struct` after all.

- revise all functions such as `display()` that declare `Time` parameters or that return `Time` structs or pointers.

- hunt for and modify statements that assign or use component values in `Time` structs. These statements must be revised to use the new date and time format.

In a large program with thousands of lines of code, the prospect of tracking down every use of `Time` is disheartening. The change to the data structure forces you to revise code that has already been tested, thus requiring new debugging sessions and wasting valuable development time.

The old rule of thumb is: "Let the data structure the code".

In the real world, however, data specifications are likely to change during a program's development making extra work for programmers. With conventional programming, the data representations you choose early in a program's development limit your freedom to make changes later.

We'll see that with object-oriented programming, data may change with only limited effects on the code. C++ classes can restrict a program's access to internal data storage details. So changes to data representations do not necessarily affect every use of that information.

C++ Classes

A C++ class is a kind of structure that encapsulates data and functions into a handy package. Below we rewrite the C `Time struct` as a C++ class and encapsulate data and functions:

```cpp
// time1.cpp

#include <iostream>

using namespace std;

class Time1 {
  public:
      int  month;
      int  day;
      int  year;
      int  hour;
      int  min;
      int  sec;
      void display();
};


void Time1::display() {
    cout << (month < 10 ? "0" : "") << month << "/"
         << (day   < 10 ? "0" : "") << day   << "/"
         << year << "  ";
    cout << ((hour == 0 || hour == 12) ? 12 : hour % 12) << ":"
         << (min < 10 ? "0" : "")    << min << ":"
         << (sec < 10 ? "0" : "")    << sec
         << (hour < 12 ? " AM" : " PM") << endl;
}


int main() {
    Time1 appt;

    appt.month = 9;
    appt.day   = 23;
    appt.year  = 2012;
    appt.hour  = 8;
    appt.min   = 30;
    appt.sec   = 0;

    cout << "Appointment = ";
    appt.display();
}
```

The class looks very much like the `Time struct` except that it begins with the `class` keyword.

C++ Classes vs C Structs

Keep in mind that C++ classes and C structs have much in common. In fact, all of the features supported by C structs are supported by C++ classes:

- class objects can be passed to functions
- class objects can be returned by functions
- pointers can be used to access class objects
- classes can be used as array types
- classes can be nested
- memory can be allocated dynamically for class objects

But keep in mind that classes are much more powerful than C structs:

- classes have facilities for information hiding with public, private, protected, and friend declarations
- classes can have functions as members
- classes have facilities for initializing and cleaning up objects using special features called constructor and destructor functions
- classes have facilities to support inheritance with derived classes and virtual functions

Class Data Members

Data members in a class are exactly like data members in a C `struct`. They may be:

- variables
- pointers
- references
- arrays
- structures
- class objects

However, class data members may not use the storage class specifiers:

- auto
- extern
- register

Class Function Members

Class member functions such as `display()` are declared as function prototypes. The actual function statements are provided at another place in the program.

Let's look at the implementation of the member function `display()`:

```
void Time1::display() {
    cout << (month < 10 ? "0" : "") << month << "/"
         << (day   < 10 ? "0" : "") << day   << "/"
         << year << "  ";
    cout << ((hour == 0 || hour == 12) ? 12 : hour % 12) << ":"
         << (min < 10 ? "0" : "")     << min << ":"
         << (sec < 10 ? "0" : "")     << sec
         << (hour < 12 ? " AM" : " PM") << endl;
}
```

There are two main differences between the implementation of a member function and the implementation of a common function:

- The member function name is prefaced by the class name and the *scope resolution operator* (`::`). This tells the compiler that `display()` is a member of the `Time1` class. Because the class name uniquely qualifies the member function's name, the program may have other functions and other class member functions named `display()` without conflicts.

- Inside the member function, statements have direct access to the class's members without the need for pointers or even parameter passing.

Time Class main() Function

```
int main() {
    Time1 appt;

    appt.month = 9;
    appt.day   = 23;
    appt.year  = 2012;
    appt.hour  = 8;
    appt.min   = 30;
    appt.sec   = 0;

    cout << "Appointment = ";
    appt.display();
}
```

The `main()` program uses the `Time1` class as much as it would any other data type.

- It declares `appt` as an object of type `Time1`. The class `Time1` is a template that describes the format of the class members. You must create an object such as `appt` of the class type in order to use the class.

- It assigns values to `appt`'s data members, initializing a date and time.

Note that we use the dot notation to refer to a class object's public data members just as you refer to members of a `struct` variable. When such a member is a function, C++ calls the function that typically performs some action upon the class object's data.

Additional Time Objects
Note that a program might have other `Time1` objects declared as:

```
Time1 today;
Time1 yesterday;
Time1 tomorrow;
```

These objects could be assigned values and then displayed with the statements:

```
today.display();
tomorrow.display();
yesterday.display();
```

Note, however, that our program does nothing to prevent the problem with data representation mentioned earlier. All data members in the `Time1` class are public and are therefore directly accessible by other statements in the program. A change to the `Time1` class data members would require modifying all such statements.

A C++ `class` is exactly the same thing as a C++ `struct` with one key difference:

- structs default to public so all the data within them is available to programmers

- classes default to private so using classes is another way to hide data within C++

Classes are an important concept in C++; so important that when Bjorne Stroustrup created the C++ language he called it *C with classes*. It wasn't called C++ for another 3 years!

Time2 Class
We can fix this problem in the `Time1` class by declaring a new and improved `Time2` class that uses member functions to access a class object's data members:

```cpp
// time2.cpp

#include <iostream>

using namespace std;

class Time2 {
  private:
    int month;
    int day;
    int year;
    int hour;
    int min;
    int sec;
  public:
    void display();
    void setTime(int m, int d, int y, int h, int mn, int s);
    void getTime(int& m, int& d, int& y, int& h, int& mn, int& s);
};

void Time2::display() {
    cout << (month < 10 ? "0" : "") << month << "/"
         << (day   < 10 ? "0" : "") << day   << "/"
         << year << "  ";
    cout << ((hour == 0 || hour == 12) ? 12 : hour % 12) << ":"
         << (min < 10 ? "0" : "")     << min << ":"
         << (sec < 10 ? "0" : "")     << sec
         << (hour < 12 ? " AM" : " PM") << endl;
}

void Time2::setTime(int m, int d, int y, int h, int mn, int s) {
    month = m;
    day = d;
    year = y;
    hour = h;
    min = mn;
    sec = s;
}

void Time2::getTime(int& m, int& d, int& y, int& h, int& mn, int& s) {
    m = month;
    d = day;
    y = year;
    h = hour;
    mn = min;
    s = sec;
}
```

```
int main() {
    Time2 appt;
    int m, d, y, h, mn, s;

    appt.setTime(7, 23, 2012, 8, 30, 0);
    cout << "Appointment = ";
    appt.display();
    appt.getTime(m, d, y, h, mn, s);
    appt.setTime(m, d, y, ++h, mn, s);
    cout << "Appointment = ";
    appt.display();
}
```

Time2 Class Additions

The new `Time2` class has 3 additions:

- The private access specifier makes the `int` data members private to the class. Because of this change, only the class member functions may refer directly to these data members.

Rule: A classes private data members are available only to members of that class. Private members are invisible outside the class.

Note: By default, members of a class are private. Therefore, the access specifier `private:` is not needed. For clarity, however, it's a good idea to mark private members with an explicit private access specifier.

- The two other additions to `Time2` are the `setTime()` and `getTime()` member functions:

  `setTime()` is used to assign values to the `appt` class object

  `getTime()` copies `appt`'s data member values to local variables which are passed back to `getTime()` with the hours incremented.

Along with `display()` the three member functions limit access to the class's private parts. The only legitimate way to use `Time2`'s private data members is to call one of the class's public member functions. Therefore, as a direct result of encapsulating data and functions in a class, if the data representation changes, only the class's member functions need to be rewritten.

Also, if a problem develops, debugging can begin with an examination of the class member functions because no other functions or statements in the program can have direct access to `Time2`'s private members. Keep in mind that functions in a class have access to one another, i.e., two class functions may call on each other.

`setTime()` and `getTime()` use the scope resolution operator with the class name `Time2::` informing the compiler that these functions belong to the `Time2` class.

The program could declare other `setTime()` and `getTime()` functions and member functions in other classes without conflicting with those that belong to `Time2`.

Inside the member function implementations, statements have direct access to the class's data members (`month, day, year, hour, minute, second`).

Note: In the `main()` function, because `Time2`'s data members are private to the class, it is no longer possible to assign values directly to a class object. In other words, statements like:

```
appt.month = 9;
```

would be illegal in `main()` and would be flagged by the compiler because `month` is private to the `Time2` class and is therefore directly usable only by members of the same class.

Do classes waste memory?

We have probably given the impression that each object created from a class contains separate copies of that class's data and member functions. This might be a good first guess, since it emphasizes that objects are complete self-contained entities designed using the class specifier.

However, things are not quite so simple. It's true that each object has its own separate data items. But all objects in a given class use the same member functions which are created and placed in memory only once when they are defined by the class specifier.

The data items, on the other hand, will hold different values. So there must be a separate instance of each member data in each object. Space for data is therefore created in memory when each object is instantiated or defined.

Organizing Your Code

Notice that all this code fits inside a single file. But when you work on larger projects, odds are that the code will be divided into several files. Your code will be a lot more maintainable and much easier to rebuild if you spread it out over many files.

Both C and C++ use a concept called header files. A header file has declarations that are used in more than one place. You include the header file anyplace you want to use the names in that file, both when the names are defined and when they are used.

In C++, a header can include any of the following declarations:

- class declarations
- class name declarations
- inline functions
- constant values
- function declarations

Note that none of the information in a C++ header allocates storage. Since a header file is included in several files that are later linked, allocating storage in a header file would cause multiple definitions across the files where the header was included. The linker wouldn't like this.

Header File Rules

The first rule for header files is:

- they can only include declarations

The second rule is:

- include header files anywhere the class name is used.

Since you are forced to declare a class anyplace an object of that class is used, and anyplace member functions of that class are defined, you are basically forced to put the class declaration in a header file in C++.

Note: the interface to the class is defined in the header file (.h) but the implementation is defined in the definition file (.cpp).

Time3 - Separation of interface and implementation.

```cpp
// time3.h

#ifndef _TIME3_H
#define _TIME3_H

class Time3 {
  private:
      int month;
      int day;
      int year;
      int hour;
      int min;
      int sec;
  public:
      void display();
      void setTime(int m, int d, int y, int h, int mn, int s);
      void getTime(int& m, int& d, int& y, int& h, int& mn, int& s);
};

#endif


// time3.cpp

#include <iostream>
#include "time3.h"

using namespace std;

void Time3::display() {
      cout << (month < 10 ? "0" : "") << month << "/"
           << (day   < 10 ? "0" : "") << day   << "/"
           << year << "  ";
      cout << ((hour == 0 || hour == 12) ? 12 : hour % 12) << ":"
           << (min < 10 ? "0" : "")     << min << ":"
           << (sec < 10 ? "0" : "")     << sec
           << (hour < 12 ? " AM" : " PM") << endl;
}

void Time3::setTime(int m, int d, int y, int h, int mn, int s) {
      month = m;
      day = d;
      year = y;
      hour = h;
      min = mn;
      sec = s;
}
```

```
void Time3::getTime(int& m, int& d, int& y, int& h, int& mn, int& s) {
    m  = month;
    d  = day;
    y  = year;
    h  = hour;
    mn = min;
    s  = sec;
}




// time3drv.cpp

#include <iostream>
#include "time3.h"

using namespace std;

int main() {
    Time3 appt;
    int m, d, y, h, mn, s;

    appt.setTime(7, 29, 1993, 8, 31, 30);
    cout << "Appointment = ";
    appt.display();
    appt.getTime(m, d, y, h, mn, s);
    appt.setTime(m, d, y, ++h, mn, s);
    cout << "Appointment = ";
    appt.display();
}
```

Note the two preprocessor commands would prevent the Time3 class header file from being included more than once in a program. Storing the Time3 class in a header file makes the class readily available to other modules.

Note that the program directly references no class data members. In all cases, the program calls member functions to perform actions on the appt object.

Preprocessor Commands

Just what are those preprocessor commands #ifndef, #define, and #endif anyway? Glad you asked!

We've already seen the necessity for breaking up our code into header files and definition files. Class declarations (the interface to a class) are always placed in the header file (.h) while the class definitions (the implementation to a class) are always placed in the definition file (.cpp).

Suppose we have a String class defined in String.h and String.cpp. Now suppose we have code in another source file foo.h and foo.cpp that wants to create and manipulate objects of the String class. The only way for foo.cpp to compile (and recognize just what a String object is) would be to:

```
#include "String.h"
```

Perhaps in the same program we're developing we also have need for faa.h and faa.cpp and the faa code also needs to create and manipulate String objects. Again, faa.cpp would need to:

```
#include "String.h"
```

to compile correctly. So far so good. Every .cpp file can be individually compiled.

Now it's time to link together all the object codes that were created through the individual compilation of each of the .cpp files. Unfortunately, the linker will find the declaration for the String class in each of the object codes and the linker will complain. It doesn't want to see more than one String declaration!

We can avoid this situation by using preprocessor commands that will tell the linker to only look at one String declaration. To do this, every header file should begin with:

```
#ifndef _NameOfHeaderFile
#define _NameOfHeaderFile
```

The code for the header file should go here. The header file should end with:

```
#endif
```

When the linker sees the String declaration for the first time it will recognize that `_NameOfHeaderFile` has not yet been defined and will then define it and look through the header file code. It will look at all the code until it reached the `#endif` preprocessor command.

If the linker sees another String declaration, it will have already defined `_NameOfHeaderFile` and will therefore ignore all declarations until the `#endif` statement.

Problem solved.

Distance Objects

We can use objects to represent distances measured in the English system. We'll create a `Distance` class with 2 data items:

```
int feet;
float inches;
```

and 3 member functions:

```
setDist() - uses function arguments to set feet and inches

getDist() - get values for feet and inches from the user during program
            execution

showDist() - displays the distance in feet and inches format
```

Thus, the value of an object of class `Distance` can be set in either of two ways using `setDist()` or `getDist()`.

```cpp
// dist1.h

#ifndef _DIST1_H
#define _DIST1_H

class Distance {
    private:
        int feet;
        float inches;
    public:
        void setDist(int ft, float in);
        void getDist();
        void showDist();
};

#endif
```

```
// dist1.cpp

#include <iostream>

#include "dist1.h"

using namespace std;

void Distance::setDist(int ft, float in) {
    feet = ft;
    inches = in;
}

void Distance::getDist() {
    cout << "feet: ";
    cin >> feet;
    cout << "inches: ";
    cin >> inches;
}

void Distance::showDist() {
    cout << feet << "' " << inches << "\"";
}


// distdrv1.cpp

#include <iostream>

#include "dist1.h"

using namespace std;

int main() {
    Distance d1, d2;

    d1.setDist(11, 6.25);
    d2.getDist();

    cout << endl;
    cout << "d1 = ";
    d1.showDist();
    cout << endl << endl;

    cout << "d2 = ";
    d2.showDist();
    cout << endl << endl;
}
```

Adding Two Distance Objects

Now, let's add a new member function:

```
AddDist()
```

that adds two Distance objects together.

```
// dist2.h

#ifndef _DIST2_H
#define _DIST2_H

class Distance {
    private:
        int feet;
        float inches;
    public:
        void setDist(int ft, float in);
        void getDist();
        void showDist();
        void addDist(Distance d1, Distance d2);
};

#endif
```

```
// dist2.cpp

#include <iostream>
#include "dist2.h"

using namespace std;

void Distance::setDist(int ft, float in) {
    feet = ft;
    inches = in;
}

void Distance::getDist() {
    cout << "feet: ";
    cin >> feet;
    cout << "inches: ";
    cin >> inches;
}

void Distance::showDist() {
    cout << feet << "' " << inches << "\"";
}

void Distance::addDist(Distance d1, Distance d2) {
    inches = d1.inches + d2.inches;
    feet = d1.feet + d2.feet;
    if (inches >= 12.0) {
        inches -= 12.0;
        ++feet;
    }
}
```

```
// distdrv2.cpp

#include <iostream>
#include "dist2.h"

using namespace std;

int main() {
    Distance d1, d2, d3;

    d1.getDist();
    d2.getDist();
    d3.addDist(d1, d2);

    cout << endl;

    cout << "d1 = ";
    d1.showDist();
    cout << endl << endl;

    cout << "d2 = ";
    d2.showDist();
    cout << endl << endl;

    cout << "d3 = ";
    d3.showDist();
    cout << endl << endl;
}
```

The two distances to be added, d1 and d2, are supplied as arguments. Note that the syntax for arguments that are objects is the same as that for arguments that are simple data types like int. The object name is supplied as the argument.

Since addDist() is a member function of class Distance, it can access private data in any Distance object supplied as an argument. Thus, it has access to the data members of d1 and d2.

When a member function is called, with or without an argument, it is given direct access to only one object, the object that called it. In our example, we called:

```
d3.addDist(d1, d2);
```

Therefore, when variables `inches` and `feet` are referenced, they really refer to `d3.inches` and `d3.feet`.

Note that the result is not returned by the function. The return type of `addDist()` is `void`. The result is stored automatically in the `d3` object.

Function That Returns an Object

Here we have a function that returns a `Distance` object:

```
Distance Distance::addDist(Distance d);
```

Previously with `d3.addDist(d1, d2);`, we passed 2 arguments of class `Distance` and the result was stored in the object which actually called `addDist()`. Now, the function has only one `Distance` object passed and returns a `Distance` object as a function return value.

```
// dist3.h

#ifndef _DIST3_H
#define _DIST3_H

class Distance {
    private:
        int feet;
        float inches;
    public:
        void setDist(int ft, float in);
        void getDist();
        void showDist();
        void addDist(Distance d1, Distance d2);
        Distance addDist(Distance d);
};

#endif
```

```cpp
// dist3.cpp

#include <iostream>
#include "dist3.h"

using namespace std;

void Distance::setDist(int ft, float in) {
    feet = ft;
    inches = in; }

void Distance::getDist() {
    cout << "feet: ";
    cin >> feet;
    cout << "inches: ";
    cin >> inches;
}

void Distance::showDist() {
    cout << feet << "' " << inches << "\"";
}

void Distance::addDist(Distance d1, Distance d2) {
    inches = d1.inches + d2.inches;
    feet = d1.feet + d2.feet;
    if (inches >= 12.0) {
        inches -= 12.0;
        feet++;
    }
}

Distance Distance::addDist(Distance d) {
    Distance temp;

    temp.inches = inches + d.inches;
    temp.feet = feet + d.feet;
    if (temp.inches >= 12.0) {
        temp.inches -= 12.0;
        temp.feet += 1;
    }
    return temp;
}
```

```
// distdrv3.cpp

#include <iostream>
#include "dist3.h"

using namespace std;

int main() {
    Distance d1, d2, d3;

    d1.getDist();
    d2.getDist();
    d3 = d1.addDist(d2);

    cout << endl;
    cout << "d1 = ";
    d1.showDist();
    cout << endl << endl;

    cout << "d2 = ";
    d2.showDist();
    cout << endl << endl;

    cout << "d3 = ";
    d3.showDist();
    cout << endl << endl;
}
```

Note that d1 and d2 are added together using addDist() and the object returned is then assigned to d3.

Note also that only one Distance object is passed. Within the member function, we create a temporary Distance object that holds the sum until it can be returned to the calling program. The sum is calculated by adding two distances:

- the first is the object which calls addDist(), i.e., d1

- the second is passed as an argument

Finally, note that the standalone references to inches and feet are really references to the data members of the object d1, the object that called addDist().

Our newest version is more natural looking, since the assignment operator = is used in a natural way:

```
d3 = d1.addDist(d2);
```

Later we'll see how to use the arithmetic + operator to achieve the even more obvious expression:

```
d3 = d1 + d2;
```

❖ Classes Lab (Poker)

For this assignment you will be given the code for two classes (`Card` and `Deck`) representing a set of playing cards. Using these classes, you will write a program, `game1.h` and `game1.cpp`, that calculates the frequency of occurrence of certain poker hands.

Note that while I will give a simple description of each of the classes below, it will not be necessary to understand the code (and at this point I don't expect you to) for the classes in order to use the classes in your program. You will only need to understand the interface (member functions) to each of the classes. You should be able to compile these classes without modification. In fact, I do not want you to modify the classes, unless your `include` statement requires a change from `iostream` to `iostream.h` and your compiler does not require a `namespace` declaration.

The `Card` class is defined in `card.h` and `card.cpp`. These files define a class (data type) for representing individual playing cards. A card can be *initialized*, *inspected*, and *printed* using the following methods:

```
getSuit()
getValue()
printSuit()
printValue()
printCard())
```

Note that the `Card` class defines enumerated data types to represent the suit and value. You can use these types in your program, but since they are defined within the `Card` class, you have to use the scope resolution operator to access them. That is, if your program needs to define a variable to contain a card value, it would be declared:

```
Card::Value cv;
```

If you want to assign the value `ACE` to this variable, you would write:

```
cv = Card::ACE;
```

Of course, you can also define variables of type `Card`. Since this class is defined in the "global" scope, you would simply write:

```
Card theCard;
```

to declare a variable called `theCard`.

Look in the `.cpp` and `.h` files for other examples of the declaration, initialization, and assignment of `Card` objects. Any file that uses a `Card` object should have the directive:

```
#include "card.h"
```

The `Deck` class is defined in `deck.h` and `deck.cpp`. As you can see, a deck consists of 52 cards (although this can be changed), and a deck of cards can be *initialized, shuffled, dealt* (only from the top), and *added to* using the following methods:

```
shuffle()
getCard()
addCard()
totalCards()
```

A "fresh" deck has its cards in order, and you can use the `shuffle()` member function to randomize the cards. The `shuffle()` function takes an integer parameter that determines how many shuffles take place. A shuffle consists only of exchanging two cards at random, so a reasonably large number (say, 100) is appropriate.

Once you've entered the `Card` and `Deck` files, write a program that deals out 10,000 five-card hands and counts how many hands contain a pair and how many contain a flush. Recall that a flush is a hand in which all five cards are the same suit. For the purposes of this assignment, consider a pair to be any hand in which *at least* two cards have the same value. Note that it is not necessary to examine every card in a hand to determine the presence of a pair or the absence of a flush. Your program should deal out a hand, check it for a pair or flush, put the cards back in the deck, and reshuffle.

Each trial that you run should gather statistics on 10,000 hands dealt and output in tabular format the following:

- trial number
- number of hands dealt
- number of hands with at least a pair
- number of hands with a flush
- percentage of hands with at least a pair (to two decimal places)
- percentage of hands with a flush (to two decimal places)

Your program should run a total of 10 trials and at the end of the 10 trials should output the average percentage of pairs and average percentage of flushes (to two decimal places).

Once again, don't worry if it's not completely clear to you how the `Card` and `Deck` classes work. You should not need to change them to make the program work. Note carefully that class declarations always end with a semicolon, but function definitions don't need them. The program that you create should make use of these predefined classes and each class should be contained within its own source code files. Don't forget to modularize your code and link together all `.cpp` files.

Be sure to include a call (only once) to the randomize function:

```
srand((unsigned)time(NULL));
```

in `main()` to properly seed the random number generator used in the `Deck::shuffle()` member function. You will need to include both the `stdlib.h` and `time.h` header files.

Be sure to modularize your code using functions to perform operations such as checking for pairs and flushes, etc. Be sure to comment all functions.

Poker Lab - Part 2

Next, create a new program, `game2.h` and `game2.cpp`, using the `Card` and `Deck` classes outlined above (again without making any changes in either of these two classes.) However, instead of placing all of the code for a poker hand in `game1()` as you did in the first part of the lab, the code for a poker hand will now reside within its own class, `Hand`.

All of the code that you wrote to manipulate the poker hand (check for pair, check for flush, deal hand, return cards to the deck at the end of each hand, etc.) will now become the member functions of this new `Hand` class. The private data members of the `Hand` class will contain the cards in the hand (i.e., an array of `Card` objects). This new class should reside in its own source files, `hand.h` and `hand.cpp`.

Keep in mind that as you write the member functions of the `Hand` class, you will also need to rewrite `main()` to accommodate these changes. You should find that there is much less code in `game2.cpp` than `game1.cpp`.

Hint: There is no need for a declaration of a `Deck` in the `Hand` class. However, a `Deck` object could be passed to a member function of the `Hand` class.

Finally, be sure to turn in separate runs for both the `game1` and `game2` programs. Also make sure to use separate output file for each run: `csis1.txt` for `game1` and `csis2.txt` for `game2`.

```
// card.h

#ifndef _CARD_H
#define _CARD_H

class Card {
    public:
        // Define types for the suit and value
        enum Suit { Diamonds, Hearts, Clubs, Spades };
        enum Value { NullCard, Ace, Two, Three, Four, Five, Six,
                     Seven, Eight, Nine, Ten, Jack, Queen, King };
    private:
        static const char *snames[4];
        static const char *vnames[14];
        Suit s;
        Value v;
    public:
        // Constructors initialize a card
        Card();
        Card(Suit newSuit, Value newValue);
        Suit getSuit();             // Returns a card's suit.
        Value getValue();           // Returns a card's value.
        void printSuit();           // Print a card's suit.
        void printValue();          // Print a card's value.
        void printCard();           // Print a card's suit and value.
};

// Return the next suit or card value in succession

Card::Suit nextSuit(Card::Suit);
Card::Value nextValue(Card::Value);

#endif
```

```cpp
// card.cpp

#include <iostream>
#include "card.h"

using namespace std;

const char * Card::snames[4] = {"Diamonds", "Hearts", "Clubs",
                                "Spades" };
const char * Card::vnames[14] = {"Bad Card", "Ace", "2", "3", "4", "5",
                                 "6", "7", "8", "9", "10", "Jack",
                                 "Queen", "King" };
Card::Card() {
    v = NullCard;
}

Card::Card(Suit newSuit, Value newValue) {
    s = newSuit;
    v = newValue;
}

Card::Suit Card::getSuit() {
    return s;
}

Card::Value Card::getValue() {
    return v;
}

void Card::printSuit() {
    cout << snames[s];
}

void Card::printValue() {
    cout << vnames[v];
}

void Card::printCard() {
    printValue();
    cout << " of ";
    printSuit();
}

Card::Suit nextSuit(Card::Suit s) {
    return(s+1 > Card::Spades) ? Card::Diamonds : (Card::Suit) (s+1);
}

Card::Value nextValue(Card::Value v) {
    return(v+1 > Card::King) ? Card::Ace : (Card::Value) (v+1);
}
```

```
// deck.h

#ifndef _DECK_H
#define _DECK_H

#include "card.h"

const int DECKSIZE = 52;

class Deck {
    private:
        Card inDeck[DECKSIZE]; // These are private data members and
        int nextCard;          // can be used only by member functions.
    public:
        Deck();                    // Initialization. Called automatically
                                   // when a Deck variable is declared.
        void shuffle(int);     // Exchange random pairs of cards.
        Card getCard();        // Returns top card from the deck.
        void addCard(Card);    // Put named card in the deck.
        int totalCards();      // Returns number of cards left in deck.
};

#endif
```

```
// deck.cpp

#include <stdlib.h>
#include "deck.h"

Deck::Deck() {
    int i;
    Card::Suit curSuit;
    Card::Value curValue;

    nextCard = 0;
    for(i = 0, curSuit = Card::Diamonds, curValue = Card::Ace;
                                        i < DECKSIZE; i++) {
        inDeck[i] = Card(curSuit, curValue);
        curValue = nextValue(curValue);
        if (curValue == Card::Ace)
            curSuit = nextSuit(curSuit);
    }
}

void Deck::shuffle(int swaps) {
    Card temp;

    for (int i = 0; i < swaps; i++) {
        int i1 = rand() % DECKSIZE;
        int i2 = rand() % DECKSIZE;
        temp = inDeck[i1];
        inDeck[i1] = inDeck[i2];
        inDeck[i2] = temp;
    }
}

Card Deck::getCard() {
    return(nextCard < DECKSIZE) ? inDeck[nextCard++] :
                                Card(Card::Clubs, Card::NullCard);
}

void Deck::addCard(Card newCard) {
    if (nextCard > 0)
        inDeck[--nextCard] = newCard;
}

int Deck::totalCards() {
    return DECKSIZE - nextCard;
}
```

Chapter 5. Constructors and Destructors

Introduction to Constructors and Destructors
Arrays of Class Objects
Constructor and Destructor Homework
Dynamic Memory Allocation
Pointer Traps
Constructor Lab (Bar Codes)
Olympic Lab

Introduction to Constructors and Destructors

The concept of creating new data types and designing new data types is the foundation of C++. What constructors do is make sure objects are properly initialized. In C, you're probably used to doing that initialization yourself. What that means though is that you actually have to remember to do the initializing.

When we first looked at types, we used an example of an elevator:

```
class Elevator {
    private:
        int floorSelected;
        int floorNumber;
    public:
        void initialize();
        void selectFloor(int floor);
        void go();
};
```

The initialize function sets the initial value of the floor selected and the current floor number:

```
void Elevator::initialize() {
    floorSelected = 0;
    floorNumber = 0;
}
```

In C, when you define a variable of a built-in type, you expect the initial value to be garbage. So you must always initialize it before you use it. However, if we forget to initialize an `int`, the garbage value it contains will still be a legitimate `int` value, and similarly for the other built-in types.

If we do not call the `Elevator::initialize()` function, however, the initial state of the elevator will probably not be legitimate:

```
Floor # 1001, -47
```

Forgetting an initialization is even more critical if the class contains pointers.

When creating user-defined types, then, it is essential to be able to ensure proper initialization. C++ has a feature to ensure proper initialization, the *constructor*. The constructor is a special member function which has the same name as the class. If we changed the `Elevator::initialize()` function to a constructor, it would look like this:

```
class Elevator {
    private:
        int floorSelected;
        int floorNumber;
    public:
        Elevator();
        void selectFloor(int floor);
        void go();
};
```

Note that a constructor has the same name as the class and has no return value (this is different than a `void` return value).

Defining a constructor follows the same rules as other member functions:

```
Elevator:: Elevator() {
    floorSelected = 0;
    floorNumber = 0;
}
```

The constructor is guaranteed to be called automatically for each object of a class when it is created. In fact, initialization now occurs at the point of definition of the object, before the programmer can even access the object.

```
int main() {
    Elevator e;

    e.selectFloor(7);
    e.go();
    return 0;
}
```

Since the constructor is automatically called, we can guarantee proper initialization of variables. This eliminates a large source of errors and ensures that the object is in a stable state when it's created.

Default Constructors

A constructor with no arguments is called the *default constructor*.

```
Elevator::Elevator();
```

If no constructor is written for the class, the compiler will supply one that does nothing. However, usually the class design overrides the compiler's default constructor by supplying an explicit default constructor. If a constructor is defined for a class, and is not a default constructor, the default constructor supplied by the compiler ceases to exist.

It's no accident that constructor functions have exactly the same name as the class of which they are members. This is how the compiler knows that they are constructors.

Note that there's no return type with constructors. Why? Since the constructor is automatically called by the system when an object is defined, there's no program to return anything to. Therefore, a return value wouldn't make any sense.

Constructor Example

```cpp
#include <iostream>
using namespace std;

class Counter {
    private:
        int count;
    public:
        Counter()          { count = 0; }
        void incCount()    { ++count; }
        int getCount()     { return count; }
};

int main() {
    Counter c1, c2;

    cout << "c1 = " << c1.getCount() << endl;
    cout << "c2 = " << c2.getCount() << endl;
    c1.incCount();
    c2.incCount();
    c2.incCount();
    cout << "c1 = " << c1.getCount() << endl;
    cout << "c2 = " << c2.getCount() << endl;
    return 0;
}
```

When an object of type `Counter` is first created, we want its count to be initialized to 0. We could provide a `setCount()` function to do this and call it with an argument of 0, or we could provide a `zeroCount()` function which would always set count to 0. However, such functions would need to be executed every time we created a `Counter` object:

```
Counter c1;
c1.zeroCount();
```

To prove the constructor does its job, let's modify the class:

```cpp
#include <iostream>

using namespace std;

class Counter {
    private:
        int count;
    public:
        Counter()           { count = 0; cout << "I'm here. " << endl; }
        void incCount()     { ++count; }
        int getCount()      { return count; }
};

int main() {
    Counter c1, c2;

    cout << "c1 = " << c1.getCount() << endl;
    cout << "c2 = " << c2.getCount() << endl;
    c1.incCount();
    c2.incCount();
    c2.incCount();
    cout << "c1 = " << c1.getCount() << endl;
    cout << "c2 = " << c2.getCount() << endl;
    return 0;
}
```

Output:
```
I'm here.
I'm here.
c1 = 0;
c2 = 0;
c1 = 1;
c2 = 2;
```

One Argument Constructor

Let's now modify the `Counter` class to allow us to initialize its value to anything we want:

```cpp
#include <iostream>

using namespace std;

class Counter {
    private:
        int count;
    public:
        Counter(int num)   { count = num; }
        void incCount()    { ++count; }
        int getCount()     { return count; }
};

int main() {
    Counter c1(10);
    c1.incCount();
    cout << c1.getCount() << endl;

    // Note that c1(10); is not a call to a function named c1().
    // Instead it calls the constructor Counter() with the
    // specified arguments.

    Counter c2 = 20;
    c2.incCount();
    cout << c2.getCount() << endl;

    // Here is another way for constructor to be called.
    // If a constructor has a single argument, a one-arg constructor,
    // we don't have to use the function call syntax. We can use the
    // = operator for initialization.

    Counter c3 = Counter(30);
    c3.incCount();
    cout << c3.getCount() << endl;
    return 0;
}
```

Output:
```
11
21
31
```

In this last example, the constructor is called explicitly and initializes `c3`. Although you can't specify a return value, a constructor, in a sense, does have one. It returns a reference to the object being created.

In our example, `Counter()` returns a reference to an object created and is initialized with the arguments specified.

Objects created by calling a constructor aren't given a name. However, a name is assigned to the object when the assignment for `c3` executes.

Overloading Constructors

You can even have more than one constructor for a class. Suppose you wanted to use both the default constructor (initializes to 0), along with a constructor that allows you to initialize the counter to any value you wish.

Including both constructors in the class declaration is called *overloaded constructors*. Overloaded constructors allow us to provide several versions of a constructor for a single class. Each constructor can have a different number and type of arguments. This is useful for objects that have more than one kind of external representation. For example, we might want to initialize a Time object using just seconds or we might want to use month, day, year, hour, minute, second.

```cpp
#include <iostream>

using namespace std;

class Counter {
    private:
        int count;
    public:
        Counter()           { count = 0; }
        Counter(int num)    { count = num; }
        void incCount()     { ++count; }
        int getCount()      { return count; }
};

int main() {
    Counter c1;
    Counter c2(10);

    cout << c1.getCount() << endl;;
    cout << c2.getCount() << endl;
    return 0;
}

Output:
0
10
```

We can now modify our `Elevator` class as follows:

```cpp
class Elevator {
    private:
        int floorSelected;
        int floorNumber;
    public:
        Elevator() {
            floorSelected = 0;
            floorNumber = 0;
        }
        Elevator (int selected, int number) {
            floorSelected = selected;
            floorNumber = number;
        }
        void selectFloor(int floor) {
            floorSelected = floor;
        }
        void go() {
        }
};

int main() {
    Elevator a;
    Elevator b(7, 2);

    a.selectFloor(7);
    a.go();
    return 0;
}
```

Remember that the constructor with no arguments is again called the default constructor and is called by the compiler when no arguments are available. Calling a constructor with arguments is simply a matter of adding an argument list after the object name.

Note: once you create a constructor taking arguments, you must also explicitly declare and define the default constructor if you want to declare any class instances that would employ the default constructor. Remember, the automatically supplied default constructor disappears. Later we'll see that constructors also automatically perform certain other initializations.

Constructors are pretty amazing when you think about it. Whoever writes language compilers (for C, C++, Java, etc) must execute the equivalent of a constructor when the user defines a variable. If you define an `int`, for example, somewhere there's a constructor allocating 4 bytes of memory for it. If we can write our own constructors, we can start to take over some of the tasks of a

compiler writer. This is one step on the path to creating our own data types, as we'll see later.

Destructors

But we need destructors too because once we finish with an object, we might want to do some cleanup.

If we continue with our `Elevator` class example, cleanup might involve bring all our elevators back to floor zero when we shut down the elevator system. That's what a destructor can do!

A destructor is represented as a class name with a tilde (~) in front of it. The definition of the destructor is just like the other member functions. Here's what it looks like:

```
class Elevator {
    private:
        int floorSelected;
        int floorNumber;
    public:
        Elevator() {
            floorSelected = 0;
            floorNumber = 0;
        }
        Elevator (int selected, int number) {
            floorSelected = selected;
            floorNumber = number;
        }
        ~Elevator() {
            floorSelected = 0;
            go();
        }
        void selectFloor(int floor) {
            floorSelected = floor;
        }
        void go() {
        }
};

int main() {
    Elevator a;
    Elevator b(7, 2);

    a.selectFloor(7);
    a.go();
    return 0;
}
```

Note that you can only have one destructor per class and the destructor has no return value. Destructors have no arguments because it's assumed that there's only one way to destroy an object. Therefore destructors cannot be overloaded.

The destructor is automatically called when an object goes out of scope or is deleted. Like the constructor, the destructor calls are not made by you but by the compiler.

The destructor in the `Elevator` class will cause both elevators to move to the ground floor when the program shuts down.

Actually, the most common use of destructors is to deallocate memory that was allocated for the object by the constructor.

So with constructors and destructors, we can create objects with proper initialization and destroy them with proper cleanup. That means we can avoid a whole class of errors.

String Class Example
There are some problems with strings as normally used in C and C++. For one thing, we can't use this perfectly reasonable looking expression to assign one string to another:

```
s1 = s2;
```

On the other hand, if we define our own string type, using a C++ class, we can use just such assignment statements. In fact, many other string operations can be simplified as well, but we'll have to wait until we've learned about overloaded operators.

For a vision of possibilities, we'll create a `String` class, one with limited functionality.

```
// String Class

const int MAX=80;

#include <iostream>
#include <string.h>

using namespace std;

class String {
    private:
        char str[MAX];
    public:
        String();
        String(char *s);
        void display();
        void concat(String s);
};

String::String() {
    str[0] = '\0';
}

String::String(char *s) {
    strcpy(str, s);
}

void String::display() {
    cout << str << endl;;
}

void String::concat(String s) {
    if (strlen(str) + strlen(s.str) < MAX)
        strcat(str, s.str);
    else
        cout << "String too long";
}

int main() {
    String s1, s2("Hello, world."), s3("Samantha");

    s1 = s2;
    s3.concat(s2);
    s1.display();
    s2.display();
    s3.display();
    return 0;
}
```

Our newly defined String class may seem just like the original definition of a string, an array of characters. But by making it a class, we gain some benefits.

Since an object can be assigned the value of another object of the same class, we can use statements like:

```
s1 = s2;
```

to assign one string to another.

The first constructor:

```
String::String();
```

is the default constructor that takes no arguments. It places a `'\0'` in the first array element to create an empty string. It is invoked by:

```
String s1;
```

The second constructor creates the String object `s2` and copies the characters in the constant into the data member `str`. It is invoked by:

```
String s2("Hello, world.");
```

Two alternate formats for calling this one-arg constructor are:

```
String s2 = "Hello, world.";
String s2 = String("Hello, world.");
```

Whichever format is used, this constructor effectively converts a string to a String, i.e., a normal string constant to an object of class String.

The member function `concat()` appends one string to another. It can be called as follows:

```
s3.concat(s2);
```

Here, the variable `str` is the data member of `s3`, the object which invoked the member function. The argument passed to the member function is an object of class String. Using the `strcat()` function, `s2.str` will be appended to `s3.str`.

Therefore the call from `main()`:

```
s3.concat(s2);
```

causes `s2` to be appended to the existing `s3`.

Note that `s1 = s2` effectively copies the contents of the data members from object `s2` to the object `s1`. Thus, we have created an assignment facility for strings. This assignment facility is already built in because we defined the class `String`. It is not due to any member function. We'll enhance the `String` class later.

Arrays of Class Objects

Just as we can declare arrays of common data types, we can declare arrays of class objects. For example, we can declare an array of 5 `Elevator` objects:

```
// e1.cpp

#include <iostream>
#include "elevator.h"

using namespace std;

int main() {
    Elevator e[5];
    ...
    ...
}
```

Storage for the entire array is allocated on the stack as shown below:

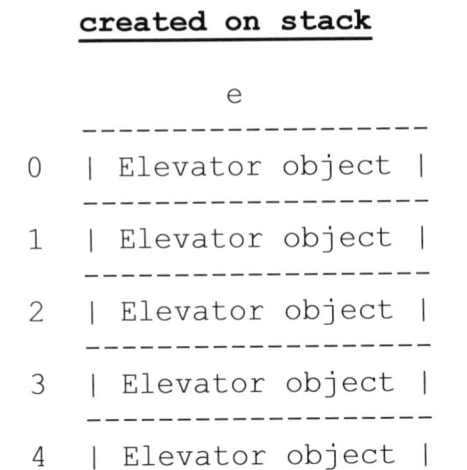

```
            created on stack

                    e
        _____
    0  | Elevator object |
        _____
    1  | Elevator object |
        _____
    2  | Elevator object |
        _____
    3  | Elevator object |
        _____
    4  | Elevator object |
        _____
```

There's only one hard and fast rule to remember: class objects to be stored within arrays must have default constructors. The default constructors could either be provided by the compiler or the programmer. However, if the programmer declares a constructor and does not provide a default constructor, the default constructor provided by the compiler is gone.

When an array of class objects comes into being, C++ calls the class default constructor once for each object in the array. In the case of the `Elevator` declaration above, C++ calls `Elevator`'s default constructor 5 times, thus initializing objects `e[0]` through `e[4]` to the correct values.

If you want C++ to call a different constructor, you must explicitly initialize each array element directly in the array's declaration:

```cpp
// e2.cpp

#include <iostream>
#include "elevator.h"

using namespace std;

int main() {
    Elevator e[5] = {
        Elevator(),
        Elevator(),
        Elevator(2, 5),
        Elevator(7, 3),
        Elevator()
    };
    . . .
    . . .
}
```

Note that sometimes braces are used as if we were initializing fields of a `struct` but some compilers are ok without them.

There is no easy way to initialize some arrayed class objects using the default constructor and some using an alternate constructor. If you need to do that, it's probably best to use the default constructor for the entire array and then reinitialize selected objects as needed. Except for small arrays, this is probably too unwieldy to be of much practical use.

Dynamic Arrays

Arrays of class objects may also be stored on the heap (free store) and addressed with pointers. The following program declares an array of pointers to `Elevator` objects. Storage for the array of pointers is allocated on the stack while storage for each of the actual `Elevator` objects is allocated on the heap.

```
// e3.cpp

#include <iostream>
#include "elevator.h"

using namespace std;

int main() {
    int i;
    Elevator *e[5];

    for (i = 0; i < 5; i++)
        e[i] = new Elevator;
    for (i = 0; i < 5; i++)
        cout << e[i]->getFloor() << endl;
    for (i = 0; i < 5; i++)
        delete e[i];
}
```

Below is a representation of the dynamic array created in the preceding program:

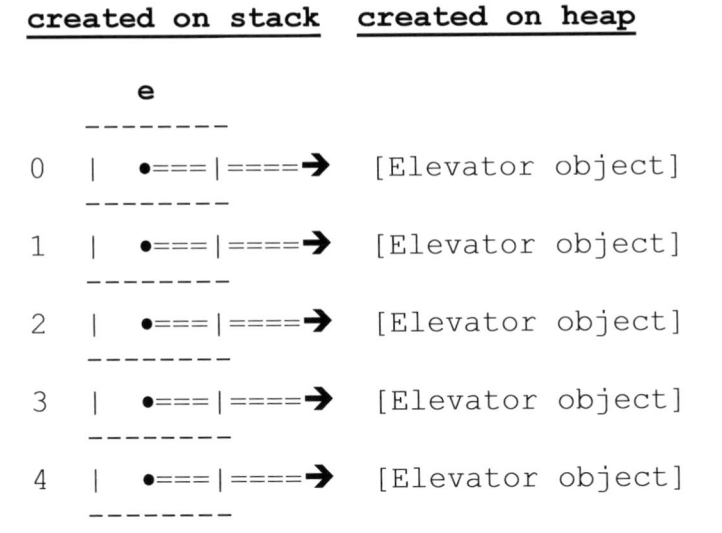

An alternate approach to dynamically create an array of class objects on the heap is as follows:

```cpp
// e4.cpp

#include <iostream>
#include "elevator.h"

using namespace std;

int main() {
    Elevator *eptr;

    eptr = new Elevator[5];
    for (int i = 0; i < 5; i++)
        cout << eptr[i].getFloor() << endl;
    delete [] eptr;
}
```

As shown in the diagram below, the only storage that is allocated on the stack is the storage for the pointer to the Elevator array. Storage for the array itself (i.e., storage for each Elevator object) is allocated on the heap.

created on stack   **created on heap**

```
                          --------------------
    eptr=========➔     0 | Elevator object |
                          --------------------
                       1 | Elevator object |
                          --------------------
                       2 | Elevator object |
                          --------------------
                       3 | Elevator object |
                          --------------------
                       4 | Elevator object |
                          --------------------
```

In the above program, note that we use the direct component selector (.) operator rather than the indirect component selector (->) operator to access the member functions. This is due to the fact that we do not have an array of pointers to class objects this time, but rather the class objects themselves are contained within the array.

As with other dynamic objects, we should delete dynamic class object arrays after we are done using them. However, deleting such an array leads to a problem. Due to the C and C++ inability to distinguish between pointers and arrays, programs must use a special form of delete (`[]`) to ensure that when deleting a dynamic array of class objects any destructor for those objects is properly called.

If we had simply written:

```
delete eptr;
```

the compiler would not know whether `eptr` addresses a single `Elevator` class object or an array of `Elevator` class objects. To inform the compiler that `eptr` addresses an array, we add empty brackets to delete:

```
delete [] eptr;
```

This ensures that the class destructor is called for each object in the array. Note that older compilers required the following:

```
delete [5] eptr;
```

Our next example also creates a dynamic array of class objects on the heap. This time, however, not only is an array of pointers created on the heap, but the Elevator objects that the pointers point to are also created on the heap.

```
// e5.cpp

#include <iostream>
#include "elevator.h"

using namespace std;

int main() {
    int i;
    Elevator **eptr;

    eptr = new Elevator *[5];
    for (i = 0; i < 5; i++)
        eptr[i] = new Elevator;
    for (i = 0; i < 5; i++)
        cout << eptr[i]->getFloor() << endl;
    for (i = 0; i < 5; i++)
        delete eptr[i];
    delete [] eptr;
}
```

Below is a representation of the dynamic array created in the above program:

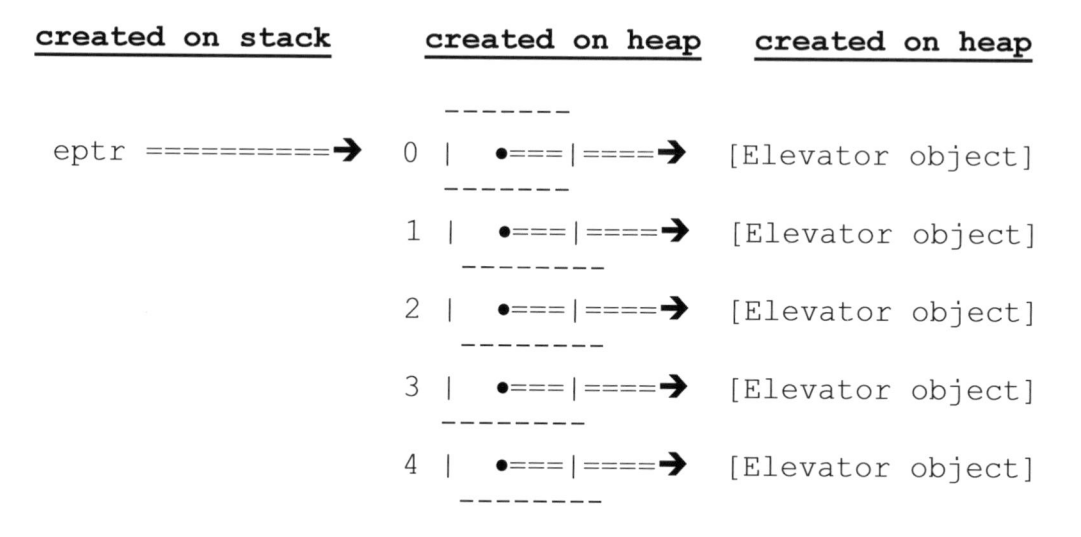

Dynamic Memory Allocation

Like the `malloc()` function in C, the C++ keyword `new` is used to allocate contiguous memory at execution time. In its simplest form, we use the keyword `new` followed by whatever type of data you want to allocate.

```
char *p = new char;          // allocates a single char

int *p = new int;            // allocates an int

double *p = new double;      // allocates a single double
```

Unlike `malloc()`, you no longer need to use the `sizeof` keyword to specify the exact number of bytes needed. Instead, you merely specify a particular type, either fundamental or user-defined. The fact that different types occupy different amounts of storage is handled automatically by the compiler.

When allocating an object of type `T` with the `new` operator, a first argument of `sizeof(T)` is automatically supplied to this function. If `operator new()` cannot allocate the requested amount of heap space, it returns a pointer whose value is zero. If `operator new()` successfully allocates storage on the heap, then it returns the address of this space which is converted to a pointer to the type being allocated.

Because it's so important to be able to initialize variables and objects in C++, fundamental types allocated from the heap via `new` can be initialized with some value by enclosing the value within parentheses immediately after the type name.

```
char *p = new char ('A');    // allocates a single char

int *p = new int(65);        // allocates an int

double *p = new double(5.2); // allocates a single double
```

Memory Allocation for an Array of a Fundamental Type
To allocate memory for an array of some fundamental type, we use the keyword `new`, the type name, and the number of array elements enclosed within square brackets. This number can be either a constant or some expression whose value

is determined at execution time. The return type is identical to a single instance, and the actual value returned is a pointer to the first element of the array.

For example:

```
int max 5;

char *p = new char[1];         // an array of one character

float* p = new float[max];     // an array of 5 floats

char** p = new char*[max];     // an array of 5 pointers to char
```

The `delete` Keyword

The `delete` keyword in C++ is used to release the space that was reserved by `new`. It is analogous to the function `free()` in C, which takes as its one argument a pointer to the memory to be released. The content of the pointer itself is not modified by `delete`.

Note: You should not commingle C and C++ styles of dynamic memory allocation. That is, if you `malloc()` some storage, then you should `free()` it. Similarly, if you `new` some storage, then you should `delete` it. Do not use `delete` with `malloc()`, nor `free()` with `new`.

To delete a single instance from the heap (including a pointer), we use the keyword `delete` followed by the name of the pointer that points to the heap space.

For example:

```
int* ptr1 = new int;
int** ptr2 = new int*;
...
...
delete ptr1;
delete ptr2;
```

To delete an array of instances from the heap (including an array of pointers), we use the keyword `delete` followed by a pair of empty brackets and the name of the pointer variable.

```
int* ptr1 = new int[5];
int** ptr2 = new int*[8];
...
...
delete [] ptr1;
delete [] ptr2;
```

The brackets are required to tell the compiler to expect an array of objects on the heap. This has important ramifications when you need to delete an array of user-defined objects.

So the rule is this: If you used brackets in `new`, *then use brackets in* `delete`. *If you didn't use brackets in* `new`, *then don't use brackets in* `delete`.

Note: If the contents of a pointer given to `delete` *contains the value zero or* `null`, *it is perfectly permissible since this situation is guaranteed not to do anything harmful.*

In the following example, the user is asked to enter string data from the keyboard. The input is captured into a buffer area of some fixed length, after which it is copied onto the heap. Note that exactly the right amount of heap space is allocated to store each string. After all strings have been entered, they are printed and their heap space released via `delete`.

The key data element is `pptr`, which is a pointer-to-pointer so that the array of pointers needed to keep track of the strings can grow dynamically on the heap.

```
// Dynamic Memory Allocation

#include <iostream>
#include <string.h>

using namespace std;

void input(char**& pptr, int& counter);
void output(char** pptr, int counter);
void release(char** pptr, int counter);
```

```
int main() {
    char** pptr = 0;
    int counter = 0;
    input(pptr, counter);
    output(pptr, counter);
    release(pptr, counter);
    return 0;
}

void input(char**& pptr, int& counter) {
    const int MAX = 256;

    cout << "Enter string: ";
    char buf[MAX];
    // End input with Ctrl-D
    while (!(cin >> buf).eof()) {
        // Get exact amount of space
        char* temp = new char[strlen(buf) + 1];
        strcpy(temp, buf);
        // Array of pointers grows dynamically
        char** temp_pptr = new char*[counter+1];
        for (int i = 0; i < counter; ++i)
            temp_pptr[i] = pptr[i];
        temp_pptr[counter++] = temp;
        // Release old space for array of pointers
        delete [] pptr;
        // Update pointer to current array of pointers
        pptr = temp_pptr;
        cout << "Next string: ";
    }
}

void output(char** pptr, int counter) {
    const char QUOTE = '"';

    if (counter > 0) {
        cout << "\nThe strings:\n";
        for (int i = 0; i < counter; ++i)
            cout << QUOTE << pptr[i] << QUOTE << '\n';
    }
    else
        cout << "Empty list\n";
}

void release(char** pptr, int counter) {
    for (int i = counter - 1; i >= 0; --i)
        delete [] pptr[i];
    delete [] pptr;
}
```

The output is as follows:

```
Enter string: All
Next string: you
Next string: need
Next string: is
Next string: love.
Next string:
The strings:
"All"
"you"
"need"
"is"
"love."
```

Pointer Traps

Pointer variables give the programmer great flexibility, but they can lead to difficult to find and correct errors in your programs. We will examine the common problems and give hints about how to avoid them.

Let's assume the following type and variable declarations:

```
int *first = NULL;
int *second;
int *third;
int num = 4;
```

We also assume that the following code has been executed:

```
second = new int;
*second = 2;
```

Dereferencing NULL
It is an error to dereference NULL. This is the safest error to make, since the system will usually catch the error at run time.

```
cout << *first << endl;        // Error, first is NULL
```

It is just as much a problem to dereference NULL with the -> operator in the case that your type is a pointer to a class or struct. A NULL pointer does not point to anything. Dereferencing obtains the thing the pointer points to.

Dangling Pointer
It is an error to dereference a pointer that has not been correctly given a value. This error will probably cause an eventual crash, but usually after running incorrectly for a while.

```
cout << *third << endl;        // Error, third has no value
```

Interestingly, something will be printed most likely. If third happens to have a value that looks to the running program like a valid pointer we will get the contents of a few memory locations interpreted as an `int`, independently of what they actually contain.

It is far worse to say:

```
*third = 3;         // Serious error - changing a random memory location
```

A pointer can get a valid value by assigning it `NULL`, or another pointer known to be valid. You can assign it the result of calling operator `new`. You can also get a valid pointer using the "address of" operator `&`.

```
third = NULL;
third = second;      // second was valid, so third is also valid
third = new int;     // OK, but see Losing Cells, below
third = &fixed;      // Third points to fixed. *third IS fixed (alias)
```

Aliasing Problems

Having more than one name for the same cell can add flexibility to our programs, but there are some dangers. For example, after executing:

```
third = new int;
```

we will eventually want to execute:

```
delete third;
```

to reclaim the storage that third occupies. However it is an error to delete a cell not allocated with `new`:

```
third = &fixed;
delete third;        // Error, third does not point to the free store
```

It is also an error to dereference a pointer after deleting any of its aliases.

```
third = second;
delete second;
cout << *third << endl;        // Error, third is dangling
```

The above code will probably print out a 2, actually, since third still points to the same memory location and those locations have not been changed yet. They soon will be however. The following may produce something very different:

```
third = second;
delete second;
int *fourth = new int;
*fourth = 4;
cout << *third << endl;        // Error, third is dangling
```

On some systems 4 will be produced, since the storage that *second (and *third) occupied previous to the delete may be reallocated to *fourth.

Losing Cells

It is an error to give a new value to a pointer that references a free store cell unless you either first:

- create an alias of the cell, or
- delete the cell

This is an error that is not caught by the system. It simply results in free store cells that cannot be recycled for later use in the program. A long running program that allocates but does not free cells will possibly crash when an attempt to allocate a cell cannot be fulfilled.

```
second = third;     // Error, second already references the free store
```

You can make this safe by first deleting second or assigning another pointer the value of second.

```
delete second;
second = third;              //OK

or

first = second;
second = third;              //OK
```

Note that giving `second` a value in other ways (`NULL`, `new`...) is just as much an error.

Modern systems throw an exception when the allocator cannot allocate the requested cell. Older systems set the pointer to `NULL` when you call `new` and `new` has no cells to give. It can be tested for, but many programmers improperly ignore the tests.

Deleting Without Allocating

We covered one of these problems above: deleting a pointer not allocated with `new`. Note that as in the last sentence, we usually say "delete a pointer". *Actually, though, it is not the pointer that is deleted. It is the thing that it points to.* It is useful to keep this distinction in mind.

It is a logical error to delete a `NULL` pointer. In C++ this is OK though, and the system will ignore it. Doing so sometimes indicates an error in your logic, however. C may not be so friendly if you try to free a `NULL` pointer. It is an error to delete a dangling pointer. This one is deadly. You might actually damage the free store mechanism itself. It might write garbage data into the cells that the free store uses for its own internal accounting.

```
delete third;      // Serious Error, Potential free store damage
```

See if you can (cautiously) write a program to test the effect of these errors. Note that some of them (dereferencing dangling pointers) can be very dangerous. If you restrict yourself to retrieving data from such dangling pointers, however, you are extremely unlikely to do any damage.

❖Constructor and Destructor Homework

1. Examine and execute the files below. Show the output produced for each. Note that the source code is shown on the following pages and can also be found on Canvas.

- `elevatr1.cpp` - initialization function
- `elevatr2.cpp` - constructor
- `elevatr3.cpp` - overloading constructors
- `elevatr4.cpp` - destructors

2. Copy `elevatr4.cpp` to `elevatr5.cpp`. Verify that the constructor and destructor are called for `Elevator` variables by putting `cout` statements in the constructor and destructor member functions, as well as before and after the object instantiations in `main()`. Show the new code as well as the resultant output.

3. Create a new program, `order1.cpp`, with a class, `Order`, that contains only a constructor and a destructor. Determine the order of constructor and destructor calls for variables by putting `cout` statements inside the constructor and destructor member functions and creating several variables inside `main()`. To generate a unique identifier for each variable, use the keyword `this` inside the `cout` statement. `this` is the address of the current object, and is a pointer. Show the new code as well as the resultant output.

4. Copy `order1.cpp` to `order2.cpp`. Examine the scoping of variables and constructor and destructor calls using the class in `order2.cpp`. Put several scopes inside `main()`, and define one or more `order` objects at the beginning of each scope. Show the new code as well as the resultant output.

```cpp
// elevatr1.cpp - initialization function

#include <iostream>
#define   FLOORS 10

using namespace std;

class elevator {
    private:
        int floorSelected;
        int floorNumber;
    public:
        void initialize();
        void selectFloor(int floor);
        void go();
};

void elevator::initialize() {
    floorSelected = 0;
    floorNumber = 0;
}

void elevator::selectFloor(int floor) {
    if(floor < 0 || floor >= FLOORS)
        return;
    floorSelected = floor;
}

void elevator::go() {
    if(floorNumber < floorSelected)
        while(++floorNumber < floorSelected)
            cout << "floor: " << floorNumber << "; going up!" << endl;
    else if(floorNumber > floorSelected)
        while(--floorNumber > floorSelected)
            cout << "floor: " << floorNumber << "; going down!" << endl;
    cout << "floor: " << floorNumber << "; stopping." << endl;
}

int main() {
    elevator A;
    A.initialize();
    A.selectFloor(7);
    A.go();
    return 0;
}
```

```cpp
// elevatr2.cpp - constructor

#include <iostream>
#define   FLOORS 10

using namespace std;

class elevator {
    private:
        int floorSelected;
        int floorNumber;
    public:
        elevator();     // constructor has class name
        void selectFloor(int floor);
        void go();
};

elevator::elevator() {
    floorSelected = 0;
    floorNumber = 0;
}

void elevator::selectFloor(int floor) {
    if(floor < 0 || floor >= FLOORS)
        return;
    floorSelected = floor;
}

void elevator::go() {
    if(floorNumber < floorSelected)
        while(++floorNumber < floorSelected)
            cout << "floor: " << floorNumber << "; going up!" << endl;
    else if(floorNumber > floorSelected)
        while(--floorNumber > floorSelected)
            cout << "floor: " << floorNumber << "; going down!" << endl;
    cout << "floor: " << floorNumber << "; stopping." << endl;
}

int main() {
    elevator A;        // constructor called here
    A.selectFloor(7);
    A.go();
    return 0;
}
```

```
// elevatr3.cpp - overloading constructors

#include <iostream>
#define  FLOORS 10

using namespace std;

class elevator {
    private:
        int floorSelected;
        int floorNumber;
    public:
        elevator();
        elevator(int selected, int number);
        void selectFloor(int floor);
        void go();
};

elevator::elevator() {
    floorSelected = 0;
    floorNumber = 0;
}

elevator::elevator(int selected, int number) {
    floorSelected = selected;
    floorNumber = number;
}

void elevator::selectFloor(int floor) {
    if(floor < 0 || floor >= FLOORS)
        return;
    floorSelected = floor;
}

void elevator::go() {
    if(floorNumber < floorSelected)
        while(++floorNumber < floorSelected)
            cout << "floor: " << floorNumber << "; going up!" << endl;
    else if(floorNumber > floorSelected)
        while(--floorNumber > floorSelected)
            cout << "floor: " << floorNumber << "; going down!" << endl;
    cout << "floor: " << floorNumber << "; stopping." << endl;
}

int main() {
    elevator A;            // constructor called here
    elevator B(7,2);    // constructor with arguments
    A.selectFloor(7);
    A.go();
    return 0;
}
```

```
// elevatr4.cpp - destructors

#include <iostream>

#define  FLOORS 10

using namespace std;

class elevator {
    private:
        int floorSelected;
        int floorNumber;
    public:
        elevator();
        elevator(int selected, int number);
        ~elevator();
        void selectFloor(int floor);
        void go();
};

elevator::elevator() {
        floorSelected = 0;
        floorNumber = 0;
}

elevator::elevator(int selected, int number) {
        floorSelected = selected;
        floorNumber = number;
}

elevator::~elevator() {
        selectFloor(0);
        go();
}

void elevator::selectFloor(int floor) {
        if(floor < 0 || floor >= FLOORS)
            return;
        floorSelected = floor;
}

void elevator::go() {
    if(floorNumber < floorSelected)
        while(++floorNumber < floorSelected)
            cout << "floor: " << floorNumber << "; going up!" << endl;
    else if(floorNumber > floorSelected)
        while(--floorNumber > floorSelected)
            cout << "floor: " << floorNumber << "; going down!" << endl;
    cout << "floor: " << floorNumber << "; stopping." << endl;
}
```

```
int main() {
    elevator A;
    elevator B(7,2);
    A.selectFloor(7);
    A.go();
    return 0;
}
```

❖ Constructors Lab (Bar Codes)

Prior to 2009 the bar code on the envelope used by the U.S. Postal Service represented a five (or more) digit zip code using a format called POSTNET. The bar code consists of long and short bars as shown below:

For this program we will represent the bar code as a string of digits. The digit 1 represents a long bar, and the digit 0 represents a short bar. Therefore, the bar code shown above would be represented in our program as:

11010010100010101100001001 1

The first and last digits of the bar code are always 1. Removing these leave 25 digits. If these 25 digits are split into groups of five digits each then we have:

10100 10100 01010 11000 01001

Next, consider each group of five digits. There will always be exactly two 1's in each group of digits. Each digit stands for a number. From left to right the digits encode the values 7, 4, 2, 1, and 0. *Multiply the corresponding value with the digit and compute the sum to get the final encoded digit for the zip code.* The table below shows the encoding for 10100.

Bar Code Digits	1	0	1	0	0
Value	7	4	2	1	0
Product of Digit x Value	7	0	2	0	0

Zip Code Digit = 7 + 0 + 2 + 0 + 0 = 9

Repeat this for each group of five digits and concatenate to get the complete zip code. Note that there is one special value. If the sum of a group of five digits is 11, then this represents the digit 0 (this is necessary because with two 1's per group it is not possible to represent zero). The zip code for the sample bar code

above decodes to 99504. While the POSTNET scheme may seem unnecessarily complex, its design allows machines to detect if errors have been made in scanning the zip code.

Your assignment will be to write a `ZipCode` class, `ZipCode.h` and `ZipCode.cpp`, that encodes and decodes five digit bar codes used by the U.S. Postal Service on envelopes. The class should have two constructors. The first constructor should input the zip code as an integer and the second constructor should input the zip code as a bar code string consisting of 0's and 1's as described above. **Although you have two ways to input the zip code, internally the class should only store the zip code using one format (you may choose to store it as a bar code string consisting of 0's and 1's or as a zip code number.)** The class should also have at least two `public` member functions, one to return the zip code as an integer, and the other to return the zip code in bar code format as a string consisting of 0's and 1's. All helper member functions (functions not called within `main()` but used internally by the `ZipCode` class to perform calculations) should be declared `private`.

Use the following driver program to test your class `ZipCode` class. You should be able to compile and link your code to it without making any changes. However, you will eventually have to make changes to `main()` to capture your program's output to an output file, `csis.txt`. No other changes should be made to `main()`.

```cpp
// main.cpp

#include <iostream>
#include <iomanip>

#include "ZipCode.h"

using namespace std;

int main() {
    ZipCode zip1(99504);
    ZipCode zip2(12345);
    ZipCode zip3(67890);
    ZipCode zip4("100101010011100001100110001");
    ZipCode zip5("110100001011100001100010011");
    ZipCode zip6("100011000110101000011100101");

    cout << "Digits" << "        " << "Bar Code" << endl;
    cout << zip1.getZipCode() << setw(35) << zip1.getBarCode() << endl;
    cout << zip2.getZipCode() << setw(35) << zip2.getBarCode() << endl;
    cout << zip3.getZipCode() << setw(35) << zip3.getBarCode() << endl;
    cout << endl;
    cout << zip4.getZipCode() << setw(35) << zip4.getBarCode() << endl;
    cout << zip5.getZipCode() << setw(35) << zip5.getBarCode() << endl;
    cout << zip6.getZipCode() << setw(35) << zip6.getBarCode() << endl;
    return 0;
}
```

❖ Olympic Lab

The International Olympic Committee needs your help. The computer program that figures out who won races based on the timing data from each lane needs to be rewritten in object-oriented style. You have been selected for the job. Fortunately (or perhaps not), some of the preliminary design work has been done, and the main program that inputs the data and prints it out has already been written. Your job is to write two C++ classes, `Competitor` and `Ranker`, both declaration (`.h`) and implementation (`.cpp`), to store information on each competitor in a race and to rank these competitors from lowest to highest finishing time. This code should be divided into four files: `competitor.h`, `competitor.cpp`, `ranker.h`, and `ranker.cpp`.

Objects of the `Competitor` class are initialized with the person's name and lane assignment, specified as a character string and an integer. The person's name should be copied into the `Competitor` object's own memory; that is, don't just record the pointer that was passed in. Note that there should be no predefined size for the person's name. Also note that you should not use any C++ predefined `String` class for this lab. Since the person's time is not known at the outset, this value is set with a separate member function called `setTime()`, which accepts a double floating-point value. There should also be member functions to retrieve the name, lane, and time, as well as a member function to print out the object as a single line of text. Finishing times should be output to one decimal place. Finally, don't forget to have the class destructor deallocate any storage when an object of this class goes out of scope.

Objects of the `Ranker` class are initialized with the number of items (`Competitor` objects) to be ranked. Pointers to `Competitor` objects are passed to an `addList()` member function one by one. If a `NULL` pointer is passed in, or if the total number of objects passed in exceeds the pre-defined number of items, `addList()` should return 0; otherwise, it returns the total number of `Competitor` objects passed in. Items are retrieved (but not removed) from the list using two member functions: `getLane()` and `getFinish()`. Each accepts an integer from 1 to the number of items in the list. `getLane()` returns a pointer to the `Competitor` object in the specified lane, while `getFinish()` returns a pointer to the competitor who finished at the specified position (first, second, etc.). Finally, when a `Ranker` object is destroyed, it frees any memory that it allocated for itself, but it does not delete the memory for the `Competitor` objects (since it has no way of knowing how these were created).

Use the following driver program to test your classes. You should be able to
compile and link your code to it without making any changes:

```
// olympic.h

#ifndef _OLYMPIC_H
#define _OLYMPIC_H

#include "competitor.h"
#include "ranker.h"

#endif
```

```
// olympic.cpp

#include <iostream>
#include <fstream>
#include "olympic.h"

using namespace std;

ofstream csis;

int main() {
    const int lanes = 4;
    Ranker rank(lanes);

    csis.open("csis.txt");

    // First make a list of names and lane assignments
    Competitor* starters[lanes];

    starters[0] = new Competitor("Emmylou Harris", 1);
    starters[1] = new Competitor("Nanci Griffith", 2);
    starters[2] = new Competitor("Bonnie Raitt",   3);
    starters[3] = new Competitor("Joni Mitchell",  4);

    // The race is run; now assign a time to each person
    starters[0]->setTime((float)12.0);
    starters[1]->setTime((float)12.8);
    starters[2]->setTime((float)11.0);
    starters[3]->setTime((float)10.3);
```

```cpp
    // Put everyone into the ranker
    for (int i = 0; i < lanes; i++)
        rank.addList(starters[i]);

    // Now print out the list to make sure it's right
    cout << "Competitors by lane are:" << endl;
    csis << "Competitors by lane are:" << endl;
    for (int i = 1; i <= lanes; i++)
        rank.getLane(i)->print();

    // Finally, show how they finished
    cout << "Rankings by finish are:"  << endl;
    csis << "Rankings by finish are:"  << endl;
    for(int i = 1; i <= lanes; i++)
        rank.getFinish(i)->print();
    for(int i = 0; i < lanes; i++)
        delete starters[i];

    csis.close();
} // myset's destructor will be called automatically here
```

Chapter 6. Inline & Const

Inline Functions
Const Modifier

Inline Functions

C programmers commonly use the preprocessor to make code easier to write and read with macros and value substitution. We'll see how this process invites errors because:

- a macro looks like a function but doesn't act like one
- value substitution looks like an expression but doesn't act like one

Then we'll see how C++ rectifies the problem with inline functions and consts.

Functions or Subprograms
The purpose or advantages of the function or subprogram are numerous:

- makes programs modular
- easier to debug programs
- easier to maintain programs
- usually less utilization of memory

Functions save memory space because all the calls to the function cause the same code to be executed. Therefore the function body need not be duplicated in memory.

Keep in mind, however, that there are drawbacks to using functions. One of the sad truths of programming languages such as C and C++ is that function calls waste time. To call a function, the compiled program has to:

- save the return address of the current location on the stack
- push onto the stack any argument values needed by the function
- save register contents
- execute an assembly language call instruction to invoke the function
- return some function value
- clear the stack and return to the starting place
- restore the registers

All of these instructions slow down the program. While these sequence of events may save memory space, they take some extra time. Although a single function call takes place rather quickly, numerous function calls can shave points off of a program's performance.

However, avoiding functions is not an acceptable solution to this problem! Functions make programs modular and easier to maintain and debug. Without functions, it is extremely difficult, perhaps impossible, to write even medium-sized programs that run correctly. In fact, if a function is very short, the instructions necessary to call it may even take up as much space as the instructions within the function body itself. In this case, there is not only a time penalty, but a space penalty as well.

Macros

One possible solution for this function overhead is to use the preprocessor and define macros that use code substitution. For example, here is a function that checks the range of its argument and calls `error()` if the argument is out of range:

```
void test(int a) {
    if (a < 0 || a > 10)
        error(a);
}
```

Now, here is a corresponding macro to perform the same range checking:

```
#define TEST(a)   if (a < 0 || a > 10) error(a)
```

The problem here is that a preprocessor macro looks like a function but doesn't act like one! If we make the following rather innocent calls, we get a number of unexpected results:

```
int main() {
    int x = 0;
    int y = 0;
    int u = 10;
    int v = 10;

    test(--x);          // error: -1 out of range
    TEST(--y);          // error: -2 out of range
    test(++u);          // error: 11 out of range
    TEST(++v);          // error: 13 out of range
}
```

We get different results depending upon whether the function or macro is called and also depending on the value of the arguments. The problem occurs because the macros are expanded like this:

```
• if (--y < 0 || --y > 10) error(--y);

• if (++v < 0 || ++v > 10) error(++v);
```

The execution of the first macro decrements `y` for both tests in the `if` statement as well as the call to `error()`. The execution of the second macro only increments `v` for the first test which evaluates to true so the second test is short-circuited, but it also increments `v` for the call to `error()`.

In C, the only way to write small, function-like entities that have no function overhead is to use the preprocessor. In C++, preprocessor macros were considered problematic enough to add a new feature to the language, *inline functions*.

Inline Functions

`inline` is a C++ keyword which tells the compiler to treat a function as a macro, somewhat. When the C++ compiler encounters an inline function, it doesn't allocate storage for the function body as it would with a normal function. Instead, the compiler creates space in its symbol table and stores the prototype of the function and the function body in the symbol table. Thus, the function definition generates no object code at the point it is defined!

When you call an `inline` function, the compiler inserts the function code directly inline, giving you the efficiency of a macro. Everything else about an inline function is exactly the same as a normal function:

• You get type checking on the arguments and return values.
• There are no surprise non-function-like interactions.

Thus, there is no confusion about the behavior of an inline function since it is exactly the same as a regular function. The only difference between an `inline` function and a non-inline function is the use of the `inline` keyword.

```cpp
// inline.cpp

#include <iostream>

using namespace std;

int max(int a, int b);

int main() {
    int x = 5, y = 3, z;

    z = max(x, y);
    cout << z;
}

inline int max(int a, int b) {
    if (a > b)
        return a;
    else
        return b;
}
```

We use inline functions the same way we use normal functions. The source code calls a `max()` function from within `main()`. However, in the compiled code the `max()` function is not called. Instead, the compiler inserts the functions statements directly into the program effectively compiling the function call as if it had been written:

```cpp
if (x > y)
    z = x;
else
    z = y;
```

You can also declare a member function within its class and define it elsewhere as `inline`, or you can both declare and define a member function within its class, in which case it is automatically an `inline` function.

For example, we'll define a C++ class `Point` that includes both a point's location and a member function to set the location.

```cpp
class Point {
    private:
        int x;
        int y;
    public:
        void setxy(int newX, int newY);
};

inline void Point::setxy(int newX, int newY) {
    x = newX;
    y = newY;
}
```

Here the member function is coded `inline`. As stated above, we can also both declare and define the member function within the class itself:

```cpp
class Point {
    private:
        int x;
        int y;
    public:
        void setxy(int newX, int newY) {
            x = newX;
            y = newY;
        }
};
```

Member functions that are defined in the body of the class are treated as if `inline` was requested. Note that the `inline` keyword is actually just a polite request to the compiler. The compiler may choose to ignore the request and compile the function as a normal function. Some functions may be too long to be `inline`. Any short function is a candidate to be an `inline` function.

So `inline` functions operate like macros in the sense that they can be used to eliminate the overhead of function calls. However, they are different from macros in that:

• type checking is performed with `inline` functions just as it is for any function having a prototype

• you don't get the side effects that can occur with macros

In fact, there's almost no place a preprocessor macro is used that an `inline` isn't better. The `inline` function gives you the efficiency of preprocessor macro combined with the error checking of ordinary functions.

Because an `inline` function duplicates the code for every function call, you might think it automatically increases code space. But for small functions (which inlines were designed for), this isn't necessarily true! Keep in mind that a function call requires code to pass arguments, make the call, and handle the return value. This code isn't present for an `inline` function.

If your `inline` function turns out to be smaller than the amount of code necessary for the ordinary call, you are actually saving space. In addition, if the `inline` function is never called, no code is ever generated. With an ordinary function, code for that function is there (only once) whether you call it or not.

The speed benefit of `inline` functions tend to diminish as the function grows in size. At some point, the overhead of the function call becomes small compared to the execution of the function body, and the benefit is lost.

Const Modifier

const and ANSI C

Value substitution is useful because it allows us to take a name and substitute it everywhere we would normally place a value:

```
// value1.c

#include <stdio.h>

#define TOP      100
#define BOTTOM  47
#define SIZE     TOP - BOTTOM

int main(void) {
    int i, a[SIZE];

    for (i = 0; i < SIZE; i++)
        a[i] = 0;
    printf("SIZE = %d\n", SIZE);
    return 0;
}
```

The output produced is:

```
SIZE = 53
```

This is a good programming practice. The program is easier to understand and you can easily change SIZE. A problem occurs in value substitution because the preprocessor does textual substitution and doesn't follow the normal rules of the compiler. Thus, an expression may look like a normal expression but it may not act like one. This can cause errors that are difficult to find.

Suppose you make an innocent modification to the above program:

```
// value2.c

#include <stdio.h>

#define TOP      100
#define BOTTOM   47
#define SIZE     TOP - BOTTOM

int main(void) {
    int i, a[SIZE * 2];
    for (i = 0; i < SIZE; i++)
        a[i] = 0;
    printf("SIZE * 2 = %d\n", SIZE * 2);
    return 0;
}
```

All we've done is to multiply SIZE by 2, but the results are unexpected. The output produced is:

```
SIZE * 2 = 6
```

If SIZE were a C variable, this wouldn't happen since the operator precedence would occur as you would expect. This problem can certainly be avoided by the correct use of parentheses. But if code like this gets buried and you aren't looking for it, it can be very difficult to find.

You might think that you can solve this problem by making SIZE a const. But ANSI C treats a const differently than a preprocessor name. A C const is a variable that cannot be changed, but the C compiler cannot evaluate an expression that contains a const within array brackets. So if SIZE is a const, you cannot say this in ANSI C:

```
int a[SIZE * 2];
```

Therefore in ANSI C, const cannot replace the use of the preprocessor.

C++ and const

In C++ the old style of declaring program constants with a `#define` is considered obsolete. `const` should be used instead. In C++, `const` has a distinctly different meaning than in ANSI C. `const` allows the compiler to perform type-checking because it knows more about the items being handled.

```
const int TOP = 100;
const int BOTTOM = 47;
const int SIZE = TOP - BOTTOM;
```

Here, C++ knows that the `const SIZE` is an `int` value. The compiler doesn't know anything about textual substitution.

The compiler is also smart enough to avoid allocating storage to a `const` when unnecessary. Instead, the compiler will evaluate the expression and store the result in the symbol table. Then it can use that result after performing type checking. In other words, the `const` is an identifier that the compiler replaces with actual values at compile time.

There are occasions, however, in which storage is actually allocated to a `const`:

```
const int& num = 9;
```

This statement contains a constant reference and allocates storage to it. The identifier `num` is in the symbol table and behaves just like any other reference variable with the understanding that it cannot be modified. More on references later!

Thus in C++ we can use a const just like a preprocessor value:

```
// const.cpp

#include <iostream>

using namespace std;

const int TOP = 100;
const int BOTTOM = 47;
const int SIZE = TOP - BOTTOM;

int main() {
    int a[SIZE * 2];

    for (int i = 0; i < SIZE; i++)
        a[i] = 0;
    cout << "SIZE * 2 = " << SIZE * 2 << endl;
}
```

The constant expression above:

```
const int SIZE = TOP - BOTTOM;
```

is evaluated at compile time so it can be used in the array definition. The output produced here is:

```
SIZE * 2 = 106
```

In C++, const has all the benefits of using the preprocessor. There is no situation where it is better to use #define for variable substitution. You always get much better benefits from using const.

const Variables

Actually, the const identifier is a C++ innovation adopted by ANSI C. Declaring a variable to be a const in effect makes the variable read-only. It may be assigned an initial value, but other statements cannot change it.

For example:

```
const int w = 10;  // declares w as a read-only variable

int x = w;         // legal - just accessing constant w

w = 25;            // illegal - cannot change value to const w

const y = 5;       // legal - can omit the int keyword

const int z;       // illegal - const must be initialized when defined
```

The reason to declare an object of a built-in type as `const` is so the compiler will ensure that it isn't changed. You can also tell the compiler that an object of a user-defined type is a `const`, although it isn't stored in the symbol table. Storage is actually allocated for the object! More on this later.

`const` and Pointers (just get a feel for this part)

Some C++ constructs frequently seem to be the source of confusion. One is the use of `const` with pointers. Sometimes it isn't clear whether the item referenced, the pointer, or both, are constant. We'll take a look at:

- pointers to constants
- constant pointers
- constant pointers to constants

When using a pointer, two things are involved:

- the pointer itself
- the object pointed to

Either can be declared as `const` and therefore cannot be assigned to. In a pointer declaration, the position of `const` indicates whether it is the pointer or the object being pointed to that cannot be changed.

Pointer to `const`

"Prefixing" a declaration of a pointer with `const` makes the object pointed to, but not the pointer, a constant.

```
syntax:        const <type> *pc;
```

Here, `pc` points to a constant value of type `<type>`. For example:

```
const char *pc = "Haley";
```

defines a pointer to a constant, or more specifically, a pointer to a character constant, i.e., a pointer to read-only memory. These types of declarations are actually better read right to left to get the meaning: *pc is a pointer to `char`s that are constant.* In other words, `pc` is a pointer to a constant object of type `char` and therefore the value of the object addressed by `pc` cannot be modified through `pc`:

```
pc[3] = 'x';        // error
*pc = 'x';          // error
```

Each of the examples above are invalid because the character that `pc` addresses cannot be changed. But, `pc` itself is not a constant and therefore `pc` can be changed to address a different variable of type `char` as seen below:

```
pc = "Samantha";   // ok
```

Here, `pc` is being changed to address another character.

Declaring something as a pointer to `const` is a statement about the pointer, not about the space being pointed to. The compiler will not ensure that the data being pointed at will not change. It ensures only that it will not change through that pointer. The pointed at object can still be changed by some other means. So by declaring a pointer to `const`, you are declaring that the pointer cannot be used to change the pointed-at object.

Constant Pointer

To declare a pointer itself to be a constant, rather than the thing being pointed to, `*const` is used:

```
syntax:     <type> *const cp;
```

Here, `cp` is a constant pointer to an object of type `<type>`. For example:

```
char *const cp = "Haley";
```

defines a constant pointer, or more specifically a constant pointer to a character, i.e., a read-only pointer to memory. We say that `cp` is a constant pointer to characters. The constant pointer `cp` always points to the same piece of memory and therefore must be initialized at the point of declaration:

```
cp = "Haley"            // error
```

Although a constant pointer cannot be modified, the object to which it points can be:

```
cp[3] = 'x';        // ok
*cp = 'x';          // ok
```

Here we are only modifying the characters `cp` addresses.

Another example:

```
int errNum;                      // possible error states of program

int *const curErr = &errNum;   // constant pointer
```

In the declarations above, `curErr` is a constant pointer to an object of type `int`. Therefore we can modify the value of the object `curErr` addresses:

```
if (*curErr) {
    errorHandler();
    *curErr = 0;
}
```

but cannot modify the address that `curErr` contains:

```
curErr = &myError;      // illegal
```

Constant Pointer to `const`

To make both the pointer and the object being pointed to constant, they must both be declared `const`.

```
syntax:      const <type> *const cpc;
```

Here, `cpc` is a constant pointer to a constant value of type `<type>`.

For example:

```
const char *const cpc = "Haley";
```

defines a constant pointer to a constant. `cpc` is a constant pointer to characters that are constant. In this case neither the value of the object addressed by `cpc` nor the address itself may be changed, i.e., `cpc` can't change and the thing it points to can't change:

```
cpc[3] = 'x';        // error
*cpc = 'x';          // error
cpc = "Samantha";    // error
```

Passing `const` Parameters to a Function

C++ allows you to declare a function parameter to be a `const`:

```
void foo(const int num);
```

Why would you want to declare a function in this manner? Because the function receives a copy of the actual parameter, the caller of the function really doesn't care whether the function modifies its copy. This is called pass by value.

But what if you pass a pointer to something solely for efficiency purposes? You may want to ensure that the called function doesn't use the pointer to modify the entity referenced. By declaring a pointer argument `const`, a function is prohibited from modifying the parameter pointed to.

For example:

```
void foo(const int *num);
void foo(const char *buf);
void foo(const bigStruct *bs);
char *strcpy(char *p, const char *q);
void foo(const int& num);
```

It is also legal to declare a function that returns a `const`:

```
const int foo();
```

Thus, the value returned can be used only as you would use a `const int`. For example:

```
// return.cpp

#include <iostream>

using namespace std;

int main() {
    const char *pc = getName();
    cout << pc << endl;
}

const char *getName() {
    return "Haley";
}
```

The variable `pc` points to a constant string. Any attempt to modify the constant string through `pc` causes a compile-time error.

Constant Arrays

To prevent changes to array elements, we can preface the array declaration with `const`:

```
const int num[] = { 0, 1, 2, 3, 4 };
```

This can be especially useful when programs are maintained by different people who may be tempted to subvert a module's design rules. Note that because of the `const` modifier, any statement that attempts to change an array value does not compile:

```
num[3]++;           // illegal
```

You can pass an array by value to ensure that the members of the array cannot be changed. Pass the array using a pointer argument that is defined with the `const` keyword:

```
int foo(const int *num) {
    num[i] = 25;           // illegal
    return num[i];         // ok
}
```

The statement:

```
num[i] = 25;
```

is illegal because the argument `num` was declared to be a pointer to an array of constant integers which cannot be modified.

Constant Class Members and the Constructor Initializer List

You can make a member of a class `const`, but the meaning reverts to that of C. Storage is always allocated for a `const` data member so a `const` occupies space inside a class. The `const` must still be initialized at the point it is defined. But storage isn't allocated for an object until the object is created, and that is the point where the `const` must be initialized. Therefore the meaning of `const` for class members is constant for that object for its lifetime.

Accordingly, the initialization of a `const` must occur in the constructor. It's a special action and must happen in a special way so that its value is guaranteed to be set at all times. This is performed in the *constructor initializer list* that occurs after the constructor's argument list but before its body. This indicates that the code is executed before the constructor body is entered. For example:

```
class Counter {
    private:
        int count;
        const int MAX;
    public:
        Counter (int init, int size) : MAX(size) {
            count = init;
            ...
        }
};

void main() {
    Counter A(0, 10);
    ...
}
```

The statement in the constructor initializer list:

```
MAX(size)
```

performs the initialization. Its meaning is simply assignment equivalent to:

```
MAX = size;
```

Note that it would have been illegal to say:

```
const int MAX = size;
```

because we cannot initialize a `const` class member here! It would also have been illegal to say:

```
MAX = size;
```

within the body of the constructor!

However, it would have been legal (and is usually preferred) to initialize all the data members in the constructor initializer list as follows:

```
Counter (int init, int size) : count(init), MAX(size) {}
```

True Constants Inside Classes

The treatment of `const` inside classes creates an inconvenient situation when you're creating an array inside a class. When dealing with ordinary arrays (not inside classes) the best programming practice is to use a named constant to define the size of the array:

```
const MAX = 10;
int num[MAX];
```

This way, any code that refers to the size of the array uses MAX and if you need to change the value of MAX, you only change it in one place, at the `const` definition.

However, you cannot use a `const` data inside an array definition within a class because the compiler must know the size of the array when it is defined and because a `const` inside a class always allocates storage (the compiler cannot know the contents of a storage location). So the following is illegal:

```
class Array {
    private:
        const int MAX;
        int num[MAX];
};
```

Fortunately, there is a convenient way to work around the problem. The enumerated data type `enum` is designed to associate names with integral numbers. Normally `enum` is used to distinguish a set of names by letting the compiler automatically assign numbers to them. However, you can force a name to be associated with a particular number by:

```
enum {MAX = 10};
```

This introduces a name called MAX that has the integral value 10. Storage is never allocated for enumerated names so the compiler always has the values available. This provides a technique, called the enum hack, to solve the problem of using names for array sizes without reverting back to the use of the preprocessor:

```
class Array {
    private:
        enum {MAX = 10};
        int num[MAX];
};
```

Constant Member Functions

Constant member functions are used to support constant objects. A constant object is an object whose data members cannot be changed once they are initialized. All data members must be initialized within the constructor. This definition introduces some interesting consequences for an object's member functions.

Let's look at a simple Point class to see what's involved in making a constant Point object:

```
// point.cpp

#include <iostream>

using namespace std;

class Point {
    private:
        int x, y;
    public:
        Point (int a, int b)       { x = a; y = b; }
        int xPos()                 { return x; }
        int yPos()                 { return y; }
        void move (int dx, int dy) { x += dx; y += dy; }
};
```

```
int main() {
    const Point p(40, 12);

    cout << p.xPos() <<"," << p.yPos() << endl;
    p.move(1, 1);
    cout << p.xPos() << "," << p.yPos() << endl;
}
```

We've declared p to be a constant Point object. However, the compiler will complain that we are calling the move() function. Why? Because p was declared as a constant and therefore cannot be modified!

But the compiler will also complain that we are calling the Point member functions xPos() and yPos(). Why? Since xPos() and yPos() don't actually change any numbers, we should be able to use them. Unfortunately the compiler won't let us. The compiler can't tell if the functions might be modifying the members x and y of object p which would be illegal since p is a constant.

How can we get around this problem? We can declare xPos() and yPos() to be constant member functions. A member function is declared constant by attaching the keyword const after the function arguments, but before the function body:

```
class Point {
    private:
        int x, y;
    public:
        Point (int a, int b)        { x = a; y = b; }
        int xPos() const            { return x; }
        int yPos() const            { return y; }
        void move (int dx, int dy)  { x += dx; y += dy; }
};
```

By placing the const keyword before a function body, we are stating that this function is guaranteed not to modify any data members of the class. The compiler will then allow us to call the function even if the function is for an object declared as a constant. *The only member functions that can be called from a constant object are constant member functions.* Therefore, p.xPos() and p.yPos() can be called, but not p.move().

It would be illegal to declare `move()` to be a constant member function. For any constant member function, the hidden this pointer is treated as a pointer to a constant object and thus any attempted updates to the object's data are flagged as errors. The following code will not compile due to the `const` keyword added to the `move()` member function:

```
class Point {
    private:
        int x, y;
    public:
        Point (int a, int b)            { x = a; y = b; }
        int xPos() const                { return x; }
        int yPos() const                { return y; }
        void move (int dx, int dy) const { x += dx; y += dy; }
};
```

A `const` member function defined outside the class body must specify the `const` keyword in both its declaration and definition:

```
class Point {
    private:
        int x, y;
    public:
        Point (int a, int b)        { x = a; y = b; }
        int xPos() const;
        int yPos() const;
        void move (int dx, int dy)  { x += dx; y += dy; }
};

int Point::xPos() const             { return x; }
int Point::yPos() const             { return y; }
```

You must also be careful when calling other functions inside a constant member function. These other functions cannot be allowed to modify any of the data members. The compiler ensures that when inside a constant member function, each data member is typed as though it were a constant! Thus, the only legal way to pass the members to another function is either by value, or by using pointers to constants, or by using references to constants.

We can introduce a new member function `modify()`, which calls a `random()` function to assign the coordinates random values:

```
class Point {
    private:
        int x, y;
    public:
        Point (int a, int b)       { x = a; y = b; }
        int xPos() const           { return x; }
        int yPos() const           { return y; }
        void move (int dx, int dy) { x += dx; y += dy; }
        void modify() const        { random(&x); random(&y); }
};

void random(int *num) {
    // set num to some random value
}
```

We've incorrectly declared `modify()` to be a constant function. The code won't compile because the calls to `random()` require pointers to integers when, in fact, `x` and `y` are typed as constants.

We can change `random()` to be acceptable to the compiler by changing its argument type:

```
void random(const int *num) {
    // can't really change num here
}
```

This prevents `random()` from changing the value of the coordinate.

The same rule about calling other functions inside constant functions applies to calling other member functions. A constant member function can only call other constant member functions. It can't call standard member functions since they might update the object's data.

Note that you could find a clever way to get around the restrictions for constant member functions by using clever programming techniques! You could use type casts and pointers to trick the compiler into letting you change the constants. There is no guarantee that a constant member function won't change the data members. C++ doesn't always stop you from violating the rules. It only prevents accidental misuse.

Also note that constructors and destructors are not required to be declared `const` in order to be applied to constant class objects.

Chapter 7. Function Overloading & Default Arguments

Function Overloading
Default Arguments
Function Overloading & Default Arguments Homework

Function Overloading

Suppose you want to create a `print()` function and you want to give it various types of arguments:

```
char, char*, int, double
```

In C, you might come up with something like this:

```
void print(int);
void printChar(char);
void printString(char*);
void printDouble(double);
```

Although it's the same concept, printing, you must create a different name for each implementation of the concept. Not only must you create these names, you must explain them to the user of your library.

Now suppose the user has mastered this group of print functions and then decides to use another library that has a way to print an `Elevator` object:

```
void print(struct Elevator);
```

If the function declarations are all made in the same file, the C compiler will give an error message about a type mismatch in the redeclaration of `print()`. If the colliding declarations are not made in the same file, the C compiler will not catch the error. However, the linker will detect a multiple definition error.

To solve the problem of name collisions between libraries in C, library authors must not only create function with different names to differentiate the type of arguments, but libraries must also be differentiated from one another:

```
void abcLibPrint(int);
void abcLibPrintChar(char);
void abcLibPrintString(char*);
void abcLibPrintDouble(double);
void defLibPrint(struct Elevator);
```

In C++, multiple functions may have the same name as long as all the functions have distinct argument lists. In this way, the same abstract operation can be implemented on different argument types without concocting artificial names to differentiate the functions. In C++ we can therefore create the following functions:

```
void print(int);
void print(char);
void print(char*);
void print(double);
void print(struct Elevator);
```

The functions are overloaded, because, though named the same, they perform multiple jobs. The function name is overloaded because it can be interpreted in more than one way. The concept of printing is the same for all the functions, but its use is different!

Overloaded functions must have a different number of arguments or different types of arguments, or both.

```
void print(int);
void print(int, int);
void print(int, float);
```

Note that each of the functions must have different argument lists, even if the difference is just in the ordering of the lists:

```
void print(float, char*);
void print(char*, float);
```

Overloaded function can have any return value, but two overloaded functions cannot differ by only their return value:

```
void print(float);      // invalid
int print(float);       // invalid
```

The only time two function declarations will clash is if they have identical argument lists, which is a logically ambiguous situation. Note that the const modifier does make a difference in the argument list:

```
int foo(int);           // ok
int foo(const int);     // ok

int i = 5;
int j = foo(i);         // int foo(int) called
int k = foo(5);         // int foo(const int) called
```

Name Mangling

How does an overloaded function know what to do? It performs one operation on one kind of data, but another operation on a different kind. Overloading is like the joke about the famous scientist who insisted that the thermos bottle was the greatest invention of all time. Why? It's a miracle device. It keeps hot things hot, but cold things cold. How does it know?

When a C++ module is compiled, the compiler generates function names that include an encoding of the function's argument types. This process is called name mangling and makes overloaded functions possible. In C++, the function name that is in the source code is different from the function name generated by the C++ system.

In C, if you make the function call:

```
void foo(int);

int main() {
    foo(5);
    return 0;
}
```

a compiler might generate an assembler-language call :

```
call _foo
```

To generate a name, the C compiler simply adds an underscore to the function name (there's no information about the argument types that would make the function name unique).

A C++ compiler generates unique names by combining the name of the function as given in the source code with information about its argument types. This encoding is sometimes called *name mangling*, because the function name and its argument types are mangled together to create the name used in the assembly language. For example, the following C++ code:

```
void foo(int);
void foo(char*);
void foo(double);

int main() {
    foo(5);
    foo("Haley");
    foo(5.5);
}
```

might generate the following calls:

```
call    ?foo@@YAXH@Z
call    ?foo@@YAXPAD@Z
call    ?foo@@YAXN@Z
```

Notice that extra characters have been added. The characters after the @ signs contain argument type information. Thus, name encoding generates unique names by combining the name of the function with its argument types, so you can overload the same function name using different argument lists.

Name encoding is also used to generate the names of class member functions, by combining the function name and argument types with the class name. For example:

```
class button {
    private:
        int pressed;
    public:
        void push();
};
```

```
int main() {
    button b;
    b.push();
}
```

Here, the call to `push()` generates the assembly-language call:

```
call    ?push@button@@QAEXXZ
```

This shows that name encoding also puts class member function names inside their own scope, so even if a class member function has the same name and argument list as a global function or a member function from another class, there will be no name clash.

Overloading Member Functions

We can also overload member functions:

```
class Point {
    private:
        int x, y;
    public:
        void set(int xp, int yp) {
            x = xp;
            y = yp;
        }
        void set(Point p) {
            x = p.x;
            y = p.y;
        }
};
```

We can call these two member functions as follows:

```
int main() {
    Point p, q;

    p.set(42, 14);
    q.set(p);
}
```

Class constructors are often overloaded to accommodate different formats of class object initialization:

```cpp
class Name {
    private:
        char first[80];
        char middle[80];
        char last[80];
    public:
        Name() { first[0] = middle[0] = last[0] = '\0'; }
        Name(char*, char*, char*);
};
```

A constant member function can be overloaded with a non-`const` instance that defines the same function arguments:

```cpp
class Screen {
    public:
        char get(int x, int y);
        char get(int x, int y) const;
};
```

In this case, the `const`ness of the class object determines which of the functions is involved:

```cpp
int main() {
    const Screen sc;
    Screen s;

    char ch = sc.get(0, 0);    // const member
    ch = s.get(0, 0);          // non-const member
}
```

Default Arguments

In C, if a function requires two arguments, both arguments must be passed when the function is called. For example, suppose we have a function to position the cursor (which has an x and y location on the screen), within a window:

```
void move(int a, int b) {
    int x, y;

    x = a;
    y = b;
    :
    :
}
```

Whenever you want to position the cursor at the home position (0, 0), you would call the move() function and pass it the zero values:

```
move(0, 0);
```

Because it is common to reposition the cursor to the home position, which is always (0, 0), C++ allows us to provide default argument values as follows:

```
void move(int a = 0, int b = 0) {
    int x, y;

    x = a;
    y = b;
    :
    :
}
```

This function can be called in 3 ways:

```
move(35, 70);    // position cursor at (35, 70)
move(10);        // position cursor at (10, 0)
move();          // position cursor at (0, 0)
```

Note that variable a was never defaulted when a function call was made unless variable b was first defaulted. It's impossible to provide a value for the last parameter and take any default values for previous parameters.

Restrictions on Default Arguments

If default arguments are used, the right-most value must be given a default value before arguments to its left are defaulted. Only trailing arguments can be defaulted.

After the first default argument, all subsequent arguments must also include default arguments.

```
void foo(int width, float v = 3.14159, char q = '$')'
```

In this example, width could not be declared anywhere else, i.e., it could not be placed between v and q or at the end of the argument list. When calling foo(), a program must specify a width. You always have to supply argument values for a function's normal parameters

• Default arguments appear only in the function prototype (declaration), not in the function definition header.

• Default arguments can only be declared once.

• Default arguments must be included in header files that are included in all files that use the defaulted notation.

• Default arguments must be passed by value, not reference or pointer.

• Default argument values may be literal values or constant definitions. They may not be variables:

```
void foo(int x, int y = n);   // invalid
```

So default arguments can be useful when function calls typically need to supply only some, but not all arguments.

Default arguments look a lot like function overloading since you can use the same name and different arguments to call the function. However, there is no overloading when default arguments are used. The compiler is performing argument substitution and calling the same function.

You can use function overloading and default arguments together as long as no ambiguities are created:

```
void foo();              // ambiguous
void foo(int a = 10);    // ambiguous
```

❖ Function Overloading & Default Arguments Homework

We've seen how C++ lets us create functions that can be called in different ways. We can create a single function name for a concept, and implement that name in many convenient ways. This means we have less to teach the end user, and the code is more readable. C++ supports this feature in two ways:

Function overloading allows us to create more than one function with the same name, as long as all the functions have distinct argument lists. Function overloading also prevents name clashes when using multiple libraries. Here are some overloaded examples of functions:

```
void foo(int);
void foo(int, char);
void foo(double);
```

Default arguments are used with a single function when you want some of the arguments to be automatically inserted by the compiler instead of writing them all out yourself, every time you call the function. Here's a function with default arguments:

```
int foo(float f, float f2 = 1.1, char x = 'i');
```

You can declare a function more than once, but you may only give default arguments once. Only trailing arguments may be given default values, and once you start giving default values, all the rest of the arguments in the list must have defaults.

1. Name encoding schemes vary from compiler to compiler, and they are up to the implementer of the compiler. In addition, encoded names are not intended to be used directly by the programmer (which would mean, for instance, that an encoding scheme might legitimately be changed by a compiler vendor, which would require recompilation of all code but no other changes.) See if you can find out what encoded names will be generated by the compiler for the file `array5.cpp`.

```
// array5.cpp - friend function
// This program will compile, but it won't link without the set()
definition

#include <iostream>

using namespace std;

#define SIZE 10

class intArray {
private:
    int a[SIZE];
public:
    void setVal(int index, int value);
    friend void print(intArray);
};

void print(intArray x) {
    int i;
    for(i = 0; i < SIZE; i++)
        cout << "a[" << i << "] = " << x.a[i] << endl;
}

int main() {
    intArray num;
    int i;

    for(i = 0; i < SIZE; i++)
        num.setVal(i,i);
    print(num);
    return 0;
}
```

An assembler listing shows the low-level assembler language that C++ is translated into before it's translated into the 0's and 1's of machine language.

Instructions to generate assembler code for Visual Studio and Xcode can be found at this Canvas location:

Files I C++ Compilers

2. Create a `Point` class that has integer `x`, `y`, `z` coordinates. Add an `initialize()` member function to the class which contains default arguments for the three coordinates. The member function should be defined outside of the class.

3. Prove to yourself that default arguments use the same function. See if you can generate an assembler listing for the file `send.cpp` and find the three function calls to see that they are actually calling the same function.

```
// send.cpp

void send(char * msg = "test", int times = 1);
void send(char *, int);

int main() {
    send();
    send("hello");
    send("hello", 3);
    return 0;
}
```

Chapter 8. Static Data Members and Static Member Functions

Static Data Members
Static Member Functions
Static Data Members and Static Member Functions Homework

Static Data Members

It is sometimes necessary that all the objects of a particular class have access to the same variable. This may be some condition flag or counter related to the class that changes dynamically in the course of a program.

For example, perhaps a count is needed of how many objects of a class exist at any one point in a program. Sometimes it is simply more efficient to provide one variable for all the objects of one class rather than having each object maintain its own copy. This variable may be a pointer to an error-handling routine for that class or a pointer to the heap for the class.

For these cases, a static data member can provide a solution. A static data member allows you to share information among objects of the class without revealing it to the rest of the world, such as a global class variable.

For example, say that I have a basket of apples, and there is a variable that defines ripeness. Now when an apple reaches a certain state, we'll say that it has gone bad. Once this happens, all the other apples should know about it so they can also go bad. In this case, one bad apple really does spoil the whole bunch.

Suppose you want an object which holds a single word. Then you can turn a text file into a group of these objects. Also suppose you want any word to be able to know how many total word objects there are. For example:

```
// word1.h

#ifndef _WORD1_H
#define _WORD1_H

class Word {
  private:
      char *buf;
  public:
      Word(char *s);
};

#endif
```

```
// word1.cpp

#include <string.h>
#include "word1.h"

int wordCount = 0;              // global variable

Word::Word(char *s) {
    buf = strdup(s);
    wordCount++;                // increment global variable
}
```

Now, any word object can access `wordCount` to find out how many words there are. But there are two problems with this approach:

- Since `wordCount` is global, anyone can modify it. There is no safety.

- `wordCount` is in the global name space, which means if you bring in some other library which uses the same name, there will be a clash.

You may think to solve this problem by bringing `wordCount` inside class Word as follows:

```
// word2.h

#ifndef _WORD2_H
#define _WORD2_H

class Word {
    private:
        char *buf;
        int wordCount;
    public:
        Word(char *s);
};

#endif
```

```
// word2.cpp

#include <string.h>
#include "word2.h"

Word::Word(char *s) {
    buf = strdup(s);
    wordCount++;
}
```

The name `wordCount` is no longer global and the data is protected so only members of the class `Word` can read or modify it. However, it doesn't produce the desired effect because storage for `wordCount` is different for each object. C++ solves these problems by adding yet another meaning to the keyword `static`.

In C, `static` has two meanings:

• When used to define a variable outside all functions, it means the variable has internal linkage, i.e., the variable can only be accessed within the file in which it is declared. If the same variable is defined as `static` or global within another file, there will be no clash and different storage will be reserved for both variables.

• The second meaning for `static` in C occurs when it is used to define local variables in a function. A `static` variable is not an automatic or local variable. It is not created on the stack but instead in a special static data area. This means that its value is held, i.e. static, between calls to the function.

In C++, a `static` data member inside a class is similar to this second definition. A `static` data member is kept in the static data area and there is only one variable allocated for all the objects in a class. Each object in the class shares the same variable. This is exactly what we need for class `Word`:

```
// word3.h

#ifndef _WORD3_H
#define _WORD3_H

class Word {
    private:
        char *buf;
        static int wordCount;
    public:
        Word(char *s);
};

#endif


// word3.cpp
#include <string.h>
#include "word3.h"

Word::Word(char *s) {
    buf = strdup(s);
    wordCount++;
}
```

So it's not 100% correct to say that each object contains its own separate data. If a data item in a class is defined as `static`, then only one such item in a class is created for the entire class, no matter how many objects of the same class must share a common item of information. There is only one instance of a static data member of a class and it is in scope whenever the class is in scope, i.e., whenever the class definition is accessible. So a `static` data member is a single shared variable accessible to all objects of its class. It acts as a global variable for its class.

There are two advantages of a `static` data member over the use of a global variable:

- Information hiding can be enforced. A `static` member can be made non-public; a global variable cannot.

- A `static` member is not entered into the program's global name space, thus removing the possibility of an accidental conflict of names.

Initialization of Static Data Member

Since all the objects of a class have access to the static member, who is responsible for reserving storage and initializing it? You cannot do it in every constructor since every object created would then reinitialize the static member. There is a special way to create and initialize static data members. You must create a definition outside of all functions:

```cpp
// word4.h

#ifndef _WORD4_H
#define _WORD4_H

class Word {
  private:
     char *buf;
     static int wordCount;
  public:
     Word(char *s);
};

#endif



// word4.cpp

#include <string.h>
#include "word4.h"

int Word::wordCount = 0;

Word::Word(char *s) {
    buf = strdup(s);
    wordCount++;
}
```

This looks somewhat unusual because you must not only give the variable name, but also its class using the scope resolution operator. So a static data member is initialized outside of the class definition in the same manner as a nonmember variable using the scope resolution operator. It's important to note that only one initialization of a static data member can occur within a program.

Note that when initializing a static member, don't include the keyword `static` again. If you do, you will encounter a compiler error. Typically the allocation of the `static` member is placed in the same source file as the implementation of the class (the `.cpp` source file). Keep in mind that class declarations typically go into header files. If the inclusion of a static member declaration inside the class

were to also allocate that member, then in effect we would be allocating data in a header file, something that should never be done.

Accessing Static Data Members

Now suppose you want to read or modify a static data member. You can access a static data member just like any other data member:

```cpp
// word5.h

#ifndef _WORD5_H
#define _WORD5_H

class Word {
  private:
     char *buf;
     static int wordCount;
  public:
     Word(char *s);
     int getCount();
};

#endif


// word5.cpp

#include <string.h>
#include "word5.h"

int Word::wordCount = 0;

Word::Word(char *s) {
    buf = strdup(s);
    wordCount++;
}

int Word::getCount() {
    return wordCount;
}
```

```
// word5drv.cpp

#include <iostream>
#include "word5.h"

using namespace std;

int main() {
    Word w1("Red");
    Word w2("White");
    Word w3("Blue");

    cout << w1.getCount() << endl;
    cout << w2.getCount() << endl;
    cout << w3.getCount() << endl;
    return 0;
}


// word5 output

3
3
3
```

Had `wordCount` been an ordinary automatic or local variable as opposed to a static data member, each constructor would have incremented its own private copy of `wordCount` once, and the output would have been 1 for all three objects, instead of 3.

Another Example

Suppose an object needed to know how many other objects of its class there were in the program. For example, in a road-racing game, a race car might want to know how many other cars were still in the race. In this case a `static` variable count could be included as a member of the class:

- all objects would have access to this variable
- it would be the same variable for all of the objects
- each object would all see the same count variable

Each object has access to the information in the `static` data member so you can use the `static` data members to allow objects to communicate with each other. So `static` data members can be thought of as a way to communicate among objects of a class.

Static Member Functions

Suppose you want to read or modify a `static` data member. You can access a `static` data member just like any other member by using a normal member function like this:

```cpp
// word6.h
#ifndef _WORD6_H
#define _WORD6_H

class Word {
    private:
        char *buf;
        static int wordCount;
    public:
        Word(char *s);
        int getCount();
};

#endif


// word6.cpp
#include <string.h>
#include "word6.h"

int Word::wordCount = 0;

Word::Word(char *s) {
    buf = strdup(s);
    wordCount++;
}

int Word::getCount() {
    return wordCount;
}
```

```
// word6drv.cpp
#include <iostream>
#include "word6.h"

using namespace std;

int main() {
    Word w1("Red");

    cout << w1.getCount() << endl;
return 0;
}

// word6 output
1
```

However, when you call `getCount()`, you must then call it for an object, i.e., send a `getCount()` message to a particular object and `this`, the address of the object, will be secretly passed to the member function.

While the extra overhead of the `this` pointer may not be a problem, there are some cases where the reduction in the efficiency is unacceptable. This is true not only for functions which only access `static` data members, but for any action which acts for the class as a whole and doesn't need to access members of a particular object.

We can eliminate the extra overhead of passing in the `this` pointer by making the `getCount()` function a friend function:

```
// word7.h
#ifndef _WORD7_H
#define _WORD7_H

class Word {
    private:
        char *buf;
        static int wordCount;
    public:
        Word(char *s);
        friend int getCount();
};

#endif
```

```
// word7.cpp
#include <string.h>
#include "word7.h"

int Word::wordCount = 0;

Word::Word(char *s) {
    buf = strdup(s);
    wordCount++;
}

int getCount() {
    return Word::wordCount;
}


// word7drv.cpp
#include <iostream>
#include "word7.h"

using namespace std;

int main() {
    Word w1("Red");
    Word w2("White");
    Word w3("Blue");
    Return 0;

    cout << getCount() << endl;
}


// word7 output
3
```

We've now defined getCount() outside the class as:

```
int getCount() {
    return Word::wordCount;
}
```

Now getCount() is a global function which has access to static private data, but it doesn't have the extra overhead. However the function is no longer in the scope of class Word. So if someone introduces the name getCount() by bringing in another library, for example, there will be a clash.

C++ allows you to surmount these problems in an elegant fashion. You can define a member function as `static`! This meaning is close to the one for `static` data members, i.e., the function is for the whole class and not just one object of the class.

The key is that the `this` pointer is not secretly passed into a `static` member function. Thus you don't get the extra overhead. However, you also cannot access anything but `static` data members and `static` member functions.

Since a function is a member of the class, its name is hidden inside the scope of the class and will not collide with global names. In addition, you have the notational convenience of associating the name of the class, so it's more obvious what the function is supposed to do and that it may be affected by changes in the class.

```cpp
// word8.h

#ifndef _WORD8_H
#define _WORD8_H

class Word {
    private:
        char *buf;
        static int wordCount;
  public:
        Word(char *s);
        static int getCount();
};

#endif


// word8.cpp

#include <string.h>
#include "word8.h"

int Word::wordCount = 0;

Word::Word(char *s) {
    buf = strdup(s);
    wordCount++;
}

int Word::getCount() {
    return wordCount;
}
```

Calling a Static Member Function

You can call a static member function in two ways:

```cpp
// word9.h

#ifndef _WORD9_H
#define _WORD9_H

class Word {
    private:
        char *buf;
        static int wordCount;

public:
    Word(char *s);
    static int getCount();
};

#endif


// word9.cpp

#include <string.h>
#include "word9.h"

int Word::wordCount = 0;

Word::Word(char *s) {
    buf = strdup(s);
    wordCount++;
}

int Word::getCount() {
    return wordCount;
}

// word9drv.cpp

#include <iostream>
#include "word9.h"

int main() {
    Word w1("Red");
    Word w2("White");
    Word w3("Blue");

    cout << w1.getCount() << endl;
    cout << Word::getCount() << endl;
    return 0;
}
```

```
// word9 output

3
3
```

Example 1 is just like any other member function. Example 2 is unique to `static` member functions by specifying the class using the scope resolution operator. This can be an important way to call the static function because it doesn't require a reference to an object. Calling a static member function for an object when it doesn't actually use or modify that particular object can be confusing to the reader of the code.

Remember, the address of the object, `this`, is not passed to a static member function; therefore the static member function cannot access non-static data members, i.e., it cannot access the member data of the object used to invoke the static function. The only data member that can be accessed by a static member function are static data members.

If a static member function needs access to the member data of a class, it needs to be passed an explicit `this` pointer:

```
// word10.h

#ifndef _WORD10_H
#define _WORD10_H

class Word {
    private:
        char *buf;
        static int wordCount;
  public:
    Word(char *s);
    static int getCount(Word *w);
};

#endif
```

```
// word10.cpp

#include <iostream>
#include <string.h>
#include "word10.h"

using namespace std;

int Word::wordCount = 0;

Word::Word(char *s) {
    buf = strdup(s);
    wordCount++;
}

int Word::getCount(Word *w) {
    cout << w->buf << endl;
    return wordCount;
}
```

```
// word10dr.cpp

#include <iostream>
#include "word10.h"

using namespace std;

int main() {
    Word w1("Red");
    Word w2("White");
    Word w3("Blue");

    cout << w1.getCount(&w1) << endl;
    cout << w2.getCount(&w2) << endl;
    cout << w3.getCount(&w3) << endl;
    return 0;
}
```

```
// word10 output

Red
3
White
3
Blue
3
```

In this example we passed an explicit object pointer in the function call. The next example shows that it would also have been appropriate to use references for this to happen.

```
// word11.h

#ifndef _WORD11_H
#define _WORD11_H

class Word {
    private:
        char *buf;
        static int wordCount;
    public:
        Word(char *s);
        static int getCount(Word &w);
};

#endif


// word11.cpp

#include <iostream>
#include <string.h>
#include "word11.h"

using namespace std;

int Word::wordCount = 0;

Word::Word(char *s) {
    buf = strdup(s);
    wordCount++;
}

int Word::getCount(Word &w) {
    cout << w.buf << endl;
    return wordCount;
}
```

```
// word11dr.cpp

#include <iostream>
#include "word11.h"

using namespace std;

int main() {
    Word w1("Red");
    Word w2("White");
    Word w3("Blue");

    cout << w1.getCount(w1) << endl;
    cout << w2.getCount(w2) << endl;
    cout << Word::getCount(w3) << endl;
    return 0;
}


// word11 output

Red
3
White
3
Blue
3
```

Note that you cannot declare constructors to be static. If a constructor was allowed to be static, it could be invoked without a this pointer and therefore would not be able to build objects correctly.

Remember that when you see a static member function in a class, the designer intended that function to be conceptually associated with the class as a whole. That function will have the faster calling time of an ordinary, global function, but its name will be visible only within the class so it won't clash with global function names.

❖ Static Homework

1. Add a destructor to the `Word5` class below which decrements `wordCount`. Create a driver for your code and show that your destructor works.

```
// word5.h

#ifndef _WORD5_H
#define _WORD5_H

class Word {
  private:
    char *buf;
    static int wordCount;
  public:
    Word(char *s);
    int getCount();
};

#endif


// word5.cpp

#include <string.h>
#include "word5.h"

int Word::wordCount = 0;

Word::Word(char *s) {
    buf = strdup(s);
    wordCount++;
}

int Word::getCount() {
    return wordCount;
}
```

2. Write some code that shows the `this` pointer is successfully passed into normal member functions but cannot be accessed within static member functions.

3a. Create a class that contains a normal member function and a static member function.

3b. Create an object of that class and, inside a `for` loop, call both functions multiple times.

3c. Compile the program and see if you can check the assembler listing to see where the arguments are being passed and verify that `this` is only passed in one situation.

Chapter 9. This Pointer & Friends

This Pointer
Friends

This Pointer

At this point you might be wondering how member functions are implemented in C++. Consider the following class:

```
class Alpha {
    private:
        int a, b, c;
    public:
        void foo(int x, int y, int z);
};

int main() {
    Alpha a1, a2;
    :
    :
}
```

An ordinary member function is invoked as:

```
a1.foo(i, j, k);
a2.foo(i, j, k);
```

The code for the member function is kept within the class `Alpha`. When `foo()` is invoked by object `a1`, the member function `foo()` operates on the private data of object `a1`. Likewise, when `foo()` is invoked by object `a2`, the member function `foo()` operates on the private data of object `a2`. How does the member function know what object it's dealing with, and which private data to manipulate?

Let's explore the invocation of the member function `foo()`:

```
a1.foo(i, j, k);
```

How many arguments are passed to the function? The answer may surprise you. It is 7!

- The explicit argument list contains 3 arguments: `i`, `j`, `k`.

- The implicit argument list contains 3 arguments consisting of the member data of the object: a, b, c.

- There is one additional implicit argument called the this pointer, which is a pointer to the object from which the member function is invoked.

C++ automatically declares a local variable named this in every (non-static) member function of a class and initializes it to point to the instance (or object) of the class to which the member function was applied.

So the this pointer is implicitly passed as a hidden parameter to every non-static member function in a class and has the same type as a declared pointer to the class. Actually, we'll see that the implicit arguments can be thought of as a list of arguments accessible through the this pointer.

So member functions of every object have access to the this pointer which points to (or contains the address of) the object itself. Therefore, any member function can find out the address of the object of which it is a member.

```cpp
class Alpha {
    private:
        int a, b, c;
    public:
        void foo(int x, int y, int z);
        void display() { cout << this; }
};

int main() {
    Alpha a1, a2;

    a1.display();
    a2.display();
    :
    :
}
```

So when you call a member function, it comes into scope with the value of this set to the address of the object for which it was called. The this pointer can be treated like any other pointer to an object, and can thus be used to access the data in the object it points to.

```
class Alpha {
    private:
        int a, b, c;
    public:
        void foo(int x, int y, int z) {
            this->a = x;
            this->b = y;
            this->c = z;
            cout << this->a << this->b << this->c;
        }
        void display() {
            cout << this;
        }
};

int main() {
    Alpha a1, a2;

    a1.foo(2, 4, 6);
    a2.foo(1, 3, 5);
    :
    :
}
```

Note that the syntax works but there is no reason for it except to show that `this` does indeed point to the object. The `this` pointer is actually the reason that we can refer to member data as `a, b, c` instead of `a1.a, a1.b, a1.c` in our objects. The compiler adds the this prefix to all member data and functions.

Note the following distinctions:

- `this` refers to the address of the object being pointed at

- `this->memberData` refers to a member data of the object which called the member function

- `*this` refers to the actual object and, depending on context, can be an lvalue or an rvalue

Another example:

```cpp
// this.cpp

#include <iostream>
#include <iomanip>

using namespace std;

class Alpha {
private:
    char ch;
public:
    void init(char c) {
        ch = c;
    }
    Alpha increment() {
        ch++;
        return *this;
    }
    Alpha* whereAmI() {
        return this;
    }
    void print() {
        cout << ch;
    }
};

int main() {
    Alpha a, b, c;

    a.init('a');
    b.init('b');
    c.init('c');
    a.print();
    cout << " is at " << a.whereAmI() << endl;
    b.print();
    cout << " is at " << b.whereAmI() << endl;
    c.print();
    cout << " is at " << c.whereAmI() << endl;
    c.increment().print();
    cout << " is at " << c.whereAmI() << endl;
    return 0;
}

The program output is:

a is at 0x0064FDF4
b is at 0x0064FDF0
c is at 0x0064FDEC
d is at 0x0064FDEC
```

The member function `increment()` uses the implicitly provided `this` pointer to return the newly incremented value of `ch`. The member function `whereAmI()` returns the address of the given object. The `this` keyword provides for a built-in self-referential pointer. It is as if `Alpha` implicitly declared the private member:

```
Alpha* const this
```

Note that the `this` pointer is always defined as non-modifiable.

`this` is most often used to:

- pass the object (or a pointer to it) to another function

- return the object (or a pointer to it) from the member function.

For example:

```
#include <stdio.h>
#include <ctype.h>

class String {
    private:
        char *buf;
    public:
        :
        :
        String upperCase();
    };

String String::upperCase() {
    for (char* cp = buf; *cp; ++cp)
        if (islower(*cp))
            *cp -= 32;
    return *this;
}

int main() {
    String s("hello");
    s.uppercase().print();
    return 0;
}

output:
HELLO
```

Friends

One of object-oriented programming's main gifts is the encapsulation of data in classes. We've seen many examples of this data-hiding concept and we've considered its benefits of reduced maintenance and easier debugging.

Like many rules in life and programming, those of data-hiding may be broken. In C++, you can break encapsulation's rules by using friends, although you do so at some risk to your program's welfare. The friend mechanism gives non-members of the class access to the private members of the class.

C++ classes can declare two kinds of friends:

- An entire class may be a friend of another class.
- A single function may be declared as a friend.

Note: If friends have a counterpart in other languages, it's the `goto` statement. Like `goto`, a friend enables you to break the rules intended to help you write reliable code. Don't interpret the following as an endorsement of friends. Top C++ programmers avoid using friends unless absolutely necessary. Needless to say, the use of friend functions in C++ is controversial.

Friend Classes

A class may declare another class as a friend. The first class (the one that declares the friend) gives another class (the friend) permission to access all private members of the first class. Note that public members are always accessible, so you don't need to declare a class a friend to give it access to public members.

Typically, friend classes are used when two unrelated classes require access to the class's private parts. For example:

```
class Alpha {
    private:
        double value;
    public:
        Alpha() { value = 3.14159; }
};
```

Now, suppose you declare another class that contains an object of `Alpha` as a data member:

```
class Beta {
    private:
        Alpha anObject;
    public:
        void showValue() {
            cout << anObject.value;
        }
};
```

Member `anObject` of type `Alpha` is private to `Beta`. Member function `showValue()` attempts to display `anObject` value. However, the declaration doesn't compile because `value` is private to `Alpha`, and only member functions in `Alpha` may access `value`.

The solution is to make the `Beta` class a friend of the `Alpha` class. Objects of the `Beta` class can then access value and any other private members of class `Alpha`. But statements outside the two classes are still prevented from entering `Alpha`'s restricted areas. To make this change, we use the `friend` keyword inside the class to which the other class needs access. For example:

```
class Alpha {
    private:
        double value;
    public:
        Alpha() { value = 3.14159; }
        friend class Beta;        // Beta is a friend of Alpha

};
```

Here, class `Beta` must access the private value member inside class `Alpha`. So to give class `Beta` permission to use that private data, class `Alpha` can declare class `Beta` as a friend. Now, `Beta` has access to `Alpha`'s private members.

Note that other statements in other classes and in the `main()` program still can't use `Alpha`'s private declarations. In this manner, you may declare any number of classes as friends. The only restriction is that the `friend` keyword must appear inside a class declaration.

Note the following:

- A class must name all of its friends in advance

- The class containing the private data is the one that declares another class to be a friend, thus giving that friend special access to the normally hidden members of the declaring class. A class can never declare itself to be a friend of another class.

- The friend class may be declared before or after the class that declares the friend. The order of the declarations is unimportant, but the friend class is typically declared last so the friend's inline member functions can refer to the other class's private parts.

Friend Functions

A friend function is similar to a friend class. Declaring a function as a friend of a class gives that function access to private members in class objects. The friend function may be a common C++ function or a class member function.

A friend function declares parameters of classes to which the function owes its friendship, which gives the friend function access to the private members of the class objects.

A friend function has the same access privileges as a member function, but isn't associated with an object of the host class.

The host class has control over granting friend privileges to other functions, so you always know who has the ability to change your private data.

There are times when the program design just won't work out right. You can't always make everything fit neatly into one class. Sometimes other functions must have access to private elements of your class for everything to work together harmoniously. You could make some elements public, but this is a bad idea unless you really want the client programmer to change the data.

```
#include <iostream>

using namespace std;

class Beta;    // incomplete class declaration allows class Alpha
               // to refer to class Beta

class Alpha {
    private:
        char *s1;
    public:
        Alpha() {
            s1 = new char(strlen("Testing")+1);
            strcpy(s1, "Testing");
        }
        friend void show(Alpha a, Beta b);
};

class Beta {
    private:
        char* s2;
    public:
        Beta() {
            s2 = new char(strlen("one, two, three")+1);
            strcpy(s2, "one, two, three");
        }
    friend void show(Alpha a, Beta b);
};

void show(Alpha a, Beta b) {
    cout << a.s1 << b.s2;
}

int main() {
    Alpha a;
    Beta b;
    show(a, b);
    return 0;
}
```

Note that a friend function declaration must appear inside the class declaration to which it is a friend. The function is prefaced by the keyword `friend` and can appear in either the public or private part of the class without affecting its meaning.

Friend functions are not member functions and therefore do not implicitly have the `this` pointer passed as their first argument.

Friend functions are not scoped within the class but have the same scope as any normal function declared in C++. However, these functions have access into all access regions of the class just like member functions.

A big difference between member functions and friend functions is how the function gets the address of the object it will read from or write to. In friend functions, we must explicitly pass the object of interest or a reference to the object. We'll see this later with operator overloading.

Note the keyword `friend` is with the function declaration within the class but not with the function definition. Only a class can say who its friends are. If any function could declare itself a friend of a particular class, that function could access the private data without going through the class interface.

Friend Member Functions

A member function of one class can be a friend of another class. For example, a class can declare a member function of another class as a friend. The friend member function can access the declaring class's private members.

```cpp
#include <iostream>

using namespace std;

class Alpha;

class Beta {
    private:
        char* s2;
    public:
        Beta() {
            s2 = new char(strlen("one, two, three")+1);
            strcpy(s2, "one, two, three");
        }
        void show(Alpha a);
};

class Alpha {
    private:
        char *s1;
    public:
        Alpha() {
            s1 = new char(strlen("Testing")+1);
            strcpy(s1, "Testing");
        }
        friend void Beta::show(Alpha a);
};
```

```
void Beta::show(Alpha a) {
    cout << a.s1 << s2;
}


int main() {
    Alpha a;
    Beta b;
    b.show(a);
    return 0;
}
```

Note that the order of the two classes is reversed from the previous example because the class that prototypes the member function must be declared before the class that specifies the member function as a friend. For example, in order for Alpha to declare Beta::show() as a friend of the class, Beta's declaration must have been seen by the compiler.

Another difference is the way show() refers to private data in the two classes. The member function declares only one parameter, Alpha a. Because show() is a member of class Beta, it can access all members of Beta directly.

The expression a.s1 is allowed because show() is a friend of class Alpha of which s1 is a private data member.

show() can refer to s2 directly because this member function belongs to class Beta, the same class that declares show().

Because show() is a member function of class Beta, it now has a this pointer that addresses the class object for which the member function was called. Consequently, the program must define an object of class Beta and then call show() for that object.

Friends for Functional Notation

Sometimes a friend allows a more obvious syntax for calling a function than does a member function. For example, suppose we want a function which will double a `Distance` object. Using a member function, our class becomes:

```cpp
#include <iostream>

using namespace std;

class Distance {
    private:
        int feet;
        float inches;
    public:
        void setDist(int ft, float in);
        void getDist();
        void showDist();
        void addDist(Distance d2, Distance d3);
        Distance addDist(Distance d);
        Distance doubleDist();
};

Distance::Distance(int ft, float in) {
    feet = ft;
    inches = in;
}

void Distance::setDist(int ft, float in) {
    feet = ft;
    inches = in;
}

void Distance::getDist() {
    cout << "feet: ";
    cin >> feet;
    cout << "inches: ";
    cin >> inches;
}

void Distance::showDist() {
    cout << feet << "' " << inches << "\"";
}
```

```
void Distance::addDist(Distance d2, Distance d3) {
    inches = d2.inches + d3.inches;
    feet = d2.feet + d3.feet;
    if (inches >= 12.0) {
        inches -= 12.0;
        feet++;
    }
}

Distance Distance::addDist(Distance d) {
    Distance temp;

    temp.inches = inches + d.inches;
    temp.feet = feet + d.feet;
    if (temp.inches >= 12.0) {
        temp.inches -= 12.0;
        temp.feet += 1;
    }
    return temp;
}

Distance Distance::doubleDist() {
    int ft = feet * 2;
    float in = inches * 2;
    if (in >= 12.0) {
        ++ft;
        in -= 12.0;
    }
    return Distance(ft, in);
}

int main() {
    Distance d1(3, 6.0), d2;
    d2 = d1.doubleDist();
    cout << "d2 = ";
    d2.showDist();
}
```

Note that:

```
d2 = d1.doubleDist();
```

works all right but we might want to work with `Distance` objects using the same syntax that we use with ordinary numbers:

```
d2 = doubleDist(d1);
```

We can achieve this by making `doubleDist()` a friend of the `Distance` class:

```cpp
#include <iostream>

using namespace std;

class Distance {
    private:
        int feet;
        float inches;
    public:
        void setDist(int ft, float in);
        void getDist();
        void showDist();
        void addDist(Distance d2, Distance d3);
        Distance addDist(Distance d);
        friend Distance doubleDist(Distance d);
};

Distance::Distance(int ft, float in) {
    feet = ft;
    inches = in;
}

void Distance::setDist(int ft, float in) {
    feet = ft;
    inches = in;
}

void Distance::getDist() {
    cout << "feet: ";
    cin >> feet;
    cout << "inches: ";
    cin >> inches;
}

void Distance::showDist() {
    cout << feet << "' " << inches << "\"";
}

void Distance::addDist(Distance d2, Distance d3) {
    inches = d2.inches + d3.inches;
    feet = d2.feet + d3.feet;
    if (inches >= 12.0) {
        inches -= 12.0;
        feet++;
    }
}
```

```
Distance Distance::addDist(Distance d) {
    Distance temp;

    temp.inches = inches + d.inches;
    temp.feet = feet + d.feet;
    if (temp.inches >= 12.0) {
        temp.inches -= 12.0;
        temp.feet += 1;
    }
    return temp;
}

Distance doubleDist(Distance d) {
    int ft = d.feet * 2;
    float in = d.inches * 2;
    if (in >= 12.0) {
        ++ft;
        in -= 12.0;
    }
    return Distance(ft, in);
}

int main() {
    Distance d1(3, 6.0), d2;
    d2 = doubleDist(d1);
    cout << "d2 = ";
    d2.showDist();
}
```

In `main()`, we now say:

```
d2 = doubleDist(d1);
```

Note: where `doubleDist()`, as a member function, takes no arguments, it takes one as a friend.

In general, the friend version of a function requires one more argument than when the function is a member.

Chapter 10. References

Reference Variables
Reference Arguments
Returning References
References Homework

Reference Variables

We know how to create pointers as aliases that address other variables:

```
int num = 5;
int *p = &num;
```

declares an `int` variable `num` initialized to 5 and an `int` pointer `p` to the address of `num`.

```
cout << num << *p << endl;          // 5   5
```

To use a pointer's addressed value requires dereferencing the pointer leading to many asterisks that could confuse code. C++ offers an alternative to pointer aliases called *references*. There are three main uses for references:

- as variables
- as function parameters
- as function results

Reference Variables

A reference variable in C++ is simply an alias for another variable. An alias is just another name for another variable. This means that whenever an alias is used, the action is really taking place on the variable to which the alias refers.

For example:

```
int num = 5;
int& val = num;
```

This sets up a reference variable `val` and assigns it to the existing variable `num`. We say that `val` is a reference to an `int`. A reference variable contains the address of another variable, like a pointer. So `val` has the address of `num`.

At this point, the referenced location has two names associated with it: `num` and `val`.

```
cout << num << val << endl;          // 5    5
```

Because both variables point to the same location in memory, they are in fact the same variable and therefore output the same value. Note that unlike pointers, reference variables are automatically dereferenced when they are used. There is therefore no need for the asterisk *.

```
++num;
cout << num << val << endl;   //     6      6
++val;
cout << num << val << endl;   //     7      7
```

Any assignment made to num is reflected through val and any assignment made to val is reflected through num. Any operation on val has precisely the same effect as an operation on num.

```
int z = 100;
val = z;
cout << num << val << endl;   //     100    100
```

This does not cause val to refer to z since the compiler always dereferences the address for you. The above code will copy the value of z into num.

Note: Once a reference variable is initialized to some existing value, it cannot be changed or reassigned to some other variables as long as the reference variable remains in scope. However, the value the reference variable points to can always change.

Rule of References

When a reference variable is created, it must always be initialized with the name of some other existing variable. That means that references must be initialized in their declarations:

```
int& iref;   // error
```

Unlike a pointer that can be declared in an uninitialized state, or set to NULL if the pointer does not address valid data, a reference must always refer to an object. There is no equivalent of a NULL pointer for a reference.

References do not have to be pre-initialized in four special cases:

- when declared extern
- when used as function arguments
- when used as function return types
- in class declarations

Some examples:

```
int&    - reference to an int
int*&   - reference to a pointer to an int

int a[10];
int& last = a[9]; // last is an alias for a[9];
```

Often the reason for giving an object a name is that it doesn't already have one. This example attaches the name last to the last element of the array a. One might do this if one were about to use a[9] several times. Using a reference would save program text and might save computation time as well.

Reference to a Reference

If a reference variable is an alias for another reference variable, it really refers to the variable for which the first reference variable is as alias. For example, the two reference variables are aliases for the same variable in memory.

```
int num1 = 1;
int& num2 = num1; // num2 is an alias for num1
int& num3 = num2; // num3 is an alias for num2 which
                  // really means that num3 is an
                  // alias for num1
```

Address of a Reference Variable

We can show that reference variables are not separate entities in memory with their own addresses. If you take the address of a reference variable, all you will get is the address of the variable for which it is an alias.

For example:

```
#include <cstdio>

int main() {
    int num1;
    int& num2 = num1;
    printf("%p    %p", &num1, &num2); // FFFFFFF4 FFFFFFF4
    return 0;
}
```

Therefore a reference variable does not have its own unique address.

Reference Variables are Useless!

Finally, reference variables are generally useless!!!

If you already have a variable, all a reference to the variable will provide is complexity. Now you'll have two ways of changing the same variable. You'll gain nothing from reference variables. It's as messy as it sounds and you should think hard before you decide you must use a reference variable.

References were really designed for argument passing and return values and reference variables were only included for syntactic completeness, not necessarily because they were a good idea. References are usually necessary only in the context of a user-defined class type.

Reference Arguments

The declaration of reference variables up to now has not really served a very useful purpose. After all, if a variable and an alias for that variable are both defined within the same scope, then why bother to use the alias? Why not just use the name of the variable itself?

The real usefulness of a reference variable occurs when it is created in a scope that is different than the scope of the variable for which it has an alias.

For example, a function sets up a brand new scope whenever it is created. If a reference variable is used as a formal parameter, then it merely becomes an alias for the corresponding actual parameter or argument to the function call. This means that whenever the formal parameter is used within the body of the function, the actual parameter in the calling function is the one that is really being manipulated!

Simulate Call By Reference

```cpp
void swap(int *a, int *b) {
    int temp = *a;
    *a = *b;
    *b = temp;
}

int main() {
    int x = 4, y = 6;
    swap(&x, &y);
    return 0;
}
```

The `swap()` function:

- declares the function parameter to be a pointer
- uses the dereferenced pointer in the function body
- passes an address as an argument when the function is called

Call By Reference

```
void swap(int& a, int& b) {
    int temp = a;
    int a = b;
    b = temp;
}

int main() {
    int x = 4, y = 6;
    swap(x, y);
    return 0;
}
```

Note that there is no need to dereference `a` and `b`. They are automatically dereferenced for us!

Reference Arguments

Behind the scenes, all references are translated to pointers according to the following two rules:

- whenever a reference variable `p` is used, replace `p` by `*p`
- whenever a reference variable is initialized with `a`, replace `a` by `&a`

The machine code that's generated is exactly the same. We don't have to manually generate the address of a variable. We don't have to create a pointer variable in which to receive the address. This allows the programmer, not the user, to take responsibility for the way the arguments are passed to functions. A reference in C++ is like a pointer because it contains an address, and like an ordinary variable because you don't have to dereference it.

Constant Reference Parameters

Reference parameters that are used to prevent argument copying (pass by value) and are not intended to change arguments can be declared `const` to ensure that they do not.

For example:

```
int foo(const bigStruct &bs);
```

This gives you the best of both worlds:

- you can pass arguments efficiently
- you don't have to worry about an argument being changed accidently

The `const` prevents the actual parameters from being altered within the function.

Is it always more advantageous to use a pass-by-reference instead of pass-by-value? No. It's just as efficient to pass the actual value of a primitive type as it is to create an alias in the function. When a primitive type variable needs to be used in a function, and no changes need to be made to that variable, pass it by value.

Inexact Matches Between Actual and Formal Parameters

Consider the following example:

```
int main() {
    int num = 1;
    foo(num);
    cout << "num = " << num << endl;
    return 0;
}

void foo(float& val) {
    ++val;
    cout << "val = " << val << endl;
}

output:
val = 2.00
num = 1
```

Huh???

Integer `num` is being passed by reference into function `foo()`, but is received as a reference-to-float, not a reference-to-int. Some compilers will give you a warning message that you may not get the correct results while other compilers will just give a fatal error.

Note that the value of `num` did not get changed by the function. In order to pass an argument by reference and have the function modify the actual value, there can be no temporary variables created in the process.

For example, if you pass an `int` by reference and create a `float` reference to it, then the compiler will convert the `int` into a temporary `float` variable and then make the reference to this temporary `float` via the formal argument. Therefore, if you attempt to modify this `float`, you would only be modifying the temporary `float` instead of the actual `int` argument itself.

Avoiding a Double Reference by Using Reference Variables

Sometimes in C the address of a pointer variable needs to be passed into a function. In this case the function argument needs to be declared as a pointer to a pointer in order to access the original variable that was being pointed at in the first place.

For example:

```
int main() {
    char *ptr;
    foo(&ptr);
    cout << "ptr = " << ptr << endl;
    return 0;
}

void foo(char **pptr) {
    *pptr = "Hello C++";
}

output:
ptr = Hello C++
```

A better solution is to create a reference variable for this pointer. For example, since the type of the pointer variable `ptr` is `char*`, then a reference to it would be type `char*&`. Therefore, in order to use the actual value of the variable `ptr` from within the function, no dereferences are required:

```
int main() {
    char *ptr;
    foo(ptr);
    cout << "ptr = " << ptr << endl;
    return 0;
}
```

```
void foo(char*& pptr) {
    pptr = "Hello C++";
}

output:
ptr = Hello C++
```

Why References Are Essential

Let's look at this class that uses an overloaded addition operator:

```
class Plus {
    private:
        int i;
    public:
        Plus(int x = 0)    {i = x; }
        Plus operator+(Plus arg) { return Plus(i + arg.i); }
        void print(char *msg) { cout << msg << endl;
};

int main() {
    Plus a(13), b(34);
    Plus c;
    c = a + b;
    ...
    ...
    return 0;
}
```

The `operator+()` takes a single argument, which is the value to the right of the + sign. Note that the function is invoked for object on the left of the + sign.

The `operator+()` function creates a new `Plus` object by calling the `Plus` constructor. The argument to the `Plus` constructor is the sum of the private data parts of the two objects. The `Plus` constructor creates a new (temporary) `Plus` object which is copied out of the function as the return value of the operator. Note that all objects are passed and returned by value!

However, there are many occasions when you need to pass the address of an object. For example, objects may be too large to pass by value without destroying the efficiency of the program.

Here is the same class using pointers:

```
class Plus {
    private:
        int i;
    public:
        Plus(int x = 0)    {i = x; }
        Plus operator+(Plus *arg) { return Plus(i + arg->i); }
        void print(char *msg) { cout << msg << endl;
};

int main() {
    Plus a(13), b(34);
    Plus c;
    c = a + &b;
    return 0;
}
```

The program works but the syntax:

```
a + &b
```

is confusing to say the least! As a last ditch, you could attempt to make the syntax homogeneous:

```
&a + &b
```

This won't work with a member function so you might think it could work with a friend operator+() function instead of a member function. But it doesn't! The C++ compiler must have some way of determining that you want to call the operator+() for a particular class. If you say, &a + &b, the compiler's best guess is that you are adding two pointers!

References allow you to pass addresses to overloaded operator functions and still retain the clear notation of a + b:

```
class Plus {
    private:
        int i;
    public:
        Plus(int x = 0)    {i = x; }
        Plus operator+(Plus &arg) { return Plus(i + arg.i); }
        void print(char *msg) { cout << msg << endl;
};

int main() {
    Plus a(13), b(34);
    Plus c;
    c = a + b;
    return 0;
}
```

With references, you can use operator overloading and still pass addresses when using large objects. This is the reason for the new feature!

You should pass arguments to a function via reference when:

• you want the function to modify the arguments

• the function is not going to modify the arguments but you want to avoid copying the objects to the stack. In this case, use the const qualifier to indicate that the argument being passed by reference will not be altered by the function.

Returning References

There are numerous reasons for functions to return references:

The first common use for functions to return references is for assignment operators:

```
class Alpha {
    public:
        Alpha& operator=(const Alpha&) {
            ...
            ...
            return *this;
        }
};
```

The usual practice is for such operators to return `*this` as a reference. This makes it possible to chain assignments:

```
Alpha a1, a2, a3, a4;
...
...
a1 = a2 = a3 = a4;
```

This is also possible if operator = returned an `Alpha` instead of an `Alpha&`:

```
class Alpha {
    public:
        Alpha operator=(const Alpha&) {
            ...
            ...
            return *this;
        }
};
```

But doing this would not be a good idea. Such an assignment operator would always form a copy of the object being assigned, even though that copy would eventually be thrown away.

The second common use for functions that return references is to allow member function calls to be chained easily.

Consider a class `Text` whose objects represent strings of text. Such objects might have an associated font, size, etc.

```
class Text {
    public:
        Text(const String&);
        Text& setSize(int);
        Text& setFont(const String&);
};
```

If the `setSize()` and `setFont()` members each return `*this` as a reference, that allows expressions such as:

```
Text t("The quick brown fox");
t.setSize(12).setFont("Arial");
```

Otherwise, it would be necessary to write:

```
t.setSize(12);
t.setFont("Arial");
```

The gain here seems relatively minor but we gain great advantage with the extraction (`<<`) operator. By convention, `<<` produces a human-readable representation of an object. Its left operand is an `ostream&` and its right operand is an object of the relevant type:

```
cout << "x = " << x << ", y = " << y << endl;
```

This example works only because `ostream::operator<<()` returns `*this` as a reference.

Each call to operator $<<$ therefore passes on its left argument as its result which becomes the left argument of the next call to operator $<<$.

The statement is really interpreted as:

```
(((( cout << "x = ") << x) << ", y = " )<< y) << endl;
```

It's important for operator $<<$ to return an `ostream&` so that several uses of it can be chained together. So for it to work, the value of `cout << x` must be a reference to `cout`!

References become even more important for input. Without them, it would be impossible for:

```
cin >> x;
```

to change the value of `x`.

For this reason, the right argument of operator $>>$ is always a reference.

Usually the functions we've seen tell us what value will be returned. But now, by allowing a function to return a reference, the function is allowed to be the target of an assignment, i.e., used as an lvalue (on the left-hand side of an = sign).

Here's a trivial example:

```
int& foo(int &x) {
    return x;
}
```

This function simply returns a reference to its argument. We can now write:

```
int a;
foo(a) = 7;        // sets a to 7
```

Another example:

```
int array[100];                  // static storage

int& getItem(int index) {
    return array[index];
}

int main() {
    getItem(4) = 5;              // not allowed in C
    return 0;
}
```

The function call is now an lvalue and `array[4]` is now = 5. Function `getItem()` will tell you at which location a value will be assigned. For example, function `getItem()` returns a reference to the element of the array index and sets it to 5.

Just as passing arguments by reference prevents unnecessary copying to the stack, you can return an object by reference to avoid copying the returned object.

Check what happens when you return an object by value. For example, suppose you return a `String` object from a function `addString()` that returns the concatenation of two strings:

```
String addString(const String& s1) {
    String s;

    // append s1 and s2 to a
    . . .
    . . .
    return s;
}
```

In this case, the return statement has to copy the `String s` to an area of memory provided by the calling program. To do so properly, the return statement will call a special constructor, the *copy constructor*, which can create and initialize a new instance of an object from an existing one. After creating the copy, return calls the destructor of the `String` class to destroy `String s` before returning to the calling program. This is a good example of work done behind the scenes by the C++ compiler.

Although returning by reference would save the time spent in copying the object, you cannot return a reference to `String` s because s is a temporary object that exists only within the function `addString()` and is destroyed when the function returns. Note: because a reference is like a pointer, you cannot return a reference to anything that is temporary!! *You should never return a reference to a local variable because the local variable will disappear as soon as the function returns and the reference will lead nowhere. This is called a dangling reference. Be sure that if you do return a reference, it refers to an object that will still be around after the function returns.*

Note that there is nothing wrong with returning a formal parameter as a reference provided that the formal parameter is itself a reference. For example:

```
// don't do this
int& foo(int x) {
    return x;
}

// but this is OK
int& foo(int& x) {
    return x;
}
```

Here the parameter x is just an alternate name for something that existed before `foo()` was called and will presumably still exist after `foo()` has returned.

There would have been no problem returning a reference to a global or static variable or to a formal parameter that is itself a reference.

Create a function that takes a `char&` argument and modifies that argument. In `main()`, print out the value of a `char` variable, call your function for that variable, and print it out again to prove to yourself that it has been changed. Be sure to show all output. How does this affect program readability?

Chapter 11. Operator Overloading

Introduction to Operator Overloading
Overloading Unary Operators
Distinguishing Prefix and Postfix for ++ and − Operators
Overloading Binary Operators
Overloading Arithmetic Assignment Operators
Selecting Friend or Member Functions
Concatenating String with + Operator
Overloading the Comparison == Operator
Overloading the Insertion << and Extraction >> Operators
Updated Distance Class
Overloading the Array Subscript Operator []
Operator Overloading Homework
Operator Overloading Lab (IntArray)
Operator Overloading Lab (IntArray) Driver

Introduction to Operator Overloading

Operator overloading is one of the most exciting features of object-oriented programming. It can transform complex, obscure programs into intuitively obvious ones. For example, a statement like:

```
d3.addObjects(d1, d2);
```

can be changed to the much more readable:

```
d3 = d1 + d2;
```

The term operator overloading refers to giving the normal C++ operators additional meanings when they are applied to user-defined types. We're already familiar with operator overloading:

```
c = a + b;
```

When you use the plus sign with `int` values, one specific set of machine language instructions is involved. When used with `float` values, a completely different set of machine language instructions is invoked. Attempting to apply + and = when `a`, `b`, `c` are objects of a user-defined class will cause complaints from the compiler. However, this statement would become legal with operator overloading. In effect, operator overloading gives us the opportunity to redefine the C++ language.

You can overload the following operators:

• arithmetic	+, -, *, /, %
• logical	&&, ||, !
• bitwise	&, |, ~, ^, <<, >>
• comparison	<, >, <=, >=

```
• equality                    ==,  !=

• assignment                  =,  +=,  -=,  *=,  /=,  %=,  &=,
                              |=,  ^=,  <<=,  >>=

• comma                       ,

• increment & decrement       ++,  --

• subscript                   []

• function call               ()

• class pointer               ->

• member pointer selector     ->*

• free store allocator        new

• recycling                   delete
```

You cannot overload these operators:

```
• member                      .

• member object selector      .*

• ternary conditional         ?:

• sizeof                      sizeof

• scope resolution            ::
```

Note the following:

• You can only overload existing operators. That is, you cannot design new operators, tempting as it may be:

```
:=,    <>,    |x|,    **
```

- You cannot change operator precedence.

- You cannot change the arity of an operator.

  - no postfix !
  - no binary ~
  - no unary ^

- You cannot define a new action for an intrinsic operator for a native C++ data type. That is, you cannot change the way the + operator works on two `int`s.

- Overloaded operators may not have default arguments.

To overload an operator, you define a function for the compiler to call when that operator is used with the appropriate data types. Whenever the compiler sees those data types used with the operator, it calls the function.

Overloading Unary Operators

Let's start off by overloading a unary operator. Unary operators act on only one operand (`++`, `--`, `-`).

We'll look at the `Counter` class we previously defined. Objects of that class were incremented by calling a member function:

```
c1.incCount();
```

That did the job, but the listing would have been more readable if we could have used the increment operator ++ and simply said:

```
++c1;
```

Here's our newest version of the `Counter` class, rewritten to use an overloaded increment variable:

```
class Counter {
    private:
        int count;
    public:
        Counter()              { count = 0;      }
        int getCount()         { return count; }
        void operator++()      { ++count;        }
};

int main() {
    Counter c1, c2;
    cout << "c1 = " << c1.getCount() << endl;
    cout << "c2 = " << c2.getCount() << endl;
    ++c1;
    ++c2;
    cout << "c1 = " << c1.getCount() << endl;
    cout << "c2 = " << c2.getCount() << endl;
    return 0;
}
```

The keyword `operator` is used to overload the ++ operator in the class. The overloaded ++ operator function is made up of four components:

- return type: `void`
- keyword operator: `operator`
- operator to be overloaded ++
- argument list in parentheses `()`

The compiler calls this member function whenever the ++ operator is encountered, provided the operand is of type `Counter`.

Note: as seen before, the compiler distinguishes between overloaded functions by the type and number of their arguments. Similarly, the compiler distinguishes between overloaded operators by the types of their operands.

In `main()`, the ++ operator is applied to a specific object as shown. Yet, `operator++()` takes no arguments. So what does this operator increment? It increments the count data member in the object `c1`, the object which used the operator ++. Since member functions can always access the particular object for which they've been called, this operator requires no arguments.

The `operator++()` function has a subtle defect which you'll see if you to execute this statement:

```
c1 = ++c2;
```

The compiler will complain. Why?

We've defined the ++ operator to have a return type of `void` in the `operator++()` function. But in the statement:

```
c1 = ++c2;
```

the function is asked to return a value of type `Counter`. The compiler is being asked to return whatever value `c2` has after being operated on by the ++ operator, and then assign this value to `c1`. So we can't use our ++ operator to increment `Counter` objects in expressions.

To make it possible to use our `operator++()` in expressions, we must provide a way for it to return a value:

```
class Counter {
    private:
        int count;
    public:
        Counter()                    { count = 0;      }
        int getCount()               { return count; }
        Counter operator++() {
            ++count;
            Counter temp;
            temp.count = count;
            return temp;
        }
};
```

In this version of the class, our `operator++()` function returns a value of class `Counter`.

- it increments the count data in its own object as before.

- the function creates a new object of type `Counter`, called `temp`, to use as a return value

- it assigns `count` in the temporary object the same value as in itself

- it returns the `temp` object

These changes now allow for the following style of expression in `main()`:

```
c2 = ++ c1;
++c1.getCount()    // displays itself using the
                   // getCount() member function.
```

There are more convenient ways to return objects from functions and overloaded operators. Previously, we created a temporary object to return a function value that took three lines of code. We can actually do it with a simple statement:

```
Counter operator++() {
      ++count;
      return Counter(count);
}
```

This code:

• creates an object of type `Counter` that has no name (it won't be around long enough to need one)

• this unnamed object is initialized to the value provided by the argument `count`.

• C++ treats this as a constructor call for a temporary object with no name and a very brief lifetime. However, the code does require a constructor that takes one argument:

```
Counter(int c) { count = c; }
```

• Once the unnamed object is initialized to the value of `count`, it can then be returned

We can do the same with decrement operators:

```
Counter operator-() {
      --count;
      return Counter(count);            // or return *this;
}
```

Distinguishing Prefix and Postfix for ++ and -- Operators

In earlier versions of C++, it was impossible to define separate overloaded operations for postfix and prefix ++ and -- operators. However, the compiler can now distinguish between the prefix and postfix versions of these operators by calling the operator function with no arguments or with one `int` argument, respectively.

```
class Counter {
    private:
        int count;
    public:
        Counter()                   { count = 0;    }
        Counter(int x)              { count = x;    }
        int getCount()              { return count; }
        Counter operator++() {
            ++count;
            return Counter(count);      // or return *this;
        }
        Counter operator++(int) {
            int temp = count;
            ++count;
            return Counter(temp);
        }
};
```

Member functions `operator++(int)` and `operator--(int)` defines a postfix increment and decrement operator for a `Counter` object. C++ assigns 0 to the single `int` parameter.

For an object `c1` of type `Counter`, the expression `++c1` calls the overloaded prefix ++ operator, in effect executing the statement:

```
c1.operator++();
```

The expression `c1++` calls the overloaded postfix ++ operator and executes as though written:

```
c1.operator++(0);
```

The decrement operator works similarly.

Because older versions of C++ did not make the distinction between prefix and postfix, some newer compilers might allow both calls to be made with only the prefix definition of `operator++()` or `operator--()`, although they usually issue a warning.

Notice that the argument `int` has no identifier, so it is not used in the function. It is only there to generate a different function signature. Because there is no identifier, the compiler will not complain that a variable has been "created but not used".

Our main concern when creating prefix and postfix operators is to decide what incrementing and decrementing means for an object of this class. In most cases, there's a specific data member in the class that you'll want to increment or decrement.

Overloading Binary Operators

Our previous `Distance` class was defined as follows:

```
class Distance {
    private:
        int feet;
        float inches;
    public:
        Distance() {} // default constructor
        Distance (int ft, float in) { feet = ft; inches = in; }
        void addDist(Distance, Distance);
};

void Distance::addDist(Distance d2, Distance d3) {
    feet = d2.feet + d3.feet;
    inches = d2.inches + d3.inches;
    if (inches >= 12.0) {
        inches -= 12.0;
        ++feet;
    }
}
```

Using this `Distance` class, we would add two `Distance` objects as:

```
d3.addDist(d1, d2);
```

We will now overload the + operator to perform the addition of two `Distance` objects as:

```
d3 = d1 + d2;
```

We can now rewrite the class as:

```
class Distance {
    private:
        int feet;
        float inches;
    public:
        Distance() {} // default constructor
        Distance (int ft, float in) { feet = ft; inches = in; }
        Distance operator+(Distance d);
};

Distance Distance::operator+(Distance d) {
    int ft = feet + d.feet;
    float in = inches + d.inches;
    if (in >= 12.0) {
        in -= 12.0;
        ++ft;
    }
    return Distance(ft, in);
}
```

When the compiler sees an expression like:

```
d3 = d1 + d2;
```

it realizes it must use the `Distance` member function `operator+()`. But what does this function use as its argument, d1 or d2? And why doesn't it need two arguments, since two numbers are added?

The argument on the left side of the operator, d1, is the object that really calls the operator member function. The object on the right side of the operator, d2, must be furnished as an argument to the operator.

In the `operator+()` function, data members left of the operand are accessed directly as feet and inches, since this is the object which really called the member function. The right operand is accessed as the function's argument, as `d2.feet` and `d2.inches`.

Note: *Overloaded operators always require one less argument than the number of operands.* This is true since one operand is the object that called the member function. That is also why unary operators require no arguments.

Note: If a member function overloads the addition operator, the expression:

```
c = a + b;
```

is semantically equivalent to the expression:

```
c = a.operator+(b);
```

Overloading Arithmetic Assignment Operators

This operator combines assignment and addition in one step. We'll use this operator to add one `Distance` object to a second `Distance` object, leaving the result in the first. This is similar to the previous `Distance` program which provided an overloaded + operator with a subtle difference.

```
void Distance::operator+=(Distance d) {
    feet += d.feet;
    inches += d.inches;
    if (inches >= 12.0) {
        inches -= 12.0;
        ++feet;
    }
}
```

The previous version of the overloaded + operator returned a `Distance` object, while here we return nothing. In the previous `Distance` program we created the overloaded + operator and it allowed us to add two `Distance` objects and assign the value to a third using:

```
d3 = d1 + d2;
```

This version permits:

```
d1 += d2;
```

The object `d2` maps onto `d1`. The object on the left side of the += operator is the object that effectively calls the `operator+=()` function. Therefore, it has direct access to `feet` and `inches`.

Note: Because the function alters `d1` directly, adding `d2` into it, there is no need to return a value.

Note: If you wanted to use this operator in more complex expressions, like:

```
d3 = d1 += d2;
```

you would need to return a `Distance` object. This could be done by ending the `operator+=()` function with a statement such as:

```
return Distance(feet, inches);
```

We would have to change the return value to a `Distance` object rather than returning a `void`. This would create a nameless object, initialize it with values, and return it.

Selecting Friend or Member Functions

In many situations you get equivalent results by using either a friend function or a member function when you overload an operator. A friend function simply contains an extra argument. The friend function must have both objects passed to it, while the member function only needs a single argument. However, sometimes friend functions are too convenient to avoid. One example is when friends are used to increase the versatility of overloaded operators.

For example, the program below shows a limitation in the use of overloaded operators when friends are not used:

```cpp
#include <iostream.h>

class Distance {
    private:
        int feet;
        float inches;
    public:
        Distance() {                              // default constructor
            feet = 0;
            inches = 0.0;
        }
        Distance(float fFeet) {                   // conversion constructor
            feet = int(fFeet);                    // feet is integer part
            inches = 12 * (fFeet - feet);         // inches is what's left
        }
        Distance (int ft, float in) {
            feet = ft;
            inches = in;
        }
        void showDist() {
            cout << feet << "\'-" << inches << "\"|;
        }
        Distance operator+(Distance d);
};

Distance Distance::operator+(Distance d) {
    int ft = feet + d.feet;
    float in = inches + d.inches;
    if (in >= 12.0) {
        in -= 12.0;
        ++ft;
    }
    return Distance(ft, in);
}
```

```
int main() {
    Distance d1 = 2.5;
    Distance d2 = 1.25;
    Distance d3;

    cout << "d1 = " << d1.showDist();
    cout << "d2 = " << d2.showDist();

    d3 = d1 + 10.0;
    cout << "d3 = " << d3.showDist();

    d3 = 10.0 + d1;
    cout << "d3 = " << d3.showDist();
    return 0;
}
```

Note that the line:

```
d3 = 10.0 + d1;
```

produces a compiler error. Why?

In this program, the + operator is overloaded to add two objects of type Distance. There's also a one-argument constructor that converts a value of type float (representing feet or decimal fractions of feet) into a Distance value. That is, converts 10.25 feet into 10 feet and 3 inches. When such a constructor exists, you can make statements like this in main():

```
d3 = d1 + 10.0;
```

The overloaded + operator is looking for objects of type Distance both on its left and on its right. But if the argument on the right is type float, the compiler will use the one-argument constructor to convert the float to a Distance value and then carry out the addition.

But why doesn't this work?

```
d3 = 10.0 + d1;
```

Because the object of which the overloaded + operator is a member must be the variable to the left of the operator. When we place a variable of a different type there, or a constant, there is no object of which the overloaded + operator can be a member and so there is no constructor to convert the type `float` to type `Distance`. The compiler simply cannot handle this situation.

A friend function can allow us to write statements that have non-member data types to the left of the operator. So we can change the declaration inside the `Distance` class to read:

```
friend Distance operator+(Distance, Distance);
```

We would now redefine the `operator+()` function as follows:

```
Distance operator+(Distance d1, Distance d2) {
    int ft = d1.feet + d2.feet;
    float in = d1.inches + d2.inches;
    if (in >= 12.0) {
        in -= 12.0;
        ++ft;
    }
    return Distance(ft, in);
}
```

We can now say the following from within `main()` without a compiler error message:

```
d3 = d1 + 10.0;
d3 = 10.0 + d1;
```

Notice that, while the overloaded + operator took one argument as a member function, it takes two as a friend function. In a member function, one of the objects on which a + operates is the object of which it was a member, and the second is an argument. In a friend, both objects must be arguments.

A friend function has one argument for a unary operator and two arguments for a binary operator, while a member function has zero arguments for a unary operator and one argument for a binary operator. This is true because a member function is automatically dealing with one variable already, the object it was called for.

Concatenating String with + Operator

Consider the following `String` class:

```cpp
const int SIZE = 100;

#include <iostream>
#include <string.h>

using namespace std;

class String {
    private:
        char buf[SIZE];
    public:
        String() { buf[0] = '\0'; }
        String(char s[]) { strcpy(buf, s); }
        void display() { cout << buf; }
        String operator+(String s);
};

String String::operator+(String s) {
    if (strlen(buf) + strlen(s.buf) < SIZE) {
        String temp;
        strcpy(temp.buf, buf);
        strcat(temp.buf, s.buf);
        return temp;
    }
    else
        return String();   // returns NULL String
}

int main() {
    String s1 = "Hello ", s2 = "world! ", s3;
    s3 = s1 + s2;
    s3.display();
    return 0;
}
```

output:
Hello world!

The + operator takes one argument of type `String` and returns an object of type `String`.

Note that with the `Counter` and `Distance` classes, we were able to simplify the `operator+()` function with the statement:

```
return Counter(count);
```

However, here we cannot use:

```
return String(String);
```

to create a nameless temporary `String` because we need access to the temporary `String` for several steps. We must be careful that we don't overflow the fixed-length strings used in the `String` class. We must check that the combined length of the two strings will not exceed the maximum string length. If it does, an overflow error occurs.

Also note that you could put the two overloaded + operator functions together in the same program:

- \+ operator for `String` objects

- \+ operator for `Distance` objects

and C++ would still know how to interpret the + operator for each context. The compiler would select the correct function to carry out the "addition" based on the type of operand (either `Distance` or `String`).

Overloading the Comparison == Operator

We can also explore the use of overloaded relational operators to create a comparison operator == to be used with our `String` class to compare two `String` objects. The function returns true if the `String` objects are the same and false is they're different.

```cpp
#include <iostream.h>
#include <iomanip.h>
#include <string.h>

const int SIZE = 100;

using namespace std;

class String {
    private:
        char buf[SIZE];
    public:
        String() { buf[0] = '\0'; }
        String(char s[]) { strcpy(buf, s); }
        void getString() { cin.get(buf, SIZE); }
        void display() { cout << buf; }
        String operator+(String s);
        int operator==(String s);
};

String String::operator+(String s) {
    if (strlen(buf) + strlen(s.buf) < SIZE) {
        String temp;
        strcpy(temp.buf, buf);
        strcat(temp.buf, s.buf);
        return temp;
    }
    else
        return String();   // returns NULL String
}

int String::operator==(String s) {
    return !strcmp(buf, s.buf);
}

int main() {
    String s1, s2 = "YES";
    cout << "Enter YES / NO: ";
    s1.getString();
    if (s2 == s1)
        cout << "You typed YES." << endl;
    return 0;
}
```

The approach here is similar to overloading the + operator in our previous program with one main difference, the return value. Here we return an `int` value. Previously we returned an object of class `String` or an object of class `Distance`.

Overloading the Insertion << and Extraction >> Operators

Normally, input and output streams handle only simple data types (int, long, float, double, char*, etc). By overloading the iostream input and output stream operators (>>, <<) you can easily add your own classes to the data types that I/O stream statements are designed to use.

For example, we can treat I/O for user-defined data types in the same way as built-in data types. This is a powerful feature of C++. It lets you say:

```
MyClass foo;
    cout << foo;
```

The operator<<() function has already been overloaded for the iostream class, but you can easily overload it for your new class. Let's overload the insertion and extraction operators for the Distance class.

```cpp
class Distance {
    private:
        int feet;
        float inches;
    public:
        Distance () { feet = 0; inches = 0; }
        Distance (int ft, float in) {
            feet = ft;
            inches = in;
        }
        friend istream& operator>>(istream& is, Distance& d);
        friend ostream& operator<<(ostream& os, const Distance& d);
};

istream& operator>>(istream& is, Distance& d) {
    cout << "Enter feet: "; is >> d.feet;
    cout << "Enter inches: "; is >> d.inches;
    return is;
}

ostream& operator<<(ostream& os, const Distance& d) {
    os << d.feet << "\' " << d.inches << '\"';
    return os;
}
```

```
int main() {
    Distance d1, d2;

    cout << "Enter two Distance values: ";
    cin >> d1 >> d2;
    Distance d3(11, 6.25);
    cout << "d1 = " << d1 << endl;
    cout << "d2 = " << d2 << endl;
    cout << "d3 = " << d3 << endl;
    return 0;
}
```

This program asks for two distance values from the user, and then prints out these values and another value that was initialized in the program. Note that arguments are passed by reference for efficiency.

The input and output operators require, respectively, an `istream` and `ostream` object as their left operand. Both return the object upon which they operate. This allows successive input or output operators to be concatenated.

For example:

```
((((( cout << "d1 = ") << d1) << "d2 = ") << d2) << endl;)
```

Here, each parenthetical sub-expression returns the `ostream` object `cout` which becomes the left operand of the next outermost expression.

Because of the design of `iostream`s, an overloaded `operator<<()` or `operator>>()` function must be a friend function and it must take arguments of an `istream` object (either `istream` or `ostream`, depending on the operator) followed by an object of your user-defined type. The function must return the same stream object as it takes as an argument.

Note, it's important that the stream object be passed into and out of the function so you can have expressions of the form:

```
cout << obj1 << obj2;
```

In effect, each object is passed to the stream and then the stream is passed down the line.

Note, why couldn't we define the overloaded operator function as a member function of the `Distance` class? If we did, here would be the declaration of the insertion output operator as a member function of `Distance`:

```
class Distance {
    ...
    ...
    ostream& operator<<(ostream&);
    ...
    ...
};
```

The left operand of every member function is an object or pointer to an object of its class. This is why the member function instance of the output operator declares only the one `ostream` object. A call of this instance takes the following form:

```
d << cout;
```

It would be very confusing both to the programmer and the human readers of the program to provide this instance. Therefore it's better to use friend functions.

```
// dist4.h

#ifndef _DIST4_H
#define _DIST4_H

#include <iostream.h>

using namespace std;

class Distance {
    private:
        int feet;
        float inches;
        void normalize();
    public:
        Distance();
        Distance(float f);
        Distance(int ft, float in);
        Distance operator+=(const Distance&);
        friend Distance operator+(const Distance&, const Distance&);
        friend istream& operator>>(istream&, Distance&);
        friend ostream& operator<<(ostream&, const Distance&);
};

#endif


// dist4.cpp

#include <iostream.h>
#include "dist4.h"

Distance::Distance() {
    feet = 0;
    inches = 0.0;
}

Distance::Distance(float f) {
    feet = int (f);
    inches = 12 * (f - feet);
}

Distance::Distance(int ft, float in) {
    feet = ft;
    inches = in;
    normalize();
}
```

```
void Distance::normalize() {
    while(inches >= 12.0) {
        inches -= 12.0;
        feet++;
    }
}

Distance Distance::operator+=(const Distance& d) {
    feet += d.feet;
    inches += d.inches;
    return Distance(feet, inches);
}

Distance operator+(const Distance& d1, const Distance& d2) {
    int ft = d1.feet + d2.feet;
    float in = d1.inches + d2.inches;
    return Distance(ft, in);
}

istream& operator>>(istream& is, Distance& d) {
    cout << "Enter feet: ";   is >> d.feet;
    cout << "Enter inches: "; is >> d.inches;
    d.normalize();
    return is;
}

ostream& operator<<(ostream& os, const Distance& d) {
    os << d.feet << "\' " << d.inches << "\"";
    return os;
}
```

```
// distdrv4.cpp

#include <iostream>
#include "dist4.h"

using namespace std;

int main() {
    Distance d1(5, 22), d2(8.25), d3, d4, d5;
    cout << "d1: " << d1 << endl;
    cout << "d2: " << d2 << endl;
    cout << "d3: " << d3 << endl;
    cout << "d4: " << d4 << endl;
    cout << "d5: " << d5 << endl;
    d3 = d1 + d2;
    cout << "d3 = d1 + d2: " << d3 << endl;
    d4 = d3 + 10;
    cout << "d4 = d3 + 10: " << d4 << endl;
    cin >> d5;
    cout << "d5: " << d5 << endl;
    return 0;
}


// Output

d1: 6' 10"
d2: 8' 3"
d3: 0' 0"
d4: 0' 0"
d5: 0' 0"
d3 = d1 + d2: 15' 1"
d4 = d3 + 10: 25' 1"
Enter feet: 8
Enter inches: 31
d5: 10' 7"
```

Overloading the Array Subscript Operator []

The subscript operator [] is normally used to index arrays. In fact, the subscript operator is a binary operator that requires two arguments. For example:

```
p = x[i];

[] : operator
x  : first argument
i  : second argument
```

The subscript operator actually performs a useful function, it hides pointer arithmetic for us! For example, if we have the following array:

```
char name[20];
```

and we execute a statement such as:

```
ch = name[12];
```

The [] operator directs the assignment statement to add 12 to the base address of the array name to locate the data stored in this memory location.

```
ch = name[12] implements pointer arithmetic: *(name + 12)
```

You can overload the array subscript operator [] to provide array-like access to a class's data members, even though that data might be stored as individual members.

Let's overload the [] operator for a class that stores four integer values as separate data members.

```
// parray.h

#ifndef _PARRAY_H
#define _PARRAY_H

class PseudoArray {
    private:
        int value0, value1, value2, value3;
    public:
        PseudoArray(int v0, int v1, int v2, int v3);
        int getInt(int i);
        int operator[](int i);
};

#endif
```

```
// parray.cpp

#include "parray.h"

PseudoArray::PseudoArray(int v0, int v1, int v2, int v3) {
    value0 = v0;
    value1 = v1;
    value2 = v2;
    value3 = v3;
}

int PseudoArray::getInt(int i) {
    switch(i) {
        case 0 : return value0;
        case 1 : return value1;
        case 2 : return value2;
        case 3 : return value3;
        default: return value0;
    }
}

int PseudoArray::operator[](int i) {
    return getInt(i);
}
```

```
// parraydr.cpp

#include <iostream>
#include "parray.h"

using namespace std;

int main() {
    PseudoArray pa(10, 20, 30, 40);

    for (int i = 0; i < 4; i++)
        cout << "pa[" << i << "] == " << pa[i] << endl;
    return 0;
}


// parray output

pa[0] == 10
pa[1] == 20
pa[2] == 30
pa[3] == 40
```

To get a data element we might have previously said:

```
cout << pa.getInt(i);
```

We can now say:

```
cout << pa[i];
```

We are using object pa as if it were an array! Overloading the subscript operator makes it possible to use array indexing rather than call member functions for an object of a class.

The overloaded subscript operator function can only be a class member:

```
pa[i]        →        pa.operator[](i)
```

Note that only one explicit argument is needed because the `this` pointer is provided.

Another String Example

Suppose we have a `String` class that represents strings of different lengths and we wish to provide bounds-checking for array subscripting operations on the strings.

In our example, we'll support strings that are statically allocated and can hold a maximum of 255 characters:

```cpp
#include <iostream>

using namespace std;

const int SIZE = 255;

int main() {
    String255 a(10);
    char c;
    a[5] = 17;              // a.operator[](5) = 17;
    c = a[7];               // c = a.operator[](7);
    c = a[26];              // c = a.operator[](26); error
    return 0;
}

class String255 {
    private:
        char buf[SIZE];         // maximum size
        int size;               // actual size
    public:
        String255(int num) { size = num; }
        char& operator[](int i);
};

char& String255::operator[](int i) {
    if (i < 0 || i >= SIZE) {
        cout << "subscript out of bounds" << endl;
        exit(1);
    }
    else
        return buf[i];          // the normal [] operator used here
}
```

In each case, the hidden first operand is a pointer to the object being subscripted, the object `a`.

Note that the `operator[]()` function passes back a reference to a character. We do this so that we can use it on both sides of an assignment statement.

Note also that the subscript operator must be able to appear on both the right and the left hand side of an expression. In order to appear on the left-hand side, its return value must be an lvalue. This is achieved by specifying the return value as a reference type:

```
char& String255::operator[](int i) {
    return buf[i];
}
```

The return value of the subscript operator is the lvalue of the indexed element. That's why it can appear as the target off an assignment.

C++ has reference declarations, and such type modifiers produce lvalue's (stands for location value).

Note: on the right side of an assignment expression, an lvalue is automatically dereferenced. On the left side of an expression, an lvalue specifies where an appropriate value is to be stored.

One Last Example

```
// phone.h

#ifndef _PHONE_H
#define _PHONE_H

typedef struct {
    char *name;
    long number;
} ENTRY;
```

```
class PhoneBook {
    private:
        int length;
        int count;
        ENTRY *listing;
    public:
        PhoneBook(int);
        ~PhoneBook();
        void addEntry(char*, long);
        void displayPhoneBook();
        long operator[](char *);
        char *operator[](long);
};

#endif


// phone.cpp

#include <iostream>
#include <iomanip>
#include <string.h>
#include "phone.h"

using namespace std;

PhoneBook::PhoneBook(int size) {      // Create phone book of
    listing = new ENTRY[size];        // specified size
    length = size;
    count = 0;
}

PhoneBook::~PhoneBook() {
    for (int i = 0; i < count; i++)
        delete [] listing[i].name;
    delete [] listing;
}

void PhoneBook::displayPhoneBook() {
    for (int i = 0; i < count; i++)
        cout << setw(10) << listing[i].name
             << setw(10) << listing[i].number << endl;
}
void PhoneBook::addEntry(char *Name, long Number) {
    listing[count].name = new char[strlen(Name) + 1];
    strcpy(listing[count].name, Name);
    listing[count++].number = Number;
}
```

```cpp
long PhoneBook::operator[](char *Name) {    // look up persons number
    for (int i = 0; i < count; i++)
        if (strcmp(listing[i].name, Name) == 0)
            return (listing[i].number);
    return 0;
}

char *PhoneBook::operator[](long Number) {   // look up person's name
    for (int i = 0; i < count; i++)
        if (listing[i].number == Number)
            return listing[i].name;
    return "Person not found.";
}

// pbook.cpp

#include <iostream>
#include "phone.h"

using namespace std;

int main() {
    PhoneBook pb(100);
    long number;
    char *name;

    pb.addEntry("Haley"   , 4861144);
    pb.addEntry("Shari"   , 4933684);
    pb.addEntry("Samantha", 4861199);
    cout << "Display phone book:" << endl << endl;
    pb.displayPhoneBook();
    cout << endl;
    number = pb["Shari"];
    cout << "number = " << number << endl;
    name = pb[4861199];
    cout << "  name = " << name << endl;
    return 0;
}
```

❖ Operator Overloading Homework

1. What is the difference between a binary operator and a function?

2. When overloading operators, give reasons for choosing **member** functions or **friend** functions.

3. When the stream **input** and **output** operators are overloaded to work with user-defined classes, are these operator functions usually **members** or **friends**? Why?

4. Is it possible using operator overloading to change the behavior of + on integers? Why or why not?

5. Which operators cannot be overloaded?

6. What restrictions apply to overloading the following operators?

```
=, (), [], ->
```

7. Why can we not overload << or >> as member operators?

8. The function below overloads the + operator and adds two `Distance` objects. It assumes the existence of a `Distance` class that contains two private data members:

```
int feet;
float inches;.
```

Given the function below, answer the following questions as specifically as you can.

```
Distance operator+(const Distance &a, const Distance &b) {
        Distance sum;

        sum.feet = a.feet + b.feet;
        sum.inches = a.inches + b.inches;
        if (sum.inches >= 12.0) {
                sum.inches -= 12.0;
                sum.feet++;
        }
        return sum;
}
```

a) Is this function most likely to be a **member** function, a **friend** function, or an **ordinary** function? How can you tell?

b) The `&` symbols in the parameter declarations make `a` and `b` reference parameters. What does that mean?

c) Would the function work correctly without the `&` symbols? If not, why? If so, why are they there?

d) What would be the effect of adding an `&` after `Distance` in the return type specification?

❖ Operator Overloading Lab (IntArray)

The array construct in C is very efficient but also very dangerous for the unwary. For example, many novice programmers fall into the trap of declaring an array of 100 elements and then try to access the element with index 100. Not only is this an error in C, but the language won't even alert the user when the mistake is made. C++ allows programmers to define safer and more flexible array constructs if they are willing to sacrifice some of C's runtime efficiency. The purpose of this lab is to see how this is done and to gain some experience in overloading operators.

The heart of this assignment is a class that you will define called `IntArray`. With it, the user will be able to declare integer arrays of any size with automatic range checking of indices. The upper and lower indices can be any integer, positive or negative, rather than the fixed limits of 0 to $SIZE-1$. It will also be possible to assign entire arrays to each other, compare two arrays for equality and inequality, add two arrays, and output arrays using the overloaded $<<$ operator. For example:

```
#include <iostream>
#include "IntArray.h"

using namespace std;

int main() {
    IntArray a(10), w(10);          // Ten elements, indexed 0 to 9
    IntArray b(-3, 6);              // Ten elements, indexed -3 to 6
    IntArray c(6, 8);               // Three elements, indexed 6 to 8
    IntArray d(5, 5);               // Single element array, indexed at 5
    IntArray z;                     // Ten elements, indexed 0 to 9

    // high() and low() return largest and smallest legal indices
    for (int i = a.low(); i <= a.high(); ++i)
        a[i] = i * 10;              // Access just like normal array

    // Output array contents. Note that you will overload
    // the << operator to do this.
    cout << a << endl;

    // The code for the overloaded << operator will look
    // something like this:
    for (int i = a.low(); i <= a.high(); i++)
        cout << "a[" << i << "] = " << a[i] << " ";
```

```
    // Similar code initializes b, c, and d.
    // Note that we will not be able to mimic the C style
    // aggregate initializer: int a[3] = { 1, 2, 3 };

    IntArray e(c);                  // e is a copy of c

    b = a;                          // b holds copy of a's elements,
                                    // but indices run from -3 to 6

    w = a + b;                      // sum two arrays into third array

    w += a;                         // sum two arrays into first array

    a[10] = 1;                      // Runtime error: illegal index
                                    // Print diagnostic and halt.

    b = c;                          // Runtime error: lengths different
                                    // Print diagnostic and simulate
                                    // halt.

    if (a == d)                     // Elements compared.
                                    // No run-time error
        cout << "TRUE" << endl;     // if lengths don't match. != also
                                    // works.
    else
        cout << "FALSE" << endl;
}
```

It's up to you to define the `IntArray` class, but it should fulfill the following requirements:

• An `IntArray` object may be declared in any of four ways:

> 1) With a single integer giving the number of elements, in which case the indices will run from 0 to one less than the number of elements.
>
> 2) With two integers, which will be taken as the lower and upper indices. It is an error for the first index to be greater than the second index, but having both indices the same specifies a valid, single element array.
>
> 3) With no integers, which creates a ten-element array whose indices run from 0 to 9.
>
> 4) With another object of type `IntArray`. The new object will be a copy of the old one, with each entry and the index range duplicated.

• Except in case 4, the initial array elements are undefined. Once an `IntArray` object has been created, its size cannot be changed, and each object should use just enough storage to hold its elements, plus a fixed amount of space for bookkeeping information.

• Elements of an `IntArray` are accessed using the standard C style array indexing notation. If the index falls outside the legal range, an error message should be printed, and the program should simulate a halt.

• `IntArray` objects may be assigned to each other if they have the same number of elements. The indices need not match, and the indices of the target object will not change; only the elements will be copied. If the sizes of objects being assigned don't match, an error message should be printed, and the program should simulate a halt.

• Two `IntArray` objects can be compared using the `==` and `!=` operators. If the arrays have the same number of elements and each of the corresponding elements match, the objects are equal and the comparison operator (`==`) should return a non-zero value, even if their indices are different. Comparing `IntArray` objects with different number of elements is legal, and the result is always zero. The `!=` operator should return zero whenever `==` would return non-zero, and vice versa.

• Class `IntArray` should also overload the operators below. An error should be generated if the arrays are of a different size.

`[]`	- allows index range checking
`+`	- allows the sum of two arrays to be assigned to a third array
`+=`	- allows the sum of two arrays to be assigned to the first array
`<<`	- allows the contents of an array to be output

• The `IntArray` class should have `high()` and `low()` member functions that return the maximum and minimum legal index for a given array as defined by the driver (user).

• The `IntArray` class should also contain a private data member that will hold the name of the `IntArray` object when instantiated. This data member should be initialized with a call to a `setName()` member function.

I will supply a driver program, `iadrv.cpp`, that demonstrates the use of the `IntArray` class. The driver will demonstrate all operation and error condition features separately. Each test will note what feature is being demonstrated and output each array name, index, and corresponding element using the overloaded `<<` operator.

Note that for each array test data, the driver will simply multiply the array index by 10 immediately after each array is initialized or modified and output its contents. When your program encounters a run-time error, you should "simulate" a halt with appropriate diagnostics rather than actually halting the program. Be sure that all error messages sent to the terminal window are also sent to the csis.txt output file.

When elements of an array are output, the array name should be output, a left bracket, an index value, a right bracket, an equal sign, and finally a value. For example:

```
a[3] = 30
```

Note that the driver program outputs array elements vertically down the page. This behavior should not be changed. Be sure that you include a destructor for the IntArray class. The driver program will include the following tests:

1. Test array declared with single integer:

```
IntArray a(10);
```

2. Test array declared with two integers, one of which is negative:

```
IntArray b(-3, 6);
```

3. Test array declared with two non-negative integers:

```
IntArray c(6, 8);
```

4. Test array declared with two identical integers:

```
IntArray d(5, 5);
```

5. Test array declared with no integers:

```
IntArray z;
```

6. Test array declared with another object of type `IntArray`:

```
IntArray c(6, 8);
IntArray e(c);
```

7. Test array assigned to another array with different indices:

```
IntArray f(1, 4);
IntArray g(5, 8);
f = g;
```

8. Test multiple array assignment with different indices:

```
IntArray j(3, 6);
IntArray k(6, 9);
IntArray l(1, 4);
j = k = l;
```

9. Test overloaded equality operator (identical elements):

```
IntArray m(3, 7);
IntArray n(1, 5);
m == n
```

10. Test overloaded equality operator (different elements):

```
IntArray o(3, 7);
IntArray p(1, 5);
o == p
```

11. Test overloaded equality operator (different size arrays):

```
IntArray q(1, 3);
IntArray r(1, 4);
q == r
```

12. Test overloaded inequality operator (identical elements):

```
IntArray s(3, 7);
IntArray t(1, 5);
s != t
```

13. Test overloaded inequality operator (different elements):

```
IntArray u(3, 7);
IntArray v(1, 5);
u != v
```

14. Test overloaded inequality operator (different size arrays):

```
IntArray w(1, 3);
IntArray x(1, 4);
w != x
```

15. Test sum of two arrays assigned to third array (note that corresponding array elements get added to each other and assigned to the corresponding element in the third array):

```
IntArray a(1, 5);
IntArray b(4, 8);
IntArray c = a+b;
```

Note that `IntArray c` should take on the indices of the invoking `IntArray` object, in this case `a`.

16. Test sum of two arrays assigned to first array (note that corresponding array elements get added to each other and assigned to the corresponding element in the first array):

```
IntArray d(10,13);
IntArray e(30,33);
d += e;
```

17. Test array declared with illegal array bounds:

```
IntArray f(5, 2);
```

18. Test array for index out of range:

```
IntArray g(10);
num = g[10];
```

19. Test arrays for length mismatch:

```
IntArray m(1, 4);
IntArray n(3, 7);
m = n;
```

20. Test array subscript operator as an lvalue and rvalue:

```
IntArray o(7, 8);
o[7] = 25;
o[8] = o[7];
```

```
// iadrv.h

#ifndef _IADRV_H
#define _IADRV_H

#include "intarray.h"

int main();
void test1();
void test2();
void test3();
void test4();
void test5();
void test6();
void test7();
void test8();
void test9();
void test10();
void test11();
void test12();
void test13();
void test14();
void test15();
void test16();
void test17();
void test18();
void test19();
void test20();
void wait();

#endif
```

```
// iadrv.cpp - driver program for testing IntArray class

#include <iostream>
#include <iomanip>
#include <fstream>
#include <stdlib.h>
#include "iadrv.h"

using namespace std;

ofstream csis;

int main() {
    csis.open("csis.txt");
    test1();
    test2();
    test3();
    test4();
    test5();
    test6();
    test7();
    test8();
```

```
        test9();
        test10();
        test11();
        test12();
        test13();
        test14();
        test15();
        test16();
        test17();
        test18();
        test19();
        test20();
        csis.close();
}

void test1() {
        cout << "1. Array declared with single integer: IntArray a(10);" << endl << endl;
        csis << "1. Array declared with single integer: IntArray a(10);" << endl << endl;
        IntArray a(10);
        for(int i = a.low(); i <= a.high(); i++)
            a[i] = i * 10;
        a.setName("a");
        cout << a << endl;
        csis << a << endl;
        wait();
}

void test2() {
        cout << "2. Array declared with two integers: IntArray b(-3, 6);" << endl << endl;
        csis << "2. Array declared with two integers: IntArray b(-3, 6);" << endl << endl;
        IntArray b(-3, 6);
        for(int i = b.low(); i <= b.high(); i++)
            b[i] = i * 10;
        b.setName("b");
        cout << b << endl;
        csis << b << endl;
        wait();
}

void test3() {
        cout << "3. Array declared with two integers: IntArray c(6, 8);" << endl << endl;
        csis << "3. Array declared with two integers: IntArray c(6, 8);" << endl << endl;
        IntArray c(6, 8);
        for(int i = c.low(); i <= c.high(); i++)
            c[i] = i * 10;
        c.setName("c");
        cout << c << endl;
        csis << c << endl;
        wait();
}

void test4() {
        cout << "4. Array declared with two identical integers: IntArray d(5, 5);"
             << endl << endl;
        csis << "4. Array declared with two identical integers: IntArray d(5, 5);"
             << endl << endl;
        IntArray d(5, 5);
        for(int i = d.low(); i <= d.high(); i++)
            d[i] = i * 10;
        d.setName("d");
        cout << d << endl;
        csis << d << endl;
        wait();
}
```

```
void test5() {
    cout << "5. Array declared with no integers: IntArray z;" << endl << endl;
    csis << "5. Array declared with no integers: IntArray z;" << endl << endl;
    IntArray z;
    for(int i = z.low(); i <= z.high(); i++)
        z[i] = i * 10;
    z.setName("z");
    cout << z << endl;
    csis << z << endl;
    wait();
}

void test6() {
    cout << "6. Array declared with another object of type IntArray: IntArray c(6, 8);"
        << endl;
    cout << "                                                        Intarray e(c);"
        << endl << endl;
    csis << "6. Array declared with another object of type IntArray: IntArray c(6, 8);"
        << endl;
    csis << "                                                        Intarray e(c);"
        << endl << endl;
    IntArray c(6, 8);
    for(int i = c.low(); i <= c.high(); i++)
        c[i] = i * 10;
    c.setName("c");
    cout << c << endl;
    csis << c << endl;
    IntArray e(c);
    e.setName("e");
    cout << e << endl;
    csis << e << endl;
    wait();
}

void test7() {
    cout << "7. Array assigned to another array w/ different indices: IntArray f(1, 4);"
        << endl;
    cout << "                                                        IntArray g(5, 8);"
        << endl;
    cout << "                                                            f = g;"
        << endl << endl;
    csis << "7. Array assigned to another array w/ different indices: IntArray f(1, 4);"
        << endl;
    csis << "                                                        IntArray g(5, 8);"
        << endl;
    csis << "                                                            f = g;"
        << endl << endl;
    IntArray f(1, 4);
    for(int i = f.low(); i <= f.high(); i++)
        f[i] = i * 10;
    f.setName("f");
    cout << f << endl;
    csis << f << endl;
    IntArray g(5, 8);
    for(int i = g.low(); i <= g.high(); i++)
        g[i] = i * 10;
    g.setName("g");
    cout << g << endl;
    csis << g << endl;
    wait();
    f = g;
    cout << f << endl;
    cout << g << endl;
    csis << f << endl;
    csis << g << endl;
    wait();
}
```

```
void test8() {
    cout << "8. Multiple array assignment with different indices: IntArray j(3, 6);"
        << endl;
    cout << "                                                        IntArray k(6, 9);"
        << endl;
    cout << "                                                        IntArray l(1, 4);"
        << endl;
    cout << "                                                        j = k = l;"
        << endl << endl;
    csis << "8. Multiple array assignment with different indices: IntArray j(3, 6);"
        << endl;
    csis << "                                                        IntArray k(6, 9);"
        << endl;
    csis << "                                                        IntArray l(1, 4);"
        << endl;
    csis << "                                                        j = k = l;"
        << endl << endl;
    IntArray j(3, 6);
    for(int i = j.low(); i <= j.high(); i++)
        j[i] = i * 10;
    j.setName("j");
    cout << j << endl;
    csis << j << endl;
    IntArray k(6, 9);
    for(int i = k.low(); i <= k.high(); i++)
        k[i] = i * 10;
    k.setName("k");
    cout << k << endl;
    csis << k << endl;
    IntArray l(1, 4);
    for(int i = l.low(); i <= l.high(); i++)
        l[i] = i * 10;
    l.setName("l");
    cout << l << endl;
    csis << l << endl;
    wait();
    j = k = l;
    cout << j << endl;
    cout << k << endl;
    cout << l << endl;
    csis << j << endl;
    csis << k << endl;
    csis << l << endl;
    wait();
}
```

```
void test9() {
    cout << "9. Overloaded equality operator (identical elements): IntArray m(3, 7);"
        << endl;
    cout << "                                                      IntArray n(1, 5);"
        << endl;
    cout << "                                                              m == n"
        << endl << endl;
    csis << "9. Overloaded equality operator (identical elements): IntArray m(3, 7);"
        << endl;
    csis << "                                                      IntArray n(1, 5);"
        << endl;
    csis << "                                                              m == n"
        << endl << endl;
    IntArray m(3, 7);
    for(int i = m.low(); i <= m.high(); i++)
        m[i] = i * 10;
    m.setName("m");
    cout << m << endl;
    csis << m << endl;
    IntArray n(1, 5);
    for(int i = n.low(); i <= n.high(); i++)
        n[i] = i * 10;
    n.setName("n");
    cout << n << endl;
    csis << n << endl;
    wait();
    m = n;
    cout << m << endl;
    cout << n << endl;
    cout << "Returns " << (m == n ? "True." : "False.") << endl << endl;
    csis << m << endl;
    csis << n << endl;
    csis << "Returns " << (m == n ? "True." : "False.") << endl << endl;
    wait();
}

void test10() {
    cout << "10. Overloaded equality operator (different elements): IntArray o(3, 7);"
        << endl;
    cout << "                                                       IntArray p(1, 5);"
        << endl;
    cout << "                                                               o == p"
        << endl << endl;
    csis << "10. Overloaded equality operator (different elements): IntArray o(3, 7);"
        << endl;
    csis << "                                                       IntArray p(1, 5);"
        << endl;
    csis << "                                                               o == p"
        << endl << endl;
    IntArray o(3, 7);
    for(int i = o.low(); i <= o.high(); i++)
        o[i] = i * 10;
    o.setName("o");
    cout << o << endl;
    csis << o << endl;
    IntArray p(1, 5);
    for(int i = p.low(); i <= p.high(); i++)
        p[i] = i * 10;
    p.setName("p");
    cout << p << endl;
    cout << "Returns " << (o == p ? "True." : "False.") << endl << endl;
    csis << p << endl;
    csis << "Returns " << (o == p ? "True." : "False.") << endl << endl;
    wait();
}
```

```
void test11() {
    cout << "11. Overloaded equality operator (different size arrays): IntArray q(1, 3);"
        << endl;
    cout << "                                                          IntArray r(1, 4);"
        << endl;
    cout << "                                                          q == r;"
        << endl << endl;
    csis << "11. Overloaded equality operator (different size arrays): IntArray q(1, 3);"
        << endl;
    csis << "                                                          IntArray r(1, 4);"
        << endl;
    csis << "                                                          q == r;"
        << endl << endl;
    IntArray q(1, 3);
    for(int i = q.low(); i <= q.high(); i++)
        q[i] = i * 10;
    q.setName("q");
    cout << q << endl;
    csis << q << endl;
    IntArray r(1, 4);
    for(int i = r.low(); i <= r.high(); i++)
        r[i] = i * 10;
    r.setName("r");
    cout << r << endl;
    cout << "Returns " << (q == r ? "True." : "False.") << endl << endl;
    csis << r << endl;
    csis << "Returns " << (q == r ? "True." : "False.") << endl << endl;
    wait();
}

void test12() {
    cout << "12. Overloaded inequality operator (identical elements): IntArray s(3, 7);"
        << endl;
    cout << "                                                         IntArray t(1, 5);"
        << endl;
    cout << "                                                         s != t;"
        << endl << endl;
    csis << "12. Overloaded inequality operator (identical elements): IntArray s(3, 7);"
        << endl;
    csis << "                                                         IntArray t(1, 5);"
        << endl;
    csis << "                                                         s != t;"
        << endl << endl;
    IntArray s(3, 7);
    for(int i = s.low(); i <= s.high(); i++)
        s[i] = i * 10;
    s.setName("s");
    cout << s << endl;
    csis << s << endl;
    IntArray t(1, 5);
    for(int i = t.low(); i <= t.high(); i++)
        t[i] = i * 10;
    t.setName("t");
    cout << t << endl;
    csis << t << endl;
    wait();
    s = t;
    cout << s << endl;
    cout << t << endl;
    cout << "Returns " << (s != t ? "True." : "False.") << endl << endl;
    csis << s << endl;
    csis << t << endl;
    csis << "Returns " << (s != t ? "True." : "False.") << endl << endl;
    wait();
}
```

```
void test13() {
    cout << "13. Overloaded inequality operator (different elements): IntArray u(3, 7);"
        << endl;
    cout << "                                                          IntArray v(1, 5);"
        << endl;
    cout << "                                                          u != v;"
        << endl << endl;
    csis << "13. Overloaded inequality operator (different elements): IntArray u(3, 7);"
        << endl;
    csis << "                                                          IntArray v(1, 5);"
        << endl;
    csis << "                                                          u != v;"
        << endl << endl;
    IntArray u(3, 7);
    for(int i = u.low(); i <= u.high(); i++)
        u[i] = i * 10;
    u.setName("u");
    cout << u << endl;
    csis << u << endl;
    IntArray v(1, 5);
    for(int i = v.low(); i <= v.high(); i++)
        v[i] = i * 10;
    v.setName("v");
    cout << v << endl;
    cout << "Returns " << (u != v ? "True." : "False.") << endl << endl;
    csis << v << endl;
    csis << "Returns " << (u != v ? "True." : "False.") << endl << endl;
    wait();
}

void test14() {
    cout << "14. Overloaded inequality operator (different size arrays):
                                                            IntArray w(1, 3);" << endl;
    cout << "                                                          IntArray x(1, 4);" << endl;
    cout << "                                                          w != x;"           << endl
        << endl;
    csis << "14. Overloaded inequality operator (different size arrays):
                                                            IntArray w(1, 3);" << endl;
    csis << "                                                          IntArray x(1, 4);" << endl;
    csis << "                                                          w != x;"           << endl
        << endl;
    IntArray w(1, 3);
    for(int i = w.low(); i <= w.high(); i++)
        w[i] = i * 10;
    w.setName("w");
    cout << w << endl;
    csis << w << endl;
    IntArray x(1, 4);
    for(int i = x.low(); i <= x.high(); i++)
        x[i] = i * 10;
    x.setName("x");
    cout << x << endl;
    cout << "Returns " << (w != x ? "True." : "False.") << endl << endl;
    csis << x << endl;
    csis << "Returns " << (w != x ? "True." : "False.") << endl << endl;
    wait();
}
```

```
void test15() {
    cout << "15. Sum of two arrays assigned to third array: IntArray a(1, 5);"   << endl;
    cout << "                                                 IntArray b(4, 8);"   << endl;
    cout << "                                                 IntArray c = a + b;" << endl
         << endl;
    csis << "15. Sum of two arrays assigned to third array: IntArray a(1, 5);"   << endl;
    csis << "                                                 IntArray b(4, 8);"   << endl;
    csis << "                                                 IntArray c = a + b;" << endl
         << endl;
    IntArray a(1, 5);
    for(int i = a.low(); i <= a.high(); i++)
        a[i] = i * 10;
    a.setName("a");
    cout << a << endl;
    csis << a << endl;
    IntArray b(4, 8);
    for(int i = b.low(); i <= b.high(); i++)
        b[i] = i * 10;
    b.setName("b");
    cout << b << endl;
    csis << b << endl;
    wait();
    IntArray c = a + b;
    c.setName("c");
    cout << c << endl;
    csis << c << endl;
    wait();
}

void test16() {
    cout << "16. Sum of two arrays assigned to first array: IntArray d(10, 13);" << endl;
    cout << "                                                IntArray e(30, 33);" << endl;
    cout << "                                                d += e;"             << endl
         << endl;
    csis << "16. Sum of two arrays assigned to first array: IntArray d(10, 13);" << endl;
    csis << "                                                IntArray e(30, 33);" << endl;
    csis << "                                                d += e;"             << endl
         << endl;
    IntArray d(10, 13);
    for(int i = d.low(); i <= d.high(); i++)
        d[i] = i * 10;
    d.setName("d");
    cout << d << endl;
    csis << d << endl;
    IntArray e(30, 33);
    for(int i = e.low(); i <= e.high(); i++)
        e[i] = i * 10;
    e.setName("e");
    cout << e << endl;
    csis << e << endl;
    d += e;
    cout << d << endl;
    csis << d << endl;
    wait();
}
```

```
void test17() {
    cout << "17. Array declared with illegal array bounds: IntArray f(5, 2);"
         << endl << endl;
    csis << "17. Array declared with illegal array bounds: IntArray f(5, 2);"
         << endl << endl;
    IntArray f(5, 2);
    for(int i = f.low(); i <= f.high(); i++)
        f[i] = i * 10;
    f.setName("f");
    cout << f << endl;
    csis << f << endl;
    wait();
}


void test18() {
    cout << "18. Array with index out of range: IntArray g(10);"    << endl;
    cout << "                                   g[10] = 1;"          << endl << endl;
    csis << "18. Array with index out of range: IntArray g(10);"    << endl;
    csis << "                                   g[10] = 1;"          << endl << endl;
    IntArray g(10);
    for(int i = g.low(); i <= g.high(); i++)
        g[i] = i * 10;
    g.setName("g");
    cout << g << endl;
    csis << g << endl;
    g[10] = 1;
    wait();
}

void test19() {
    cout << "19. Arrays with length mismatch: IntArray m(1, 4);" << endl;
    cout << "                                 IntArray n(2, 4);" << endl;
    cout << "                                 m = n;"            << endl << endl;
    csis << "19. Arrays with length mismatch: IntArray m(1, 4);" << endl;
    csis << "                                 IntArray n(2, 4);" << endl;
    csis << "                                 m = n;"            << endl << endl;
    IntArray m(1, 4);
    for(int i = m.low(); i <= m.high(); i++)
        m[i] = i * 10;
    m.setName("m");
    cout << m << endl;
    csis << m << endl;
    IntArray n(2, 4);
    for(int i = n.low(); i <= n.high(); i++)
        n[i] = i * 10;
    n.setName("n");
    cout << n << endl;
    csis << n << endl;
    wait();
    m = n;
    cout << m << endl;
    cout << n << endl;
    csis << m << endl;
    csis << n << endl;
    wait();
}
```

```
void test20() {
    cout << "20. Array subscript operator: IntArray o(7, 8);" << endl;
    cout << "                                 o[7] = 25;"        << endl;
    cout << "                                 o[8] = o[7];"      << endl << endl;
    csis << "20. Array subscript operator: IntArray o(7, 8);" << endl;
    csis << "                                 o[7] = 25;"        << endl;
    csis << "                                 o[8] = o[7];"      << endl << endl;
    IntArray o(7, 8);
    for(int i = o.low(); i <= o.high(); i++)
        o[i] = i * 10;
    o.setName("o");
    cout << o << endl;
    csis << o << endl;
    o[7] = 25;
    o[8] = o[7];
    cout << o << endl;
    csis << o << endl;
    wait();
}

void wait() {
    char buf;

    cout << "Press any key to continue." << endl;
    cin.get(buf);
}
```

Chapter 12. Assignment & Initialization

Introduction to Assignment & Initialization
Memberwise Initialization
Memberwise Assignment

Introduction to Assignment & Initialization

The C++ compiler is always busy on our behalf, doing things we can't be bothered to do. If we take charge, it will defer to our judgment. Otherwise, it will do things its own way! Two important examples of this process are the *assignment operator* and the *copy constructor*.

We've used the assignment operator many times, probably without thinking too much about it. For example, suppose d1 and d2 are two Distance objects. Unless you tell the compiler otherwise, the statement:

```
d2 = d1;
```

will cause the compiler to copy the data from d1, member-by-member, into d2. This is the default activation of the assignment operator, =.

We're also familiar with initializing variables. Initializing an object with another object, such as:

```
Distance d2(d1);
```

causes a similar action. The compiler creates a new object, d2, and copies the data from d1, member-by-member, into d2. This is the default action of the copy constructor.

Both of these activities are provided, free of charge, by the compiler. If member-by-member copying is what you want, you need take no further action. However, if you want assignment or initialization to do something more complex, then you can override these default functions.

Memberwise Initialization

When an object is used to initialize a new object of a class, C++ copies each data member from the existing object to the new one. For example, a simple class is shown below with no pointer data members:

```
class Alpha {
    private:
        int i;
        double d;
    public:
        Alpha() { i = 0; d = 0.0; }
        Alpha(int ii, double dd) { i = ii; d = dd; }
};
```

Note that in practice, `Alpha` needs other member functions to access its private data members `i` and `d`, but this example simply illustrates how object copying works.

We can now declare `Alpha` objects `a1` and `a2` as follows:

```
Alpha a1;
Alpha a2(100, 3.14159);
```

We can also use an existing `Alpha` object such as `a2` to initialize a new object `a3`:

```
Alpha a3(a2);
```

This copies `a2` to `a3` by performing a memberwise initialization of `a2`'s data members. That is, `a2`'s data members are copied one by one to `a3`'s data members as though we had written:

```
Alpha a3;
A3.i = a2.i;
A3.d = a2.d;
```

Note that these assignment statements are for illustration purposes only. Because the class data members are private, these statements do not compile. Memberwise initialization of a new class object occurs in three contexts:

- when an object is used to initialize a newly constructed object, i.e., as part of a declaration

- when passing an object to a function by value

- when a function returns a class object

If any of these actions cause a pointer data member to be duplicated, a serious bug might arise!

Let's redefine class Alpha to include a pointer data member:

```
#include <string.h>

class Alpha {
    private:
        char *buf;
        int length;
    public:
        Alpha();         // default constructor
        Alpha(char *);   // constructor accepts null-terminated
                         // character
                         // array and creates an Alpha object from it
        ~Alpha();
};

Alpha::Alpha() {
    length = 100;
    buf = new char[length + 1];
    buf[0] = '\0';
}

Alpha::Alpha(char *s) {
    length = strlen(s);
    buf = new char[length + 1];
    strcpy(buf, s);
}

Alpha::~Alpha() {
    delete [] buf;
}
```

Now that our `Alpha` class contains a pointer data member, let's see what happens when we pass an `Alpha` object by value. For example, suppose we decide to pass an `Alpha` object by value to a function `addSpace()` prototyped below:

```
void addSpace(Alpha a);
```

We'll assume that `addSpace()` is a friend function to the `Alpha` class and, therefore, has access to the private data members of `Alpha` objects. The purpose of `addSpace()` is to add a space to the end of the private data member `buf` in the `Alpha` object passed as a formal parameter. Now suppose we execute the following code:

```
Alpha a("Hello");
AddSpace(a);
```

To implement the call to `addSpace()`, the C++ compiler has to make a copy of the `Alpha` object a on the stack:

Original instance of `Alpha` object a:

```
 -------------------             -------------------------
|     char *buf      | =========➤  | h | e | l | l | o | \0 |
 -------------------   =======➤    -------------------------
|  int length = 5    |   ||
 -------------------     ||
                        ||
                        ||
              =====================||
                                    ||
                                    ||
Copy of Alpha object a on the stack: ||
                                    ||
 -------------------                 ||
|     char *buf      | =======================➤
 -------------------
|  int length = 5    |
 -------------------
```

Note that both the copy (`str.buf`) and the original private member (`a.buf`) point to the same null-terminated string because the character pointers are identical!

Now, what could be the problem with duplicate pointers addressing the same locations in memory?

Deleting one of those pointers, perhaps within a destructor, might cause the other pointer to address this freed memory (a dangling pointer). Worse yet, deleting the same space more than once in a class destructor might corrupt the heap!

C++ provides two mechanisms for ensuring that classes with pointer members are copied and assigned safely:

• memberwise initialization
• memberwise assignment

To create a complete copy of an Alpha object, we must define a constructor for the Alpha class that takes an Alpha reference as an argument, i.e., a constructor that can create a copy of an Alpha object, a copy constructor:

```
Alpha::Alpha(const Alpha& a) {
    length = a.length;
    buf = new char[length + 1];
    strcpy(buf, a.buf);
}
```

Note that the copy constructor allocates room for the null-terminating string buf and copies into it the string that was passed to it as an argument.

When the copy constructor is used, you get a complete copy of the Alpha object:

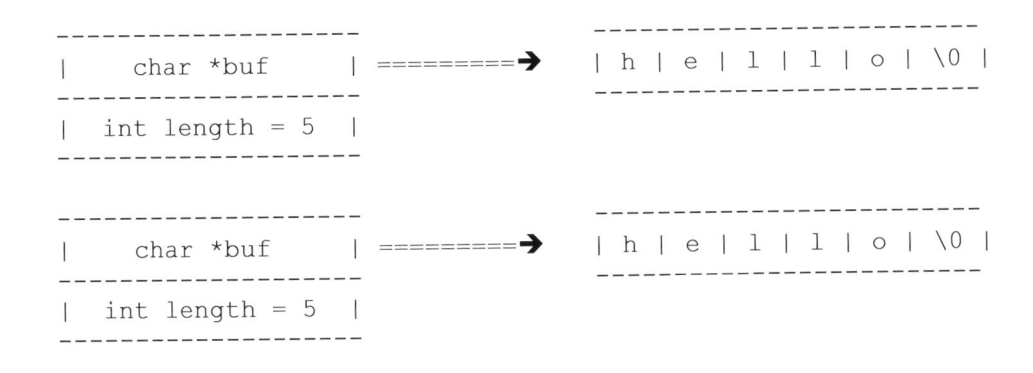

Note also that the argument to the copy constructor is a reference to an `Alpha` object and not an Alpha object itself. Since the whole point of a copy constructor is to enable the implementer to define the semantics of copying an object, it allows the constructor to gain access to the `Alpha` object without copying it.

The copy constructor also comes into play when you write:

```
• Alpha a1 = "Hello";      // string first gets converted into an
                           // Alpha object and then the copy
                           // constructor is called

• Alpha a2 = a1;           // the compiler has to initialize a2 with
                           // the value of another Alpha object a1.

• Alpha a3(a2);            // the compiler has to initialize a3 with
                           // the value of another Alpha object a2.
```

See if you can give the output of the following program before reading on:

```
#include <iostream>
#include <conio.h>
#include <string.h>

using namespace std;

class Alpha {
    private:
        char *buf;
    public:
        Alpha() {
            cout << "Default constructor" << endl;
            buf = new char[100];
            buf[0] = '\0';
        }
        Alpha(const Alpha& a) {
            cout << "Copy constructor" << endl;
            buf = new char[strlen(a.buf)+1];
            strcpy(buf, a.buf);
        }
        Alpha(const char* s) {
            cout << "One arg constructor" << endl;
            buf = new char[strlen(s)+1];
            strcpy(buf, s);
        }
        ~Alpha() {
            cout << "destructor called" << endl;
        }
```

```
};
```

```
int main() {
    Alpha a;
    Alpha b = "Hello";
    Alpha c(b);
    Alpha d = c;
    Alpha e("Hello");
    return 0;
}
```

The output produced is:

```
default constructor
one arg constructor
copy constructor
copy constructor
one arg constructor
destructor called
destructor called
destructor called
destructor called
destructor called
```

When to Provide a Copy Constructor

If your class does not include any pointers that have to be properly initialized, you do not have to provide a copy constructor. The C++ compiler will use a default copy constructor that performs a memberwise copy of one class instance to another.

You must provide a copy constructor only for classes with a pointer variable. Otherwise both copies will hold pointers to a single array of characters.

Even if you can live with the memberwise copy, you will run into another problem because of the destructor. If an Alpha object created by a memberwise copy operation is no longer needed, the Alpha destructor will be called and will free the storage whose address is in the character pointer of that Alpha object.

This leaves the remaining copy of Alpha with a dangling pointer, a pointer that does not point to any valid block of memory. When that remaining copy of Alpha has to be destroyed, the delete operator will be called with the address of memory that has already been freed. Worse yet, if that memory has been allocated to some other object, the destructor will inadvertently free memory belonging to some other object. To avoid such problems, you should always

provide a copy constructor for any class that includes dynamically allocated members.

So a copy constructor is a member function of a class that is executed to create an object of that class. Because such a constructor is usually written as taking a single argument of type reference to the class of which it is a member, it is usually referred to as:

```
X(X&)  (pronounced X of X ref)
```

Our `Alpha` copy constructor tells how to construct an `Alpha` object with an initial value that is a copy of some other `Alpha`:

```
Alpha(const Alpha&);
```

So initializations are always expressed as copy constructors that are capable of creating a new object as a replica of an existing object.

Initializations are always accomplished through constructors. A class of this kind will usually have at least three constructors:

• default constructor - `Alpha::Alpha()`, that determines what value to give to an `Alpha` object that is not explicitly initialized

• conversion constructor – a constructor from some other type such as `char*` to make it possible to put values into `Alpha` objects that do not originate from other `Alpha` objects

• copy constructor, `Alpha::Alpha(const Alpha&)`, that tells how to construct an `Alpha` object with an initial value that is a copy of some other `Alpha` object.

Memberwise Assignment

Consider the following class declarations:

```
class Alpha {
    private:
        char *buf;
        int i;
        double d;
    public:
        // appropriate constructors etc.
};

int main() {
    Alpha a;                        // 1
    Alpha b(5, 12.2, "Hello");      // 2
    Alpha c(b);                     // 3
    Alpha d = b;                    // 4
    a = d;                          // 5
    return 0;
}
```

Statement 1 creates a new `Alpha` object `a` by a call to the default constructor.

Statement 2 creates a new `Alpha` object `b` by a call to the three arg constructor.

Statement 3 creates and initializes a new `Alpha` object `c` by a call to the copy constructor.

Statement 4 creates and initializes a new `Alpha` object `d` by a call to the copy constructor.

However, statement 5 has an operation that is fundamentally different from the previous 4 operations, even though the same symbol = is used to represent it. This statement says to set the value of `Alpha` object `a` equal to the value of `Alpha` object `d`, i.e., an assignment statement copies `d` into `a`. *This act of giving a new value to an object that already exists is called assignment.*

This assignment involves (potentially) changing the value of `Alpha` object `a` but does not create any new `Alpha` objects. Because no new `Alpha` objects are created, no class constructors are called. Not even the class copy constructor is called!

The copy constructor is not invoked here because the objects invoked have already been constructed. C++ calls constructors only to initialize new objects, not when assigning one object to another.

An assignment occurs whenever an object that has already been created receives a value through an assignment statement:

```
a = d;
```

An initialization occurs when an object receives a value at the time it is declared:

```
Alpha d = b;
```

If you are ever uncertain whether a piece of C++ code involves assignment or initialization, ask yourself the following question:

"Is a new object being created here?"

If the answer is yes, a constructor is involved. If the answer is no, no constructors are involved. Initializations are performed by constructors and assignments are performed by overloading the assignment operator, =.

In C++, an initialization occurs when a new object is created. An assignment changes the value of an existing object without creating a new object. It therefore always obliterates a previous value. Why is this distinction important?

We've previously seen that if we don't write a copy constructor for our class, the compiler will automatically provide one for us. But we've seen that if our class has pointers, the provided copy constructor will not correctly make a copy of our object.

Like the default copy constructor, the C++ compiler will provide an assignment operator for our class if one is not explicitly written, and the default assignment operator will assign each of the data members. But, as with initialization, if the class declares any pointer data members, the copied pointers will now address the same location in memory, almost certainly causing a major bug if that memory is deleted in a destructor. This is a dangerous situation to be avoided at all costs.

We can solve this problem by overloading the assignment operator, =. We use a declaration of the form:

```
Alpha& operator=(const Alpha&);
```

So let's now add an overloaded assignment operator function to the Alpha class:

```
class Alpha {
    private:
        char *buf;
        int i;
        double d;
    public:
        // appropriate constructors etc.
        Alpha& operator=(const Alpha&);
};
```

The overloaded assignment operator specifies a constant reference parameter of type Alpha. The parameter refers to the object being copied.

Returning a reference to the assigned-to object allows users to write code like:

```
Alpha x, y;
x = y = Alpha(num);     // note that assignment is right associative
```

Assigning Alpha Objects

Assigning one Alpha object to another requires several steps:

- freeing the memory belonging to the pointer being obliterated
- allocating memory to contain a copy of the source string pointer
- making that memory contain a copy of the string
- copy all other data members

Such an operation might look like this:

```
Alpha& Alpha::operator=(const Alpha& a) {
    delete [] buf;
    buf = new char[strlen(a.buf) + 1];
    strcpy(buf, a.buf);
    i = a.i;
    d = a.d;
    return *this;
}
```

We return *this to go along with the usual C++ convention that an assignment should return a reference to its left-hand side.

However, this code has a serious bug. Try and find it before you read on.

...

...

It fails if you try to assign an object to itself:

```
a = a;
```

To fix this bug, we need a conditional statement in the code to guard against:

```
String s;
s = s;
```

Without this test, an assignment of an Alpha object to itself would delete the data member buf of that Alpha object and then try to copy the just-deleted data.

Once we are sure that we are dealing with two different objects, the assignment operator (unlike the copy constructor) should release any resources that held the old value. After that, the assignment operator usually does the same thing as the copy constructor, except that the assignment operator returns a value.

Here's the Fix
We must check for assigning something to itself as a special case and do nothing at all if the check succeeds:

```
Alpha& Alpha::operator=(const Alpha& a) {
    if (this != &a) {    // or if (*this != a)
        delete [] buf;
        buf = new char[strlen(a.buf) + 1];
        strcpy(buf, a.buf);
        i = a.i;
        d = a.d;
        return *this;
}
```

The `if` statement compares the `this` pointer to the address of the a reference parameter. It then catches an accidental assignment of the same object to itself. Similar precautions are not necessary for copy constructors. Why not?

Note that we cannot declare `operator=()` outside of a class. It must be declared as a member function and cannot be a friend function.

Copying of Objects
So in order to control copying of objects of a class `X`, it is generally necessary to define 4 things:

• a default constructor `X::X()` to establish a default value for `X` objects

• a copy constructor `X::X(const X&)` to control initialization and argument passing and return

• a destructor `X::~X()`

• the overloaded assignment operator, `X& X::operator=(const X&)`

Note that the argument to the assignment operator is an `X&` and not an `X`. This is to forestall copying the object again on its way to being assigned.

Assignment Operators vs Copy Constructors

Assignment operators and copy constructors have a lot of logic in common. However, here are three main differences:

• the assignment operator must work correctly when an object is assigned to itself

• each call to the assignment operator overwrites an existing value. If that value was using resources external to the object, those resources may have to be freed.

• the assignment operator can return a value.

Chapter 13. Container Relationships & Inheritance

Container Relationships
Introduction to Inheritance
Derivation
Counter Class
Protected Access Specifier
Employee Class
Class Hierarchies
Accessing Mechanisms
Constructors and Destructors with Derived Classes
Constructor Initializer List
Base and Derived Class Destructors
Overloading Copy Constructors
Overloading the Assignment Operator
Casting from Derived into Base
Composition & Inheritance Homework
Inheritance Lab (String Class)
Inheritance Lab (String Class) Driver

Container Relationships

Up to now, the only type of data members within a class have been mostly primitive types: `int`, `float`, `char`, etc. But there is nothing that forbids us from having a class definition that contains an instance of some other class.

The relationship between a class and an instance of another class is called a *container relationship*, because one class literally contains an instance of the other class. In object-oriented programming terminology, this is also referred to as a *has-a relationship*, because the containing class has-a(n) instance of the other class.

Consider the following program:

```cpp
// contain1.cpp

#include <iostream>
#include <string.h>

using namespace std;

class String {
    private:
        char *buf;
    public:
        String() {
            cout << "String default constructor" << endl;
            buf = new char[strlen("Shari")+1];
            strcpy(buf, "Shari");
        }
        String(char *s) {
            cout << "String one arg constructor" << endl;
            buf = new char[strlen(s)+1];
            strcpy(buf, s);
        }
        ~String() {
            delete [] buf;
        }
        void printString() {
            cout << buf << endl;
        }
};
```

```
class Employee {
    private:
        String name;
        int id;
    public:
        Employee() {
            cout << "Employee default constructor" << endl;
            name = "Haley";
            id = 108;
        }
        Employee(char *s, int num) {
            cout << "Employee two arg constructor" << endl;
            name = s;
            id = num;
        }
        void printEmployee() {
            name.printString();
            cout << id << endl;
        }
};

int main() {
    Employee e("Sally", 123);
    e.printEmployee();
    return 0;
}
```

Note that the private members of the contained instance are still inaccessible to the member function of the containing class. But the public functions of the contained instance may be accessed by the function of the containing class simply by writing the instance name and the dot operator:

```
name.printString();
```

Before reading on, see if you can figure out the output that gets generated from this program:

I did say BEFORE READING ON!!!

Here's the output produced:

```
String default constructor
Employee two arg constructor
String one arg constructor
Sally
123
```

Why the strange output? Why the extra call to the `String` default constructor?

Well, here's a C++ rule: *By default, before the body of the a constructor is executed, if the class contains a data member which is itself a class object, its default constructor will automatically be called first.*

So when an `Employee` object is created, the name member will first be initialized using the default `String` constructor, and then its value will be changed by the assignment in the body of the `Employee` constructor. This is two operations when one would have sufficed! Very inefficient!

Therefore, when a data member of a class is itself a class object, we must be careful or else we'll invoke the object's default constructor when it's not needed. The constructor will produce the correct results, but it is slower than it needs to be. Thus, for reasons of efficiency, we must use the *constructor initializer list* to initialize member variables. The constructor initializer list can override this default behavior.

Below we explicitly initialize the name data member with the correct value by using the constructor initializer list:

```cpp
Employee() : name("Haley"), id(108) {
    cout << "Employee default constructor" << endl;
}

Employee(char *s, int num) : name(s), id(num) {
    cout << "Employee two arg constructor" << endl;
}
```

When you write a constructor definition, stop after the `()` that encloses the formal argument list, supply a colon followed by a list of initializers, and see how many of the members of the class can be initialized using this constructor initializer syntax. Note that each initializer consists of a data member name and an argument list that specifies the arguments to the constructor for that member.

We'll often find that all of the members of the class can be so initialized, and by the time we get to the body of the constructor, there is nothing left to do. This is a good sign. Your constructor has initialized all of its members with the correct initial values.

Also note that the initialization syntax allows us to initialize data members that are not class objects. Since built-in types have no constructor, initializing the integer member id using the initialization syntax is no faster than using assignment. But the code is easier to read if we initialize all of the data members in the same way.

Here is the complete program using the constructor initializer lists:

```cpp
// contain2.cpp

#include <iostream>
#include <string.h>

using namespace std;

class String {
    private:
        char *buf;
    public:
        String() {
            cout << "String default constructor" << endl;
            buf = new char[strlen("Shari")+1];
            strcpy(buf, "Shari");
        }
        String(char *s) {
            cout << "String one arg constructor" << endl;
            buf = new char[strlen(s)+1];
            strcpy(buf, s);
        }
        ~String() {
            delete [] buf;
        }
        void printString() {
            cout << buf << endl;
        }
};
```

```
class Employee {
    private:
        String name;
        int id;
    public:
        Employee() : name("Haley"), id(108) {
            cout << "Employee default constructor" << endl;
        }
        Employee(char *s, int num) : name(s), id(num) {
            cout << "Employee two arg constructor" << endl;
        }
        void printEmployee() {
            name.printString();
            cout << id << endl;
        }
};

int main() {
    Employee e("Sally", 123);
    e.printEmployee();
    return 0;
}
```

The output produced now looks like this:

```
String one arg constructor
Employee two arg constructor
Sally
123
```

Note that this constructor does one `String` operation (an initialization), while the original version did two `String` operations, an initialization followed by an assignment.

Even if we were to ignore the efficiency of the constructor initializer lists, on some occasions we'll find that they must be used to properly create an instance of a class. This occurs when the class in question has one of the following:

- a `const` data member

```
class Employee {
    private:
        const long ssnum;
    public:
        Employee(long n) : ssnum(n) {}
};
```

• a reference data member

```
class Employee{
    private:
        const long ssnum;
        int& value;
    public:
        Employee(long n, int val) : ssnum(n), value(val) {}
};
```

Introduction to Inheritance

Inheritance is one of the best things about C++. After classes, inheritance is probably the most powerful feature of object-oriented programming. It's the feature that sets object-oriented programming apart from traditional programming because it allows us to extend and reuse existing code without having to rewrite the code.

So far all the classes we've created have been built from scratch, which is probably the way you've been working when you want to add new features to a program, since it's usually just as easy to write new code as it is to change existing code, debug it, etc. What we want to do is to be able to add new features to a program without changing the original code, and that's what inheritance lets you do!

In C, if you have a library of code and its source, which isn't always the case, you can make changes. But first you have to have a copy of the source code. As you make modifications, you need to understand the internals of the functions that someone else may have written. As you do this, it is a given that you're going to introduce some bugs, which may or may not be easy to find. So you end up with 2 bodies of code

> • the old debugged body

> • and your new body of code which contains the old code in some form, plus additions and changes, and of course new bugs.

Future changes mean more copies and the amount of code you need to maintain increases, but without an equivalent increase in functionality.

Creating new data types or classes is the fundamental activity in C++, since we use types or classes to model the problem being solved. When we think about changing or extending code in C++, it's natural that we think not in terms of data or functions, but in terms of classes. Thus, reusing code in C++ means reusing classes.

It's very common that an existing class that you or someone else has written or debugged doesn't quite do what you want. In C++, you can make a new class from an old one using inheritance. When you inherit, you say:

"This new type or class is just like that old one plus some data and functions."

With inheritance, you don't make a copy of the existing member functions. You just use the ones that are already there:

• you can add new data members and new member functions

• you can also change the meaning of existing functions by redefining them so they have a different behavior for your new class

You don't even need the source code for the member function definitions, just the header files containing the class definitions which must always be available to use the class in any case. This means you can inherit from any vendors class library, whether they provide source code or not!!

Derivation

C++ implements inheritance through the mechanism of derivation. A derived class is derived from a base class and inherits all the data members and member functions of the base class.

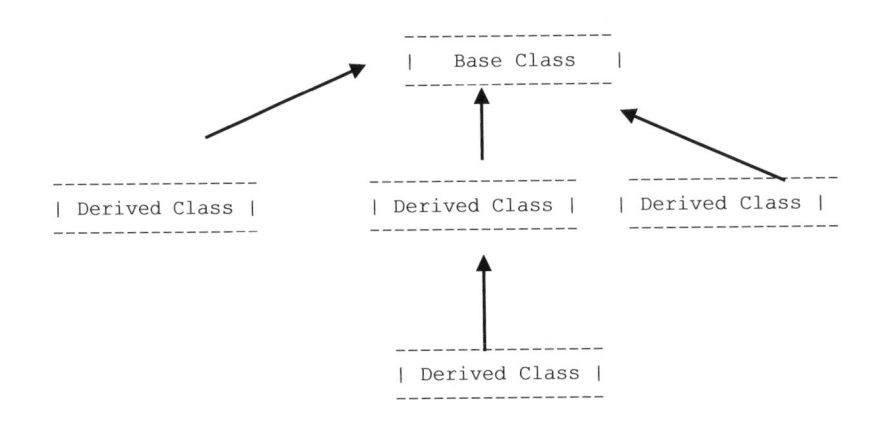

• Many classes might be derived from a single base class.

• A derived class may itself be a base class from which additional classes are derived.

Thus, we can build up a class hierarchy where each class can serve as a parent or root for a new set of classes.

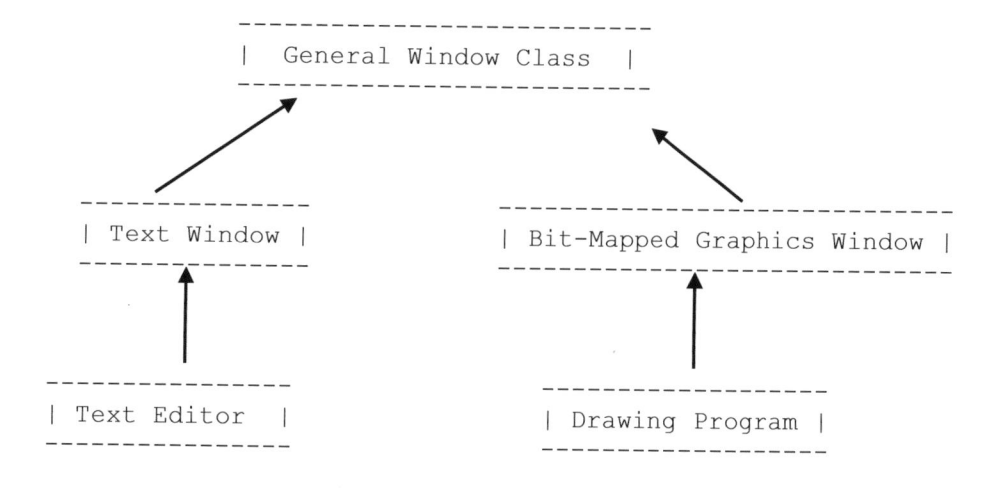

This is all single inheritance because all derived classes possess only one parent each. Note that there is no set limit on the number of classes that may be derived from one another.

So inheritance allows you to collect related classes into a hierarchy with the classes at the top serving as abstractions for those below. This implies that a derived class is a specialization of its parent class, i.e., the derived class "is a" type of base class, but with more detail added. For this reason, the relationship between a derived class and its base class is called an *is-a relationship*.

The base class remains unchanged by this process of inheritance regardless of the number of classes that are derived from it. Therefore a derived class is always at least as big as its parent class so that an instance of the derived class always contains all of the members of its parent class.

Note that you cannot subtract or remove anything from a parent class. However, accessing the inherited members is a different matter we'll look at later, i.e., just because you happen to derive a class does not mean that you are automatically granted complete and unlimited access privileges to the members of the base class.

Inheritance is commonly used in two ways:

• in one role it can be thought of as a *reuse mechanism*, a way to create a new class that strongly resembles an existing one.

• in another role, it can be thought of as an *abstraction mechanism*, a tool to organize classes into hierarchies of specializations

In both cases, you can think of it as a way for one class to subcontract or delegate some of its tasks to another.

Counter Class

Previously we defined a `Counter` class that included a general purpose counter variable, `count`, that could be:

- initialized to 0 with the default constructor
- initialized to some other value with an one-arg constructor
- incremented with the ++ operator
- returned with the `getCount()` operator

The class is shown below:

```
class Counter {
    private:
        int count;
    public:
        Counter()          { count = 0; }
        Counter (int c)  { count = c; }
        int getCount()    { return count; }
        Counter operator++() {
            ++count;
            return *this;
        }
};
```

Suppose we've worked long and hard to make the `Counter` class operate just the way we want and we're pleased with the results, except for one thing. We really need a way to decrement the count. Perhaps we're counting people entering a bank and we want to increment the count when they come in and decrement the count when they go out. So the count represents the number of people in the bank at any moment.

We could insert a decrement routine directly into the source code of the `Counter` class, but there are several reasons why we might not want to do this:

- The `Counter` class works very well and has undergone many hours of testing and debugging. (While this is an exaggeration in this case, it might be true in a larger and more complex class.)

- If we start fooling around with the source code for `Counter`, the testing process will need to be carried out again and we may foul something up and spend hours debugging code that worked fine before we modified it.

• Another reason for not modifying the Counter class is that we might not have access to its source code, especially if it has been distributed as part of a class library.

To avoid these problems, we can use inheritance to create a new class based on Counter, without modifying Counter itself. We can derive a new class, CountDn, which adds a decrement operator to the Counter class:

```
class Counter {
    protected:
        int count;
    public:
        Counter()        { count = 0; }
        Counter (int c)  { count = c; }
        int getCount()   { return count; }
        Counter operator++() {
            ++count;
            return *this;
        }
};

class CountDn : public Counter {
    public:
        CountDn operator--() {
            --count;
            return *this;
        }
};
```

Here's a breakdown of the syntax:

• The private access specifier in the Counter class has been replaced with the protected access specifier. More on this later.

• CountDn is the derived class name

• : specifies that CountDn is derived from Counter

• public refers to the type of inheritance which may be public, private, or protected. We'll look at this at a later time.

• Counter is the name of the base class

```
int main() {
    CountDn c;
    ++c;
    ++c;
    cout << c.getCount() << endl;
    --c;
    cout << c.getCount() << endl;
    return 0;
}
```

The output produced is:

```
2
1
```

Note that the new CountDn class inherits all the features of the Counter class. CountDn doesn't need a constructor, it doesn't need the getCount() function, and it doesn't need the operator++() function because they already exist in the base class Counter. Keep in mind that the public members of a public base class remain public in the derived class.

Accessing Base Class Members

An important part to inheritance is knowing when a member function in the base class can be used by objects of the derived class. This is called *accessibility*. For example, in main(), we created an object of class CountDn:

```
CountDn c;
```

which caused c to be created as an object of class CountDn and initialized to 0. But wait! How is this possible? There is no constructor in the CountDn class specifier. So what entity carries out the initialization?

It turns out that under certain circumstances, if you don't specify a constructor, the derived class will use an appropriate constructor from the base class. Here, there's no constructor in CountDn, so the compiler uses the default constructor from the base class, Count.

Note that object c of the CountDn class also uses the operator++() and getCount() member functions from the Counter class. If the compiler doesn't

find these functions in the class of which `c` is a member, it uses member functions from the base class.

Derived Class Constructors

There's a potential glitch in our program. What if we want to initialize a `CountDn` object to a value? Can the one-arg constructor in `Counter` be used? No it cannot!

Although the compiler will substitute a default constructor from the base class, it draws the line at more complex constructors. Therefore we must write a new set of constructors for the derived class:

```
class CountDn : public Counter {
    public:
        CountDn() : Counter() {}
        CountDn(int c) : Counter(c) {}
        CountDn operator--() {
            --count;
            return *this;
        }
};

int main() {
    CountDn c1, c2(100);
    return 0;
}
```

Here,

```
CountDn() : Counter() {}
```

causes the `CountDn` default constructor to call the `Counter` default constructor in the base class and

```
CountDn(int c) : Counter(c)
```

causes the `CountDn` one-arg constructor to call the `Counter` one-arg constructor in the base class.

Note that the `CountDn` constructor could add additional statements of its own, but in this case it doesn't need to so the braces are empty. But if it defined a new data member, it could initialize it here and then call the base class constructor to initialize the rest of the object. More on this later.

Protected Access Specifier

• Member functions of a class can access both private and public members.

• Objects of a class can access only public members.

• Private members are private.

This is all we need to know if we don't use inheritance. With inheritance, there are new possibilities. The question that concerns us at the moment is this:

Can member functions of the derived class access data members of the base class? For example, can `operator--()` in `CountDn` access `count` in `Counter`?

The answer is that member functions can access members of the base class if the members are `public` or `protected`. They can't access `private` members.

We don't want to make count `public` since that would allow it to be accessed by any function anywhere in the program and eliminate the advantages of data-hiding. A `protected` member, on the other hand, can be accessed by member functions in its own class, or in any class derived from its own class. It can't be accessed from functions outside the class, such as `main()` which is just what we want!

So a `protected` class member is a cross between a `private` and a `public` member.

• Like `private` members, `protected` members are accessible only to other class member functions. Outside of the class, `protected` members are invisible.

• Like `public` members, `protected` members are inherited by the derived class and are accessible to member functions in the derived class.

Keep the following rules in mind when deciding whether to make members
`private`, `protected`, or `public`:

• A `private` member is accessible only to members of the class in which the
`private` member is declared.

• A `protected` member is accessible to members of its own class and to any
members in a derived class.

• A `public` member is accessible to the class's own members, to a derived
class's members, and to all other users of the class.

Access Specifier	Accessible From Own Class	Accessible From Derived Class	Accessible From Objects Outside Class
private	yes	no	no
protected	yes	yes	no
public	yes	yes	yes

The moral is that if you are writing a class that you suspect might be used, at any
point in the future, as a base class for other classes, then any data that the
derived class might need to access should be made `protected` rather than
`private`. This ensures that the class is "inheritance ready", to adapt a phrase
from TV set advertising.

Note that inheritance doesn't work in reverse. The base class and its objects
don't know anything about any classes derived from the base class. For instance,
in the declaration:

```
Counter c;
```

object `c` cannot use the `operator--()` function defined in `CountDn`. If you
want a counter that you can decrement, it must be of class `CountDn` and not
class `Counter`.

Employee Class

Consider the `Employee` base class below:

```cpp
// employee.h

#ifndef _EMPLOYEE_H
#define _EMPLOYEE_H

using namespace std;

class Employee {
    protected:
        char *name;
        char *address;
        int age;
        int dependent;
        int jobClass;
    public:
        Employee(char *, char *, int, int, int);
        ~Employee();
        void display();
        int getAge()        {return age;}
        int getDependent() {return dependent;}
        int getJobClass()   {return jobClass;}
};

#endif


// employee.cpp

#include <iostream>
#include <string.h>
#include "employee.h"

Employee::Employee(char *nm, char *ad, int ag, int dp, int jc) {
    name = new char[strlen(nm)+1];
    strcpy(name, nm);
    address = new char[strlen(ad)+1];
    strcpy(address, ad);
    age = ag;
    dependent = dp;
    jobClass = jc;
}

Employee::~Employee() {
    delete [] name;
    delete [] address;
}
```

```
void Employee::display() {
    cout << name << endl;
    cout << address << endl;
}
```

The base class `Employee` is set up like any other class. However, because we intend that this class have a class derived from it, we put the data members in the `protected` access region because we want the derived class member functions to have direct access to these data. Note that this class has a constructor, a destructor, output functions, and three access functions.

Occasionally very little is done with a base class. We may not even declare any objects of a base class because a base class object is not required to declare a derived object. Base classes can be created to provide data members and member functions that will be shared by a variety of different derived class definitions. The derived class looks like this:

```
// workdata.h

#ifndef _WORKDATA_H
#define _WORKDATA_H

#include "employee.h"

class Workdata : public Employee {
    private:
        char *occupation;
    public:
        Workdata(char *, char *, char *, int, int, int);
        ~Workdata() {delete [] occupation;}
        char* getOccupation() {return occupation;}
        void display();
        int getAge() {return age;}
};

#endif
```

```
// workdata.cpp

#include <iostream>
#include <string.h>
#include "workdata.h"

using namespace std;

Workdata::Workdata(char *nm, char *ad, char *oc, int ag, int dp,
                        int jc) : Employee(nm, ad, ag, dp, jc) {
    occupation = new char[strlen(oc)+1];
    strcpy(occupation, oc);
}

void Workdata::display() {
    Employee::display();
    cout << occupation << endl;
}
```

Class `Workdata` is a derived class that inherits from the public base class `Employee`. The base class for this derived class can either be `private`, `protected`, or `public`. More on this later.

A derived class will inherit all the data members and member functions of its base class and can declare some of its own. The member functions of the derived class can see the `public` and `protected` data members of the base class, but not the `private` data members.

The base class member functions cannot see the data members of any of its derived classes. How could it? The base class has no idea what classes may be derived from it. The derived class can only indirectly access the private data members of the base class by using the `protected` or `public` member functions of the base class.

Note that the derived class has a constructor with 6 arguments, although the class has only declared one data member of its own:

```
char *occupation;
```

One argument is used to give `occupation` a value, while the remaining five arguments are used to initialize the memory inherited from the base class. More on this later when we discuss base class initializations. So the constructor for

`Workdata` is responsible for initializing the base class portion of a derived class object as well as its own data member.

The derived class inherits the ability to use the following functions of the base class:

```
display()
getAge()
getDependent()
getJobClass()
```

In addition, the derived class has added a new function called:

```
getOccupation()
```

and redefines the functions:

```
display()
getAge()
```

for the derived class.

Note that these redefined functions are not examples of function name overloading because both the base class and derived class functions have identical prototypes. The derived class can actually use both `display()` functions and both `getAge()` functions, i.e., those defined in `Employee` and the others defined in `Workdata`.

If a derived class object invokes a `display()` or `getAge()` function, it gets the local version by using the following calls:

```
Workdata obj("Bill", "EB2", "Trainer", 49, 1, 57);
obj.display();
int x = obj.getAge();
```

The derived class can access the member functions of the base class by using the form:

```
int x = obj.getDependent();
```

or by using the scope resolution operator:

```
obj.Employee::display();
```

Note the scope resolution operator is required when functions in base and derived classes have the same name. Also note that a derived class can act as a base class for classes derived from it allowing you to create a hierarchy of derived classes.

Class Hierarchies

In the examples so far, inheritance has been used to add functionality to an existing class. Let's look at an example where inheritance is used for a different purpose, as part of the original design of a program.

Our example models a database of employees of a widget company. In this company, there are three kinds of employees:

- Managers – manage other employees
- Scientists - perform research to develop better widgets
- Laborers - operate the dangerous widget-stamping presses.

The database stores a name and an employee id number for all employees, no matter what category they are.

- For Managers, it also stores their title and golf club dues.
- For Scientists it stores the number of scholarly articles they have published.
- Laborers need no additional data beyond their names and id numbers.

So our program design will start with a base class `Employee` that handles the employee's last name and id number. From this class, three other classes will be derived: `Manager`, `Scientist`, and `Laborer`.

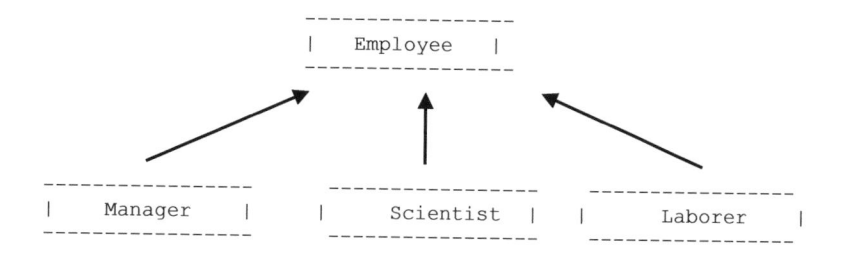

The `Manager` and `Scientist` classes contain additional information about these categories of `Employee` as well as member functions to handle this information.

```
// hier.cpp

#include <iostream>

class Employee {
    private:
        char name[81];
        long id;
    public:
        void getData() {
            cout << "Enter last name: "; cin >> name;
            cout << "Enter ID: "; cin >> id;
        }
        void putData() {
            cout << "Name: " << name << endl;
            cout << "ID:   " << id   << endl;
        }
};

class Manager : public Employee {
    private:
        char title[81];
        double dues;
    public:
        void getData() {
            Employee::getData();
            cout << "Title: " ; cin >> title;
            cout << "Golf Club Dues: " ; cin >> dues;
        }
        void putData() {
            Employee::putData();
            cout << "Title: " << title << endl;
            cout << "Golf Club dues: " << dues << endl;
        }
};

class Scientist : public Employee {
    private:
        int pubs;
    public:
        void getData() {
            Employee::getData();
            cout << "Number of Publications: " ; cin >> pubs;
        }
        void putData() {
            Employee::putData();
            cout << "Number of Publications: " << pubs << endl;
        }
};

class Laborer : public Employee {
};
```

Note that the `Laborer` class operates identically to the `Employee` class since it contains no additional data or functions. It may seem that the `Laborer` class is unnecessary, but by making it a separate class we emphasize that all classes are derived from the same source, `Employee`.

Another reason for creating an apparently empty `Laborer` class concerns the issues of maintenance. In the future, if we decide to modify the `Laborer` class, we would not need to change the specifier for `Employee`.

Classes used only for deriving other classes, i.e. `Employee`, are sometimes called *abstract classes*, meaning that no actual instances (objects) of this class are created. The term abstract has a more precise definition connected with polymorphism and virtual functions that we'll come across later.

Levels of Inheritance

Classes can be derived from classes that are themselves derived classes. Consider the class hierarchy below:

```
class A {
};

class B : public A {
};

class C : public B {
};
```

Here, class `B` is derived from class `A`, and class `C` is derived from class `B`. This process can be extended to an arbitrary number of levels.

In this way, we can extend our `Employee` class program to add a special kind of laborer called a foreman. Since a foreman is a kind of laborer, the `Foreman` class will be derived from the `Laborer` class.

Foremen oversee the widget-stamping operation, supervising groups of laborers. They are responsible for the widget production quota for their group. A foreman's ability is measured by the percentage of production quotas successfully met. So we'll have a quota data item in the `Foreman` class that represents this percentage:

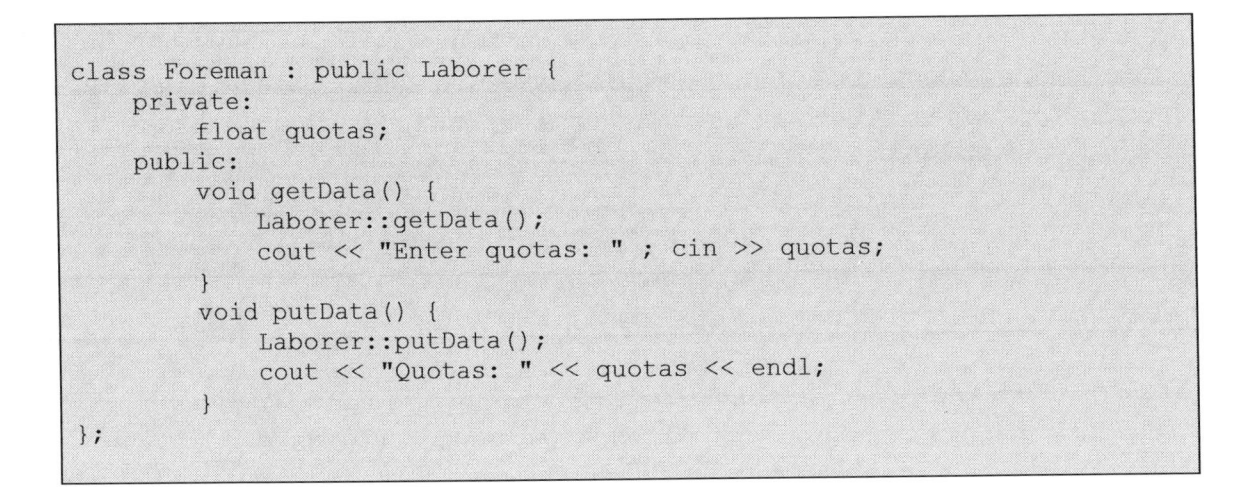

```
class Foreman : public Laborer {
    private:
        float quotas;
    public:
        void getData() {
            Laborer::getData();
            cout << "Enter quotas: " ; cin >> quotas;
        }
        void putData() {
            Laborer::putData();
            cout << "Quotas: " << quotas << endl;
        }
};
```

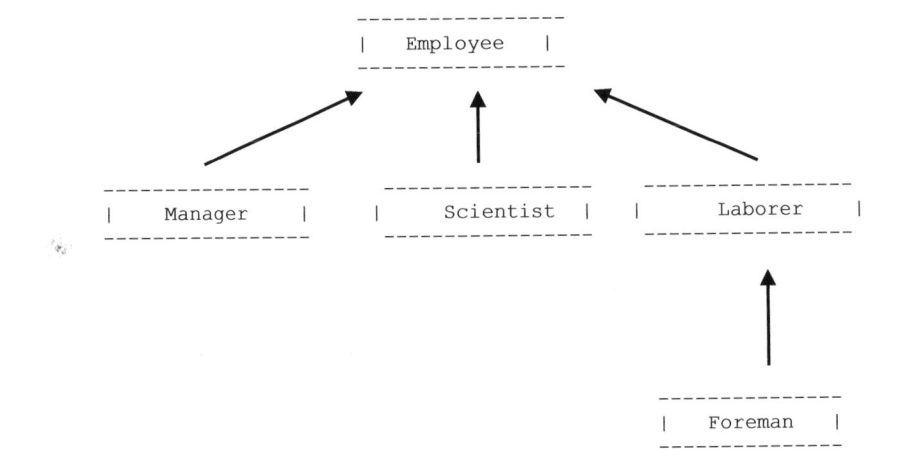

Notice that a class hierarchy is not the same as an organizational chart. An organization chart shows lines of commands. A class hierarchy results from generalizing common characteristics. The more general the class, the higher it is on the chart! Thus, a Laborer is more general than a Foreman, who is a specialized kind of Laborer. So Laborer is shown above Foreman in the class hierarchy.

Accessing Mechanisms

Accessing Base Class Members Using Derived Class Instance

An instance of a derived class has complete access to the public members (usually functions) of the base class. Assuming that a member with the same name does not exist within the scope of the derived class, the member from the base class will automatically be used.

```
// derived.cpp

class Base {
    protected:
        int num;
    public:
        Base (int n = 0) : num(n) {}
        int getNum() const { return num; }
};

class Derived : public Base {
};

int main() {
    Derived d;

    cout << d.getNum();        // first checks class Derived
                               // and then class Base
    cout << d.Base::getNum();  // goes directly to class Base
    return 0;
};
```

In the first call, the compiler will look at the scope of class `Derived` to see if a function called `getNumber()` exists. If not, it will then look at the scope of the class `Base`. In the second call, the compiler will look directly at the scope of the `Base` class for `getNum()`.

Override Base Class Data Member Names

In addition to creating new members whose names are different than those inherited from the base class, a derived class may selectively override or hide existing names. This is done by declaring a member in the derived class with exactly the same name as that of a member from the base class.

That is, if an instance or member function of the derived class refers to a data member of the base class that has been overridden, the member from the derived class will be used by default. Since the derived class contains its base

class, we can use the scope resolution operator to refer to the member of the base class.

```cpp
// base.cpp

class Base {
    protected:
        int num;
    public:
        Base (int n = 0) : num(n) {}
        int getNum() { return num; }
};

class Derived : public Base {
    private:
        int num;
    public:
        Derived(int n = 1) : num(n) {}
        int getNum() { return num + 1; }
};

int main() {
    Derived d;

    cout << d.getNum() << endl;
    cout << d.Base::getNum() << endl;
    return 0;
};
```

Overriding Member Functions

We can also override base class member functions. We can use member functions in a derived class that have the same name as those in the base class. You might want to do this so that calls in your program work the same way for objects of both base and derived classes.

Consider the following `Stack` class:

```cpp
// stack.cpp

const int MAX = 100;

class Stack {
    protected:
        int s[MAX];
        int top;
    public:
        Stack() { top = -1; }
        void push(int num) { s[++top] = num; }
        int pop() { return s[top--]; }
};
```

There is a potential flaw here. There is no check for overflow or underflow.

To cure these defects, we'll create a new class, Stack2, derived from Stack. Objects of Stack2 behave in exactly the same way as those of Stack except that you will be warned if you attempt to push too many items on the stack, or if you try to pop an item from an empty stack.

```cpp
class Stack2 : public Stack {
    public:
        void push(int num) {
            if (top < MAX-1)
                Stack::push(num);
            else {
                cout << "Stack overflow. ";
                exit(1);
            }
        }
        int pop() {
            if (top > 0)
                return Stack::pop();
            else {
                cout << "Stack underflow. ";
                exit(1);
            }
        }
};
```

```
int main() {
    Stack2 s2;

    s2.push(5);
    s2.push(6);
    s2.push(7);
    cout << s2.pop();
    cout << s2.pop();
    cout << s2.pop();
    cout << s2.pop();
    return 0;
}
```

Note that functions `push()` and `pop()` in `Stack2` have the same names, arguments, and return types are functions in class `Stack`. When we call these functions in `main()`, how does the compiler know which of the two functions to use?

Rule: When the same function exists in both the base class and the derived class, the function in the derived class will be executed. The derived class functions override the base class functions.

Hiding Overloaded Functions

You cannot overload (or hide) a base class function by redefining it in a derived class with a different argument list. The rationale behind this design is that if you override a function inherited from the base class, there should be no danger of accidentally calling the base class version through the process of argument matching.

In other words, if a base class function expects an `int` as an argument, and you hide it by redefining it in a derived class to accept a `char*`, then calling the function using a derived instance with an `int` actual argument should cause a compilation error.

```
// hide1.cpp

class Base {
    public:
        void foo(int n) { cout << n; }
};

class Derived : public Base {
    public:
        void foo(double n) { cout << n * n; }
};

int main() {
    Derived d;

    d.foo(5);
    return 0;
}
```

This shows that a function inherited from the base class is not examined under the rules of argument matching. It appears that the function call using an `int` as the actual argument would be a better match for the `Base` class function that expects an `int`, rather than the `Derived` class function that expects a `double`. But the `Derived` class function is the one that gets called because it is the only one that the compiler knows about.

Even if there is no conceivable match within the scope of the `Derived` class, the inherited members from the `Base` class will not be examined.

```
// hide2.cpp
class Base {
    public:
        void foo(char *ptr) { cout << ptr; }
};

class Derived : public Base {
    public:
        void foo(double n) { cout << n; }
};

int main() {
   Derived d;

   d.foo("A");            // error - cannot convert char * to double
   d.Base::foo("A");      // ok
   return 0;
}
```

Compiler Lookup Rules

Using derived instances to access member functions without the base class name and scope resolution operator, the compiler resolves the address by asking:

• Does the member function have scope in the derived class? If so, use it. If not, generate a compilation error.

• Does the member function have scope in the parent class? If so, use it. If not, generate a compilation error.

• Does the member function have scope higher up in the hierarchy? If so, use it. If not, generate a compilation error.

Note that functions with the same name at file scope are not considered.

Similarly, a function in a derived class resolves the address of a data member by asking:

• Does the data member have scope in the function class itself? If so, use it. If not, generate a compilation error.

• Does the data member have derived class scope? If so, use it. If not, generate a compilation error.

• Does the data member have scope in the parent class that is accessible? If so, use it. If not, generate a compilation error.

• Does the data member have accessible scope up the hierarchy? If so, use it. If not, generate a compilation error.

• Does the data member exist at file scope? If so, use it. If not, generate a compilation error.

Constructors and Destructors with Derived Classes

Recall that an instance of a derived class always contains the data members inherited from its base class. Therefore it logically follows that whenever a derived class instance gets created, two constructor functions are needed to ensure the proper initialization of all of the data members of the instance.

So the constructor for the base class must somehow get control to do the initialization of base class members, and the constructor for the derived class must get control to initialize any new derived class data members. Fortunately, the compiler takes care of this problem, and will automatically call both constructors.

This happens because as soon as the derived class constructor gets control and establishes its formal arguments, the base class constructor will be called immediately. The only question now is, which base class constructor will be called?

The answer is this: If you fail to tell the compiler which one to call, it will attempt to call the default constructor. Therefore, it's always a good idea to have a default constructor present in a base class.

On the other hand, you may explicitly tell the compiler which base class constructor to call by using the syntax in the derived class's *constructor initialization list*. The constructor initialization list appears after the derived constructor header. It is preceded by a colon, and contains:

• the base class name (which is the name of the constructor function)

• a list of arguments between parentheses, as in any normal function call

To explicitly invoke the base class default constructor, we leave the argument list empty.

To repeat:

Because the initialization of any new derived class members may be dependent upon the proper initialization of the base class members, the base class constructor would always get called first.

As you would expect, the destructor functions will get called in reverse order, i.e., derived class first, and base class last.

Below is a program to show how an argument can be passed into a base class constructor function and traces the order of the calls to the constructor and destructor functions of both classes.

```cpp
// constructor.cpp

#include <iostream.h>

class Base {
    protected:
        int bNum;
    public:
        Base(int n = 0) : bNum(n) {
            cout << "Base constructor." << endl;
        }
        ~Base() {
            cout << "Base destructor." << endl;
        }
        int getBaseNum() {
            return bNum;
        }
};

class Derived : public Base {
    private:
        int dNum;
    public:
        Derived(int a = 0, int b = 0) : Base(a), dNum(b) {
            cout << "Derived constructor." << endl;
        }
        ~Derived() {
            cout << "Derived destructor." << endl;
        }
        int getDerivedNum() const {
            return dNum;
        }
};

int main() {
    Derived d(1,2);

    cout << d.getBaseNum() << " " << d.getDerivedNum() << endl;
    return 0;
}
```

Here's the program's output:

```
base constructor
derived constructor
1 2
derived destructor
base destructor
```

Note that the derived instance d is being created with two integers:

• the first of these integers, a, is needed by the base class constructor to initialize bNum
• the second integer, b, is needed by the derived class in order to initialize dNum

Caution:

When you initialize base class data members, do not attempt to list them individually in the member initialization list of the derived class constructor. Their initialization is not the responsibility of the derived class, and will cause a compilation error because the only items that can legally be written in a derived class initialization list are the derived class data member names and base class constructor names.

Note that the base class constructor and destructor functions are unique in the sense that they are not inherited by the derived class. If you create a derived class instance, then the proper constructor to handle that particular type of creation must exist within the derived class, not the base class.

Constructor Initializer List

One of the easiest ways to improve a C++ program's efficiency is to specify an appropriate base class constructor function in the colon list of a derived class constructor function. What happens without a constructor initialization list?

• To initialize an object of a particular class, your program calls the appropriate constructor for that class.

• To initialize a derived class's base object, your program implicitly calls the appropriate base class constructor function.

Unfortunately, that implicit call to the base class constructor function makes it easy to initialize the base class's data members twice! See the example below:

```cpp
// shape1.cpp

#include <iostream>

uses namespace std;

class Shape {
    private:
        int color;
        int sideCount;
    public:
        Shape() {
            cout << "Shape default constructor" << endl;
            SetColor(0);
            SetSideCount(0);
        }
        Shape(int c, int sc) {
            cout << "Shape 2 arg constructor" << endl;
            SetColor(c);
            SetSideCount(sc);
        }
        void display() {
            cout << "color = " << color << endl;
            cout << "sides = " << sideCount << endl;
        }
        void SetColor(int c) {
            cout << "set color to " << c << endl;
            color = c;
        }
        void SetSideCount(int sc) {
            cout << "set side count to " << sc << endl;
            sideCount = sc;
        }
};
```

```cpp
class Square : public Shape {
    public:
        Square() {
            cout << "Square default constructor" << endl;
            SetColor(0);
            SetSideCount(4);
        }
        Square(int c) {
            cout << "Square one arg constructor " << endl;
            SetColor(c);
            SetSideCount(4);
        }
};

int main() {
    Square s(10);

    s.display();
    return 0;
}
```

We derived the Square class from the Shape class to show how the compiler calls constructors for embedded classes.

```
Output:
Shape default constructor
Set Color to 0
Set Side Count to 0
Square integer constructor
Set Color to 10
Set Sidr Count to 4
Color = 10
Sides = 4
```

Note that the program doesn't act as you'd expect! It contains two Square (derived) class constructors which initialize the color and side data members of the Shape (base) class. But the Square (derived) class constructors initialize these data members a second time.

How are these data members first initialized? Remember, that derived class constructors execute the base class constructor. Therefore, the constructors for both the Square (derived) and the Shape (base) classes initialize Shape's data members, color and side.

To correct the extra calls to the initializing functions, we can simply use a colon list in the `Square` (derived) class constructor declaration. By specifying the proper `Shape` (base) class constructor, the `Square` (derived) constructor allows the `Shape` (base) class constructor to perform the initializations.

```cpp
// shape2.cpp

#include <iostream.h>

class Shape {
    private:
        int color;
        int sideCount;
    public:
        Shape() {
            cout << "Shape default constructor" << endl;
            SetColor(0);
            SetSideCount(0);
        }
        Shape(int c, int sc) {
            cout << "Shape 2 arg constructor " << endl;
            SetColor(c);
            SetSideCount(sc);
        }
        void display() {
            cout << "color = " << color << endl;
            cout << "sides = " << sideCount << endl;
        }
        void SetColor(int c) {
            cout << "set color to " << c << endl;
            color = c;
        }
        void SetSideCount(int sc) {
            cout << "set side count to " << sc << endl;
            sideCount = sc;
        }
};

class Square : public Shape {
    public:
        Square() : Shape(0, 4) {
            cout << "Square default constructor" << endl;
        }
        Square(int c) : Shape(c, 4) {
            cout << "Square one arg constructor" << endl;
        }
};
```

```
int main() {
    Square s(10);
    s.display();
    return 0;
}
```

We only had to specify the appropriate Shape (base) class constructor in the colon list of the Square (derived) class constructor. The Square (derived) class now specifies which Shape (base) class to call. By doing so, the Square class allows the Shape class constructor to initialize its data members.

```
output:
Shape 2 arg constructor
Set Color to 10
Set Side Count to 4
Square one arg constructor
Color = 10
Sides = 4
```

Base and Derived Class Destructors

A derived class typically provides a constructor if its base class has one, and a base class constructor must be called in order to initialize any data members in the base class. Destructors, on the other hand, do not require such strict handling. A derived class needs to supply a destructor only if the derived class has any members that require deleting when an object of the derived class goes out of scope.

```
// dtor.cpp

#include <iostream>
#include <string.h>

uses namespace std;

class Base {
    private:
        char *buf;
    public:
        Base(char *s) {
            buf = new char[strlen(s) + 1];
            strcpy(buf, s);
        }
        ~Base() {
            delete [] buf;
        }
        char *getStr() {
            return buf;
        }
};
```

Note that the constructor copies a string argument to the heap, assigning the address of the allocated memory to `buf`. The destructor deletes this memory block when an object of type `Base` goes out of scope. Consider what happens in a new class derived from `Base`:

```
class Derived : public Base {
    private:
        char *upper;
    public:
        Derived(char *s) : Base(s) {
            upper = new char[strlen(s) + 1];
            strcpy(upper, strupr(s));
        }
        ~Derived() {
            delete [] upper;
        }
        char *getUpper() {
            return upper;
        }
};

int main() {
    Derived d("Haley Laurel");

    cout << d.getStr() << endl;
    cout << d.getUpper() << endl;
    return 0;
}

output:
Haley Laurel
HALEY LAUREL
```

Note that the `Derived` constructor calls the `Base` constructor, which copies the string addressed by s to the heap. The `Derived` class constructor makes another copy of the string, and converts its characters to upper case, and assigns to uppercase the address of the copied string. If you now create an object of type `Derived`, you create two copies of a string, one changed and one in uppercase.

A derived class object must carefully delete any memory that it allocates. When the object goes out of scope, its destructor must delete this memory and return it to the heap for reuse. The `Base` class destructor must also delete the memory addressed by buf.

Unlike constructors, a destructor in a derived class does not explicitly call the base class destructor. C++ calls destructors automatically when objects of a class go out of scope. When `Derived` object d goes out of scope, C++ first calls the `Derived` class destructor, and then calls the `Base` class destructor. Note that the two strings are deleted in the reverse order in which they were constructed.

Overloading Copy Constructors

The initialization of a derived instance with an existing derived instance automatically invokes the copy constructor function in the derived class. Remember, if you fail to write one, the compiler will automatically supply one for you.

Note that even if you write a copy constructor for the base class, and someone derives a new class from that base class and writes a copy constructor for the derived class, then a potential problem exists because the compiler will not automatically invoke the copy constructor in the base class when the copy constructor in the derived class is invoked. Instead, the default copy constructor in the base class will get called!

The solution to this problem is to make an explicit call from the derived class copy constructor to the base class copy constructor. You do this using the constructor initialization list to write the name of the base class copy constructor and, within parentheses, the same argument that the derived class copy constructor received.

```
// copyctor.cpp

#include <iostream>

uses namespace std;

class Base {
    protected:
        int bnum;
    public:
        Base(int x=0) : bnum(x) {
            cout << "base constructor" << endl;
        }
        Base(const Base& b) : bnum(b.bnum) {
            cout << "base copy constructor" << endl;
        }
        ~Base() {
            cout << "base destructor" << endl;
        }
        int getBnum() const {
            return bnum;
        }
};
```

```
class Derived : public Base {
    private:
        int dnum;
    public:
        Derived(int x=0, int y=0) : Base(x), dnum(y) {
            cout << "derived constructor" << endl;
        }
        Derived(const Derived& d) : Base(d), dnum(d.dnum) {
            cout << "derived copy constructor" << endl;
        }
        ~Derived() {
            cout << "derived destructor" << endl;
        }
        int getDnum() const {
            return dnum;
        }
};

int main() {
    Derived d1(1, 2);
    Derived d2 = d1;      // or Derived d2(d1);

    cout << "d2 = " << d2.getBnum() << ", " << d2.getDnum() << endl;
    return 0;
}

output:
base constructor
derived constructor
base copy constructor
derived copy constructor
d2 = 1 2
derived destructor
base destructor
derived destructor
base destructor
```

While making the call to the base class copy constructor, note that object d is simply an alias for object d1 (the existing instance), and that a cast occurs from a *reference-to-derived* into a *reference-to-base*. This is perfectly acceptable and normal code since the formal argument b in the base class copy constructor now is an alias for only the base class portion of the d1 instance.

Overloading the Assignment Operator

The situation for the overloaded assignment operator in a derived class works in a fashion similar to that of the overloaded copy constructor. When you need to write an overloaded `operator=()` function in a derived class, it is your responsibility to make an explicit call to the overloaded `operator=()` function in the base class.

Unlike overloading a copy constructor function, you have complete control over when this call is made, although it makes sense to do it immediately so that the base portion of the derived object gets its data members assigned first, followed by the derived portion.

Remember, if the base class does not have an overloaded assignment operator, the compiler will supply one (that we may not like).

```cpp
// assign.cpp

#include <iostream>

uses namespace std;

class Base {
    protected:
        int bnum;
    public:
        Base(int x=0) : bnum(x) {
            cout << "base constructor" << endl;
        }
        Base(const Base& b) : bnum(b.bnum) {
            cout << "base copy constructor" << endl;
        }
        Base& operator=(const Base& b) {
            cout << "base operator=()" << endl;
            if (&b != this)
                bnum = b.bnum;
            return *this;
        }
        ~Base() {
            cout << "base destructor" << endl;
        }
        int getBnum() const {
            return bnum;
        }
};
```

```
class Derived : public Base {
    private:
        int dnum;
    public:
        Derived(int x=0, int y=0) : Base(x), dnum(y) {
            cout << "derived constructor" << endl;
        }
        Derived(const Derived& d) : Base(d), dnum(d.dnum) {
            cout << "derived copy constructor" << endl;
        }
        Derived& operator=(const Derived& d) {
            cout << "derived operator=()" << endl;
            if (&d != this) {
                Base::operator=(d);
                dnum = d.dnum;
            }
            return *this;
        }
        ~Derived() {
            cout << "derived destructor" << endl;
        }
        int getDnum() const {
            return dnum;
        }
};

int main() {
    Derived d1(1, 2);
    Derived d2;

    d2 = d1;
    cout << "d2 = " << d2.getBnum() << ", " << d2.getDnum() << endl;
    return 0;
}
```

output:
```
base constructor
derived constructor
base constructor
derived constructor
derived operator=()
base operator=()
d2 = 1 2
derived destructor
base destructor
derived destructor
base destructor
```

Note that the explicit call to the functions `Base::operator=()` was written using functional notation. If you want to use infix notation, you may use either of the following two methods:

- `(base)this = d;`

- `((base&)*this) = d;`

The first statement casts the `Derived` pointer into a `Base` pointer, and then dereferences the result to produce the `Base` class portion of the invoking instance. Therefore, the `Base` class `operator=()` will be called.

The second statement achieved the same result, but in a more direct fashion, with the invoking instance itself now cast into a reference to the `Base` class.

Casting from Derived into Base

Casting from a derived class instance, pointer, or reference, into a base class object of the same type presents no problem to the compiler. We've seen previously how a reference-to-derived in the derived class copy constructor was passed into the base class copy constructor and received as a reference-to-base.

Since a derived class instance always contains a base class instance, when the cast occurs the compiler simply ignores the derived class portion of the instance that is being used, pointer at, or referred to. For example:

```
Derived d(1, 2);
```

creates an object of the derived class.

```
Base b = d;
```

creates an object of the base class. Here, however, a derived instance is used to initialize a base instance. Object b now only contains the Base class portion of object d.

```
Base *pbase = new Derived;
```

In this example, new Derived returns a pointer to a Derived object and is used to initialize a pointer to a Base object. In other words, a pointer to Derived is used to initialize a pointer to Base. So pBase "sees" only the Base class portion of the Derived object on the heap.

Remember, since a Derived class instance always contains a Base class instance, when the cast occurs, the compiler simply ignores the Derived class portion of the instance that is being pointed at.

Casting from Base into Derived

On the other hand, if you attempt to cast from a base class instance, pointer, or reference into a derived class object of the same type, the compiler will complain

because what you are trying to do now is put a smaller object (the base class) into a larger object (the derived class). The compiler will not let you do this automatically. We'll not worry about this at this time.

Pointers to Base and Derived Class Instances

As we've already seen, pointers to base and derived class instances can be declared just like any other type of pointer. In terms of assignment, you can always assign a derived class pointer into a base class pointer, because then the base class pointer will be pointing (only) to the base class portion of the derived object.

- A base class pointer can always point to a base class instance,
- A derived class pointer can always point to a derived class instance.

When a pointer of type `Base` or `Derived` class is used to call a member function, and the function has been defined in both the base and derived classes, there is no ambiguity as to which version of the function will be called. The governing factor here is the type of the pointer itself.

```
Class Base {
    public:
        void print() { cout << "Base class print()" << endl; }
};

class Derived : public Base {
    public:
        void print() { cout << "Derived class print()" << endl; }
};

int main() {
    Base* pBase = new Base;              // base class instance and pointer
    Derived* pDerived = new Derived; // derived class instance & pointer

    pBase->print();             // calls Base::print()
    pDerived->print()           // calls Derived::print()
    pDerived->Base::print();    // calls Base::print()
    delete pBase;
    delete pDerived;
    return 0;
}

output:
Base class print
Derived class print
Base class print
```

Note the call `pBase->Derived::print()` is not allowed.
Two pointers are declared and each one is used to call the `print()` function in its respective class. In addition, the `Derived` class pointer may be used to call the `print()` function in the `Base` class provided that the function name is qualified with the `Base` class name using the scope resolution operator.

Base Class Pointer Pointing to a Derived Class Object

Things become a lot more interesting when a base class pointer is used to store the address of a derived class object. This is no problem for the compiler because the pointer is, in reality, just pointing to the base class portion of the derived instance. (Remember a derived class instance always contains a base class instance).

Now suppose that the base class pointer, which points to a derived class object, is used to invoke a member function that is defined in the `Base` class, and overridden in the `Derived` class. Which function will be called?

The answer is that the member function of the `Base` class will be used, even though the pointer may be pointing to a `Derived` class instance!

Why?

Because the compiler must generate an offset address for the function member at compilation time, even though the address of the instance stored in the pointer may be changed at execution time. Therefore, the compiler must say:

" I know that the type of the pointer is base class so I will generate an address relative to the start of some base class instance."

"If at execution time, the instance to which the pointer points happens to be of type `Derived` class, that's too bad because I will still call the function in the `Base` class and not the overridden function that is in the derived class."

```
Class Base {
    public:
        void print() { cout << "Base Class Print()" << endl; }
};

class Derived : public Base {
    public:
        void print() { cout << "Derived class Print()" << endl; }
};
```

```
int main() {
    Base* pBase = new Derived;
    pBase->print();  // calls Base::print()
    delete pBase;
    return 0;
}

output:
Base class print()
```

If this result troubles you, then your instincts are correct, because it serves no useful purpose to be "locked" into the base class world, even though the base class pointer can legitimately point to a variety of derived class objects.

This is especially troubling if you consider any kind of container object, such as an array, linked list, queue, stack, tree, etc. that might hold base class pointers as its data.

For example, suppose we are running a used car lot and we have declared a base class `Vehicle` and have derived multiple classes from the `Vehicle` class (`Lexus, Ford, BMW`). So we've also created an array of `Base` class pointers, i.e., pointers to `Vehicle` objects, and have initialized each base class pointer to point to a dynamically created derived object (`Lexus, Ford, BMW`). So our inventory consists of a collection of "pointers" to the various cars on our lot. Someone then asks us to provide an inventory of all the cars present. So we then proceed to scan our collection of pointers and send some kind of identify yourself message to each car, rightfully expecting that each car will identify itself in some unique fashion. Unfortunately, as things now stand, the only type of response that we'll get back from all cars is, in effect, `I AM A VEHICLE`. The message only gets sent to the base class `identify()` function and never reaches the `identify()` functions in the derived classes. Not very useful!!

How we are going to free ourselves from this situation is the topic of polymorphism and virtual functions.

1. Create two classes, A and B, with default constructors that announce themselves. Inherit a new class called C from A, and create a member object of B in C, but do not create a constructor for C. In main(), create an object of class C and explain the results.

2. Create a three-level hierarchy of classes with default constructors, along with destructors, both of which announce themselves to **cout**. Verify that for an object of the most derived type, all three constructors and destructors are automatically called. Explain the order in which the calls are made.

3. As part of a military video game, a designer has created a vehicle class hierarchy. The base class is Vehicle, and it has AirVehicle, LandVehicle, and SeaVehicle as derived classes. The class SeaPlane inherits from both AirVehicle and SeaVehicle. Specifically, what **design issues** had to be considered in developing the SeaPlane class?

4. Without executing the code below, give the output for the program.

```
class A {
    protected:
        int fa;
    public:
        A(int va = 0) : fa(va) {
            cout << "A constructor" << endl;
        }
        A(const A& va) : fa(va.fa) {
            cout << "A copy constructor " << endl;
        }
        ~A() {
            cout << "A destructor" << endl;
        }
        A& operator=(const A& va) {
            cout << "A = operator" << endl;
            fa = va.fa;
            return *this;
        }
};
```

```cpp
class B : public A {
    protected:
        int fb;
    public:
        B(int va = 0, int vb = 0) : fb(vb), A(va) {
            cout << "B constructor " << endl;
        }
        B(const B& vb) : fb(vb.fb), A(vb) {
            cout << "B copy constructor " << endl;
        }
        ~B() {
            cout << "B destructor" << endl;
        }
        B& operator=(const B& vb) {
            cout << "B = operator" << endl;
            A::operator=(vb);
            fb = vb.fb;
            return *this;
        }
};

int main() {
    B b1(1, 2);
    B b2;
    b2 = b1;
    B b3(b1);
}
```

❖ Inheritance Lab (String Class)

One of the most popular classes to define in C++ is the one that emulates a string literal in C. The reason is that a string is not a built-in type in C, and yet it is used all the time. Thus, to perform the various manipulations on strings, you must call upon functions in the library `string.h` that perform operations such as copy, duplicate, compare, etc. In C++, this need is obviated by the use of function and operator overloading. In order to represent the state of a `String` object, you will use two data members:

• the length, `length`, of the string as an `int`

• a pointer, `buf`, to the string literal of type `char*`

A pointer to a character (`char*`) is used instead of a fixed-length character array so that memory can be allocated at run-time (via the operator `new`) and, therefore, have no restrictions placed upon the length of the string (too big or too small). The length field is not strictly mandatory since it can always be obtained from the string pointer via the `strlen()` function. However, it's probably a good idea to have it to avoid repeated calls to `strlen()`.

Overloaded Constructor Functions

The creation of each `String` instance involves allocation of space on the heap, copying the string literal, and setting the length data member accordingly. Because there are many different ways in which a `String` object can be initialized, the constructor functions for the `String` class must be overloaded to accommodate all of the various situations.

First, if a `String` object is created with no explicit initialization, then the default constructor (the one that can be called with no actual arguments) will be called. It makes sense that we should have the invoking instance point to an empty string literal whose length is zero.

```
String();
```

Second, because there is such a close relationship between a string literal (which is type `char*`) and the `String` class, it should be easy to create a `String` object based upon that literal. In this case a separate copy of the literal must be made so that it can be encapsulated as part of the class.

```
String(const char*);
```

Third, it might be useful to be able to create a `String` object from a single character.

```
String(char);
```

Fourth, here is a constructor that creates a `String` object from an `int` value. This implies allocating `int` bytes from the heap, and setting the first byte to the `NULL` character. Note that a value less than zero should be interpreted as zero.

```
String(int);
```

Recall from our discussion on copy constructors that the compiler automatically supplies one for you if you fail to write one. This is fine if a class contains no pointer data members, but will cause a big problem if at least one pointer data member is present. The reason for this is that two or more pointer members may not point to the same space on the heap, and this is exactly what would happen if you used the default copy constructor. Therefore, it is now necessary to create one for the `String` class.

```
String(const String&);
```

Finally, it might be useful to create and initialize a `String` object with a repeated character `int` number of times.

```
String (char, int);
```

Destructor Function
When the `String` object goes out of scope, the destructor function must release the space on the heap. Otherwise, we will have a severe case of "memory leakage".

```
~String();
```

Assigning One String Instance To Another String Instance
Another function that we need in the `String` class is one that can assign one `String` instance to another instance (every C programmers dream!). In order to do this successfully, you must overload the assignment operator for the `String` class. Why is this necessary since the compiler automatically supplies a default assignment operator that is used to copy all of the data members? For much the same reason that we had to write an explicit copy constructor, i.e., without one the address in the pointer would get copied into the corresponding pointer data member of the receiving instance and, as a result, the pointer would then be pointing to the same space on the heap. This could cause trouble when the destructor function attempts to release this space. Therefore, the default assignment operator that the compiler supplies in this situation obviously is not good enough and therefore needs to be overloaded. Remember, this must be done as a binary member function.

```
String& operator=(const String&);
```

Implicit Type Conversion
In addition to assigning one `String` instance to another `String` instance, it makes sense to have the capability to assign a string literal `(char*)` to a `String` instance.

```
String& operator=(const char*);
```

String Concatenation

The next step in the process of overloading functions for the `String` class is to add two `Strings` together. For our purposes, we will assume that this implies a `String` concatenation operation. Since addition typically involves the binary + operator, it seems natural to overload it for the `String` class. There are, however, some design considerations to be made here.

First, should the function be a member or a friend? The better choice would be to use a friend function. This has the advantage of allowing the left-hand argument to be a string literal (type `char*`) or a character (type `char`) so that an implicit type conversion can occur. In addition, if you coded it as a member function, then one of the `String` instances would have to be the invoking instance, and the other would have to be the explicit argument. This doesn't make sense because the arguments in an add operation are, in essence, equal. Therefore, both of them should be explicit arguments, thereby implying a friend function.

Second, should the function literally emulate what `strcat()` does and modify the first argument? In this case the better answer would be no, because normally an add operation implies that neither argument is modified. Instead, they are simply combined to produce a brand new object. Here is one way to do it.

```
friend String operator+(const String&, const String&);
```

Note that the return type of the function is a `String` object by value. A reference cannot be used because you should never make an alias to a local (automatic) object, which is what the result of the function really is. Also, this gives us a chance to use the constructor taking an `int`.

You can also overload the `operator+()` four more times in order to achieve an exact match under all possible concatenation circumstances.

```
friend String operator+(const String&, const char*);
friend String operator+(const char*, const String&);
friend String operator+(const String&, char);
friend String operator+(char, const String&);
```

String Concatenation - Another Approach

You know that in C the operator += has the effect of adding the two operands together and then storing the results into the left-hand argument. If your intent in doing String concatenation is to modify the left-hand argument, then this operator would be a natural choice for you to use. Obviously, you must allocate new space on the heap for the resulting literal to avoid an overflow situation.

```
String& operator+=(const String&);
```

Overloading An Operator As Both Unary And Binary

To demonstrate an operator being overloaded in both its unary and binary modes, you'll overload the + operator as a unary member operator. Of course, what the operator actually does is completely arbitrary, so let's just assume that its purpose is to ensure that all letters of the string literal are upper case. Note that this function does not modify the String object itself, but instead returns a new String object that contains the modification. This is somewhat consistent with how unary operators behave. If you really do want to modify the invoking object, then it's just a matter of storing the result back into the invoking object.

```
String operator+() const;
```

Comparing Two String Instances

The next operations you might want to perform on instances of the String class would be to compare them against each other. This implies overloading the six relational operators. All comparisons will be case sensitive, although you could easily make them non-case sensitive. Once again, note that the functions are all binary friends to allow the left-hand operand (which could be type char* or type char) to be implicitly cast into a temporary String object.

```
friend int operator==(const String&, const String&);
friend int operator!=(const String&, const String&);
friend int operator< (const String&, const String&);
friend int operator<=(const String&, const String&);
friend int operator> (const String&, const String&);
friend int operator>=(const String&, const String&);
```

Overloading `Operator[]()`

As you know, in C the operator `[]` is typically used to represent indexing. That is, the value between the square brackets specifies which element of the designated array is to be referenced. Of course, if you were to attempt to apply the bracket operator to an instance of some user-defined class, the compiler would complain because it has no idea what you are trying to do. However, you may give it meaning for your class by overloading it. Typically this would be done for a class that contains an array or a pointer as a data member. The big advantage here is that having your own function allows you to do a validity check on the index value. Recall that `operator[]()` must be a binary member (not friend) function of the class. In other words, there must be one explicit argument which is the index value itself. In the case of the `String` class, the `[]` operator function will be used to fetch a particular character from the array of characters.

```
char& operator[](int);
```

Note that the character is being returned by reference. The validity of the index value should be checked. If it turns out to be invalid, the function should output an error message and return a `NULL` value.

Bracket vs. Pointer Notation

The use of brackets in C is just a convenience for you that avoids your having to use pointer dereferencing when accessing an array element. In other words, for some array `s`, the following three constructs should all yield the same result since they are equivalent:

```
s[i]          *(s + i)          *(i + s)
```

Thus, it is incumbent upon you to allow the user of the `String` class to be able to write code using the latter two notations. As things stand now, the compiler would complain because it has no idea how to add a `String` object and an `int`. Therefore, the + operator must be overloaded again to accommodate both syntax forms.

```
friend char* operator+(const String&, int);
friend char* operator+(int, const String&);
```

Overloading the Increment and Decrement Operators

In C, the operators ++ (increment by 1) and -- (decrement by 1) are unique in that they may appear on either side of the variable to which they apply. If the operator appears on the left, it implies prefix notation while on the right it implies postfix notation.

In older C++ compilers, the distinction between prefix and postfix notation for user-defined types was lost. That is, for the `String` class and an instance called s, the expressions s++ and ++s cannot be distinguished from each other because the compiler will generate exactly the same code for each one. However with more modern compilers, any postfix usage automatically generates an extra argument consisting of an integer 0. The actual value of this integer, of course, is meaningless, so that a formal argument name does not need to be written when the function is defined. Nevertheless, this extra argument allows both prefix and postfix versions of the operator to be present.

```
• String& operator++();       // prefix form - add 1 to each character
                              // in the string literal pointed at by
                              // buf and return the resulting String

• String& operator--();       // prefix form - subtract 1 from each
                              // character in the string literal
                              // pointed at by buf and return the
                              // resulting String

• String operator++(int);     // postfix form - add 1 to each character
                              // in the string literal pointed at by
                              // buf and return the original String

• String operator--(int);     // postfix form - subtract 1 from each
                              // character in the string literal
                              // pointed at by buf and return the
                              // original String
```

Return String Length

It will be useful in this exercise to define a function to return the length of the string literal pointed to by `buf` (excluding the null character).

```
int getLength();
```

Substring Function

It is often useful to be able to access substrings of `String` objects. We'll need to extract and return the substring of a `String` object starting at a particular index in the buffer and of a specified length.

```
String substr(int, int);
```

Output Function

In order to display the state of any `String` object, we have to write a print function. This function outputs the literal itself between double quotation marks and outputs the length of the string literal on the same line.

```
void print();
```

It is also useful to overload the `operator<<()` function to output just the contents of the string pointer itself.

```
friend ostream& operator<<(ostream&, const String&);
```

Class Inheritance

Inheritance is the process by which a derived class is created as the result of extending, or enhancing, an existing class. Stated differently, the derived class is inherited from a parent (or base) class. This capability saves you from having to "reinvent the wheel", because if a new class greatly resembles an existing class, then all of the properties of the existing class can be inherited into the new class, and any new members can then be added.

Inheritance allows you to collect related classes into a hierarchy with the classes at the top serving as abstractions for those below. This implies that a derived class is a specialization of its parent class. In other words, the derived class is a type of base class, but with more detail added. For this reason, the relationship between a derived class and its base class is called an is-a relationship.

We've seen how a `String` class could be created and then utilize the capabilities of operator function overloading to make the class interface more

intuitive. Now suppose you were to put this class into a library and announce its existence so that users could take advantage of it. But someone then says to you, "The class is fine, except that I meant to overload the unary ~ as a member function so that it reverses the sequence of letters in the String and returns a new instance of the class. What should I do?". Your answer in this case would be for the user to derive a new class from the String class and add the new function. Then, when instantiations are made from the newly created class, all of the capabilities of the original class will still be there, plus the added capability of the new function. Since this newly created class is the String class with more features, the new class really is a kind of String class.

Write a class called ReverseString that is publicly derived from the String class, with the following member functions:

• a default constructor function

• a copy constructor function

• a constructor function that takes a const char* as its one argument

• an overloaded assignment operator

• an overloaded unary operator~() function that reverses the letters of the string itself, and returns a new ReverseString instance by value (note that the actual contents of the string literal pointed to by buf are never modified)

Write a class called CaseString, that is publicly derived from the String class, with the following data members:

• a pointer, upper, of type char*, which holds the contents of the string literal in upper case

• a pointer, lower, of type char*, which holds the contents of the string literal in lower case

---

Also include the following member functions in the `CaseString` class:

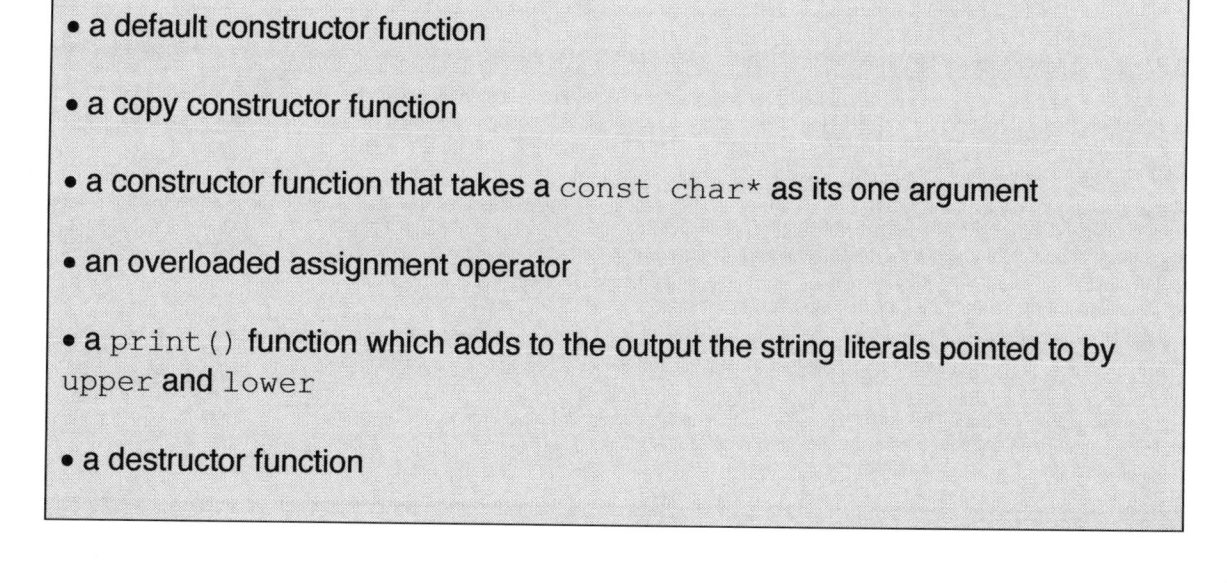

- a default constructor function

- a copy constructor function

- a constructor function that takes a `const char*` as its one argument

- an overloaded assignment operator

- a `print()` function which adds to the output the string literals pointed to by `upper` and `lower`

- a destructor function

Use the `StringDriver` code provided to test your classes.

```cpp
// StringDriver.h

#ifndef _StringDriver_H
#define _StringDriver_H

#include "String.h"
#include "ReverseString.h"
#include "CaseString.h"

int main();
void test1();
void test2();
void test3();
void test4();
void test5();
void test6();
void test7();
void test8();
void test9();
void test10();
void test11();
void test12();
void test13();
void test14();
void test15();
void test16();
void test17();
void test18();
void test19();
void test20();
void wait();

#endif


// StringDriver.cpp

#include <iostream>
#include <fstream>
#include <stdlib.h>
#include "StringDriver.h"

using namespace std;

ofstream csis;
```

```
int main() {
    csis.open("csis.txt");
    test1();
    test2();
    test3();
    test4();
    test5();
    test6();
    test7();
    test8();
    test9();
    test10();
    test11();
    test12();
    test13();
    test14();
    test15();
    test16();
    test17();
    test18();
    test19();
    test20();
    csis.close();
}

void test1() {
    cout << "1. Testing S1: String default constructor."
        << endl << endl;
    csis << "1. Testing S1: String default constructor."
        << endl << endl;
    String s1;
    s1.print();
    wait();
}

void test2() {
    cout << "2. Testing S2: String one arg (char *) constructor."
        << endl << endl;
    csis << "2. Testing S2: String one arg (char *) constructor."
        << endl << endl;
    String s2("ABC");
    s2.print();
    wait();
}
```

```
void test3() {
    cout << "3. Testing S3: String one arg (char) constructor."
        << endl << endl;
    csis << "3. Testing S3: String one arg (char) constructor."
        << endl << endl;
    String s3('Z');
    s3.print();
    wait();
}


void test4() {
    cout << "4. Testing S4: String one arg (int) constructor."
        << endl << endl;
    csis << "4. Testing S4: String one arg (int) constructor."
        << endl << endl;
    String s4(10);
    s4.print();
    wait();
}

void test5() {
    cout << "5. Testing S5, T5: String copy constructor."
        << endl << endl;
    csis << "5. Testing S5, T5: String copy constructor."
        << endl << endl;
    String s5("Purple Rain");
    s5.print();
    String t5(s5);
    t5.print();
    wait();
}

void test6() {
    cout << "6. Testing S6: String two arg (char, int) constructor."
        << endl << endl;
    csis << "6. Testing S6: String two arg (char, int) constructor."
        << endl << endl;
    String s6('*', 10);
    s6.print();
    wait();
}
```

```
void test7() {
    cout << "7. Testing S7, T7, U7: String assignment."
        << endl << endl;
    csis << "7. Testing S7, T7, U7: String assignment."
        << endl << endl;
    String s7("Sally Ride"), t7, u7;
    t7 = u7 = s7;
    s7.print();
    t7.print();
    u7.print();
    wait();
}

void test8() {
    cout << "8. Testing S8: String assignment."
        << endl << endl;
    csis << "8. Testing S8: String assignment."
        << endl << endl;
    String s8("ABC");
    s8 = s8;
    s8.print();
    wait();
}

void test9() {
    cout << "9. Testing S9: Implicit type conversion."
        << endl << endl;
    csis << "9. Testing S9: Implicit type conversion."
        << endl << endl;
    String s9;
    s9 = "ABC";
    s9.print();
    wait();
}

void test10() {
    cout << "10. Testing S10, T10, U10: String concatenation."
        << endl << endl;
    csis << "10. Testing S10, T10, U10: String concatenation."
        << endl << endl;
    String s10("DEF");
    String t10('H');
      String u10("ABC" + s10 + "G" + t10 + 'I');
    u10.print();
      String v10('X' + u10);
    v10.print();
    wait();
}
```

```
void test11() {
    cout << "11. Testing S11, T11: String concatenation."
        << endl << endl;
    csis << "11. Testing S11, T11: String concatenation."
        << endl << endl;
    String s11('A');
    String t11("BC");
    s11 += s11 += t11 += 'D';
    s11.print();
    t11.print();
    wait();
}

void test12() {
    cout << "12. Testing S12, T12: String unary operator."
        << endl << endl;
    csis << "12. Testing S12, T12: String unary operator."
        << endl << endl;
    String s12("Unary +");
    String t12(+s12);
    s12.print();
    t12.print();
    s12 = +s12;
    s12.print();
    wait();
}

void test13() {
    cout << "13. Testing S13, T13: String comparison operators."
        << endl << endl;
    csis << "13. Testing S13, T13: String comparison operators."
        << endl << endl;
    String s13("ABC"), t13("ABCD");
    s13.print();
    t13.print();
    cout << endl;
    cout << "== " << (s13 == t13 ? "True" : "False") << endl;
    cout << "!= " << (s13 != t13 ? "True" : "False") << endl;
    cout << "<  " << (s13 <  t13 ? "True" : "False") << endl;
    cout << "<= " << (s13 <= t13 ? "True" : "False") << endl;
    cout << ">  " << (s13 >  t13 ? "True" : "False") << endl;
    cout << ">= " << (s13 >= t13 ? "True" : "False") << endl;
    csis << endl;
    csis << "== " << (s13 == t13 ? "True" : "False") << endl;
    csis << "!= " << (s13 != t13 ? "True" : "False") << endl;
    csis << "<  " << (s13 <  t13 ? "True" : "False") << endl;
    csis << "<= " << (s13 <= t13 ? "True" : "False") << endl;
    csis << ">  " << (s13 >  t13 ? "True" : "False") << endl;
    csis << ">= " << (s13 >= t13 ? "True" : "False") << endl;
    wait();
}
```

```
void test14() {
    cout << "14. Testing S14: Overloaded subscript operator."
        << endl << endl;
    csis << "14. Testing S14: Overloaded subscript operator."
        << endl << endl;
    String s14("C++ is fun.");
    for (int i = -1; i <= s14.getLength(); i++) {
        char& ch = s14[i];
        if (ch != '\0')
            ++ch;
    }
    s14.print();
    wait();
}

void test15() {
    cout << "15. Testing S15: Pointer notation."
        << endl << endl;
    csis << "15. Testing S15: Pointer notation."
        << endl << endl;
    String s15("ABCDE");
    for(int i = 0; i < s15.getLength(); i++)
        ++(*(s15+i));
    for (int j = 0; j < s15.getLength(); j++) {
        cout << *(j + s15);
        csis << *(j + s15);
    }
    cout << endl;
    csis << endl;
    wait();
}
```

```
void test16() {
    cout << "16. Testing S16, T16, U16, V16, W16, X16, Y16, Z16: "
        <<  "Increment and decrement operators."
        << endl << endl;
    csis << "16. Testing S16, T16, U16, V16, W16, X16, Y16, Z16: "
        << "Increment and decrement operators."
        << endl << endl;
    String s16("ABC");
    String t16(++s16);
    s16.print();
    t16.print();

    String u16("ABC");
    String v16(u16++);
    u16.print();
    v16.print();

    String w16("ABC");
    String x16(--w16);
    w16.print();
    x16.print();

    String y16("ABC");
    String z16(y16--);
    y16.print();
    z16.print();
    wait();
}

void test17() {
    cout << "17. Testing S17, T17: Substr function."
        << endl << endl;
    csis << "17. Testing S17, T17: Substr function."
        << endl << endl;
    String s17("All You Need Is Love"), t17;
    t17 = s17.substr(4, 8);
    s17.print();
    t17.print();
    wait();
}
```

```
void test18() {
    cout << "18. Testing S18, T18: Output function."
        << endl << endl;
    csis << "18. Testing S18, T18: Output function."
        << endl << endl;
    String s18("Red-");
    String t18("Green-");
    String u18("Blue");
    cout << s18 << t18 << u18;
    csis << s18 << t18 << u18;
    cout << endl;
    csis << endl;
    wait();
}

void test19() {
    cout << "19. Testing S19, T19, U19: ReverseString class."
        << endl << endl;
    csis << "19. Testing S19, T19, U19: ReverseString class."
        << endl << endl;
    ReverseString s19("Computer");
    ReverseString t19;
    t19 = ~s19;
    s19.print();
    t19.print();
    ReverseString u19(~~s19);
    u19.print();
    wait();
}

void test20() {
    cout << "20. Testing S20, T20, U20: CaseString class."
        << endl << endl;
    csis << "20. Testing S20, T20, U20: CaseString class."
        << endl << endl;
    CaseString s20("BaLLooN");
    CaseString t20;
    t20 = s20;
    s20.print();
    t20.print();
    CaseString u20(s20);
    u20.print();
    wait();
}

void wait() {
    char buf;

    cout << endl << "Press any key to continue." << endl;
    csis << endl << endl;
    cin.get(buf);
}
```

Chapter 14. Polymorphism and Virtual Functions

Introduction to Polymorphism
Static and Dynamic Binding
Virtual Functions
Abstract Classes
Virtual Destructors
Virtual Function Mechanism
How to Leak Memory in C++
Polymorphism Homework
Polymorphism Lab (Library Holdings)

Introduction to Polymorphism

For any computer language to call itself an object-oriented programming language, it must contain three essential components:

· *Data Abstraction and Encapsulation*

This is the process of defining some new data type or class. This involves specifying the internal representation of that type along with the functions used to manipulate the type. C++ supports data abstraction and encapsulation through the class mechanism.

· *Inheritance*

Inheritance involves the creation of a new type from an existing type in some hierarchical fashion. C++ supports inheritance through class derivation.

· *Polymorphism and Dynamic Binding*

Polymorphism comes from the Greek words: poly, or many, and morphos, or form. It refers to the quality of having more than one form, of having many shapes. In the context of object-oriented programming, polymorphism describes the capability of C++ code of behaving differently depending on runtime situations. It refers to the ability to have different objects derived from a common class respond to the same message differently. A single operation can have different behavior in different objects. C++ supports polymorphism and dynamic binding through virtual functions.

Static and Dynamic Binding

In most traditional programming languages the compiler calls fixed function identifiers based on the source code. The linker then takes these identifiers and replaces them with a physical address.

This is called *early or static binding* because function identifiers are associated with physical addresses before runtime during the process of compilation and linkage. The problem with early binding is that the programmer must predict what objects will be used in all function calls in every situation. This is not only limiting, but its sometimes impossible.

The best thing about early binding is that it is fast. The only runtime overhead incurred is in:

• passing arguments
• performing a function call
• and cleaning up the stack

In C++, a polymorphic function can become associated with one of many possible functions only when an actual object is passed to the polymorphic function. In other words, the source code itself doesn't always tell you how a section of code is going to execute. A function call is only indicated in the source code without specifying the exact function to call. This is called *late or dynamic binding.*

The problem with late binding is its runtime efficiency. The code itself must deduce at runtime which function to invoke, and then invoke it. Some languages such as Smalltalk use late binding exclusively, which results in a powerful language but one that suffers from certain speed penalties. On the other hand, ANSI C uses early binding exclusively, resulting in high speed but a lack of flexibility.

C++ makes polymorphism possible with the application of a rule we've seen before:

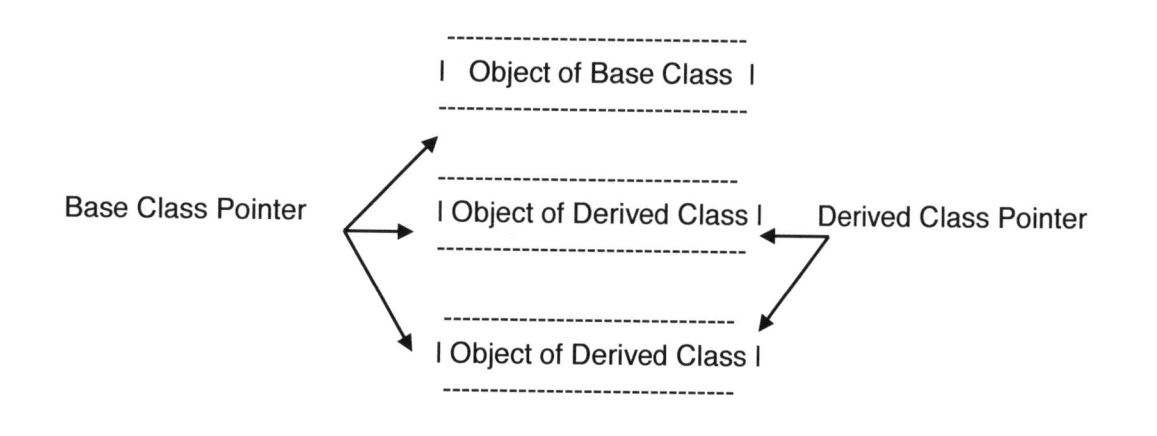

A base class pointer may address objects of the base class or of any derived class. However, this rule is a one-way street. *A pointer to a derived class object may not address an object of a base class.*

Recall, however, that this presented some difficulties. When we used pointers of type `Base` class, the only member function (or data) that could be referenced were those belonging to the `Base` class, even if this pointer pointed to a `Derived` class object!

This is not polymorphism because the object being pointed at never received the message being sent! The reason is that the compiler only knows about the type of the pointer itself, and nothing about the object that the pointer will eventually point to during execution time.

The goal of an object-oriented programming language is to send a message to whatever object happens to be pointed at by the pointer, as opposed to the type of the pointer itself.

Why is this important? Because then you can control the content (or address) of the pointer at execution time, whereas the type of the pointer variable itself cannot be changed once it is hard-coded. In this fashion, the object itself can "figure out" what to do with the message and thereby implement the concept of polymorphism.

Virtual Functions

Accessing Member Functions

Let's see what happens when a base class and derived classes all have functions with the same name, and we access these functions using pointers, but without virtual functions. Given the following class:

```cpp
class Base {
    protected:
        int value;
    public:
        Base (int x)    { value = x;    }
        int getValue() { return value; }
};
```

We can now declare and use a `Base` class object:

```cpp
int main() {
    Base b(10);

    cout << "value = " << b.getValue() << endl;
    return 0;
}
```

The above code generates the output:

```
value = 10
```

We will now derive a class from `Base` that multiplies the `Base` class value by a specified amount:

```cpp
class Derived : public Base {
    protected:
        int mult;
    public:
        Derived(int x, int y) : Base(x), mult(y) {}
        int getValue() { return value * mult; }
};
```

We can now declare and use a `Derived` class object:

```
int main() {
    Derived d(10, 2);

    cout << "value = " << d.getValue() << endl;
    return 0;
}
```

Note that the `getValue()` member function from the `Derived` class gets called here and returns the product of the two values used to initialize the object. The above code generates the output:

```
value = 20
```

Accessing Member Function with Pointers

Let's now declare a `Base` class pointer and call the `getValue()` member function for the addressed object:

```
int main() {
    Base *bptr;

    bptr = new Base(10);
    cout << "value = " << bptr->getValue() << endl;
    return 0;
}
```

The output generated by the above code is:

```
value = 10
```

which is as expected.

We can now declare a pointer to a `Derived` class object and call the replacement `getValue()` member function defined in the `Derived` class.

```
int main() {
    Derived *dptr;

    dptr = new Derived(10, 2);
    cout << "value = " << dptr->getValue() << endl;
    return 0;
}
```

The output generated by the above code is:

```
value = 20
```

which is also as expected.

Applying the C++ rule about pointers and derived classes, we can also declare a pointer to the `Base` class and assign the address of a `Derived` class object to that pointer:

```
int main() {
    Base *bptr;

    bptr = new Derived(10, 2);
    cout << "value = " << bptr->getValue() << endl;
    return 0;
}
```

Note that `bptr` was declared to address an object of class `Base` and we created a `Derived` object and assigned the object's address to `bptr`. But look at the output generated by this code:

```
value = 10
```

The code compiles but it does not produce the expected result which is:

```
value = 20
```

As far as C++ knows, `bptr` addresses a `Base` class object. Therefore, C++ calls the `getValue()` member function in the `Base` class! *The C++ rule is that it's the type of the pointer itself that determines which member function gets invoked. The type of the object being pointed to is never even taken into account!*

Virtual Functions

But we really want C++ to call the `getValue()` member function for the `Derived` object that the pointer actually addresses. In order for this to happen, we must declare `getValue()` as a *virtual member function*. Calls to virtual functions are linked at runtime (late binding). Virtual member functions and late binding make it possible for objects to determine their own behavior at runtime, which is the chief characteristic of polymorphism. In terms of C++, what is needed is a method to delay the choice of which member function gets executed until execution time.

Having such a method would imply that by using a single pointer and member function name, the same function call (message) can be sent to the object that has its address stored in the pointer. If these objects implement the same member function differently, then different results can be obtained! Such a method is called late or dynamic binding because the decision as to which member function to invoke is no longer made at compile time but is made at runtime. The C++ rule is as follows:

"If a function in a base class definition is declared to be virtual, and is declared exactly the same way (including the return type) in one or more derived classes, then all calls to that function using pointers or references of type `Base` class will invoke the function that is specified by the object being pointed at, and not by the type of the pointer itself."

The following code demonstrates this rule:

```
// virtual1.cpp

#include <iostream>
using namespace std;

class Base {
    protected:
        int value;
    public:
        Base (int x) { value = x; }
        virtual int getValue() { return value; }
};
```

```
class Derived : public Base {
    protected:
        int mult;
    public:
        Derived(int x, int y) : Base(x), mult(y) {}
        virtual int getValue() { return value * mult; }
};


int main() {
    Base *bptr;

    bptr = new Base(10);
    cout << "value = " << bptr->getValue() << endl;
    delete bptr;

    bptr = new Derived(10, 2);
    cout << "value = " << bptr->getValue() << endl;
    delete bptr;
    return 0;
}
```

The keyword `virtual` informs the compiler that calls to `getValue()` are to be linked at runtime. The addresses of the two `getValue()` virtual member functions are stored in an internal table. When statements call virtual member functions, C++ looks up the correct function address from this table. We'll talk more about this later. But if you think this lookup action takes time, you are correct. Using virtual member functions might reduce the program's performance (two extra lookups) although in practice serious slowdowns are rare.

The output produced by the above program is:

```
value = 10
value = 20
```

Now compare the two output statements above. They are exactly identical! So how can they produce different output? We now have enough information to answer this question. The addressed objects determine for themselves which of the two virtual `getValue()` member functions to call. The decision is made by the program at runtime and not at compile time.

Note that once we declare a function to be `virtual` in a class, any subclass that declares the same function with the same prototype is also considered a `virtual` function. Keep in mind that late binding makes sense in C++ only for

objects that are part of a class hierarchy. Declaring a function `virtual` for a class that is not used as a base class is syntactically correct but results in unnecessary runtime overhead.

It may be helpful to think of polymorphism as a response to a message. In classical object-oriented programming terms, an expression such as:

```
bptr->getValue()
```

sends a message named `getValue()` to the object addressed by `bptr`. Upon receiving this message, the object responds in some predetermined way. More importantly, that same message could be sent to other objects that determine their own responses or they might ignore the message. Regardless of the type of object that `bptr` addresses, the object itself responds in its programmed fashion to the `getValue()` message.

Here's another example:

```cpp
// virtual2.cpp

#include <iostream>
using namespace std;

class Base {
    public:
        void show() { cout << "Base" << endl; }
};

class Derived1 : public Base {
    public:
        void show() { cout << "Derived1" << endl; }
};

class Derived2 : public Base {
    public:
        void show() { cout << "Derived2" << endl; }
};
```

```
int main() {
    Derived1 dv1;
    Derived2 dv2;
    Base *bptr;

    bptr = &dv1;
    bptr->show();

    bptr = &dv2;
    bptr->show();
    return 0;
}
```

Now, when we execute the lines:

```
bptr = &dv1;
bptr->show();
```

which member function gets called:

```
Base::show()
```

or

```
Derived1::show()
```

Or when we execute the lines:

```
bptr = &dv2;
bptr->show();
```

which member function gets called:

```
Base::show()
```

or

```
Derived2::show()
```

In each case the member function that gets invoked is:

```
Base::show()
```

producing the following output:

```
Base
Base
```

The compiler simply ignores the contents of the pointer, `bptr`, and chooses the member function that matches the type of the pointer itself.

We can again change this behavior by placing the keyword `virtual` in front of the declaration for the `show()` member function in the `Base` class, and optionally in the derived classes.

```cpp
// virtual3.cpp

#include <iostream>

using namespace std;

class Base {
    public:
        virtual void show() { cout << "Base" << endl; }
};

class Derived1 : public Base {
    public:
        virtual void show() { cout << "Derived1" << endl; }
};

class Derived2 : public Base {
    public:
        virtual void show() { cout << "Derived2" << endl; }
};
```

```
int main() {
    Derived1 dv1;
    Derived2 dv2;
    Base *bptr;

    bptr = &dv1;
    bptr->show();

    bptr = &dv2;
    bptr->show();
    return 0;
}
```

The output generated is now:

```
Derived1
Derived2
```

Now the member functions of the derived classes and not the `Base` class are executed. So the function call:

```
bptr->show();
```

executes different functions depending on the contents of the pointer itself. Once again, the rule is that the function called is based on the contents of the pointer rather than on the type of the pointer, as before.

Pure Virtual Functions
In our previous example, `Base::show()` is never executed. This is a common situation. There is often no need for the base class version of a particular function. We only use the versions of the function in the derived classes. When this is true, the body of the virtual function in the base class can be removed and the notation =0 added to the function declaration:

```
class Base {
    public:
        virtual void show() = 0;
};
```

The = sign has nothing to do with assignment and the value 0 is not assigned to anything. The =0 syntax is simply how we tell the compiler that a function will be pure, i.e., will have no body. This is called a *pure virtual function*.

```
// pvirtual.cpp

#include <iostream>

using namespace std;

class Base {
    public:
        virtual void show() = 0;
};

class Derived1 : public Base {
    public:
        virtual void show() {
            cout << "Derived1" << endl;
        }
};


class Derived2 : public Base {
    public:
        virtual void show() {
            cout << "Derived2" << endl;
        }
};

int main() {
    Base *list[2];
    Derived1 dv1;
    Derived2 dv2;
    list[0] = &dv1;
    list[1] = &dv2;
    for (int i=0; i<2; i++)
        list[i]->show();
}
```

The output produced is:

```
Derived1
Derived2
```

You might wonder if we can remove the body of the virtual function in the base class, why can't we remove the function altogether! That would be even cleaner, but it wouldn't work. Without a function `show()` in the `Base` class, statements like:

```
list[0]->show();
```

would not be valid because the pointers in the `list[]` array must point to members of class `Base`.

Here's another example.

```cpp
// student.cpp

#include <iostream>

using namespace std;

class Person {
    protected:
        char name[40];
    public:
        void setName() {
            cout << "Enter name: ";
            cin >> name;
        }
        void getName() {
            cout << "Name is: " << name << endl;
        }
        virtual int isOutstanding() = 0;
};
```

```cpp
class Student : public Person {
    private:
        float gpa;
    public:
        void setGpa() {
            cout << "Enter students gpa: ";
            cin >> gpa;
        }
        virtual int isOutstanding() {
            return gpa > 3.5;
        }
};

class Professor : public Person {
    private:
        int numPubs;
    public:
        void setNumPubs() {
            cout << "Enter number publications ";
            cin >> numPubs;
        }
        virtual int isOutstanding() {
            return numPubs > 100;
        }
};
```

```
int main() {
    Person *persPtr[100];
    Student *studPtr;
    Professor *profPtr;
    int n=0;
    char choice;

    do {
        cout << "Enter student or professor (s/p): ";
        cin >> choice;
        if (choice == 's') {
            studPtr = new Student;
            studPtr->setName();
            studPtr->setGpa();
            persPtr[n++] = studPtr;
        }
        else if (choice == 'p') {
            profPtr = new Professor;
            profPtr->setName();
            profPtr->setNumPubs();
            persPtr[n++] = profPtr;
        }
        else
            cout << "Invalid choice." << endl;
        cout << "Enter another (y/n): ";
        cin >> choice;
    }
    while (choice == 'y');

    for (int i=0; i<n; i++) {
        persPtr[i]->getName();
        if (persPtr[i]->isOutstanding())
            cout << "This person is outstanding!" << endl;
    }
}
```

Note that we have only one call to `isOutstanding()` in `main()` and it will dynamically bind to the correct version at runtime. The decision about which version of the `isOutstanding()` member function to use is delayed until runtime. It has to be delayed since the array of pointers to `Person` objects won't have any pointer values until the program is run, and it's these pointer values which determine which version is ultimately called.

Abstract Classes

While it is true that all objects are represented by a class, the converse is not true. All classes do not necessarily represent objects. It is possible, and often desirable, for a class to be insufficient to completely represent an object. Such classes are called *abstract classes*. Like an abstract idea, an abstract class describes an unrealized concept. It's a kind of schematic from which you are expected to build one or more derived classes.

An abstract base class is generally used to declare an interface without declaring a full set of implementations for that interface. That interface specifies the abstract operations supported by all objects derived from that class and it is up to the derived classes to supply implementations for those abstract operations.

So an abstract class is a class that does not fully represent an object, but represents a broad range of different classes of objects. However, it's important to realize that this representation extends only to the features those classes of objects have in common. Thus, an abstract class provides only a partial description of its objects.

Because abstract classes do not fully represent an object, they cannot be instantiated. At first it may seem a little odd that a class is incapable of having any instances, but everyday life is full of such classes. For example, we may describe a particular animal as belonging to the class of all Mammals, but we will never see an instance of the class Mammal. Every animal belonging to the class Mammal must also belong to a class that is subordinate to a Mammal, such as a Mouse, Dog, Human, or Platypus. That is because the class Mammal does not fully represent any animal.

In object-oriented design, we will never see an instance of an abstract class, unless it is also an instance of a subordinate class. Again, this is because the abstract class does not fully represent any object. An abstract class is written with the expectation that its subclasses will add to its structure and behavior.

In C++, any class that declares or inherits a pure virtual function is an abstract class, and exists only to act as a base class for other classes. In fact, it is illegal in C++ to create an instance or an object of an abstract class!

Abstract Base Class Example

Consider two classes, Window and Door:

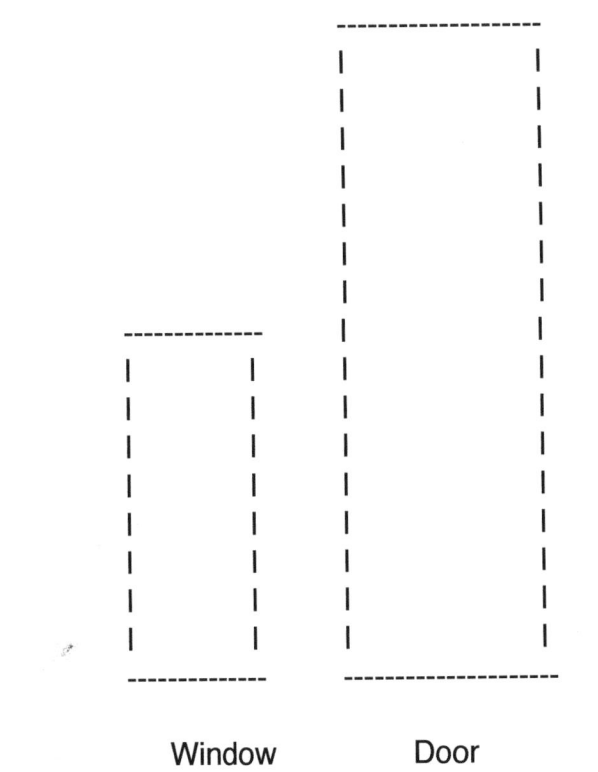

Window   Door

At first glance these classes may not seem to have anything to do with each other. But both share some interesting common features:

- They are both holes in the wall.
- They both have particular locations and sizes with respect to the wall
- They both may exist in one of 3 states: open, closed, locked
- They can be sent similar messages: open, close, lock

The code below shows the declarations of the two classes:

```
// window.cpp

class Window {
    public:
        enum windowState {opened, closed, locked};
    private:
        windowState state;
        int xPos, yPos, width, height;
    public:
        void open();
        void close();
        void lock();
        void getState() const;
        void getXpos()  const;
        void getYpos()  const;
        int getHeight() const;
        int getWidth()  const;
};

class Door {
    public:
        enum doorState {opened, closed, locked};
    private:
        doorState state;
        int xPos, yPos, width, height;
    public:
        void open();
        void close();
        void lock();
        void getState() const;
        void getXpos()  const;
        void getYpos()  const;
        int getHeight() const;
        int getWidth()  const;
};
```

These two classes are horribly redundant and cry for some form of unification. That unification is supplied by creating an abstract base class. We can describe both classes as inheriting from a common abstract class, Portal, which contains all the common features of their interface.

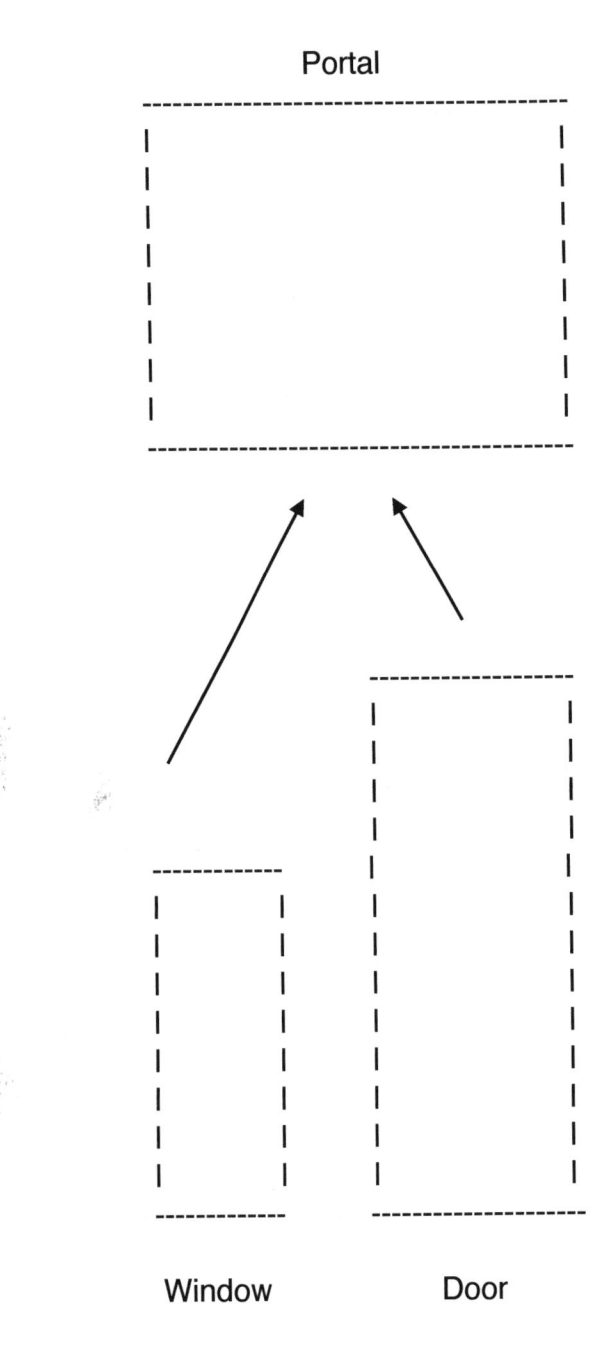

Portal

Window   Door

The code below shows the declarations of the `Portal` abstract base class, along with the `Window` and `Door` derived classes:

```cpp
// portal.cpp

class Portal {
    public:
        enum portalState {opened, closed, locked};
    private:
        portalState state;
        int xPos, yPos, width, height;
    public:
        virtual void open()  = 0;
        virtual void close() = 0;
        virtual void lock()  = 0;
        void getState() const;
        void getXpos()  const;
        void getYpos()  const;
        int getHeight() const;
        int getWidth()  const;
};

class Window : public Portal {
    public:
        virtual void open();
        virtual void close();
        virtual void lock();
};

class Door : public Portal {
    public:
        virtual void open();
        virtual void close();
        virtual void lock();
};
```

Notice how this allows `Window` and `Door` objects to be stereotyped. They can both be referred to as `Portal`s. While this is an incomplete description of both objects, it is nonetheless accurate and useful.

This ability to stereotype an object is a powerful design tool. It allows us to bundle all the common aspects of a set of objects together into an abstract class. That's why we say: "An abstract class provides a common public interface for the entire class hierarchy."

The efficiency that we gained is obvious.

- Several of the `Portal` functions can be reused
- The declarations for `Window` and `Door` are terse and understandable in terms of the functionality of `Portal`
- Both `Window` and `Door` can be stereotyped as a `Portal` when convenient.

For example:

```
for (i = 0; i < max; ++i)
    portallist[i] -> open();
```

Clearly this method for opening all `Window`s and `Door`s is superior to handling each type separately. As the application matures, more types of `Portal`s will probably be added. But since the code deals with all types of `Portal`s, it will not have to change. Thus, the polymorphic behavior of the abstract class `Portal` provides a more maintainable and robust design!

Note that forcing `Window` and `Door` to inherit from `Portal` is not necessary to the proper functioning of the application. The application could be designed without the abstract class. However, creating the abstract `Portal` class results in a superior design that promotes polymorphism and code reuse.

Virtual Destructors

Destructors can be virtual, as well as member functions, but not constructors. A virtual destructor is typically used when one class needs to delete objects of a derived class that are addressed by pointers to a base class.

Example without using virtual destructors:

```
// vdtor1.cpp

#include <iostream>
#include <string.h>

using namespace std;

class Base {
    private:
        char *buf1;
    public:
        Base(const char* s) {
            cout << "Base 1 arg constructor" << endl;
            buf1 = new char[strlen(s) + 1];
            strcpy(buf1, s);
        }
        ~Base() {
            cout << "Base destructor" << endl;
            delete [] buf1;
        }
};

class Derived : public Base {
    private:
        char *buf2;
    public:
        Derived(const char *s1, const char *s2) : Base(s1) {
            cout << "Derived 2 arg constructor " << endl;
            buf2 = new char[strlen(s2) + 1];
            strcpy(buf2, s2);
        }
        ~Derived() {
            cout << "Derived destructor" << endl;
            delete [] buf2;
        }
};
```

```
int main() {
    Base *pbase = new Derived("Haley", "Laurel");

    delete pbase;
    return 0;
}
```

The `Base` class constructor allocates storage for a string and stores the new string in `buf1`. The `Base` class destructor is set up to delete this storage and return it to the heap when an object of type base goes out of scope.

Likewise, the `Derived` class constructor calls the `Base` class constructor, passing one string to the `Base` class and then allocates storage for a string and stores the new string in `buf2`. The `Derived` class destructor is also set up to delete this storage and return it to the heap when an object of type `Derived` goes out of scope. When a `Derived` object goes out of scope it's important that both strings be deleted.

Suppose you declare a pointer to `Base` but assign it the address of a `Derived` object. That's OK since a `Base` class pointer may address an object of that class or of any `Derived` class. This is shown below:

```
int main() {
    Base* pBase = new Derived ("Haley", "Laurel");

    delete pBase;
    return 0;
}
output:
Base destructor
```

Note that the `Derived` destructor is not called! Why? Because `pBase` is a pointer to the `Base` class and therefore the `Base` class destructor is linked to the call, i.e., only `Base` class destructor is available at compile time. This is *static binding*.

In the example below, let's make the destructors all virtual and see what happens:

```cpp
// vdtor2.cpp

#include <iostream>
#include <string.h>

using namespace std;

class Base {
    private:
        char *buf1;
    public:
        Base(const char* s) {
            cout << "Base 1 arg constructor " << endl;
            buf1 = new char[strlen(s) + 1];
            strcpy(buf1, s);
        }
        virtual ~Base() {
            cout << "Base destructor" << endl;
            delete [] buf1;
        }
};

class Derived : public Base {
    private:
        char *buf2;
    public:
        Derived(const char *s1, const char *s2) : Base(s1) {
            cout << "Derived 2 arg constructor " << endl;
            buf2 = new char[strlen(s2) + 1];
            strcpy(buf2, s2);
        }
        virtual ~Derived() {
            cout << "Derived destructor" << endl;
            delete [] buf2;
        }
};

int main() {
    Base *pbase = new Derived("Haley", "Laurel");

    delete pbase;
    return 0;
}
```

Output:
```
Derived destructor
Base destructor
```

Because the destructors are now declared to be `virtual`, calls to them are linked at runtime, and the object itself determines which destructor should be called! Therefore, the `Derived` destructor is called which then implicitly invokes the `Base` class destructor! Therefore, the entire two strings are deleted.

Rule: If you allocate memory in a Base class, you should make the destructor virtual.

Note, that virtual destructors violate the rule that states that virtual functions must have the same name and destructors obviously do not have the same name because they are based upon their respective class names. Therefore the compiler takes this into account and simply ignores the name of the destructor.

Virtual Function Mechanism

It is easier to understand how the compiler locates virtual functions at runtime by considering the data structures involved and the memory layout used with class objects using virtual functions. Let's do an experiment and look at the sizes of some objects:

```
class Alpha {
    private:
        int i;
    public:
        void foo() {}
        void faa() {}
};

class Beta {
    private:
        int i;
    public:
        virtual void foo() {}
        void faa() {}
};

class Gamma {
    private:
        int i;
    public:
        virtual void foo() {}
        virtual void faa() {}
};

int main() {
    Alpha a;
    Beta  b;
    Gamma g;

    cout << sizeof(Alpha) << endl;       // 4
    cout << sizeof(Beta)  << endl;       // 8
    cout << sizeof(Gamma) << endl;       // 8

    cout << sizeof(a) << endl;           // 4
    cout << sizeof(b) << endl;           // 8
    cout << sizeof(g) << endl;           // 8

    void *voidPointer;
    cout << sizeof(voidPointer);         // 4
    return 0;
}
```

Looking at the size of various classes and objects, we find a difference in size when one or more of the functions are `virtual`. Whenever a class has one or more virtual functions, the `sizeof()` that class is increased by 4 bytes or one pointer size. This extra pointer data member that has secretly been inserted into the class by the compiler is called the `vpointer` or `vptr`.

The `vptr` contains the address of an array of function pointers called the `vtable`. The `vtable` contains the addresses of (pointers to) all the functions that have been declared virtual in the class or in any inherited class. Every object of a class containing a virtual function is provided with a `vptr` to the `vtable` for the class.

Let's take another look at our `Gamma` class:

```
class Gamma {
    private:
        int i;
    public:
        virtual void foo() {}
        virtual void faa() {}
};
```

Here is a picture of what the data structures look like:

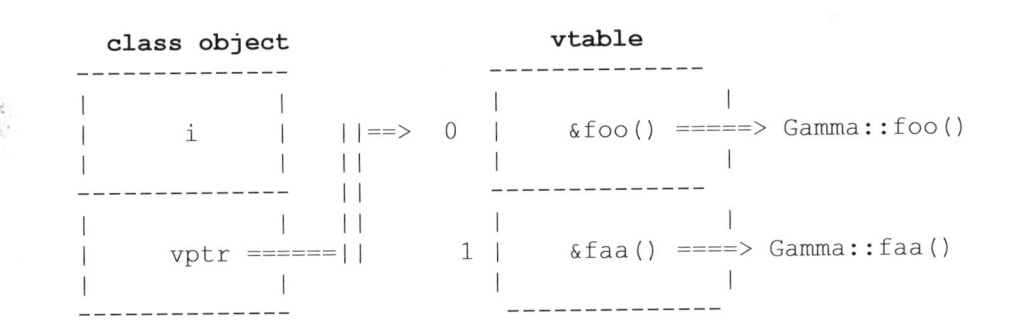

When an object is created, the appropriate `vtable` address is inserted into the objects `vptr`. The `vptr` is initialized to the starting address of the `vtable` in the constructor, so there's no chance of making a virtual function call before the `vptr` has been initialized. This is accomplished by secret code inserted into the constructor by the compiler. Note that if a `vptr` wasn't initialized, you'd have a disaster the first time you made a function call. This is one essential reason why the compiler automatically creates a default constructor if you don't explicitly create any constructors. One job of the generated default constructor is to initialize the `vptr`.

When you make a call to a virtual function, the compiler doesn't execute a normal function call that is an absolute call to some address. This would be early or static binding that is what a C compiler always does. We don't want that here! Instead, after the compiler performs the usual type checking for a function call, the compiler inserts a special bit of code that says:

- go to the address pointed to by the vptr

- index into this array by x places and call that function

So the exact function call is determined at runtime by which vtable address is in the vptr. Note that the order of functions in the vtable is crucial since the compiler calls the function by indexing into the table. Therefore, all the vtables must have the same order of functions. This fixed index remains associated with the particular virtual function throughout the inheritance hierarchy. Fortunately, this is all taken care of for us by the compiler!

Keep in mind that each derived class has its own vtable. For each derived class, the vtable of the base class is duplicated and if a virtual function is redefined in the derived class, then its address replaces the address of the corresponding function from the base class. Any new virtual functions defined in the derived class are appended to the end of the vtable. For example, shown below is a base class with three virtual functions declared:

```
class Base {
    private:
        int i;
    public:
        virtual void foo();
        virtual void faa();
        virtual void laa();
};
```

The internal data structures that get created are shown below:

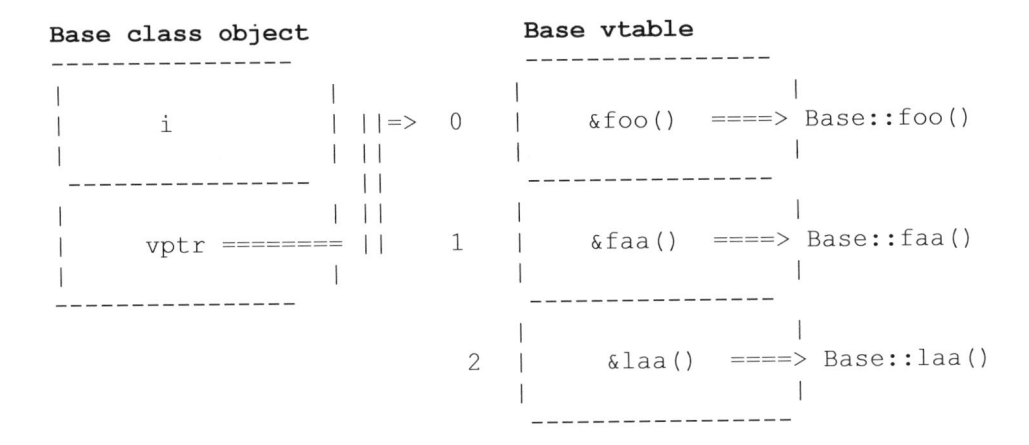

```
Base class object               Base vtable
----------------                ----------------
|              |     |          |              |
|      i       |     | ||=>  0  |   &foo()  ====> Base::foo()
|              |     | ||       |              |
----------------     ||         ----------------
|              |     | ||       |              |
|    vptr =======    ||      1  |   &faa()  ====> Base::faa()
|              |     |          |              |
----------------                ----------------
                                |              |
                             2  |   &laa()  ====> Base::laa()
                                |              |
                                ----------------
```

A derived class is declared as follows:

```
class Derived : public Base {
    public:
        virtual void faa();
        virtual void taa();
};
```

Its internal data structures are shown below:

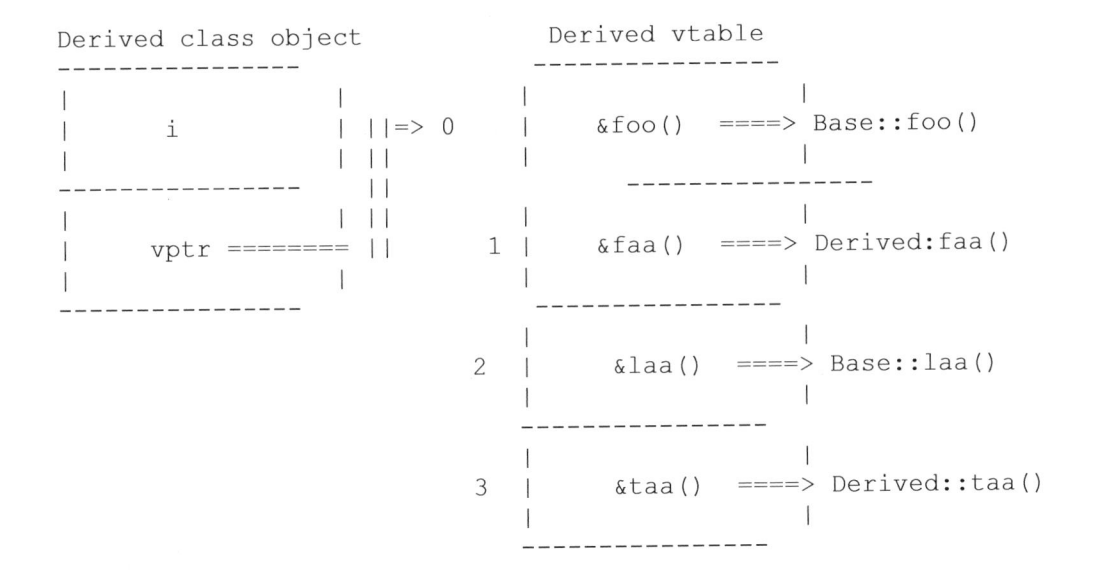

```
Derived class object               Derived vtable
----------------                   ----------------
|              |     |             |              |
|      i       |     | ||=> 0      |   &foo()  ====> Base::foo()
|              |     | ||          |              |
----------------     ||            ----------------
|              |     | ||          |              |
|    vptr =======    ||      1     |   &faa()  ====> Derived:faa()
|              |     |             |              |
----------------                   ----------------
                                   |              |
                             2     |   &laa()  ====> Base::laa()
                                   |              |
                                   ----------------
                                   |              |
                             3     |   &taa()  ====> Derived::taa()
                                   |              |
                                   ----------------
```

So each time you derive a new class, the compiler creates a new `vtable` for that derived class. There are three possibilities that can occur in a derived class:

- The derived class can inherit the virtual function declared within the base class. The address of that virtual function is copied into the associated slot in the derived class's virtual table.

- The derived class can override the virtual function declared within the base class with one of its own virtual functions. In this case, the address of derived class's virtual function is placed within the associated slot.

- The derived class can introduce a new virtual function not present in the base class. In this case, the virtual table is increased in size one slot and the address of the virtual function is placed within that slot. Note that there's always a virtual function address in every slot of the virtual table.

The virtual function calling scheme is as follows:

- A pointer to the object's `vtable` is retrieved from `vptr`.

- Remember that the index of each virtual function in the `vtable` is a constant value for each `vtable` in the class hierarchy established at compile time. Therefore, the `vtable` pointer will be offset by the index of the virtual function in question

- The `vtable` contains an entry (a pointer) to the desired virtual function.

- The retrieved function pointer is dereferenced and the function call is made.

Thus, given the following declaration:

```
Base *bptr = new Derived;
```

the compiler would translate the statement:

```
bptr->faa();
```

as follows:

```
*(bptr->vptr)[1]();
```

We can interpret this statement as follows:

- `vptr` is a pointer in the `Derived` object that points to the `Derived` class's virtual table

- `bptr->vptr` is the address of the `Derived` class's virtual table

- `(bptr->vptr)[1]` is the second entry in the `Derived` class's virtual table which contains the address of the virtual function `Derived::faa()`.

- `*(bptr->vptr)[1])()` dereferences the pointer to the function and invokes the function `Derived::faa()`.

How To Leak Memory in C++

This article appeared in the March 1997 issue of the C/C++ Users Journal (www.cuj.com) and appears here with the permission of Mark Briand, the managing editor. The article was written by Rodney B. Doe, P.E. who has a B.S. in Mechanical Engineering from the University of Washington in Seattle. He is employed as a Senior Software Engineer with Micro Encoder. Inc., in Kirkland. WA. He can be reached via e-mail at rodneyd@microen.com.

Warning: These are untrained professionals at work here. Do not try this at home.

It's common practice in software engineering to dynamically allocate memory from the operating system for use by a process. C and C++ do not require a process to release this memory when it is no longer needed. When a process neglects to return memory to the operating system, the utility of that memory is lost. Losing the use of memory in this manner is known colloquially as memory leakage.

The purpose of this article is to describe to the practicing software engineer how to achieve memory leakage using the C++ language. The tactics presented here are tried and true and have been proven in actual industrial practice. They are trivially easy to implement. Furthermore, these techniques are portable - they can be applied to any operating system. If used rigorously and diligently, they will provide an adequately debilitating stream of memory leaks for your application.

Tactic No. 1: Just Don't Delete It
If you like to keep things simple, then this is the memory leakage tactic for you. To use it, simply omit the applicable `delete` operator on memory that you have dynamically allocated with the `new` operator. Bear in mind, though, that using this technique will brand you as an amateur. Any decent memory error detection tool will flag it. For more insidious memory leakage tactics, read on.

Tactic No. 2: Use Multiple Return Statements
Remember when some dotty old chalk-covered college professor told you to use only one `return` statement in a function? Well, clearly, that professor knew nothing about how to leak memory. Using multiple `return`s actually allows you to exit a function before dynamically allocated memory is freed.

Here is an example:

```
// Memory leak

bool AdjustTemperature(HTHING hThing) {
    if (!hThing)
        return(false);

    // Instantiate a Thing using a handle
    Thing *pThing = new Thing(hThing);
    if (!pThing)
        return(false);

    // Get Thing's color and adjust temperature accordingly
    switch(pThing->Color()) {
        case BLUE: AdjustTemperature(5);  return(true);
        case RED:  AdjustTemperature(-5); return(true);
        default:   break;
    }

    // For what colors does this get executed?
    delete pThing;
    return( false );
}
```

As you can see in the above code fragment, the programmer saved a few clock cycles by not using a local Boolean variable to keep track of success or failure. He didn't need a local variable because whenever his code encountered an error it just returned to the caller. This code does return the correct result to its caller. However, if the color of the `Thing` is `red` or `blue`, the memory that `pThing` points to never gets deleted. Note that if you must implement memory leakage in legacy code, this technique can also be implemented in good old C by using `malloc` and `free` instead of `new` and `delete`.

Tactic No. 3: Use Numerous Pointers with the Same Name

C++ allows and encourages a programmer to declare an automatic variable close to the point of usage. So, you can declare an automatic variable that will point to a quantity of allocated memory right at the point of allocation. This capability provides the memory leaker with a unique opportunity. Declare a pointer at the start of a function (in the traditional manner), and declare another pointer of the same name within an inner scope, at the point of instantiation. This will create a fairly confusing memory leak. Here is an example:

```
void ChangeColor(HTHING hThing, int color) {
    Thing *pThing  = NULL;
    if (IsValid(hThing) != false) {
        Thing *pThing = new Thing(hThing);
        pThing->Color(color);
    }

    // Which pThing is this?
    delete pThing;
}
```

In this snippet, two `Thing` pointers called `pThing` are declared. The one inside the `if` statement points to an instance of a Thing object which is never deleted. The pThing that does get deleted is `NULL` at the time of deletion. C++ guarantees that deleting a null pointer is harmless. Thus, a memory error detection tool may not flag this deletion as an error because it is a valid operation.

A quick glance at the above function makes you think `pThing` gets deleted, so there seems to be no problem. This is a nice technique to slip past code reviewers.

Tactic No. 4: Use delete on an Array

To use this tactic, allocate an array with the array `new` operator (`new[]`) then release the memory with plain old `delete`, as in:

```
char *pBuf;
pBuf - new char[100];
delete pBuf;        // NOT delete[] pBuf;
```

Probably more than a few readers are suddenly getting a nervous feeling in the pits of their stomachs. If you use the `delete` operator (and not the `delete[]` operator) to release an object that was allocated with `new[]`, the resulting behavior is currently undefined. Even if you get away with this operation now it could bite you later. Or. it may simply leak the memory. Go ahead, put some peril in your life and use it liberally.

Tactic No. 5: Practice Careless Polymorphism

This tactic takes advantage of the fact that destructors are not inherited. In C++, it is possible and in fact highly desirable to store the address of a derived-class object as a pointer to its base class. You get polymorphic behavior by referencing this derived class object through the base class pointer. The following program

instantiates a derived-class object and stores the address as a pointer to its base class:

```cpp
#include <iostream>

using namespace std;

class Base {
    public:
        Base();
        ~Base();
};

class Derived : public Base {
    protected:
        char *pBuf;
    public:
        Derived();
        -Derived();
};

Base::8ase() {
    cout << "Base::Base()" << endl;
}

Base::~Base() {
    cout << "Base::~Base()" << endl;
}

Derived::Derived() : Base() {
    pBuf = new char[100];
    cout << "Derived::Derived()" << endl;
}

Derived::~Derived() {
    delete[] pBuf;
    cout << "Derived::~Derived()" << endl;
}

int main() {
    Base *pObject = new Derived();
    delete pObject;
    return 0;
}
```

Output:
```
Base::Base()
Derived::Derived()
Base::~Base()
```

Note that the destructor for the `Derived` class did not get called. The memory leaker can thus leak the buffer `pBuf` because it was allocated in the `Derived` constructor. To fix this, simply declare the destructor of the `Base` class as a virtual function, as in:

```
class Base {
    public:
        Base();
        virtual ~Base();
};
```

As modified, the program output will be:

```
Base::Base()
Derived::Derived()
Derived::~Derived()
Base::~Base()
```

Now the `Derived` class destructor is called, and `Derived::pBuf` is deleted.

Conclusion

Now you have seen a number of handy tricks of the trade that varsity-level C++ programmers use to implement memory leakage. If you use the tactics described above, you will not have to invent your own. Code reuse is, after all, what C++ is all about. A cautionary note: these memory leak scenarios are made possible by improper use of C++. They are not intrinsic to the language. As such, when your application runs out of memory at a customer site on the other side of the planet, you may not be successful in casting the blame on C++.

❖ Polymorphism Homework

Note: These were all questions that appeared on a recent final exam.

1. How does polymorphism contribute to the design of object-oriented systems?

2. Explain the role of the `vptr` and `vtable` in the implementation of virtual functions.

3. What is a pure virtual function and why would you ever use one?

4. What is the purpose of a virtual destructor?

5. Describe a situation where it makes sense for a class to have a destructor, but no constructor.

6. Without executing the code below, give the output of the following program.

```cpp
class Base {
    public:
        void f() {
            cout << "Base::f()" << endl;
        }
        virtual void g() {
            cout << "Base::g()" << endl;
        }
};

class Derived : public Base {
    public:
        void f() {
            cout << "Derived::f()" << endl;
        }
        virtual void g() {
            cout << "Derived::g()" << endl;
        }
};
```

```
int main() {
    Base b;
    Derived d;

    Base* bp = &b;
    Base* bq = &d;
    Derived* dp = &d;
    bp->f();
    bp->g();
    bq->f();
    bq->g();
    dp->f();
    dp->g();
    return 0;
}
```

❖ Polymorphism Lab (Library Holdings)

In this assignment, you will learn how to use inheritance and virtual functions by designing a set of classes to represent library holdings. Our library has two kinds of items: books and recordings. Each item has a title and an integer call number. In addition, books have authors, and recordings have performers and formats (MP3, WAV, AIFF). In addition to constructors and destructors, each holding should have a print() function that displays its data to an ostream object (such as cout) that the user specifies.

You should begin by designing a Holding class that contains all the data common to books and recordings. It should also have a pure virtual print() function. Then design the Book and Recording classes to add the details necessary for those types of items and to implement the print() function. Note that Holding will be an abstract base class, since it will have a pure virtual function. This makes sense, since there are no generic holdings, only books and recordings.

All three classes should have both copy constructors and constructors that accept the individual data fields as arguments. Although it may seem natural to have a zero-argument default constructor that queries the user and reads in the data fields, doing I/O from within a constructor is probably unwise. If the constructor did I/O, the class implementation would be tied to a particular style of getting data. Applications that use this class may read input from a file, in which case the prompt messages would be inappropriate, or they may use a window-based interface that does not support the cin and cout I/O stream objects. Since constructors are such a basic part of the operation of a class, it is best to let the user decide where their input data comes from.

Once you have written the code to implement the Holding, Book, and Recording classes, you should write a simple program to test them. This program should declare an array of five pointers to Holding objects and then repeatedly call a function that creates a new Holding. This function asks the user which kind of Holding is to be entered and then inputs the data needed to create either a Book or a Recording. A new object of the appropriate class is created with this data, and the resulting pointer is returned.

Note that even though the function always returns either a Book pointer or a Recording pointer, the function can be declared to return a Holding pointer, and the pointer to the newly created object can be returned without conversion. This pointer is then stored in the array and the function is called again until all

five objects have been created (see data below). Now there is an array to five pointers to `Holding` objects, and each of these objects "knows" whether it is a `Book` or a `Recording`, although the program has no direct way of querying them to get this information. However, it can run through the array and call the virtual `print()` function for each object, which will then use the appropriate virtual function for printing itself out. A sample session is shown below, with user responses shown in bold type for clarity.

```
Enter holdings to be stored in a list:

Enter B for book, R for recording: B
Enter book title: Harry Potter
Enter book author: Rowling, J.K.
Enter call number: 101
Enter B for book, R for recording: R
Enter recording title: Abbey Road
Enter performer: The Beatles
Enter format: (M)P3, (W)AV, (A)IFF: W
Enter call number: 401
Enter B for book, R for recording: R
Enter recording title: Lone Star State of Mind
Enter performer: Nanci Griffith
Enter format: (M)P3, (W)AV, (A)IFF: A
Enter call number: 402
Enter B for book, R for recording: B
Enter book title: The Hitchhiker's Guide to the Galaxy
Enter book author: Adams, Douglas
Enter call number: 102
Enter B for book, R for recording: R
Enter recording title: Tapestry
Enter performer: Carole King
Enter format: (M)P3, (W)AV, (A)IFF: M
Enter call number: 103

Here are the holdings:

BOOK: Rowling, J.K. "Harry Potter" 101
RECORDING: "Abbey Road" The Beatles (WAV) 401
RECORDING: "Lone Star State of Mind" Nanci Griffith (AIFF) 402
BOOK: Adams, Douglas "The Hitchhiker's Guide to the Galaxy" 102
RECORDING: "Tapestry" Carole King (MP3) 103
```

Turn in all the code for the three classes and for the test program, as well as a sample run showing the input of Books and Recordings interspersed with each other, as is done in the sample above. Please use my data above for the lab. Output should be to the terminal window as well as the `csis.txt` output file. Please do not use the C++ String class to create any String objects in your constructors.

Chapter 15. Multiple Inheritance

Introduction to Multiple Inheritance
Multiple Base Class Constructors
Virtual Base Classes
Final Comment on Multiple Inheritance
Multiple Inheritance Homework

Introduction to Multiple Inheritance

So far, derived and base classes have been related by single inheritance. Multiple inheritance describes a more complex relationship in which a single class inherits the data members and member functions from more than one base class.

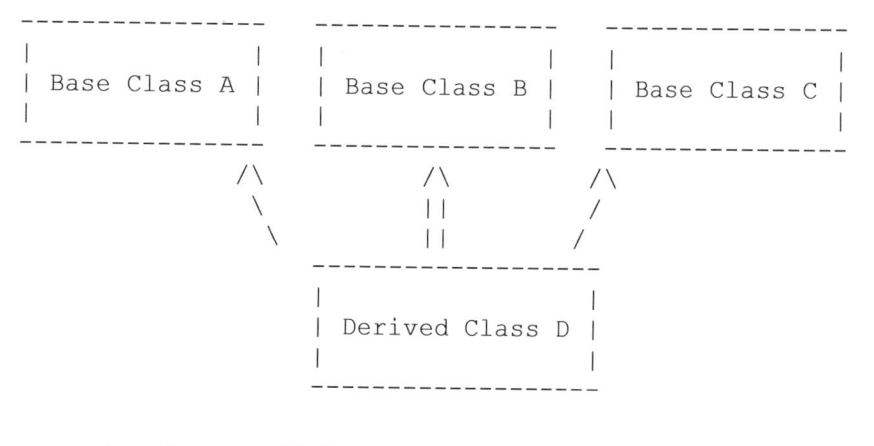

To derive a new class from multiple base classes, we simply list the base class names after the new derived class:

```
class D : public A, public B, public C {
};
```

Here, class D inherits from classes A, B, and C.

The basic need for multiple inheritance occurs because in C++ it's possible to grow inheritance hierarchies from scratch. This means you may purchase a class library from vendor A and one from vendor B, each with its own inheritance tree. Because you may want to mix the characteristics of two or more independent class hierarchies, you need to be able to inherit from more than one class at a time.

Class 1	+	Class 2	=	Derived Class 3
Boat	+	Plane	=	SeaPlane
Date	+	Time	=	DateTime
Painter	+	Artist	=	PortraitPainter
Text Window	+	Scrollable Window	=	ScrollableTextWindow

In multiple inheritance, a derived class receives all the inheritable properties from all of the listed base classes. If a name conflict occurs, such as the same name appearing in more than one base class (i.e., two base classes declaring a member function or data member with the same name), and the name is not overridden in a derived class, it's use would be an error, even if members are private or have different sets of parameters.

The fix would be to use the scope resolution operator to specify which member you intend to use:

```cpp
class A {
    public:
        int a;
        void foo();
};

class B {
    private:
        int a;
        void foo(int);
};

class C : public A, public B {
};

int main() {
    C myObj;

    myObj.foo();        // error - ambiguous
    myObj.foo(5);       // error - ambiguous
    myObj.a = 3;        // error - ambiguous
    myObj.A::foo();     // OK
    myObj.A::a = 3;     // OK
    return 0;
}
```

Multiple Base Class Constructors

A class derived from multiple base classes might also need to call multiple base class constructors:

```
class D : public A, public B, public C {
    public :
        D() : A(), B(), C() {}
};
```

This assumes that A, B, and C are classes with default constructors and class D can call those constructors.

Note that the constructors are called in the order in which their base classes are declared. However, it is generally unwise to devise classes that depend on other class objects to be constructed in a particular order.

As a rule, classes should operate as independently of each other as possible.

Virtual Base Classes

When using multiple inheritance, a base class may not be specified more than once in a derived class: This declaration is therefore not legal:

```
class Base {};

class Derived : public Base, public Base{};
```

However, a base class can legally be indirectly passed to the derived class more than once:

```
class X : public Base {
    public:
        int num;
        void getNum();
};

class Y : public Base {
    public:
        int num;
        void getNum();
};

class Z : public X, public Y {
    public:
        int num;
        void getNum() { return num; };
};
```

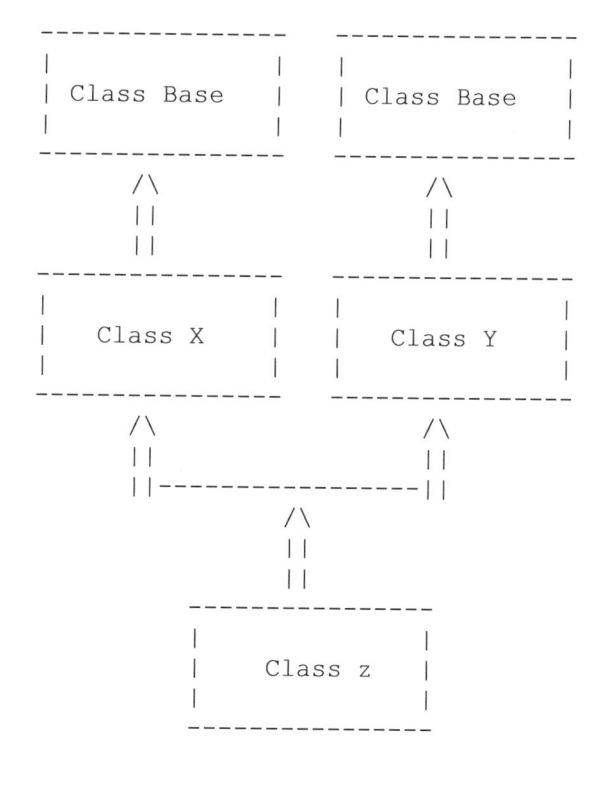

Note that the above inheritance tree is shown with the same `Base` class appearing twice. In this case, each object of class `Z` will have two sub-objects of class `Base`, each with its own data!

Also note that the function `getNum()` is now ambiguous. The compiler doesn't know which `num` is being referred to! The scope resolution operator must therefore be applied to the function:

```
int Z::getNum() {
    return X::num;
}
```

Having multiple copies of the same base class in an instance hierarchy is not only confusing, it wastes storage! The solution is to declare the base class `virtual`. This forces the compiler to allow only one copy of a given base class in the declaration of a derived class:

```
class base {
    public:
        int num;
};

class X : public virtual Base {};

class Y : public virtual Base {};

class Z : public X, public Y {
    public:
        int getNum { return num; }
};
```

Note that we no longer need to use the scope resolution operator in the return
statement of the `getNum()` member function in class Z. Base is now a virtual
base class and class Z only has one object of class Base.

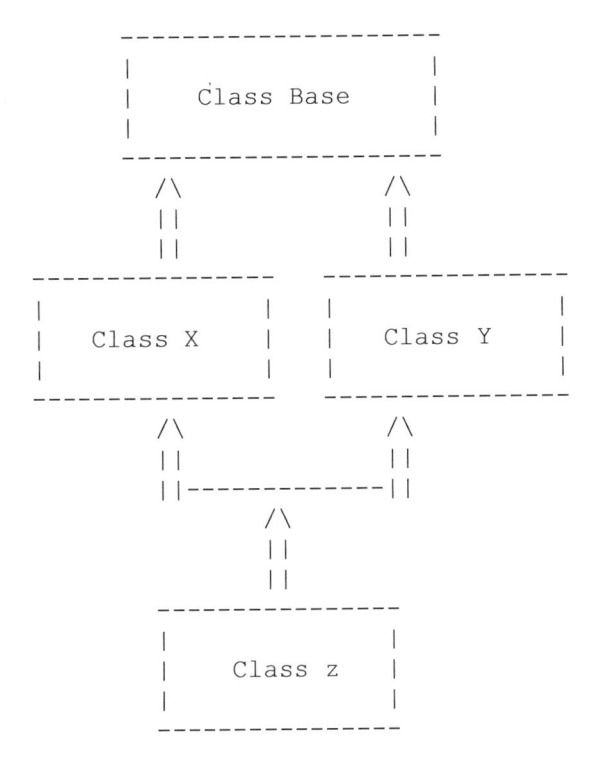

Note that virtual base classes are used only in the context of multiple inheritance
and have nothing to do with abstract base classes (which contain a pure virtual
function).

Final Comment on Multiple Inheritance

Multiple inheritance is a feature that was added to the C++ language after its initial release amidst much arguing for and against. The object-oriented programming language gurus are still arguing over whether multiple inheritance is a bug or a feature of C++ because the presence of multiple inheritance and virtual functions make it possible to design inadvertent ambiguities into your program that can't be detected until runtime.

Multiple inheritance has been considered the `goto`'s of the 90's and is something that you should know about, but generally avoid.

1. As part of a military video game, a designer has created a vehicle class hierarchy. The base class is `Vehicle`, and it has `AirVehicle`, `LandVehicle`, and `SeaVehicle` as derived classes. The class `SeaPlane` inherits from both `AirVehicle` and `SeaVehicle`. Specifically, what **design issues** had to be considered in developing the `SeaPlane` class?

2. What is the distinction between an **abstract base class** and a **virtual base class**?

3. Give the output for the following program without executing the code.

```cpp
class A {
    protected:
        int fa;
    public:
        A(int va = 0) : fa(va) {
            cout << "A constructor " << endl;
        }
        A(const A& va) : fa(va.fa) {
            cout << "A copy constructor " << endl;
        }
        ~A() {
            cout << "A destructor" << endl;
        }
        A& operator=(const A& va) {
            cout << "A = operator" << endl;
            fa = va.fa;
            return *this;
        }
};
```

```
class B : public A {
    protected:
        int fb;
    public:
        B(int va = 0, int vb = 0)  : fb(vb), A(va) {
            cout << "B constructor " << endl;
        }
    B(const B& vb)  : fb(vb.fb), A(vb) {
        cout << "B copy constructor " << endl;
    }
    ~B() {
        cout << "B destructor" << endl;
    }
    B& operator=(const B& vb) {
        cout << "B = operator" << endl;
        A::operator=(vb);
        fb = vb.fb;
        return *this;
    }
};

int main() {
    B b1(1, 2);
    B b2;
    b2 = b1;
    B b3(b1);
    return 0;
}
```

Chapter 16. Templates

Introduction to Templates
Function Templates
Class Templates

Introduction to Templates

Just as a class is a schematic for building objects, a template is a schematic for building functions and classes. They provide specifications for functions and classes, but not the actual implementation details.

With templates, you can develop generic functions and classes that support the same operations but for different data types. You can write template functions that sort, search, and swap various data types. You can create template classes that model generic data structures, such as arrays, matrices, lists, trees, hash tables, and graphs.

Templates can be used to implement data structures and algorithms that are largely independent of the type of the objects they operate on. For instance, a stack template might describe how to implement a stack of arbitrary objects. Once that template has been defined, users can write code that uses stacks of integers, stacks of strings, etc.

The compiler will figure out which kinds of stacks have been used and automatically generate the implementations of those stack classes.

Function Templates

Function templates provide you with the capability to write a single function that is a skeleton, or template, for a family of functions. They provide a mechanism to create a generic function, a single function that can simultaneously support multiple data types for its parameter or parameters.

In this function, at least one formal argument is generic. This function template becomes a real function with real types when and only when it needs to be invoked.

But What's Wrong with Function Overloading?

We've seen a great deal about how to overload functions in C++ and the benefits we may achieve from doing so.

The chief advantage is that function overloading relieves someone who is using your functions from having to know about different names for various functions that essentially do the same task. Unfortunately, overloaded functions, while a vast improvement over what's available in C aren't the ultimate solution to the problem that is inherent when writing functions that are similar in their behavior.

To see why, here's a function called `max()` that returns the greater of its two input arguments:

```
int max(int x, int y)          { return x > y ? x : y; };

long max(long x, long y)       { return x > y ? x : y; };

double max(double x, double y) { return x > y ? x : y; };

char max(char x, char y)       { return x > y ? x : y; };
```

We let C++'s overloaded function feature sort out which function to call depending on the data type!

Note, however, that `max()` has to be continually overloaded in order to accommodate the various types of arguments that can conceivably be used to invoke it. That involves too much repetitious coding. Each function is doing essentially the same thing, returning the greater of its two arguments.

Instead of your having to write many similar `max()` functions, wouldn't it be nice to write just one `max()` function that returns the greater of its two arguments and can accommodate almost any type of argument?

Why Not Use A Macro

You can write a macro to solve this problem but you lose the ability to ensure that you are not comparing arguments of different types. For example, if your intent is to guarantee that both arguments are the same type, then a macro does not work:

```
#define max (x, y) (((x) > (y)) ? (x) : (y))

int main() {
    cout << max(1, 2.98)  << endl;
    cout << max('A', 32)  << endl;
    cout << max(++a, b++) << endl;
    return 0;
}
```

There is no type checking here and the arguments could be evaluated twice when using prefix or postfix notation.

Template Functions

Template functions provide a mechanism by which the programmer can take advantage of the compactness of the macro solution without forfeiting any of the benefits of a strongly-typed language.

A function template solves the problem by allowing you to write just one function that serves as a skeleton, or template, for a family of functions whose tasks are all similar. At the same time, full type checking is carried out.

You should consider using a function template when you find yourself writing a sequence of identical, overloaded functions. By using templates, you can define a pattern for a family of related overloaded functions by letting the data type itself be a parameter!

```
template <class T>
T max (T a, T b) {
    return a > b ? a : b;
}

int main() {
    cout << max(15, 16)    << endl;
    cout << max('A', 'B') << endl;
    return 0;
}
```

The `template` keyword indicates to the compiler that a function template is present. The `T` symbol is a placeholder symbol indicating that the compiler can substitute the appropriate data type at compilation time.

When the compiler sees a template function, it stores the template code for future reference. The template code itself is not compiled. To employ the template function, you need to associate it with a data type. The compiler then instantiates a copy of the template function for the data type you specify.

A template function is typically declared in a header file:

```
// minmax.h

#ifndef _MINMAX_H
#define _MINMAX_H

template <class T>
T max (T a, T b) {
    return a > b ? a : b;
}

template <class T>
T min (T a, T b) {
    return a < b ? a : b;
}

#endif
```

In this example, `template <class T>` states that `T` is an unspecified type, which a template function `max()` returns. In addition, `max()` and `min()` require two `T` arguments. The function's statements are schematics for the real statements that are generated later when `T`'s actual type is specified.

```
int main() {
    int i1 = 100, i2 = 200;
    double d1 = 3.14, d2 = 9.876;
    char c1 = 'A', c2 = 'z';

    cout << max(i1, i2) << endl;
    cout << max(d1, d2) << endl;
    cout << max(c1, c2) << endl;
    return 0;
}
```

Each `max()` function returns a different data type and requires pairs of arguments of the same types.

Keep in mind that you could also use more than one placeholder. For example, you could declare `max()` with two different possible types:

```
template <class T1, class T2>
T1 max(T1 a, T2 b);
```

This returns a value of type `T1` and requires two arguments, one of type `T1` and one of type `T2`.

Class Templates

You already know how classes can help you reuse code and recycle your code for more efficient programming. Templates raise the level of abstraction one level and enable you to automatically create individual classes that work with different types of data.

Template classes are even more powerful than template functions! You can use a template to create one class definition to describe a class that works with integers, a class that works with characters, and a class that works with floats, all within one class definition and without writing any new code.

A class template specifies how individual classes can be constructed, much as a class declaration specifies how individual objects can be constructed.

For example, a user typically needs a vector of elements of some type unknown to the writer of the vector class. Therefore the vector class should be expressed in such a way that it takes the element type as an argument.

```cpp
// vector1.cpp

void error(char *);

class String {};
class Distance {};

template <class T>
class Vector {
    private:
        T *v;
        int size;
    public:
        Vector(int s) {
            if (s <= 0)
                error("Bad vector size.");
                size = s;
                v = new T[size];    // allocate an array of s T's
        }
        ~Vector() { delete [] v; }
        int getsize() { return size; }
        T& operator[](int i) { return v[i]; }
};
```

The `template <class T>` prefix to the otherwise ordinary class declaration specifies that a template that requires a type argument is being declared. This defines a class template that is a specification of how to make a family of related classes. In this case, the vector template specifies how the vector classes can be made for individual element types.

Vectors of specified types can now be defined and used:

```
int main() {
    Vector <int>       v1(100);     // generate a vector of 100 ints
    Vector <char>      v2(200);     // generate a vector of 200 chars
    Vector <char*>     v3(300);     // generate a vector of 300 char *
    Vector <String>    v4(400);     // generate a vector of 400 objects
                                    // of the String class
    Vector <Distance> v5(500);      // generate a vector of 500 objects
                                    // of the Distance class

    return 0;
}
```

Note that there is no run-time overhead compared with a class where all types involved are specified directly.

Class Template for the Stack Class

```
// stack1.cpp

template <class T>
class Stack {
    private:
        int size;
        int top;
        T *s;
    public:
        Stack(int max = 100);
        ~Stack();
        void push(T);
        T pop();
        T Top();
        int empty();
        int full();
        void clear();
};
```

```
template <class T>
Stack<T>::Stack(int max) {
    size = max;
    top = -1;
    s = new T[size];
}

template <class T>
Stack<T>::~Stack() {
    delete [] s;
}

template <class T>
void Stack<T>::push(T d) {
    s[++top] = d;
}

template <class T>
T Stack<T>::pop() {
    return s[top--];
}

template <class T>
T Stack<T>::Top() {
    return s[top];
}

template <class T>
int Stack<T>::empty() {
    return top== -1;
}

template <class T>
int Stack<T>::full() {
    return top== size-1;
}

template <class T>
void Stack<T>::clear() {
    top = -1;
}
```

The `Stack` class template defines a family of classes when the `Stack` template is used to declare an object. Note that a type must be supplied to occupy the place of the dummy type `T` in the template declaration.

We can instantiate `Stack` objects as follows:

```
int main() {
    Stack <char>   sc(100);    // declares an object sc which is a stack
                               // that stores up to 100 character values
    Stack <int>    si(200);    // declares an object si which is a stack
                               // that stores up to 200 integer values
    Stack <char*> sp(300);     // declares an object sp which is a stack
                               // that stores up to 300 char* values
    Stack <string> ss(400);    // declares an object ss which is a stack
                               // that stores up to 400 String values
    return 0;
}
```

Except for the special syntax of its name, `Stack <char>` works exactly as if `char` had been built into the definition. This is the same for the instantiations of all the `Stack` objects. Note that this mechanism saves our rewriting class declarations when the only variation would be type declarations.

The principle motivation for the addition of templates to C++ is their support for general-purpose collection classes. Not only can stacks of integers and characters be created, but stacks of floats, stacks of strings, etc, can be created. We can create stacks of any user-defined type!

```
// Record Class Example - record1.cpp

class String {};

template <class K, class D>
class Record {
    private:
        K key;
        D data;
    public:
        Record() {}
        Record(K kx, D dx) {
            key = kx;
            data = dx;
        }
        K getkey() { return key; }
        D getdata() { return data; }
};
```

```
template <class K, class D>
Record<K, D> :: Record(K kx, D dx) {
    key = kx;
    data = dx;
}
```

Remember, members defined outside the template class definition need to be templates themselves. Class scope identifiers must contain template arguments:

```
// record2.cpp

class String {};

template <class K, class D>
class Record {
private:
    K key;
    D data;
public:
    Record() {}
    Record(K kx, D dx);
    K getkey() { return key; }
    D getdata() { return data; }
};

template <class K, class D>
Record<K, D>::Record(K kx, D dx) {
    key = kx;
    data = dx;
}

int main() {
    Record <int, float> r1;
    Record <long, char> r2(50000, 'A');
    Record <int, String> r3;
    return 0;
}
```

```
int main() {
    Record <int, float> r1;
    Record <long, char> r2(50000, 'A');
    Record <int, String> r3;
    return 0;
}
```

566575766R00247